Nitrification

Nitrification

EDITED BY

Bess B. Ward
Daniel J. Arp
Martin G. Klotz

ASM PRESS WASHINGTON, DC

Cover image: *Nitrobacter winogradskyi* Nb255 (courtesy of William Hickey; see Fig. 2 in Chapter 11)

ASM Press
American Society for Microbiology
1752 N St., N.W.
Washington, DC 20036-2904

Library of Congress Cataloging-in-Publication Data

Nitrification / edited by Bess B. Ward, Daniel J. Arp, and Martin G. Klotz.
p. ; cm.
Includes bibliographical references and index.
ISBN-13: 978-1-55581-481-6 (hardcover : alk. paper)
ISBN-10: 1-55581-481-6 (hardcover : alk. paper) 1. Nitrification. 2. Nitrogen—Fixation. I. Ward, Bess B. II. Arp, D. J. III. Klotz, Martin G. IV. American Society for Microbiology.
[DNLM: 1. Nitrogen Fixation—physiology. 2. Ammonia—metabolism. 3. Archaea—metabolism. 4. Bacteria—metabolism. 5. Ecological and Environmental Phenomena. 6. Nitrates—physiology. QU 70]

QR89.7.N542 2011
572'.545—dc22

2010041260

10 9 8 7 6 5 4 3 2 1

Address editorial correspondence to ASM Press, 1752 N St., N.W., Washington, DC 20036-2904, USA

Send orders to ASM Press, P.O. Box 605, Herndon, VA 20172, USA
Phone: 800-546-2416; 703-661-1593
Fax: 703-661-1501
E-mail: books@asmusa.org
Online: http://estore.asm.org

CONTENTS

CONTRIBUTORS

Wiebe R. Abma
Paques B.V., Balk 8560AB-52, The Netherlands

Yoshiteru Aoi
Waseda Institute for Advanced Study, Tokyo 169-8050, Japan

Daniel J. Arp
Department of Botany and Plant Pathology, Oregon State University, Corvallis, OR 97331

Annette Bollmann
Department of Microbiology, Miami University, Oxford, OH 45056

Peter J. Bottomley
Departments of Microbiology and Crop and Soil Science, Oregon State University, Corvallis, OR 97331-3804

Holger Daims
Department of Microbial Ecology, University of Vienna, 1090 Vienna, Austria

Pia Engström
Civil and Environmental Engineering,
Chalmers University of Technology, SE-412 96 Göteborg, Sweden

Mike S. M. Jetten
Department of Microbiology, Faculty of Science,
Radboud University of Nijmegen, Nijmegen, The Netherlands

Boran Kartal
Department of Microbiology, Faculty of Science,
Radboud University of Nijmegen, Nijmegen, The Netherlands

Jan T. Keltjens
Department of Microbiology, Faculty of Science,
Radboud University of Nijmegen, Nijmegen, The Netherlands

Martin G. Klotz
Department of Biology, University of Louisville, Louisville, KY 40292

Hendrikus J. Laanbroek
Department of Microbial Ecology,
Netherlands Institute of Ecology (NIOO-KNAW), Nieuwersluis, The Netherlands

Sven Leininger
Sars International Centre for Marine Molecular Biology,
Thormøhlensgate 55, N-5008 Bergen, Norway

Denis Le Paslier
CEA/Genoscope, CNRS UMR8030, Evry, France

Sebastian Lücker
Department of Microbial Ecology, University of Vienna, 1090 Vienna, Austria

Willm Martens-Habbena
Department of Civil and Environmental Engineering,
University of Washington, Seattle, WA 98195-5014

Graeme W. Nicol
Institute of Biological & Environmental Sciences,
University of Aberdeen, Aberdeen, AB24 3UU, United Kingdom

Jeanette M. Norton
Department of Plants, Soils and Climate, Utah State University, Logan, UT 84322

Satoshi Okabe
Department of Urban and Environmental Engineering, Graduate School of Engineering,
Hokkaido University, Sapporo 060-8628, Japan

James I. Prosser
Institute of Biological and Environmental Sciences, University of Aberdeen,
Aberdeen AB24 3UU, United Kingdom

Hisashi Satoh
Department of Urban and Environmental Engineering, Graduate School of Engineering,
Hokkaido University, Sapporo 060-8628, Japan

Luis A. Sayavedra-Soto
Department of Botany and Plant Pathology, Oregon State University, Corvallis, OR 97331

Christa Schleper
Department of Genetics in Ecology, University of Vienna, Althanstrasse 14,
A-1090 Vienna, Austria

Eva Spieck
Universität Hamburg, Biozentrum Klein Flottbek, Mikrobiologie & Biotechnologie,
Ohnhorststraße 18, D-22609 Hamburg, Germany

David A. Stahl
Department of Civil and Environmental Engineering,
University of Washington, Seattle, WA 98195-5014

Shawn R. Starkenburg
Life Technologies, 29851 Willow Creek Rd., Eugene, OR 97402-9132

Lisa Y. Stein
Department of Biological Sciences, University of Alberta,
Edmonton, Alberta T6G 2E9, Canada

Yuichi Suwa
Faculty of Science and Engineering, Chuo University, Tokyo 112-8551, Japan

Mark Trimmer
School of Biological and Chemical Sciences, Queen Mary University of London,
London E1 4NS, United Kingdom

Hidetoshi Urakawa
Department of Civil and Environmental Engineering,
University of Washington, Seattle, WA 98195-5014

Wouter R. L. van der Star
Department of Geo-Engineering, Deltares, Delft 2600MH-177, The Netherlands

Mark C. M. van Loosdrecht
Department of Biotechnology, Delft University of Technology,
Delft 2628BC-67, The Netherlands

Michael Wagner
Department of Microbial Ecology, University of Vienna, 1090 Vienna, Austria

Bess B. Ward
Department of Geosciences, Princeton University, Princeton, NJ 08544

PREFACE

Nitrification, the oxidation of reduced forms of nitrogen to nitrite and nitrate, is an essential link in the nitrogen cycle of natural, industrial, and agricultural systems. Nitrate, the end product of nitrification, is the major bioavailable form of nitrogen in seawater and an important factor limiting primary production. Nitrification in agricultural systems can lead to fertilizer loss and to nitrogen pollution in receiving waters. In wastewater treatment, nitrification is a key step in nitrogen removal, linking to denitrification and fixed nitrogen loss.

Autotrophic nitrifying bacteria were discovered near the end of the 19th century and for about 100 years, were considered the only organisms capable of the oxidation of ammonium to nitrite and then nitrate. Two recent discoveries have transformed our understanding of nitrification. Although the anaerobic oxidation of ammonia to N_2 using nitrite is thermodynamically favorable, organisms capable of anaerobic ammonia oxidation, or anammox, were first described only in the 1990s. Even more recently, strong evidence has been presented that many, if not the vast majority of, ammonia-oxidizing microbes in the ocean and many terrestrial environments are archaea, rather than bacteria. This discovery has many still unexplored ramifications for regulation of ammonia oxidation and the magnitude of autotrophic carbon fixation in the deep sea, as well as for control of nitrification in terrestrial and aquatic environments. In addition, it has become clear that the overlap of lithotrophic ammonia oxidation and methane cycling have significant ecological ramifications for methane and nitrous oxide emissions. These discoveries in the last 15 years radically changed our understanding of the N cycle and nitrification, in particular, and have dramatically increased the interest in nitrification and the N cycle.

The last monograph to review nitrification was published in 1986 (J. I. Prosser, editor), before the discoveries of anammox and ammonia-oxidizing archaea and before the revolution in molecular biology and genomics was applied to the genetics and biochemistry of nitrification. Since that time, the level of interest in the research community and the number of publications on nitrification have risen dramatically. Thus, it is timely again to review the state of the field and collect the current knowledge in one comprehensive volume.

This book is focused around the microbes that perform the expanding repertoire of nitrification reactions and pathways. Four sections cover the main groups of microbes involved: the conventional aerobic bacterial ammonia oxidizers, the recently discovered aerobic archaeal ammonia oxidizers, the anaerobic ammonia-oxidizing planctomycetes, and the nitrite-oxidizing bacteria. For each group, comprehensive information and referencing is provided on phylogeny, distributions, biochemistry, and genomics. Nitrification in its narrowest sense is liked tightly to several other processes in the nitrogen cycle, and in its broadest sense, even includes parts of those processes. Thus, we include topics such as nitrifier denitrification and anammox but do not cover denitrification thoroughly and do not include the closely related process of methane oxidation. The last section places nitrification in the ecological context of major ecosystems: oceans, terrestrial (including agriculture), freshwater, and wastewater treatment.

The idea for the book arose from discussions among the founding members of the Nitrification Network (http://nitrificationnetwork.org/). It is our intention to provide a thorough resource for all things nitrification that will be of use to students and practitioners, researchers, and teachers. We hope that it will provide the background and state-of-the-art knowledge for the next generation of nitrification researchers, recognizing that the pace of advancement in this field means that another review will most likely be necessary in much less time than has elapsed since Prosser's 1986 volume.

We thank all of our chapter authors for their time and expertise and willingness to share their work with us. We also thank the many other nitrification researchers who reviewed the chapters and provided valuable input and feedback. We look forward to the next exciting decade of nitrification research.

B. B. WARD
D. J. ARP
M. G. KLOTZ

OVERVIEW

I

NITRIFICATION: AN INTRODUCTION AND OVERVIEW OF THE STATE OF THE FIELD

Bess B. Ward

1

NITRIFICATION IN THE NITROGEN CYCLE

Dinitrogen gas (N_2) makes up most of the nitrogen (N) in the atmosphere and in natural waters. Other gaseous forms of N occur either transiently or at trace levels in the environment. Fixed N, the ionic and organic forms, comprises a very small fraction of the total N inventory on earth, but these are the forms that are most important to biogeochemical processes and to the sustenance of life on earth. N is an essential element for life, a major component of proteins and nucleic acids. In addition to its role as a nutrient, N occurs in a range of oxidation states from +5 (nitrate) to −3 (ammonium and amino-nitrogen) and thus serves as an electron donor or acceptor in a variety of microbially mediated transformations. These transformations ensure that the fluxes of nitrogen are large, while the pool sizes are often small compared to biological demand, leading to rapid nitrogen cycling (see Fig. 1 in Chapter 13).

In oxic environments, such as rivers, lakes, and the ocean, nitrate is the stable and most abundant form of fixed N, and it tends to accumulate in aphotic environments where plants and algae are not able to live photosynthetically. In anoxic environments, such as waterlogged soils, subsurface sediments, and wastewater, ammonium usually dominates the N inventory. Nitrification links the most reduced and most oxidized components of the N cycle: the oxidation of ammonium occurs in two steps, first to nitrite and then to nitrate. Where nitrate is supplied to oxygen-depleted environments, conventional denitrification, in which nitrate is used as a respiratory substrate instead of oxygen, catalyzes the return of fixed N to the atmospheric reservoir of N_2. Nitrification is generally an aerobic process, while conventional denitrification mainly occurs in the absence of oxygen, so the two processes are often linked across oxic/anoxic interfaces, such as across the sediment/water interface or the surfaces of aggregates in soil. The linkage between nitrification and denitrification is of interest in agriculture because it leads to loss of fixed N, which is often applied at great expense as fertilizer. On the other hand, excess fixed nitrogen from agriculture that accumulates coastal or inland waters can cause nuisance algal or weed blooms, and denitrification is the process by which this accumulation is limited. Thus, linked nitrification/denitrification controls the N inventory of natural and managed systems.

Bess B. Ward, Department of Geosciences, Princeton University, Princeton, NJ 08544.

Nitrification, Edited by Bess B. Ward, Daniel J. Arp, and Martin G. Klotz

Replenishment of the fixed N pool is accomplished by biological and industrial nitrogen fixation. Industrial N fixation for fertilizer use adds about as much fixed N to terrestrial systems as is fixed by naturally occurring microbes in the ocean or on land; all three fluxes occur on the scale of about 100 Tg year^{-1} (Galloway et al., 2004). Nitrification is not directly responsible for changes in the fixed N inventory, but it is tightly linked to two processes that do contribute to fixed N loss: nitrification can reduce loss of ammonia that might be volatilized from agricultural systems, and it supplies the primary substrate for denitrification, the main biological loss term.

NITRIFYING MICROORGANISMS

Bacterial chemoautotrophy was discovered by Winogradsky (1890) in the course of his study on nitrifying bacteria. While the process of nitrification in soils had been known for some time, Winogradsky isolated both ammonia-oxidizing bacteria (AOB) and nitrite-oxidizing bacteria (NOB) and proved quantitatively that they were autotrophs. Bacterial nitrifiers were assumed to be the only microbes capable of autotrophic nitrification for over a century. Cultivated AOB and NOB provided the basis for investigations into the physiology and biochemistry of nitrification for decades and supported the ecological inferences obtained from field studies. Characteristics of the general physiology of nitrifiers, such as an obligate requirement for oxygen to oxidize ammonia, but tolerance of very low oxygen and sensitivity to inhibition by light, were observed in natural systems and verified in culture. Thus, the most important development in the study of nitrification in a century came as a surprise: ammonia-oxidizing archaea (AOA) are much more abundant than AOB in most of the environments in which they have been investigated since their discovery in 2004 (Venter et al., 2004; Konneke et al., 2005; Schleper et al., 2005; Treusch et al., 2005).

Among the bacteria, the capacity for ammonia and nitrite oxidation is apparently limited to a small number of genera, most of which descended from a photosynthetic proteobacterial ancestor (Teske et al., 1994). With the advent of molecular ecological tools for the investigation of diversity and distribution of microbes, the nitrifiers became the poster children of the approach because their physiology was strongly coherent with their phylogeny (Kowalchuk and Stephen, 2001). PCR amplification techniques led to the discovery of much greater diversity in functional genes, and thus by inference in physiology, of nitrifiers than had been suspected from culture-based research. Most of this work was done on AOB, and even now, the NOB are much less studied in the environment. The focus on the genes (*amoABC*) that encode the first enzyme in the bacterial oxidation of ammonia, ammonia monooxygenase, also led to the discovery of the AOA. A homologue of the bacterial *amoA* was discovered in archaeal scaffolds of metagenomic libraries from the ocean (Venter et al., 2004) and soil (Treusch et al., 2005). The archaeal *amoA* gene was then rapidly reported from a variety of environments (Francis et al., 2005), and a major shift in our understanding of nitrifiers began. Although AOA share the first enzyme in the ammonia-oxidizing pathway, the rest of the pathway to nitrite is still speculative in AOA. Thus, there are many unanswered questions at this writing, including the biochemical pathways involved in archaeal nitrification, the question of whether AOA produce nitrous oxide, the extent to which AOA are autotrophic, and the extent to which AOA contribute to nitrification in natural and managed ecosystems.

While conventional nitrification is obligately aerobic, the thermodynamics of anaerobic ammonia oxidation had long suggested that oxidation of ammonia at the expense of nitrite or nitrate was a viable way for microbes to make a living, and profiles suggestive of this link had been noted (Richards, 1965). The discovery of anaerobic ammonia-oxidizing (anammox) bacteria in 1995 (van de Graaf et al., 1995) was thus both expected and surprising. The stoichiometry of anammox was soon verified as the 1:1 consumption of ammonium and nitrite leading to N_2 gas, and the organisms involved were identified as unusual autotrophs in the *Planctomycetales* (Strous et al., 1999). Unlike aerobic ammonia oxidation, N_2 as the end product of anammox makes this process a form of denitrification (Kartal et al., 2006), leading to the loss of fixed N rather than its oxidation. Over the succeeding decade after its discovery in wastewater treatment systems, anammox was discovered in sediments and seawater environments, where it was shown to be more prevalent and to occur at greater rates than conventional denitrification (for a review, see Dalsgaard et al., 2005). In some sites, no denitrification at all was detected while anammox was almost ubiquitously found. While it makes little difference to the overall inventory of fixed N whether N_2 is produced by conventional denitrification or anammox, there are internal mass balance questions about the supply of ammonium and nitrite for anammox, if these compounds are not supplied by denitrification. Thus, the relative contribution of anammox and denitrification to fixed N loss remains a topic of research and debate.

ADVANCES IN NITRIFICATION IN THE LAST 25 YEARS

The discovery of novel organisms and novel pathways are the most important findings to be documented in the field of nitrification since the publication of the last monograph in 1986. But just as important, and absolutely critical to these discoveries, are the changes in the study and methodology of nitrification. Like all of microbiology, the molecular biology revolution has completely changed both the questions and the answers in the field of nitrification. PCR was first reported by Saiki et al. (1985) only 1 year before the previous monograph on nitrification was published (Prosser, 1986). Its first environmental applications were in amplification of 16S rRNA genes, which immediately opened our eyes to an immense and previously hidden microbial world of diversity (Pace, 1997). The narrow phylogeny of AOB made this group amenable to investigation by 16S rRNA PCR (Head et al., 1993), and interest in nitrifiers and nitrification grew rapidly. The current state of knowledge on the diversity, distribution, and biogeography of AOB, largely derived from 16S rRNA and *amoA* sequence data, are reviewed by Norton (see Chapter 3). NOB are reviewed in the chapter by Daims et al. (see Chapter 12), and AOA are reviewed by Nicol et al. (see Chapter 7).

Beyond PCR for investigation of diversity and biogeography based on single genes, the sequencing and analysis of complete genomes and metagenomes has contributed greatly to our knowledge of the biochemistry of AOB (Arp et al., 2007; Stein et al., 2007; Norton et al., 2008), NOB (Starkenburg et al., 2006, 2008), and anammox (Strous et al., 2006). Both independently and in parallel with these advances in the molecular biology of nitrification, major advances in understanding their biochemistry and regulation have also occurred in the last 25 years. Genomics and metabolism of AOB are reviewed in Chapters 4, 2, and 5, respectively, by Klotz and Stein, Sayavedra-Soto and Arp, and Stein; NOB are reviewed by Starkenburg et al. (see Chapter 11), and anammox is reviewed by Kartal et al. (see Chapter 8). Environmental metagenomics was directly responsible for the discovery of AOA, and only 5 years after the first cultivation of AOA, a complete genome of this first free-living AOA has now been completed (Walker et al., 2010). Major insights about AOA genomics and metabolic capabilities are discussed in the chapter by Urakawa et al. (see Chapter 6).

Microbial ecology has been transformed into molecular ecology, so great has been the impact of molecular biological methods in the study of microbes in natural and managed systems. Ribosomal RNA and functional gene sequence data are now the standard for investigation of microbial diversity, distribution, and activity in the environment. These methods have made it possible to investigate environmental control of nitrification, regulation in response to changing conditions, the discovery of great uncultured diversity, and an understanding of succession and biogeography among functionally similar types. In several chapters of this volume, the ecology of major environments on earth in terms of the role of nitrification are reviewed, including terrestrial ecosystems (see Chapter 14), estuarine and freshwater (see Chapter 15), oceans (see Chapters 9 and 13), and wastewater (see Chapters 10 and 16).

THE FUTURE OF NITRIFICATION RESEARCH

This volume was conceived in 2005, at a time when it was clear that paradigms in nitrification were shifting rapidly. Now, at the time of publication, 5 years later, major unresolved questions remain.

1. Who are the main nitrifiers in the environment? How does community composition vary with environmental conditions, and what are the metabolic capabilities and characteristics of the environmentally important groups? The best answers to these questions now are in the form of sequence data from clone libraries and metagenomes, and these answers make it clear that the major players in the environment are not represented in our culture collections. For example, the phylotypes of AOB most abundantly found in both clone libraries from both terrestrial and aquatic systems are *Nitrosospira* like and are not in culture. The recently cultivated AOA *Nitrosopumilus maritimus* has some attributes of an open ocean organism (high substrate affinity), but it is not found at some sites in the open ocean (N. J. Bouskill and B. B. Ward, unpublished data), and its pH and temperature optima make it an unlikely candidate to represent the abundant AOA documented in deep cold ocean water. Thus, with only a few signature genes to work with, our understanding of the autecology of nitrifiers in the environment is limited. This is an area of current progress, however, and genes in addition to *amoA*, such as *hao* (encoding hydroxylamine oxidoreductase and its homologues in AOB and anammox organisms) and *nxr* (encoding nitrite oxidase in nitrite oxidizers), and others, will facilitate this research.

2. What is the metabolism of AOA? What is the pathway by which ammonia is oxidized to nitrite? Because the AOA do not possess a hydroxylamine oxidoreductase, the enzyme that performs the second step in the ammonia oxidation pathway of AOB, this must be something quite novel (see Chapter 6). Do AOA produce nitrous oxide (as do AOB), and if so, by what pathway and under what environmental conditions?

3. What is the relationship between anammox and denitrification? While not strictly related to conventional nitrification, this question has major implications for N cycling in both terrestrial and aquatic environments and for regulation of the global N inventory of fixed N. Nevertheless, anammox bacteria, although constrained to an anaerobic metabolism, may compete with aerobic AOA and AOB for ammonium in microaerobic environments (Lam et al., 2009), thus forming a pivot in the N cycle between N oxidation and fixed N loss.

4. How will nitrification and the N cycle respond to human-driven N enrichment of the environment? The environmental and economic relevance of nitrification has never been more appreciated than in the nitrogen-enriched modern world (Galloway et al., 2008). Conventional nitrification is essential in ameliorating the process of eutrophication, caused by excess N loading in estuarine and coastal waters. Substrate concentration is often identified as a controlling variable in determining AOB and AOA community composition; increasing N loads in natural waters may affect a change in the resident assemblages. Resilience of natural systems may rely heavily on the redundancy or resistance of nitrifier assemblages for this ecosystem service. Wastewater treatment relies heavily on conventional nitrification, denitrification, and anammox to minimize the impact of human, agricultural, and industrial waste on receiving waters. Europe and Japan are taking the lead on harnessing the metabolic potential of microbes for wastewater treatment (see Chapters 10 and 16), and it may be that only by following their lead to reduce N inputs to the ocean will it be possible to avert major shifts in ocean chemistry (Duce et al., 2008). Conventional nitrification, by producing the substrates for conventional denitrification, and anammox, by combining the two processes in one organism, both play essential roles in regulating the global fixed N inventory. Nitrifier denitrification (see Chapter 5) and conventional denitrification both contribute to the production of nitrous oxide, a potent greenhouse gas, and this flux may be enhanced by modern agricultural practices. How are these removal processes affected by the increased nitrogen load in the environment? While the world is not likely to run out of fixed nitrogen any time soon, the marine research community continues to debate the state of the oceanic N balance. Large errors accompany such estimates, but rates of removal processes (denitrification plus anammox) are usually estimated to exceed input processes (biological nitrogen fixation and terrestrial inputs). Conversely, excess N loading from terrestrial systems has caused major changes in many coastal systems. Intensely cultivated agricultural systems and forests suffer the effects of N saturation and lead to excess loading in inland waterways.

Thus, while much has been learned since the last monograph on the topic, *Nitrification*, was published (Prosser, 1986), it is clear that compelling practical and basic research questions remain. We hope that this book will provide the state of the art in 2010 and the background for future research progress on the same scale that has occurred over the intervening quarter of a century.

REFERENCES

Arp, D. J., P. S. G. Chain, and M. G. Klotz. 2007. The impact of genome analyses on our understanding of ammonia-oxidizing bacteria. *Annu. Rev. Microbiol.* **61:**503–528.

Dalsgaard, T., B. Thamdrup, and D. E. Canfield. 2005. Anaerobic ammonium oxidation (anammox) in the marine environment. *Res. Microbiol.* **156:**457–464.

Duce, R. A., J. LaRoche, K. Altieri, K. R. Arrigo, A. R. Baker, D. G. Capone, S. Cornell, F. Dentener, J. Galloway, R. S. Ganeshram, R. J. Geider, T. Jickells, M. M. Kuypers, R. Langlois, P. S. Liss, S. M. Liu, J. J. Middelburg, C. M. Moore, S. Nickovic, A. Oschlies, T. Pedersen, J. Prospero, R. Schlitzer, S. Seitzinger, L. L. Sorensen, M. Uematsu, O. Ulloa, M. Voss, B. Ward, and L. Zamora. 2008. Impacts of atmospheric anthropogenic nitrogen on the open ocean. *Science* **320:**893–897.

Francis, C. A., K. J. Roberts, M. J. Beman, A. E. Santoro, and B. B. Oakley. 2005. Ubiquity and diversity of ammonia-oxidizing archaea in water columns and sediments of the ocean. *Proc. Natl. Acad. Sci. USA* **102:**14683–14688.

Galloway, J. N., F. J. Dentener, D. G. Capone, E. W. Boyer, R. W. Howarth, S. P. Seitzinger, G. P. Asner, C. C. Cleveland, P. A. Green, E. A. Holland, D. M. Karl, A. F. Michaels, J. H. Porter, A. R. Townsend, and C. J. Vorosmarty. 2004. Nitrogen cycles: past, present, and future. *Biogeochemistry* **70:**153–226.

Galloway, J. N., A. R. Townsend, J. W. Erisman, M. Bekunda, Z. C. Cai, J. R. Freney, L. A. Martinelli, S. P. Seitzinger, and M. A. Sutton. 2008. Transformation of the nitrogen cycle: recent trends, questions, and potential solutions. *Science* **320:**889–892.

Head, I. M., W. D. Hiorns, T. M. Embley, A. J. McCarthy, and J. R. Saunders. 1993. The phylogeny of autotrophic ammonia-oxidizing bacteria as determined by analysis of 16S ribosomal-RNA gene-sequences. *J. Gen. Microbiol.* **139:**1147–1153.

Kartal, B., M. M. Kuypers, G. Lavik, J. Schalk, H. J. M. Op den Camp, M. S. M. Jetten, and M. Strous. 2006. Anammox bacteria disguised as denitrifiers: nitrate reduction to dinitrogen gas via nitrite an ammonium. *Environ. Microbiol.* doi:10.1111/j.1462-2920.2006.01183x.

Konneke, M., A. E. Berhnard, J. R. de la Torre, C. B. Walker, J. B. Waterbury, and D. A. Stahl. 2005. Isolation of an autotrophic ammonia-oxidizing marine archaeon. *Nature* **437:**543–546.

Kowalchuk, G. A., and J. R. Stephen. 2001. Ammonia-oxidizing bacteria: a model for molecular microbial ecology. *Annu. Rev. Microbiol.* **55:**485–529.

Lam, P., G. Lavik, M. M. Jensen, J. van de Vossenberg, M. Schmid, D. Woebken, G. Dimitri, R. Amann, M. S. M. Jetten, and M. M. M. Kuypers. 2009. Revising the nitrogen cycle in the Peruvian oxygen minimum zone. *Proc. Natl. Acad. Sci. USA* **106:**4752–4757.

Norton, J. M., M. G. Klotz, L. Y. Stein, D. J. Arp, P. J. Bottomley, P. S. G. Chain, L. J. Hauser, M. L. Land, F. W. Larimer, M. W. Shin, and S. R. Starkenburg. 2008. Complete genome sequence of *Nitrosospira multiformis*, an ammonia-oxidizing bacterium from the soil environment. *Appl. Environ. Microbiol.* **74:**3559–3572.

Pace, N. R. 1997. A molecular view of microbial diversity and the biosphere. *Science* **276:**734–740.

Prosser, J. I. 1986. *Nitrification.* IRL Press, Oxford, United Kingdom.

Richards, F. A. 1965. Anoxic basins and fjords, p. 611–645. *In* J. P. Riley and G. Skirrow (ed.), *Chemical Oceanography*, vol. 1. Academic Press, London, United Kingdom.

Saiki, R. K., S. Scharf, F. Faloona, K. B. Mullis, G. T. Horn, H. A. Erlich, and N. Arnheim. 1985. Enzymatic amplification of beta-globin genomic sequences and restriction site analysis for diagnosis of sickle-cell anemia. *Science* **230:**1350–1354.

Schleper, C., G. Jurgens, and M. Jonuscheit. 2005. Genomic studies of uncultivated Archaea. *Nat. Rev. Microbiol.* **3**:479–488.

Starkenburg, S. R., P. S. G. Chain, L. A. Sayavedra-Soto, L. Hauser, M. L. Land, F. W. Larimer, S. A. Malfatti, M. G. Klotz, P. J. Bottomley, D. J. Arp, and W. J. Hickey. 2006. Genome sequence of the chemolithoautotrophic nitrite-oxidizing bacterium *Nitrobacter winogradskyi* Nb-255. *Appl. Environ. Microbiol.* **72:**2050–2063.

Starkenburg, S. R., F. W. Larimer, L. Y. Stein, M. G. Klotz, P. S. G. Chain, L. A. Sayavedra-Soto, A. T. Poret-Peterson, M. E. Gentry, D. J. Arp, B. Ward, and P. J. Bottomley. 2008. Complete genome sequence of *Nitrobacter hamburgensis* X14 and comparative genomic analysis of species within the genus *Nitrobacter*. *Appl. Environ. Microbiol.* **74:**2852–2863.

Stein, L. Y., D. J. Arp, P. M. Berube, P. S. G. Chain, L. Hauser, M. S. M. Jetten, M. G. Klotz, F. W. Larimer, J. M. Norton, H. den Camp, M. Shin, and X. M. Wei. 2007. Whole-genome analysis of the ammonia-oxidizing bacterium, *Nitrosomonas eutropha* C91: implications for niche adaptation. *Environ. Microbiol.* **9:**2993–3007.

Strous, M., J. A. Fuerst, E. H. M. Kramer, S. Logemann, G. Muyzer, K. T. van de Pas-Schoonen, R. Webb, J. G. Kuenen, and M. S. M. Jetten. 1999. Missing lithotroph identified as new planctomycete. *Nature* **400:**446–449.

Strous, M., E. Pelletier, S. Mangenot, T. Rattei, A. Lehner, M. W. Taylor, M. Horn, H. Daims, D. Bartol-Mavel, P. Wincker, V. Barbe, N. Fonknechten, D. Vallenet, B. Segurens, C. Schenowitz-Truong, C. Medigue, A. Collingro, B. Snel, B. E. Dutilh, H. J. M. Op den Camp, C. van der Drift, I. Cirpus, K. T. van de Pas-Schoonen, H. R. Harhangi, L. van Niftrik, M. Schmid, J. Keltjens, J. van de Vossenberg, B. Kartal, H. Meier, D. Frishman, M. A. Huynen, H. W. Mewes, J. Weissenbach, M. S. M. Jetten, M. Wagner, and D. Le Paslier. 2006. Deciphering the evolution and metabolism of an anammox bacterium from a community genome. *Nature* **440:**790–794.

Teske, A., E. Alm, J. M. Regan, S. Toze, B. E. Rittmann, and D. A. Stahl. 1994. Evolutionary relationships among ammonia- and nitrite-oxidizing bacteria. *J. Bacteriol.* **176:**6623–6630.

Treusch, A. H., S. Leininger, A. Kletzin, S. C. Schuster, H. P. Klenk, and C. Schleper. 2005. Novel genes for nitrite reductase and Amo-related proteins indicate a role of uncultivated mesophilic crenarchaeota in nitrogen cycling. *Environ. Microbiol.* **7:**1985–1995.

van de Graaf, A. A., A. Mulder, P. Debruijn, M. S. M. Jetten, L. A. Robertson, and J. G. Kuenen. 1995. Anaerobic oxidation of ammonium is a biologically mediated process. *Appl. Environ. Microbiol.* **61:**1246–1251.

Venter, C. J., K. Remington, J. G. Heidelberg, A. L. Halpern, D. Rusch, J. A. Eisen, D. Wu, I. Paulsen, K. E. Nelson, W. Nelson, D. E. Fouts, S. Levy, A. H. Knap, M. W. Lomas, K. Nealson, O. White, J. Peterson, J. Hoffman, R. Parsons, H. Baden-Tillson, C. Pfannkoch, J.-H. Rogers, and H. O. Smith. 2004. Environmental genome shotgun sequencing of the Sargasso Sea. *Science* **304:**66–74.

Walker, C. B., J. R. de la Torre, M. G. Klotz, H. Urakawa, N. Pinel, D. J. Arp, C. Brochier-Armanet, P. S. G. Chain, P. P. Chan, A. Gollabgir, J. Hemp, M. Hügler, E. A. Karr, M. Könneke, M. Shin, T. J. Lawton, T. Lowe, W. Martens-Habbena, L. A. Sayavedra-Soto, D. Lang, S. M. Sievert, A. C. Rosenzweig, G. Manning, and D. A. Stahl. 2010. *Nitrosopumilis maritimus* genome reveals unique mechanisms for nitrification and autotrophy in globally distributed marine crenarchaea. *Proc. Natl. Acad. Sci. USA* **107:**8818–8823.

Winogradsky, S. 1890. Recherches sur les organismes de la nitrification. *Ann. Inst. Pasteur* **4:**213–231, 258–275, 760–771.

AMMONIA-OXIDIZING BACTERIA

II

AMMONIA-OXIDIZING BACTERIA: THEIR BIOCHEMISTRY AND MOLECULAR BIOLOGY

Luis A. Sayavedra-Soto and Daniel J. Arp

2

INTRODUCTION

Ammonia and Ammonia-Oxidizing Bacteria

Ammonia (NH_3) is an important molecule in the biogeochemical nitrogen (N) cycle (see Chapter 1) (Mancinelli and McKay, 1988). Ammonia is produced and consumed in diverse ecosystems predominantly by microorganisms. Ammonia is released into the environment mainly from the decay of organic matter or from the use of NH_3-based fertilizers in agriculture and serves as an N supply to plants and microorganisms. Ammonia-oxidizing bacteria (AOB) (Arp and Stein, 2003), ammonia-oxidizing archaea (AOA) (Francis et al., 2007), and anaerobic ammonia-oxidizing (anammox) bacteria (Jetten et al., 2005) can derive energy for growth from the oxidation of NH_3. This chapter covers our current understanding of the biochemical and genetic underpinnings relevant to ammonia oxidation by aerobic bacteria. The AOA and anammox bacteria are covered in Sections III and IV, respectively.

AOB are predominantly chemolithoautotrophs (i.e., use NH_3 for energy and reductant and CO_2 as their carbon source). Some are known to have limited heterotrophic capability (i.e., can take up and assimilate simple organic compounds [Arp and Bottomley, 2006]) although, under oxic conditions, none are known to use organic compounds as a sole energy source. Given the thermodynamically low energy yield ($\Delta G^{0\prime} = -271$ kJ mol^{-1}) produced in the oxidation of NH_3 (Wood, 1986), the obligate dependence of all AOB on NH_3 for growth is enigmatic (see below). Nonetheless, AOB are able to derive sufficient energy from the oxidation of NH_3 to perform all necessary metabolic processes, including assimilation of CO_2 (Hooper et al., 1997; Arp and Bottomley, 2006).

Much of what is known about the molecular biology, physiology, and biochemistry of AOB was derived from studies on *Nitrosomonas europaea* (Fig. 1). This bacterium has the advantages of growing relatively rapid (7 to 8 h doubling times) for an AOB and being able to tolerate high concentrations of ammonium (up to 100 mM) and nitrite (which can accumulate up to 25 mM in batch cultures). This AOB can be grown in batch cultures, chemostats, and retentostats and as individual colonies on agar plates. *N. europaea* was also used to construct the first AOB mutants. An AOB with similar properties, *Nitrosomonas* sp. strain ENI-11, has also been used in a number

Luis A. Sayavedra-Soto and Daniel J. Arp, Department of Botany and Plant Pathology, Oregon State University, Corvallis, OR 97331.

Nitrification, Edited by Bess B. Ward, Daniel J. Arp, and Martin G. Klotz

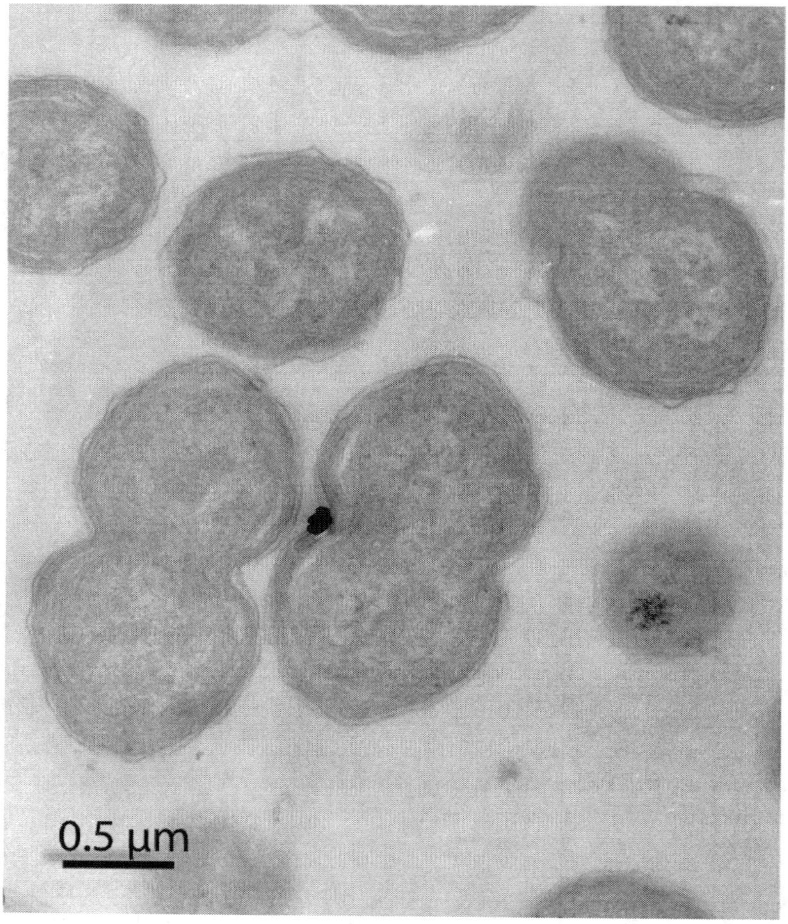

FIGURE 1 Electron microscopy picture of thin sections of cells of *N. europaea,* some of which are dividing. Note the ICM in the periphery of the cells.

of studies (Yamagata et al., 2000; Hirota et al., 2006). Studies of other AOB, including *Nitrosococcus oceani*, *Nitrosomonas eutropha*, and *Nitrosospira* sp. strain NpAV, have also added substantially to our understanding of the AOB. Although *N. europaea* is the most widely used model AOB, *N. europaea* is not widely distributed in all the environments in which AOB flourish (see Chapter 3). *N. europaea* is a common inhabitant of sediments and wastewater treatment communities where ammonia may be present in relatively high concentrations, but it is not typical of soil or marine environments. Therefore, it is important to continue to study AOB isolates representative of other habitats.

Bioinformatics and Ammonia-Oxidizing Bacteria

When the first nitrification treatise was published in 1986 (Prosser, 1986), no genes from any AOB that were associated with ammonia metabolism had been sequenced. Today, we have complete genomes from AOB representative of several ecotypes. The genomes from *Nitrosomonadaceae* in the Betaproteobacteria (e.g., *N. europaea*, *N. eutropha*, and *Nitrosospira multiformis*) and *Chromatiaceae* in the Gammaproteobacteria (e.g., *N. oceani*) provide the basis for a more complete understanding of the metabolic and cellular functions performed by AOB (Arp et al., 2007).

AOB have relatively small genomes (average, 3 Mb), a characteristic often associated with microorganisms found in specialized niches (i.e., petroleum-degrading bacteria, obligate methanotrophs, and microorganisms living in extreme environments). The genomes of AOB reveal genes for the biosynthesis of all the necessary cellular constituents from inorganic nutrients (Arp et al., 2007). The genomes also revealed the scant number of genes encoding enzymes in pathways for the degradation of organic compounds in AOB. For example, genes encoding enzymes in the degradation pathways of most amino acids, carbohydrates, phospholipids, and purines are not present in the known genomes of AOB. Similarly, systems for the uptake of organic molecules are few in AOB.

The genomes are beginning to provide insights to niche differentiation among the four major AOB ecotypes proposed, namely freshwater sediments, sewage/wastewater, soils, and marine (see Chapter 3). In freshwater sediments, facultative aerobes compete with AOBs for O_2 (Koops and Pommerening-Roser, 2001; Kowalchuk and Stephen, 2001). In this environment O_2 is often in low supply, but *Nitrosomonadaceae* (AOB commonly found in this environment) have the necessary genetic composition to coexist. For example, *N. eutropha* has the genes for a *cbb*$_3$-type terminal oxidase usually implicated in microaerophilic respiration (Stein et al., 2007; Norton et al., 2008). In the ecosystems where *Nitrosomonadaceae* are found soluble, iron can be present at extremely low concentrations (10^{-18} M at pH 7), and the sequenced AOB genomes have genes to acquire iron efficiently. *N. multiformis*, an AOB commonly found in soils, has genes for urea catabolism that may give *Nitrosospira* a competitive edge (Norton et al., 2008). The capacity to use urea for growth may be an evolutionary niche adaptation for acid soils (Norton et al., 2008). In addition to the high salt concentrations in marine environments, ammonium is consistently found in low concentrations. Marine AOB in the genus *Nitrosococcus* have genes to express multiple proton- and sodium-dependent ATPase and NDH-1 complexes that may help the cells to derive benefit from their environment (Klotz et al., 2006).

The genomes of AOB also show that all encode four specialized proteins to perform the oxidation of NH_3: ammonia monooxygenase (AMO), hydroxylamine oxidoreductase (HAO), and cytochromes c_{554} (cyt c_{554}) and c_{m552} (cyt c_{m552}). The genes encoding these proteins (i.e., *amo*, *hao*, *cycA*, and *cycB*, respectively) are present in nearly identical multiple copies, albeit in different number and chromosome location, in the sequenced Betaproteobacteria AOB genomes but singly in the sequenced Gammaproteobacterium (nitrosococcus) genome (Arp et al., 2007). Conserved open reading frames (ORFs) flanking the gene clusters encoding AMO are present among the different AOB, but the roles of these unknown ORFs in ammonia oxidation have not been clearly established.

Overall, the composition of the sequenced AOB genomes is consistent with their highly specialized chemolithotrophic growth (Arp et al., 2007). Nonetheless, some unexpected aspects of AOB became evident from sequencing efforts. For example, *N. europaea* has a gene inventory consistent with the ability to completely oxidize some organic compounds (see below) (Chain et al., 2003; Hommes et al., 2003). The genome sequences also showed that among the known AOB genomes, only *N. multiformis* has genes to encode a NiFe hydrogenase, which raises the possibility that this AOB can derive energy from H_2 in addition to ammonia (Norton et al., 2008). Hydrogen might alternate with ammonia as the sole source of reductant, or it might supplement the energy budget while cells are oxidizing ammonia (Norton et al., 2008). The reported growth of *N. europaea* and *N. eutropha* using H_2 in oxygen-limited and ammonia-free conditions (Bock, 1995) remains enigmatic, as there are no genes with similarity to characterized hydrogenase genes in their genomes (Stein et al., 2007).

AMMONIA AS AN ENERGY SOURCE

Conversion of Ammonia to Nitrite

Under oxic conditions, AOB derive all the energy (reductant) required for their metabolism from the oxidation of NH_3 to nitrite (NO_2^-) in a two-step process (Fig. 2). AOB first use the membrane-bound enzyme AMO to catalyze the oxidation of NH_3 to hydroxylamine (NH_2OH) and then, in the periplasmic space, use HAO to catalyze the oxidation of NH_2OH to NO_2^-. The oxidation of NH_3 to NH_2OH requires O_2, two protons, and two electrons: one O is inserted into NH_3 to form NH_2OH, and the other O is combined with the two protons and two electrons to form H_2O (Wood, 1986; Hooper et al., 1997; Poughon et al., 2001).

In the oxidation of NH_2OH to NO_2^-, four electrons are released and channeled through the tetraheme cytochrome c_{554} located in the periplasm, and then likely through a second membrane-bound cytochrome, tetraheme cytochrome c_{m552}, to the ubiquinone pool (see Fig. 3 and below). The membrane-bound cytochrome c_{m552} serves as a quinone reductase (Hooper, 1989). Electrons are then partitioned at the level of the periplasmic ubiquinone pool; two electrons go to support further ammonia oxidation by AMO, and two electrons pass through the electron transport chain to generate a proton gradient for ATP generation and to provide reductant for other cellular processes (i.e., the assimilation of inorganic nutrients). This model is reinforced by the observation that tetra- and trimethylhydroquinols support ammonia oxidation in vitro (Shears and Wood, 1986). The electron transport chain of *N. europaea* has the same major electron transfer complexes as the electron transport chain of mitochondria. However, there are some significant differences in the flow of electrons through these complexes (Wood, 1986; Whittaker et al., 2000; Poughon et al., 2001). Most importantly, electrons released from NH_2OH oxidation are not expected to have a forward flow (i.e., toward more positive reduction potential) through Complex I (NADH oxidoreductase). Electrons from the oxidation of NH_2OH enter the electron transport chain at about +127 mV, much too positive for the direct reduction of $NAD(P)^+$ to NAD(P)H ($E°' = -320$ mV). Inhibitors of electron transfer through cytochrome$_{bc1}$ block ammonia utilization (Arp and Stein, 2003), indicating that in AOB electrons derived from ammonia oxidation flow through this cytochrome complex (Suzuki and Kwok, 1970). Generation of NAD(P)H requires transfer of electrons "uphill" (i.e., to a more negative reduction potential) from the potential at which they are generated and is therefore referred to as reverse electron flow.

The overall oxidation of NH_4^+ to NO_2^- with O_2 as the terminal electron acceptor results in the release of two protons ($NH_4^+ + 1.5\ O_2 \rightarrow NO_2^- + H_2O + 2H^+$). Therefore, ammonia oxidation can cause the acidification of the growth medium or environment and thereby shift the NH_3/NH_4^+ equilibrium toward NH_4^+ (pK_a of 9.25 at 25°C). Because AMO uses NH_3, not NH_4^+, lowering the pH lowers the concentration of NH_3 that is available for growth (Suzuki et al., 1974). To illustrate, for *N. europaea* the K_s for NH_4^+ is about 1.3 mM (Keener and Arp, 1993), corresponding to an available NH_3 concentration of about 46 μM at pH 7.7. However, it is known that some AOB are able to grow in environments where the bulk pH is relatively acidic. Often, the AOB take advantage of microenvironments where the pH is higher and more

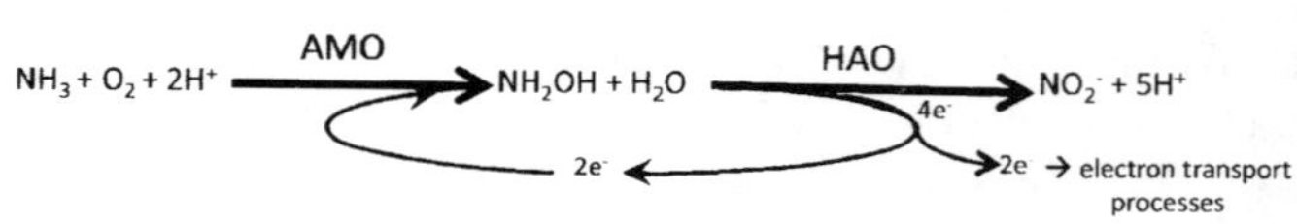

FIGURE 2 Catabolism of ammonia: proteins involved, product and flow of electrons.

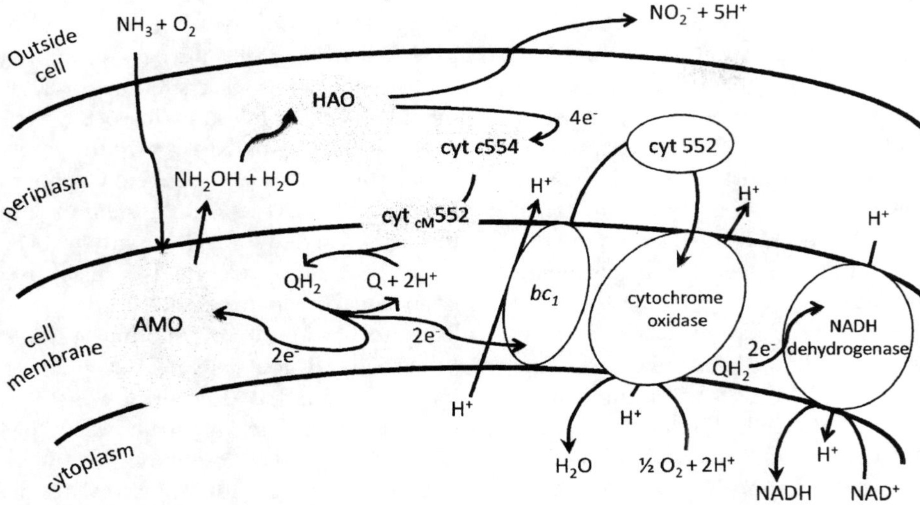

FIGURE 3 Model for the oxidation of ammonia and the proteins involved. *bc*$_1$, complex III; QH_2, quinol. (Adapted from Arp and Stein [2003] and Hooper et al. [1997] with permission.)

of the available N is in the form of NH_3, as when AOB aggregate or form a biofilm (De Boer et al., 1991; Gieseke et al., 2005). There is also evidence that some ammonia oxidizers can remain highly active at low pH (Tarre and Green, 2004); this finding suggests an active transport mechanism must be facilitating the uptake of NH_3 (Weidinger et al., 2007) (see below).

To understand ammonia catabolism by AOB, it is necessary to appreciate the bioenergetic challenge these bacteria face. First, the overall energy yield from the aerobic oxidation of NH_3 to NO_2^- is $\Delta G°' = -271$ kJ mol^{-1} (Wood, 1986), a modest energy yield given the change in the formal oxidation state of the N from -3 to $+3$. In comparison, the aerobic oxidation of a mole of carbon in glucose to CO_2, where the formal oxidation state of the carbon changes from 0 to +4, yields 480 kJ mol^{-1}. Second, the pathway of ammonia catabolism places limits on the steps where energy released in redox reactions is captured in an electrochemical potential gradient. The oxidation of NH_3 to NH_2OH occurs at a calculated midpoint potential of +800 to +900 mV (Wood, 1986; Poughon et al., 2001) and does not produce but rather consumes reductant. The oxidation of NH_2OH to NO_2^- occurs at a midpoint potential of +127 mV, providing some energy as electrons flow through the electron transport chain to O_2 ($E°' = +820$ mV for O_2/H_2O couple). But the amount of energy is much less than in systems where NADH serves as electron donor ($E°' = 320$ mV) to the electron transport chain (e.g., mitochondria). Recall that of the electrons released in the oxidation of NH_2OH to NO_2^-, half are used to sustain further ammonia oxidation, leaving the other half to fill the remaining reductant needs of the cell, including biosynthesis and generation of a proton motive force (Fig. 2). Reductant available for the initial oxidation of NH_3 must pass through the NO_2^-/NH_2OH couple ($E°' = +120$ mV). Therefore, NADH is unlikely to serve as a source of reductant for ammonia oxidation in AOB given the low midpoint potential of the NAD^+/NADH couple. The more likely source of reductant is the ubiquinone pool where midpoint potentials of the ubiquinone/ubiquinol couple are +50 to +100 mV.

The major product of ammonia oxidation, NO_2^- has been shown to have a variety of effects on AOB. For example, NO_2^- inactivates AMO activity in an unknown mechanism where NH_3 protects it from inactivation (Stein and Arp, 1998b). Interestingly, after starvation periods, NO_2^- can stimulate ammonia oxidation in AOB (Laanbroek et al., 2002). Minor products produced during the oxidation of NH_3 by AOB include trace amounts of nitrous oxide, nitric oxide, and N (Arp and Stein, 2003). Production of these trace gases is discussed in the chapter by Stein (see Chapter 5).

Some AOBs produce extensive intracytoplasmic membranes (ICM). For example, *Nitrosomonas* has ICM in stacks running along the periphery of the cells (Fig. 1). *Nitrosococcus* has ICM flattened and centrally located in the cell. In the pleomorphic lobes of *Nitrosolobus*, ICM are divided into cell compartments by the cytomembrane (Watson et al., 1971). Although the exact function of ICM in AOBs is unknown, ICM would increase the surface area available for the enzymes required for the metabolism of NH_3. Electron microscopy studies also show high concentrations of AMO protein associated with the ICM in these bacteria (Fiencke and Bock, 2006). However, *Nitrosospira* and *Nitrosovibrio* have no ICM though their cell shape has high surface area that might compensate for not having ICM.

AMO

COMPOSITION, STRUCTURE, AND METAL CONTENT

AMO is an integral membrane enzyme catalyzing the oxidation NH_3 to NH_2OH that has not yet been purified to homogeneity with activity. Therefore, many details of the structure and catalytic mechanism of this enzyme remain to be elucidated. Much has been deduced to date from a variety of experimental approaches, and by comparison to particulate methane monooxygenase (pMMO), which is structurally and catalytically similar and is evolutionarily related to AMO. Studies of the structure of pMMO are currently more advanced than those of AMO.

AMO consists of three subunits: AmoA or α (27 kDa), AmoB or β (38 kDa), and AmoC or γ (31.4 kDa). The primary protein amino acid sequences of each subunit reveal several membrane-spanning α-helices. Studies with the mechanism-based inactivator acetylene, which binds AmoA (see below), led to the suggestion that the AmoA subunit contains the active site. By analogy with pMMO, for which a crystal structure is available (Hakemian and Rosenzweig, 2007), the enzyme is likely to have an $\alpha_3\beta_3\gamma_3$ subunit composition. Inhibitor and activity studies support a role for Cu in catalysis (Ensign et al., 1993), and Cu was present in the structure of pMMO (Hakemian and Rosenzweig, 2007). The requirement of Cu for in vivo AMO activity was shown through Cu-binding compounds (e.g., allylthiourea, xanthanes, carbon disufide, α,α'-dipyridyl, or cyanide) or in vitro, by recovering AMO activity temporarily in cell extracts upon the addition of Cu (Ensign et al., 1993). A role for Fe has also been suggested in both AMO and pMMO. Although Fe was not found in the crystal structure for pMMO, the preparations were also inactive. Recent work with pMMO has identified a role for Fe in a di-iron center similar to that found in soluble methane monooxygenase (Martinho et al., 2007). A similar role for Fe in AMO is yet to be determined but seems likely given the similarities between AMO and pMMO.

Other indirect evidence of a role for Cu in AMO is that when *N. europaea* cells are exposed to intense light they rapidly lose AMO activity. Shears and Wood (1985) proposed a catalytic cycle for AMO in which O_2 is reduced at a binuclear copper site on the enzyme. The photosensitive state of an oxygenated AMO then would be similar to the photosensitive state of the copper enzyme tyrosinase. Determination of the exact composition of copper in AMO will require purification of the enzyme to homogeneity with activity.

AMO readily loses activity upon cell breakage (Suzuki et al., 1981; Ensign et al., 1993). Cell extracts with activity were obtained after the addition of animal serum albumins, spermine, or Mg^{2+} as stabilizing agents (Ensign et al., 1993; Juliette et al., 1995). However, activity was readily lost in these preparations after short periods of storage (hours). The addition of fractions containing HAO or soluble cytochromes was also found to partially restore the activity of AMO (Suzuki and Kwok, 1981; Suzuki et al., 1981). The stimulation of AMO activity in vitro by the addition of Cu suggested that the loss of copper upon cell lysis might be at least part of the cause of the lack of activity in cell extracts (Ensign et al., 1993). Cell breakage might disrupt the coupling integrity between the enzyme and the accessory proteins for electron transfer downstream (Ensign et al., 1993). The accumulation of free fatty acids in cell extracts also caused loss of AMO activity in the preparations during storage. The addition of bovine serum albumin (BSA) during cell breakage had a stabilizing effect on AMO activity attributed to the inhibition of lipolysis (Juliette et al., 1995). Although BSA is not essential for activity, it stabilized AMO activity in cell-free preparations. Correlation of palmitoleic acid generation in cell extracts and loss of activity was demonstrated (Juliette et al., 1995). It is thought that the stabilizing effect of BSA is more due to the inhibition of lypolysis than the interaction with the free fatty acids that are released upon cell breakage. The increase in activity in vitro upon addition of exogenous copper suggests that a pool of copper-deficient AMO is produced upon cell lysis (Ensign et al., 1993). Divalent metals such as zinc, nickel, or iron cannot substitute for copper in restoring activity; rather, they compete with copper in restoring activity. Interestingly, divalent copper and divalent mercury also might help to maintain activity because they are known inhibitors of lipolysis in cell extracts in other systems (Juliette et al., 1995).

A soluble form of AMO was recently isolated as an alpha-beta-gamma trimer (molecular mass, 238 kDa) where the gamma subunit was not AmoC but rather heme cytochrome c_1 (Gilch et al., 2009a). This soluble form contained Cu, Fe, and tentatively Zn (Gilch et al., 2009a). The soluble form of AMO, like the particulate form, lost activity readily upon cell disruption. In intact cells, the soluble form could be labeled with radioactive acetylene, suggesting that it was catalytically viable (Gilch et al., 2009a). The role of this soluble AMO in the catabolism of NH_3 is not yet known.

In contrast to AMO, pMMO has been isolated with activity, although the success in the preparation of active cell extract of pMMO has been mixed and reported specific activities are lower than needed to support the activities observed in intact cells (Hakemian and Rosenzweig, 2007). Still, the work on pMMO has helped to guide our understanding of AMO. pMMO and AMO are similar in putative subunit composition, catalytic properties, metal content, and the sequence of nucleotides of the encoding genes (see below). The active preparations suggest that pMMO is formed of three subunits with approximate molecular masses of 47, 27, and 25 kDa. In some instances, the isolation of pMMO resulted in an enzyme formed of only the two larger subunits; however, these preparations had no enzymatic activity. It is accepted that pMMO contains Cu, but the stoichiometry varies with the preparations from 4 to 59 copper ions per 100 kDa. Electron paramagnetic resonance spectra confirm the presence of redox-active copper atoms in pMMO (Zahn et al., 1996). Iron has also been found in some preparations of membrane-bound and purified pMMO. The stoichiometry ranged from 0.5 to 2.5 Fe atoms per 100 kDa of purified *Methylococcus capsulatus* pMMO (Lieberman and Rosenzweig, 2005; Hakemian and Rosenzweig, 2007). Recent results provide evidence of a di-iron center in pMMO that seems correlated with activity (Martinho et al., 2007).

Although there is not as yet evidence of the binding sites for the metal cofactors in AMO, the analysis of the pMMO crystal structure from *M. capsulatus* (Bath) has yielded some interesting possibilities (Hakemian and Rosenzweig, 2007). For example, an encoded motif that includes four His and one Gln residues in AmoB and an encoded motif that includes one Glu, one Asp, and two His in AmoA could be the binding site of the Cu atom(s).

In addition, putative metal-binding motifs can be inferred in AmoB (a dinuclear copper center and a mononuclear copper center) and in AmoC and AmoA (a mononuclear metal center; Zn during crystallization; perhaps Cu or Fe in vivo), suggesting that, as for pMMO, Cu and probably Fe are necessary for catalysis (Hakemian and Rosenzweig, 2007). Among the many AOB for which the nucleotides of genes for AMO have been sequenced, the putative metal-binding amino acids are highly conserved.

SUBSTRATES AND INHIBITORS OF AMO

As with many monooxygenases, AMO has broad substrate specificity (Fig. 4). AMO can catalyze the oxidation of diverse alkanes, alkenes, aromatic hydrocarbons, and ethers, all by inserting one O atom into the molecules (Hoffman and Lee, 1953; Hyman and Wood, 1983; Vanelli and Hooper, 1995; Keener and Arp, 1994). AMO can also catalyze dehydrogenation of ethylbenzene and reductive dehalogenation of organic compounds such as nitrapyrin (Arp and Stein, 2003). The broad substrate specificity also extends to many chlorinated hydrocarbons, including vinyl chloride, trichloroethylene, chloroform, and chlorobenzene. A common characteristic of all AMO substrates is that they are mostly uncharged and of low polarity, which suggests a hydrophobic substrate-binding active site (Arp and Stein, 2003).

Studies of inhibitors of ammonia oxidation activity have provided many insights to the mechanism of ammonia oxidation and the possible pathways for electron transfer. The inhibition of AMO activity can be competitive, noncompetitive, or mechanism based. Competitive inhibitors include methane, ethylene, and carbon monoxide (Hooper and Terry, 1973; Keener and Arp, 1993). Noncompetitive inhibitors include ethane, chloroethane, and thiourea. Among the known

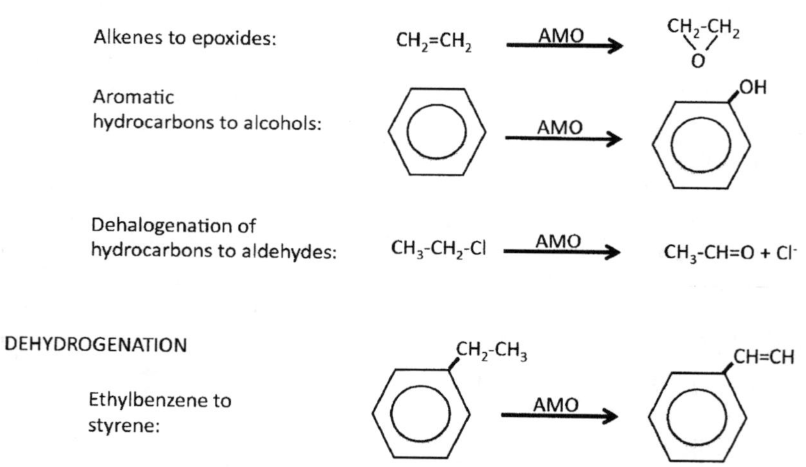

FIG. 4 Reactions catalyzed by AMO are broad in substrate specificity and include oxidation and dehydrogenation (Hoffman and Lee, 1953; Hyman and Wood, 1983; Vanelli and Hooper, 1995; Keener and Arp, 1994).

inhibitors of AMO activity, the natural nitrification byproduct, NO_2^-, inhibits ammonia oxidation in the presence of O_2 by an as yet unknown mechanism (Stein and Arp, 1998b). Interestingly, NH_3 itself and short alkanes can protect AMO from NO_2^- inhibition.

Nitrapyrin is an inhibitor of ammonia oxidation that is marketed under the trade name of Nserve. This inhibitor is used in some croplands to slow the conversion of ammonia-based fertilizers to nitrate, thereby reducing the losses of these fertilizers due to leaching and denitrification. As mentioned above, nitrapyrin is a substrate for AMO and undergoes the unusual reaction of reductive elimination (Vannelli and Hooper, 1992).

Diphenyliodonium (DPI), a well-characterized flavoprotein inhibitor, was used to investigate the electron transfer pathway to pMMO and AMO (Shiemke et al., 2004). At low concentrations (K_i, 5 μM), DPI interferes with electron flow from NADH to pMMO in methanotrophs by inactivating a type-2 NADH:quinone oxidoreductase that mediates electron flow from NADH to the quinone pool. At higher concentrations (K_i, 100 μM), DPI inhibits pMMO and AMO activities directly, apparently by blocking electron flow from the ubiquinone pool to the monooxygenase. Consistent with this mechanism, genes encoding type-2 NADH:quinone oxidoreductase were not identified in *N. europaea*, *N. eutropha*, *N. multiformis*, or *N. oceani*. Cosubstrates did not protect the enzyme from the inhibition by DPI (Shiemke et al., 2004). DPI did not affect the electron transfer pathway from HAO to the terminal oxidase.

Among the mechanism-based inhibitors of AMO, acetylene (C_2H_2) has been most useful. When active preparations of AMO are incubated in the presence of $^{14}C_2H_2$, the 27-kDa AmoA subunit is labeled, demonstrating that this subunit contains the catalytic site (Hyman and Arp, 1992). The residue His-191 in AmoA of *N. europaea* was modified by acetylene, suggesting that this residue is part of or in close association with the acetylene-binding site (Gilch et al., 2009b). Other mechanism-based inactivators include longer alkynes (up to octyne) and allylsulfide (Hyman et al., 1988; Juliette et al., 1993).

MOLECULAR BIOLOGY

AMO is encoded from a gene cluster (*amoCAB*) present in one to three copies in the genomes of different AOB (Arp et al., 2007). Alignment of the encoded amino acids for AMO against the NCBI database results in significant matches only to AMO of other AOB (>85% identities) and to pMMO of the methanotrophs [i.e., *M. capsulatus* (Bath), or *Methylosinus trichosporium* with >80% identities, and with most of the divergence occurring at the N terminus] (Hakemian and Rosenzweig, 2007). In *N. europaea*, there are two gene copies of *amoCAB* with their DNA sequences differing in *amoA* by a single nucleotide that results in only one amino acid change (Hommes et al., 1998). Similarly, in the genomes of other AOB that have multiple AMO gene copies, all copies are almost identical within an organism. The only Gamma-AOB examined, *N. oceani*, contains a single copy of the AMO operon (Alzerreca et al., 1999; Klotz et al., 2006). The genes encoding pMMO (*pmoCAB*) in methanotrophs occur in the same order as their homologues in AOB (*amoCAB*) as part of an operon (Hakemian and Rosenzweig, 2007).

Tandem promoter nucleotide sequences similar to sigma 70-type *Escherichia coli* promoters are associated with the AMO operon of *N. europaea* (Hommes et al., 2001). These promoters are differentially expressed upon exposure to a new supply of NH_3, suggesting specific copy expression for different growth conditions (Hommes et al., 2001). The role of the multiple promoters associated with *amoC* (the first gene in the AMO operon) or the reason for the multiple copies of the *amo* operon (Sayavedra-Soto et al., 1998; Hommes et al., 2001; Berube et al., 2007) are not apparent. There is another possible sigma 70-type promoter in front of *amoA* in *N. europaea*, whose function is unknown (Hommes et al., 2001). This *amoA* promoter is also present in *Nitrosospira* sp. strain NpAV and can drive the expression of *amoA* in

E. coli, suggesting that it is a viable promoter, at least in these two nitrifiers (Klotz and Norton, 1995). Interestingly, the putative promoters of *amo*, *hao*, and *cycA* do not share common elements (Hommes et al., 2001) (see below).

Most of the sequenced AOB genomes exhibit an extra copy of *amoC*. This version of the gene is not found in association with *amoAB* and is slightly dissimilar to the other two versions. For example, *amoC3* in *N. europaea* has 67.5% identity, 81.4% similarity to *amoC1* or *amoC2* (Sayavedra-Soto et al., 1998). The function of *amoC3* is not known but may have a role in recovery from starvation. The transcript level of *amoC3* was raised during recovery after long periods of starvation (Berube et al., 2007). However, a deletion mutant of *amoC3* in *N. europaea* did not show a different phenotype from the wild-type strain either while growing or during recovery from substrate deprivation experiments (Berube et al., 2007).

In *N. europaea* and *Nitrosospira* sp. strain NpAV, the genes for AMO are transcribed as a 3.5-kb mRNA forming a polycistronic transcript (Sayavedra-Soto et al., 1998). Transcript analysis by northern hybridizations revealed three mRNAs in these two nitrifiers, one derived from the whole operon *amoCAB* and two others derived from *amoAB* and *amoC*, perhaps as a result of processing of the full mRNA. Of the three fragments, *amoC* is the most stable, probably as a result of stem-loop structures that can be predicted by computer modeling (Sayavedra-Soto et al., 1998). The stability of the mRNA of *amoC* is similar in Gamma- and Beta-AOB. The role for the *amoC* mRNA stability and $AmoC_{1,2}$ function remains to be determined.

In *N. europaea*, both copies of AMO are functional and each is sufficient for growth, but there is evidence for differential regulation of the two copies (Hommes et al., 1998; Stein et al., 2000). In mutational studies, inactivation of one of the copies (*amoA1*) slowed growth by about 25%, while inactivation of the other (*amoA2*) did not have a negative effect on growth. If the mutant cells were exposed to fresh medium, the *N. europaea* strain lacking the *amoA1* or *amoB1* copies responded more slowly than the mutant strain lacking the *amoA2* or *amoB2* copies (Stein et al., 2000). The mutants also synthesized a smaller amount of AMO polypeptides and recovered slightly more slowly after AMO inactivation than the wild type (Stein et al., 2000). The mechanism for the differences in growth of cells lacking one or the other copy of *amoAB* is not apparent, as the DNA sequence of the putative AMO promoters are identical in *N. europaea*. In the single *amo* operon of *N. oceani*, three promoters were identified and were differentially expressed depending on the available ammonia. This interesting observation suggests a different regulation mechanism for AMO expression in Gamma-AOB from that in Beta-AOB (El Sheikh and Klotz, 2008). In *N. oceani*, the gene *amoR* was identified in front of *amoCAB* and was expressed in cells that were exposed to ammonia. In addition, an adjacent downstream ORF named *amoD* was cotranscribed in the same mRNA (*amoCABD*) at higher concentrations of ammonium (5 mM) (El Sheikh et al., 2008). The transcription of the *amo* gene cluster in *N. oceani* showed more resemblance to the transcription of *M. capsulatus* (a methanotroph) than to that of the *amo* gene cluster in *N. europaea* (a Beta-AOB).

In *N. europaea*, the activity of AMO is regulated by ammonia at three levels: transcriptional, translational, and posttranslational (Hyman and Arp, 1992; Sayavedra-Soto et al., 1996; Stein et al., 1997; Geets et al., 2006). Different regulatory mechanisms might be at play depending on the environmental conditions. Among prokaryotes, the regulation of mRNA degradation can help cells to respond to starvation and to recover readily when new substrate becomes available. Stable mRNAs would mean that energy would not be utilized to synthesize new RNA pools to produce key enzymes. *N. europaea* has mechanisms at the transcriptional and translational level to cope with lack of O_2 and ammonia (Geets et al., 2006). In the stored cells, the potential AMO activity decreased by 85% within 24 h; however, the HAO potential activity remained unaffected (Stein and Arp,

1998a). In *N. europaea* cells kept suspended in ammonia-free medium at low cell density, *amo* mRNAs were detected for up to 4 days (Berube et al., 2007). This suggests that the mRNAs levels could be maintained by a transcriptional control mechanism that either prevents their degradation or allows its continued synthesis of mRNAs at a steady level. The ability to respond rapidly after ammonia deprivation under different physiological conditions is thought to be an important survival tool for AOB in environments where there is competition for available ammonia. For example, cells presented with a new supply of NH_3 showed a twofold spike of AMO activity, which then returned to the initial level of activity (Stein et al., 1997). This response was accompanied by an increase in both mRNA synthesis and AMO peptides. Increased concentration of ammonia in the medium resulted in higher concentration of AMO in the cell. Similarly, the mRNAs for AMO and HAO were also present at higher concentrations when cells were incubated in ammonia-rich medium (Sayavedra-Soto et al., 1996). When *Nitrosospira briensis* was starved 10 days in a batch culture, the *amoA* mRNA concentration decreased, and a relatively small change in total soluble proteins concentration was observed (Bollmann et al., 2005). These cells readily synthesized new *amoA* mRNA upon transfer to fresh ammonia medium. De novo *amoA* mRNA synthesis, while preserving protein levels, might be an adaptation of AOA to tolerate fluctuations of ammonia availability (Bollmann et al., 2005).

HAO

STRUCTURE AND METAL CONTENT

The second enzyme in the catabolism of NH_3, HAO, catalyzes the oxidation of NH_2OH to NO_2^- and is considered the link to the respiratory chain in AOB. HAO is a complex heme-containing enzyme in an α_3-oligomeric state. Each of the three subunits contains a modified, high-spin, five-coordinated c-type heme designated heme P460 that is the catalytic site. This heme P460 is unique to HAO. Seven additional c-type hemes in each subunit participate in electron transfer from the catalytic site (Arciero et al., 1993). Heme P460 in HAO derives its name from a ferrous Soret peak maximum at 460 nm (Andersson et al., 1991). A second P460 chromophore has also been identified in the AOB and resides in a small soluble periplasmic protein, cytochrome P460, of unknown function (Pearson et al., 2007). Cytochrome P460 has a single high-spin, five-coordinate P460 heme per 18.8-kDa polypeptide and has no structural similarity to HAO. Cytochrome P460 binds hydroxylamine, hydrazine, and cyanide in the ferric form and CO in the ferrous form and exhibits a weak hydroxylamine oxidation/cytochrome c oxidoreductase activity (Numata et al., 1990).

The X-ray crystallographic structure of HAO of *N. europaea* at 2.8 Å (Fig. 5) revealed the oligomeric nature of HAO composed of three identical subunits (Igarashi et al., 1997). The crystal structure shows a 100 Å pear-shaped structure with a candidate cavity (30 Å wide and 8 Å deep) where cytochrome c_{554} (cyt c_{554}) could bind (Igarashi et al., 1997). Each subunit is folded into two distinct domains in addition to a flexible hydrophobic C terminal. The first 269 amino acids form a short, two-stranded beta-sheet and contain 5 *c*-type hemes and the heme P460. The central domain between amino acids 270 and 499 contains two *c*-type hemes and 10 α-helices. The 24 hemes in HAO, eight per subunit, are located in the thicker bottom half of the molecule (Igarashi et al., 1997). The *c*-type hemes have octahedral coordination of the Fe atom completed by two His as axial ligands; each has a different redox potential (Igarashi et al., 1997), and they strongly interact with each other (Hendrich et al., 2001). The unique redox potential of each *c*-type heme is determined by its surrounding environment. The midpoint potentials of the *c*-type hemes determined by Mössbauer and electron paramagnetic resonance (EPR) studies at pH 7 against a normal hydrogen electrode range from −412 to +288 mV, while the midpoint potential for heme P460 is −260 mV (Kurnikov et al., 2005). Analysis of the crystal

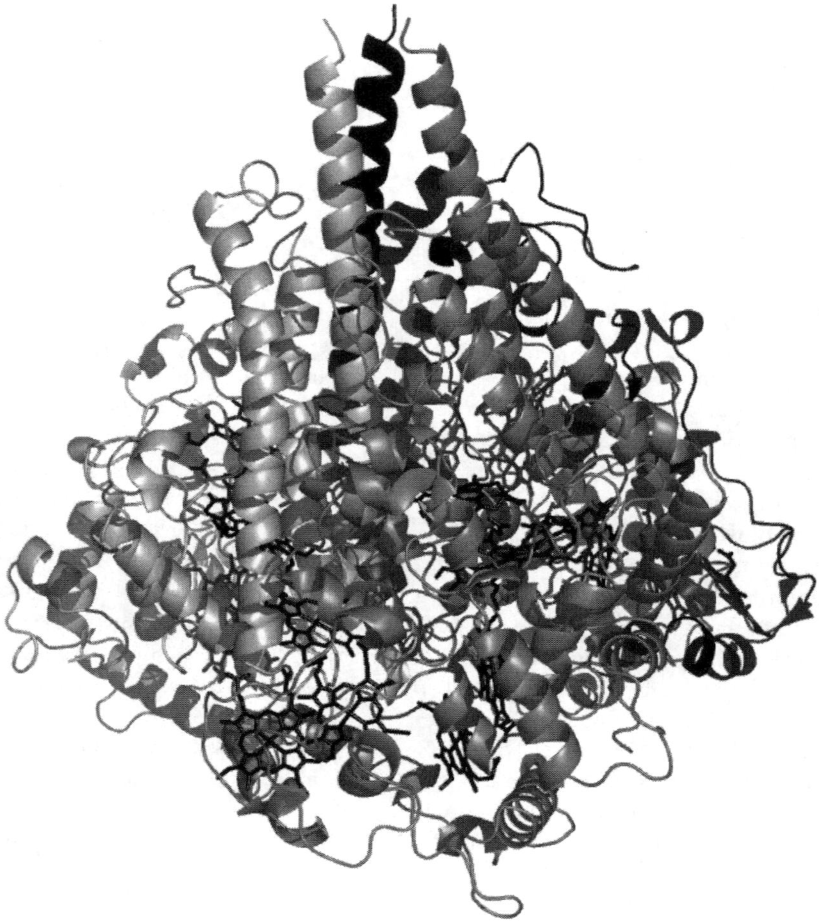

FIGURE 5 Three-dimensional X-ray crystal structure of hydroxylamine oxidoreductase from *N. europaea*. Each subunit is shown in ribbon form of a different shade. The heme molecules are shown as stick structures. The figure was derived from file PDB ID 1FGJ (www.pdb.org) (Igarashi et al., 1997) and MacPyMOLsoftware (wwwpymol.org).

structure of HAO revealed the hemes organized in four distinct entities: a cluster consisting of P460 and two *c*-type clusters, two double-heme clusters, and a single-heme cluster (Fig. 5). The P460 heme is localized at the catalytic pocket and is held by a typical heme-binding motif (Cys-X-X-Cys-His) that, in addition, is uniquely covalently linked through a tyrosine residue to an adjacent subunit. The linkage is required for the trimerization and is considered essential for stabilization of the molecule and for catalysis. The planes of the Tyr and the heme ring are perpendicular (Igarashi et al., 1997; Pearson et al., 2007). The Fe in the P460 heme has one of the six available coordination positions available to bind NH_2OH. The proximity and circular arrangement of the heme clusters enable HAO to transfer electrons efficiently over a relatively large distance (Igarashi et al., 1997; Kurnikov et al., 2005).

To initiate the oxidation of NH_2OH, HAO presumably withdraws two electrons simultaneously from NH_2OH, forming HNO as an enzyme-bound intermediate. To prevent the formation of N_2O or NO from HNO, the oxidation must occur in a continuous way to remove two more electrons. However, the exact mechanism is not yet known. Fully oxidized HAO at 1 atm of NO produced stable $[FeNO]^6$ species (K_{eq}, ~10^5 M^{-1} or higher) (Hendrich et al., 2002). This finding has significance in that NO may not be released easily, thus allowing complete oxidation of NH_2OH by HAO. Other possible reaction intermediates include Fe^{III}-NH_2OH, Fe^{III}-HNO, and $[FeNO]^7$ (Fernandez et al., 2008).

The difference between the redox potential of the solvent-exposed heme P460 and heme 2 can be enough to hold two electrons produced from the oxidation of hydroxylamine. The electrons then could be released when the attachment of cyt c_{554} shifts the potential of P460 to a more positive value (Kurnikov et al., 2005). The heme that is found singly is located between subunits, and it may serve to redirect excess electrons to another available oxidized heme in a neighboring subunit. The electrons from one of the other two double hemes in HAO are transferred in succession to the periplasmic abundant cyt c_{554}. The C terminal of HAO is flexible and highly hydrophobic, features that may be involved in the association of the enzyme with the membrane or with respiratory chain enzymes that are membrane bound. The iron atom in the P460 heme is high spin and probably pentacoordinate (5c) in the resting enzyme, though the presence of water at the sixth position cannot be ruled out (Igarashi et al., 1997; Arciero et al., 1998; Hendrich et al., 2001). The sixth vacant coordination site is available to bind hydroxylamine. The remaining *c*-type hemes are in the low-spin ferric state, hexacoordinated (6c), favoring the electron transfer down the electron transport chain to provide energy for all metabolic processes.

HAO can catalyze in vitro the reduction of NO to NH_3 in the presence of reduced methyl viologen (Kostera et al., 2008). In this reaction, NO is sequentially reduced to NH_2OH rapidly and then, at a 10-fold slower rate than the first step, to NH_3. This reduction may have some physiological relevance in low O_2 or anoxic conditions by preventing the accumulation of NO. A likely redox partner for HAO-catalyzed reduction of NO to ammonia is cyt c_{554}. cyt c_{554} has four hemes (see below) of which two have midpoint potentials of +47 mV (Arciero et al., 1991a; Upadhyay et al., 2003). The reduction potential for the NO/ammonia couple is about +339 mV; thus, reduced cyt c_{554} will yield a favorable cell potential (+292 mV) for the reduction of NO to ammonia (Kostera et al., 2008). HAO can also oxidize hydrazine to dinitrogen gas in a reaction similar to that performed by hydrazine oxidoreductase (Klotz et al., 2008; Jetten et al., 2009) (see Section IV).

MOLECULAR BIOLOGY

Among Beta-AOB, the genes encoding HAO are in multiple copies. The genes for a putative membrane protein (*orf2*), cyt c_{554} (*cycA*), and cyt c_{m552} (*cycB*) are adjacent to *hao* and in similar organization among all known AOBs (Arp et al., 2007). However, the chromosomal distances between these nearly identical copies differ by organism. The gene *orf2* follows *hao* in all the sequenced AOB genomes, as well as in the genome of the Gamma-MOB *M. capsulatus* and in a plasmid of the sulfur oxidizer *Silicibacter pomeroyi* (Klotz et al., 2008). One of the three gene copies of *hao* in *N. europaea* and *N. eutropha* does not have a copy of *cycB* associated with it. This change in gene structure is attributed to evolutionary divergences among the nitrosomonads (Purkhold et al., 2000, 2003). There are HAO-like genes in the genomes of *M. capsulatus* (Bath), *S. pomeroyi*, *Magnetococcus* sp. strain Mc-1, *Desulfovibrio desulfuricans* G20, *Geobacter metallireducens* GS15, and *Methanococcoides burtoni*, organisms that do not catalyze ammonia oxidation (Bergmann et al., 2005; Klotz et al., 2008). Whether these non-AOB produce functional HAO proteins is not known. The HAO-like genes have ~30% similarity to any of the HAO genes from AOB. The low similarity is attributed to the gaps in nucleotide sequence in those HAO-like genes. In *M. capsulatus* (Bath), the HAO-like gene, along with the gene (*orf2*) located immediately downstream, was transcribed in response to ammonia, thereby supporting the presence of a functional HAO in this methane-oxidizing bacterium (Poret-Peterson et al., 2008).

The HAO primary protein amino acid sequences from *N. europaea* and *N. multiformis* are 68% identical, both autotrophic Betaproteobacteria, and have somewhat less similarity (~50%) to Gammaproteobacteria and to the known HAO-like proteins in non-ammonia oxidizers (see Chapter 4). Other characterized multi-heme-containing proteins such as cytochrome *c* nitrite reductase (Einsle et al., 1999) and a tetraheme cytochrome *c* (Leys et al., 2002), although unrelated to HAO, have similar spatial heme arrangements. Fumarate dehydrogenase has a similarity to HAO in arrangement of three of the heme groups, although there is no significant amino acid sequence conservation between the two proteins (Taylor et al., 1999). An anammox HAO with enzymatic properties different than HAO from AOB and to the anammox hydrazine oxidoreductase was isolated from an anoxic sludge where the anammox bacterium strain KSU-1 was dominant (Shimamura et al., 2008). This HAO had a P468 chromophore reminiscent of the P460 chromophore.

Only one start of transcription was detected for each of the copies of *hao* in *N. europaea* (Sayavedra-Soto et al., 1994), which suggests that the multiple copies of *hao* might be transcribed simultaneously. Contrary to what was observed in *N. europaea*, *hao-3* in *Nitrosomonas* ENI-11 was transcribed from two promoters (Hirota et al., 2006). The expression of the multicopy *hao* was studied through transcriptional fusions (Hirota et al., 2006) and by gene inactivation in *Nitrosomonas* sp. strain ENI-11 (Yamagata et al., 2000) and by gene inactivation in *N. europaea* (Hommes et al., 1996, 2002). None of the copies in either strain was essential for growth. While the *N. europaea* strains with a single *hao* copy disrupted grew similarly to the wild type, in ENI-11 a single inactivation of any of the copies of *hao* led to ~30% lower growth than the wild type (Yamagata et al., 2000). In ENI-11, *hao-3* showed the highest expression. The promoters of *hao-1* and *hao-2* are almost the same, while the promoter of *hao-3* is different (Hirota et al., 2006). In spite of the similarity of the promoters of *hao-1* and *hao-2*, *hao-2* was expressed at higher levels than *hao-1* in ENI-11 (Hirota et al., 2006). In *N. europaea*, the double mutants had about half the in vitro activity of wild-type cells and were reflected in the mRNA levels but showed no decreases in either observed growth rates or in vivo HAO activity (Hommes et al., 2002). These results suggest that cells can lose substantial HAO activity without becoming limiting for growth. Single-copy gene inactivation of HAO genes did not produce a discernible phenotype (no effect on growth rate or ammonia- or hydroxylamine-dependent O_2 uptake rates). In ENI-11, it was suggested that *hao3* has a role in recovery from energy-depleted conditions, as it increased in expression considerably more than the other two copies of *hao* after ammonia addition. The transcription of the three copies of *hao* in *Nitrosomonas* sp. strain ENI11 showed that the copies were transcribed differentially in response to the supply of energy to the cell (Hirota et al., 2006).

Electron Transport from HAO

CYTOCHROME c_{554}

Physical evidence that the electrons flow from HAO to cytochrome c_{554} (cyt c_{554}) is suggested in its crystal structure. The structure shows an area where there could be interaction between HAO and cyt c_{554} for efficient transfer of electrons (Iverson et al., 1998, 2001). Cytochrome c_{554} is a 25-kDa monomeric protein with no amino acid sequence similarity to other known proteins. Cytochrome c_{554} contains four c-type hemes covalently linked through typical cheme-binding motifs: two Cys thioether linkages in the sequence -Cys-X-Tyr-Cys-His-. Cytochrome c_{554} has one heme five-coordinate with an axial His ligand and three hemes with bis-His axial coordination. Despite the dissimilar primary sequence of amino acids, the hemes have a conserved structural arrangement that is also observed in other bacterial multiheme c-cytochromes such as HAO, cytochrome *c*, nitrite reductase, fumarate reductase,

NapB, and split-Soret cytochrome (Upadhyay et al., 2003). However, the cyt c_{554} motif has no resemblance to other characterized tetraheme cytochrome c3 proteins (Iverson et al., 2001). The UV-visible spectrum of cyt c_{554} has a broad Soret band with a maximum at 407 nm attributed to high- and low-spin hemes. Upon reduction, an α-band is observed at 554 nm, hence the name of the protein. The oxidized cyt c_{554} shows all of its iron in the ferric state by Mössbauer spectroscopy. Ligand-binding experiments indicate that this cytochrome has no other function than electron transport. However, a possible alternative role of cyt c_{554} is in detoxification by the reduction of NO as it can both accept electrons from HAO and catalytically donate electrons to nitric oxide (Upadhyay et al., 2006). Biochemical experiments in vitro demonstrated that cyt c_{554} can accept electrons from HAO (Yamanaka and Shinra, 1974). The reduction potentials of the four hemes in cyt c_{554} have been calculated in vitro at pH 7: 47 mV for the high-spin heme and 47, −147, and 276 mV for the remaining three hemes, respectively (Upadhyay et al., 2003). The reduction of the hemes has a rate constant of 250 to 300 s^{-1} for one of the electrons transferred to cyt c_{554} from HAO, while the second has a rate constant of 25 to 30 s^{-1} (Arciero et al., 1991b).

Although the gene of cytochrome c_{554} (*cycA*) is likely transcribed independently from *hao*, probably through a sigma 70-type promoter, its proximity to *hao* (Arp et al., 2007) suggests that they act in concert.

CYTOCHROME c_{m552}

The membrane location of cytochrome c_{m552} (cyt c_{m552}) makes it a good candidate as the intermediate for the transfer of electrons between cyt c_{554} and the ubiquinone pool (Kim et al., 2008), though this remains to be verified experimentally. Recently, cyt c_{m552} was purified from *N. europaea* cell membranes and tended to form dimers that were attributed to transmembrane motifs (Kim et al., 2008). Based on this evidence, it was suggested that a dimeric or multimeric state is necessary for function. UV-visible spectrophotometric characterization of purified preparations indicated features found in cytochromes belonging to the NapC/NRH family and the presence of a high-spin heme. Cytochrome c_{m552} at pH 7.8 has a characteristic absorbance maximum at 408 nm for the Soret γ-band and a broad peak at 532 nm with a weak shoulder at 550 nm in the Q-band region (the α-band region of the pyridine ferrochrome spectrum [Kim et al., 2008]). Furthermore, Mössbauer spectroscopy of the reduced ^{57}Fe-enriched protein suggested features consistent with several low-spin or high-spin Fe(III) heme species in a 1:3 ratio. EPR spectra of purified cyt c_{m552} also suggested an interacting high-spin/low-spin pair of hemes. Ancestry similarities to the nitrite-reducing protein suggest that cyt c_{m552} may directly accept electrons from HAO, though this also has not been shown experimentally.

From the encoded amino acid sequences, the core tetraheme of cyt c_{m552} shows common ancestry to the NapC/NrfH/NirT/TorC family of tetra and pentaheme quinol dehydrogenases (Bergmann et al., 2005). These dehydrogenases are present in facultative anaerobes and function to transfer electrons from the ubiquinone pool to alternative electron acceptors (Bergmann et al., 2005). Based on its amino acid sequence homology, it has been proposed that cyt c_{m552} may have a quinol oxidoreductase function (Kim et al., 2008). Cytochrome c_{m552} has a predicted molecular mass of 27.1 kDa.

THE QUINONE POOL

The following step in the transfer of electrons from cyt c_{m552} is to the quinone pool. Ubiquinone is the predominant quinone in aerobic nitrifying bacteria. Cells contain ubiquinone-8 (Hooper et al., 1972) in 13-fold excess relative to HAO. Genes to produce proteins that can interact with the ubiquinone pool (Q/QH2) are present in the genomes of known AOB. The terminal oxidase of the cytochrome aa3 family (DiSpirito et al., 1986) and ubiquinone-8 (Hooper et al., 1972) were purified from *N. europaea*.

NITROSOCYANIN

Nitrosocyanin is a small mononuclear copper protein that is unique to AOB. Although its function is not yet known, it is included in this section dealing with electron transfer proteins because of its similarity to another electron transfer protein, plastocyanin. One characteristic that differentiates nitrosocyanin from other blue copper proteins is the cupric absorption band at 390 nm, which gives rise to a characteristic brilliant red color, in contrast to the 450 and 600 nm bands of blue copper proteins. A second distinguishing characteristic of nitrosocyanin is a redox potential of +85 mV, lower than those of blue copper proteins, which range from +184 to +680 mV. Nitrosocyanin is found in the same proportion as other components of the ammonia-oxidizing system, and its gene nucleotide sequence suggests that it is located in the periplasm (Arciero et al., 2002). Although its exact role remains to be determined, its properties and abundance suggest an important physiological role. Although an electron transfer role for nitrosocyanin is possible, its crystal structure has features that are not consistent with this role (Lieberman et al., 2001). The EPR characteristics are type 2 tetragonal copper centers often associated with a catalytic role rather than with electron transfer (Arciero et al., 2002). The presence of water coordinated to the copper delineates an open coordination site for substrate binding, and a cavity in the oxidized form of nitrosocyanin capable of binding a substrate reinforces a catalytic role (Lieberman et al., 2001).

CENTRAL CARBON METABOLISM

Autotrophy

Many studies have shown that AOB could take up and assimilate small amounts of organic carbon, mostly in anoxic conditions (Clark and Schmidt, 1966; Wallace et al., 1970; Krümmel and Harms, 1982; Martiny and Koops, 1982; Schmidt, 2009). However, the amounts of carbon taken up in oxic conditions are not sufficient to satisfy the carbon needs of the cells, and the majority of the cellular carbon comes from carbon dioxide/bicarbonate. The same phenomenon was observed in other, although not all, lithotrophs. A theory was developed that obligate autotrophy was due to an incomplete tricarboxylic acid (TCA) cycle. In early studies in *N. europaea*, α-ketoglutarate dehydrogenase activity was not detected and was considered the basis of the dependence on autotrophy (Hooper, 1969). Furthermore, an oxidative TCA cycle is incompatible with use of ammonia as an energy source. In the facultative methylotroph *Methylobacterium extorquens* AM1, the inactivation of the gene for α-ketoglutarate dehydrogenase resulted in a mutant unable to grow on substrates other than C_1, adding evidence to the hypothesis for obligate autotrophy (Van Dien et al., 2003). Based on all the aforementioned evidence then, AOB came to be known as obligate autotrophs.

With the completion of the *N. europaea* genome, obligate autotrophy was examined in another way. The gene profiles were consistent with the complete metabolism of some sugars (e.g., fructose) and organic acids (e.g., pyruvate) (Chain et al., 2003). This analysis led to experiments that demonstrated lithoheterotrophic growth of *N. europaea* with either fructose or pyruvate as the carbon source (Fig. 6), though ammonia was still required as the energy source (Hommes et al., 2006). Complete removal of CO_2 was required to demonstrate heterotrophy and resulted in slow growth, indicating that autotrophy is the preferred growth mode. In the known genomes of AOB, the genes *sucA*, *sucB*, and *lpd* (encoding α-ketoglutarate dehydrogenase) are present. The expression of the mRNA for this enzyme was corroborated in *N. europaea* showing that the genes are functional (Hommes et al., 2006). Nonetheless, activity could not be detected under oxic conditions. An *N. europaea sucA* deletion mutant grew as well as the wild type with ammonia and fructose or pyruvate and indicated that an incomplete TCA cycle still is compatible with heterotrophy in AOB. An incomplete TCA cycle does not prohibit carbon from either CO_2 or organic compounds from serving as the carbon source. A branched TCA cycle can

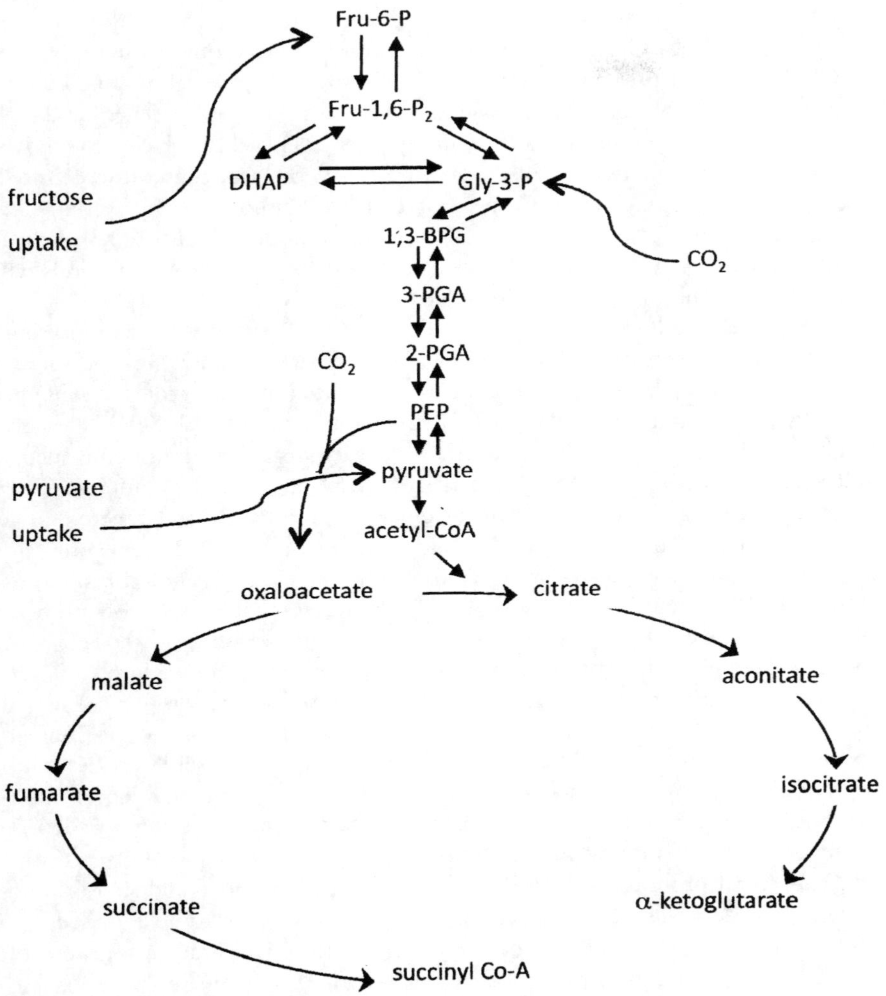

FIGURE 6 Central carbon metabolism in *N. europaea* under oxic conditions.

provide the necessary carbon backbones for biosynthesis. However, an incomplete TCA cycle cannot support organotrophy. The lack of α-ketoglutarate dehydrogenase activity does not explain obligate autotrophy but is consistent with the obligate lithotrophy observed in AOB grown under oxic conditions. Transcriptional studies showed higher levels of the mRNA for *sucA* in the late stationary phase, suggesting that a role for α-ketoglutarate dehydrogenase in AOB might be in assisting the cell to cope with a short supply of ammonia (Hommes et al., 2006). Under anaerobic growth on pyruvate as electron donor and nitrite as electron acceptor, a complete TCA cycle, including α-ketoglutarate dehydrogenase activity, would provide a likely mechanism to complete the oxidation of pyruvate to CO_2.

CO_2 Assimilation

In AOB, CO_2 assimilation takes place via the Calvin-Benson-Bassham (CBB) cycle, where the carboxylation reaction is catalyzed by ribulose-1,5-bisphosphate carboxylase/oxygenase (RubisCO). All enzymes for the CBB cycle are present in sequenced AOB genomes, with the exception of sedoheptulose-1,7-bisphosphatase. In AOB, fructose-1,6-bisphosphate dehydrogenase may be the enzyme performing the hydrolysis of sedoheptulose-1,7-bisphosphate in the CBB cycle, rather than for fructose-1,6-bisphosphate hydrolysis in gluconeogenesis (Yoo and Bowien, 1995). There are four recognized forms of RubisCO (Form I to Form IV) (Ezaki et al., 1999; Maeda et al., 1999; Watson et al., 1999; Utaker et al., 2002; Tabita et al., 2008). The available genomes indicate that RubisCO in AOB is predominantly Form I. For example, *N. europaea* and *N. eutropha* have Form IA (green-like) RubisCO, while *N. multiformis*, and *N. oceani* have a Form IC (red-like) RubisCO (Stein et al., 2007; Norton et al., 2008). Among AOB, RubisCO has >80% identity in the sequence of amino acids.

The only AOB isolate with the capacity to produce carboxysomes is *N. eutropha*. The carboxysomes are similar to those observed in other unrelated autotrophs, such as *Thiobacillus denitrificans* (Beller et al., 2006). The carboxysome genes in *N. eutropha* include those encoding structural proteins, carbon dioxide-concentrating proteins, and shell proteins, all of which are characteristic of the carboxysomes of other unrelated autotrophs (Stein et al., 2007).

Glycogen and Sucrose

Analysis of the AOB genomes showed genes to produce and metabolize the carbohydrates glycogen and sucrose (Arp et al., 2007). Genes for glycogen biosynthesis and degradation are present in *N. europaea* and are concentrated in two gene clusters with additional genes present at other loci (Chain et al., 2003). Glycogen is a carbon and energy reserve commonly found in animals and sometimes also in prokaryotes (Ball and Morell, 2003; Lodwig et al., 2005). Cellular stress can lead to accumulation of glycogen (Sherman et al., 1983). *N. europaea* contains approximately 10 to 20 ng of glycogen/mg protein (detected as glucose after α-amylase hydrolysis) when grown under standard laboratory conditions (Vajrala et al., 2010) (Fig. 7). The disruption of the genes encoding glycogen synthase (NE2264) in *N. europaea* caused the cells to be less resistant to ammonia deprivation (Arp laboratory, unpublished). Thus, AOB likely use glycogen to help them through periods when ammonia is in low supply.

The pathways for transmembrane transport and degradation of sucrose are documented in prokaryotes (Monchois et al., 1999; Ajdic and Pham, 2007). However, reports of microorganisms producing sucrose are limited (Arp et al., 2007; Lunn, 2002). Sucrose production has been detected in cyanobacteria (e.g., Lunn, 2002), but only a few proteobacteria have genes for sucrose biosynthesis. Sucrose probably serves more to protect the cell against osmotic shock than as an energy or carbon reserve (Lunn, 2002). Two genes for sucrose production are conserved in the four sequenced AOB genomes (Arp et al., 2007). In *N. europaea*, sucrose was detected (0.15 to 1.0 μg/mg of protein), demonstrating the functionality of the genes for sucrose production (Vajrala et al., 2010). In the AOB, sucrose phosphatase synthase and sucrose phosphate phosphatase appear to be encoded in one gene. The conserved residues associated with the haloacid dehalogenase phosphatase superfamily are encoded in the C-terminus extension of the sucrose phosphatase synthase. The gene for sucrose synthase is adjacent to the sucrose phosphate synthase in the four AOB sequenced (Arp et al., 2007). An *N. europaea* mutant with the genes for sucrose synthesis deleted did not produce detectable levels of sucrose (Vajrala et al., 2010).

BIOSYNTHESIS AND TRANSPORT

Ammonia Assimilation and Transport

On the basis of gene profiles of sequenced AOB, ammonia appears to be assimilated via glutamate dehydrogenase. Pathways for the

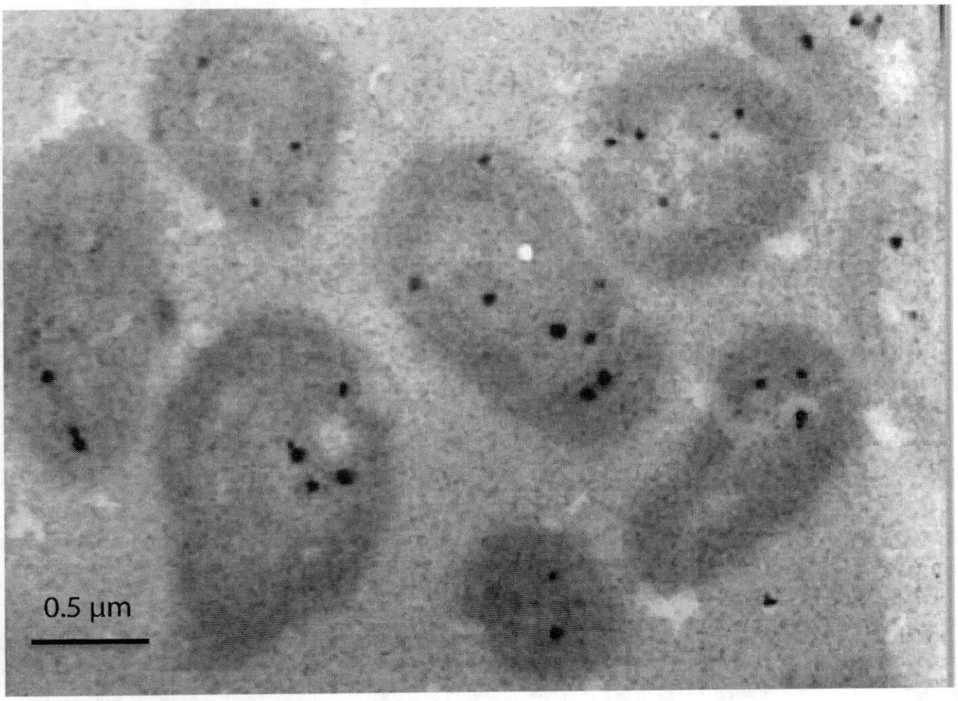

FIGURE 7 Electron microscopy picture of thin sections of *N. europaea* treated with acid-thiosemicarbazide-osmium tetroxide to visualize glycogen granules in the cells (dark spots).

synthesis of amino acids and other N-containing compounds could be identified and are consistent with established pathways in other organisms (Arp et al., 2007).

A gene encoding an Rh (Rhesus)-type transporter similar to the Amt-B ammonium transport proteins common in other organisms was identified in the genome of *N. europaea* (Weidinger et al., 2007). The Amt-B proteins function as channels for ammonia/ammonium and have been identified in bacteria, fungi, and plants (Winkler, 2006). The protein is capable of ammonium transport (Weidinger et al., 2007). The relative expression of the *rh1* decreased in *N. europaea* cells under denitrifying conditions and increased when the bacteria were transferred to oxic conditions. However, the transcription of *rh1* did not change with changes in ammonia concentration. An effective inhibition of ^{14}C-labeled methylammonium uptake by ammonium was observed indicating that the same transport system was involved for both. The transport of methylammonium was independent of pH, suggesting that an uncharged molecule, such as NH_3, is transported. However, an *N. europaea* mutant with the gene encoding Rh1 disrupted grew as well as the wild type over a range of ammonium concentrations (Vajrala et al., 2010).

The X-ray crystal structure of the *N. europaea* ammonium transport Rh protein at 1.8 Å (Li et al., 2007) and at 1.3 Å (Lupo et al., 2007) resolution has been determined. The protein is an α_3 homotrimer generated by a crystallographic threefold axis with the first 24 to 27 amino acids missing, likely cleaved by a signal peptidase. A C-terminal extension suggests an interaction with an unknown cytoplasmic partner. The structure of the *N. europaea* Rh protein in comparison to other known Amt proteins is consistent with the transport of NH_3 or CO_2 (Li et al., 2007; Lupo et al., 2007), although there is currently no evidence for CO_2 transport by this protein.

Iron

In AOB, Fe is essential for the many cytochromes, heme-containing enzymes, and other iron-containing enzymes required for ammonia oxidation and cell growth and maintenance. Indeed, the AOB genomes revealed numerous genes for Fe uptake (Arp et al., 2007). Among AOB, *N. europaea* has the most genes for Fe accumulation, with nearly 100 genes dedicated to iron uptake, although, interestingly, none for siderophore biosynthesis. In the genome of *N. multiformis*, 29 genes were identified for active transport of Fe and 4 putative genes for the production of the siderophore pyoverdin (Norton et al., 2008). *N. oceani* and *N. eutropha*, on the other hand, have 22 and 28 genes, respectively, dedicated to iron uptake and 2 putative genes each for the production of siderophores. The many siderophore transducers/receptors encoded by *N. europaea*, *N. eutropha*, *N. multiformis*, and *N. oceani* genomes are probably involved in uptake of Fe-loaded siderophores produced by other organisms (e.g., Fe-loaded siderophores ferrichrome, desferrioxamine, coprogen, pyoverdin, and catechol/catecholate type) (Arp et al., 2007).

N. europaea grown in Fe-replete medium (10 μM Fe) has high cellular Fe concentration (i.e., 16.3 mM, 80-fold higher than in *E. coli*). *N. europaea* can also grow moderately well at Fe concentrations as low as 0.2 μM, even when exogenous siderophores are not provided (Wei et al., 2006). Growth at this low Fe concentration is extraordinary given the high requirement of *N. europaea* for Fe and its inability to produce siderophores. Other microorganisms with a much lower requirement for Fe rely on siderophores to grow at concentrations below 1 μM Fe.

Uptake of iron siderophores requires Fe ABC transporters in addition to siderophore transducers/receptors, and, in several microorganisms, these genes are often found associated with each other. For example, *E. coli* has three sets of genes, and *Pseudomonas aeruginosa* has four sets of genes (Andrews et al., 2003). In addition to numerous siderophore transducer/receptor genes, *N. europaea* has genes encoding one complete set of the Fe ABC transporter set, although these genes are not associated with any of the receptor genes. The *N. europaea* Fe ABC transporter was shown to be specific for hydroxamate-type siderophores and the mixed chelating-type siderophore pyoverdin (Vajrala et al., 2010). The genes for the siderophore transporter specific for catecholate-type siderophores or the mixed chelating-type siderophore aerobactin remain unknown. *N. eutropha* has no discernible Fe ABC transporter; however, it does have a major facilitator superfamily (MFS) family transporter that could potentially import loaded siderophores (Stein et al., 2007).

When the intracellular Fe concentration becomes low, upregulation of Fe uptake genes generally occurs through the release of gene repression imposed by ferric uptake regulator (Fur) (Andrews et al., 2003). A putative *fur* gene (NE0616) was disrupted in *N. europaea* and resulted in the upregulation of Fe uptake genes (Arp laboratory, unpublished). Several Fe-containing proteins in *N. europaea* were present at lower levels when *N. europaea* grew in Fe-limited medium (Wei et al., 2006). These observations suggest that *N. europaea* maintains a delicate balancing act between iron uptake and Fe consumption, because the very enzymes that permit *N. europaea* to derive energy from NH_3 are also those that have high Fe content (Hooper, 1989; Arp et al., 2002).

N. europaea has putative genes that encode high-affinity, siderophore-independent Fe uptake systems. For example, it has a putative siderophore-independent Fe^{3+} transporter (encoded by NE0294, a cytochrome *c*-type protein) that is similar to the yeast Ftr1 Fe^{3+} transporter with a characteristic Glu-X-X-Glu Fe-binding site (Stearman et al., 1996; Severance et al., 2004). Though a Fe^{3+} transporter, Ftr1 also supports high-affinity Fe^{2+} transport because it is coupled with a multicopper oxidase that oxidizes Fe^{2+} to Fe^{3+}, which is then transported by Ftr1 into the cytoplasm. There are at least seven putative multicopper oxidase genes in *N. europaea*. Multicopper oxidases were shown to be involved in Fe acquisition in bacteria (Herbik et al., 2002; Huston et al., 2002). A gene with relatively low similarity to a Fe^{2+} transporter is also present (NE1286, *feoB*), but in an aerobic growth environment Fe^{2+} would be scarce. The function of FeoB and the level of contribution of Fe^{2+} to *N. europaea* Fe nutrition remain to be characterized. Other genes related to Fe nutrition include the Fe-storage protein bacterioferritin and bacterioferritin comigratory protein.

PERSPECTIVES

When the first treatise on nitrification was published in 1986 (Prosser, 1986), the basic pathway of ammonia catabolism had been determined. However, the study of the enzymes and genes involved in the process was in its infancy. In the intervening 23 years, our knowledge of the enzymes that are central to ammonia catabolism has grown considerably, as has our knowledge of the genes encoding these proteins. Several proteins and enzymes have been purified and biochemically characterized, and crystal structures for a number of key proteins in the oxidation of ammonia have been determined. Perhaps most significant was the elucidation of the structure of HAO, the remarkably complex trimer of octaheme subunits. Structures of cytochromes unique to AOB were determined along with some additional proteins, including nitrosocyanin and AmtB. The work described above has greatly enhanced our knowledge of how ammonia is oxidized. Nonetheless, a number of questions remain. Most notably, the enzyme that initiates the entire process, AMO, has not yet been purified to homogeneity with activity. As a consequence, we do not yet have a clear picture of the metal content in AMO, or the mechanism of ammonia oxidation. Studies with inhibitors, alternative substrates, and comparisons to pMMO have provided insights, but deeper characterizations await active preparations. On the other hand, our understanding of the catalytic versatility, or lack of specificity, of this enzyme has grown considerably. Other aspects of the pathway also remain to be determined. For example, the precise pathway of electron transfer is not yet known, though well-supported working models have been presented. One of the fascinating aspects of ammonia oxidizers is the need to produce reductant [NAD(P)H] for biosynthesis from reductants that have more positive potentials (i.e., through "reverse electron flow").

There have been fewer studies on the biochemistry, physiology, or molecular biology dealing with carbon metabolism than ammonia metabolism in AOB. Several gene nucleotide sequences for RubisCOs have been determined, but only scant attention has been focused on the biochemical properties of the protein. Most attention to carbon metabolism has come from gene profiles deduced from sequenced genomes. Analysis of the gene profiles led to the first demonstration of chemolithoheterotrophic growth of an AOB under oxic conditions. But heterotrophic growth was slower than growth on carbon dioxide and was limited to just a few carbon sources. So, questions remain about how the carbon metabolism of these bacteria is tuned to work best when assimilating carbon dioxide. And we still do not fully understand the obligate lithotrophic nature of the AOB when under oxic conditions. The genes for a complete TCA cycle are present and expressed, but the activities through some of the steps are below detection. The gene profiles of the AOB also point to a need to further understand central carbon metabolism. In particular, the interconversion of fructose-1,6-bisphosphate and fructose-6-phosphate, a critical control point between glycolysis and gluconeogenesis, is not well characterized. A reversible pyrophosphate-dependent enzyme has been suggested, but experimental proof has not been provided. The gene profiles revealed pathways for synthesis of essential biomolecules as expected for this autotroph. On the other hand, pathways for scavenging organic molecules have not been identified. While this absence of recycling capacity is consistent with the predominantly autotrophic nature of the AOB, it also raises the question of the fate of amino acids, nucleotides, and phospholipids as proteins, RNA, and lipids are turned over. The role of sucrose synthesis and degradation also remains to be determined.

An area that has advanced dramatically since the last treatise (Prosser, 1986) is the knowledge of gene nucleotide sequences and gene profiles present in the AOB. We have advanced from having no nucleotide sequences for the genes involved in ammonia catabolism to complete genome sequences for several AOB, with more genome sequences in progress. These nucleotide sequences are allowing in-depth consideration of the core genes required for AOB, as distinct from the genes required for autotrophy, cell division, maintenance, etc.

The AOB are of profound importance to the movement of N through various ecosystems, including natural, managed (e.g., croplands), and engineered (e.g., wastewater treatment). And it is this importance that sustains our general interest in this group of bacteria. But the unique lifestyle of these bacteria, namely their ability to derive energy from ammonia, and seemingly at the exclusion of other energy sources, has and will continue to attract the attention of biochemists and others interested in the enzymes and pathways that have evolved to give these bacteria the capacity to fill a unique niche.

ACKNOWLEDGMENTS

We acknowledge the work of the many researchers who have contributed to our understanding of the biochemistry, molecular biology, and metabolism of ammonia-oxidizing bacteria over the last four decades. While we have attempted to capture the major advances, especially in the last 20 years, we were not able to cite all relevant publications given space limitations. We thank three anonymous reviewers and our chapter editor, Martin Klotz, for their careful and critical reading of the chapter.

REFERENCES

Ajdic, D., and V. T. Pham. 2007. Global transcriptional analysis of *Streptococcus mutans* sugar transporters using microarrays. *J. Bacteriol.* **189:**5049–5059.

Alzerreca, J. J., J. M. Norton, and M. G. Klotz. 1999. The *amo* operon in marine, ammonia-oxidizing gamma-proteobacteria. *FEMS Microbiol. Lett.* **180:**21–29.

Andersson, K. K., G. T. Babcock, and A. B. Hooper. 1991. P460 of hydroxylamine oxidoreductase of *Nitrosomonas europaea:* Soret resonance raman evidence for a novel heme-like structure. *Biochem. Biophys. Res. Commun.* **174:**358–363.

Andrews, S. C., A. K. Robinson, and F. Rodriguez-Quinones. 2003. Bacterial iron homeostasis. *FEMS Microbiol. Rev.* **27:**215–237.

Arciero, D. M., C. Balny, and A. B. Hooper. 1991a. Spectroscopic and rapid kinetic studies of reduction of cytochrome *c*554 by hydroxylamine oxidoreductase from *Nitrosomonas europaea. Biochemistry* **30:**11466–11472.

Arciero, D. M., M. J. Collins, J. Haladjian, P. Bianco, and A. B. Hooper. 1991b. Resolution of the four hemes of cytochrome-c_{554} from *Nitrosomonas europaea* by redox potentiometry and optical spectroscopy. *Biochemistry* **30:**11459–11465.

Arciero, D. M., A. B. Hooper, M. Cai, and R. Timkovich. 1993. Evidence for the structure of the active site heme P460 in hydroxylamine oxidoreductase of *Nitrosomonas. Biochemistry* **32:**9370–9378.

Arciero, D. M., A. Golombek, M. P. Hendrich, and A. B. Hooper. 1998. Correlation of optical and EPR signals with the P460 heme of hydroxylamine oxidoreductase from *Nitrosomonas europaea. Biochemistry* **37:**523–529.

Arciero, D. M., B. S. Pierce, M. P. Hendrich, and A. B. Hooper. 2002. Nitrosocyanin, a red cupredoxin-like protein from *Nitrosomonas europaea. Biochemistry* **41:**1703–1709.

Arp, D. J., and P. J. Bottomley. 2006. Nitrifiers: more than 100 years from isolation to genome sequences. *Microbe* **1:**229–234.

Arp, D. J., and L. Y. Stein. 2003. Metabolism of inorganic N compounds by ammonia-oxidizing bacteria. *Crit. Rev. Biochem. Mol. Biol.* **38:**471–495.

Arp, D. J., L. A. Sayavedra-Soto, and N. G. Hommes. 2002. Molecular biology and biochemistry of ammonia oxidation by *Nitrosomonas europaea. Arch. Microbiol.* **178:**250–255.

Arp, D. J., P. S. Chain, and M. G. Klotz. 2007. The impact of genome analyses on our understanding of ammonia-oxidizing bacteria. *Annu. Rev. Microbiol.* **61:**503–528.

Ball, S. G., and M. K. Morell. 2003. From bacterial glycogen to starch: understanding the biogenesis of the plant starch granule. *Annu. Rev. Plant Biol.* **54:**207–233.

Beller, H. R., P. S. G. Chain, T. E. Letain, A. Chakicherla, F. W. Larimer, P. M. Richardson, M. A. Coleman, A. P. Wood, and D. P. Kelly. 2006. The genome sequence of the obligately chemolithoautotrophic, facultatively anaerobic bacterium *Thiobacillus denitrificans*. *J. Bacteriol.* **188:**1473–1488.

Bergmann, D. J., A. B. Hooper, and M. G. Klotz. 2005. Structure and sequence conservation of hao cluster genes of autotrophic ammonia-oxidizing bacteria: evidence for their evolutionary history. *Appl. Environ. Microbiol.* **71:**5371–5382.

Berube, P. M., R. Samudrala, and D. A. Stahl. 2007. Transcription of all *amoC* copies is associated with recovery of *Nitrosomonas europaea* from ammonia starvation. *J. Bacteriol.* **189:**3935–3944.

Bock, E. 1995. Nitrogen loss caused by denitrifying *Nitrosomonas* cells using ammonium or hydrogen as electron donors and nitrite as electron acceptor. *Arch. Microbiol.* **163:**16–20.

Bollmann, A., I. Schmidt, A. M. Saunders, and M. H. Nicolaisen. 2005. Influence of starvation on potential ammonia-oxidizing activity and *amoA* mRNA levels of *Nitrosospira briensis*. *Appl. Environ. Microbiol.* **71:**1276–1282.

Chain, P., J. Lamerdin, F. Larimer, W. Regala, V. Lao, M. Land, L. Hauser, A. Hooper, M. Klotz, J. Norton, L. Sayavedra-Soto, D. Arciero, N. Hommes, M. Whittaker, and D. Arp. 2003. Complete genome sequence of the ammonia-oxidizing bacterium and obligate chemolithoautotroph *Nitrosomonas europaea*. *J. Bacteriol.* **185:**2759–2773.

Clark, C. and E. L. Schmidt. 1966. Effect of mixed culture on *Nitrosomonas europaea* simulated by uptake and utilization of pyruvate. *J. Bacteriol.* **91:**367–373.

De Boer, W., P. J. A. K. Gunnewiek, M. Veenhuis, E. Bock, and H. J. Laanbroek. 1991. Nitrification at low pH by aggregated chemolithotrophic bacteria. *Appl. Environ. Microbiol.* **57:**3600–3604.

DiSpirito, A. A., J. D. Lipscomb, and A. B. Hooper. 1986. Cytochrome aa3 from *Nitrosomonas europaea*. *J. Biol. Chem.* **261:**17048–17056.

Einsle, O., A. Messerschmidt, P. Stach, G. P. Bourenkov, H. D. Bartunik, R. Huber, and P. M. Kroneck. 1999. Structure of cytochrome c nitrite reductase. *Nature* **400:**476–480.

El Sheikh, A. F., and M. G. Klotz. 2008. Ammonia-dependent differential regulation of the gene cluster that encodes ammonia monooxygenase in *Nitrosococcus oceani* ATCC 19707. *Environ. Microbiol.* **10:**3026–3035.

El Sheikh, A. F., A. T. Poret-Peterson, and M. G. Klotz. 2008. Characterization of two new genes, *amoR* and *amoD*, in the *amo* operon of the marine ammonia oxidizer *Nitrosococcus oceani* ATCC 19707. *Appl. Environ. Microbiol.* **74:**312–318.

Ensign, S. A., M. R. Hyman, and D. J. Arp. 1993. In vitro activation of ammonia monooxygenase from *Nitrosomonas europaea* by copper. *J. Bacteriol.* **175:**1971–1980.

Ezaki, S., N. Maeda, T. Kishimoto, H. Atomi, and T. Imanaka. 1999. Presence of a structurally novel type ribulose-bisphosphate carboxylase/oxygenase in the hyperthermophilic archaeon, *Pyrococcus kodakaraensis* KOD1. *J. Biol. Chem.* **274:**5078–5082.

Fernandez, M. L., D. A. Estrin, and S. E. Bari. 2008. Theoretical insight into the hydroxylamine oxidoreductase mechanism. *J. Inorg. Biochem.* **102:**1523–1530.

Fiencke, C., and E. Bock. 2006. Immunocytochemical localization of membrane-bound ammonia monooxygenase in cells of ammonia oxidizing bacteria. *Arch. Microbiol.* **185:**99–106.

Francis, C. A., J. M. Beman, and M. M. Kuypers. 2007. New processes and players in the nitrogen cycle, the microbial ecology of anaerobic and archeal ammonia oxidation. *ISME J.* **1:**19–27.

Geets, J., N. Boon, and W. Verstraete. 2006. Strategies of aerobic ammonia-oxidizing bacteria for coping with nutrient and oxygen fluctuations. *FEMS Microbiol. Ecol.* **58:**1–13.

Gieseke, A., J. L. Nielsen, R. Amann, P. H. Nielsen, and D. de Beer. 2005. In situ substrate conversion and assimilation by nitrifying bacteria in a model biofilm. *Environ. Microbiol.* **7:**1392–1404.

Gilch, S., O. Meyer, and I. Schmidt. 2009a. A soluble form of ammonia monooxygenase in *Nitrosomonas europaea*. *Biol. Chem.* **390:**863–873.

Gilch, S., M. Vogel, M. W. Lorenz, O. Meyer, and I. Schmidt. 2009b. Interaction of the mechanism-based inactivator acetylene with ammonia monooxygenase of *Nitrosomonas europaea*. *Microbiology* **155:**279–284.

Hakemian, A. S., and A. C. Rosenzweig. 2007. The biochemistry of methane oxidation. *Annu. Rev. Biochem.* **76:**223–241.

Hendrich, M. P., D. Petasis, D. M. Arciero, and A. B. Hooper. 2001. Correlations of structure and electronic properties from EPR spectroscopy of hydroxylamine oxidoreductase. *J. Am. Chem. Soc.* **123:**2997–3005.

Hendrich, M. P., A. K. Upadhyay, J. Riga, D. M. Arciero, and A. B. Hooper. 2002. Spectroscopic characterization of the NO adduct of hydroxylamine oxidoreductase. *Biochemistry* **41:**4603–4611.

Herbik, A., C. Bolling, and T. J. Buckhout. 2002. The involvement of a multicopper oxidase in iron uptake by the green algae *Chlamydomonas reinhardtii*. *Plant Physiol.* **130:**2039–2048.

Hirota, R., A. Kuroda, T. Ikeda, N. Takiguchi, H. Ohtake, and J. Kato. 2006. Transcriptional analysis of the multicopy hao gene coding for hydroxylamine oxidoreductase in *Nitrosomonas* sp. strain ENI-11. *Biosci. Biotechnol. Biochem.* **70:**1875–1881.

Hommes, N. G., L. A. Sayavedra-Soto, and D. J. Arp. 1996. Mutagenesis of hydroxylamine oxidoreductase in *Nitrosomonas europaea* by transformation and recombination. *J. Bacteriol.* **178:**3710–3714.

Hommes, N. G., L. A. Sayavedra-Soto, and D. J. Arp. 1998. Mutagenesis and expression of *amo*, which codes for ammonia monooxygenase in *Nitrosomonas europaea. J. Bacteriol.* **180:**3353–3359.

Hommes, N. G., L. A. Sayavedra-Soto, and D. J. Arp. 2001. Transcript analysis of multiple copies of *amo* (encoding ammonia monooxygenase) and *hao* (encoding hydroxylamine oxidoreductase) in *Nitrosomonas europaea. J. Bacteriol.* **183:**1096–1100.

Hommes, N. G., L. A. Sayavedra-Soto, and D. J. Arp. 2002. The roles of the three gene copies encoding hydroxylamine oxidoreductase in *Nitrosomonas europaea. Arch. Microbiol.* **178:**471–476.

Hommes, N. G., L. A. Sayavedra-Soto, and D. J. Arp. 2003. Chemolithoorganotrophic growth of *Nitrosomonas europaea* on fructose. *J. Bacteriol.* **185:**6809–6814.

Hommes, N. G., E. G. Kurth, L. A. Sayavedra-Soto, and D. J. Arp. 2006. Disruption of *sucA*, which encodes the E1 subunit of alpha-ketoglutarate dehydrogenase, affects the survival of *Nitrosomonas europaea* in stationary phase. *J. Bacteriol.* **188:**343–347.

Hooper, A. B. 1969. Biochemical basis of obligate autotrophy in *Nitrosomonas europaea. J Bacteriol.* **97:**776–779.

Hooper, A. B. 1989. Biochemistry of the nitrifying lithoautotrophic bacteria, p. 239–281. *In* H. G. Schlegel and B. Bowien (ed.), *Autotrophic Bacteria.* Science Tech Publishers, Madison, WI.

Hooper, A. B., and K. R. Terry. 1973. Specific inhibitors of ammonia oxidation in *Nitrosomonas. J. Bacteriol.* **115:**480–485.

Hooper, A. B., R. H. Erickson, and K. R. Terry. 1972. Electron transport systems of *Nitrosomonas*: isolation of a membrane-envelope fraction. *J. Bacteriol.* **110:**430–438.

Hooper, A. B., T. Vannelli, D. J. Bergmann, and D. M. Arciero. 1997. Enzymology of the oxidation of ammonia to nitrite by bacteria. *Antonie van Leeuwenhoek* **71:**59–67.

Huston, W. M., M. P. Jennings, and A. G. McEwan. 2002. The multicopper oxidase of *Pseudomonas aeruginosa* is a ferroxidase with a central role in iron acquisition. *Mol. Microbiol.* **45:**1741–1750.

Hyman, M. R., and D. J. Arp. 1992. $^{14}C_2H_2$- and $^{14}CO_2$-labeling studies of the *de novo* synthesis of polypeptides by *Nitrosomonas europaea* during recovery from acetylene and light inactivation of ammonia monooxygenase. *J. Biol. Chem.* **267:**1534–1545.

Hyman, M. R., I. B. Murton, and D. J. Arp. 1988. Interaction of ammonia monooxygenase from *Nitrosomonas europaea* with alkanes, alkenes, and alkynes. *Appl. Environ. Microbiol.* **54:**3187–3190.

Igarashi, N., H. Moriyama, T. Fujiwara, Y. Fukumori, and N. Tanaka. 1997. The 2.8 Å structure of hydroxylamine oxidoreductase from a nitrifying chemoautotrophic bacterium, *Nitrosomonas europaea. Nat. Struct. Biol.* **4:**276–284.

Iverson, T. M., D. M. Arciero, B. T. Hsu, M. S. Logan, A. B. Hooper, and D. C. Rees. 1998. Heme packing motifs revealed by the crystal structure of the tetra-heme cytochrome c554 from *Nitrosomonas europaea. Nat. Struct. Biol.* **5:**1005–1012.

Iverson, T. M., D. M. Arciero, A. B. Hooper, and D. C. Rees. 2001. High-resolution structures of the oxidized and reduced states of cytochrome c554 from *Nitrosomonas europaea. J. Biol. Inorg. Chem.* **6:**390–397.

Jetten, M. S., I. Cirpus, B. Kartal, L. van Niftrik, K. T. van de Pas-Schoonen, O. Sliekers, S. Haaijer, W. van der Star, M. Schmid, J. van de Vossenberg, I. Schmidt, H. Harhangi, M. van Loosdrecht, J. Gijs Kuenen, H. Op den Camp, and M. Strous. 2005. 1994–2004: 10 years of research on the anaerobic oxidation of ammonium. *Biochem. Soc. Trans.* **33:**119–123.

Jetten, M. S., L. V. Niftrik, M. Strous, B. Kartal, J. T. Keltjens, and H. J. Op den Camp. 2009. Biochemistry and molecular biology of anammox bacteria. *Crit. Rev. Biochem. Mol. Biol.* **44:**65–84.

Juliette, L. Y., M. R. Hyman, and D. J. Arp. 1993. Mechanism-based inactivation of ammonia monooxygenase in *Nitrosomonas europaea* by allylsulfide. *Appl. Environ. Microbiol.* **59:**3728–3735.

Juliette, L. Y., M. R. Hyman, and D. J. Arp. 1995. Roles of bovine serum albumin and copper in the assay and stability of ammonia monooxygenase activity in vitro. *J. Bacteriol.* **177:**4908–4913.

Keener, W. K., and D. J. Arp. 1993. Kinetic studies of ammonia monooxygenase inhibition in *Nitrosomonas europaea* by hydrocarbons and halogenated hydrocarbons in an optimized whole-cell assay. *Appl. Environ. Microbiol.* **59:**2501–2510.

Kim, H. J., A. Zatsman, A. K. Upadhyay, M. Whittaker, D. Bergmann, M. P. Hendrich, and A. B. Hooper. 2008. Membrane tetraheme cytochrome cm552 of the ammonia-oxidizing *Nitrosomonas europaea*: a ubiquinone reductase. *Biochemistry* **47:**6539–6551

Klotz, M. G., and J. M. Norton. 1995. Sequence of an ammonia monooxygenase subunit A-encoding gene from *Nitrosospira* sp. NpAV. *Gene* **163:**159–160.

Klotz, M. G., D. J. Arp, P. S. Chain, A. F. El-Sheikh, L. J. Hauser, N. G. Hommes, F. W. Larimer, S. A. Malfatti, J. M. Norton, A. T. Poret-Peterson, L. M. Vergez, and B. B. Ward. 2006. Complete genome sequence of the marine, chemolithoautotrophic, ammonia-oxidizing bacterium *Nitrosococcus oceani* ATCC 19707. *Appl. Environ. Microbiol.* **72:**6299–6315.

Klotz, M. G., M. C. Schmid, M. Strous, H. J. op den Camp, M. S. Jetten, and A. B. Hooper. 2008. Evolution of an octahaem cytochrome c protein family that is key to aerobic and anaerobic ammonia oxidation by bacteria. *Environ. Microbiol.* **10:**3150–3163.

Koops, H. P., and A. Pommerening-Roser. 2001. Distribution and ecophysiology of the nitrifying bacteria emphasizing culture species. *FEMS Microbiol.* **37:**1–9.

Kostera, J., M. D. Youngblut, J. M. Slosarczyk, and A. A. Pacheco. 2008. Kinetic and product distribution analysis of NO* reductase activity in *Nitrosomonas europaea* hydroxylamine oxidoreductase. *J. Biol. Inorg. Chem.* **13:**1073–1083.

Kowalchuk, G. A., and J. R. Stephen. 2001. Ammonia-oxidizing bacteria: a model for molecular microbial ecology. *Annu. Rev. Microbiol.* **55:**485–529.

Krümmel, A., and H. Harms. 1982. Effect of organic matter on growth and cell yield of ammonia-oxidizing bacteria. *Arch. Microbiol.* **133:**50–54.

Kurnikov, I. V., M. A. Ratner, and A. A. Pacheco. 2005. Redox equilibria in hydroxylamine oxidoreductase. Electrostatic control of electron redistribution in multielectron oxidative processes. *Biochemistry* **44:**1856–1863.

Laanbroek, H. J., M. J. Bar-Gilissen, and H. L. Hoogveld. 2002. Nitrite as a stimulus for ammonia-starved *Nitrosomonas europaea. Appl. Environ. Microbiol.* **68:**1454–1457.

Leys, D., T. E. Meyer, A. S. Tsapin, K. H. Nealson, M. A. Cusanovich, and J. J. Van Beeumen. 2002. Crystal structures at atomic resolution reveal the novel concept of "electron-harvesting" as a role for the small tetraheme cytochrome c. *J. Biol. Chem.* **277:**35703–35711.

Li, X., S. Jayachandran, H. H. Nguyen, and M. K. Chan. 2007. Structure of the *Nitrosomonas europaea* Rh protein. *Proc. Natl. Acad. Sci. USA* **104:**19279–19284.

Lieberman, R. L., and A. C. Rosenzweig. 2005. Crystal structure of a membrane-bound metalloenzyme that catalyses the biological oxidation of methane. *Nature* **434:**177–182.

Lieberman, R. L., D. M. Arciero, A. B. Hooper, and A. C. Rosenzweig. 2001. Crystal structure of a novel red copper protein from *Nitrosomonas europaea. Biochemistry* **40:**5674–5681.

Lodwig, E. M., M. Leonard, S. Marroqui, T. R. Wheeler, K. Findlay, J. A. Downie, and P. S. Poole. 2005. Role of polyhydroxybutyrate and glycogen as carbon storage compounds in pea and bean bacteroids. *Mol. Plant Microbe. Interact.* **18:**67–74.

Lunn, J. E. 2002. Evolution of sucrose synthesis. *Plant Physiol.* **128:**1490–1500.

Lupo, D., X. D. Li, A. Durand, T. Tomizaki, B. Cherif-Zahar, G. Matassi, M. Merrick, and F. K. Winkler. 2007. The 1.3-Å resolution structure of *Nitrosomonas europaea* Rh50 and mechanistic implications for NH_3 transport by Rhesus family proteins. *Proc. Natl. Acad. Sci. USA* **104:**19303–19308.

Maeda, N., K. Kitano, T. Fukui, S. Ezaki, H. Atomi, K. Miki, and T. Imanaka. 1999. Ribulose bisphosphate carboxylase/oxygenase from the hyperthermophilic archaeon *Pyrococcus kodakaraensis* KOD1 is composed solely of large subunits and forms a pentagonal structure. *J. Mol. Biol.* **293:**57–66.

Mancinelli, R. L., and C. P. McKay. 1988. The evolution of nitrogen cycling. *Orig. Life Evol. Biosph.* **18**:311–325.

Martinho, M., D. W. Choi, A. A. Dispirito, W. E. Antholine, J. D. Semrau, and E. Munck. 2007. Mossbauer studies of the membrane-associated methane monooxygenase from *Methylococcus capsulatus* Bath: evidence for a diiron center. *J. Am. Chem. Soc.* **129:**15783–15785.

Martiny, H., and H. P. Koops. 1982. Incorporation of organic compounds into cell protein by lithotrophic, ammonia-oxidizing bacteria. *Antonie Van Leeuwenhoek* **48:**327–336.

Monchois, V., R. M. Willemot, and P. Monsan. 1999. Glucansucrases: mechanism of action and structure-function relationships. *FEMS Microbiol. Rev.* **23:**131–151.

Norton, J. M., M. G. Klotz, L. Y. Stein, D. J. Arp, P. J. Bottomley, P. S. Chain, L. J. Hauser, M. L. Land, F. W. Larimer, M. W. Shin, and S. R. Starkenburg. 2008. Complete genome sequence of *Nitrosospira multiformis*, an ammonia-oxidizing bacterium from the soil environment. *Appl. Environ. Microbiol.* **74:**3559–3572.

Numata, M., T. Saito, T. Yamazaki, Y. Fukumori, and T. Yamanaka. 1990. Cytochrome P-460 of *Nitrosomonas europaea:* further purification and further characterization. *J. Biochem.* **108:**1016–1021.

Pearson, A. R., B. O. Elmore, C. Yang, J. D. Ferrara, A. B. Hooper, and C. M. Wilmot. 2007. The crystal structure of cytochrome P460 of *Nitrosomonas europaea* reveals a novel cytochrome fold and heme-protein cross-link. *Biochemistry* **46:**8340–8349.

Poret-Peterson, A. T., J. E. Graham, J. Gulledge, and M. G. Klotz. 2008. Transcription of nitrification genes by the methane-oxidizing bacterium, *Methylococcus capsulatus* strain Bath. *ISME J.* **2:**1213–1220.

Poughon, L., C. G. Dussap, and J. B. Gros. 2001. Energy model and metabolic flux analysis for autotrophic nitrifiers. *Biotechnol. Bioeng.* **72:**416–433.

Prosser, J. I. 1986. *Nitrification*, p. 217. IRL Press, Oxford, United Kingdom.

Purkhold, U., A. Pommerening-Roser, S. Juretschko, M. C. Schmid, H. P. Koops, and M. Wagner. 2000. Phylogeny of all recognized species of ammonia oxidizers based on comparative 16S rRNA and *amoA* sequence analysis: implications for molecular diversity surveys. *Appl. Environ. Microbiol.* **66:**5368–5382.

Purkhold, U., M. Wagner, G. Timmermann, A. Pommerening-Roser, and H.-P. Koops. 2003. 16S rRNA and *amoA*-based phylogeny of 12 novel betaproteobacterial ammonia-oxidizing isolates: extension of the dataset and proposal of a new lineage within the nitrosomonads. *Int. J. Syst. Evol. Microbiol.* **53:**1485–1494.

Sayavedra-Soto, L. A., N. G. Hommes, and D. J. Arp. 1994. Characterization of the gene encoding hydroxylamine oxidoreductase in *Nitrosomonas europaea. J. Bacteriol.* **176:**504–510.

Sayavedra-Soto, L. A., N. G. Hommes, S. A. Russell, and D. J. Arp. 1996. Induction of ammonia monooxygenase and hydroxylamine oxidoreductase mRNAs by ammonium in *Nitrosomonas europaea. Mol. Microbiol.* **20:**541–548.

Sayavedra-Soto, L. A., N. G. Hommes, J. J. Alzerreca, D. J. Arp, J. M. Norton, and M. G. Klotz. 1998. Transcription of the *amoC*, *amoA*, and *amoB* genes in *Nitrosomonas europaea* and *Nitrosospira* sp. NpAV. *FEMS Microbiol. Lett.* **167:**81–88.

Schmidt, I. 2009. Chemoorganoheterotrophic growth of *Nitrosomonas europaea* and *Nitrosomonas eutropha. Curr. Microbiol.* **59:**130–138.

Severance, S., S. Chakraborty, and D. J. Kosman. 2004. The Ftr1p iron permease in the yeast plasma membrane: orientation, topology and structure-function relationships. *Biochem. J.* **380:**487–496.

Shears, J. H., and P. M. Wood. 1985. Spectroscopic evidence for a photosensitive oxygenated state of ammonia mono-oxygenase. *Biochem. J.* **226:**499–507.

Shears, J. H., and P. M. Wood. 1986. Tri- and tetramethylhydroquinone as electron donors for ammonia monooxygenase in whole cells of *Nitrosomonas europaea. FEMS Microbiol. Lett.* **33:**281–284.

Sherman, W. M., D. L. Costill, W. J. Fink, F. C. Hagerman, L. E. Armstrong, and T. F. Murray. 1983. Effect of a 42.2-km footrace and subsequent rest or exercise on muscle glycogen and enzymes. *J. Appl. Physiol.* **55:**1219–1224.

Shiemke, A. K., D. J. Arp, and L. A. Sayavedra-Soto. 2004. Inhibition of membrane-bound methane monooxygenase and ammonia monooxygenase by diphenyliodonium: implications for electron transfer. *J. Bacteriol.* **186:**928–937.

Shimamura, M., T. Nishiyama, K. Shinya, Y. Kawahara, K. Furukawa, and T. Fujii. 2008. Another multiheme protein, hydroxylamine oxidoreductase, abundantly produced in an anammox bacterium besides the hydrazine-oxidizing enzyme. *J. Biosci. Bioeng.* **105:**243–248.

Stearman, R., D. S. Yuan, Y. Yamaguchi-Iwai, R. D. Klausner, and A. Dancis. 1996. A permease-oxidase complex involved in high-affinity iron uptake in yeast. *Science* **271:**1552–1557.

Stein, L. Y., and D. J. Arp. 1998a. Ammonium limitation results in the loss of ammonia-oxidizing activity in *Nitrosomonas europaea. Appl. Environ. Microbiol.* **64:**1514–1521.

Stein, L. Y., and D. J. Arp. 1998b. Loss of ammonia monooxygenase activity in *Nitrosomonas europaea* upon exposure to nitrite. *Appl. Environ. Microbiol.* **64:**4098–4102.

Stein, L. Y., D. J. Arp, and M. R. Hyman. 1997. Regulation of the synthesis and activity of ammonia monooxygenase in *Nitrosomonas europaea* by altering pH to affect NH_3 availability. *Appl. Environ. Microbiol.* **63:**4588–4592.

Stein, L. Y., L. A. Sayavedra-Soto, N. G. Hommes, and D. J. Arp. 2000. Differential regulation of *amoA* and *amoB* gene copies in *Nitrosomonas europaea. FEMS Microbiol. Lett.* **192:**163–168.

Stein, L. Y., D. J. Arp, P. M. Berube, P. S. Chain, L. Hauser, M. S. Jetten, M. G. Klotz, F. W. Larimer, J. M. Norton, H. J. Op den Camp, M. Shin, and X. Wei. 2007. Whole-genome analysis of the ammonia-oxidizing bacterium, *Nitrosomonas eutropha* C91: implications for niche adaptation. *Environ. Microbiol.* **9:**2993–3007.

Suzuki, I., and S. C. Kwok. 1970. Cell-free ammonia oxidation by *Nitrosomonas europaea* extracts: effects of polyamines, Mg^{2+} and albumin. *Biochem. Biophys. Res. Commun.* **39:**950–955.

Suzuki, I., and S.-C. Kwok. 1981. A partial resolution and reconstitution of the ammonia-oxidizing system of *Nitrosomonas europaea*: role of cytochrome c_{554}. *Can. J. Biochem.* **59:**484–488.

Suzuki, I., U. Dular, and S. C. Kwok. 1974. Ammonia or ammonium ion as substrate for oxidation by *Nitrosomonas europaea* cells and extracts. *J. Bacteriol.* **120:**556–558.

Suzuki, I., S.-C. Kwok, U. Dular, and D. C. Y. Tsang. 1981. Cell-free ammonia-oxidizing system of *Nitrosomonas europaea*: general conditions and properties. *Can. J. Biochem.* **59:**477–483.

Tabita, F. R., T. E. Hanson, S. Satagopan, B. H. Witte, and N. E. Kreel. 2008. Phylogenetic and evolutionary relationships of RubisCO and the RubisCO-like proteins and the functional lessons provided by diverse molecular forms. *Philos. Trans. R. Soc. Lond. B. Biol. Sci.* **363:**2629–2640.

Tarre, S., and M. Green. 2004. High-rate nitrification at low pH in suspended- and attached-biomass reactors. *Appl. Environ. Microbiol.* **70:**6481–6487.

Taylor, P., S. L. Pealing, G. A. Reid, S. K. Chapman, and M. D. Walkinshaw. 1999. Structural and mechanistic mapping of a unique fumarate reductase. *Nat. Struct. Biol.* **6:**1108–1112.

Upadhyay, A. K., D. T. Petasis, D. M. Arciero, A. B. Hooper, and M. P. Hendrich. 2003. Spectroscopic characterization and assignment of reduction potentials in the tetraheme cytochrome c_{554} from *Nitrosomonas europaea*. *J. Am. Chem. Soc.* **125:**1738–1747.

Upadhyay, A. K., A. B. Hooper, and M. P. Hendrich. 2006. NO reductase activity of the tetraheme cytochrome c_{554} of *Nitrosomonas europaea*. *J. Am. Chem. Soc.* **128:**4330–4337.

Utaker, J. B., K. Andersen, A. Aakra, B. Moen, and I. F. Nes. 2002. Phylogeny and functional expression of ribulose 1,5-bisphosphate carboxylase/oxygenase from the autotrophic ammonia-oxidizing bacterium *Nitrosospira* sp. isolate 40KI. *J. Bacteriol.* **184:**468–478.

Vajrala, N., L. A. Sayavedra-Soto, P. J. Bottomley, and D. J. Arp. 2010. Role of *Nitrosomonas europaea* NitABC iron transporter in the uptake of Fe^{3+}-siderophore complexes. *Arch. Microbiol.* **192:**899–908.

Van Dien, S. J., Y. Okubo, M. T. Hough, N. Korotkova, T. Taitano, and M. E. Lidstrom. 2003. Reconstruction of C(3) and C(4) metabolism in *Methylobacterium extorquens* AM1 using transposon mutagenesis. *Microbiology* **149:**601–609.

Vannelli, T., and A. B. Hooper. 1992. Oxidation of nitrapyrin to 6-chloropicolinic acid by the ammonia-oxidizing bacterium *Nitrosomonas europaea*. *Appl. Environ. Microbiol.* **58:**2321–2325.

Wallace, W., S. E. Knowles, and D. J. Nicholas. 1970. Intermediary metabolism of carbon compounds by nitrifying bacteria. *Arch. Mikrobiol.* **70:**26–42.

Watson, G. M., J. P. Yu, and F. R. Tabita. 1999. Unusual ribulose 1,5-bisphosphate carboxylase/oxygenase of anoxic Archaea. *J. Bacteriol.* **181:**1569–1575.

Watson, S. W., L. B. Graham, C. C. Remsen, and F. W. Valois. 1971. A lobular, ammonia-oxidizing bacterium, *Nitrosolobus multiformis* nov. gen. nov. sp. *Arch. Mikrobiol.* **76:**183–203.

Wei, X., N. Vajrala, L. Hauser, L. A. Sayavedra-Soto, and D. J. Arp. 2006. Iron nutrition and physiological responses to iron stress in *Nitrosomonas europaea*. *Arch. Microbiol.* **186:**107–118.

Weidinger, K., B. Neuhauser, S. Gilch, U. Ludewig, O. Meyer, and I. Schmidt. 2007. Functional and physiological evidence for a rhesus-type ammonia transporter in *Nitrosomonas europaea*. *FEMS Microbiol. Lett.* **273:**260–267.

Whittaker, M., D. Bergmann, D. Arciero, and A. B. Hooper. 2000. Electron transfer during the oxidation of ammonia by the chemolithotrophic bacterium *Nitrosomonas europaea*. *Biochim. Biophys. Acta.* **1459:**346–355.

Winkler, F. K. 2006. Amt/MEP/Rh proteins conduct ammonia. *Pflugers Arch.* **451:**701–707.

Wood, P. M. 1986. Nitrification as a bacterial energy source, p. 39–62. *In* J. I. Prosser (ed.), *Nitrification*. IRL Press, Oxford, United Kingdom.

Yamagata, A., R. Hirota, J. Kato, A. Kuroda, T. Ikeda, N. Takiguchi, and H. Ohtake. 2000. Mutational analysis of the multicopy *hao* gene coding for hydroxylamine oxidoreductase in *Nitrosomonas* sp. strain ENI-11. *Biosci. Biotechnol. Biochem.* **64:**1754–1757.

Yamanaka, T., and M. Shinra. 1974. Cytochrome *c*-552 and cytochrome *c*-554 derived from *Nitrosomonas europaea*. Purification, properties, and their function in hydroxylamine oxidation. *J. Biochem.* **75:**1265–1273.

Yoo, J. G., and B. Bowien. 1995. Analysis of the *cbbF* genes from *Alcaligenes eutrophus* that encode fructose-1,6-/sedoheptulose-1,7-bisphosphatase. *Curr. Microbiol.* **31:**55–61.

Zahn, J. A., D. M. Arciero, A. B. Hooper, and A. A. DiSpirito. 1996. Evidence for an iron center in the ammonia monooxygenase from *Nitrosomonas europaea*. *FEBS Lett.* **397:**35–38.

DIVERSITY AND ENVIRONMENTAL DISTRIBUTION OF AMMONIA-OXIDIZING BACTERIA

Jeanette M. Norton

3

INTRODUCTION

Delineating a taxonomic group of bacteria as responsible for an environmental process is often a compromise that must later be retracted, and yet the goal remains enticing. From 1890 through the 1980s, the nitrifying bacteria were grouped in the family *Nitrobacteraceae*, with genera distinguished by their function of ammonia or nitrite oxidation and their cell morphology (Bock et al., 1986). As early as 1971, Stanley Watson (Watson, 1971b) recognized the need for a thorough reconsideration of the family, but restructuring based on phylogeny began in earnest only with the availability of 16S ribosomal RNA oligonucleotide catalogs (Woese et al., 1984, 1985; Head et al., 1993). As additional ribosomal sequencing progressed, it became apparent that the nitrifiers, as a group, were not derived from any ancestral nitrifying phenotype but that these lines of descent arose multiple times from distinct photosynthetic ancestors (Teske et al., 1994) and the ammonia-oxidizing inventory was likely acquired by the ancestor of the family *Nitrosomonadaceae* of the *Betaproteobacteria* through lateral transfer (Klotz and Stein, 2007) (see Chapter 4). More recently, a group of *Archaea* abundant in both terrestrial and marine environments was identified as having genes related to those encoding ammonia monooxygenase (*amo*) in *Bacteria* (Treusch et al., 2005), and isolates have ammonia-oxidizing metabolism (Konneke et al., 2005) (see Section III). Archaeal ammonia oxidation and the anaerobic ammonia oxidation (anammox) processes have important roles in the global nitrogen cycle (Francis et al., 2007) and are discussed in the following chapters. This chapter reviews the diversity, distribution, and biogeography of a subset of the ammonia-oxidizing prokaryotes, the aerobic chemolithotrophic ammonia-oxidizing bacteria (AOB).

PHYLOGENY AND SYSTEMATICS OF AEROBIC AOB

Taxonomic Outline

Current taxonomy of the aerobic chemolithotrophic AOB is based upon ribosomal sequences and comparative genomics (see Chapter 4), while some older retained taxonomic designations are based on cell shape and the arrangement of intracytoplasmic membranes. The terrestrial AOB are generally restricted to the *Betaproteobacteria*, while

Jeanette M. Norton, Department of Plants, Soils and Climate, Utah State University, Logan, UT 84322.

Nitrification, Edited by Bess B. Ward, Daniel J. Arp, and Martin G. Klotz

TABLE 1. Outline of the taxonomy of chemolithotrophic AOB for selected pure culture isolates and strains[a]

Genus and species	Strain	Cluster[b]	Typical habitat(s)	Cell shape	GenBank 16S rRNA	GenBank *amo*	Reference(s)
Betaproteobacteria							
Nitrosomonas europaea	ATCC 19718	7	Soil, water, sewage	Straight rods	Genome: AL954747		Winogradsky, 1892; Chain et al., 2003
N. mobilis[c]	Nc2	7	Brackish water	Coccus	AJ298701	AF037108	Koops et al., 1976
N. communis	Nm2	8	Soil	Rods	AF272417	AF272399	Koops et al., 1991
N. eutropha	C-91 (Nm57)	7	Sewage	Rods to pear	Genome: CP000450 chromosome 1, CP000451 and CP000452 plasmids 1 and 2		Koops et al., 1991; Stein et al., 2007
N. halophila	Nm1 (Nc4)		Salt or soda lakes or lagoons	Short rods	AF272413	AF272398	Koops et al., 1991
N. marina	Nm22	6B	Marine	Straight rods	AF272418	AF272405	Koops et al., 1991
N. nitrosa	Nm90	8	Eutrophic waters, sewage	Coccus or rods	AF272425	AF272404	Koops et al., 1991
N. oligotropha	Nm 45	6A	Freshwater, soil	Straight rods	AF272422	AF272406	Koops et al., 1991
N. ureae	Nm 10	6B	Soils, fresh water	Rods	AF272414	AF272403	Koops et al., 1991
N. aestuarii	Nm36	6B	Brackish water	Rods	AJ298734	AF272400	Koops et al., 1991
N. cryotolerans	NW430 (Nm55)		Marine	Straight rods	AJ298738	AF314753	Jones et al., 1988
Nitrosomonas sp.	Nm143		Marine estuary	Rods	AY123794	AY123816	Purkhold et al., 2003
Nitrosospira sp.	40KI	0	Soil	Spiral tight coils	X84656	AJ298687	Utaker et al., 1995; Jiang and Bakken, 1999
Nitrosospira	No cultures	1	Marine waters or sediments	ND	AY461519 LD2-2 clone	Unknown	Freitag and Prosser, 2004
Nitrosospira sp.	AHB1	2	Acid soils	Spiral	X90820	X90821	DeBoer et al., 1991; Rotthauwe et al., 1995
Nitrosospira briensis	C-76 (Nsp10)	3	Soil	Spirals	AY123800	AY123821	Watson, 1971a; Purkhold et al., 2003
N. briensis	C-128 (Nsp4)	3	Soil	Spiral	L35505 M96396	U76553	Watson, 1971a; Norton et al., 2002
Nitrosospira multiformis[d]	C-71	3	Soil	Lobate	Genome NC_007614 (chromosome), NC_ 007615, NC_007616, NC_ 007617 (plasmids 1, 2, 3)		Watson et al., 1971; Norton et al., 2008
Nitrosospira tenuis[e]	Nv1	3	Soil	Curved rods, vibroid	AY123803	AY123824	Harms et al., 1976; Purkhold et al., 2003

Nitrosospira sp.	AF	3	Acid soil	Vibroid	X84658	AJ298689	Utaker et al., 1995; Jiang and Bakken, 1999
Nitrosospira sp.	KA3	4	Soil	Spirals	AY123806	AY123827	Jiang and Bakken, 1999
Nitrosospira sp.	Nsp57	--	Masonry	Spirals	AY123791	AY123835	Purkhold et al., 2003
Nitrosospira sp.	Nsp65		Masonry	Spirals	AY123813	AY123838	Purkhold et al., 2003
Gammaproteobacteria							
N. oceani	C-107		Marine	Coccus	Genome NC_007484 (chromosome) and NC_007483 (plasmid A)		Watson, 1965; Alzerreca et al., 1999; Klotz et al., 2006
N. oceani	C-27		Marine	Coccus			
N. oceani	AFC27		Marine	Coccus	AF508988	AF509001	Ward and O'Mullan, 2002
N. halophilus	Nc4		Saline ponds	Coccus	AF287298	AJ555509	Koops et al., 1990; Purkhold et al., 2000
Nitrosococcus "watsonii"	C-113		Marine	Coccus	AF153343	AF153344	Alzerreca et al., 1999; M. G. Klotz, personal communication

[a]For strains with a genome sequence available, the accession for the genome is given rather than those for individual genes.
[b]See Purkhold et al. (2003).
[c]Should be reclassified to the genus *Nitrosomonas*.
[d]Previously known as *Nitrosolobus multiformis* (Head et al., 1993).
[e]Formerly *Nitrosovibrio tenuis* (Head et al., 1993).

the marine organisms are found both in the *Betaproteobacteria* and the *Gammaproteobacteria*. The genus *Nitrosococcus* (class *Gammaproteobacteria*, order *Chromatiales*, family *Chromatiaceae*) is widely distributed in marine systems (Ward and O'Mullan, 2002). The known pure cultures and species of the AOB were reviewed (Purkhold et al., 2000), and the phylogenetic lineages or clusters have been outlined (Purkhold et al., 2000; Kowalchuk and Stephen, 2001). A guide tree for reference to cluster designations in the proteobacterial AOB is shown in Fig. 1, and Table 1 gives an updated version of selected cultured AOB with appropriate primary references and GenBank accessions.

The AOB of the *Betaproteobacteria* (*Bacteria*, *Proteobacteria*, *Betaproteobacteria*, *Nitrosomonadales*, *Nitrosomonadaceae*)

Nitrosomonas and *Nitrosospira* are the currently accepted genera comprising the betaproteobacterial AOB. *Nitrosococcus mobilis* is not a validly published name for a betaproteobacterial AOB but has not yet been officially reclassified to the genus *Nitrosomonas*. Phylogeny inference based on the current data set does not support that all nitrosomonads are more closely related to each other than to members of the *Nitrosospira* lineage (Purkhold et al., 2000); especially problematic is the placement of *Nitrosomonas cryotolerans* and *Nitrosomonas* sp. strain Nm143 (Fig. 1). Therefore, the classical divisions within the *Nitrosomonadaceae* into two or more genera may not be retained through revisions based on phylogenomic approaches to taxonomy. While *Nitrosolobus* and *Nitrosovibrio* have been superseded by *Nitrosospira*, this decision remains controversial, especially for "*Nitrosolobus*." Current genome sequencing within these groups may help to delineate the boundaries for the genera or lineages.

NITROSOMONAS

There are at least six lines of descent within the genus *Nitrosomonas* as currently defined (Pommerening-Roser et al., 1996). The lineages are defined by 16S rRNA gene sequences, *amoA* sequences, and ecophysiological characteristics (Purkhold et al., 2000; Koops and Pommerening-Roser, 2001). Unfortunately, many species designations within these groups are in danger of removal from the valid lists since deposit in two different collections in different countries has not been documented (Euzeby and Tindall, 2004) and several type strains are not publicly accessible. Within the *Nitrosomonas*, the lineages and/or species are generally distributed in distinct environments (Koops and Pommerening-Roser, 2001). The crucial environmental characteristics include salinity, ammonia concentration, and pH (see the chapters in Section VI). The cluster 6A represented by *Nitrosomonas oligotropha* and other ammonia-sensitive strains is found primarily in freshwaters but also in estuaries and in terrestrial systems (Coci et al., 2005, 2008; Fierer et al., 2009). Members of the closely related cluster 6B, including *Nitrosomonas aestuarii* and *Nitrosomonas marina*, have tolerance for higher salt concentrations, including marine systems. Cluster 7 includes *Nitrosomonas europaea*, *N. mobilis* strains, and *Nitrosomonas eutropha*, which are capable of tolerating high-ammonia concentrations. Representatives of this group have been isolated from a wide variety of environments, including sewage and wastewaters and also aquatic and terrestrial environments. *Nitrosomonas communis* and *Nitrosomonas nitrosa* and related strains (sometimes referred to as cluster 8) have diverse ecophysiological characteristics and sources. The lineages represented by *Nitrosomonas* sp. strain NM 143 and *N. cryotolerans* are both deep branching marine groups.

NITROSOSPIRA

Stable clusters based on 16S rRNA phylogeny are problematic within *Nitrosospira* because of the overall high levels of identity of the 16S rRNA (>97%). A finer resolution within the group may be achieved by using additional markers such as the 16S-23S rRNA intergenic spacer region (Aakra et al., 2001) or full-length *amoA* genes (Norton et al., 2002).

Additional genome sequencing of *Nitrosospira* spp. and DNA homology value determinations will give further insights. Current clusters with representative isolates include clusters 0, 2, 3, and 4 (Fig. 1 and Table 1); cluster 1 still has no pure culture representative. Soils are often dominated by *Nitrosospira* spp., while marine and freshwater systems often have mixtures of the genera of AOB present. The distribution of *Nitrosospira* clusters is related to ecophysiological traits including pH tolerance, urease activity, pH optimum for ureolysis, and salt tolerance (De Boer and Kowalchuk, 2001; Koops and Pommerening-Roser, 2001; Pommerening-Roser and Koops, 2005). Sequences grouping with cluster 0 have been found in soils and freshwater environments, while cluster 1 sequences are recovered predominantly from marine waters or sediments. Clusters 2, 3, and 4 are found in a variety of environments including soils, freshwater, and marine systems, with cluster 2 sequences often recovered from acidic soils (Kowalchuk and Stephen, 2001). Cluster 3 sequences remain the most commonly recovered from terrestrial environments, particularly from agricultural, grassland, or turfgrass systems (Kowalchuk and Stephen, 2001; Webster et al., 2002; Dell et al., 2008; Le Roux et al., 2008; Norton, 2008). Further divisions of cluster 3 (i.e., into 3A and 3B) have been suggested to facilitate groups with characteristic kinetics or growth parameters (Avrahami et al., 2003; Webster et al., 2005).

The AOB of the *Gammaproteobacteria* (*Bacteria*, *Proteobacteria*, *Gammaproteobacteria*, *Chromatiales*, *Chromatiaceae*, *Nitrosococcus*)

Currently, all AOB in the *Gammaproteobacteria* belong in the genus *Nitrosococcus*. The original type strain for the genus *Nitrosococcus* is *Nitrosococcus winogradskyi* 1892 (Winogradsky, 1892), which has been lost; however, *Nitrosococcus oceani* ATCC 19707 (C-107) (Watson, 1965) is deemed to be very similar to the strain described earlier. *N. oceani* belongs in the family *Chromatiaceae* (ectothiospiraceae branch), also known as the purple sulfur bacteria, Currently, *N. oceani* and *Nitrosococcus halophilus* are the only recognized species of gammaproteobacterial AOB, although an additional strain, *Nitrosococcus watsoni* C-113, has been described. *N. oceani* has been found to be widespread in marine environments by immunofluorescence and detection of both 16S rRNA gene and *amo* sequences from DNA extracted from natural seawater (Ward and O'Mullan, 2002; O'Mullan and Ward, 2005). In addition to the truly marine environment, *Nitrosococcus* was detected in the saline waters of permanently ice-covered lakes in Antarctica (Voytek et al., 1999). *N. halophilus* has been isolated only from saline ponds (Koops et al., 1990). The limited diversity of the gammaproteobacterial AOB detected in environmental samples based on 16S rRNA gene sequences may be partially explained by primer selectivity, but this observation has been confirmed by approaches using *amoA* (O'Mullan and Ward, 2005). A listing of selected *Nitrosococcus* strains is given in Table 1, and their relationships are shown in Fig.1B.

ENVIRONMENTAL DISTRIBUTION AND BIOGEOGRAPHY OF THE AEROBIC AOB

Marine (See also Chapters 7 and 13)

AOB from both the *Gammaproteobacteria* and the *Betaproteobacteria* are found in marine systems, although the archaeal ammonia oxidizers are found in larger numbers in most marine environments examined to date (Francis et al., 2005; Wuchter et al., 2006). Strains of *N. oceani* have been detected in seawater globally and from a permanently ice-covered Antarctic saline lake (Ward and O'Mullan, 2002; O'Mullan and Ward, 2005). *N. cryotolerans*, *N. marina*, *Nitrosomonas* sp. strain Nm143, and some salt-tolerant cluster 6 representatives such as *Nitrosomonas ureae* have been isolated or detected from surface waters and sediments

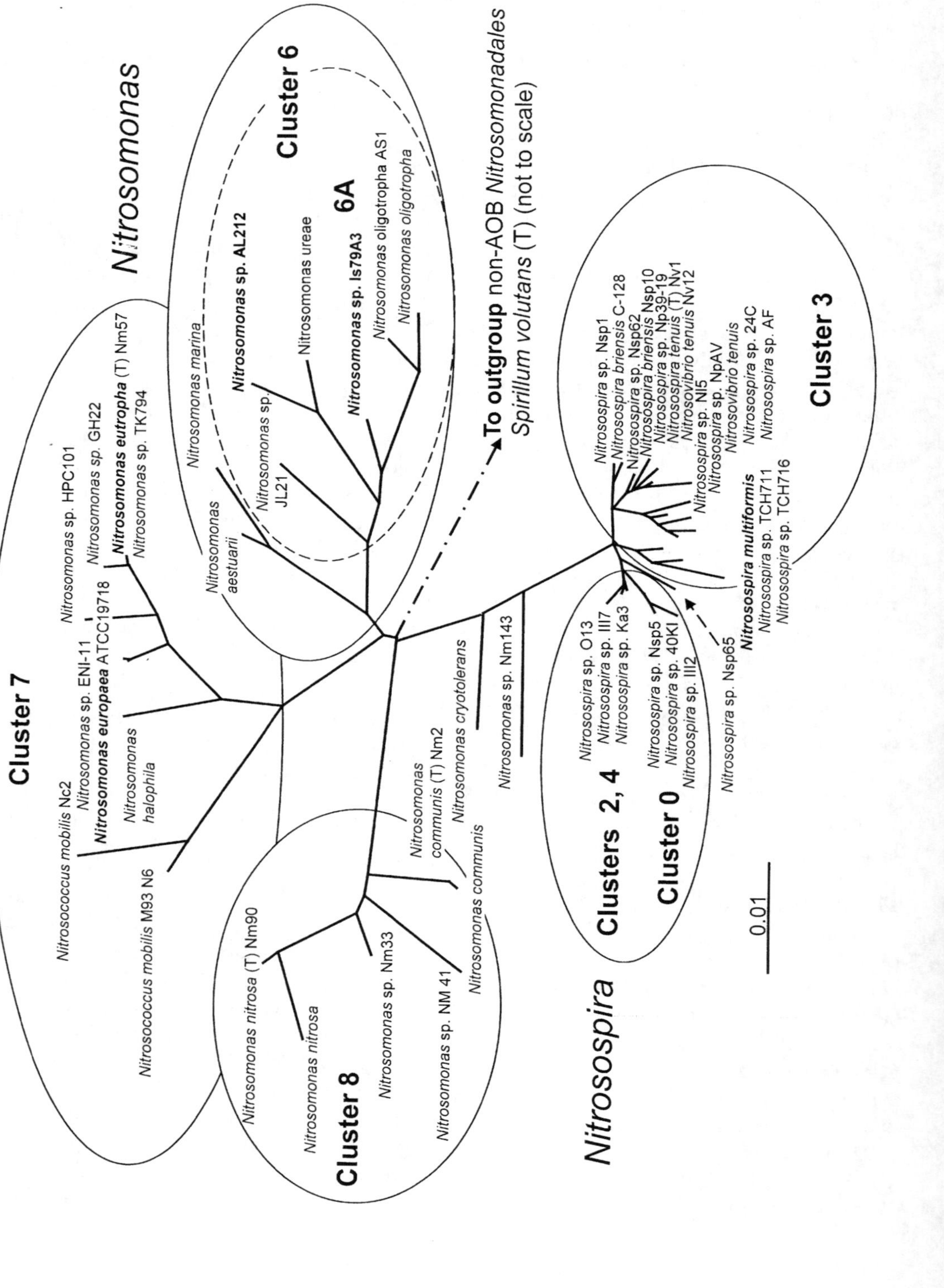
Nitrosomonas
Cluster 6
6A
Nitrosomonas marina
Nitrosomonas sp. AL212
Nitrosomonas ureae
Nitrosomonas sp. Is79A3
Nitrosomonas oligotropha AS1
Nitrosomonas oligotropha
Nitrosomonas sp. JL21
Nitrosomonas aestuarii
To outgroup non-AOB Nitrosomonadales
Spirillum volutans (T) (not to scale)
Cluster 7
Nitrosomonas eutropha (T) Nm57
Nitrosomonas sp. GH22
Nitrosomonas sp. TK794
Nitrosomonas sp. HPC101
Nitrosomonas sp. ENI-11
Nitrosomonas europaea ATCC19718
Nitrosomonas halophila
Nitrosococcus mobilis Nc2
Nitrosococcus mobilis M93 N6
Cluster 8
Nitrosomonas nitrosa (T) Nm90
Nitrosomonas nitrosa
Nitrosomonas sp. Nm33
Nitrosomonas sp. NM 41
Nitrosomonas communis
Nitrosomonas communis (T) Nm2
Nitrosomonas cryotolerans
Nitrosomonas sp. Nm143
Nitrosospira
Clusters 2, 4
Cluster 0
Nitrosospira sp. O13
Nitrosospira sp. III7
Nitrosospira sp. Ka3
Nitrosospira sp. Nsp5
Nitrosospira sp. 40KI
Nitrosospira sp. III2
Nitrosospira sp. Nsp65
Cluster 3
Nitrosospira sp. Nsp1
Nitrosospira briensis C-128
Nitrosospira sp. Nsp62
Nitrosospira briensis Nsp10
Nitrosospira sp. Np39-19
Nitrosospira tenuis (T) Nv1
Nitrosovibrio tenuis Nv12
Nitrosospira sp. Nl5
Nitrosospira sp. NpAV
Nitrosovibrio tenuis
Nitrosospira sp. 24C
Nitrosospira sp. AF
Nitrosospira multiformis
Nitrosospira sp. TCH711
Nitrosospira sp. TCH716
0.01

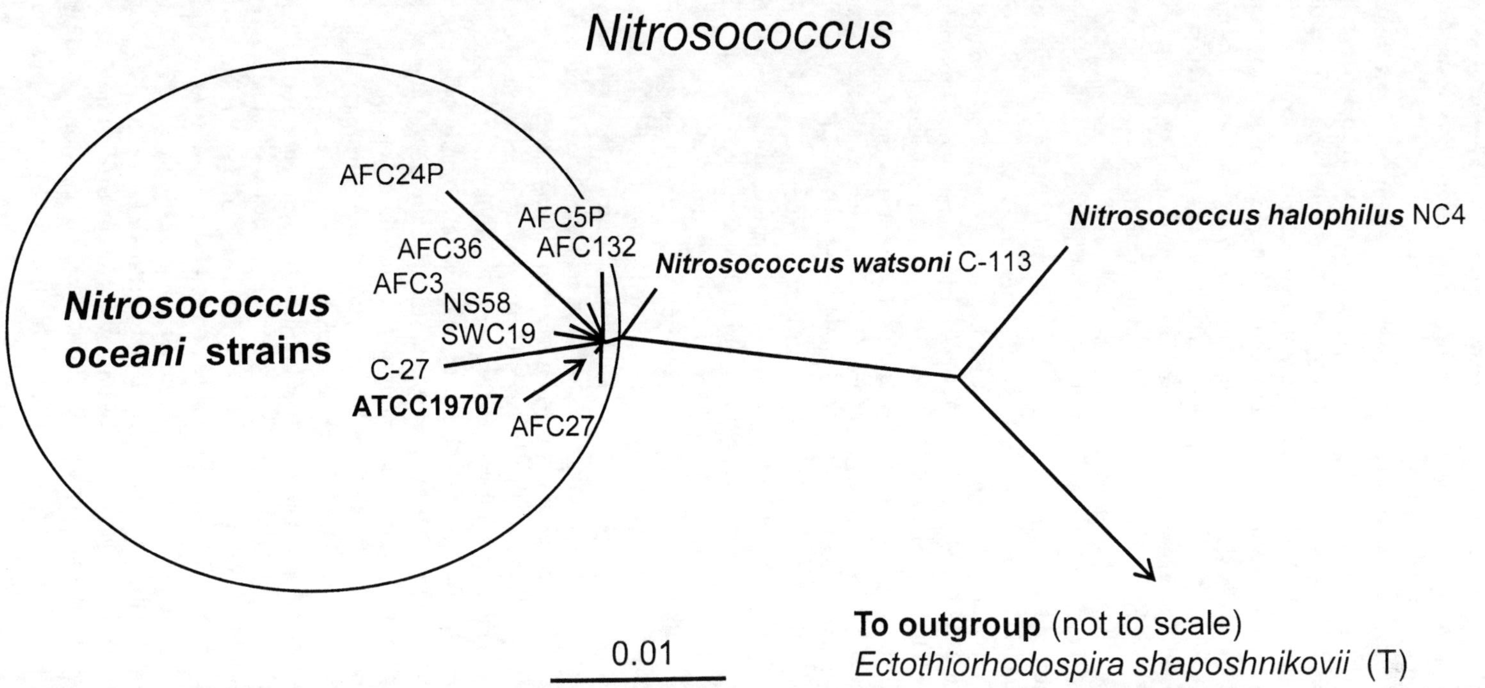

FIGURE 1 Two 16S ribosomal RNA guide trees for the clusters of beta-proteobacterial (top) and gamma-proteobacterial (bottom) AOB based on high-quality sequences (>1,200 bp) from isolates. Sequence data retrieval and analysis was preformed with RDP version 10 database functions (Cole et al., 2009). Several *Nitrosococcus* 16S rRNA gene sequences were from ongoing genomic sequencing projects (M. G. Klotz, personal communication). The scale is substitution per site. Strain selection and cluster designations are based on those of Purkhold et al. (2000, 2003), Kowalchuk and Stephen (2001), and Ward and O'Mullan (2002).

of marine systems (Purkhold et al., 2003; O'Mullan and Ward, 2005; Ward et al., 2007). *Nitrosospira* cluster 1 and cluster 3 sequences have been detected in marine systems (Bano and Hollibaugh, 2000; Freitag and Prosser, 2004; O'Mullan and Ward, 2005). Tolerance of salinity and temperature appear to be strong selective factors on community composition in these environments (Ward et al., 2007).

Estuarine and Freshwater Systems (See Chapters 7 and 15)

The community of AOB has been examined along estuarine gradient in systems from several continents, with the most intensive studies being those in the Chesapeake Bay, USA (Ward et al., 2007); the lower Seine River, France (Cebron et al., 2003, 2004); Plum Island Sound, Massachusetts, USA (Bernhard et al., 2005, 2007); the Ythan Estuary, on the east coast of Scotland, United Kingdom (Freitag et al., 2006); and the Schelde Estuary, The Netherlands, and Belgium (de Bie et al., 2001; Bollmann and Laanbroek, 2002; Coci et al., 2005). Most studies did not evaluate the role or community composition of archaeal ammonia oxidizers, which may have importance in these systems, particularly toward the marine end of salinity gradients (Ward et al., 2007). Observations from several studies show changes in the AOB community with salinity along the gradient from fresh to marine waters. Commonly observed groups include *Nitrosomonas* similar to strain Nm143, *Nitrosomonas* cluster 6A, and nitrosospiras related to others found in marine systems (Bernhard et al., 2005; Freitag et al., 2006; Ward et al., 2007).

Freshwater lakes and streams vary from oligotrophic to eutrophic in character, often as a result of N inputs from wastewater treatment or agriculture. The AOB communities in these systems reflect the altered N status (Whitby et al., 2001; Caffrey et al., 2003; Cebron et al., 2004; Coci et al., 2008). Commonly observed AOB include those related to *N. oligotropha* (cluster 6A) and nitrosospiras in sediments and in epiphytic niches (Coci et al., 2008).

Wastewater and Other Engineered Nitrogen Treatment Systems (See Chapter 16)

Wastewater treatment plants are highly managed environments with specific goals for the treatment of ammonia/ammonium levels. Additional stages of treatment are often used to promote nitrification and to retain nitrifying biomass (Viessman and Hammer, 2004), and secondary or industrial high-ammonia wastes are often treated separately to remove excess ammonia. The AOB have been studied extensively in these systems and are assumed to be a rate-limiting factor, although the focus is often on nitrification kinetics and process characteristics rather than the organisms involved. Both cultivation-dependent and non-cultivation-dependent approaches have found various *Nitrosomonas* to be common AOB in wastewater treatment systems (Wagner et al., 1996; Juretschko et al., 1998; Kelly et al., 2005; Wells et al., 2009), although *Nitrosospira* have also been observed (Park et al., 2002) and may be favored under cooler temperatures and higher dissolved oxygen (Wells et al., 2009). The type strains of *N. eutropha* and *N. nitrosa*, both tolerant of high ammonia levels, were isolated from sewage (Table 1) (Koops et al., 1991). Constructed wetlands for wastewater treatment are often colonized by both *Nitrosospira* and *Nitrosomonas* (Ibekwe et al., 2003; Gorra et al., 2007; Ruiz-Rueda et al., 2009) with controlling factors related to plant species and waste strength. While it is often assumed that wastewater systems have high ammonia/ammonium levels, well-managed, mature systems often maintain high nitrification rates (i.e., high fluxes) through a rather low ammonium/ammonia pool. Therefore, it is not surprising that molecular surveys often find *Nitrosomonas* cluster 6A related to *N. oligotropha* as the most numerous AOB (Park et al., 2002; Siripong and Rittmann, 2007) and that related strains have been isolated from sewage using low-ammonia media (Suwa et al., 1994). Recently, archaeal ammonia-oxidizer *amoA* sequences have been found in some but not all wastewater systems surveyed; their functional importance in these systems remains a topic of current research (Park et al., 2006; Wells et al., 2009; Zhang et al., 2009).

Terrestrial Systems and Soils (See Chapter 14)

Molecular surveys in terrestrial environments commonly find *Nitrosospira* clusters 3, 2, and 4 as the most common types of AOB, while less commonly, the *Nitrosomonas* clusters 6a and 7 have also been observed (Kowalchuk and Stephen, 2001; Prosser and Embley, 2002; Avrahami and Conrad, 2005; Norton, 2008; Fierer et al., 2009). Overall, cluster 3 *Nitrosospira* are the most commonly observed AOB, but this may reflect the large numbers of observations from agricultural and grassland systems worldwide (see the "Biogeography" section). There are some noted correlations with ecophysiological traits and phylogenetic clusters; for example, *Nitrosospira* cluster 2 is associated with acid soils (De Boer and Kowalchuk, 2001; Nugroho et al., 2007), and *Nitrosospira* cluster 4 is more common in native, never-tilled soils (Bruns et al., 1999; Kowalchuk et al., 2000a, 2000b). Generalizations are problematic given the difficulty of differentiating among terrestrial *Nitrosospira* clusters adequately based solely on the 16S rRNA gene signatures. Differences in the genes encoding ammonia monooxygenase, urease, and nitrite reductase and their respective activities may be helpful for further delineating functional traits and ecotypes of *Nitrosospira* (Koper et al., 2004; Avrahami and Conrad, 2005; Pommerening-Roser and Koops, 2005; Webster et al., 2005; Avrahami and Bohannan, 2007; Cantera and Stein, 2007; Garbeva et al., 2007; Le Roux et al., 2008).

AOB COMMUNITIES DURING PRIMARY AND SECONDARY SUCCESSION

Primary succession occurs on newly exposed or deposited substrates such as lava flows, sand dunes, and glacial till and requires the input of generally wind- or water-dispersed propagules from outside the site (Chapin et al., 2002). Ecologists have often investigated nitrogen cycling during succession since nitrogen availability frequently limits plant establishment and growth. While the activity of nitrifiers has been examined (Vitousek et al., 1989; Merila et al., 2002), few studies have specifically examined the ability of AOB to colonize new primary substrates. AOB have been detected in intercontinental airborne dust (Polymenakou et al., 2008) and in recently deglaciated glacier forefields (Nemergut et al., 2007) through the use of molecular methods. Colonization of new substrates is often governed by local site conditions and proximity to sources of inocula (Sigler and Zeyer, 2002; Gomez-Alvarez et al., 2007). Based on inference from pool sizes of inorganic nitrogen and activity measurements, nitrification is often established several decades to centuries after primary succession begins (Kitayama, 1996; Merila et al., 2002; King, 2003; Gomez-Alvarez et al., 2007; Nemergut et al., 2007).

During secondary succession, microbial communities develop on the soils that are often depleted in the number and diversity of microorganisms. Some studies that have examined nitrification during secondary succession include those in postfire forest systems (Smithwick et al., 2005; Turner et al., 2007) and in shifting sand dunes (Kowalchuk et al., 1997) and several during reversion of former agricultural sites (Bruns et al., 1999; Kowalchuk et al., 2000a). Inorganic nitrogen availability and turnover were examined after the severe stand-replacing wildfires of the Yellowstone ecosystem in 2000 (Turner et al., 2007). Soil inorganic N pools (mostly ammonium) were elevated postfire and then rapidly declined. Nitrate and nitrification rates increased annually during the 4 years postfire (Turner et al., 2007). It would be interesting to examine changes in the nitrifier community during this progression. Across a transect of shifting sand dunes spanning approximately 200 years, sequences belonging to the marine clusters *Nitrosomonas* and *Nitrosospira* were recovered from the youngest dunes adjacent to the ocean, while the landward sites were dominanted by *Nitrosospira* from clusters 3, 4, and 2 (Kowalchuk et al., 1997). In calcareous grasslands in the Netherlands, soils in which fertilization was recently halted in early stages of secondary succession were dominated by *Nitrosospira* of cluster 3 shifting toward cluster 4 in older fields that had spent decades without fertilization (Kowalchuk et al., 2000a, 2000b). Similarly, soils that had been tilled and fertilized for 100 years were dominated by *Nitrosospira* cluster 3, while the adjacent native soils also contained sequences from clusters 4 and 2 (Bruns et al., 1999). In pasture soils differing in fertility and plant species management for 10 to 20 years, the AOB community from the improved fertilized site was less diverse, and cluster 3 and 2 *Nitrosospira* were common. In the unimproved site, diverse representatives from clusters 3, 7, 2, and a novel group were detected (Webster et al., 2002, 2005). Further discussion on the communities of soil nitrifiers and their ecophysiological niches is found in Chapter 14.

Environmental and Geographic Limits for AOB (see Section VI)

The AOB have been investigated for more than a century with isolation techniques based on their chemolithotrophic lifestyle (Winogradsky, 1892; Koops and Pommerening-Roser, 2001) and more recently using molecular surveys based on 16S rRNA sequences and genes encoding a key enzyme, ammonia monooxygenase (Kowalchuk and Stephen, 2001; Prosser and Embley, 2002). There have been relatively few examples of environments that exhibited no detectable molecular signatures for the AOB (Bano and Hollibaugh, 2000; Hatzenpichler et al., 2008), although many have noted a need for nested PCR approaches for consistent detection of low-abundance populations (Ward et al., 1997; Hastings et al., 1998; Phillips et al., 1999; Whitby et al., 2001; Fierer et al., 2009). AOB have been detected on all continents (Kowalchuk and Stephen, 2001; Yergeau et al., 2007) and oceans (Ward and O'Mullan, 2002). Activity of *N. cryotolerans* has been detected at well below freezing (Miteva et al., 2007), and AOB have been detected in moderately thermophilic environments (Lebedeva et al., 2005) although ammonia-oxidizing archaea may be the dominant ammonia oxidizers in most high-temperature environments (Zhang et al., 2008). AOB have been detected and isolated in acidic (De Boer and Kowalchuk, 2001) and extreme alkaline (Sorokin et al., 2001) habitats. While the isolation and detection of AOB has been accomplished from a wide variety of environments that supply their basic metabolic needs of ammonia and oxygen, their successful cultivation in the laboratory remains a challenge and requires a careful match with the conditions found in their habitat of origin.

PERSPECTIVES ON BIOGEOGRAPHY OF THE AEROBIC AOB

Speciation, extinction, and dispersal generate the observable distribution of microbes globally (Ramette and Tiedje, 2007a). Microbial biogeography and distributions in the environment are the result of both environmental determination and schochastic dispersal and colonization processes (Martiny et al., 2006; Green et al., 2008). The ability of bacteria to survive extended periods of dormancy under conditions unfavorable for growth promotes their dispersal across ecosystems barriers. The overall soil bacterial community diversity has been compared in terms of phylotype diversity and richness summary variables (Fierer and Jackson, 2006; Horner-Devine and Bohannan, 2006). Soil pH was found to be the best predictor of overall bacterial diversity (Fierer and Jackson, 2006), and this factor may be important for the AOB as well (see Chapter 14). The aerobic AOB have been used as a model for molecular ecology (Kowalchuk and Stephen, 2001) and have been a group of great interest to ecologists and biogeochemists. For these reasons, it is one of the few coherent groups with a known function that has sufficient depth of characterization to complete a biogeographic comparison on the continental scale (Ramette and Tiedje, 2007a, 2007b). For this analysis, *Nitrosospira* 16S rRNA gene sequences

(>440 bp) originating in surface grassland and agricultural soils within the near-neutral pH range (pH 5.8 to 8) but from different continents were selected for comparison (493 total sequences). The selected 16S rRNA gene sequences include those from isolates as well as those from environmental clone libraries from studies worldwide (Koops and Harms, 1985; Utaker et al., 1995; Stephen et al., 1996; Bruns et al., 1999; Mendum et al., 1999; Phillips et al., 2000; Purkhold et al., 2000; Oved et al., 2001; Mendum and Hirsch, 2002; Ida et al., 2005; Nejidat, 2005; Mertens et al., 2006; Song et al., 2007; Dell et al., 2008; Le Roux et al., 2008) as available in the Ribosomal Database Project (Cole et al., 2009). The goal of this analysis was to inquire about biogeography at a finer scale of taxonomic resolution and to explore the issue of whether the *Nitrosospira* spp. are endemic to their locations or cosmopolitan in their distribution (Ramette and Tiedje, 2007a). The matrices of genetic distance and geographic distance were significantly correlated ($r = 0.22$, $P = 0.001$) as determined by the Mantel r test (Dray and Dufour, 2007). The genetic distance between pairs of 16S rRNA gene sequences was greater when the strains were from locations further apart than for those geographically closer (Fig. 2). Our results indicate the existence of a spatial structure in the genus *Nitrosospira* over large geographic distances but at a very fine level of genetic resolution (less than 3% divergence total) (Fig. 2). Understanding the biogeography of functional differentiation within the *Nitrosospira* may require finer-scale tools, and still it may be that a large amount of the observed genetic variability will be unexplained and related to ecologically neutral processes rather than niche differentiation (Ramette and Tiedje, 2007b).

Recently, the AOB communities for 23 soils from a range of ecosystem types (exclusive of agriculture) within North America have been compared (Fierer et al., 2009). At a 97% 16S rRNA gene sequence similarity cutoff level, there were only 24 AOB phylotypes observed (602 total sequences examined); 80% of these were within the *Nitrosospira*. Although there were differences in diversity among sites, the observed spatial patterns were not clearly related to ecosystem type or site characteristics. Overall, the site mean annual temperature was the best predictor of AOB community relatedness (Fierer et al., 2009). The importance of temperature as a selective factor is supported by more detailed ecophysiological investigations on the impact of temperature on AOB communities in grassland and agricultural soils

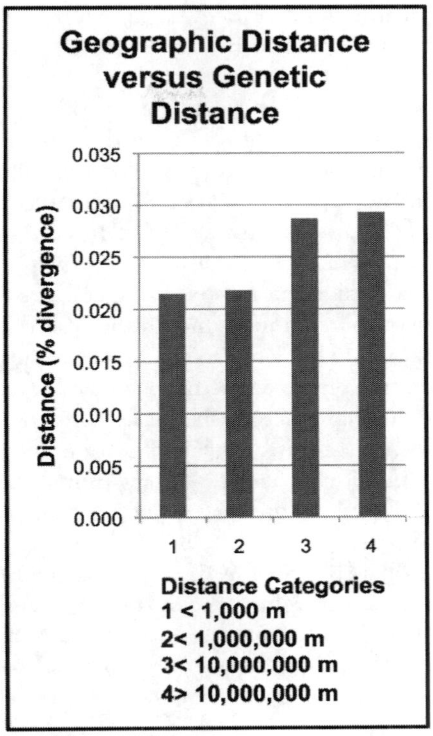

FIG. 2 Pairwise comparisons of 493 16S rRNA gene sequences of *Nitrosospira* spp. for genetic distance (percentage of divergence) and geographic distance (m) between sources. Sequence data retrieval and analysis was performed with RDP version 10 database functions (Cole et al., 2009). Geographic distances calculated in ArcGIS (version 9.1; Environmental Systems Research Institute, Redlands, CA) were transformed into four categories, as displayed. Statistical analysis on the untransformed geographic and sequence distance matrices was performed with the Mantel r test (Ade4 version 1.4-11 (Dray and Dufour, 2007) and indicated that DNA distance increased as geographic distance increased ($r = 0.22$, $P = 0.001$).

(Avrahami et al., 2003; Avrahami and Conrad, 2005; Avrahami and Bohannan, 2007).

On a global basis, it may remain difficult to disentangle the biogeography of AOB in terrestrial habitats given the complex interactions of the soil-forming factors of climate, biota, and topography acting on parent materials over time (Jenny, 1941). In marine habitats, *Nitrosococcus*, *Nitrosomonas*, and *Nitrosospira* coexist with ammonia-oxidizing archaea; their relative contributions and diversity are only beginning to be delineated. Current anthropogenic disruptions of the nitrogen cycle further alter the ecological niche space for nitrifiers. For investigations of local adaptation and distinct functional characteristics within phylogenetically narrow groups, polyphasic taxonomic or genomic approaches will be essential. The functional cohort of AOB, ammonia-oxidizing archaea, and the nitrite-oxidizing prokaryotes will persist as important model organisms for linking the process of nitrification to microbial diversity and biogeography.

ACKNOWLEDGMENTS
This research was supported by the Utah Agricultural Experiment Station and Utah State University and was approved as journal paper number 8141.

REFERENCES

Aakra, A., J. B. Utaker, A. Pommerening-Roser, H. P. Koops, and I. F. Nes. 2001. Detailed phylogeny of ammonia-oxidizing bacteria determined by rDNA sequences and DNA homology values. *Int. J. Syst. Evol. Microbiol.* **51:**2021–2030.

Alzerreca, J. J., J. M. Norton, and M. G. Klotz. 1999. The *amo* operon in marine, ammonia-oxidizing gamma-proteobacteria. *FEMS Microbiol. Lett.* **180:**21–29.

Avrahami, S., and B. J. A. Bohannan. 2007. Response of *Nitrosospira* sp. strain AF-like ammonia oxidizers to changes in temperature, soil moisture content, and fertilizer concentration. *Appl. Environ. Microbiol.* **73:**1166–1173.

Avrahami, S., and R. Conrad. 2005. Cold-temperate climate: a factor for selection of ammonia oxidizers in upland soil? *Can. J. Microbiol.* **51:**709–714.

Avrahami, S., W. Liesack, and R. Conrad. 2003. Effects of temperature and fertilizer on activity and community structure of soil ammonia oxidizers. *Environ. Microbiol.* **5:**691–705.

Bano, N., and J. T. Hollibaugh. 2000. Diversity and distribution of DNA sequences with affinity to ammonia-oxidizing bacteria of the beta subdivision of the class Proteobacteria in the Arctic Ocean. *Appl. Environ. Microbiol.* **66:**1960–1969.

Bernhard, A. E., T. Donn, A. E. Giblin, and D. A. Stahl. 2005. Loss of diversity of ammonia-oxidizing bacteria correlates with increasing salinity in an estuary system. *Environ. Microbiol.* **7:**1289–1297.

Bernhard, A. E., J. Tucker, A. E. Giblin, and D. A. Stahl. 2007. Functionally distinct communities of ammonia-oxidizing bacteria along an estuarine salinity gradient. *Environ. Microbiol.* **9:**1439–1447.

Bock, E., H. P. Koops, and H. Harms. 1986. Cell biology of nitrifying bacteria, p. 17–38. *In* J. I. Prosser (ed.), *Nitrification*. IRL Press, Oxford, United Kingdom.

Bollmann, A., and H. J. Laanbroek. 2002. Influence of oxygen partial pressure and salinity on the community composition of ammonia-oxidizing bacteria in the Schelde Estuary. *Aquat. Microbiol. Ecol.* **28:**239–247.

Bruns, M. A., J. R. Stephen, G. A. Kowalchuk, J. I. Prosser, and E.A. Paul. 1999. Comparative diversity of ammonia oxidizer 16S rRNA gene sequences in native, tilled, and successional soils. *Appl. Environ. Microbiol.* **65:**2994–3000.

Caffrey, J. M., N. Harrington, I. Solem, and B. B. Ward. 2003. Biogeochemical processes in a small California estuary. 2. Nitrification activity, community structure and role in nitrogen budgets. *Mar. Ecol. Prog. Ser.* **248:**27–40.

Cantera, J. J. L., and L. Y. Stein. 2007. Molecular diversity of nitrite reductase genes (*nirK*) in nitrifying bacteria. *Environ. Microbiol.* **9:**765–776.

Cebron, A., T. Berthe, and J. Garnier. 2003. Nitrification and nitrifying bacteria in the lower Seine River and Estuary (France). *Appl. Environ. Microbiol.* **69:**7091–7100.

Cebron, A., M. Coci, J. Garnier, and H. J. Laanbroek. 2004. Denaturing gradient gel electrophoretic analysis of ammonia-oxidizing bacterial community structure in the lower Seine River: impact of Paris wastewater effluents. *Appl. Environ. Microbiol.* **70:**6726–6737.

Chain, P., J. Lamerdin, F. Larimer, W. Regala, V. Lao, M. Land, L. Hauser, A. Hooper, M. Klotz, J. Norton, L. Sayavedra-Soto, D. Arciero, N. Hommes, M. Whittaker, and D. Arp. 2003. Complete genome sequence of the ammonia-oxidizing bacterium and obligate chemolithoautotroph *Nitrosomonas europaea*. *J. Bacteriol.* **185:**2759–2773.

Chapin, F. S., P. A. Matson, and H. A. Mooney. 2002. *Principles of Terrestrial Ecosystem Ecology*. Springer, New York, NY.

Coci, M., D. Riechmann, P. L. E. Bodelier, S. Stefani, G. Zwart, and H. J. Laanbroek. 2005.

Effect of salinity on temporal and spatial dynamics of ammonia-oxidising bacteria from intertidal freshwater sediment. *FEMS Microbiol. Ecol.* **53:**359–368.

Coci, M., P. L. E. Bodelier, and H. J. Laanbroek. 2008. Epiphyton as a niche for ammonia-oxidizing bacteria: detailed comparison with benthic and pelagic compartments in shallow freshwater lakes. *Appl. Environ. Microbiol.* **74:**1963–1971.

Cole, J. R., Q. Wang, E. Cardenas, J. Fish, B. Chai, R. J. Farris, A. S. Kulam-Syed-Mohideen, D. M. McGarrell, T. Marsh, G. M. Garrity, and J. M. Tiedje. 2009. The Ribosomal Database Project: improved alignments and new tools for rRNA analysis. *Nucl. Acids Res.* **37:**D141–D145.

de Bie, M. J. M., A. Speksnijder, G. A. Kowalchuk, T. Schuurman, G. Zwart, J. R. Stephen, O. E. Diekmann, and H. J. Laanbroek. 2001. Shifts in the dominant populations of ammonia-oxidizing beta-subclass Proteobacteria along the eutrophic Schelde Estuary. *Aquat. Microbiol. Ecol.* **23:**225–236.

De Boer, W., and G. A. Kowalchuk. 2001. Nitrification in acid soils: micro-organisms and mechanisms. *Soil Biol. Biochem.* **33:**853–866.

De Boer, W., P. Gunnewiek, M. Veenhuis, E. Bock, and H. J. Laanbroek. 1991. Nitrification at low pH by aggregated chemolithotrophic bacteria. *Appl. Environ. Microbiol.* **57:**3600–3604.

Dell, E. A., D. Bowman, T. Rufty, and W. Shi. 2008. Intensive management affects composition of betaproteobacterial ammonia oxidizers in turfgrass systems. *Microb. Ecol.* **56:**178–190.

Dray, S., and A. B. Dufour. 2007. The ade4 package: implementing the duality diagram for ecologists. *J. Stat. Softw.* **22:**1–20.

Euzeby, J. P. and B. J. Tindall. 2004. Status of strains that contravene Rules 27(3) and 30 of the Bacteriological Code. Request for an Opinion. *Int. J. Syst. Evol. Microbiol.* **54:**293–301.

Fierer, N., and R. B. Jackson. 2006. The diversity and biogeography of soil bacterial communities. *Proc. Natl. Acad. Sci. USA* **103:**626–631.

Fierer, N., K. M. Carney, M. C. Horner-Devine, and J. P. Megonigal. 2009. The biogeography of ammonia-oxidizing bacterial communities in soil. *Microbiol. Ecol.* **58:**435–445.

Francis, C. A., K. J. Roberts, J. M. Beman, A. E. Santoro, and B. B. Oakley. 2005. Ubiquity and diversity of ammonia-oxidizing archaea in water columns and sediments of the ocean. *Proc. Natl. Acad. Sci. USA* **102:**14683–14688.

Francis, C. A., J. M. Beman, and M. M. Kuypers. 2007. New processes and players in the nitrogen cycle: the microbial ecology of anaerobic and archaeal ammonia oxidation. *ISME J.* **1:**19–27.

Freitag, T. E., and J. I. Prosser. 2004. Differences between betaproteobacterial ammonia-oxidizing communities in marine sediments and those in overlying water. *Appl. Environ. Microbiol.* **70:**3789–3793.

Freitag, T. E., L. Chang, and J. I. Prosser. 2006. Changes in the community structure and activity of betaproteobacterial ammonia-oxidizing sediment bacteria along a freshwater-marine gradient. *Environ. Microbiol.* **8:**684–696.

Garbeva, P., E. M. Baggs, and J. I. Prosser. 2007. Phylogeny of nitrite reductase (*nirK*) and nitric oxide reductase (*norB*) genes from *Nitrosospira* species isolated from soil. *FEMS Microbiol. Lett.* **266:**83–89.

Gomez-Alvarez, V., G. M. King, and K. Nusslein. 2007. Comparative bacterial diversity in recent Hawaiian volcanic deposits of different ages. *FEMS Microbiol. Ecol.* **60:**60–73.

Gorra, R., M. Coci, R. Ambrosoli, and H. J. Laanbroek. 2007. Effects of substratum on the diversity and stability of ammonia-oxidizing communities in a constructed wetland used for wastewater treatment. *J. Appl. Microbiol.* **103:**1442–1452.

Green, J. L., B. J. M. Bohannan, and R. J. Whitaker. 2008. Microbial biogeography: from taxonomy to traits. *Science* **320:**1039–1043.

Harms, H., H. P. Koops, and H. Wehrmann. 1976. An ammonia-oxidizing bacterium, *Nitrosovibrio tenuis* nov. gen. nov. sp. *Arch. Microbiol.* **108:**105–111.

Hastings, R. C., J. R. Saunders, G. H. Hall, R. W. Pickup, and A. J. McCarthy. 1998. Application of molecular biological techniques to a seasonal study of ammonia oxidation in a eutrophic freshwater lake. *Appl. Environ. Microbiol.* **64:**3674–3682.

Hatzenpichler, R., E. V. Lebecleva, E. Spieck, K. Stoecker, A. Richter, H. Daims, and M. Wagner. 2008. A moderately thermophilic ammonia-oxidizing crenarchaeote from a hot spring. *Proc. Natl. Acad. Sci. USA* **105:**2134–2139.

Head, I. M., W. D. Hiorns, T. M. Embley, A. J. McCarthy, and J. R. Saunders. 1993. The phylogeny of autotrophic ammonia-oxidizing bacteria as determined by analysis of 16S ribosomal-RNA sequences *J. Gen. Microbiol.* **139:**1147–1153.

Horner-Devine, M. C., and B. J. M. Bohannan. 2006. Phylogenetic clustering and overdispersion in bacterial communities. *Ecology* **87:**S100–S108.

Ibekwe, A. M., C. M. Grieve, and S. R. Lyon. 2003. Characterization of microbial communities and composition in constructed dairy wetland wastewater effluent. *Appl. Environ. Microbiol.* **69:**5060–5069.

Ida, T., M. Kugimiya, M. Kogure, R. Takahashi, and T. Tokuyama. 2005. Phylogenetic relationships among ammonia-oxidizing bacteria as revealed by gene sequences of glyceraldehyde

3-phosphate dehydrogenase and phosphoglycerate kinase. *J. Biosci. Bioeng.* **99:**569–576.

Jenny, H. 1941. *Factors of Soil Formation; a System of Quantitative Pedology*. McGraw Hill, New York, NY.

Jiang, Q. Q., and L. R. Bakken. 1999. Comparison of *Nitrosospira* strains isolated from terrestrial environments. *FEMS Microbiol. Ecol.* **30:**171–186.

Jones, R. D., R. Y. Morita, H. P. Koops, and S. W. Watson. 1988. A new marine ammonium-oxidizing bacterium, *Nitrosomonas cryotolerans* sp.-nov. *Can. J. Microbiol.* **34:**1122–1128.

Juretschko, S., G. Timmermann, M. Schmid, K. H. Schleifer, A. Pommerening-Roser, H. P. Koops, and M. Wagner. 1998. Combined molecular and conventional analyses of nitrifying bacterium diversity in activated sludge: *Nitrosococcus mobilis* and *Nitrospira*-like bacteria as dominant populations. *Appl. Environ. Microbiol.* **64:**3042–3051.

Kelly, J. J., S. Siripong, J. McCormack, L. R. Janus, H. Urakawa, S. El Fantroussi, P. A. Noble, L. Sappelsa, B. E. Rittmann, and D. A. Stahl. 2005. DNA microarray detection of nitrifying bacterial 16S rRNA in wastewater treatment plant samples. *Water Res.* **39:**3229–3238.

King, G. M. 2003. Contributions of atmospheric CO and hydrogen uptake to microbial dynamics on recent Hawaiian volcanic deposits. *Appl. Environ. Microbiol.* **69:**4067-4075.

Kitayama, K. 1996. Soil nitrogen dynamics along a gradient of long-term soil development in a Hawaiian wet montane rainforest. *Plant Soil* **183:**253–262.

Klotz, M. G., and L. Y. Stein. 2007. Nitrifier genomics and evolution of the N-cycle. *FEMS Microbiol. Lett.* **278:**146–156.

Klotz, M. G., D. J. Arp, P. S. G. Chain, A. F. El-Sheikh, L. J. Hauser, N. G. Hommes, F. W. Larimer, S. A. Malfatti, J. M. Norton, A. T. Poret-Peterson, L. M. Vergez, and B. B. Ward. 2006. Complete genome sequence of the marine, chemolithoautotrophic, ammonia-oxidizing bacterium *Nitrosococcus oceani* ATCC 19707. *Appl. Environ. Microbiol.* **72:**6299–6315.

Konneke, M., A. E. Bernhard, J. R. de la Torre, C. B. Walker, J. B. Waterbury, and D. A. Stahl. 2005. Isolation of an autotrophic ammonia-oxidizing marine archaeon. *Nature* **437:**543–546.

Koops, H. P., and H. Harms. 1985. Deoxyribonucleic-acid homologies among 96 strains of ammonia-oxidizing bacteria. *Arch. Microbiol.* **141:**214–218.

Koops, H. P., and A. Pommerening-Roser. 2001. Distribution and ecophysiology of the nitrifying bacteria emphasizing cultured species. *FEMS Microbiol. Ecol.* **37:**1–9.

Koops, H. P., H. Harms, and H. Wehrmann. 1976. Isolation of a moderate halophilic ammonia-oxidizing bacterium, *Nitrosococcus mobilis* nov. sp. *Arch. Microbiol.* **107:**277–282.

Koops, H. P., B. Bottcher, U. C. Moller, A. Pommereningroser, and G. Stehr. 1990. Description of a new species of *Nitrosococcus. Arch. Microbiol.* **154:**244–248.

Koops, H.P ., B. Bottcher, U. C. Moller, A. Pommerening-Roser, and G. Stehr. 1991. Classification of eight new species of ammonia-oxidizing bacteria: *Nitrosomonas communis* sp. nov., *Nitrosomonas ureae* sp. nov., *Nitrosomonas aestuarii* sp. nov., *Nitrosomonas marina* sp. nov., *Nitrosomonas nitrosa* sp. nov., *Nitrosomonas oligotropha* sp. nov., *Nitrosomonas halophila* sp. nov. *J. Gen. Microbiol.* **137:**1689–1699.

Koper, T. E., A. F. El-Sheikh, J. M. Norton, and M. G. Klotz. 2004. Urease-encoding genes in ammonia-oxidizing bacteria. *Appl. Environ. Microbiol.* **70:**2342–2348.

Kowalchuk, G. A., and J. R. Stephen. 2001. Ammonia-oxidizing bacteria: a model for molecular microbial ecology. *Annu. Rev. Microbiol.* **55:**485–529.

Kowalchuk, G. A., J. R. Stephen, W. DeBoer, J. I. Prosser, T. M. Embley, and J. W. Woldendorp. 1997. Analysis of ammonia-oxidizing bacteria of the beta subdivision of the class Proteobacteria in coastal sand dunes by denaturing gradient gel electrophoresis and sequencing of PCR-amplified 16S ribosomal DNA fragments. *Appl. Environ. Microbiol.* **63:**1489–1497.

Kowalchuk, G. A., A. W. Stienstra, G. H. J. Heilig, J. R. Stephen, and J. W. Woldendorp. 2000a. Changes in the community structure of ammonia-oxidizing bacteria during secondary succession of calcareous grasslands. *Environ. Microbiol.* **2:**99–110.

Kowalchuk, G. A., A. W. Stienstra, G. H. J. Heilig, J. R. Stephen, and J. W. Woldendorp. 2000b. Molecular analysis of ammonia-oxidising bacteria in soil of successional grasslands of the Drentsche A (The Netherlands). *FEMS Microbiol. Ecol.* **31:**207–215.

Le Roux, X., F. Poly, P. Currey, C. Commeaux, B. Hai, G. W. Nicol, J. I. Prosser, M. Schloter, E. Attard, and K. Klumpp. 2008. Effects of aboveground grazing on coupling among nitrifier activity, abundance and community structure. *ISME J.* **2:**221–232.

Lebedeva, E. V., M. Alawi, C. Fiencke, B. Namsaraev, E. Bock, and E. Spieck. 2005. Moderately thermophilic nitrifying bacteria from a hot spring of the Baikal rift zone. *FEMS Microbiol. Ecol.* **54:**297–306.

Martiny, J. B. H., B. J. M. Bohannan, J. H. Brown, R. K. Colwell, J. A. Fuhrman, J. L. Green, M. C. Horner-Devine, M. Kane, J. A. Krumins, C. R. Kuske, P. J. Morin, S. Naeem, L. Ovreas, A. L. Reysenbach, V. H. Smith, and J. T. Staley. 2006. Microbial biogeography: putting microorganisms on the map. *Nat. Rev. Microbiol.* **4:**102–112.

Mendum, T. A., and P. R. Hirsch. 2002. Changes in the population structure of beta-group autotrophic ammonia oxidising bacteria in arable soils in response to agricultural practice. *Soil Biol. Biochem.* **34:**1479–1485.

Mendum, T. A., R. E. Sockett, and P. R. Hirsch. 1999. Use of molecular and isotopic techniques to monitor the response of autotrophic ammonia-oxidizing populations of the beta subdivision of the class Proteobacteria in arable soils to nitrogen fertilizer. *Appl. Environ. Microbiol.* **65:**4155–4162.

Merila, P., A. Smolander, and R. Strommer. 2002. Soil nitrogen transformations along a primary succession transect on the land-uplift coast in western Finland. *Soil Biol. Biochem.* **34:**373–385.

Mertens, J., D. Springael, I. De Troyer, K. Cheyns, P. Wattiau, and E. Smolders. 2006. Long-term exposure to elevated zinc concentrations induced structural changes and zinc tolerance of the nitrifying community in soil. *Environ. Microbiol.* **8:**2170–2178.

Miteva, V., T. Sowers, and J. Brenchley. 2007. Production of N_2O by ammonia oxidizing bacteria at subfreezing temperatures as a model for assessing the N_2O anomalies in the vostok ice core. *Geomicrobiol. J.* **24:**451–459.

Nejidat, A. 2005. Nitrification and occurrence of salt-tolerant nitrifying bacteria in the Negev desert soils. *FEMS Microbiol. Ecol.* **52:**21–29.

Nemergut, D. R., S. P. Anderson, C. C. Cleveland, A. P. Martin, A. E. Miller, A. Seimon, and S. K. Schmidt. 2007. Microbial community succession in an unvegetated, recently deglaciated soil. *Microb. Ecol.* **53:**110–122.

Norton, J. M. 2008. Nitrification in agricultural soils, p. 173–199. *In* J. S. Schepers and W. R. Raun (ed.), *Nitrogen in Agricultural Systems.* American Society of Agronomy, Inc.; Crop Science Society of America, Inc.; Soil Science Society of America, Inc., Madison, WI.

Norton, J. M., J. J. Alzerreca, Y. Suwa, and M. G. Klotz. 2002. Diversity of ammonia monooxygenase operon in autotrophic ammonia-oxidizing bacteria. *Arch. Microbiol.* **177:**139–149.

Norton, J. M., M. G. Klotz, L. Y. Stein, D. J. Arp, P. J. Bottomley, P. S. G. Chain, L. J. Hauser, M. L. Land, F. W. Larimer, M. W. Shin, and S. R. Starkenburg. 2008. Complete genome sequence of *Nitrosospira multiformis*, an ammonia-oxidizing bacterium from the soil environment. *Appl. Environ. Microbiol.* **74:**3559–3572.

Nugroho, R. A., W. F. M. Roling, A. M. Laverman, and H. A. Verhoef. 2007. Low nitrification rates in acid scots pine forest soils are due to pH-related factors. *Microbiol. Ecol.* **53:**89–97.

O'Mullan, G. D., and B. B. Ward. 2005. Relationship of temporal and spatial variabilities of ammonia-oxidizing bacteria to nitrification rates in Monterey Bay, California. *Appl. Environ. Microbiol.* **71:**697–705.

Oved, T., A. Shaviv, T. Goldrath, R. T. Mandelbaum, and D. Minz. 2001. Influence of effluent irrigation on community composition and function of ammonia-oxidizing bacteria in soil. *Appl. Environ. Microbiol.* **67:**3426–3433.

Park, H. D., J. M. Regan, and D. R. Noguera. 2002. Molecular analysis of ammonia-oxidizing bacterial populations in aerated-anoxic Orbal processes. *Water Sci. Tech.* **46:**273–280.

Park, H. D., G. F. Wells, H. Bae, C. S. Criddle, and C. A. Francis. 2006. Occurrence of ammonia-oxidizing archaea in wastewater treatment plant bioreactors. *Appl. Environ. Microbiol.* **72:**5643–5647.

Phillips, C. J., Z. Smith, T. M. Embley, and J. I. Prosser. 1999. Phylogenetic differences between particle-associated and planktonic ammonia-oxidizing bacteria of the beta subdivision of the class Proteobacteria in the northwestern Mediterranean Sea. *Appl. Environ. Microbiol.* **65:**779–786.

Phillips, C. J., D. Harris, S. L. Dollhopf, K. L. Gross, J. I. Prosser, and E. A. Paul. 2000. Effects of agronomic treatments on structure and function of ammonia-oxidizing communities. *Appl. Environ. Microbiol.* **66:**5410–5418.

Polymenakou, P. N., M. Mandalakis, E. G. Stephanou, and A. Tselepides. 2008. Particle size distribution of airborne microorganisms and pathogens during an intense African dust event in the eastern Mediterranean. *Environ. Health Perspect.* **116:**292–296.

Pommerening-Roser, A., and H. P. Koops. 2005. Environmental pH as an important factor for the distribution of urease positive ammonia-oxidizing bacteria. *Microbiol. Res.* **160:**27–35.

Pommerening-Roser, A., G. Rath, and H. P. Koops. 1996. Phylogenetic diversity within the genus *Nitrosomonas*. *Syst. Appl. Microbiol.* **19:**344–351.

Prosser, J. I., and T. M. Embley. 2002. Cultivation-based and molecular approaches to characterisation of terrestrial and aquatic nitrifiers. *Antonie Van Leewenhoek Int. J. Gen. Mol. Microbiol.* **81:**165–179.

Purkhold, U., A. Pommerening-Roser, S. Juretschko, M. C. Schmid, H. P. Koops, and M. Wagner. 2000. Phylogeny of all recognized species of ammonia oxidizers based on comparative 16S rRNA and *amoA* sequence analysis: Implications for molecular diversity surveys. *Appl. Environ. Microbiol.* **66:**5368–5382.

Purkhold, U., M. Wagner, G. Timmermann, A. Pommerening-Roser, and H. P. Koops. 2003. 16S rRNA and amoA-based phylogeny of 12 novel betaproteobacterial ammonia-oxidizing isolates: extension of the dataset and proposal of a new lineage within the nitrosomonads. *Int. J. Syst. Evol. Microbiol.* **53:**1485–1494.

Ramette, A., and J. M. Tiedje. 2007a. Biogeography: An emerging cornerstone for understanding prokaryotic diversity, ecology, and evolution. *Microbiol. Ecol.* **53:**197–207.

Ramette, A., and J. M. Tiedje. 2007b. Multiscale responses of microbial life to spatial distance and environmental heterogeneity in a patchy ecosystem. *Proc. Natl. Acad. Sci. USA* **104:**2761–2766.

Rotthauwe, J. H., W. de Boer, and W. Liesak. 1995. Comparative analysis of gene sequences encoding for ammonia monooxygenase of *Nitrosospira* sp. AHB1 and *Nitrosolobus multiformis* C-71. *FEMS Microbiol. Lett.* **133:**133–135.

Ruiz-Rueda, O., S. Hallin, and L. Baneras. 2009. Structure and function of denitrifying and nitrifying bacterial communities in relation to the plant species in a constructed wetland. *FEMS Microbiol. Ecol.* **67:**308–319.

Sigler, W. V., and J. Zeyer. 2002. Microbial diversity and activity along the forefields of two receding glaciers. *Microbiol. Ecol.* **43:**397–407.

Siripong, S., and B. E. Rittmann. 2007. Diversity study of nitrifying bacteria in full-scale municipal wastewater treatment plants. *Water Res.* **41:**1110–1120.

Smithwick, E. A. H., M. G. Turner, M. C. Mack, and F. S. Chapin. 2005. Postfire soil N cycling in northern conifer forests affected by severe, stand-replacing wildfires. *Ecosystems* **8:**163–181.

Song, Y. N., P. Marschner, L. Li, X. G. Bao, J. H. Sun, and F. S. Zhang. 2007. Community composition of ammonia-oxidizing bacteria in the rhizosphere of intercropped wheat (*Triticum aestivum* L.), maize (*Zea mays* L.), and faba bean (*Vicia faba* L.). *Biol. Fertil. Soils* **44:**307–314.

Sorokin, D., T. Tourova, M. C. Schmid, M. Wagner, H. P. Koops, J. G. Kuenen, and M. Jetten. 2001. Isolation and properties of obligately chemolithoautotrophic and extremely alkali-tolerant ammonia-oxidizing bacteria from Mongolian soda lakes. *Arch. Microbiol.* **176:**170–177.

Stein, L. Y., D. J. Arp, P. M. Berube, P. S. G. Chain, L. Hauser, M. S. M. Jetten, M. G. Klotz, F. W. Larimer, J. M. Norton, H. J. M. O. d. Camp, M. Shin, and X. Wei. 2007. Whole-genome analysis of the ammonia-oxidzing bacterium, *Nitrosomonas eutropha* C91: implications for niche adaptation. *Environ. Microbiol.* **9:**2993–3007.

Stephen, J. R., A. E. McCaig, Z. Smith, J. I. Prosser, and T. M. Embley. 1996. Molecular diversity of soil and marine 16S rRNA gene sequences related to beta-subgroup ammonia-oxidizing bacteria. *Appl. Environ. Microbiol.* **62:**4147–4154.

Suwa, Y., Y. Imamura, T. Suzuki, T. Tashiro, and Y. Urushigawa. 1994. Ammonia-oxidizing bacteria with different sensitivities to $(NH_4)_2SO_4$ in activated sludges. *Water Res.* **28:**1523–1532.

Teske, A., E. Alm, J. M. Regan, S. Toze, B. E. Rittmann, and D. A. Stahl. 1994. Evolutionary relationships among ammonia-oxidizing and nitrite-oxidizing bacteria. *J. Bacteriol.* **176:**6623–6630.

Treusch, A. H., S. Leininger, A. Kletzin, S. C. Schuster, H. P. Klenk, and C. Schleper. 2005. Novel genes for nitrite reductase and Amo-related proteins indicate a role of uncultivated mesophilic crenarchaeota in nitrogen cycling. *Environ. Microbiol.* **7:**1985–1995.

Turner, M. G., E. A. H. Smithwick, K. L. Metzger, D. B. Tinker, and W. H. Romme. 2007. Inorganic nitrogen availability after severe stand-replacing fire in the Greater Yellowstone Ecosystem. *Proc. Natl. Acad. Sci. USA* **104:**4782–4789.

Utaker, J. B., L. Bakken, L. J. Jiang, and I. F. Nes. 1995. Phylogenetic analysis of seven new isolates of ammonia-oxidizing bacteria based on 16S rRNA gene sequences. *Syst. Appl. Microbiol.* **18:**549–559.

Viessman, W., and M. J. Hammer. 2004. *Water Supply and Pollution Control.* Pearson Prentice Hall, Upper Saddle River, NJ.

Vitousek, P. M., P. A. Matson, and K. Vancleve. 1989. Nitrogen availability and nitrification during succession-primary, secondary and old-field seres. *Plant Soil* **115:**229–239.

Voytek, M. A., J. C. Priscu, and B. B. Ward. 1999. The distribution and relative abundance of ammonia-oxidizing bacteria in lakes of the McMurdo Dry Valley, Antarctica. *Hydrobiologia* **401:**113–130.

Wagner, M., G. Rath, H. P. Koops, J. Flood, and R. Amann. 1996. In situ analysis of nitrifying bacteria in sewage treatment plants. *Water Sci. Tech.* **34:**237–244.

Ward, B. B., and G. D. O'Mullan. 2002. Worldwide distribution of *Nitrosococcus oceani*, a marine ammonia-oxidizing gamma-proteobacterium, detected by PCR and sequencing of 16S rRNA and amoA genes. *Appl. Environ. Microbiol.* **68:**4153–4157.

Ward, B. B., M. A. Voytek, and R. P. Witzel. 1997. Phylogenetic diversity of natural populations of ammonia oxidizers investigated by specific PCR amplification. *Microbiol. Ecol.* **33:**87–96.

Ward, B. B., D. Eveillard, J. D. Kirshtein, J. D. Nelson, M. A. Voytek, and G. A. Jackson. 2007. Ammonia-oxidizing bacterial community composition in estuarine and oceanic environments assessed using a functional gene microarray. *Environ. Microbiol.* **9:**2522–2538.

Watson, S. W. 1965. Characteristics of a marine nitrifying bacterium, *Nitrosocystis oceanus* sp. nov. *Limnol. Oceanogr.* **10:**R247–R289.

Watson, S. W. 1971a. Reisolation of *Nitrosospira briensis* S. Winogradsky and H. Winogradsky 1933. *Arch Mikrobiol* **75:**179–188.

Watson, S.W. 1971b. Taxonomic considerations of the family *Nitrobacteraceae* Buchanan. *Int. J. Syst. Bacteriol.* **21:**254–270.

Watson, S. W., L. B. Graham, C. C. Remsen, and F. W. Valois. 1971. A lobular, ammonia-oxidizing bacterium, *Nitrosolobus multiuformis* nov. gen. nov. sp. *Arch. Microbiol.* **76:**183–203.

Webster, G., T. M. Embley, and J. I. Prosser. 2002. Grassland management regimens reduce small-scale heterogeneity and species diversity of beta-proteobacterial ammonia oxidizer populations. *Appl. Environ. Microbiol.* **68:**20–30.

Webster, G., T. M. Embley, T. E. Freitag, Z. Smith, and J. I. Prosser. 2005. Links between ammonia oxidizer species composition, functional diversity and nitrification kinetics in grassland soils. *Environ. Microbiol.* **7:**676–684.

Wells, G. F., H. D. Park, C. H. Yeung, B. Eggleston, C. A. Francis, and C. S. Criddle. 2009. Ammonia-oxidizing communities in a highly aerated full-scale activated sludge bioreactor: betaproteobacterial dynamics and low relative abundance of *Crenarchaea*. *Environ. Microbiol.* **11:**2310–2328.

Whitby, C. B., J. R. Saunders, R. W. Pickup, and A. J. McCarthy. 2001. A comparison of ammonia-oxidiser populations in eutrophic and oligotrophic basins of a large freshwater lake. *Antonie Van Leewenhoek Int. J. Gen. Mol. Microbiol.* **79:**179–188.

Winogradsky, S. 1892. Contributions a la morphologie des organismes de la nitrification. *Arch. Sci. Biol. St. Peterburg.* **1:**86–137

Woese, C. R., W. G. Weisburg, B. J. Paster, C. M. Hahn, R. S. Tanner, N. R. Krieg, H. P. Koops, H. Harms, and E. Stackebrandt. 1984. The phylogeny of the purple bacteria—the Beta-subdivision. *Syst. Appl. Microbiol.* **5:**327–336.

Woese, C. R., W. G. Weisburg, C. M. Hahn, B. J. Paster, L. B. Zablen, B. J. Lewis, T. J. Macke, W. Ludwig, and E. Stackebrandt. 1985. The phylogeny of the purple bacteria-the Gamma-subdivision. *Syst. Appl. Microbiol.* **6:**25–33.

Wuchter, C., B. Abbas, M. J. L. Coolen, L. Herfort, J. van Bleijswijk, P. Timmers, M. Strous, E. Teira, G. J. Herndl, J. J. Middelburg, S. Schouten, and J. S. S. Damste. 2006. Archaeal nitrification in the ocean. *Proc. Natl. Acad. Sci. USA* **103:**12317–12322.

Yergeau, E., S. Kang, Z. He, J. Zhou, and G. A. Kowalchuk. 2007. Functional microarray analysis of nitrogen and carbon cycling genes across an Antarctic latitudinal transect. *ISME J.* **1:**163–179.

Zhang, C. L., Q. Ye, Z. Y. Huang, W. J. Li, J. Q. Chen, Z. Q. Song, W. D. Zhao, C. Bagwell, W. P. Inskeep, C. Ross, L. Gao, J. Wiegel, C. S. Romanek, E. L. Shock, and B. P. Hedlund. 2008. Global occurrence of archaeal *amoA* genes in terrestrial hot springs. *Appl. Environ. Microbiol.* **74:**6417–6426.

Zhang, T., T. Jin, Q. Yan, M. Shao, G. Wells, C. Criddle, and H. H. P. Fang. 2009. Occurrence of ammonia-oxidizing *Archaea* in activated sludges of a laboratory scale reactor and two wastewater treatment plants. *J. Appl. Microbiol.* **107:**970–977.

GENOMICS OF AMMONIA-OXIDIZING BACTERIA AND INSIGHTS INTO THEIR EVOLUTION

Martin G. Klotz and Lisa Y. Stein

4

INTRODUCTION

Nitrification is defined as the aerobic oxidation of ammonia (oxidation state of −3) to nitrite (oxidation state of +3) followed by the aerobic oxidation of nitrite to nitrate (oxidation state of +5). Together with assimilatory and dissimilatory nitrate reduction, assimilatory and respiratory ammonification, and denitrifying ammonia oxidation, nitrification represents one of the key transformation processes between different fixed nitrogen intermediates (Fig. 1) (Lin and Stewart, 1998; Moreno-Vivian et al., 1999; Allen et al., 2001; Potter et al., 2001; Simon, 2002; Butler and Richardson, 2005; Ferguson and Richardson, 2005; Jepson et al., 2006; Tavares et al., 2006; Brandes et al., 2007; Smith et al., 2007; Klotz and Stein, 2008).

Microorganisms Implicated in Nitrification

Although considerable ecophysiological information describing the lifestyles of nitrifying bacteria has been produced over the last 100 years (Winogradsky, 1892; Prosser, 1989; Bock et al., 1991; Arp and Bottomley, 2006), our knowledge of these organisms at the molecular level was quite limited prior to obtaining whole genome sequences and did not expand beyond sequences of genes encoding ribosomal RNA and a few key enzymes involved in nitrogen transformations (Arp et al., 2007, and references therein). In addition, we learned in the last 5 years that our knowledge on the taxonomic diversity of nitrifying organisms was fairly limited. The discovery of broadly distributed Archaea that aerobically oxidize ammonia to nitrite (Könneke et al., 2005; Hallam et al., 2006; Leininger et al., 2006; Nicol and Schleper, 2006; de la Torre et al., 2008; Hatzenpichler et al., 2008; Prosser and Nicol, 2008; Martens-Habbena et al., 2009) significantly extended the taxonomic range of nitrifying organisms (see the chapters in Section III). Yet another discovery along with detailed studies over the last 15 years led to fundamental changes in our understanding of nitrification at the process and molecular levels: significant amounts of ammonia are metabolized under anoxic conditions (Dalsgaard et al., 2005; Jetten et al., 2005; Kuenen, 2008, and references therein). The obligatorily anaerobic ammonia oxidation process (anammox) directly yields and releases N_2 (Kartal et al., 2007; Jetten et al., 2009) in a distinctly different way than classical denitrification (Zumft, 1997; Zumft and Kro-

Martin G. Klotz, Department of Biology, University of Louisville, Louisville, KY 40292. *Lisa Y. Stein,* Department of Biological Sciences, University of Alberta, Edmonton, Alberta T6G 2E9, Canada.

Nitrification, Edited by Bess B. Ward, Daniel J. Arp, and Martin G. Klotz

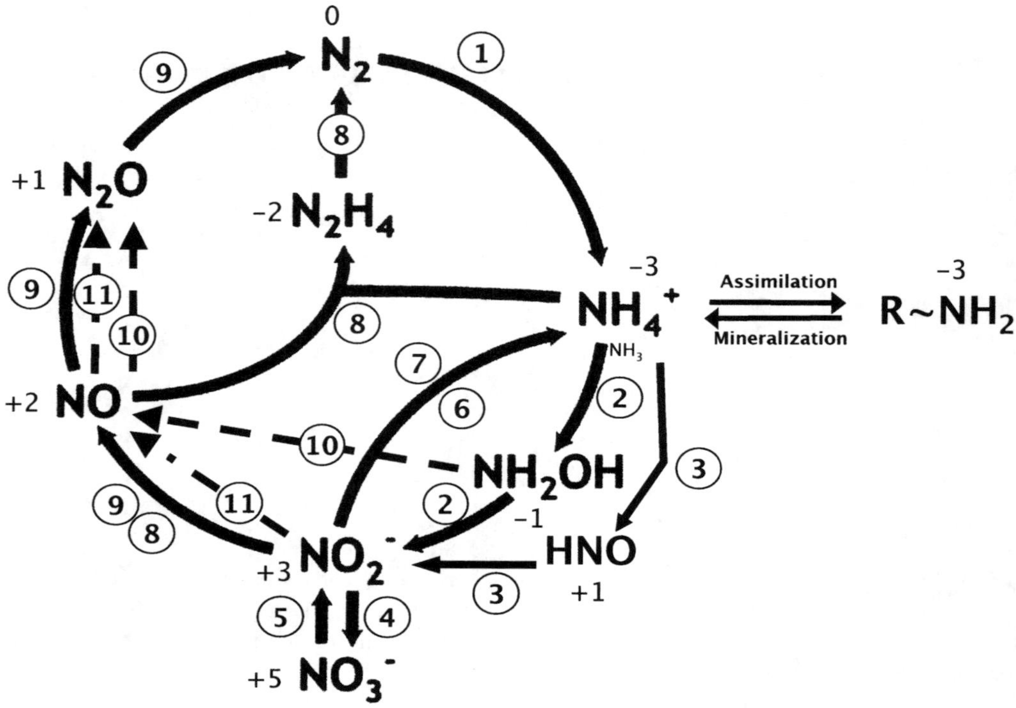

FIGURE 1 Processes in the microbial nitrogen cycle. Oxidation states of each intermediate are indicated (Klotz, 2008; Klotz and Stein, 2008); the pathway for archaeal ammonia oxidation is putative (Walker et al., 2010). 1, Dinitrogen fixation; 2, aerobic dissimilatory ammonia oxidation to nitrite by bacteria; 3, aerobic dissimilatory ammonia oxidation to nitrite by archaea; 4, aerobic dissimilatory nitrite oxidation to nitrate by bacteria; 5, assimilatory or dissimilatory nitrate reduction to nitrite by microbes; 6, respiratory ammonification as the second step of dissimilatory nitrate reduction of ammonia (DNRA, 5 and 6); 7, assimilatory ammonification as the second step of assimilatory nitrate reduction of ammonia (ANRA, 5 and 7); 8, denitrifying anaerobic ammonia oxidation (anammox, typified by ANAOB); 9, classic (anaerobic) denitrification by mixotrophs and heterotrophs; 10, aerobic oxidation of hydroxylamine to nitrous oxide by AOB and ANB; 11, aerobic denitrification by AOB and ANB.

neck, 2006). Anammox is thus best described with the term "denitrifying ammonia oxidation." The discovery of nitrifying archaea and anaerobically ammonia-oxidizing (anammox) bacteria (Jetten et al., 1998, 2005, 2009; Strous et al., 1999; Kuenen, 2008) (see Section IV) prompted the need for differentiating between processes such as ammonia oxidation or nitrification and terms used to describe organisms involved in these processes, such as "ammonia-oxidizing bacteria," "ammonia-oxidizing archaea," or "nitrifying bacteria."

Aside from chemolithotrophic bacteria and archaea, we learned recently at the molecular and physiological levels that cohorts of taxonomically diverse methanotrophic and heterotrophic bacteria have the ability to aerobically oxidize ammonia to nitrite (Poret-Peterson et al., 2008; Nyerges and Stein, 2009). These bacteria are also capable of nonclassical (aerobic) denitrification because they respiratorily produce and release NITRIC OXIDE (NO) and N_2O in the presence of oxygen (see below and Chapter 5). In addition, the recently described anaerobic methane-oxidizing bacterium (MOB) *Methylomirabilis oxyfera* in the phylum NC10 has the potential to oxidize ammonia anaerobically. While the anaerobic oxidation of methane by NC10 is coupled to denitrification (Ettwig et al., 2008, 2009,

TABLE 1 New nomenclature for ammonia-oxidizing microorganisms

Name	Acronym	Physiological description
Aerobic ammonia-oxidizing bacteria	AOB	Obligate chemolithotrophs that support growth with energy and reductant gained solely from the oxidation of ammonia
Aerobic ammonia-oxidizing archaea	AOA	Same as AOB
Aerobic nitrite-oxidizing bacteria	NOB	Obligate chemolithotrophs that support growth with energy and reductant gained solely from the oxidation of nitrite
Aerobic ammonia-oxidizing nonlithotrophic bacteria	ANB	(C1)-Organotrophs capable of co-oxidizing ammonia to nitrite (nitrification) that cannot support growth from this activity
Aerobic ammonia-oxidizing nonlithotrophic archaea	ANA	Same as ANB
Aerobic *amoA*-encoding archaea	AEA	Archaea that encode the *amoA* signature gene but whose ability to nitrify has not been demonstrated
Anaerobic ammonia-oxidizing bacteria	ANAOB	Obligate anaerobic chemolithotrophs that use the oxidation of ammonia for energy conservation and couple it with the reduction of nitrite to make N_2
Anaerobic ammonia-oxidizing nonlithotrophic bacteria	ANANB	Anaerobic bacteria that couple methane oxidation to the reduction of nitrate/nitrite to make N_2 and oxidize ammonia to nitrite via cometabolism

2010), it is not yet clear whether the oxidation of ammonia is coupled to nitrite reduction.

To describe the genetics and evolution of ammonia-oxidizing bacteria, an extended classification scheme is proposed in this chapter that captures the metabolic context of ammonia oxidation (Table 1). This scheme is also extended to nitrite-oxidizing bacteria (NOB). Taxonomic and ecological information about each of these cohorts are detailed in other chapters of this book.

The ammonium-oxidizing bacteria (AOB) that oxidize ammonia aerobically as their sole source of energy and reductant (Table 1) belong taxonomically to two monophyletic groups in different proteobacterial classes (phylogenetic tree in Chapter 3). The majority of cultured AOB identified in soils, freshwater, wastewater, and marine environments belong to the family *Nitrosomonadaceae* in the class Betaproteobacteria, whereas the AOB in the *Nitrosococcus* genus are purple sulfur bacteria (family *Chromatiaceae*) in the class Gammaproteobacteria that are restricted to marine environments (Teske et al., 1994; Utaker et al., 1995; Purkhold et al., 2000, 2003; Koops and Pommerening-Roser, 2001; Ward and O'Mullan, 2002). The family *Nitrosomonadaceae* exclusively includes AOB that are classified in the genera *Nitrosomonas*, *Nitrosospira*, and *Nitrosovibrio* (Teske et al., 1994). Phylogenetic analysis revealed that the genus *Nitrosomonas* is further divided into several lineages that correlate with particular growth conditions (Purkhold et al., 2000, 2003). In contrast, the family *Chromatiaceae* contains numerous non-ammonia oxidizers, most of which are strict anaerobes (for further details, refer to Chapter 3). The ammonium-oxidizing archaea (AOA) that oxidize ammonia aerobically as their sole source of energy and reductant are presently represented by two lineages in group I.1a of the Crenarchaeota, the mesophilic genera *Nitrosopumilus* and *Cenarchaeum*, as well as by the thermophilic genus *Nitrosocaldus* and the genus *Nitrososphaera*. Representatives of these genera have been cultured as isolates or communities of nitrifying chemolithoautotrophs (Könneke et al., 2005; Hallam et al., 2006; Wuchter et al., 2006; de la Torre et al., 2008; Hatzenpichler et al., 2008) (see Section III). The ammonia-oxidizing nonlithotrophic bacteria (ANB) that aerobically oxidize ammonia to nitrite with a modest gain of reductant but without any gain of energy (Table 1) include obligate methanotrophic *Methylococcaceae* in the class Gammaproteobacteria (Trotsenko and Murrell, 2008, and references therein) and methanotrophic *Methylacidiphilaceae* in the phylum Verrucomicrobia (Pol et al., 2007; Islam et al., 2008; Op den Camp et al., 2009, and references therein). Interestingly, although some isolated obligate (*Methylocystaceae*) and facultative (*Beijerinckiaceae*) methanotrophic Alphaproteobacteria (Trotsenko and Murrell, 2008, and references therein) contribute to the nitrification process, no known nitrite-producing inventory has been identified so far. In addition, there are numerous reports on the abundant presence of Crenarchaea, whose genomes contain one or more copies of the *amoA* signature gene associated with aerobic ammonia oxidation, for which growth-physiological or biochemical data are not yet available and which should thus be addressed as *amoA*-encoding archaea (AEA) (Francis et al., 2005; Treusch et al., 2005; Leininger et al., 2006; Nicol and Schleper, 2006; Agogue et al., 2008; Dang et al., 2008, 2009, 2010; Prosser and Nicol, 2008; Reigstad et al., 2008; Schleper, 2008; Tourna et al., 2008) (see the chapters in Section III). Pending further functional characterization, these AEA may later be classified as obligate (AOA), ammonia-co-oxidizing mixotrophs, or chemoorganotrophs with nonfunctional *amo* genes in their genomes. The aerobic NOB, which associate functionally with AOB and AOA, are found in the classes of Alphaproteobacteria, Gammaproteobacteria, and Deltaproteobacteria as well as the phylum Nitrospirae (Bock et al., 1991; Koops and Pommerening-Roser, 2001) (see the chapters in Section V).

As for the obligate anaerobic microbes involved in ammonia oxidation, the anammox bacteria, which are capable of reducing (analog to dissimilatory nitrate reduction to ammonia [DNRA]; reactions 5 and 6 in Fig. 1) and oxidizing nitrite anaerobically, are all classified as *Brocadiaceae* in the phylum Planctomycetes (Strous et al., 1999; Schmidt et al., 2002a, 2002b; Kuenen, 2008; Jetten et al., 2009; Op den Camp et al., 2009) (see the chapters in Section IV). According to the new classification scheme, these bacteria would be analogously addressed as anaerobic ammonia-oxidizing bacteria (ANAOB) (Klotz and Stein, 2008) (Table 1). Very recently, a novel taxonomic and functional cohort of methanotrophic bacteria was discovered that directly couples anaerobic oxidation of methane with denitrification (NC10 [Raghoebarsing et al., 2006; Ettwig et al., 2008]). Genomic data mining revealed that these bacteria have the necessary inventory for ammonia oxidation. Once better understood at the molecular level, discoveries like the latter will likely aid further efforts for clarification in our descriptions of processes and organisms involved in the nitrogen cycle. Following above introduced terminology (Klotz and Stein, 2008), the anaerobic methane-oxidizing NC10 bacteria, capable of ammonia oxidation to nitrite, could hence be designated anaerobic ammonia-co-oxidizing nitrifying bacteria (Table 1).

Pregenomic Era Gene Inventory Implicated in Nitrification

Given that AOB are chemolithotrophs, early attention was focused on genes and proteins that enabled use of ammonia as a source of energy and reductant (Vannelli et al., 1996; Hooper et al., 1997, 2005; Whittaker et al., 2000; for review, see Arp et al., 2002; Hommes et al., 2002; Norton et al., 2002). Attention was also focused on genes and proteins for carbon dioxide fixation prior to the discovery that some AOB can grow on organics (Hommes et al., 2003; Utaker et al., 2002). The availability of gene sequences encoding the functional ammonia monooxygenase (AMO) (McTavish et al., 1993a, 1993b; Klotz and Norton, 1995, 1998; Norton et al., 1996; Sayavedra-Soto et al., 1996, 1998; Klotz et al., 1997; Hommes et al., 1998, 2001; Alzerreca et al., 1999; Hirota et al., 2000) and hydroxylamine oxidoreductase (HAO) (Bergmann et al., 1994; Sayavedra-Soto et al., 1994, 1996; Hommes et al., 1996, 2002) protein complexes led to a surge in information about the distribution and abundance of AOB using molecular probes (e.g., Holmes et al., 1995; Rotthauwe et al., 1997; Purkhold et al., 2000, 2003; Gieseke et al., 2001; Kowalchuk and Stephen, 2001; Ward and O'Mullan, 2002; Zehr and Ward, 2002, and references therein) (see Chapter 3 and Section VI). Information on a few additional genes whose products were implicated in carbon assimilation or N transformations were reported together with physiological or biochemical data from a few representative isolates before genome sequences became available (i.e., cyt. *c*554 [Hommes et al., 1994], cyt. P460 [Bergmann and Hooper, 2003], NirK [Cantera and Stein, 2007a, 2007b; Casciotti and Ward, 2001; Garbeva et al., 2007], nitrosocyanin [Arciero et al., 2002], NorB [Ren et al., 2000; Braker and Tiedje, 2003; Garbeva et al., 2007], and urease [Koper et al., 2004]); however, very little was known about the regulation of gene expression in AOB (Sayavedra-Soto et al., 1996, 1998). In contrast, there was a much larger body of literature on the structure and function of individual enzymes involved in nitrification (Hooper and Nason, 1965; Hooper, 1968, 1969; Erickson and Hooper, 1972; Hooper and Terry, 1977, 1979; Terry and Hooper, 1981; Andersson et al., 1982; Hooper et al., 1990; Rasche et al., 1990; Arciero et al., 1991a, 1991b, 1993, 2002; Hyman and Arp, 1992, 1995; Arciero and Hooper, 1993, 1997; Ensign et al., 1993; Juliette et al., 1995; Stein et al., 1997; Iverson et al., 1998, 2001; Stein and Arp, 1998; Jiang and Bakken, 1999; Lontoh et al., 2000; Hendrich et al., 2002; Arp and Stein, 2003; Bergmann and Hooper, 2003) (see Chapter 2).

This situation changed dramatically with the onset of the genomic era (Fleischmann et al., 1995). The emergence of high-throughput sequencing and automated annotation at the turn of the century created tremendous opportunities for the genome inventory-based reevaluations of microbial physiology, the design of new genome-informed experimentation, and the reconstruction of the molecular evolutionary history of nitrifying bacteria. Since then, the ever-increasing genome-based information on the molecular underpinnings of nitrifying organisms (Table 2) has yielded a number of novel discoveries, which undoubtedly will facilitate future efforts to exploit the positive effects of nitrifying bacteria and mitigate their detrimental impacts. Based on present day genome analyses, the following parts of this chapter will address the inventory involved in nitrification, attempt metabolic reconstruction of N transformation processes, and provide insight into their evolution.

GENOMICS OF BACTERIA THAT AEROBICALLY OXIDIZE AMMONIA TO NITRITE

Genome Structure

Presently, the genomes of four AOB are published: three belong to the family *Nitrosomonadaceae* (Betaproteobacteria); they are *Nitrosomonas europaea* ATCC 19718, isolated from wastewater (Chain et al., 2003); *Nitrosomonas eutropha* C-91 (equal to Nm-57), isolated from sewage (Stein et al., 2007); and *Nitrosospira multiformis* ATCC 25196 (Norton et al., 2008), isolated from soil. The fourth AOB genome is that of *Nitrosococcus oceani* ATCC 19707 (equal to C-107), a marine Gammaproteobacterium in the family *Chromatiaceae* (Klotz et al., 2006). Additional AOB genome sequences will soon enter the literature, as described in Table 2. Collective knowledge regarding the four published AOB genomes was recently summarized in context with existing ecophysiological and biochemical data (Arp et al., 2007). Additional insights from the newer AOB genome projects have confirmed general trends of genome size, redundancy, repeats, acquisition, and degradation. Likewise, we can conclude that all completely sequenced and draft genomes support that AOB are obligate chemolithotrophs (Hommes et al., 2006) that have the ability to utilize selected organic carbon compounds (Hommes et al., 2003) but usually depend on autotrophic carbon assimilation (Wei et al., 2004; Arp et al., 2007; Stein et al., 2007; Norton et al., 2008).

AOB genomes are among the smallest of free-living Betaproteobacteria at approximately 3 Mb in size as a result of genome econo-

TABLE 2 Ongoing and completed whole-genome sequencing (WGS) projects involving nitrifying bacteria

Nitrifying organism	Metabolic organism	Genome size	Sequencing center (funding source)[a]	WGS accession number
Verrucomicrobia				
Methylacidiphilum infernorum V4	ANB	2,287,145	University of Hawaii	CP000975
Methylacidiphilum fumariolicum SolV	ANB	~ 2.5 Mb	Radboud University Nijmegen	In progress (embargo)
Alphaproteobacteria				
N. winogradskyi Nb-255	NOB	3,402,093	JGI-DOE	CP000115
Nitrobacter hamburgensis X-14	NOB	4,406,969	JGI-DOE	CP000319
Plasmid 1		294,831	JGI-DOE	CP000320
Plasmid 2		188,320	JGI-DOE	CP000321
Plasmid 3		121,410	JGI-DOE	CP000322
Nitrobacter sp. strain Nb-311A	NOB	> 4.1 Mb	WHOI/JCVI (GBMF)	CH672416–CH672426
Methylosinus trichosporium OB3b (type II)	ANB	> 4.8 Mb	JGI-DOE	Gi021903
Methylocystis sp. strain ATCC 49424 (II)	ANB	~4 Mb	JGI-DOE	In progress (embargo)
Gammaproteobacteria				
N. oceani C-107	AOB	3,481,691	JGI-DOE	CP000127
Plasmid 1	AOB	40,420	JGI-DOE	CP000126
N. oceani AFC27	AOB	3,471,807	UofL/JCVI (GBMF)	ABSG01
Nitrosococcus halophilus Nc4	AOB	4,079,427	UofL/DOE-JGI–NSF	NC013960
Plasmid 1	AOB	65,833	UofL/DOE-JGI–NSF	NC013958
Nitrosococcus watsoni C-113	AOB	3,328,570	UofL/DOE-JGI–NSF	NC014315
Plasmid 1	AOB	39,105	UofL/DOE-JGI–NSF	NC014316
Plasmid 2	AOB	5,611	UofL/DOE-JGI–NSF	NC014317
Nitrococcus mobilis Nb-231	NOB	~ 3 Mb	WHOI/JCVI (GBMF)	CH672427
M. capsulatus Bath (type X)	ANB	3,304,561	TIGR	AE017282
M. album BG 8 (type I)	ANB	~ 3.5 Mb	JGI-DOE	In progress (embargo)
Betaproteobacteria				
N. europaea ATCC 19718	AOB	2,812,094	JGI-DOE	AL954747
N. eutropha C-91	AOB	2,661,057	JGI-DOE	CP000450
Plasmid 1		65,132	JGI-DOE	CP000451
Plasmid 2		65,132	JGI-DOE	CP000452
N. multiformis ATCC 25196	AOB	3,184,243	JGI-DOE	CP000103
Plasmid 1		18,871	JGI-DOE	CP000451
Plasmid 2		17,036	JGI-DOE	CP000451
Plasmid 3		14,159	JGI-DOE	CP000452
Nitrosomonas sp. strain AL212	AOB	~ 3 Mb	JGI-DOE	Gi03896
Nitrosomonas sp. strain IS-79	AOB	~ 3 Mb	JGI-DOE	In progress (embargo)
Nitrosomonas marina C-113a	AOB	~ 3 Mb	UofL/DOE-JGI–NSF	In progress (embargo)

[a]JGI-DOE, Joint Genome Institute–U.S. Department of Energy; WHOI/JCVI (GBMF), Woods Hole Oceanographic Institute/J. Craig Venter Institute (Gordon and Betty Moore Foundation); UofL, University of Louisville, Louisville, KY; NSF, U.S. National Science Foundation; TIGR, The Institute for Genomic Research (now the J. Craig Venter Institute).

mization and reduction (Arp et al., 2007). Furthermore, it appears that genome reduction has contributed to niche differentiation as soil (Norton et al., 2008) and marine (B. B. Ward, K. L. Casciotti, P. S. G. Chain, S. A. Malfatti, M. A. Campbell, and M. G. Klotz, unpublished data) isolates tend to have slightly larger genomes than AOB isolates that live in stable high-nitrogen environments (Stein et al., 2007). The genomes of free-living marine AOB in the genus *Nitrosococcus* are approximately 3.5 to 4.0 Mb in size (Klotz et al., 2006; M. A. Campbell, S. A. Malfatti, P. S. G. Chain, J. F. Heidelberg, B. B. Ward, and M. G. Klotz, unpublished data), similar to that of many other environmental Gammaproteobacteria (Arp et al., 2007). The G+C content of genome-sequenced AOB is 48.5% (*Nitrosomonas*), ~50% (*Nitrosococcus*), and 53.9% (*Nitrosospira*), confirming the recent prediction that betaproteobacterial AOB (Beta-AOB) have among the lowest G+C content of the Betaproteobacteria, which, compared to that of other bacterial taxa, is relatively constrained (48.5% to 68.5%) (Arp et al., 2007; Stein et al., 2007; Norton et al., 2008; Ward et al., unpublished; Campbell et al., unpublished). In general, the number of ribosomal RNA (*rrn*) operon copies in AOB is below the average observed for each class of Proteobacteria. Even though the Gamma-AOB appear to have much longer generation times than the Beta-AOB, all studied Beta-AOB genomes harbor only one *rrn* operon copy, whereas all Gamma-AOB encode two (Arp et al., 2007; Stein et al., 2007; Norton et al., 2008; Ward et al., unpublished; Campbell et al., unpublished). Thus, it appears that the AOB disprove the hypotheses that faster growth rates and improved fitness correlates with the number of *rrn* operon copies per cell (Stevenson and Schmidt, 1998).

Before the genomic era, it was already known that Beta-AOB harbor multiple, nearly identical copies of gene clusters required to oxidize ammonia, including AMO (*amoCAB*) and HAO (*haoA*) and associated cytochromes *c*554 (*cycA*) and c_M552 (*cycB*) (McTavish et al., 1993; Sayavedra-Soto et al., 1994; Norton et al., 1996; Klotz et al., 1997), and that the Gamma-AOB contain only one copy (Alzerreca et al., 1999). Aside from these duplicated genome segments in Beta-AOB, it appears that a small number of gene duplications have arisen recently in all AOB genomes. While some nearly identical gene copies are present in all investigated genomes (i.e., two copies of the Tu elongation factor), each genome also includes unique duplications. For instance, *N. europaea* has a unique 7.5-kb tandem duplication of a region encoding key metabolic genes (Chain et al., 2003), which is missing from the genome of the closely related Beta-AOB *N. eutropha* (Stein et al., 2007). The genome of *N. eutropha* C-91 harbors two identical copies of approximately 12-kb DNA fragments of significantly higher G+C content that contain a number of plasmid- and phage-related proteins, and most of the encoded open reading frames appear unique to this isolate (Stein et al., 2007). In addition, the *N. eutropha* C-91 genome contains duplicated elements >6 kb with significantly lower G+C content, indicating that acquisition and loss of genome fragments in AOB are recent and likely driving forces in niche differentiation (Stein et al., 2007). A direct comparison of gene arrangement and G+C content between the genome sequences of *N. europaea* ATCC 19718 and *N. eutropha* C-91 revealed a significant extent of structural rearrangement between these two species of AOB that inhabit similar ecological niches (wastewater/sewage) (Stein et al., 2007). The genome of the marine AOB *N. oceani* ATCC 19707 also carries recent duplications of several functional genes (Klotz et al., 2006), which are present in other, but not all, genomes of analyzed *Nitrosococcus* strains (Campbell et al., unpublished). As was observed for the two nitrosomonad genomes, the structural arrangement of genes in nitrosococcus genomes was not conserved; however, the percentage of sequence identity and synteny between *N. oceani* ATCC 19707 and *N. oceani* AFC27 was significant relative to either *N. oceani* genome with *N. halophilus* Nc4 or *N. watsoni* C-113 genomes (Campbell et al., unpublished).

In addition to duplications of coding DNA, AOB genomes contain a number of unique duplicated insertion sequence (IS) elements (Arp et al., 2007; Stein et al., 2007; Norton et al., 2008). For instance, *N. oceani* ATCC 19707 harbors five families of IS elements repeated a total of 25 times that are not all present in other *Nitrosococcus* genomes (Campbell et al., unpublished). The genome of *N. europaea* carries eight families repeated a total of 89 times (Chain et al., 2003). Some of the *N. europaea* IS element families were preferentially found in proximity to other specific families, indicating possible cotransposition and perhaps historical acquisition of multiple different IS elements in a single event (Arp et al., 2007). The genome of *N. eutropha* C-91 harbors at least seven families of IS elements, repeated up to 22 times, two of which are related, but not identical, to IS elements found in *N. europaea* (Stein et al., 2007). The chromosome of *N. multiformis* ATCC 25196 contains eight families of IS elements, repeated from 2 to 13 times and spread randomly across the genome; two of these elements are also found on plasmids (Norton et al., 2008).

The genomes of AOB contain a number of predicted pseudogenes (113 in *N. europaea*, 90 in *N. eutropha*, 80 in *N. oceani*, and only 22 in *N. multiformis*) with IS elements as well as small insertions/deletions contributing to these inactivations (Arp et al., 2007; Stein et al., 2007; Norton et al., 2008). Numerous IS elements within genomes may also enhance recombinogenic activity, given the presence of so many near-identical sequences. Most proteobacterial genomes are, as a function of bidirectional semiconservative replication, partitioned into two replicores with a biased incorporation of guanine in the leading strand. Interestingly, the genome of *N. europaea* is asymmetrically partitioned, which may be a result of active recombination between IS elements of the same family.

Other foreign material has been identified in both *N. oceani* (175 kb in 10 regions) and *N. europaea* (also ca. 10 regions). Several phage-related regions with similarity to known phage genes were often found associated with transposase genes, recombinases, restriction modification systems, tRNAs, and clusters of small hypothetical genes. A large ca. 117-kbp genomic island with a markedly higher G+C content was found in the genome of *N. eutropha* flanked by tRNA genes and direct repeat sequences as well as a phage-related integrase. This region carries 64 genes (51%) that have no homologues within the other AOB genomes and contains coding regions predicted to confer resistance to heavy metals (Stein et al., 2007). Remarkably, this region also contains a second complete cluster of cytochrome *c* maturation (*ccm*) genes (Stein et al., 2007).

An additional surprise was the finding that AOB vary dramatically regarding the presence of extrachromosomal DNA. While *N. europaea* ATCC 19718 does not contain plasmids, other Beta-AOB do: *N. eutropha* C-91 harbors two plasmids, and *N. multiformis* ATCC 25196 has three (Arp et al., 2007; Stein et al., 2007; Norton et al., 2008). The *N. oceani* ATCC 19707, *N. watsoni* C-113, and *N. halophilus* Nc4 genomes contain plasmids of 40.4 kb, 39.1 and 5.6 kb, and 65.8 kb that comprise mostly hypothetical and conserved hypothetical genes along with a small number of phage-related genes and functions associated with replication and plasmid partitioning (Arp et al., 2007; Campbell et al., unpublished). Such cryptic plasmids, also present in other *Nitrosococcus* genomes, may contribute to the dynamic processes of adaptation, evolution, and speciation (Campbell et al., unpublished).

Although all AOB genes should have an equally high demand for iron to satisfy the requirement for cytochrome *c* protein synthesis, not all genomes are rich in genes involved in uptake and processing of iron. While *N. europaea*, *N. multiformis*, and *N. oceani* genomes are particularly rich in genes involved in iron (siderophore) transport, this function is nearly absent from the genome of *N. eutropha* (Stein et al., 2007). The genome of *N. oceani* encodes at least 22 genes involved in iron transport and is likely capable of synthesizing a hydroxamate-type siderophore. Surprisingly, *N. europaea* has more than 100 genes involved in iron transport, including a large number of FecIR two-component regulatory systems (>20 systems), yet the genome is devoid of siderophore biosynthesis genes (Chain et al., 2003). An understanding of this striking disparity in iron acquisition inventory, which cannot be reduced to a difference in oxygen and thus Fe^{2+} availability, will likely depend on the analysis of many additional AOB genomes and may also yield more clues as to niche differentiation of the AOB.

While the relatively small sizes of AOB genomes correlate with their limited catabolic versatility, the relatively large number of IS elements and inactivated pseudogenes, and the rare presence of genomic islands, phage- and plasmid-like segments may indicate that AOB genomes are currently undergoing the evolutionary process of genome degradation. Since genome economization is known to occur faster in A+T-rich genomes, in contrast to the expansion of G+C-rich genomes, it has been proposed that the variable copy number of key catabolic genes in AOB is the result of loss of copies that were not repairable by rectification. This loss must have occurred completely without leaving pseudogenes capable of producing nonfunctional and potentially deleterious enzymes (Klotz and Norton, 1998). Indeed, a detailed comparative analysis of the *N. eutropha* C-91 and *N. europaea* ATCC 19718 genomes revealed the existence of predictable breakpoints in synteny between the duplicated copies of catabolic gene-encoding DNA segments (Stein et al., 2007). Nevertheless, the inventory responsible for this rectification mechanism remains to be discovered.

Catabolic Inventory

Since the last two reviews on the genomics and evolution of AOB (Arp et al., 2007; Klotz and Stein, 2008), new insights into the ammonia-catabolic gene inventories were revealed regarding their organization, regulation of expression, and evolution. The *amoCAB* genes encoding AMO were believed necessary and sufficient for AMO synthesis and function in all AOB (Klotz et al., 1997; Alzerreca et al., 1999; Norton et al., 2002). However, *amoCAB* genes were recently found as members of a larger cluster of coregulated genes (Fig. 2), which differ in number and regulation between Beta-AOB (Berube et al., 2007) and Gamma-AOB (El Sheikh and Klotz, 2008; El Sheikh et al., 2008). Based on expression studies and in silico analysis of *Nitrosococcus* genomes, the *amo* gene cluster of Gamma-AOB consists of overlapping operons, the largest of which, *amoRCABD*, has five genes (El Sheikh and Klotz, 2008; El Sheikh et al., 2008). Involvement in AMO synthesis has been suggested for *amoR* (a gene unique to strains of *Nitrosococcus oceani* [ATCC 19707, AFC27] but absent from *N. halophilus* and *N. watsonii* [M. A. Campbell and M. G. Klotz, unpublished]) and *amoD*; however, their roles still need to be biochemically explored (El Sheikh and Klotz, 2008; El Sheikh et al., 2008). The *amoD* gene is found in tandem with a likely duplicated orthologue (*amoE*) downstream of the *amoCAB* genes in Beta-AOB; however, first expression experiments suggest that *amoED* is coregulated but not part of the same operon as *amoCAB* (Berube et al., 2007). Interestingly, orthologues of *amoD* (but not *amoE*) are also found in the genomes of aerobic methanotrophs where they reside either downstream of the gene cluster encoding particulate methane monooxygenase (pMMO) (Alphaproteobacteria) or in proximity of a gene tandem encoding copper-blue oxidases (Gammaproteobacteria). This blue copper oxidase gene tandem is also conserved in Gamma-AOB (upstream of the *amo* gene cluster) and in *N. eutropha* (Fig. 2). Beta-AOB encode nonoperonal copies (singleton) of *amoC* (Norton et al., 2002; Arp et al., 2007; Berube et al., 2007) and *amoE* (Norton et al., 2002; Arp et al., 2007) genes, and all AOB encode *amoD* singletons (El Sheikh et al., 2008). Singleton *amoA* and *amoB* genes have not yet been found in any AOB genome.

The capacity of AOB to aerobically catabolize ammonia as the sole source of energy and reductant requires two specialized protein complexes, AMO and HAO, as well as the cytochromes *c*554 and c_M552, which relay the electrons to the quinone pool (Whittaker et al., 2000; Arp et al., 2002; Hooper et al., 2005) (Fig. 3). The three-subunit AMO protein complex initiates ammonia catabolism by oxidizing ammonia to hydroxylamine when supplied with reductant from the quinone pool (Hooper et al., 2005). AMO is homologous to pMMO (Klotz and Norton, 1998; Norton et al., 2002), which initiates the oxidation of methane to methanol by methanotrophs (Hanson and Hanson, 1996; Murrell et al., 2000; Trotsenko and Murrell,

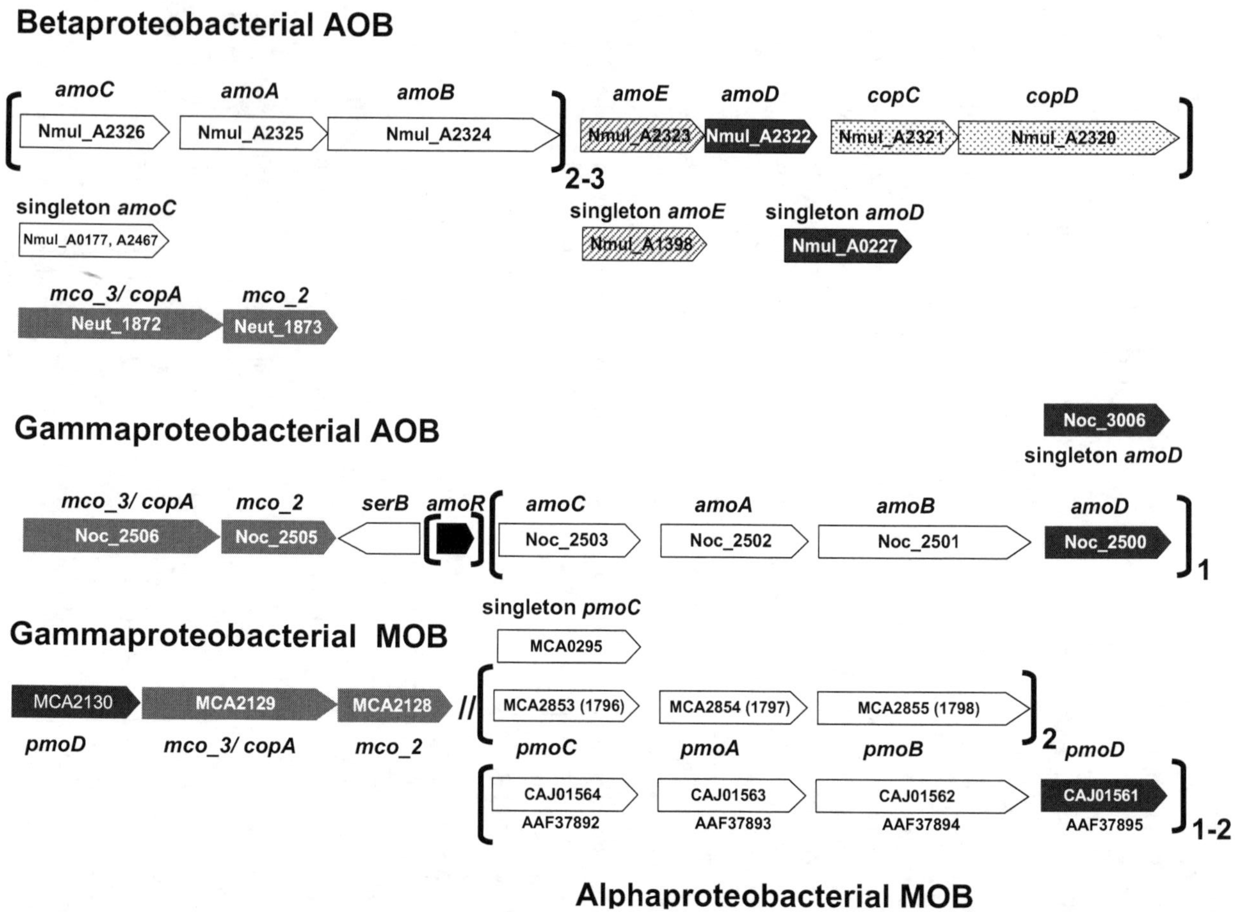

FIGURE 2 Organization of ammonia and methane monooxygenase-encoding and ancillary genes in the genomes of betaproteobacterial and gammaproteobacterial AOB and in gammaproteobacterial and alphaproteobacterial MOB. Representative protein accession numbers are provided. Multiple copies of coregulated genes with near-identical sequence are indicated by indexed parentheses. The *amoR* gene is present only in genomes of *Nitrosococcus oceani* strains ATCC 19707 and AFC-27 but absent from *N. halophilus* and *N. watsonii* (Campbell and Klotz, unpublished). The *serB* gene is conserved in all nitrosococci but not involved in nitrification.

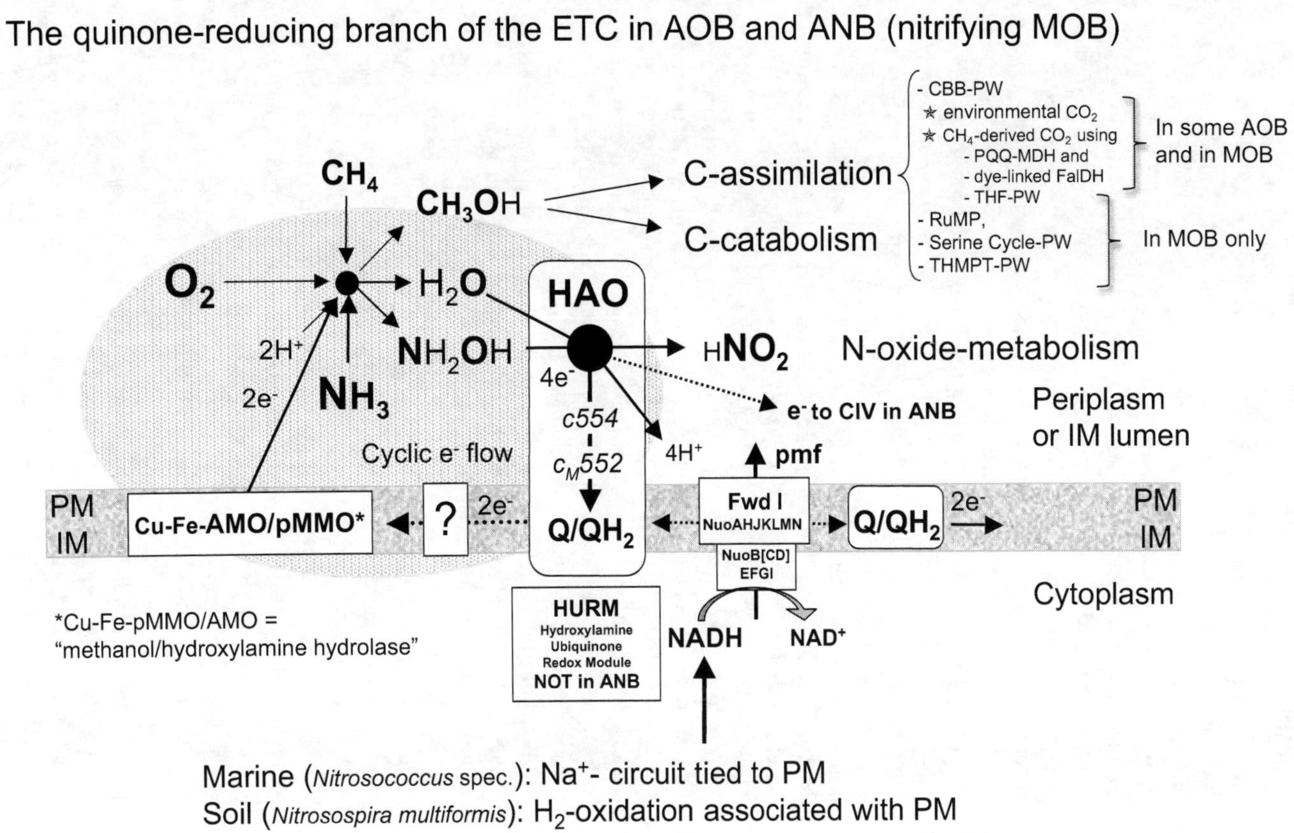

FIGURE 3 Flow of nitrogen, carbon, and electrons in the quinone-reducing branch of AOB and ANB. Q/QH_2 indicates the quinone/quinole pool in the plasma membrane (PM) and intracellular membrane (IM). The question mark indicates that a direct quinol oxidase function of AMO/pMMO has not yet been demonstrated. The stippled arrow indicates that the electrons extracted by HAO in ANB are not relayed by HURM into the Q-pool. Instead, these nitrification-borne electrons are transferred via soluble *c*552 proteins for energy conservation to pertinent terminal electron acceptors including Complex IV heme-copper oxidases that reduce oxygen or NO. The figure is modified from Klotz and Stein (2008).

2008). Based on this homology, AMO also likely facilitates catalysis with a di-iron center as was recently proposed for the oxidation of methane by pMMO (Martinho et al., 2007). The subsequent oxidation of hydroxylamine to nitrite is catalyzed by HAO, which operates in the periplasm and consists of three cross-linked HaoA protein subunits (Igarashi et al., 1997; Hooper et al., 2005). The four electrons extracted in this dehydrogenation process were proposed to enter the ubiquinone pool via a redox cascade established by the two tetraheme cytochromes *c*554 and c_M552 (Fig. 3) (Hooper et al., 2005, and references therein). The positional proximity of HAO and cytochrome c554 and c_M552 genes along with the proposed interaction of their products led to a comprehensive designation of the Hydroxylamine Ubiquinone Redox Module (HURM) (Klotz and Stein, 2008) (Fig. 3). However, the interaction between cytochromes c554 and c_M552 in AOB and the functionality of the redox chain in the absence of either cytochrome has never been experimentally established (Klotz and Stein, 2008).

The core HURM genes are encoded by a conserved gene cluster, *hao-orf2-cycAB*, in all AOB (Fig. 4) (Bergmann et al., 2005). Complete sequences of the genes encoding the HURM proteins had been published from several AOB prior to obtaining genome sequences (Arp et al., 2002; Norton et al., 2002, and references therein), and the protein structures of HAO and *c*554 have since been resolved (Igarashi et al., 1997; Iverson et al., 1998). The crystal structures of functional AMO and c_M552 protein complexes are awaiting resolution, although a threaded analysis of the *N. europaea* c_M552 structure has been presented (Kim et al., 2008). The analysis of genomes from sulfur-dependent deep-sea vent Epsilonproteobacteria recently revealed that HURM, consisting only of HAO and cytochrome c_M552, together with nitrate reductase (*napA*) and hydroxylamine reductase (*hcy*), operates as the sole pathway for nitrogen assimilation from nitrate within some *Nautiliales* (Campbell et al., 2009), thereby providing evidence for a functional redox partnership between HAO and cytochrome c_M552. Previous studies with *N. europaea* suggested that the *hao* gene and the *cycAB* genes are expressed independently, and no evidence for the transcription of *orf2* was found (Bergmann et al., 1994; Sayavedra-Soto et al., 1996). Comparative analysis of individual *hao* gene expression in *N. europaea* strain ENI-11 revealed differential regulation and identified the one *hao* gene copy not located in the vicinity of the two *amoCAB* operons as being expressed at the highest level and as the only copy transcribed in cells denied an energy source (Hirota et al., 2006). More recent experiments indicated that expression of the *hao* and *cycAB* genes is not identical in all AOB. While the *hao* and *cycAB* genes are expressed independently in *N. europaea* (Bergmann et al., 1994; Sayavedra-Soto et al., 1996), studies of the transcriptional response of the Gamma-AOB, *N. oceani* ATCC 19707, to ammonia suggested the presence of a steady-state mRNA that included all four genes; nevertheless, basal expression produced independent *hao-orf2* and *cycAB* transcripts (M. A. Campbell and M. G. Klotz, unpublished data). Ammonia also induced expression of an *hao-orf2* gene tandem in the ANB, *Methylococcus capsulatus* Bath (Poret-Peterson et al., 2008), providing the designation of the first two genes in the HURM gene cluster as *haoAB* (Campbell and Klotz, unpublished).

One of the three *haoAB-cycAB* gene clusters in *N. europaea* and *N. eutropha* lacks *cycB* (McTavish et al., 1993; Sayavedra-Soto et al., 1994; Chain et al., 2003; Stein et al., 2007), whereas it is present in all three respective gene clusters in *N. multiformis* (Norton et al., 2008). The nitrosomonad *haoAB-cycA* gene clusters lacking the *cycB* gene happen to cluster with a conserved hypothetical gene, *orfM* (NE2041, Neut_1669). The missing *cycB* gene of the nitrosomonads was likely lost by deletion in the *N. europaea*/*N. mobilis* lineage (Purkhold et al., 2000, 2003) as indicated by the presence of transposase and helicase genes flanking their *haoAB-cycA-orfM* gene clusters. The *orfM* gene is present in all AOB genomes: downstream of an *haoAB-cycAB* gene cluster in *N. multiformis* (Nmul A2658) and as an unclustered gene in the three *Nitrosococcus* genomes

(Klotz et al., 2006; Campbell and Klotz, unpublished). *orfM* was recently reported as restricted to AOB genomes (Arp et al., 2007); however, new information available from whole genome sequencing projects indicates that a variant of *orfM* also resides in non-nitrifying Proteobacteria (ABM03597, ABR71384, EDN67668), Bacteroidetes (EAQ40717, EAR12710, EAR12744), and Chloroflexi (ABX04895). Interestingly, the *orfM* gene in *Beggiotoa* sp. strain PS (EDN67668) is adjacent to *dsrC*, whose expression product, DsrC (cl01101), may be involved in the assembly, folding, or stabilization of siroheme proteins that are integral parts of enzymes such as dissimilatory sulfite reductase and assimilatory siroheme sulfite and nitrite reductases. In *Marinomonas* sp. strain MWYL1, the *orfM* gene is adjacent to a gene encoding glutathione peroxidase (EC 1.11.1.9; cd00340) and to tetrameric selenoenzymes that catalyze the reduction of a variety of hydroperoxides, including reactive oxygen species and peroxinitrite. In other genomes, the *orfM* genes are adjacent to gene clusters important to iron transport.

At the present time, the only recognized gene identified as unique to AOB encodes nitrosocyanin (*ncyA*), a novel soluble red copper protein found in equimolar quantities with HAO in the periplasm of AOB (Hooper et al., 2005). A recent paper analyzing available microbial genomes for copper transporters and cuproproteomes reported erroneously that nitrosocyanin was one of the most widespread cuproproteins in bacterial genomes (15%) and also was present in the Archaea (Ridge et al., 2008). This result was concluded from limited sequence similarity between deduced protein sequences of the red-copper protein nitrosocyanin with a domain of the blue-copper protein nitrous oxide reductase, the latter of which is, indeed, widely (15%) distributed (Zumft and Kroneck, 2006). The nitrous oxide reductase protein is a dinuclear copper protein more closely related to the CuA center in HCO (type I copper center), whereas nitrosocyanin is more closely related to the mononuclear cupredoxins such as amicyanin, azurin, pseudoazurin, plastocyanin, and rusticyanin (type II copper center).

Once ammonia oxidation has led to increases in the reduced quinone pool (Fig. 3), the oxidative branch of the respiratory electron transport chain (ETC) can be utilized for the production of ATP and NAD(P)H by ATP-synthases and NADH-(ubi)quinone oxidoreductases (NUO), respectively (Fig. 5). There are three evolutionarily independent families of NUO that functionally constitute Complex I, which extracts electrons from NADH by dehydrogenation while reducing (ubi)quinone (Kerscher et al., 2008). One of the three families, usually referred to as alternative NAD(P)H dehydrogenase (NDH-2), is encoded in all three domains of life, usually consists of one protein, and is unable to convert the redox potential difference between NADH and ubiquinone into ion translocation. In contrast, the other two NUOs pump either protons (NADH dehydrogenase NDH-I; found in all three domains of life) or sodium (Na^+-Nqr; so far only found in bacteria (Kerscher et al., 2008, and references therein). All AOB genomes encode NDH-I, a few encode Na^+-Nqr, but none encode NDH-2.

One of the main variations of NDH-I is fusion of the NuoCD subunits as found in some Gammaproteobacteria like the AOB *N. oceani* (Klotz et al., 2006; Schneider et al., 2008). The sodium-pumping Complexes I are evolutionarily unrelated functional analogues of NDH-I and are found, so far, only in bacteria (Kerscher et al., 2008, and references therein). Based on their isolation from *Vibrio* spp., *Klebsiella pneumoniae*, and *Azotobacter vinelandii*, which operate these sodium-translocating NADH:quinone oxidoreductases either as sole or alternative Complexes I, they were called Na^+-NQRs (*nqrABCDEF*) (Unemoto and Hayashi, 1993; Bertsova and Bogachev, 2004; Fadeeva et al., 2008; Tao et al., 2008). Full complements of Na^+-NQR-encoding genes have been identified in *N. europaea* (Chain et al., 2003) and *Nitrosomonas marina* C-113a (Ward et al., unpublished) but in no other AOB genome. Homologues of complete *nqr* gene sets were identified in *Rhodobacter capsulatus* as essential for nitrogen fixation activity; hence, their expression products were

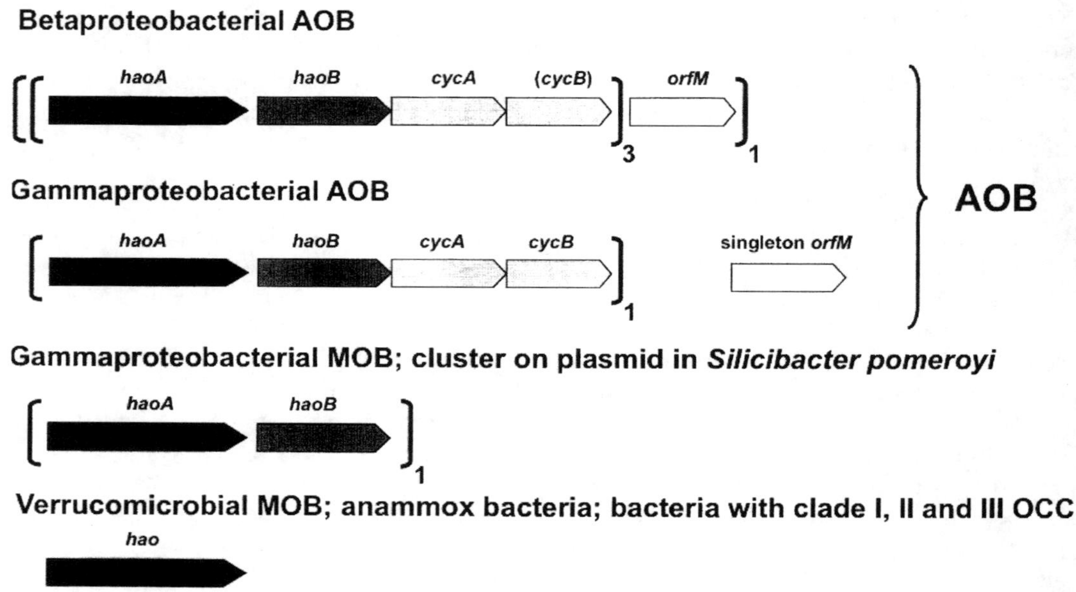

FIGURE 4 Residence and organization of genes encoding the OCC protein HAO (HaoA) and electron transfer cytochrome *c* proteins, for which catalytic activity also has been demonstrated: CycA (*c554*) – NO reductase; CycB (c_M552) – quinone reductase. Functions for putative expression products of the conserved genes *haoB* and *orfM* have not yet been elucidated. Bacteria with clade I, II, and III OCC are listed in the study by Klotz et al. (2008). The background arrow indicates that the direction of divergence on the phylogenetic tree of OCC proteins (Klotz et al., 2008) correlates with increasing co-organization of genes that encode interacting nitrification proteins.

termed "*Rhodobacter*-specific nitrogen fixation" (Rnf) proteins (Schmehl et al., 1993; Kumagai et al., 1997). The RnfABCDGE proteins are also encoded in numerous gammaproteobacterial genomes, including the methanotroph *M. capsulatus* Bath (Ward et al., 2004) and all four sequenced *Nitrosococcus* genomes (Klotz et al., 2006; Campbell and Klotz, unpublished). Interestingly, all four sequenced *Nitrosococcus* genomes have the inventory to express three functional Complexes I: two NDH-I and one Na^+-NQR (Rnf) (Fig. 5) (Klotz et al., 2006; Campbell and Klotz, unpublished). Because the genomes of these AOB also encode numerous other sodium-dependent inventory, including a sodium-pumping ATPase (Klotz et al., 2006), operation of a unique sodium circuit in addition to the proton circuit was proposed, which could allow this bacterium to use a different NDH-I complex in forward or reverse mode (Fig. 5) and discriminate between processes in the plasma membrane and internal membranes that protrude into the cytoplasm. However, all AOB genomes encode at least one proton-translocating NDH-I complex (Arp et al., 2007; Stein et al., 2007; Norton et al., 2008) that is usually used in the "reverse electron flow mode" as a quinol oxidase, which depletes proton motive force and typically facilitates synthesis of NADH in chemolithotrophs where external reductants have more positive redox potentials than that of the redox couple NAD^+/NADH (−0.32 V). Only some AOB genomes encode additional complexes I that can operate as quinone reductases and contribute to proton-motive force when supplied NADH, such as by the sodium circuit in Gamma-AOB or by hydrogenase encoded in the genome of *N. multiformis* (Fig. 5) (Norton et al., 2008) (see below also).

Interestingly, the genomes of all Gamma-AOB indicate the presence of highly redun-

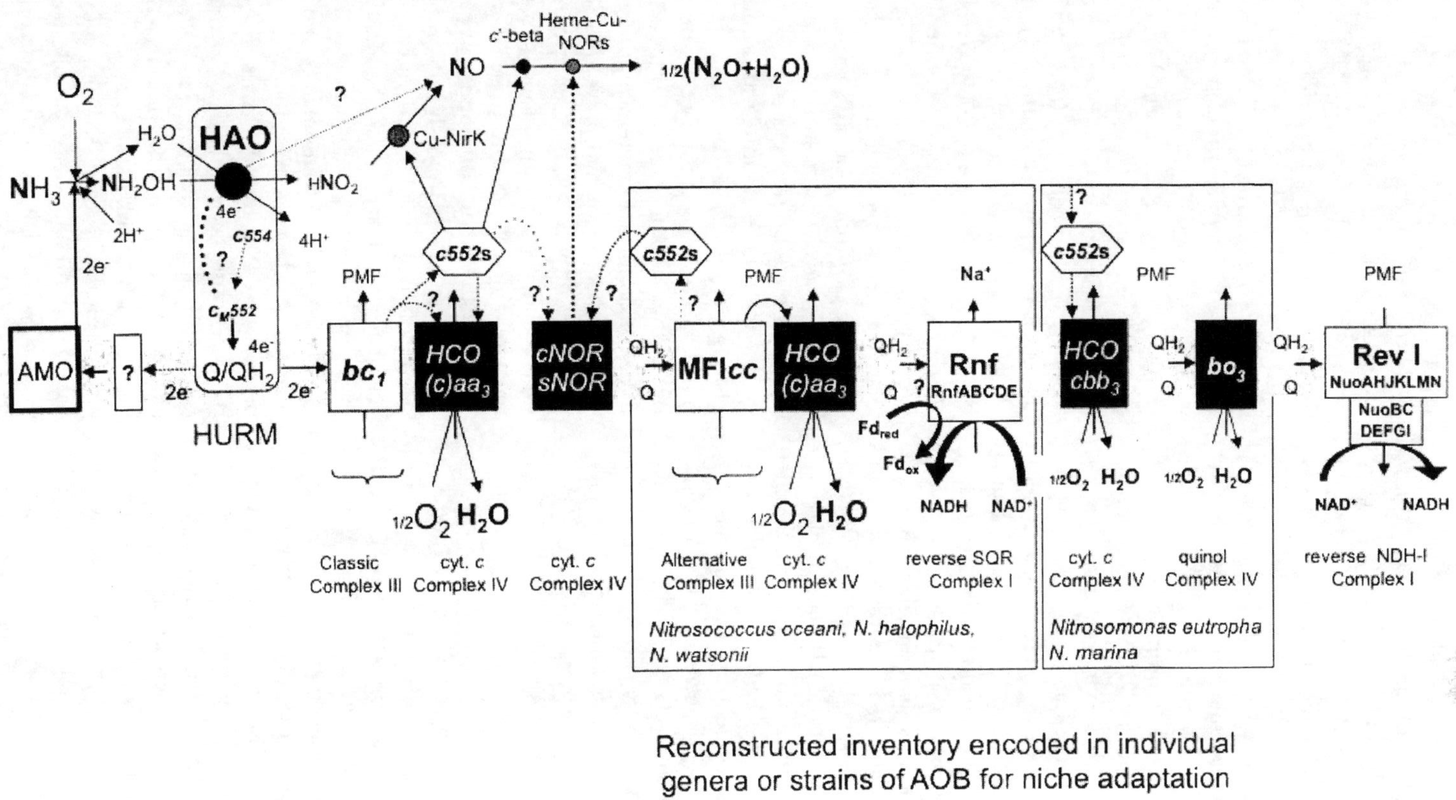

FIGURE 5 Flow of nitrogen and electrons in the quinone-reducing and quinol-oxidizing branches of the ETC of AOB. Basic inventory encoded in all AOB are shown together with reconstructed inventory encoded in individual genera or strains of AOB for niche adaptation. Abbreviations are explained in the text.

dant oxidative branches of the ETC, whereas Beta-AOB genomes usually encode only one quinol-oxidizing branch. There is a great variety of cytochrome *c*-mediated reductive pathways in the quinol-oxidizing branch in all three domains of life, all of which begin with (ubi)quinol-cytochrome *c* oxidoreductase (Complex III [Cape et al., 2006; Hunte et al., 2008, and references therein]) and usually end with soluble or membrane-bound terminal oxidases (often called "Complex IV" in sensu lato). These Complexes IV can accommodate life in habitats with varying oxygen concentrations from anoxia (anaerobic respiration) to oxygen-saturated environments. In contrast, direct quinol-oxidizing complexes include two versions of electron flow, a linear branch ending with Complex IV, such as cytochrome *bd*-type quinol oxidase, and an electron-recycling branch, exemplified by reverse operation of, for example, NDH-I from the reductive branch (Complex I) (Fig. 5). Biochemical characterization of a "novel multiheme cytochrome *bc* complex" from *Rhodothermus marinus* that lacks the classical cytochrome bc_1 Complex III (Pereira et al., 1999) triggered in silico analyses of available genomes. This novel Alternative Complex III (ACIII) is encoded by numerous bacterial genomes and assembled from a minimum of six proteins expressed from a cluster of contiguous genes (Yanyushin et al., 2005). Most genes encoding ACIII are also clustered with genes that encode a dedicated functional Complex IV (Yanyushin et al., 2005). Interestingly, of more than 2,000 sequenced bacterial genomes, only ~50 genomes, including those of *Geobacter metallireducens* GS-15, *Thermus thermophilus* HB8, *Ralstonia eutropha* JMP134, and all three *Nitrosococcus* species, contain both the classical and alternative forms of Complex III (Campbell et al., unpublished). Comparison of transcriptomes from *N. oceani* cultures denied ammonia as an energy source for 24 h and cultures stimulated with ammonia after a 24-h starvation revealed that CIII and ACIII are differentially expressed: CIII is utilized during growth in the presence of ammonia, while ACIII is expressed in starving cells to maintain electron flow (Campbell and Klotz, unpublished).

Nitric oxide and nitrous oxide are produced in small amounts by the AOB and ANB both as side products of hydroxylamine oxidation and from the reduction of nitrite. The latter process, which involves nitrite and nitric oxide reductases, is termed "nitrifier denitrification" (see Chapter 5). Hypoxia stimulates nitrifier denitrification and leads to greater losses of NH_3-N to nitrogen oxides. All examined AOB genomes encode both copper-containing nitrite reductase (*nirK*) and membrane-bound cytochrome *c* nitric oxide reductase (*norCBQD*) genes, although the diversity of these genes within the AOB is vast (Casciotti and Ward, 2001, 2005; Cantera and Stein, 2007; Garbeva et al., 2007). None of the AOB genome sequences contain homologues to nitrous oxide reductase (*nosZ*) genes, suggesting that nitrous oxide is the terminal product of NO_x reduction; however, *Nitrosomonas* spp. can produce N_2 as a main product of nitrite reduction (Schmidt et al., 2004). Furthermore, nitrosomonads can grow anaerobically by using nitrite as a terminal electron acceptor with ammonia, H_2, or organic carbon as a fuel source (Schmidt, 2009). Chemoorganoheterotrophic growth of *N. europaea* was reported previously (Hommes et al., 2003); however, the denitrifying inventory enabling the nitrosomonads to have this metabolic lifestyle remains to be characterized.

Under fully oxic conditions, NO is created in small quantities from the incomplete oxidation of hydroxylamine to nitrite by HAO (Hooper and Terry, 1979; Andersson et al., 1982; Hooper et al., 1990). As the AOB cannot prevent NO production during ammonia oxidation, they have acquired multiple lines of defense to avoid nitrosative stress. All of the AOB genomes were found to encode *c′*-beta (*cytS*), and all but *N. multiformis* encode cytochrome P460 (*cytL*), the products of which bind to NO. Although not yet examined physiologically in the AOB, both cytochromes *c′*-beta and P460 have been implicated in alleviating NO toxicity in other bacteria (Choi et al., 2006; Elmore et al., 2007; Deeudom et al., 2008). Furthermore, a four-gene cluster encoding a heme-copper nitric oxide reductase (sNOR; *norSY-senC-orf1*) Complex IV has been identified in all genomes of the AOB and a few sulfur-cycle bacteria (Stein et al., 2007; Hemp and Gennis, 2008; J. Hemp, R. B. Gennis, L. Y. Stein, and M. G. Klotz, unpublished data). The first two members of this gene cluster were upregulated in a *nirK* mutant strain of *N. europaea*, indicating involvement in the nitrosative stress response (Cho et al., 2006).

Methanotrophic bacteria that are also ANB produce nitrous oxide from both incomplete hydroxylamine oxidation and nitrifier denitrification. The genome of *M. capsulatus* Bath encodes functional cNOR (*norCB*), cytochrome *c*′-beta (*cytS*), cytochrome P460 (*cytL*), and HAO (*haoAB*) proteins. Recent experiments showed that expression of *haoA* and *cytS* was induced by NH_3 (Poret-Peterson et al., 2008), whereas expression of *norC* was induced by nitrite or sodium nitroprusside in *M. capsulatus* Bath (A. T. Poret-Peterson and M. G. Klotz, unpublished data). In contrast, expression of *cytL* was not affected by NH_3, nitrite, or sodium nitroprusside (Klotz et al., unpublished data). Exposure to hydroxylamine or NH_3 increased *haoA* mRNA levels in the Gamma-MOB, *Methylomicrobium album* (G. Nyerges and L. Y. Stein, unpublished data). Interestingly, the *norCB* gene tandem in *M. capsulatus* Bath (MCA2400-01) is in close proximity to the *cytS-c552-coxABD* (MCA2394-MCA2397) genes as well as another *c552* gene (MCA2405), which together appear to constitute a NO_x-linked electron flow gene supercluster (MCA2394 to MCA2405) that is largely flanked by hypothetical genes. It is reasonable to suggest that NO_x-linked electron flow gene superclusters were horizontally transferred in parallel with genes that generate poisonous NO_x (i.e., the *haoAB* gene tandem). Additional genome sequences are necessary to further delineate the evolution and regulation of nitrification and denitrification inventories of ANB.

The red copper cupredoxin nitrosocyanin (Arciero et al., 2002) is present at concentrations comparable to the central enzymes of ammonia catabolism such as AMO and HAO during aerobic growth of *N. europaea* (Whittaker et al., 2000) and was thus proposed to be involved in central N-oxidation pathways either catalytically or as an electron carrier (Hooper et al., 2005). It has been suggested that nitrosocyanin may participate in the recycling of electrons from the quinone pool to AMO or in the relay of electrons from hydroxylamine to O_2 (Arp et al., 2007). In unraveling the molecular underpinnings of nitrification by ANB such as Verrucomicrobia and Gamma-MOB, both of which lack *ncyA* genes and use the AMO-homologue pMMO for the oxidation of ammonia (Ward et al., 2004; Hou et al., 2008; Op den Camp et al., 2009), it appears less likely that nitrosocyanin is involved in recycling electrons from the quinone pool to AMO in AOB. On the other hand, a functional link to NirK has been suggested (Arciero et al., 2002), and a proteomics study revealed that nitrosocyanin levels greatly increased in *N. europaea* cultures exposed to exogenous NO, high levels of NO_2^-, or reduced O_2 concentration (Schmidt et al., 2004). Additional results link the expression of nitrosocyanin with response to ammonia starvation (Campbell and Klotz, unpublished). Based on its electronic structure, a potential role for NO binding and reduction by nitrosocyanin was proposed (Basumallick et al., 2005). As mentioned above, a region of the deduced nitrosocyanin protein shares significant sequence similarity with that of the binuclear copper center (CuA)-binding region of N_2O reductase (NosZ); however, physical properties of the enzyme suggest a role in electron transfer rather than catalysis. Future experiments will investigate whether nitrosocyanin has a role in ammonia catabolism and/or is a functional part of denitrification activity of AOB.

All AOB express one or more copies of soluble periplasmic monoheme and di-heme cytochrome *c552* proteins that have been experimentally or hypothetically implicated in the transfer of electrons to functional respiratory electron sinks including the soluble cytochrome *c* peroxidase, soluble nitrite, and nitric oxide reductases and membrane-bound Complex IV heme-copper oxidases that reduce O_2 or NO (Arp et al., 2007; Klotz and Stein, 2008, and references therein). Cytochrome *c552* homologues have also been similarly implicated in the sulfur-dependent catabolism of *Thiomicrospira crunogena* (Scott et al., 2006) and *Sulfurimonas denitrificans* (Sievert et al., 2008). Since increased functional protein levels of electron sinks demand an increased supply of reductant, it would be logical if expression of *c552* and its respective electron acceptors were coregulated.

It was summarized recently that the genomes of all four AOB indicate a dependence on polyphosphate as a storage compound and inorganic pyrophosphate (PP_i) in carbon and energy metabolism (Arp et al., 2007). According to genome inventory, energy could be acquired from polyphosphate as ATP at times when ammonia is limited as an energy source. In addition, the genomes indicate the capacity for hydrolysis of ATP to PP_i via the action of one of several nucleoside diphosphate (NUDIX) hydrolases, and the PP_i could then be used to generate a proton gradient via the action of a proton-translocating membrane-associated pyrophosphatase encoded in all AOB genomes (Arp et al., 2007). Hence, polyphosphate hydrolysis might contribute to motility, transport, reverse electron flow, and other cellular activities requiring a proton gradient. The genomes also encode a soluble cytoplasmic pyrophosphatase that must be regulated to avoid hydrolysis of the PP_i required for other reactions (Arp et al., 2007).

Together, apparent differences in complexity and diversity of energy flow and electron transport inventory suggest that architectural structure and organization of central metabolic networks were likely the outcome of different environmental pressures that selected for versatility to overcome specific environmental stresses. For instance, the plastic respiratory response capacity identified in genomes of the Gamma-AOB was likely achieved through highly regulated differential expression of interacting components to constitute a functional ETC. While this analysis of ETC inventory revealed an almost unprecedented richness encoded by genomes of Gamma-AOB compared with available genomes of Beta-AOB, there is presently no plausible ecological explanation for this situation considering that soil environments experience dramatic fluxes regarding environmental parameters.

Autotrophy

Since their initial isolation over 100 years ago, AOB have been viewed as obligately requiring ammonia as their sole source of energy (chemotrophy) and reductant (lithotrophy) and carbon dioxide as their sole carbon source (autotrophy) (Arp and Bottomley, 2006). Because ammonia oxidation yields only marginal useable energy, namely a maximum of two electrons per molecule (Hooper et al., 2005, and references therein), this combination of obligate ammonia oxidation at the expense of inorganic carbon assimilation seems to be unfavorable for successful selection. Complete pathways for the oxidation of a few organic compounds (e.g., pyruvate, fructose, glutamate) were identified; however, the genome sequences of all AOB revealed the absence of pathways for the uptake and catabolism of most amino acids, sugars, phospholipids, and nucleic acids (Arp et al., 2007; Stein et al., 2007; Norton et al., 2008). While analysis of all AOB genomes suggests a default mode of carbon assimilation via the Calvin-Benson-Basham cycle (Arp et al., 2007; Stein et al., 2007; Norton et al., 2008), experimentation revealed that fructose as well as pyruvate could serve as sole carbon sources for the growth of *N. europaea* (Hommes et al., 2003). Even though this observed chemolithoheterotrophic growth was slower and produced lower cell densities than when CO_2 was the available carbon source (Hommes et al., 2003), these experiments broke the 100-year-old dogma that AOB are obligate autotrophs (Arp and Bottomley, 2006). In agreement with prior hypotheses, growth of *N. europaea* on fructose or pyruvate still required ammonia as the energy and reductant source (Hommes et al., 2003), thereby upholding, for a time, the paradigm of obligate ammonia catabolism (chemolithotrophy) for AOB. Only very recently was it shown that *N. europaea* and *N. eutropha* could grow chemoheterotrophically with pyruvate, lactate, acetate, serine, succinate, alpha-ketoglutarate, or fructose as substrate and nitrite as terminal electron acceptor under anoxic conditions (Schmidt, 2009). Here, ammonium inhibited growth, showing for the first time that some previously classified AOB do not catabolize ammonia obligatorily.

Analysis of all AOB genomes suggests that the mechanism of fructose-1,6-bisphosphate and glucose-6-phosphate interconversion in gluconeogenesis and glycolysis probably occurs via a reversible, pyrophosphate-dependent phosphofructokinase (Arp et al., 2007) as proposed for the methanotroph *M. capsulatus* Bath (Ward et al., 2004). Other examples of the dependence on pyrophosphate in all examined AOB include UDP-glucose pyrophosphorylase, which catalyzes the formation of the glucosyl donor for sucrose synthesis (see below), and ADP-glucose pyrophosphorylase, which catalyzes the formation of the glucosyl donor for glycogen synthesis (Arp et al., 2007). Surprisingly, the genome of *N. multiformis* ATCC 25196 appears to lack an orthologue for a bacterial-type, ATP-independent fructose-1,6-bisphosphate aldolase (Norton et al., 2008). However, since this step of gluconeogenesis is required for autotrophic metabolism, it was suggested that this function in this strain of *N. multiformis* likely used an archaeal-type inositol monophosphatase (Norton et al., 2008).

Ecological Implications

AOB are incapable of catabolizing natural energy sources other than ammonia, with the exception of some nitrosomonads (Schmidt, 2009). As pointed out above, the majority of AOB may have evolved by losing the uptake and processing capacity for other energy sources during the process of genome reduction in concert with their adaptation to specific environmental conditions. Indeed, available ecophysiological, genetic and genomic data support the hypothesis that AOB can be grouped into four major ecotypes: (i) freshwater sediments, (ii) sewage/wastewater, (iii) soils, and (iv) marine environment (Koops and Pommerening-Roser, 2001; Kowalchuk and Stephen, 2001; Zehr and Ward, 2002). Nevertheless, several general ecophysiological features were deduced from the genomes that seemed to be characteristic to all AOB independent of their individual habitats. Most striking was the near-absence of transport systems for organic compounds, whereas a plethora of transporters for inorganic compounds were present in high redundancy (Arp et al., 2007; Stein et al., 2007; Norton et al., 2008). This imbalance is likely due to a bias in genome economization in that AOB lost most of their transporters for organics in the process of niche differentiation. Furthermore, the classical inventory to produce acyl-homoserine lactone signal molecules is absent from all AOB genomes, whereas all have the capacity to sense them with the typical receptors. Since AOB sense and respond to acyl-homoserine lactone molecules (Batchelor et al., 1997; Burton et al., 2005), an alternate pathway for their synthesis is likely present in AOB (Arp et al., 2007; Stein et al., 2007; Norton et al., 2008). This ability could be significant for the interaction and aggregation of AOB in biofilms in all but marine environments (Arp and Bottomley, 2006). A surprising finding was that all investigated AOB genomes contain two genes that code for the synthesis of sucrose (Arp et al., 2007). It was proposed that this inventory could have been horizontally acquired from cyanobacteria, with which AOB might have closely shared a niche (freshwater/sediment and marine) (Arp et al., 2007). On the other hand, sucrose-synthesizing activities are also found in some halotolerant methanotrophs (Arp et al., 2007) in the family *Methylococcaceae*, which includes several ANB and could thus also be a partner for gene transfer. Although it has not yet been demonstrated whether AOB produce sucrose and, if so, under what conditions this would occur, sucrose could provide an osmoticum to protect cells exposed to high salt concentrations (as in marine habitats) or desiccation (in fluctuating freshwater evaporation ponds and sediments).

AOB in freshwater sediments often experience NH_3/NH_4^+ and O_2 depletion due to competition with facultative anaerobes. A mode for micro-aerophilic or anaerobic respiration was proposed for the nitrosomonads based on their physiology (Schmidt et al., 2002). However, inventory for the operation of a *cbb*$_3$-type terminal oxidase usually implicated in micro-aerophilic respiration was reportedly found only in the genome of the sewage isolate *N. eutropha* (Stein et al., 2007). The sewage environment is characterized by high NH_3/NH_4^+ availability, which bears the potential for toxicity along with a stiff competition for O_2, CO_2, and iron. Hence, AOB in wastewater should have increased detoxification capacity, the ability to sequester CO_2 and perform microaerophilic respiration. With the recent discovery of anaerobic chemoheterotrophy, the wastewater AOB do appear to have additional metabolic capacity that can be advantageous in the dynamic wastewater environment. According to its genome inventory, *N. eutropha* indeed appears better able to resist toxic compounds, in particular heavy metals, and its genome contains genes for carboxysome synthesis and uniquely encodes alternative terminal oxidases, including *cbb*$_3$-type and quinol oxidases (Fig. 5) (Stein et al., 2007). The carboxysome genes were highly similar to the inventory found in the genome of *Nitrobacter winogradskyi* Nb-255, an NOB isolated from similar environments. A shared evolutionary origin of the inventory was proposed based on the known close associations between AOB and NOB in nitrification aggregates (Stein et al., 2007). On the other hand, genes encoding a similar complement of alternative terminal oxidases (*cbb*$_3$-type and quinol oxidases) were also detected in the marine isolate *N. marina* C-113a (Ward et al., unpublished), suggesting that the prediction of ecotypical genome inventory is not straightforward.

AOB in soil environments experience fluctuating NH_3/NH_4^+ availability, and stiff competition for NH_3/NH_4^+ with plants, changing resources, and acidity (pH several units below pK_a for NH_3/NH_4^+) usually results in variable growth rates. Urea hydrolysis not only adds an organic source for the main growth requirements of AOB (NH_3 and CO_2) but also allows pH manipulation of the environment. It was thus an ecological fit to find that urea-catabolic capacity was resident in most soil AOB of the genus *Nitrosospira* (Koper et al., 2004)—the genome of *N. multiformis* ATCC 25196 was found to encode both urea hydrolase and urea amidolyase activities (Norton et al., 2008)—but absent from wastewater nitrosomonads (Chain et al., 2003; Stein et al., 2007). On the other hand, the marine AOB *N. oceani*, but not *N. halophilus* Nc4, appears to have also the full complement required to access and process urea (Koper et al., 2004) including a complete urea cycle (Klotz et al., 2006). While this ureolytic capacity of soil AOB may be a major factor for their survival in acidic soils (Burton and Prosser, 2001), it is difficult to imagine a major catabolic advantage for urea hydrolysis in the oceans. On the other hand, the unique presence of hydrogenase-encoding inventory in the genome of *N. multiformis* ATCC 25196 would constitute an additional source of reductant and energy (Norton et al., 2008) and would constitute, if experimentally confirmed, an additional case of breaking the paradigm of obligate ammonia catabolism, though not that of obligate chemolithotrophy (Fig. 5).

Marine environments have a stable but low NH_3/NH_4^+ availability, and the dissolved CO_2 concentration is variable, which explains the fairly low growth rates measured for Gamma-AOB. Marine nitrifiers also experience a high salt concentration beyond the tolerance level of the three other ecotypes. The finding of inventory suited to express multiple proton- and sodium-dependent ATPase and NDH-I complexes was a first indication for marine-specific inventory as it allowed assembly of a sodium circuit in addition to the proton circuit in *N. oceani* (Fig. 5) (Klotz et al., 2006; Arp et al., 2007). Whereas this sodium circuit likely will not recruit additional energy and reductant sources (in contrast to the additional input by hydrogenase in *N. multiformis*), the ability to convert a sodium motive force into proton motive force may provide flexibility in regulating availability of proton motive force across the cytoplasmic membrane versus the intracellular membranes. The inventory necessary for such a sodium circuit has been identified in all four sequenced *Nitrosococcus* genomes (Campbell and Klotz, unpublished).

MOLECULAR EVOLUTION OF NITRIFICATION INVENTORY

Ecophysiological and taxonomic research in the pregenomic era addressed bacterial autotrophic nitrification as a functional, cooperative unit of two different cohorts of Proteobacteria, the AOB and the NOB (Prosser, 1989). The taxonomy of the NOB is complex because nitrite-oxidizing representatives are found in four of the six classes of Proteobacteria and some belong to the phylum Nitrospirae (Teske et al., 1994) (see Section V for more details). In contrast, phylogenetic inference using alignments of both 16S rRNA and *amoA* genes have placed the AOB in only two classes of the Proteobacteria (Teske et al., 1994) (see Chapter 3 for more details). Reconstructing the natural history of nitrification has focused on the molecular evolution of the subunit proteins of AMO (which is homologous to pMMO) due to the congruence of 16S rRNA and *amo* gene phylogenies, the assumption that aerobic ammonia and nitrite oxidation coevolved, and the assertion that ammonia oxidation constitutes a bottleneck in the nitrification process (Holmes et al., 1995; Rotthauwe et al., 1997; Klotz and Norton, 1998; Purkhold et al., 2000; Norton et al., 2002; Casciotti et al., 2003; Calvo and Garcia-Gil, 2004). Thus, the question was posed as to whether (i) AOB evolved ammonia-only catabolism in parallel within the Betaproteobacteria and Gammaproteobacteria, and as sister groups to pMMO-expressing methane oxidizers (MOB), via genome and functional reduction from a common, ancestral, physiologically versatile proteobacterial ammonia/methane oxidizer (holophyletic AMO/pMMO-centric model); (ii) nitrification inventory evolved in toto one time only in the ancestor of either AOB or MOB and was then distributed as a single suite of pathway genes into other taxa by lateral gene transfer (process-centric model); or (iii) individual inventory of extant aerobic ammonia oxidation and nitrite production evolved independently before Earth's oceans and atmosphere became oxygenated, was disseminated independently by lateral gene transfer and, by chance, functionally combined in the ancestors of modern nitrifiers (modular model). While there was ambiguous support for any one of the three models in the pregenomic era, recent genome-informed attempts to reconstruct the natural history of nitrification support the modular model (Klotz, 2008; Klotz et al., 2008; Klotz and Stein, 2008).

Use of the AMO/pMMO-centric and process-centric models for reconstructing the evolutionary history of nitrification was based on two premises that were incorrect. AMO and pMMO, which catalyze the first step in extant nitrification and methanotrophy, cooxidize both substrates supporting the evolution of both enzymes from a common, likely substrate-promiscuous ancestor (Holmes et al., 1995; Norton et al., 2002). The first premise assumed that all of the extant nitrifiers and methanotrophs, which are obligate aerobes and utilize aerobic respiration for energy conservation, became fit for this catabolic lifestyle once they evolved functional AMO/pMMO complexes. However, the activities of AMO and pMMO do not contribute directly to the harvest of energy and reductant during nitrification and methanotrophy, respectively; in fact, their activity drains the quinol pool (Q-pool) and merely modifies external reductants (ammonia and methane) into compounds (hydroxylamine and methanol/formaldehyde) that are more conducive to the harvest of electrons. Therefore, the evolution of these monooxygenases was reliant on the presence of inventory collectively able to provide electrons to the Q-pool by extracting electrons from intermediate metabolites and, if AMO/pMMO are not bona fide quinol oxidase (which has not yet been experimentally established), recycle electrons from the Q-pool to AMO/pMMO. In addition, these oxygen-dependent enzymes generate extremely toxic products at high throughput, hydroxylamine (by AMO) and methanol/formaldehyde (by pMMO/methanol dehydrogenase), and therefore the premise that AMO/pMMO evolved first makes little sense because efficient detoxification systems must be in place for successful natural selection (Klotz, 2008). Furthermore, AMO/pMMO and many of the respiratory key enzymes contain copper-active sites. Copper is thought to be largely non-bioavailable in the absence of oxygen, and the mainly sulfidic ocean environment of the late Archaean and Early Proterozoic did not enable copper redox processes (Anbar and Knoll, 2002; Kaufman et al., 2007; Klotz and Stein, 2008; Scott et al., 2008). Assuming that delineation of the Proteobacteria was largely completed by the time sufficient oxygen levels arose (>2 billion years ago) (Arnold et al., 2004; Kaufman et al., 2007; Scott et al., 2008; Garvin et al., 2009), the premise predicts that aerobic MOB and AOB must have evolved as unique functional cohorts after the delineation was complete. Because an anaerobic N-cycle was in place before the advent of oxygen, the evolution of AMO/pMMO was likely a late functional modular add-on to the operation of existing anaerobic pathways (Klotz, 2008) (Fig. 6).

The second premise assumed that AMO and pMMO were only functional in modern Proteobacteria, namely the gammaproteobacterial (*Chromatiales*: *Nitrosococcus*) and betaproteobacterial (*Nitrosomonadales*: *Nitrosomonas*, *Nitrosospira*) AOB and the alphaproteobacterial (*Methylocystaceae*: *Methylocystis*, *Methylosinus*) and gammaproteobacterial (*Methylococcaceae*: *Methylococcus*, *Methylomicrobium*, *Methylomonas*) MOB (Prosser, 1989; Arp and Bottomley, 2006). However, the Amo proteins from Gamma-AOB (*Nitrosococcus*) and pMmo proteins from Gamma-MOB (*Methylococcus*) are more closely related to each other than either is to respective enzymes in Beta-AOB and other pXmo proteins in certain Gamma-MOB (Purkhold et al., 2000; Norton et al., 2002; Tavormina et al., 2010). Hence, the AMO/pMMO-centric and process-centric models would predict residence of nearly identical inventory involved in ammonia/hydroxylamine and methane/methanol/formaldehyde oxidation in only one taxonomic group, or at least in closely related taxa within one proteobacterial class. Based on the current state of knowledge, present data contradict this prediction as aerobic ammonia/methane oxidation exists in organisms outside the phylum Proteobacteria (Archaea and Verrucomicrobia) and the evolutionary histories of individual nitrification inventories (Amo, Hao, pertinent electron carrier proteins) are not congruent or even identical among the organisms (Arp et al., 2007; Klotz and Stein, 2008; Op den Camp et al., 2009; Walker et al., 2010). In addition, the anaerobic oxidation of ammonia and methane involves microbes outside of the Proteobacteria that utilize some, but not all, of the inventory for aerobic ammonia/methane oxidation. These recent discoveries strongly support the modular model that best describes the evolution of nitrification, which also means that evolution of the process, exemplified by pathways, can be described adequately only once the evolution of individual inventory is understood.

Ammonification and the Evolution of HURM in an Anoxic World

Geochemists and planetary scientists agree today that the primordial atmosphere (in contrast to deep-sea vent environments) was fairly inert (N_2, CO_2, CO) and did not contain large amounts of available geothermal energy in the form of inorganic reductants (CH_4, H_2S, NH_3). A small amount of NH_3 was possibly just enough to fuel the primordial peptide and nucleotide cycles that operated strictly at S-, Fe-, and Ni/Co-containing mineral surfaces (the "ligand sphere") in anoxic space (Wachtershauser, 1994; Huber et al., 2003). These surface-bound metal centers likely served as structural templates for active site complexes in enzymes that extended the primordial cycles in the emerging cellular world

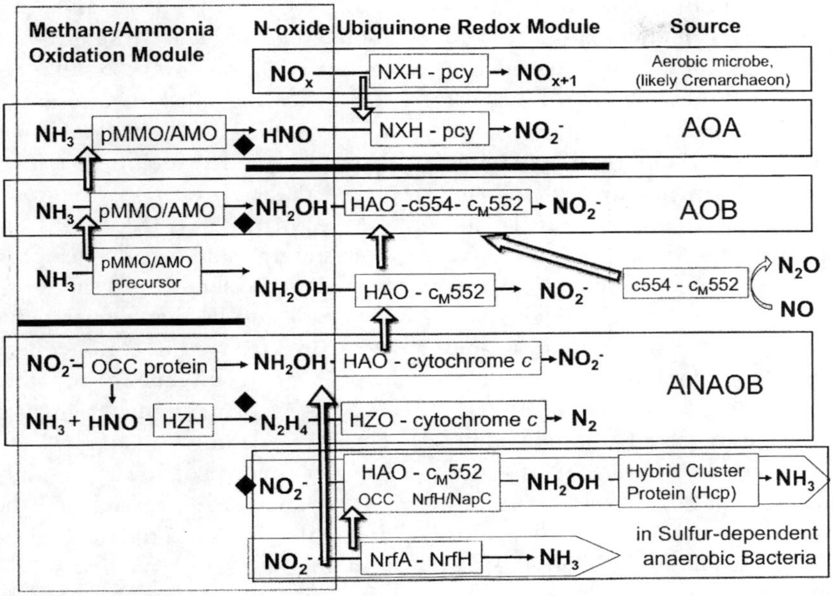

FIGURE 6 The modular concept of N-oxide transformation relevant to ammonia oxidation and nitrification. Directions of chemical and evolutionary pathways are indicated by closed and open arrows, respectively. Filled diamonds indicate the merger of modules as discussed in the text.

approximately 3.8 billion years (Gys) ago. With the evolving capacity of metal uptake, Mo, Zn, and Mn (but not Cu) were likely recruited into active sites of ancient enzymes because their oxidation state can change in the absence of oxygen (Scott et al., 2008). Hydrogen sulfide and ammonia were likely the precursors for molecules with catalytic sulfhydryl and amino groups.

The combination of hydrogen oxidation and sulfur reduction (as found in *Aquifex* and some modern archaea) was likely the chemolithotrophic beginning of cellular catabolism in largely oxygen-free and reducing microenvironments followed by the emergence of simple fermentations (substrate-level phosphorylation) and anoxygenic phototrophy (light-driven cyclic electron-flow). Further evolution of cytochromes for anoxygenic phototrophy, coupled with reactions recruited from fermentation pathways, likely led to the emergence of anaerobic chemotrophic respiration, in which external inorganic compounds served as terminal electron acceptors. Stable isotope geochemistry provides evidence for sulfur reduction at approximately 3 Gys ago; thus, anaerobic respiration involving sulfur reduction is likely an older adaptation of catabolism than phototrophy and nitrogen-based chemotrophy. Ammonification, the production of ammonium from other nitrogen compounds, likely existed within early bacteria and archaea as a consequence of simple fermentations; however, these "internal" cycles did not likely increase net NH_4^+/NH_3 availability. An early evolution of nitrogen fixation (producing reduced nitrogen) and methanogenesis (producing reduced carbon) under these anoxic conditions is imaginable but still controversially debated (Falkowski, 1997; Shen et al., 2003; Raymond et al., 2004; Canfield et al., 2006; Klotz and Stein, 2008).

Based on geophysicochemical data, it has been proposed that the early global N cycle was predominantly an atmospheric interaction of N_2, CO_2, and H_2O facilitated by lightning that generated small amounts of NO and HCN with larger amounts of NO_2^- and NO_3^- dissolved in the oceans (Mancinelli and McKay, 1988). Although free molecular oxygen was scarce, oxidized nitrogen (and sulfur) compounds accumulated predominantly in the oceans as nitrate (and sulfate) with only smaller quantities of nitrite (and sulfite) (Mancinelli and McKay, 1988). On the other hand, relatively higher ferrous iron concentrations during the Archaean might have been a strong interactive sink for nitrite due to the fast kinetics of their interaction. Because of these more or less sizable resources, nitrate and nitrite reduction (based on molybdopterin-containing and cytochrome *c* proteins) likely evolved in addition, but not parallel, to sulfate and thiosulfate reduction in the oceans (Shen et al., 2003; Arnold et al., 2004). There is a striking biochemical similarity between oxidoreductases active in the sulfur and nitrogen cycles allowing many of these enzymes to oxidize or reduce the other substrates. For instance, recent molecular-evolutionary and biochemical analyses indicated the evolutionary relatedness of pentaheme cytochrome *c* nitrite reductase (NrfA) and the octaheme cytochrome *c* (OCC) proteins, tetrathionate oxidoreductase and HAO, of which the former two can reduce both sulfur and nitrogen compounds (Einsle et al., 1999, 2000; Mowat et al., 2004; Bergmann et al., 2005; Hooper et al., 2005; Atkinson et al., 2007; Klotz et al., 2008; Lukat et al., 2008). These analyses also provided evidence for modular evolution versus the evolution of individual inventory by showing coevolution of catalytic OCC proteins with their respective redox partners in the c_M552/NrfH/NapC protein superfamily (Bergmann et al., 2005; Rodrigues et al., 2006; Kim et al., 2008; Klotz et al., 2008).

Recent studies on the evolutionary history of cytochrome *c* proteins key to the extant nitrogen cycle suggest that many of them evolved, indeed, from ancestral proteins key to the sulfur cycle (Hooper et al., 2005; Scott et al., 2006; Elmore et al., 2007; Klotz et al., 2008; Klotz and Stein, 2008; Sievert et al., 2008; M. G. Klotz and A. B. Hooper, unpublished results). Most of the nitrate and nitrite reduction inventory likely served the ever-increasing need for ammonia assimilation, in which the molybdopterin-containing (Nar, Nap) and cytochrome *c* (Nrf) proteins preceded the siroheme cytochrome proteins (NasA, NirA, NirB) (Klotz and Stein, 2008). Interestingly, some early-branching sulfur-dependent anaerobic Epsilonproteobacteria devoid of any known nitrite reductases utilize a "hydroxylamine oxidoreductase-c_M552/NapC protein" module in its pathway for the assimilation of ammonia from nitrate as the sole nitrogen source (Campbell et al., 2009). This very recent discovery strongly supports the proposal that an oxygen-independent HURM) (Fig. 3 and 6) had evolved early on from inventory that facilitates NO_x respiratory ammonification and NO_x detoxification (Arp et al., 2007; Klotz and Stein, 2008) as a reductive module that tied electron flow in anaerobic sulfur-dependent chemotrophs to N assimilation (Campbell et al., 2009). The concept of HURM was initially derived from comparative analysis of inventory encoded by genomes of AOB (Arp et al., 2007) and ANAOB (Strous et al., 2006), which identified HURM as the central oxidative module for all N-oxide-based electron flow in bacteria (Klotz and Stein, 2008). As in anoxygenic phototrophy, N-oxide-based electron flow in bacteria was initially cyclic and evolved to conserve energy in the anammox process (Klotz, 2008; Klotz and Stein, 2008). In addition to providing a quinone reductase needed for linking N-oxide chemistry with catabolic function, the high-efficiency HURM also provided high-throughput detoxification of poisonous N-oxides, thereby providing the stage for recruitment of high-throughput N-oxide-producing modules, such as the Amo and pMmo proteins (Klotz, 2008) (Fig. 6).

Aside from OCC proteins such as HAO (Klotz et al., 2008), multicopper oxidases are known to process N-oxides and both are good candidates for sources of low-level hydrazine (N_2H_4) production. Similar to the evolutionary pressure scenario that facilitated OCC protein evolution from NrfA (Klotz and Stein, 2008), disproportionation reactions facilitated by OCC proteins such as HAO that leak N-oxide intermediates (van der Star et al., 2008) might have set the evolutionary stage for hydrazine hydrolase to evolve, which freely operates as a hydrolase (and not as a synthase), and provided additional hydrazine detoxification capacity. HAO is also capable of oxidizing hydrazine (Schalk et al., 2000). It is further imaginable that exposure to hydrazine constituted the driving force for "anammoxosomeogenesis" to protect sensitive cellular structures (Klotz, 2008). Once the anammoxosome was in place and hydroxylamine/hydrazine oxidoreductase was coupled to a respective quinone reductase (establishing HURM in ANAOB), enough redox gradient was provided by the system to pull the hydrazine hydrolase into the synthase direction, in which it oxidizes ammonia using highly reactive nitroxyl (HNO) or NO as the oxidant. Using HNO to oxidize ammonia rather than using NO to oxidize NH_4^+ would avoid obligate dependence on NO, a highly reactive radical species, and make anammox more effective by distributing electrons more evenly in the process. Hence, recycled reductant from the quinone pool and a reversely operating OCC protein with nitrite reductase function (i.e., HAO) to produce HNO or NH_2OH, as in the assimilative HURM of sulfur-dependent Epsilonproteobacteria (Campbell et al., 2009), was the only inventory required to close the cycle of electron flow in the anammox process. The genomes of several species in the genera *Nautilia*, *Caminibacter*, and *Campylobacter* contain genes encoding the enzymes of the reverse-HURM pathway (Campbell et al., 2009), while the genomes of other *Epsilonproteobacteria* encode homologues of the classical NO-forming NirS/NirK, assimilatory siroheme NirA, or ammonium-forming NrfA nitrite reductases (Kern and Simon, 2009). These recent findings also suggest that HURM is bidirectional, includes at least one OCC protein and a (ubi)quinone reductase, and that the direction of operation depends on its position in either the reductive or oxidative branch of cellular electron flow (Fig. 6).

Evolution of Bacterial Inventory Involved in Aerobic, Iron-Copper-Facilitated Oxidation of Ammonia

The major evolutionary event for the vast diversification of catabolic pathways found in modern bacteria was undoubtedly the emergence of oxygenic phototrophy in the ancestors of extant cyanobacteria and prochlorophytes approximately 2.5 Gys ago, which led to a gradual increase of molecular oxygen in the atmosphere, reaching approximately 1% about 1.9 Gys ago (Falkowski, 1997; Raymond et al., 2004; Canfield et al., 2006; Kaufman et al., 2007; Garvin et al., 2009, and references therein). The most important consequences of this developing oxidizing atmosphere were the formation of an ozone layer and the coevolution of inventory allowing branched electron flow such as CIII and ACIII and the A-, B-, and C-type heme-copper oxidases that terminate electron flow by facilitating the reduction of oxygen (aerobic respiration) or NO (anaerobic respiration) (Garcia-Horsman et al., 1994; Pereira et al., 2001; Hemp and Gennis, 2008; Hemp et al., unpublished), all of which benefited from a dramatic increase in metal center diversity of enzymes (Anbar and Knoll, 2002). While ancient enzymes mostly contained Ni-, Fe-, S-redox-active sites (e.g., hydrogenase, urease, hydantoinase, etc.) and those without an oxygen requirement, or that operate best under anoxic conditions, also utilized Zn, Mn, and Mo (e.g., nitrogenase, molybdopterin-containing nitrate reductase), it was the rising oxygen levels that made copper available as an additional redox-active transition metal (Anbar and Knoll, 2002; Arnold et al., 2004). This was particularly consequential for the evolution of the nitrogen cycle (Klotz and Stein, 2008). For instance, reduction of the abiogenic and increasing biogenic nitrate pool by nitrate reductase likely increased the nitrite pool and generated strong evolutionary pressure for the emergence of numerous new variants of nitrite, NO, and nitrous oxide reductases, many of which contain copper in extant bacteria (Nakamura et al., 2004). Elevated concentrations of ammonia, nitrite, and NO are poisonous to many enzyme activities. The emergence of numerous alternate oxidoreductases (i.e., the three novel NOR families, sNOR, gNOR, and eNOR [Hemp and Gennis, 2008; Hemp et al., unpublished]) likely facilitated a continuous reduction of the nitrate pool and recycling to dinitrogen gas while maintaining a fairly small nitrite pool. This hypothesis, largely based on bioinorganic chemistry, was recently supported by the discovery of a high number of diverse multicopper oxidases in the genomes of organisms that are instrumental in the nitrogen cycle such as the ammonia and NOB (Starkenburg et al., 2006, 2008; Arp et al., 2007; Klotz and Stein, 2008).

Before the rise of oxygen, N-cycle pathway evolution likely resulted in dissimilative nitrate/nitrite reduction yielding mostly gaseous nitrogenous oxidants and thereby establishing denitrification (Mancinelli and McKay, 1988; Klotz and Stein, 2008). At this point in time, approximately 1 Gys after cellular life emerged and still approximately 1 Gy before the Earth's atmosphere became oxidizing in nature, the ancestors of emerging Proteobacteria had the metabolic opportunity to reduce oxidized nitrogen, sulfur, and carbon compounds by using them as terminal electron acceptors (Scott et al., 2008). While sulfur and nitrogen were more likely involved in anaerobic respiration, reduced (organic) carbon was the electron acceptor in fermentation processes. Early denitrification, which involved molybdopterin- and heme-iron-based redox chemistry on the path from nitrate to NO, was likely not able to progress with present-day classical denitrification inventory because cNOR and qNOR complexes are members of the heme-copper oxidase superfamily that evolved from heme-copper oxygen reductases (Hemp and Gennis, 2008; Hemp et al., unpublished). Because the latter appear to have emerged following oxygenic phototrophy, neither cNOR, qNOR, nor the blue copper protein nitrous oxide reductase was available for N-oxide reduction prior to the rise of oxygen. Hence, early denitrification likely relied on NO reduction by soluble periplasmic cytochrome *c* proteins such as *c*′-beta (*cytS*) and c554 (*cycA*) and not by membrane proteins. Today, NO-forming nitrate and nitrite reduction, singly or as part of the denitrification process and as induced by anoxia or hypoxic stress, are metabolic functions found in numerous taxonomic groups of bacteria; however, complete and efficient denitrification pathways (nitrate to dinitrogen) are almost exclusively found among the Proteobacteria that express nitrous oxide reductase (Ferguson and Richardson, 2005; Tavares et al., 2006; Zumft and Kroneck, 2006; Klotz and Stein, 2008).

The availability of oxygen as both a powerful oxidant and terminal electron acceptor allowed for the emergence of branched ETCs, a great diversity of novel copper-based electron carriers and redox-active enzymes, and coupled HURM (extracts four electrons from hydroxylamine/hydrazine by dehydrogenation) with the high-throughput oxidation of reduced nitrogen compounds. Hydrazine hydrolase and the OCC nitrite reductase are early-evolved cytochrome *c* proteins (Klotz et al., 2008; Klotz and Stein, 2008). In contrast, the genes encoding copper-containing extant pMMO and AMO homologues (Klotz and Norton, 1998; Norton et al., 2002) likely did not evolve from genes coding for pMMO in anaerobic denitrifying methanotrophic bacteria related to the NC10 clade (Raghoebarsing et al., 2006; Ettwig et al., 2008, 2009; Klotz, 2008; Tavormina et al., 2010). It is still an open question whether the (OCC protein-hydrazine hydrolase) Ammonia Oxidation Module as part of anammox was replaced by a recruited Methane/Ammonia Oxidation Module (pMMO/AMO) or whether HURM was recruited into an anaerobic bacterial genome encoding an ancestral promiscuous monooxygenase; however, it can be predicted with some certainty that the combination and functional linkage of pMMO/AMO and HURM (Fig. 6) occurred after the rise of oxygen (Klotz, 2008; Klotz and Stein, 2008).

The main advantage of this functional merger was a reduced need for recycling reductant from the Q-pool [pMMO/AMO requires two electrons to activate oxygen; (H) NO-producing OCC protein and hydrazine hydrolase together require four electrons] and, consequently, a net production of two electrons available for linear electron flow as well as the replacement of two soluble enzyme complexes with one membrane-bound complex.

This increased energetic efficiency in aerobic ammonia oxidizers resulted in less costly synthesis (reductant in the anammox process is generated by uneconomical anaerobic reoxidation of nitrite to nitrate [Strous et al., 2006; Jetten et al., 2009]) and dramatically increased growth rates as well as the global fixed N-oxide pool. It appears that the functional merger of the Methane/Ammonia Oxidation Module with HURM has had several different outcomes. Extant aerobic MOB appear to have lost the HURM quinone reductase (nitrifying MOB, the ANB) or HURM altogether, likely because the energy/reductant requirement for carbon fixation are much better met by their own unique carbon-1 metabolism than by oxidation of ammonia with reduction of CO_2. On the other hand, early HURM likely merged with another module involved in N-oxide metabolism in predecessors of extant AOB (Fig. 6): cytochrome *c554* is reported to have NO-reductase activity (Upadhyay et al., 2006), and genes encoding homologues of *c554* were found clustered together with c_M552/NapC protein- or other quinol oxidase-encoding genes in nonnitrifying bacteria, including chlorine oxide-reducing Betaproteobacteria and sulfur-dependent Epsilonproteobacteria and Deltaproteobacteria (Klotz and Hooper, unpublished). The increasing complexity of gene clusters containing *haoA* genes (Fig. 4) was congruent with the phylogenetic tree describing the evolution of OCC proteins, including HaoA from pentaheme cytochrome *c* nitrite reductase (Klotz et al., 2008).

Increased diversity in oxidoreductases and the availability of oxygen as terminal electron acceptor likely created opportunities for new redox interactions, some of which led to the reversal of electron flow (Bergmann et al., 2005; Klotz et al., 2008). Although different in their biochemical complexities, oxidation of sulfite to sulfate, for instance, is essentially the reverse of sulfate reduction, and both are performed by different groups of extant Proteobacteria. Likewise, aerobic nitrite oxidation is the reverse process of the reduction of nitrate to nitrite; thus, it is not surprising that nitrite oxidoreductase (NxrAB) of the NOB and nitrate reductase (NarGH) are evolutionarily related molybdopterin proteins (Kirstein and Bock, 1993; Starkenburg et al., 2006, 2008) (see the chapters in Section V for more details). Many of the enzymes involved in nitrification and aerobic denitrification involve copper (NirK nitrite reductase)-, iron-copper (AMO)-, or heme-copper (cNor, sNor, the Complex IV heme-copper oxidases that reduce O_2 and NO)-based redox activity. Recent physiological work with *nirK* and *norB* mutants of *N. europaea* revealed unexpected activities able to produce gaseous N-compounds that were not performed by traditional denitrification enzymes (Schmidt et al., 2004). The proteins involved have yet to be identified; however, preliminary analyses of all available nitrifier genomes uncovered several conserved multicopper oxidases predicted as alternative players in the oxidation/reduction of N-compounds. It has also become clear only recently that aerobic oxidation of ammonia by nitrifying archaea is solely facilitated by copper-based redox chemistry, with some of the inventory being unique to the AOA (Walker et al., 2010) (see the chapters in Section III for more details).

Comparison of genome inventory from all cohorts of catabolic ammonia oxidizers (AOB [Arp et al., 2007], ANAOB [Strous et al., 2006], and AOA [Walker et al., 2010]) suggests that electron flow as found in the anammox process may be the basis for all extant bacterial and archaeal ammonia oxidation mechanisms. While the evolution of all inventory essential to extant aerobic and anaerobic bacterial ammonia oxidation likely occurred before the big oxygenation event, archaeal ammonia oxidation is presently known to occur only in oxic environments. Because of this and the fact that nitrification inventories in AOB and AOA are not identical, a monophyletic origin of ammonia oxidation inventory that includes the Archaea (looking at the substrate of the pathway) and that of nitrite production (looking at the product of nitrification) is nonparsimonious. Like AOB, the AOA also utilize a Methane/Ammonia Oxidation Module; however, formation of HNO instead of NH_2OH by a modified archaeal AMO is proposed (Walker et al., 2010). This proposal is based on the absence of genes encoding HAO from AOA genomes (Walker et al., 2010) and the fact that AOA can be purged from mixed cultures by the addition of hydroxylamine. In contrast to AOB, the catabolic and respiratory electron flow inventory in the AOA is solely copper based including CIII, CIV, electron shuttles (plastocyanins instead of cytochrome *c* proteins), and, most importantly, the ubiquinone-reducing module analogous to HURM (Walker et al., 2010). Instead of a [HAO-(c554)-c_M552/NapC protein] cytochrome *c* protein module, AOA are proposed to use a (multicopper oxidase–di-copper-blue-domain membrane protein) module to relay electrons extracted from HNO to the quinone pool (Walker et al., 2010). Since AOA do not produce and utilize hydroxylamine or hydrazine as redox-active intermediates, the use of the acronym HURM to describe the quinone-reductive module is inappropriate. Instead and in extension of the HURM concept proposed previously (Klotz and Stein, 2008), use of the term N-oxide-Ubiquinone Redox Module (NURM) is proposed here to describe the general principle by which a reductant-rich N-oxide (NH_2OH, N_2H_4, HNO) can be utilized to harness energy and relay accessible reductant to the Q-pool in the membranes of obligate ammonia-oxidizing chemolithotrophs (Fig. 3 and 6).

SUMMARY AND PERSPECTIVE

Analysis of inventory encoded in ammonia-oxidizing bacteria and archaea as well as molecular evolutionary inference into the sequence and structure of nitrogen cycle inventory revealed that bacterial and archaeal ammonia oxidation pathways consist of two modules each: a (Methane) Ammonia Oxidation Module and the reductant-rich NURM, both of which functionally combined within different organisms in different geochemical backgrounds at different times during evolutionary history (Fig. 6). This is supported by the fact that NURM in aerobic and ANAOB are homologous (Klotz and Stein, 2008) but unrelated to the archaeal NURM (multicopper nitroxyl hydrolase plus copper-blue quinone reductase [Walker et al., 2010]), whereas the (Methane) Ammonia Oxidation Module is homologous in AOB and AOA (pMMO/AMO) but unrelated to the Ammonia Oxidation Module in ANAOB (equal to OCC nitrite reductase plus hydrazine hydrolase [Strous et al., 2006; Jetten et al., 2009]). Because of the highly toxic nature of the substrates for NURM, it only makes sense to propose that the emergence of a functional NURM must have preceded the functional linkage with an efficient and high-throughput (Methane) Ammonia Oxidation Module. Given that the present number of drafted and finished whole-genome projects has unraveled, so far, only one gene suspected to be unique to AOB (*ncyA*, nitrosocyanin), a few genes encoding candidate inventory with meaning for niche adaptation (Fig. 5), and the annotated bacterial genomes by far outnumber those of archaeal ammonia-oxidizers, continued isolation of ecophysiologically representative pure cultures and the sequencing and characterization of their genomes is imperative to continued progress in nitrogen cycle research. Likewise, intensified broader "omics" studies, including comparisons between isolate-based genome information and that available from the growing number of metagenome projects from environments relevant to ammonia oxidation, are needed to assess the connection between inventory and function, from a single-cell scale to physiological characterization of cohorts to a better ecological understanding.

Given the pace of discovery during that last two decades, which started with the development of ever more sophisticated applications of the primer extension method pioneered by Ray Wu in the 1970s, long before Sanger sequencing and PCR, we will likely soon see a dramatic increase in available genome sequences but highly likely also discover more biological novelty driven by bioprospecting, for instance, in extreme environments (cold, hot, saline, etc.), including the world's oceans.

REFERENCES

Agogue, H., M. Brink, J. Dinasquet, and G. J. Herndl. 2008. Major gradients in putatively nitrifying and non-nitrifying Archaea in the deep North Atlantic. *Nature* **456:**788–791.

Allen, A. E., M. G. Booth, M. E. Frischer, P. G. Verity, J. P. Zehr, and S. Zani. 2001. Diversity and detection of nitrate assimilation genes in marine bacteria. *Appl. Environ. Microbiol.* **67:**5343–5348.

Alzerreca, J. J., J. M. Norton, and M. G. Klotz. 1999. The *amo* operon in marine, ammonia-oxidizing Gammaproteobacteria. *FEMS Microbiol. Lett.* **180:**21–29.

Anbar, A. D., and A. H. Knoll. 2002. Proterozoic ocean chemistry and evolution: a bioinorganic bridge? *Science* **297:**1137–1142.

Andersson, K. K., S. B. Philson, and A. B. Hooper. 1982. ^{18}O isotope shift in ^{15}N NMR analysis of biological N-oxidations: N_2O-NO_2^- exchange in the ammonia-oxidizing bacterium *Nitrosomonas. Proc. Natl. Acad. Sci. USA* **79:**5871–5875.

Arciero, D., C. Balny, and A. B. Hooper. 1991a. Spectroscopic and rapid kinetic studies of reduction of cytochrome *c*554 by hydroxylamine reductase from *Nitrosomonas europaea. J. Biol. Chem.* **269:**11878–11886.

Arciero, D. M., and A. B. Hooper. 1993. Hydroxylamine oxidorectase is a multimer of an octa-heme subunit. *J. Biol. Chem.* **268:**14645–14654.

Arciero, D. M., and A. B. Hooper. 1997. Evidence for a crosslink between *c*-heme and a lysine residue in cytochrome P460 of *Nitrosomonas europaea. FEBS Lett.* **410:**457–460.

Arciero, D. M., M. J. Collins, J. Haladjian, P. Bianco, and A. B. Hooper. 1991b. Resolution of the four hemes of cytochrome *c*554 from *Nitrosomonas europaea* by redox potentiometry and optical spectroscopy. *Biochemistry* **30:**11459–11465.

Arciero, D. M., A. B. Hooper, M. Cai, and R. Timkovich. 1993. Evidence for the structure of the active site heme P460 in hydroxylamine oxidoreductase of *Nitrosomonas. Biochemistry* **32:**9370–9378.

Arciero, D. M., B. S. Pierce, M. P. Hendrich, and A. B. Hooper. 2002. Nitrosocyanin, a red cupredoxin-like protein from *Nitrosomonas europaea. Biochemistry* **41:**1703–1709.

Arnold, G. L., A. D. Anbar, J. Barling, and T. W. Lyons. 2004. Molybdenum isotope evidence for widespread anoxia in mid-Proterozoic oceans. *Science* **304:**87–90.

Arp, D. J., and P. J. Bottomley. 2006. Nitrifiers: more than 100 years from isolation to genome sequences. *Microbe* **1:**229–234.

Arp, D. J., and L. Y. Stein. 2003. Metabolism of inorganic N compounds by ammonia-oxidizing bacteria. *Crit. Rev. Biochem. Mol. Biol.* **38:**471–495.

Arp, D. J., L. A. Sayavedra-Soto, and N. G. Hommes. 2002. Molecular biology and biochemistry of ammonia oxidation by *Nitrosomonas europaea. Arch. Microbiol.* **178:**250–255.

Arp, D. J., P. S. G. Chain, and M. G. Klotz. 2007. The impact of genome analyses on our understanding of ammonia-oxidizing bacteria. *Ann. Rev. Microbiol.* **61:**21–58.

Atkinson, S. J., C. G. Mowat, G. A. Reid, and S. K. Chapman. 2007. An octaheme *c*-type cytochrome from *Shewanella oneidensis* can reduce nitrite and hydroxylamine. *FEBS Lett.* **581:**3805–3808.

Basumallick, L., R. Sarangi, S. DeBeerGeorge, B. Elmore, A. B. Hooper, B. Hedman, K. O. Hodgson, and E. I. Solomon. 2005. Spectroscopic and density functional studies of the red copper site in nitrosocyanin: role of the protein in determining active site geometric and electronic structure. *J. Am. Chem. Soc.* **127:**3531–3544.

Batchelor, S. E., M. Cooper, S. R. Chhabra, L. A. Glover, G. S. Stewart, P. Williams, and J. I. Prosser. 1997. Cell density-regulated recovery of starved biofilm populations of ammonia-oxidizing bacteria. *Appl. Environ. Microbiol.* **63:**2281–2286.

Bergmann, D. J., and A. B. Hooper. 2003. Cytochrome P460 of *Nitrosomonas europaea. Eur. J. Biochem.* **270:**1935–1941.

Bergmann, D. J., D. Arciero, and A. B. Hooper. 1994. Organization of the *hao* gene cluster of *Nitrosomonas europaea*: genes for two tetraheme *c* cytochromes. *J. Bacteriol.* **176:**3148–3153.

Bergmann, D. J., A. B. Hooper, and M. G. Klotz. 2005. Structure and sequence conservation of genes in the *hao* cluster of autotrophic ammonia-oxidizing bacteria: evidence for their evolutionary history. *Appl. Environ. Microbiol.* **71:**5371–5382.

Bertsova, Y. V., and A. V. Bogachev. 2004. The origin of the sodium-dependent NADH oxidation by the respiratory chain of *Klebsiella pneumoniae. FEBS Lett.* **563:**207–212.

Berube, P. M., S. C. Proll, and D. A. Stahl. 2007. Genome-wide transcriptional analysis following the recovery of *Nitrosomonas europaea* from ammonia starvation, abstr. H-105. Abstr. 107th Gen. Meet. Am. Soc. Microbiol. American Society for Microbiology, Washington, DC.

Berube, P. M., R. Samudrala, and D. A. Stahl. 2007. Transcription of all *amoC* copies is associated with recovery of *Nitrosomonas europaea* from ammonia starvation. *J. Bacteriol.* **189:**3935–3944.

Bock, E., H.-P. Koops, H. Harms, and B. Ahlers. 1991. The biochemistry of nitrifying organisms, p. 171–200. *In* J. M. Shively and L. L. Barton (ed.), *Variations in Autotrophic Life*. Academic Press Limited, San Diego, CA.

Braker, G., and J. M. Tiedje. 2003. Nitric oxide reductase (*norB*) genes from pure cultures and environmental samples. *Appl. Environ. Microbiol.* **69:**3476–3483.

Brandes, J. A., A. H. Devol, and C. Deutsch. 2007. New developments in the marine nitrogen cycle. *Chem. Rev.* **107:**577–589.

Burton, E. O., H. W. Read, M. C. Pellitteri, and W. J. Hickey. 2005. Identification of acyl-homoserine lactone signal molecules produced by *Nitrosomonas europaea* strain Schmidt. *Appl. Environ. Microbiol.* **71:**4906–4909.

Burton, S. A. Q., and J. I. Prosser. 2001. Autotrophic ammonia oxidation at low pH through urea hydrolysis. *Appl. Environ. Microbiol.* **67:**2952–2957.

Butler, C. S., and D. J. Richardson. 2005. The emerging molecular structure of the nitrogen cycle: an introduction to the proceedings of the 10th annual N-cycle meeting. *Biochem. Soc. Trans.* **33:**113–118.

Calvo, L., and L. J. Garcia-Gil. 2004. Use of *amoB* as a new molecular marker for ammonia-oxidizing bacteria. *J. Microbiol. Methods* **57:**69–78.

Campbell, B. J., J. L. Smith, T. E. Hanson, M. G. Klotz, L. Y. Stein, C. K. Lee, D. Wu, J. M. Robinson, H. M. Khouri , J. A. Eisen, and S. C. Cary. 2009. Adaptations to submarine hydrothermal environments exemplified by the genome of *Nautilia profundicola*. *PLoS Genet.* **5:**e1000362.

Canfield, D., M. Rosing, and C. Bjerrum. 2006. Early anaerobic metabolisms. *Philos. Trans. R. Soc. B Biol. Sci.* **361:**1819–1836.

Cantera, J. J., and L. Y. Stein. 2007a. Molecular diversity of nitrite reductase genes (*nirK*) in nitrifying bacteria. *Environ. Microbiol.* **9:**765–776.

Cantera, J. J., and L. Y. Stein. 2007b. Role of nitrite reductase in the ammonia-oxidizing pathway of *Nitrosomonas europaea*. *Arch. Microbiol.* **188:**349–354.

Cape, J. L., M. K. Bowman, and D. M. Kramer. 2006. Understanding the cytochrome *bc* complexes by what they don't do. The Q-cycle at 30. *Trends Plant Sci.* **11:**46–55.

Casciotti, K. L., and B. B. Ward. 2001. Dissimilatory nitrite reductase genes from autotrophic ammonia-oxidizing bacteria. *Appl. Environ. Microbiol.* **67:**2213–2221.

Casciotti, K. L., and B. B. Ward. 2005. Phylogenetic analysis of nitric oxide reductase gene homologues from aerobic ammonia-oxidizing bacteria. *FEMS Microbiol. Ecol.* **52:**197–205.

Casciotti, K., D. Sigman, and B. Ward. 2003. Linking diversity and stable isotope fractionation in ammonia-oxidizing bacteria. *Geomicrobiol. J.* **20:**335–353.

Chain, P., J. Lamerdin, F. Larimer, W. Regala, V. Lao, M. Land, L. Hauser, A. Hooper, M. Klotz, J. Norton, L. Sayavedra-Soto, D. Arciero, N. Hommes, M. Whittaker, and D. Arp. 2003. Complete genome sequence of the ammonia-oxidizing bacterium and obligate chemolithoautotroph *Nitrosomonas europaea*. *J. Bacteriol.* **185:**2759–2773.

Cho, C. M. H., T. Yan, X. Liu, L. Wu, J. Zhou, and L. Y. Stein. 2006. Transcriptome of a *Nitrosomonas europaea* mutant with a disrupted nitrite reductase gene (*nirK*). *Appl. Environ. Microbiol.* **72:**4450–4454.

Choi, P. S., Z. Naal, C. Moore, E. Casado-Rivera, H. D. Abruna, J. D. Helmann, and J. P. Shapleigh. 2006. Assessing the impact of denitrifier-produced nitric oxide on other bacteria. *Appl. Environ. Microbiol.* **72:**2200–2205.

Dalsgaard, T., B. Thamdrup, and D. E. Canfield. 2005. Anaerobic ammonium oxidation (anammox) in the marine environment. *Res. Microbiol.* **156:**457–464.

Dang, H., X. Zhang, J. Sun, T. Li, Z. Zhang, and G. Yang. 2008. Diversity and spatial distribution of sediment ammonia-oxidizing crenarchaeota in response to estuarine and environmental gradients in the Changjiang Estuary and East China Sea. *Microbiology* **154:**2084–2095.

Dang, H., J. Li, X. Zhang, T. Li, F. Tian, and W. Jin. 2009. Diversity and spatial distribution of *amoA*-encoding archaea in the deep-sea sediments of the tropical West Pacific Continental Margin. *J. Appl. Microbiol.* **106:**1482–1493.

Dang, H., X. W. Luan, R. Chen, X. Zhang, L. Guo, and M. G. Klotz. 2010. Diversity, abundance and distribution of *amoA*-encoding archaea in deep-sea methane seep sediments of the Okhotsk Sea. *FEMS Microbiol. Ecol.* **72:**370–385.

de la Torre, J. R., C. B. Walker, A. E. Ingalls, M. Könneke, D. A. Stahl. 2008. Cultivation of a thermophilic ammonia oxidizing archaeon synthesizing crenarchaeol. *Environ. Microbiol.* **10:**810–818.

Deeudom, M., M. Koomey, J. W. B. Moir. 2008. Roles of *c*-type cytochromes in respiration in *Neisseria meningitidis*. *Microbiology* **154:**2857–2864.

Einsle, O., A. Messerschmidt, P. Stach, G. P. Bourenkov, H. D. Bartunik, R. Huber, P. M. H. Kroneck. 1999. Structure of cytochrome *c* nitrite reductase. *Nature* **400:**476–480.

Einsle O., P. Stach, A. Messerschmidt, J. Simon, A. Kroeger, R. Huber, P. M. H. Kroneck. 2000. Cytochrome *c* nitrite reductase from *Wolinella succinogenes. J. Biol. Chem.* **275:**39608–39616.

El Sheikh, A. F., and M. G. Klotz. 2008. Ammonia-dependent differential regulation of the gene cluster that encodes ammonia monooxygenase in *Nitrosococcus oceani* ATCC 19707. *Environ. Microbiol.* **10:**3026–3035.

El Sheikh, A. F., A. T. Poret-Peterson, and M. G. Klotz. 2008. Characterization of two new genes, *amoR* and *amoD*, in the *amo* operon of the marine ammonia oxidizer *Nitrosococcus oceani* ATCC 19707. *Appl. Environ. Microbiol.* **74:**312–318.

Elmore, B. O., D. J. Bergmann, M. G. Klotz, and A. B. Hooper. 2007. Cytochromes P460 and *c'*-beta; A new family of high-spin cytochromes *c. FEBS Lett.* **581:**911–916.

Ensign, S. A., M. R. Hyman, and D. J. Arp. 1993. In vitro activation of ammonia monooxygenase from *Nitrosomonas europaea* by copper. *J. Bacteriol.* **175:**1971–1980.

Erickson, R. H., and A. B. Hooper. 1972. Preliminary characterization of a variant C-binding heme protein from *Nitrosomonas. Biochim. Biophys. Acta* **275:**231–244.

Ettwig, K. F., S. Shima, K. T. van de Pas-Schoonen, J. Kahnt, M. H. Medema, H. J. M. Op den Camp, M. S. M. Jetten, and M. Strous. 2008. Denitrifying bacteria anaerobically oxidize methane in the absence of Archaea. *Environ. Microbiol.* **10:**3164–3173.

Ettwig, K. F., T. van Alen, K. T. van de Pas-Schoonen, M. S. M. Jetten, M. Strous. 2009. Enrichment and molecular detection of denitrifying methanotrophic bacteria of the NC10 phylum. *Appl. Environ. Microbiol.* **75:**3656–3662.

Ettwig, K. F., M. K. Butler, D. Le Paslier, E. Pelletier, S. Mangenot, M. M. Kuypers, F. Schreiber, B. E. Dutilh, J. Zedelius, D. de Beer, J. Gloerich, H. J. Wessels, T. van Alen, F. Luesken, M. L. Wu, K. T. van de Pas-Schoonen, H. J. Op den Camp, E. M. Janssen-Megens, K. J. Francoijs, H. Stunnenberg, J. Weissenbach, M. S. Jetten, and M. Strous. 2010. Nitrite-driven anaerobic methane oxidation by oxygenic bacteria. *Nature* **464:**543–548.

Fadeeva, M. S., C. Nunez, Y. V. Bertsova, G. Espin, and A. V. Bogachev. 2008. Catalytic properties of Na^+-translocating NADH:quinone oxidoreductases from *Vibrio harveyi, Klebsiella pneumoniae,* and *Azotobacter vinelandii. FEMS Microbiol. Lett.* **279:**116–123.

Falkowski, P. G. 1997. Evolution of the nitrogen cycle and its influence on the biological sequestration of CO_2 in the ocean. *Nature* **387:**272.

Ferguson, S. J., and D. J. Richardson. 2005. The enzymes and bioenergetics of bacterial nitrate, nitrite, nitric oxide and nitrous oxide respiration, p. 169–206. *In* D. Zannoni (ed.), *Respiration in Archaea and Bacteria: Diversity of Procaryotic Respiratory Systems,* vol. 2. Springer, Dordrecht, The Netherlands.

Fleischmann, R. D., M. D. Adams, O. White, R. A. Clayton, E. F. Kirkness, A. R. Kerlavage, C. J. Bult, J.-F. Tomb, B. A. Dougherty, J. M. Merrick, K. McKenny, G. G. Sutton, W. Fitzhugh, C. Fields, J. D. Gocayne, J. Scott, R. Shirley, L.-I. Liu, A. Glodek, J. M. Kelley, J. F. Wiedman, C. A. Phillips, T. Spriggs, E. Hedblom, M. D. Cotton, T. R. Utterback, M. C. Hanna, D. T. Nquyen, D. M. Saudek, R. C. Brandon, L. D. Fine, J. L. Fritchman, J. L. Fuhrman, N. S. M. Geoghagen, C. L. Gnehm, L. A. McDonald, K. V. Small, C. M. Fraser, H. O. Smith, and J. C. Venter. 1995. Whole-genome random sequencing and assembly of *Haemophilus influenzae* Rd. *Science* **269:**496–512.

Francis, C. A., K. J. Roberts, J. M. Beman, A. E. Santoro, B. B. Oakley. 2005. Ubiquity and diversity of ammonia-oxidizing archaea in water columns and sediments of the ocean. *Proc. Natl. Acad. Sci. USA* **102:**14683–14688.

Garbeva, P., E. M. Baggs, and J. I. Prosser. 2007. Phylogeny of nitrite reductase (*nirK*) and nitric oxide reductase (*norB*) genes from *Nitrosospira* species isolated from soil. *FEMS Microbiol. Lett.* **266:**83–89.

Garcia-Horsman, J. A., B. Barquera, J. Rumbley, J. Ma, and R. B. Gennis. 1994. The superfamily of heme-copper respiratory oxidases. *J. Bacteriol.* **176:**5587–5600.

Garvin, J., R. Buick, A. D. Anbar, G. L. Arnold, and A. J. Kaufman. 2009. Isotopic evidence for an aerobic nitrogen cycle in the latest Archaean. *Science* **323:**1045–1048.

Gieseke, A., U. Purkhold, M. Wagner, R. Amann, and A. Schramm. 2001. Community structure and activity dynamics of nitrifying bacteria in a phosphate-removing biofilm. *Appl. Environ. Microbiol.* **67:**1351–1362.

Hallam, S. J., T. J. Mincer, C. Schleper, C. M. Preston, K. Roberts, P. M. Richardson, and E. F. DeLong. 2006. Pathways of carbon assimilation and ammonia oxidation suggested by environmental genomic analyses of marine Crenarchaeota. *PLoS Biol.* **4:**e95.

Hanson, R. S., and T. E. Hanson. 1996. Methanotrophic bacteria. *Microbiol. Rev.* **60:**439–471.

Hatzenpichler, R., E. V. Lebedeva, E. Spieck, K. Stoecker, A. Richter, H. Daims, and M. Wagner. 2008. A moderately thermophilic ammonia-oxidizing crenarchaeote from a hot spring. *Proc. Natl. Acad. Sci. USA* **105:**2134–2139.

Hemp, J., and R. B. Gennis. 2008. Diversity of

the heme–copper superfamily in archaea: insights from genomics and structural modeling, p. 1–31. *In* G. Schäfer and H. S. Penefsky (ed.), *Bioenergetics*. Springer, Berlin, Germany.

Hendrich, M. P., A. K. Upadhyay, J. Riga, D. M. Arciero, and A. B. Hooper. 2002. Spectroscopic characterization of the NO adduct of hydroxylamine oxidoreductase. *Biochemistry* **41:**4603–4611.

Hirota, R., A. Kuroda, T. Ikeda, N. Takiguchi, H. Ohtake, and J. Kato. 2006. Transcriptional analysis of the multicopy *hao* gene coding for hydroxylamine oxidoreductase in *Nitrosomonas* sp. strain ENI-11. *Biosci. Biotechnol. Biochem.* **70:**1875–1881.

Hirota, R., A. Yamagata, J. Kato, A. Kuroda, T. Ikeda, N. Takiguchi, and H. Ohtake. 2000. Physical map location of the multicopy genes coding for ammonia monooxygenase and hydroxylamine oxidoreductase in the ammonia-oxidizing bacterium *Nitrosomonas* sp. ENI-11. *J. Bacteriol.* **182:**825–828.

Holmes, A. J., A. Costello, M. E. Lidstrom, and J. C. Murrell. 1995. Evidence that particulate methane monooxygenase and ammonia monooxygenase may be evolutionarily related. *FEMS Microbiol. Lett.* **132:**203–208.

Hommes, N. G., L. A. Sayavedra-Soto, and D. J. Arp. 1994. Sequence of *hcy*, a gene encoding cytochrome *c*-554 from *Nitrosomonas europaea. Gene* **146:**87.

Hommes, N. G., L. A. Sayavedra-Soto, and D. J. Arp. 1996. Mutagenesis of hydroxylamine oxidoreductase in *Nitrosomonas europaea* by transformation and recombination. *J. Bacteriol.* **178:**3710–3714.

Hommes, N. G., L. A. Sayavedra-Soto, and D. J. Arp. 1998. Mutagenesis and expression of amo, which codes for ammonia monooxygenase in *Nitrosomonas europaea. J. Bacteriol.* **180:**3353–3359.

Hommes, N. G., L. A. Sayavedra-Soto, and D. J. Arp. 2001. Transcript analysis of multiple copies of *amo* (encoding ammonia monooxygenase) and *hao* (encoding hydroxylamine oxidoreductase) in *Nitrosomonas europaea. J. Bacteriol.* **183:**1096–1100.

Hommes, N. G., L. A. Sayavedra-Soto, and D. J. Arp. 2002. The roles of the three gene copies encoding hydroxylamine oxidoreductase in *Nitrosomonas europaea. Arch. Microbiol.* **178:**471–476.

Hommes, N. G., L. A. Sayavedra-Soto, and D. J. Arp. 2003. Chemolithoorganotrophic growth of *Nitrosomonas europaea* on fructose. *J. Bacteriol.* **185:**6809–6814.

Hommes, N. G., E. G. Kurth, L. A. Sayavedra-Soto, and D. J. Arp. 2006. Disruption of *sucA*, which encodes the E1 subunit of α-ketoglutarate dehydrogenase, affects the survival of *Nitrosomonas europaea* in stationary phase. *J. Bacteriol.* **188:**343–347.

Hooper, A. B. 1968. A nitrite-reducing enzyme from *Nitrosomonas europaea. Biochim. Biophys. Acta* **162:**49–65.

Hooper, A. B. 1969. Biochemical basis of obligate autotrophy in *Nitrosomonas europaea. J. Bacteriol.* **97:**776–779.

Hooper, A. B., and A. Nason. 1965. Characterization of hydroxylamine cytochrome c reductase from *Nitrosomonas europaea* and *Nitrosocystis oceanus. J. Biol. Chem.* **249:**4044–4057.

Hooper, A. B., and K. R. Terry. 1977. Hydroxylamine oxidoreductase from *Nitrosomonas*: inactivation by hydrogen peroxide. *Biochemistry* **16:**455–459.

Hooper, A. B., and K. R. Terry. 1979. Hydroxylamine oxidoreductase of *Nitrosomonas:* production of nitric-oxide from hydroxylamine. *Biochim. Biophys. Acta* **571:**12–20.

Hooper, A. B., D. M. Arciero, A. A. DiSpirito, J. Fuchs, M. Johnson, F. LaQuir, G. Mundfrom, and H. McTavish. 1990. Production of nitrite and $N_2O$by the ammonia-oxidizing nitrifiers, p. 387–391. *In* R. Gresshoff, E. Roth, G. Stacey, and W. E. Newton (ed.), *Nitrogen Fixation: Achievements and Objectives*, vol. 1. Chapman and Hall, New York, NY.

Hooper, A. B., T. Vannelli , D. J. Bergmann, and D. M. Arciero. 1997. Enzymology of the oxidation of ammonia to nitrite by bacteria. *Antonie Van Leeuwenhoek* **71:**59–67.

Hooper, A. B., D. M. Arciero, D. Bergmann, and M. P. Hendrich. 2005. The oxidation of ammonia as an energy source in bacteria in respiration, p. 121–147. *In* D. Zannoni (ed.), *Respiration in Archaea and Bacteria: Diversity of Procaryotic Respiratory Systems*, vol. 2. Springer, Dordrecht, The Netherlands.

Hou, S., K. Makarova, J. Saw, P. Senin, B. Ly, Z. Zhou, Y. Ren, J. Wang, M. Galperin, M. Omelchenko, Y. Wolf, N. Yutin, E. Koonin, M. Stott, B. Mountain, M. Crowe, A. Smirnova, P. Dunfield, L. Feng, L. Wang, and M. Alam. 2008. Complete genome sequence of the extremely acidophilic methanotroph isolate V4, *Methylacidiphilum infernorum*, a representative of the bacterial phylum Verrucomicrobia. *Biol. Direct* **3:**26–51.

Huber, C., W. Eisenreich, S. Hecht, and G. Wachtershauser. 2003. A possible primordial peptide cycle. *Science* **301:**938–940.

Hunte, C., S. Solmaz, H. Palsdóttir, and T. Wenz. 2008. A structural perspective on mechanism and function of the cytochrome bc_1 complex, p. 253–278. *In* G. Schäfer and H. S. Penefsky (ed.), *Bioenergetics*, vol. 45. Springer, Berlin, Germany.

Hyman, M. R., and D. J. Arp. 1992. $^{14}C_2H_2$- and $^{14}CO_2$-labeling studies of the *de novo* synthesis of polypeptides by *Nitrosomonas europaea* during recovery from acetylene and light inactivation of ammonia monooxygenase. *J. Biol. Chem.* **267:**1534–1545.

Hyman, M. R., and D. J. Arp. 1995. Effects of ammonia on the *de novo* synthesis of polypeptides in cells of *Nitrosomonas europaea* denied ammonia as an energy source. *J. Bacteriol.* **177:**4974–4979.

Igarashi, N., H. Moriyama, T. Fujiwara, Y. Fukumori, and N. Tanaka. 1997. The 2.8 Å structure of hydroxylamine oxidoreductase from a nitrifying chemolithotrophic bacterium, *Nitrosomonas europaea. Nat. Struct. Biol.* **4:**276–284.

Islam, T., S. Jensen, L. J. Reigstad, O. Larsen, and N.-K. Birkeland. 2008. Methane oxidation at 55°C and pH 2 by a thermoacidophilic bacterium belonging to the Verrucomicrobia phylum. *Proc. Natl. Acad. Sci. USA* **105:**300–304.

Iverson, T. M., D. M. Arciero, B. T. Hsu, M. S. P. Logan, A. B. Hooper, and D. C. Rees. 1998. Heme packing motifs revealed by the crystal structure of the tetra-heme cytochrome *c554* from *Nitrosomonas europaea. Nat. Struct. Biol.* **5:**1005–1012.

Iverson, T. M., D. M. Arciero, A. B. Hooper, and D. C. Rees. 2001. High-resolution structures of the oxidized and reduced states of cytochrome *c554* from *Nitrosomonas europaea. J. Biol. Inorg. Chem.* **6:**390–297.

Jepson, B. J. N., A. Marietou, S. Mohan, J. A. Cole, C. S. Butler, and D. J. Richardson. 2006. Evolution of the soluble nitrate reductase: defining the monomeric periplasmic nitrate reductase subgroup. *Biochem. Soc. Trans.* **34:**122–126.

Jetten, M. S., I. Cirpus, B. Kartal, L. van Niftrik, K. T. van de Pas-Schoonen, O. Sliekers, S. Haaijer, W. van der Star, M. Schmid, J. van de Vossenberg, I. Schmidt, H. Harhangi, M. van Loosdrecht, J. G. Kuenen, H. Op den Camp, and M. Strous. 2005. 1994–2004: 10 years of research on the anaerobic oxidation of ammonium. *Biochem. Soc. Trans.* **33:**119–133.

Jetten, M. S. M., M. Strous, K. T. van de Pas-Schoonen, J. Schalk, U. G. J. M. van Dongen, A. A. van de Graaf, S. Logemann, G. Muyzer, M. C. M. van Loosdrecht, and J. G. Kuenen. 1998. The anaerobic oxidation of ammonium. *FEMS Microbiol. Rev.* **22:**421–437.

Jetten, M. S. M., L. v. Niftrik, M. Strous, B. Kartal, J. T. Keltjens, and H. J. M. Op den Camp. 2009. Biochemistry and molecular biology of anammox bacteria. *Crit. Rev. Biochem. Mol. Biol.* **44:**65–84.

Jiang, Q.-Q., and L. R. Bakken. 1999. Nitrous oxide production and methane oxidation by different ammonia-oxidizing bacteria. *Appl. Environ. Microbiol.* **65:**2679–2684.

Juliette, L. Y., M. R. Hyman, and D. J. Arp. 1995. Roles of bovine serum albumin and copper in the assay and stability of ammonia monooxygenase activity in vitro. *J. Bacteriol.* **177:**4908–4913.

Kartal, B., M. M. M. Kuypers, G. Lavik, J. Schalk, H. J. M. Op den Camp, M. S. M. Jetten, and M. Strous. 2007. Anammox bacteria disguised as denitrifiers: nitrate reduction to dinitrogen gas via nitrite and ammonium. *Environ. Microbiol.* **9:**635–642.

Kaufman, A. J., D. T. Johnston, J. Farquhar, A. L. Masterson, T. W. Lyons, S. Bates, A. D. Anbar, G. L. Arnold, J. Garvin, and R. Buick. 2007. Late Archaean biospheric oxygenation and atmospheric evolution. *Science* **317:**1900–1903.

Kern, M., and J. Simon. 2009. Electron transport chains and bioenergetics of respiratory nitrogen metabolism in *Wolinella succinogenes* and other Epsilonproteobacteria. *Biochim. Biophys. Acta* **1787:**646–656.

Kerscher, S., S. Dröse, V. Zickermann, and U. Brandt. 2008. The three families of respiratory NADH dehydrogenases, p. 185–222. *In* G. Schäfer and H. S. Penefsky (ed.), *Bioenergetics*. Springer, Berlin, Germany.

Kim, H. J., A. Zatsman, A. K. Upadhyay, M. Whittaker, D. Bergmann, M. P. Hendrich, and A. B. Hooper. 2008. Membrane tetraheme cytochrome c_M552 of the ammonia-oxidizing *Nitrosomonas europaea*: a ubiquinone reductase. *Biochemistry* **47:**6539–6551.

Kirstein, K., and E. Bock. 1993. Close genetic relationship between *Nitrobacter hamburgensis* nitrite oxidoreductase and *Escherichia coli* nitrate reductases. *Arch. Microbiol.* **160:**447–453.

Klotz, M. G. 2008. Evolution of the nitrogen cycle, an ~omics perspective: evolution of the marine nitrogen cycle through time I, p. PP42A-02. Abstr. Fall meeting of the American Geophysical Union, San Francisco, CA.

Klotz, M. G., and J. M. Norton. 1995. Sequence of an ammonia monooxygenase subunit A-encoding gene from *Nitrosospira* sp. NpAV. *Gene* **163:**159–160.

Klotz, M. G., and J. M. Norton. 1998. Multiple copies of ammonia monooxygenase (*amo*) operons have evolved under biased AT/GC mutational pressure in ammonia-oxidizing autotrophic bacteria. *FEMS Microbiol. Lett.* **168:**303–311.

Klotz, M. G., and L. Y. Stein. 2008. Nitrifier genomics and evolution of the N-cycle. *FEMS Microbiol. Lett.* **278:**146–156.

Klotz, M. G., J. Alzerreca, and J. M. Norton. 1997. A gene encoding a membrane protein exists upstream of the *amoA*/*amoB* genes in ammonia oxidizing bacteria: a third member of the *amo* operon? *FEMS Microbiol. Lett.* **150:**65–73.

Klotz, M. G., D. J. Arp, P. S. G. Chain, A. F. El-Sheikh, L. J. Hauser, N. G. Hommes, F. W. Larimer, S. A. Malfatti, J. M. Norton, A. T. Poret-Peterson, L. M. Vergez, and B. B. Ward.

2006. Complete genome sequence of the marine, chemolithoautotrophic, ammonia-oxidizing bacterium *Nitrosococcus oceani* ATCC 19707. *Appl. Environ. Microbiol.* **72:**6299–6315.

Klotz, M. G., M. C. Schmid, M. Strous, H. J. M. Op den Camp, M. S. M. Jetten, and A. B. Hooper. 2008. Evolution of an octaheme cytochrome *c* protein family that is key to aerobic and anaerobic ammonia oxidation by bacteria. *Environ. Microbiol.* **10:**3150–3163.

Könneke, M., A. E. Bernhard, J. R. de la Torre, C. B. Walker, J. B. Waterbury, and D. A. Stahl. 2005. Isolation of an autotrophic ammonia-oxidizing marine archaeon. *Nature* **437:**543–546.

Koops, H.-P., and A. Pommerening-Roser. 2001. Distribution and ecophysiology of the nitrifying bacteria emphasizing cultured species. *FEMS Microbiol. Ecol.* **37:**1–9.

Koper, T. E., A. F. El-Sheikh, J. M. Norton, and M. G. Klotz. 2004. Urease-encoding genes in ammonia-oxidizing bacteria. *Appl. Environ. Microbiol.* **70:**2342–2348.

Kowalchuk, G. A., and J. R. Stephen. 2001. Ammonia-oxidizing bacteria: a model for molecular microbial ecology. *Annu. Rev. Microbiol.* **55:**485–529.

Kuenen, J. G. 2008. Anammox bacteria: from discovery to application. *Nat. Rev.* **6:**320–326.

Kumagai, H., T. Fujiwara, H. Matsubara, and K. Saeki. 1997. Membrane localization, topology, and mutual stabilization of the *rnfABC* gene products in *Rhodobacter capsulatus* and implications for a new family of energy-coupling NADH oxidoreductases. *Biochemistry* **36:**5509–5521.

Leininger, S., T. Urich, M. Schloter, L. Schwark, J. Qi, G. W. Nicol, J. I. Prosser, S. C. Schuster, and C. Schleper. 2006. Archaea predominate among ammonia-oxidizing prokaryotes in soils. *Nature* **442:**806.

Lin, J. T., and V. Stewart. 1998. Nitrate assimilation in bacteria. *Adv. Microb. Physiol.* **39:**1–30.

Lontoh, S., A. A. DiSpirito, C. L. Krema, M. R. Whittaker, A. B. Hooper, and J. D. Semrau. 2000. Differential inhibition in vivo of ammonia monooxygenase, soluble methane monooxygenase and membrane-associated methane monooxygenase by phenylacetylene. *Environ. Microbiol.* **2:**485–494.

Lukat, P., M. Rudolf, P. Stach, A. Messerschmidt, P. M. H. Kroneck, J. Simon, and O. Einsle. 2008. Binding and reduction of sulfite by cytochrome *c* nitrite reductase. *Biochemistry* **47:**2080–2086.

Mancinelli, R. L., and C. P. McKay. 1988. The evolution of nitrogen cycling. *Orig. Life Evol. Biosph.* **18:**311–325.

Martens-Habbena, W., P. M. Berube, H. Urakawa, J. R. de la Torre, and D. A. Stahl. 2009. Ammonia oxidation kinetics determine niche separation of nitrifying Archaea and Bacteria. *Nature* **461:**976–979.

Martinho, M., D.-W. Choi, A. A. DiSpirito, W. E. Antholine, J. D. Semrau, and E. Munck. 2007. Mossbauer studies of the membrane-associated methane monooxygenase from *Methylococcus capsulatus* Bath: evidence for a diiron center. *J. Am. Chem. Soc.* **129:**15783–15785.

McTavish, H., J. A. Fuchs, and A. B. Hooper. 1993a. Sequence of the gene coding for ammonia monooxygenase in *Nitrosomonas europaea*. *J. Bacteriol.* **175:**2436–2444.

McTavish, H., F. LaQuier, D. Aciero, M. Logan, G. Mundfrom, J. A. Fuchs, and A. B. Hooper. 1993b. Multiple copies of genes coding for electron transport proteins in the bacterium *Nitrosomonas europaea*. *J. Bacteriol.* **175:**2445–2447.

Moreno-Vivian, C., P. Cabello, M. Martinez-Luque, R. Blasco, and F. Castillo. 1999. Prokaryotic nitrate reduction: molecular properties and functional distinction among bacteria nitrate reductases. *J. Bacteriol.* **181:**6573–6584.

Mowat, C. G., E. Rothery, C. S. Miles, L. McIver, M. K. Doherty, K. Drewette, P. Taylor, M. D. Walkinshaw, S. K. Chapman, and G. A. Reid. 2004. Octaheme tetrathionate reductase is a respiratory enzyme with novel heme ligation. *Nat. Struct. Mol. Biol.* **11:**1023–1024.

Murrell, J. C., B. Gilbert, and I. R. McDonald. 2000. Molecular biology and regulation of methane monooxygenase. *Arch. Microbiol.* **173:**325–332.

Nakamura, K., T. Kawabata, K. Yura, and N. Go. 2004. Novel types of two-domain multi-copper oxidases: possible missing links in the evolution. *FEBS Lett.* **553:**239–244.

Nicol, G. W., and C. Schleper. 2006. Ammonia-oxidising crenarchaeota: important players in the nitrogen cycle? *Trends Microbiol.* **14:**207.

Norton, J. M., J. M. Low, and M. G. Klotz. 1996. The gene encoding ammonia monooxygenase subunit A exists in three nearly identical copies in *Nitrosospira* sp. NpAV. *FEMS Microbiol. Lett.* **139:**181–188.

Norton, J. M., J. J. Alzerreca, Y. Suwa, and M. G. Klotz. 2002. Diversity of ammonia monooxygenase operon in autotrophic ammonia-oxidizing bacteria. *Arch. Microbiol.* **177:**139–149.

Norton, J. M., M. G. Klotz, L. Y. Stein, D. J. Arp, P. J. Bottomley, P. S. G. Chain, L. J. Hauser, M. L. Land, F. W. Larimer, M. W. Shin, and S. R. Starkenburg. 2008. Complete genome sequence of *Nitrosospira multiformis*, an ammonia-oxidizing bacterium from the soil environment. *Appl. Environ. Microbiol.* **74:**3559–3572.

Nyerges, G., and L. Y. Stein. 2009. Ammonia cometabolism and product inhibition vary considerably among species of methanotrophic bacteria. *FEMS Microbiol. Lett.* **297:**131–136.

Op den Camp, H. J., T. Islam, M. B. Stott, H. R. Harhangi, A. Hynes, S. Schouten, M. S. M. Jetten, N. K. Birkeland, A. Pol, and P. F. Dunfield. 2009. Environmental, genomic, and taxonomic perspectives on methanotrophic *Verrucomicrobia*. *Environ. Microbiol. Rep.* **1:**293–306.

Pereira, M. M., J. N. Carita, and M. Teixeira. 1999. Membrane-bound electron transfer chain of the thermohalophilic bacterium *Rhodothermus marinus*: a novel multihemic cytochrome *bc*, a new Complex III. *Biochemistry* **38:**1268–1275.

Pereira, M. M., M. Santana, and M. Teixeira. 2001. A novel scenario for the evoluation of haem-copper oxygen reductases. *Biochim. Biophys. Acta* **1505:**185–208.

Pol, A., K. Heijmans, H. R. Harhangi, D. Tedesco, M. S. M. Jetten, and H. J. M. Op den Camp. 2007. Methanotrophy below pH 1 by a new Verrucomicrobia species. *Nature* **450:**874–878.

Poret-Peterson, A. T., J. E. Graham, J. Gulledge, and M. G. Klotz. 2008. Transcription of nitrification genes by the methane-oxidizing bacterium, *Methylococcus capsulatus* strain Bath. *ISME J.* **2:**1213–1220.

Potter, L., H. Angove, D. Richardson, and J. Cole. 2001. Nitrate reduction in the periplasm of gram-negative bacteria. *Adv. Microb. Physiol.* **45:**51–86.

Prosser, J. I. 1989. Autotrophic nitrification in bacteria. *Adv. Microb. Physiol.* **30:**125–181.

Prosser, J. I., and G. W. Nicol. 2008. Relative contributions of archaea and bacteria to aerobic ammonia oxidation in the environment. *Environ. Microbiol.* **10:**2931–2941.

Purkhold, U., A. Pommerening-Roser, S. Juretschko, M. C. Schmid, H.-P. Koops, and M. Wagner. 2000. Phylogeny of all recognized species of ammonia oxidizers based on comparative 16S rRNA and *amoA* sequence analysis: implications for molecular diversity surveys. *Appl. Environ. Microbiol.* **66:**5368–5382.

Purkhold, U., M. Wagner, G. Timmermann, A. Pommerening-Roser, and H.-P. Koops. 2003. 16S rRNA and *amoA*-based phylogeny of 12 novel betaproteobacterial ammonia-oxidizing isolates: extension of the dataset and proposal of a new lineage within the nitrosomonads. *Int. J. Syst. Evol. Microbiol.* **53:**1485–1494.

Raghoebarsing, A. A., A. Pol, K. T. van de Pas-Schoonen, A. J. P. Smolders, K. F. Ettwig, W. I. C. Rijpstra, S. Schouten, J. S. S. Damste, H. J. M. Op den Camp, M. S. M. Jetten, and M. Strous. 2006. A microbial consortium couples anaerobic methane oxidation to denitrification. *Nature* **440:**918–921.

Rasche, M. E., R. E. Hicks, M. R. Hyman, and D. J. Arp. 1990. Oxidation of monohalogenated ethanes and n-chlorinated alkanes by whole cells of *Nitrosomonas europaea*. *J. Bacteriol.* **172:**5368–5373.

Raymond, J., J. L. Siefert, C. R. Staples, and R. E. Blankenship. 2004. The natural history of nitrogen fixation. *Mol. Biol. Evol.* **21:**541–554.

Reigstad, L. J., A. Richter, H. Daims, T. Urich, L. Schwark, and C. Schleper. 2008. Nitrification in terrestrial hot springs of Iceland and Kamchatka. *FEMS Microbiol. Ecol.* **64:**167–174.

Ren, T., R. Roy, and R. Knowles. 2000. Production and consumption of nitric oxide by three methanotrophic bacteria. *Appl. Environ. Microbiol.* **66:**3891–3897.

Ridge, P. G., Y. Zhang, and V. N. Gladyshev. 2008. Comparative genomic analyses of copper transporters and cuproproteomes reveal evolutionary dynamics of copper utilization and its link to oxygen. *PLoS One* **3:**e1378.

Rodrigues, M. L., T. F. Oliveira, I. A. Pereira, and M. Archer. 2006. X-ray structure of the membrane-bound cytochrome *c* quinol dehydrogenase NrfH reveals novel haem coordination. *EMBO J.* **25:**5951–5960.

Rotthauwe, J. H., K. P. Witzel, and W. Liesack. 1997. The ammonia monooxygenase structural gene *amoA* as a functional marker: molecular fine-scale analysis of natural ammonia-oxidizing populations. *Appl. Environ. Microbiol.* **63:**4704–4712.

Sayavedra-Soto, L. A., N. G. Hommes, and D. J. Arp. 1994. Characterization of the gene encoding hydroxylamine oxidoreductase in *Nitrosomonas europaea*. *J. Bacteriol.* **176:**504–510.

Sayavedra-Soto, L. A., N. G. Hommes, S. A. Russel, and D. J. Arp. 1996. Induction of ammonia monooxygenase and hydroxylamine reductase mRNAs by ammonium in *Nitrosomonas europaea*. *Mol. Microbiol.* **20:**541–548.

Sayavedra-Soto, L. A., N. G. Hommes, J. J. Alzerreca, D. J. Arp, J. M. Norton, and M. G. Klotz. 1998. Transcription of the *amoC*, *amoA* and *amoB* genes in *Nitrosomonas europaea* and *Nitrosospira* sp. NpAV. *FEMS Microbiol. Lett.* **167:**81–88.

Schalk, J., S. de Vries, J. G. Kuenen, and M. S. M. Jetten. 2000. Involvement of a novel hydroxylamine oxidoreductase in anaerobic ammonium oxidation. *Biochemistry* **39:**5405–5412.

Schleper, C. 2008. Microbial ecology: metabolism of the deep. *Nature* **456:**712–714.

Schmehl, M., A. Jahn, A. Meyer zu Vilsendorf, S. Hennecke, B. Masepohl, M. Schuppler, M. Marxer, J. Oelze, and W. Klipp. 1993. Identification of a new class of nitrogen fixation genes in *Rhodobacter capsulatus*: a putative membrane complex involved in electron transport to nitrogenase. *Mol. Gen. Genet.* **241:**602–615.

Schmidt, I. 2009. Chemoorganoheterotrophic

growth of *Nitrosomonas europaea* and *Nitrosomonas eutropha*. *Curr. Microbiol.* **59:**130–138.

Schmidt, I., C. Hermelink, K. van de Pas-Schoonen, M. Strous, H. J. op den Camp, J. G. Kuenen, and M. S. M. Jetten. 2002a. Anaerobic ammonia oxidation in the presence of nitrogen oxides (NO_x) by two different lithotrophs. *Appl. Environ. Microbiol.* **68:**5351–5357.

Schmidt, I., O. Sliekers, M. Schmid, I. Cirpus, M. Strous, E. Bock, J. G. Kuenen, and M. S. M. Jetten. 2002b. Aerobic and anaerobic ammonia oxidizing bacteria—competitors or natural partners? *FEMS Microbiol. Ecol.* **39:**175–181.

Schmidt, I., P. J. M. Steenbakkers, H. J. M. op den Camp, K. Schmidt, and M. S. M. Jetten. 2004. Physiologic and proteomic evidence for a role of nitric oxide in biofilm formation by *Nitrosomonas europaea* and other ammonia oxidizers. *J. Bacteriol.* **186:**2781–2788.

Schmidt, I., R. J. M. van Spanning, and M. S. M. Jetten. 2004. Denitrification and ammonia oxidation by *Nitrosomonas europaea* wild-type, and NirK- and NorB-deficient mutants. *Microbiology* **150:**4107–4114.

Schneider, D., T. Pohl, J. Walter, K. Dörner, M. Kohlstädt, A. Berger, V. Spehr, and T. Friedrich. 2008. Assembly of the *Escherichia coli* NADH:ubiquinone oxidoreductase (complex I). *Biochim. Biophys. Acta* **1777:**735–739.

Scott, C., T. W. Lyons, A. Bekker, Y. Shen, S. W. Poulton, X. Chu, and A. D. Anbar. 2008. Tracing the stepwise oxygenation of the Proterozoic ocean. *Nature* **452:**456–459.

Scott, K. M., S. M. Sievert, F. N. Abril, L. A. Ball, C. J. Barrett, R. A. Blake, A. J. Boller, P. S. G. Chain, J. A. Clark, C. R. Davis, C. Detter, K. F. Do, K. P. Dobrinski, B. I. Faza, K. A. Fitzpatrick, S. K. Freyermuth, T. L. Harmer, L. J. Hauser, M. Uumlgler, C. A. Kerfeld, M. G. Klotz, W. W. Kong, M. Land, A. Lapidus, F. W. Larimer, D. L. Longo, S. Lucas, S. A. Malfatti, S. E. Massey, D. D. Martin, Z. McCuddin, F. Meyer, J. L. Moore, L. H. Ocampo, J. H. Paul, I. T. Paulsen, D. K. Reep, Q. Ren, R. L. Ross, P. Y. Sato, P. Thomas, L. E. Tinkham, and G. T. Zeruth. 2006. The genome of deep-sea vent chemolithoautotroph *Thiomicrospira crunogena* XCL-2. *PLoS Biol.* **4:**e383.

Shen, Y., A. H. Knoll, and M. R. Walter. 2003. Evidence for low sulphate and anoxia in a mid-Proterozoic marine basin. *Nature* **423:**632–635.

Sievert, S. M., K. M. Scott, M. G. Klotz, P. S. G. Chain, L. J. Hauser, J. Hemp, M. Hugler, M. Land, A. Lapidus, F. W. Larimer, S. Lucas, S. A. Malfatti, F. Meyer, I. T. Paulsen, Q. Ren, and J. Simon. 2008. Genome of the Epsilonproteobacterial chemolithoautotroph *Sulfurimonas denitrificans*. *Appl. Environ. Microbiol.* **74:**1145–1156.

Simon, J. 2002. Enzymology and bioenergetics of respiratory nitrite ammonification. *FEMS Microbiol. Rev.* **26:**285–309.

Smith, C. J., D. B. Nedwell, L. F. Dong, and A. M. Osborn. 2007. Diversity and abundance of nitrate reductase genes (*narG* and *napA*), nitrite reductase genes (*nirS* and *nrfA*), and their transcripts in estuarine sediments. *Appl. Environ. Microbiol.* **73:**3612–3622.

Starkenburg, S. R., P. S. G. Chain, L. A. Sayavedra-Soto, L. Hauser, M. L. Land, F. W. Larimer, S. A. Malfatti, M. G. Klotz, P. J. Bottomley, D. J. Arp, and W. J. Hickey. 2006. Genome sequence of the chemolithoautotrophic nitrite-oxidizing bacterium *Nitrobacter winogradskyi* Nb-255. *Appl. Environ. Microbiol.* **72:**2050–2063.

Starkenburg, S. R., F. W. Larimer, L. Y. Stein, M. G. Klotz, P. S. G. Chain, L. A. Sayavedra-Soto, A. T. Poret-Peterson, M. E. Gentry, D. J. Arp, B. Ward, and P. J. Bottomley. 2008. Complete genome sequence of *Nitrobacter hamburgensis* X14 and comparative genomic analysis of species within the genus *Nitrobacter*. *Appl. Environ. Microbiol.* **74:**2852–2863.

Stein, L. Y., and D. J. Arp. 1998. Ammonium limitation results in the loss of ammonia-oxidizing activity in *Nitrosomonas europaea*. *Appl. Environ. Microbiol.* **64:**1514–1521.

Stein, L. Y., and D. J. Arp. 1998. Loss of ammonia monooxygenase activity in *Nitrosomonas europaea* upon exposure to nitrite. *Appl. Environ. Microbiol.* **64:**4098–4102.

Stein, L. Y., D. J. Arp, and M. R. Hyman. 1997. Regulation of the synthesis and activity of ammonia monooxygenase in *Nitrosomonas europaea* by altering pH to affect NH_3 availability. *Appl. Environ. Microbiol.* **63:**4588–4592.

Stein, L. Y., D. J. Arp, P. M. Berube, P. S. G. Chain, L. J. Hauser, M. S. M. Jetten, M. G. Klotz, F. W. Larimer, J. M. Norton, H. J. M. Op den Camp, M. Shin, and X. Wei. 2007. Whole-genome analysis of the ammonia-oxidizing bacterium, *Nitrosomonas eutropha* C91: implications for niche adaptation. *Environ. Microbiol.* **9:**1–15.

Stevenson, B. S., and T. M. Schmidt. 1998. Growth rate-dependent accumulation of RNA from plasmid-borne rRNA operons in *Escherichia coli*. *J. Bacteriol.* **180:**1970–1972.

Strous, M., E. H. M. Kramer, S. Logemann, G. Muyzer, K. T. van De Pas-Schoonen, R. E. Webb, J. G. Kuenen, and M. S. M. Jetten. 1999. Missing lithotroph identified as new planctomycete. *Nature* **400:**446–449.

Strous, M., E. Pelletier, S. Mangenot, T. Rattei, A. Lehner, M. W. Taylor, M. Horn, H. Daims, D. Bartol-Mavel, P. Wincker, V. r. Barbe,

N. Fonknechten, D. Vallenet, B. a. Segurens, C. Schenowitz-Truong, C. Medigue, A. Collingro, B. Snel, B. E. Dutilh, H. J. M. Op den Camp, C. van der Drift, I. Cirpus, K. T. van de Pas-Schoonen, H. R. Harhangi, L. van Niftrik, M. Schmid, J. Keltjens, J. van de Vossenberg, B. Kartal, H. Meier, D. Frishman, M.A. Huynen, H.-W. Mewes, J. Weissenbach, M. S. M. Jetten, M. Wagner, and D. Le Paslier. 2006. Deciphering the evolution and metabolism of an anammox bacterium from a community genome. *Nature* **440:**790.

Tao, M., M. S. Casutt, G. N. Fritz, and J. Steuber. 2008. Oxidant-induced formation of a neutral flavosemiquinone in the Na^+-translocating NADH:quinone oxidoreductase (Na^+-NQR) from *Vibrio cholerae*. *Biochim. Biophys. Acta* **1777:**696–702.

Tavares, P., A. S. Pereira, J. J. G. Moura, and I. Moura. 2006. Metalloenzymes of the denitrification pathway. *J. Inorg. Biochem.* **100:**2087–2100.

Tavormina, P. L., V. J. Orphan, M. G. Kalyuzhnaya, M. S. M. Jetten, and M. G. Klotz. 2010. A novel family of functional operons encoding methane/ammonia monooxygenase-related proteins in gammaproteobacterial methanotrophs. *Environ. Microbiol. Rep.* (Online.) DOI: 10.1111/j.1758-2229.2010.00192.x

Terry, K., and A. B. Hooper. 1981. Hydroxylamine oxidoreductase: a 20-heme, 200,000 molecular weight cytochrome *c* with unusual denaturation properties, which forms a 63,000 molecular weight monomer after heme removal. *Biochemistry* **20:**7026–7032.

Teske, A., E. Alm, J. Regan, S. Toze, B. Rittmann, and D. Stahl. 1994. Evolutionary relationships among ammonia- and nitrite-oxidizing bacteria. *J. Bacteriol.* **176:**6623–6630.

Tourna, M., T. E. Freitag, G. W. Nicol, and J. I. Prosser. 2008. Growth, activity and temperature responses of ammonia-oxidizing archaea and bacteria in soil microcosms. *Environ. Microbiol.* **10:**1357–1364.

Treusch, A. H., S. Leininger, A. Kletzin, S. C. Schuster, H.-P. Klenk, and C. Schleper. 2005. Novel genes for nitrite reductase and Amo-related proteins indicate a role of uncultivated mesophilic crenarchaeota in nitrogen cycling. *Environ. Microbiol.* **7:**1985–1995.

Trotsenko, Y. A., and J. C. Murrell. 2008. Metabolic aspects of aerobic obligate methanotrophy. *Adv. Appl. Microbiol.* **63:**183–229.

Unemoto, T., and M. Hayashi. 1993. Na(+)-translocating NADH-quinone reductase of marine and halophilic bacteria. *J. Bioenerg. Biomembr.* **25:**385–391.

Upadhyay, A. K., A. B. Hooper, and M. P. Hendrich. 2006. NO reductase activity of the tetraheme cytochrome c_{554} of *Nitrosomonas europaea*. *J. Am. Chem. Soc.* **128:**4330–4337.

Utaker, J. B., L. Bakken, Q. Q. Jiang, and I. F. Nes. 1995. Phylogenetic analysis of seven new isolates of ammonia-oxidizing bacteria based on 16S rRNA gene sequences. *Syst. Appl. Microbiol.* **18:**549–559.

Utaker, J. B., K. Andersen, A. Aakra, B. Moen, and I. F. Nes. 2002. Phylogeny and functional expression of ribulose 1,5-bisphosphate carboxylase/oxygenase from the autotrophic ammonia-oxidizing bacterium *Nitrosospira* sp. isolate 40KI. *J. Bacteriol.* **184:**468–478.

van der Star, W. R. L., M. J. van de Graaf, B. Kartal, C. Picioreanu, M. S. M. Jetten, M. C. M. van Loosdrecht. 2008. Response of anaerobic ammonium-oxidizing bacteria to hydroxylamine. *Appl. Environ. Microbiol.* **74:**4417–4426.

Vannelli, T., D. Bergmann, D. M. Arciero, and A. B. Hooper. 1996. Mechanism of N-oxidation and electron transfer in the ammonia oxidizing autotrophs, p. 80–87. *In* M. E. Lidstrom and F. R. Tabita (ed.), *Microbial Growth on C_1 Compounds*. Kluwer Academic Publishers, Dordrecht, The Netherlands.

Wachtershauser, G. 1994. Life in a ligand sphere. *Proc. Natl. Acad. Sci. USA* **91:**4283–4287.

Walker, C. B., J. R. de la Torre, M. G. Klotz, H. Urakawa, N. Pinel, D. J. Arp, C. Brochier-Armanet, P. S. Chain, P. P. Chan, A. Gollabgir, J. Hemp, M. Hügler, E. A. Karr, M. Könneke, M. Shin, T. J. Lawton, T. Lowe, W. Martens-Habbena, L. A. Sayavedra-Soto, D. Lang, S. M. Sievert, A. C. Rosenzweig, G. Manning, and D. A. Stahl. 2010. *Nitrosopumilus maritimus* genome reveals unique mechanisms for nitrification and autotrophy in globally distributed marine crenarchaea. *Proc. Natl. Acad. Sci. USA* **107:**8818–8823.

Ward, B. B., and G. D. O'Mullan. 2002. Worldwide distribution of *Nitrosococcus oceani*, a marine ammonia-oxidizing gamma-proteobacterium, detected by PCR and sequencing of 16S rRNA and *amoA* genes. *Appl. Environ. Microbiol.* **68:**4153–4157.

Ward, N., O. Larsen, J. Sakwa, L. Bruseth, H. Khouri, A. S. Durkin, G. Dimitrov, L. Jiang, D. Scanlan, K. H. Kang, M. Lewis, K. E. Nelson, B. Methe, M. Wu, J. F. Heidelberg, I. T. Paulsen, D. Fouts, J. Ravel, H. Tettelin, Q. Ren, T. Read, R. T. DeBoy, R. Seshadri, S. L. Salzberg, H. B. Jensen, N. K. Birkeland, W. C. Nelson, R. J. Dodson, S. H. Grindhaug, I. Holt, I. Eidhammer, I. Jonasen, S. Vanaken, T. Utterback, T. V. Feldblyum, C. M. Fraser, J. R. Lillehaug, and J. A. Eisen. 2004. Genomic insights into methanotrophy: the complete genome sequence of *Methylococcus capsulatus* (Bath). *PLoS Biol.* **2:**e303.

Wei, X., L. A. Sayavedra-Soto, and D. J. Arp. 2004. The transcription of the *cbb* operon in *Nitrosomonas europaea*. *Microbiology* **150:**1869–1879.

Whittaker, M., D. Bergmann, D. Arciero, and A. B. Hooper. 2000. Electron transfer during the oxidation of ammonia by the chemolithotrophic bacterium *Nitrosomonas europaea*. *Biochim. Biophys. Acta* **1459:**346–355.

Winogradsky, S. 1892. Contributions à la morphologie des organismes de la nitrification. *Arch. Sci. Biol.* **1:**88–137.

Wuchter, C., B. Abbas, M. J. L. Coolen, L. Herfort, J. van Bleijswijk, P. Timmers, M. Strous, E. Teira, G. J. Herndl, J. J. Middelburg, S. Schouten, and J. S. Sinninghe Damste. 2006. Archaeal nitrification in the ocean. *Proc. Natl. Acad. Sci. USA* **103:**12317–12322.

Yanyushin, M. F., M. C. delRosario, D. C. Brune, and R. E. Blankenship. 2005. New class of bacterial membrane oxidoreductases. *Biochemistry* **44:**10037–10045.

Zehr, J. P., and B. B. Ward. 2002. Nitrogen cycling in the ocean: new perspectives on processes and paradigms. *Appl. Environ. Microbiol.* **68:**1015–1024.

Zumft, W. G. 1997. Cell biology and molecular basis of denitrification. *Microbiol. Mol. Biol. Rev.* **61:**522–616.

Zumft, W. G., and P. M. H. Kroneck. 2006. Respiratory transformation of nitrous oxide (N_2O) to dinitrogen by bacteria and archaea. *Adv. Microb. Physiol.* **52:**107–227.

HETEROTROPHIC NITRIFICATION AND NITRIFIER DENITRIFICATION

Lisa Y. Stein

5

INTRODUCTION

The vast combination of genome sequence, molecular microbial ecology, and physiology studies described in the proceeding chapters has greatly expanded the range of organisms known to actively participate in the biogeochemical nitrogen cycle. Aside from the nitrifying chemolithotrophic bacteria and Thaumarchaea, several genera of chemoorganotrophic bacteria and a handful of eukaryotes are capable of oxidizing ammonia, hydroxylamine, various organics, and/or nitrite in a process termed "heterotrophic nitrification." Unlike the classical definition of nitrification (i.e., the oxidation of ammonia to nitrate via nitrite), heterotrophic nitrification embraces a broadened definition to include the oxidation of any reduced form of nitrogen to a more oxidized form (Focht and Verstraete, 1977; Ralt et al., 1981; Castignetti et al., 1984; Killham, 1986; van Niel et al., 1993). Also, unlike nitrification by chemolithotrophs, heterotrophic nitrification is not necessarily coupled to energy conservation, but rather has been linked to reoxidation of NAD(P)H under hypoxic conditions in bacteria (Robertson and Kuenen, 1990), endogenous respiration in fungi (Van Gool and Schmidt, 1973), and a form of defense against competing organisms in soils (Verstraete, 1975).

Many heterotrophic nitrifiers are also capable of aerobic denitrification, such that the nitrite or nitrate produced is immediately reduced to N-oxides or dinitrogen via denitrifying enzymes. This so-named simultaneous nitrification-denitrification (SND) may have led in the past to a systematic underestimation of the heterotrophic contribution to nitrification, as oxidized intermediates did not accumulate. However, it must be noted that there remains considerable uncertainty of the types and abundances of heterotrophic nitrifiers in the environment, which equally confounds reliable measurements of their activity. The process of SND is best characterized in certain types of aerobic wastewater treatment systems from which several heterotrophic nitrifiers have been isolated (Robertson and Kuenen, 1990; Schmidt et al., 2003). Heterotrophic nitrifiers are not the only organisms capable of SND; chemolithotrophic ammonia oxidizers reduce nitrite to nitric oxide with nitrous oxide or dinitrogen as terminal products during ammonia oxidation (Poth, 1986; Wrage et al., 2001). This process termed "nitrifier denitrification" occurs aerobically but is also required for anaerobic respiration in some *Nitrosomonas*

Lisa Y. Stein, Department of Biological Sciences, University of Alberta, Edmonton, Alberta T6G 2E9, Canada.

Nitrification, Edited by Bess B. Ward, Daniel J. Arp, and Martin G. Klotz

spp. in which nitrite reduction is energetically coupled to ammonia, hydrogen, or organic carbon oxidation (Bock, 1995; I. Schmidt et al., 2004; Schmidt, 2009). Aerobic denitrification is also not restricted to organisms that nitrify as some heterotrophic bacteria and fungi simultaneously use nitrate and oxygen as terminal electron acceptors as a strategy to maximize respiration under hypoxic conditions (Robertson and Kuenen, 1990; Takaya et al., 2003; Otani et al., 2004).

The microbial activities of nitrification and denitrification (aerobic and anaerobic; autotrophic and heterotrophic) constitute the greatest source of the potent greenhouse gas, nitrous oxide, to the atmosphere (Stein and Yung, 2003). The continuing increase in atmospheric nitrous oxide is largely from conversion of land to agricultural uses, which stimulates both nitrifying and denitrifying activities by providing saturating amounts of nitrogenous fertilizers plus ample moisture (IPCC, 2006). Rates of nitrogen oxide production have significantly exceeded the threshold for ecological sustainability (Röckstrom et al., 2009); therefore, it is critical to understand the metabolic control of nitrogen oxide production, the diversity of organisms and pathways that create nitrogen oxides, and how to predict the metabolic activities leading to nitrogen oxide production and release within a given environment.

This chapter describes the physiology and biochemical pathways of heterotrophic nitrification and nitrifier (and aerobic) denitrification, a description of the genetic and organism diversity involved, and a brief description of techniques to discern one process from another. A final perspective is offered on how anthropogenic input of nitrogen affects microbial transformations of inorganic N with particular emphasis on emissions of gaseous N-oxides to the atmosphere.

HETEROTROPHIC NITRIFICATION

Biochemistry and Physiology

Ammonia-oxidizing bacteria (AOB) use ammonia monooxygenase (AMO) and the multi-heme hydroxylamine oxidoreductase (HAO) to oxidize ammonia to nitrite as their sole source of energy and reductant (see Chapter 2). Most, but not all, species of methanotrophic bacteria, organisms with functional and structural similarities to AOB, also oxidize ammonia to nitrite using similar enzymology to the AOB, but do so via cometabolism (Conrad, 1996; Nyerges and Stein, 2009). Both soluble methane monooxygenase and particulate methane monooxygenase oxidize ammonia to hydroxylamine, and particulate methane monooxygenase is evolutionarily related to AMO (Klotz and Norton, 1998; Norton et al., 2002; Hakemian and Rosenzweig, 2007). As for hydroxylamine-oxidizing activity, expression of *haoAB* genes in *Methylococcus capsulatus* Bath (Poret-Peterson et al., 2008) and *Methylomicrobium album* (Nyerges, 2008) were specifically induced by ammonia, and a purified cytochrome P460 from *M. capsulatus* Bath was shown to have hydroxylamine-oxidizing activity similar to that of cytochrome P460 isolated from *Nitrosomonas europaea* (Bergmann et al., 1998) (Table 1). Although methanotrophs are not considered heterotrophic as the majority are restricted to metabolizing single-carbon compounds, their physiological similarity to the AOB and their abundance and distribution in the environment results in substantial contributions to ammonia-oxidizing activity, particularly in oxic soils (Bodelier and Laanbroek, 2004).

Heterotrophic nitrifiers, most notably *Paracoccus pantotrophus* GB17 (formerly *Paracoccus denitrificans* GB17 and *Thiosphaera pantotropha*), can apparently share similar enzymology with the AOB; an enzyme with structural and functional similarities to AMO (Moir et al., 1996b) and a non-heme hydroxylamine oxidase (Wehrfritz et al., 1993; Moir et al., 1996a) were purified from *P. pantotrophus* GB17. However, in the absence of a complete genome sequence for this organism and further biochemical examination, the structural, functional, and evolutionary details of these enzymes remain somewhat ambiguous. Apparent hydroxylamine-oxidizing enzymes

TABLE 1 Characteristics of putative bacterial hydroxylamine-oxidizing enzymes

Parameter	Result for enzyme classification[a]:								
	Hydroxylamine oxidoreductase		Hydroxylamine oxidoreductase, non-heme	Cytochrome P460		Non-heme hydroxylamine oxidase, cyt *c* reductase, oxidoreductase			
Organism/source	*N. europaea*	Anammox enrichment	*Pseudomonas* sp. strain PB16	*N. europaea*	*M. capsulatus* Bath	*P. pantotrophus* GB17	*A. globiformis*	*A. faecalis*	*Pseudomonas* sp. strain S2.14
Native mass (kDa)	189	183	132	52	39	18.5	ND	20	19
Subunit mass (kDa)	63	58	68	17.3–18.5	16.4	18.5	ND	20	19
Subunit composition	α_3	α_3	α_2	α_3	α_2	α	ND	α	α
Metal content	24 hemes, Heme P460	26 hemes, Heme P468	Non-heme iron	Heme P460	Heme P460 and copper	Non-heme and non-iron-sulfur iron	Non-heme iron	Non-heme iron	Non-heme iron
pH optimum	ND	8.0	9.0	ND	ND	8.5	9.0	8–9	8.7
V_{max} (μmol·min^{-1}·mg^{-1})	75 (PMS)	21	0.45	ND	ND	0.99 (pseudo.) 0.13 (cyt $_{c551}$)	ND	0.031	3.6
K_m (μM)	ND	26	37	ND	ND	33 (pseudo.) 10 (cyt $_{c551}$)	ND	1500	70 (cyt *c*)
Physiological electron acceptors	cyt $_{c554}$	ND (assay: PMS and MTT)	ND (assay: ferricyanide)	cyt $_{c552}$	cyt $_{c555}$ (required PMS *in vitro*)	Pseudoazurin cyt $_{c551}$ cyt $_{c550}$	ND (assay: horse heart cyt *c*)	ND (assay: ferricyanide)	cyt $_{c551}$
Electron acceptors that did not work	ND	NAD, benzyl viologen, Wurster's blue	DCPIP, PMS, PMS+MTT, NAD, FAD	ND	cyt $_{c555}$, $_{c554}$, $_{c557}$, and cyt c' in absence of PMS	All other cyt *c* fractions from purification	ND	Heart cyt *c* Pseudoazurin	Heart cyt *c* Pseudoazurin
O_2 requirement	No	No	ND	ND	Yes	Yes	ND	ND	ND
Reference(s)	Hooper et al., 1978	Schalk et al., 2000	Jetten et al., 1997a	Erickson and Hooper, 1972; Numata et al., 1990	Zahn et al., 1994	Wehrfritz et al., 1993; Moir et al., 1996a	Kurokawa et al., 1985	Otte et al., 1999	Wehrfritz et al., 1997

[a]ND, not determined; PMS, phenazine methosulfate; MTT, methylthiazol tetrazolium bromide; DCPIP, dichlorophenol indophenol.

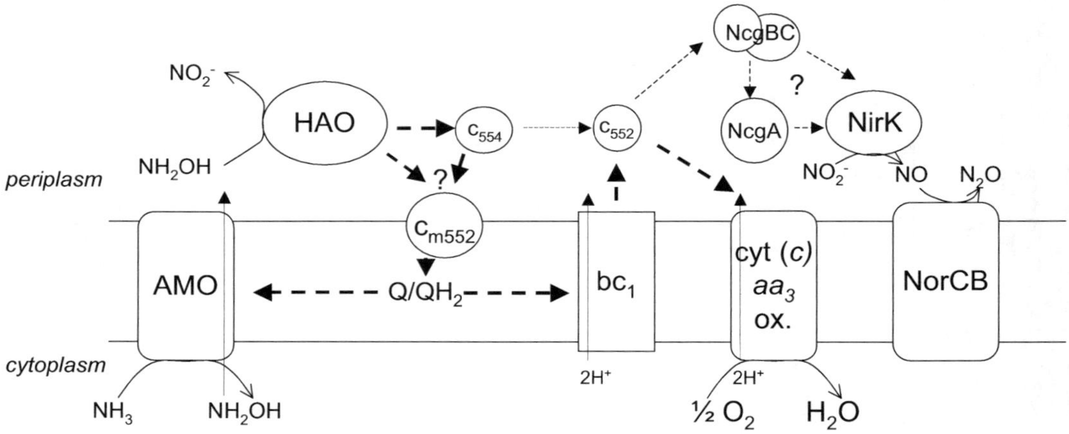

FIGURE 1 Pathways of ammonia oxidation and nitrifier denitrification in *N. europaea*. Dashed lines indicate the direction of electron flow, with thinner lines indicating less electron flow than thicker lines. The question mark above cytochrome c_{m552} indicates the uncertainty of whether electrons are delivered to this enzyme directly from HAO or via cytochrome c_{554} (Klotz and Stein, 2008). Similarly, the question mark in the middle of NcgA, NcgBC, and NirK indicates that order of electron transfer among these proteins remains uncharacterized. NorCB, nitric oxide reductase; Ncg, products of *nirK* cluster genes; Q, quinone.

have been isolated from many heterotrophic nitrifiers, but so far only the enzyme characterized from an anaerobic ammonia oxidation (anammox) enrichment was found to share similar features with purified HAO from the AOB *N. europaea* (Schalk et al., 2000) (Table 1). Two additional types of apparent hydroxylamine-oxidizing enzymes aside from HAO and cytochrome P460 have been isolated from heterotrophic nitrifiers, the most common of which is a small (ca. 20-kDa), monomeric, O_2-requiring, non-heme iron enzyme (Table 1). A completely different type of putative hydroxylamine-oxidizing enzyme was isolated from *Pseudomonas* sp. strain PB16 (Jetten et al., 1997a) but has not yet been identified in other isolates. Together, hydroxylamine-oxidizing enzymes differentiate into four distinct classes, but only HAO and cytochrome P460 have characterized biochemical, genetic, and physiological properties.

The anammox and AOB multi-heme HAO enzymes are central components of the hydroxylamine/hydrazine-ubiquinone redox module that allows efficient transfer of electrons from HAO to cytochrome *c* and then to the ubiquinone pool for generation of proton motive force and continued oxidation of ammonia (Klotz et al., 2008; Klotz and Stein, 2008) (Fig. 1) (see Chapter 4 for details). This module is central to the chemolithotrophic lifestyle of these organisms. There is no evidence that the putative non-heme hydroxylamine-oxidizing enzymes from heterotrophic microorganisms participate in hydroxylamine/hydrazine-ubiquinone redox module; however, cytochrome *c* was the most frequently reported native electron acceptor of these enzymes (Table 1).

All of the heterotrophic nitrifying bacteria from which hydroxylamine-oxidizing enzymes have been isolated are also aerobic denitrifiers, aside from *Arthrobacter globiformis*. In a model based on physiological data from *P. pantotrophus* GB17, the diversion of electrons from cytochrome *c* to the denitrification pathway relieved an electron flow bottleneck between Complexes III and IV that occurred upon a decrease in O_2 availability (Fig. 2) (Robertson et al., 1988; Wehrfritz et al., 1993). Furthermore, in the presence of both nitrate and O_2

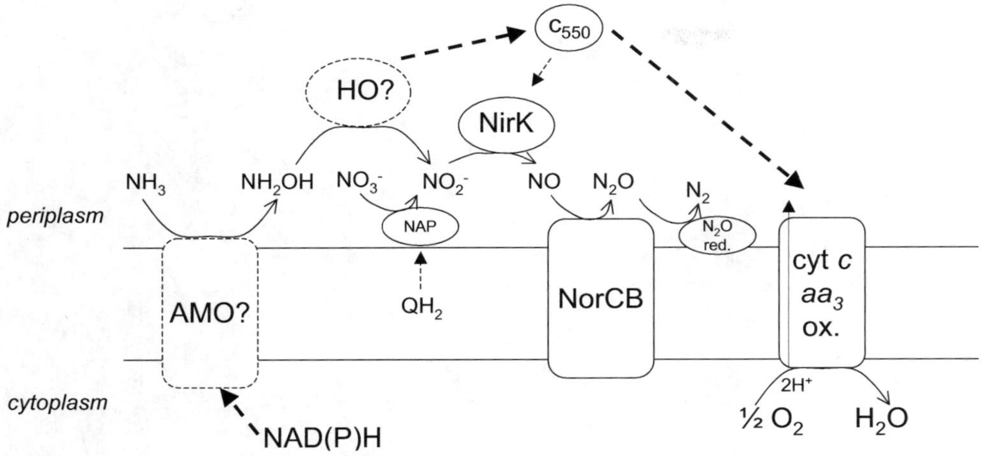

FIGURE 2 Putative pathway for heterotrophic nitrification and aerobic denitrification in *P. pantotrophus* GB17 (based on model by Stouthammer et al., 1997). Electron carriers between the putative hydroxylamine oxidase enzyme and members of the denitrification pathway remain uncharacterized. AMO, putative ammonia monooxygenase; HO, putative hydroxylamine oxidase; NorCB, nitric oxide reductase; NAP, periplasmic nitrate reductase; Q, quinone.

as electron acceptors, *P. pantotrophus* GB17 had circa four times the growth rate than with either electron acceptor alone (Robertson and Kuenen, 1990). Oxidation of ammonia by *P. pantotrophus* GB17 had the additional effect of reoxidizing NAD(P)H under low oxygen conditions. Together, the data indicated that SND stimulated the growth rate, but lowered the growth yield, of *P. pantotrophus* GB17 at low oxygen tension (Robertson et al., 1988; Robertson and Kuenen, 1990). This strategy of simultaneous nitrate and oxygen reduction to maximize respiration under hypoxic conditions is also used in the mitochondria of the fungus *Fusarium oxysporum* (Takaya et al., 2003) (Fig. 3).

Intriguingly, nitrifier denitrification by *N. europaea* is similar to SND in that a trickle of electrons to denitrifying enzymes during ammonia oxidation can speed hydroxylamine oxidation and thus increase the rate of cell growth by 10% to 20% (Fig. 1) (Beaumont et al., 2002; I. Schmidt et al., 2004; Cantera and Stein, 2007a). On the basis of these observations, the activity of denitrifying enzymes in *N. europaea* likely also function to relieve an electron flow bottleneck between Complexes III and IV, thus allowing faster hydroxylamine oxidation and electron flow to the quinone pool for ATP and reductant generation (Cantera and Stein, 2007a). In a general view, then, both heterotrophic and chemolithotrophic nitrifiers and aerobic denitrifiers apparently use denitrification simultaneously with oxygen respiration to facilitate electron flow and maximize aerobic growth. However, as with most generalizations, exceptions or modifications to these pathways are likely as they are based on published studies in a small handful of model organisms. We already know from comparing genome sequences of closely related AOB that pathway inventories, gene environments, regulatory features, and levels of protein sequence identity can be quite diverse (see Chapter 4).

Unlike the organisms discussed thus far, several heterotrophic nitrifiers have distinct enzymology from AOB and oxidize substrates other than ammonia to produce nitrite and/or nitrate. Enzymes that oxidize organic N to nitrite have been purified from bacteria and fungi. For example, pyruvic oxime dioxy-

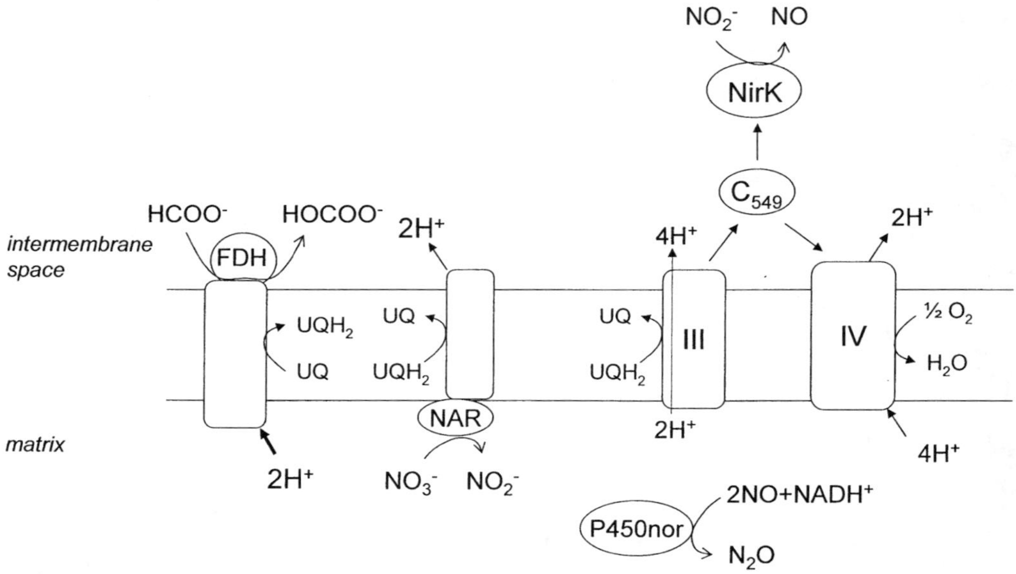

FIGURE 3 Pathway for hybrid respiration of oxygen and nitrate in *F. oxysporum*. Denitrification is linked to formate oxidation in the mitochrondria as per the model presented by Takaya et al. (2003). FDH, formate dehydrogenase; Nar, nitrate reductase; P450nor, nitric oxide reductase.

genase, a non-heme iron enzyme that requires molecular oxygen, was isolated from the heterotrophic nitrifier *Alcaligenes faecalis* (Ono et al., 1999). Although this enzyme was activated by hydroxylamine, it was not capable of directly oxidizing hydroxylamine to nitrite. Several *Pseudomonas* spp. isolated from mammalian intestines were shown to oxidize both acetohydroxamate and hydroxylamine to nitrite, although enzymes were not purified from these bacteria (Ralt et al., 1981). The fungus *Aspergillus flavus* has long been a favored model organism for the study of heterotrophic nitrification and is especially relevant to nitrate production in acidic coniferous forest soils (Killham, 1986). Studies have shown that nitrate is produced readily by *A. flavus* from the oxidation of organics like aspartate (Van Gool and Schmidt, 1973) and peptone (White and Johnson, 1982), but only when the preferred carbon and energy source for the fungus has been depleted. Thus, it was concluded from these studies that nitrification by *A. flavus* is an endogenous respiration, perhaps as a source of maintenance energy.

The oxidation of nitrite to nitrate, the second half of the classical nitrification process, is generally performed by nitrite-oxidizing chemolithotrophs typified by the genera *Nitrobacter* and *Nitrospira*. The central enzyme for nitrite oxidation by these bacteria is nitrite oxidoreductase (see Chapter 11). In contrast to the chemolithotrophs, the heterotrophic nitrite-oxidizers appear to predominantly use catalase enzymes. A nitrite-oxidizing catalase was purified from the fungi *A. flavus* (Molina and Alexander, 1972) and *Candida rugosa* IFO 0591 (Sakai et al., 1988) and from the firmicute *Bacillus badius* I-73 (Sakai et al., 2000). Based on similar isolation methods and physiological characteristics, it is most likely that catalase was the active enzyme in other nitrite-oxidizing fungi (Tachiki et al., 1988) and heterotrophic bacteria (Sakai et al., 1996) isolated at the same time as *C. rugosa* and *B. badius*, respectively. It was suggested in these studies that nitrite was detoxified by catalase and that the activity was fortuitous.

Genetics

Aside from the biochemical and physiological approaches described above, the construction of gene knock-out mutants has been another useful approach for reconstructing pathways for heterotrophic nitrification. The functionality and physical linkage of AMO- and hydroxylamine oxidase-encoding genes was demonstrated in *P. pantotrophus* GB17 by expressing both genes from a single genomic clone in a heterologous host (Crossman et al., 1997). Also, the presence of an *amoA* homologue in *Pseudomonas putida* DSMZ-1088-260 was detected by DNA hybridization to an *amoA* probe from the AOB, *N. europaea* (Daum et al., 1998). Unfortunately, further analysis of genes encoding AMO and hydroxylamine oxidases from heterotrophic bacteria has not been continued, and the ability of most model strains to nitrify has apparently been lost.

Aside from ammonia- and hydroxylamine-oxidizing enzymes, products of denitrification genes have been characterized as essential participants for the heterotrophic nitrification pathway. For instance, screening a transposon mutant library of *Pseudomonas* sp. strain M19 revealed the requirement of the nitrate reductase genes *narH*, *narJ*, and *moaE* for nitrite and nitrate production from peptone and, to a lesser extent, ammonium (Nemergut and Schmidt, 2002). An iron-containing nitrite reductase (NirS)-deficient mutant of *Burkholderia cepacia* NH-17 was unable to oxidize nitrite to nitrate or reduce nitrite to nitric oxide (Matsuzaka et al., 2003). Although a similar role of the copper-containing nitrite reductase (NirK) has not been demonstrated for heterotrophic nitrification, NirK participates in aerobic denitrification in the majority of heterotrophic nitrifier model organisms (Robertson et al., 1989) and is also the nitrite reductase that participates in aerobic denitrification by fungi (Kim et al., 2009). Similarly, a NirK-deficient mutant of the ammonia-oxidizer *N. europaea* was incapable of using nitrite as an electron acceptor (I. Schmidt et al., 2004), and NirK was implicated in reducing nitrite to nitric oxide to maintain redox balance under low-oxygen tensions by the nitrite oxidizer *Nitrobacter winogradskyi* Nb-255 (Starkenburg et al., 2008). These results offer additional insights into the tight coupling between nitrification and denitrification (i.e., SND) processes in diverse microorganisms.

Beyond this small handful of genes, no other studies have implicated additional inventory involved in heterotrophic nitrification pathways. Thus, either the genetic inventory required for this process is small or several more genes have yet to be discovered. In particular, additional observations of bona fide AMO and hydroxylamine-oxidizing enzymes are required to fully understand the nature, evolutionary history, and physiological significance of heterotrophic nitrification.

Diversity of Heterotrophic Nitrifiers

A number of bacteria and eukaryotes (mostly fungi) capable of heterotrophic nitrification have been isolated from a number of environments. General characteristics of several isolates are listed in Table 2. As discussed above, the enzymology and genes for heterotrophic nitrification are diverse and still somewhat mysterious as only a few examples of enzymes and genes have been experimentally scrutinized. Interestingly, long-term maintenance of the model heterotrophic nitrifier, *P. pantotrophus* GB17, caused a gradual loss of its heterotrophic nitrifying activity (Stouthammer et al., 1997). Thus, specific environmental conditions are obviously required to maintain this activity, especially since nitrification merely bolsters, but is not essential to, metabolic productivity of this organism. Today, *P. pantotrophus* GB17 is a model organism for the study of lithotrophic sulfur oxidation (Friedrich et al., 2001) rather than for heterotrophic nitrification. A similar observation was made with soil fungi in that they lost their ability to nitrify when inoculated into sterile soils (Schmidt, 1973). It

TABLE 2 Characteristics of select heterotrophic nitrifying microorganisms

Organism	Substrates	Aerobic denitrification	Genes/enzymes identified in pathway[a]	Environment	References
Gammaproteobacteria					
Pseudomonas chlororaphis ATCC 13985	NH_2OH, pyruvic oxime	Yes	ND	River clay	Castignetti et al., 1984
P. putida DSMZ-1088-260	Ammonia, NH_2OH, nitrite	Yes	*amoA*	Forest soil	Daum et al., 1998
Pseudomonas sp. strain M19	Peptone, ammonia	Yes	Nitrate reductase	Tundra soil	Nemergut and Schmidt, 2002
Pseudomonas sp. strain PB16	NH_2OH	Yes	HAO	Soil	Robertson et al., 1989; Jetten et al., 1997a
Pseudomonas sp. strain S2.14	NH_2OH	Yes	Hydroxylamine oxidase	Soil	Wehrfritz et al., 1997
Moraxella sp. strain S2.18	NH_2OH	No	ND	Soil	Wehrfritz et al., 1997
M. capsulatus Bath	Ammonia, NH_2OH	Yes	MMO, cyt P460, HAO	Roman Baths	Bergmann et al., 1998; Lieberman and Rosenzweig, 2004
Betaproteobacteria					
A. faecalis TUD	NH_2OH, pyruvic oxime	Yes	Pyruvic oxime dioxygenase	Soil	Ono et al., 1999; Otte et al., 1999
A. faecalis No. 4	Ammonia, NH_2OH	Yes	ND	Sewage sludge	Joo et al., 2005
Diaphorobacter nitroreducens	Ammonia	Yes	ND	Activated biomass	Anshuman et al., 2007
B. cepacia NH-17	Nitrite	Yes	*nirS*	Soil	Matsuzaka et al., 2003
Alphaproteobacteria					
P. pantotrophus GB17	Ammonia, NH_2OH	Yes	AMO, HAO	Wastewater	Robertson et al., 1988; Wehrfritz et al., 1993; Moir et al., 1996b, 1996a
Firmicutes					
B. badius I-73	Nitrite	No	Catalase	Activated sludge	Sakai et al., 2000
Bacillus sp. strain LY	Ammonia	Yes	ND	Memb. bioreactor	Lin et al., 2004
Bacillus strain MS30	Ammonia, acetate	Yes	ND	Hydrothermal vent	Mével and Prieur, 2000
Actinobacteria					
A. globiformis IFO 3062	Ammonia, NH_2OH	No	Hydroxylamine oxidase	Sewage	Verstraete and Alexander, 1972; Kurokawa et al., 1985
Arthrobacter sp. strain S2.26	NH_2OH	No	ND	Soil	Wehrfritz et al., 1997
Fungi					
A. flavus	NH_2OH, nitrite, organics	No	Catalase	Cotton seed	Molina and Alexander, 1972; Van Gool and Schmidt, 1973

(Continued)

TABLE 2 *(Continued)*

Organism	Substrates	Aerobic denitrification	Genes/enzymes identified in pathway[a]	Environment	References
A. flavus ATCC 26214	Ammonia, peptone	No	ND	Acidic forest soil	Schimel et al., 1984
C. rugosa	Nitrite	No	Catalase	Soil	Sakai et al., 1988
Absidia cylindrospora	Ammonia, organics	No	ND	Acidic forest soil	Stroo et al., 1986
Algae					
Ankistrodesmus braunii	NH_2OH, nitrite, organics	No	Catalase	Eutrophic lake	Spiller et al., 1976

[a]Verification of AMO and hydroxylamine oxidase genes and/or enzymes from heterotrophic bacteria is required to adequately compare to those found in chemolithotrophic AOB. ND, not determined.

is unclear whether nitrifying activity is equally fickle in other isolates, or whether cultivation itself causes instability of this phenotype. The ability of organisms to easily lose their ability to nitrify is an important issue to resolve to fully understand ecological implications for heterotrophic nitrifying activity.

Measuring Heterotrophic versus Chemolithotrophic Nitrification

Both the stability and the relative strength of heterotrophic nitrification depend on particular environmental conditions. Several methods have been used over the years to discriminate nitrifying activity between chemolithotrophs and heterotrophs, with varying degrees of success (Table 3) (for review, see De Boer and Kowalchuk, 2001). One issue with in situ activity measurements is that, unlike with pure culture experiments, the conditions in soils and other environments do not normally favor readily measureable (i.e., rapid) rates of heterotrophic nitrification. Limitations in the amount or accessibility of nitrogenous substrate, carbon-to-nitrogen ratios, or variations in physicochemical parameters (e.g., temperature, moisture, oxygen content, pH, etc.) all influence measurements of bulk nitrification and greatly complicate discrimination between the pathways that contribute to it. Environments are quite variable, and the majority of microbial ecology studies are single time-point snapshots of activity and/or diversity profiles in one to a small handful of samples. Furthermore, since the heterotrophic pathways are diverse and largely uncharacterized relative to those in chemolithotrophs, the application of most methods, especially inhibitors, is not an exact science. Some complications of these methods are briefly summarized in Table 3.

Although we are a long way from having a definitive understanding of how and when heterotrophic nitrification is active in complex environments like soils, engineered environments may yield answers more quickly as processes like SND are consciously encouraged by altering environmental parameters (see Chapter 16). The one soil environment where heterotrophic nitrification is consistently favored over chemolithotrophic nitrification is acidic coniferous forest soils, and the active organisms are predominantly nitrifying fungi (Schimel et al., 1984; Killham, 1986; Jordan et al., 2005). The most compelling arguments made for the success of fungal over chemolithotrophic nitrifiers in these particular soils is that fungi are not inhibited by low pH, there is an abundance of organic N for fungi to nitrify and carbon to metabolize, and the fungal biomass in these soils is absolutely massive relative to that of chemolithotrophic bacteria (and archaea) (Killham, 1986). Conversely, acidic soils not within coniferous forests sometimes

TABLE 3 Methods to discriminate between heterotrophic and chemolithotrophic nitrifiers

Name	Type of detection	Target organism	Problems	Select references
Most probable number	Enumeration	Determined by media	Media is selective	Papen and von Berg, 1998
Nitrapyrin	Selective inhibition	AOB	Some soils bind to and/or remove it	Goring, 1962
Acetylene	Selective inhibition	AOB (at low levels)	Can be degraded over time, and not equally effective on all AOB	Hynes and Knowles, 1982; Wrage et al., 2004
Chlorate	Selective inhibition	Nitrite oxidizers	Negative effects on AOB and other microbes	Belser and Mays, 1980
Cycloheximide	Selective inhibition	Fungi (eukaryotes)	Negative effects on AOB and easily degraded	Schimel et al., 1984
Gamma irradiation	Selective inhibition	All organisms	Over sterilization/ recovery of population	Ishaque and Cornfield, 1976
^{15}N pool dilution	Activity	AOB	Cannot account for NH_3 oxidation by heterotrophs; bias towards ^{14}N uptake	Barraclough and Puri, 1995
Substrate amendment	Activity	Fungi or AOB	Indirect; substrate is selective	Killham, 1986

show AOB (Killham, 1986; De Boer and Kowalchuk, 2001) or Thaumarchaea (Nicol et al., 2008) as the dominant AMO containing (ergo, ammonia-oxidizing) phylotypes. Thus, no single environmental parameter can be used to accurately predict whether heterotrophic or chemolithotrophic nitrifiers dominate any particular environment, or when heterotrophic nitrification will be a significant contributor to bulk nitrification rates.

NITRIFIER DENITRIFICATION

Nitrifier denitrification, the reduction of nitrite to nitrous oxide via nitric oxide, was originally characterized in the AOB and differs most significantly from "classical" denitrification in that it is not coupled to the oxidation of organic carbon. Furthermore, this pathway operates under aerobic conditions during ammonia oxidation but is enhanced under microaerobic conditions (Goreau et al., 1980; Lipschultz et al., 1981) and is required for growth of some nitrosomonads under anaerobic conditions (Bock, 1995; Schmidt et al., 2004; Schmidt, 2009). Early on, nitrifier denitrification was speculated to function primarily as (i) an anaerobic respiratory pathway, (ii) a mechanism to out-compete nitrite oxidizers for oxygen, or (iii) a detoxification mechanism to rid the cell of excess nitrite. However, as alluded to in the above description of heterotrophic nitrification, recent studies have also suggested that at least in *N. europaea*, nitrifier denitrification functions as an electron sink from the cytochrome pool to speed the oxidation of hydroxylamine during aerobic metabolism (Fig. 1) (Cantera and Stein, 2007a), analogous to aerobic denitrification in heterotrophic bacteria (Fig. 2) and fungi (Fig. 3). Although *N. europaea* and *N. eutropha* use nitrifier denitrification enzymes to grow anaerobically by coupling the reduction of nitrite to the oxidation of ammonia, hydrogen, and organic carbon (Bock, 1995; Schmidt and Bock, 1997; Schmidt et al., 2004; Schmidt, 2009), anaerobic respiration has not been confirmed for any other AOB genus. The only other nonheterotrophic microbes known to perform both nitrification and nitrifier denitrification are the methanotrophs, again highlighting the functional similarities between these two bacterial groups (Yoshinari, 1984; Mandernack et al., 2000; Sutka et al., 2003; Nyerges, 2008). Currently, there is no evidence for nitrifier denitrification by ammonia-oxidizing Thaumarchaea.

Nitrifier denitrification has been the subject of several review articles because of its significance to the global nitrous oxide budget (Jetten et al., 1997b; Colliver and Stephenson, 2000; Wrage et al., 2001; Arp and Stein, 2003; Stein and Yung, 2003; Klotz and Stein, 2008). Yet, the genetics and enzymology of the pathway are still poorly understood, largely due to the dearth of physiological studies beyond *Nitrosomonas* spp. For example, physiological studies have verified that *Nitrosospira* spp. produce nitrous oxide from the reduction of nitrite (Dundee and Hopkins, 2001; Shaw et al., 2005), but differences in the structure and local environment of denitrification genes in *Nitrosospira multiformis* relative to *N. europaea* and *N. eutropha* suggest that the two AOB genera acquired the genes from different lateral transfer events (Norton et al., 2008). In addition, nitrous oxide is produced readily from hydroxylamine oxidation in addition to nitrite reduction in both AOB (Fig. 4) and methanotrophs (Whittaker et al., 2000; Sutka et al., 2003; Cantera and Stein, 2007a), greatly complicating genetic and enzymatic isolation of the nitrite reduction pathway alone.

Biochemistry

There are two main activities in the nitrifier denitrification pathway: the reduction of nitrite to nitric oxide via nitrite reductase and the reduction of nitric oxide to nitrous oxide via nitric oxide reductase (Fig. 4). Evidence from anaerobically grown *Nitrosomonas* implicates NirK and NorB as the sole reductases in the nitrifier denitrification pathway, while other enzymes likely play roles in nitrous oxide production from hydroxylamine (I. Schmidt et al., 2004; Beyer et al., 2008). Only *Nitrosomonas* spp. have been directly observed to produce dinitrogen as an end product of nitrifier denitrification (Poth, 1986; Shrestha et al., 2002; I. Schmidt et al., 2004) even though homologues to nitrous oxide reductase are absent from their genomes. Nitrite reductase activity was first observed in partial protein purifications of *N. europaea* in the same fractions as hydroxylamine oxidase activity (Hooper, 1968; Ritchie and Nicholas, 1974). Therefore, these studies suggested a linkage between enzymes in the ammonia-oxidation pathway to those in the nitrifier denitrification pathway. Later studies of *N. europaea* protein extracts linked a weak nitrite reductase activity to cytochrome *c* oxidase activity (DiSpirito et al., 1985; Miller and Nicholas, 1985). Further characterization of these fractions showed the presence of blue-copper proteins, although it was fairly evident that the nitrite reductase and cytochrome *c* oxidase components had different physical properties (for review, see Arp and Stein, 2003). It was not until completion of the *N. europaea* genome sequence that the linkage between two copper enzymes, a Pan1 multicopper oxidase (i.e., the cytochrome *c* oxidase component) and a NirK nitrite reductase, became clear and that the early biochemical studies had actually identified two distinct enzymes.

The nitric oxide reductase activity of *N. europaea* was not initially characterized biochemically, but rather was observed through numerous physiological and environmental studies of nitrous oxide production by *N. europaea* and other AOB (for reviews, see Wrage et al., 2001, and Arp and Stein, 2003). The electron transfer components of the nitrifier denitrification pathway were also not resolved biochemically and remain somewhat speculative; however, genetic studies in *N. europaea* have started to clarify our view of the full pathway, at least for this organism.

Genetics

The two best characterized genes in the nitrifier denitrification pathway are the copper-containing nitrite reductase, NirK, and the membrane-bound nitric oxide reductase NorB (encoded by *norCBQD*). A diversity of both *nirK* and *norB* genes has been detected in numerous AOB species by PCR and sequencing analysis (Casciotti and Ward, 2001; Casciotti and Ward, 2005; Cantera and Stein, 2007b; Garbeva et al., 2007), but direct testing of *nirK* and *norB* function in the AOB has

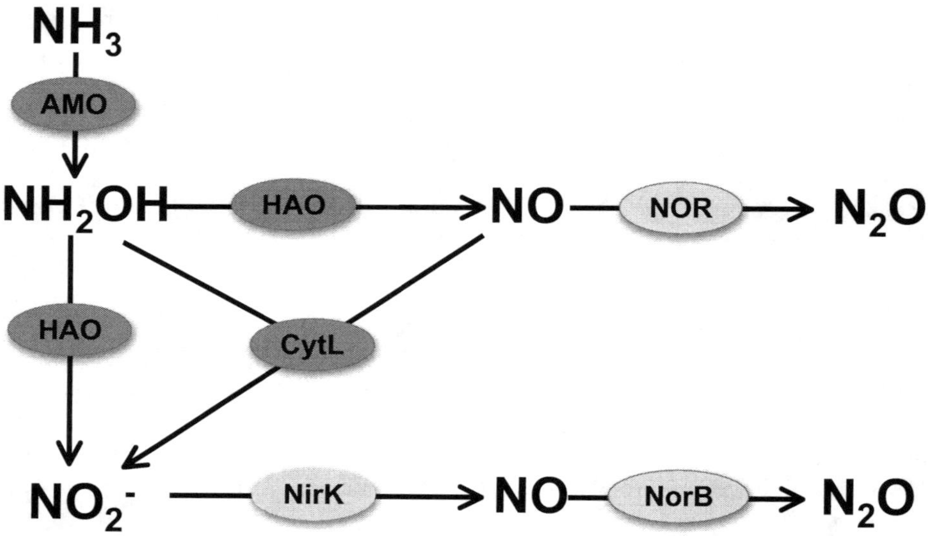

FIGURE 4 Two pathways for nitrous oxide production in *N. europaea*: hydroxylamine oxidation pathway and nitrifier denitrification pathway. Lightly shaded enzymes indicate reductases, and darkly shaded enzymes indicate oxidative processes. CytL, cytochrome P460; NorB, nitric oxide reductase implicated in denitrification pathway; NOR, generic nitric oxide reductase (descriptive of multiple enzymes in *N. europaea*).

only been accomplished thus far in *N. europaea* ATCC 19718. The blue-copper cytochrome *c* oxidase that has been purified, crystallized, and characterized (Lawton et al., 2009) is encoded by the first gene, *ncgA*, in the *nirK* gene cluster; *ncgABC-nirK* (*nirK* cluster genes [*ncg*]). The two middle genes, *ncgBC*, are small mono- and di-heme cytochrome *c* proteins, respectively. Although unusual, this particular operonic combination of genes with *nirK* has been found in the *Nitrosomonas* spp. and in members of the nitrite-oxidizing genus *Nitrobacter* (Cantera and Stein, 2007b), organisms that form tight physical associations in the environment (Mobarry et al., 1996). The translated *nirK* genes from these organisms form a distinct phylogenetic branch with few members, suggesting a unique evolutionary origin relative to the majority of other *nirK* genes (Cantera and Stein, 2007b). The majority of known *nirK* operons are also preceded by genes encoding the nitrite-regulated repressor, NsrR, and a conserved binding motif for this repressor is in the region upstream of the operon (Cantera and Stein, 2007b). Indeed, derepression of the *nirK* operon in *N. europaea* occurred in a nitrite-dependent fashion, suggesting a role in tolerance to nitrite toxicity (Beaumont et al., 2004a). The phenotype of aerobically grown *N. europaea* with a disrupted *nirK* gene showed a marked increase in both nitrous oxide production and sensitivity to nitrite (Beaumont et al., 2002). Further analysis of this NirK-deficient mutant revealed that the increase in nitrous oxide production was from the oxidation of hydroxylamine (Fig. 4) that likely accumulated due to the slower activity of HAO relative to that of AMO (Cantera and Stein, 2007a). This result not only validated the original observation of a functional linkage between HAO and nitrite reductase (via uncharacterized electron donors), as described above, but also suggested that NirK assists in the aerobic ammonia oxidation pathway by relieving an electron flow bottleneck between Complexes III and IV as described above for heterotrophic nitrification and aerobic denitrification.

Phenotypic analysis of *ncgA*, *ncgB*, and *ncgC* mutants showed that the four members of the *nirK* operon work together. Disruption of these genes caused polar effects, such that genes downstream of the disruption were also not expressed. Both the *ncgA* mutant (where none of the genes were expressed) and a *nirK*-complemented *ncgA* mutant (where only *nirK* was expressed) had the same phenotype as the *nirK* mutant, verifying that the gene products of the operon interact (Beaumont et al., 2005). The phenotypes of *ncgB* (expression of *ncgA* only) and *ncgC* (expression of *ncgA* and *ncgB*) mutants also matched that of the *nirK* mutant, again suggesting that the products of the genes necessarily interact with NirK. However, *nirK*-complemented *ncgB* (expression of *ncgA* and *nirK*) or *ncgC* (expression of *ncgA*, *ncgB*, and *nirK*) mutants created significant toxic effects to the cell during growth on ammonia, indicating that the products of the *ncgBC* genes must operate together with NcgA and NirK to prevent the accumulation of reactive nitrogen species. Either cupredoxins (Murphy et al., 2002) or cytochromes *c* (Nojiri et al., 2009) can donate electrons to NirK enzymes. Thus, either NcgA or NcgB/C is the likely electron donor to NirK as an artery from the cytochrome *c* pool (Fig. 1). This hypothetical positioning of gene products in the nitrifier denitrification pathway remains speculative as direct protein-protein interactions have yet to be tested. Both the structure and genomic context of *nirK* and *norB* genes in other AOB, like *Nitrosospira* and *Nitrosococcus*, are quite different from that in *N. europaea* and *N. eutropha* (Klotz et al., 2006; Norton et al., 2008), suggesting that nitrifier denitrification operates on different principles in these organisms. However, *N. multiformis* does produce nitrous oxide via a nitrifier denitrification pathway (Shaw et al., 2005) and awaits further physiological and genetic analysis.

Disruption of the nitric oxide reductase gene *norB* showed that production of nitrous oxide in *N. europaea* is not reliant on this enzyme alone as NorB-deficient cells produced the same amount of nitrous oxide as the wild type during ammonia oxidation (Beaumont et al., 2004b). Indeed, additional putative nitric oxide reductases, such as NorS, have been identified in the AOB (Stein et al., 2007) that are likely active under aerobic conditions (Fig. 4). Incidentally, both NirK and NorB were shown to be essential to anaerobic respiration of nitrite by *N. europaea*, indicating that they are indeed both critical members of the nitrifier denitrification pathway (I. Schmidt et al., 2004; Beyer et al., 2009).

Discrimination of Nitrous Oxide Produced by Nitrification versus Denitrification

AOB can produce nitrous oxide by two different pathways, hydroxylamine oxidation or nitrifier denitrification (Fig. 4). In pure culture studies, hydroxylamine oxidation to nitrous oxide is generally favored under high oxygen, whereas nitrifier denitrification is favored under low or no oxygen (Dundee and Hopkins, 2001; Wrage et al., 2004), although both occur with some oxygen present. Because nitrous oxide is formed readily by nitrification, nitrifier denitrification, aerobic denitrification, and classical anaerobic denitrification in the environment, tools to quantify the relative strengths of each pathway have become vital to complete our understanding of how, why, and where nitrous oxide is produced.

The technical breakthrough to discriminate nitrous oxide production from nitrification, nitrifier denitrification, and denitrification was the detection of individual nitrous oxide isotopomers using isotope ratio mass spectroscopy (Casciotti et al., 2003; Sutka et al., 2003, 2006; Shaw et al., 2005). The formation of nitrous oxide by nitric oxide reductase requires two nitric oxide molecules. It was observed that the site preference for ^{15}N in the alpha or beta position relative to the O atom in nitrous oxide ($N^{\beta}N^{\alpha}O$) depends on the catalytic mechanism of nitric oxide reductase (for review, see Stein and Yung, 2003; for comment, see H. L. Schmidt et al., 2004). Sutka et al. (2006) showed that the $\delta^{15}N$ of nitrous oxide produced from hydroxylamine oxidation was significantly more positive than that from nitrifier denitrification or denitrification. Furthermore, this study found that the site preference of ^{15}N in nitrous oxide was significantly different during nitrifier denitrification by AOB versus denitrification by two species of *Pseudomonas*. Therefore, both overall $\delta^{15}N$ values in conjunction with variation in ^{15}N placement in nitrous oxide are extraordinarily powerful measurements that can separate and quantify the relative contributions of each process to nitrous oxide flux from ecosystems. The isotopic signature of oxygen has also been used in combination with $\delta^{15}N$ values to discriminate sources of nitrous oxide, although caution must be used as O exchange between H_2O and N-oxide intermediates happens readily and can obscure metabolically derived isotopic signatures (Kool et al., 2009). It should be noted, however, that since the extent of O exchange can be quantified in both nitrification and denitrification processes, dual isotopic signatures are currently the most useful discriminatory measurement.

Isotopomer discrimination techniques are being applied more frequently and have verified significant contributions of nitrification to nitrous oxide production, particularly in marine (Charpentier et al., 2007; Yamagishi et al., 2007) and soil (Perez et al., 2006; Well et al., 2006, 2008) environments. These findings were somewhat surprising as denitrification via carbon respiration is typically thought of as the primary biological source of nitrous oxide. Nevertheless, studies are now confirming that nitrification and nitrifier denitrification can emit similar or greater amounts of nitrous oxide than anaerobic denitrification, especially from N-impacted, well-oxygenated ecosystems. As a caveat, a recent study showed that oxygen- and formate-dependent fungal denitrification contributes a significant proportion of nitrous oxide emissions from N-impacted aerated soils (Ma et al., 2008), and the contribution of bacterial aerobic denitrifiers is as yet unquantified. Furthermore, small differences in AMO enzymes were shown to change the $\delta^{15}N$ value of N_2O from different isolates of AOB (Casciotti et al., 2003). Thus, discreet quantification of individual nitrous oxide sources remains an arduous undertaking, yet isotopic measurements, particularly in combination with other molecular and microbiological techniques, have enabled more precise estimations of nitrous oxide sources than any preceding methodology.

PERSPECTIVES

This chapter has touched on largely understudied, but highly significant, processes of inorganic nitrogen metabolism (heterotrophic nitrification and nitrifier [and aerobic] denitrification) that impact the global nitrogen cycle. Many of the studies cited in this chapter suggest that these processes are strongly influenced by the availability of carbon, nitrogen, and oxygen in the environment. As soil moisture largely controls oxygen availability, it too plays a major role in governing the rates of heterotrophic nitrification and nitrifier (and aerobic) denitrification as well as other physicochemical parameters like temperature and salinity. Continued anthropogenic perturbation of the N cycle accelerates these aerobic processes as increasing N availability feeds directly into both nitrification and aerobic denitrification pathways. This acceleration is measured by the increasing amounts of nitrous oxide arising from marine and terrestrial ecosystems along with increased nitrate pollution and eutrophication. Indeed, the aerobic part of the N cycle is so far out of balance that it is considered a tell-tale feature of the Anthropocene epoch, in which human activities predominantly drive environmental change (Röckstrom et al., 2009). On a more positive note, the ability of many heterotrophic and autotrophic nitrifiers to simultaneously denitrify is being harnessed for more efficient N-removal strategies in N-impacted industrial systems like wastewater (Schmidt et al., 2003).

Given the ability of AOB, methanotrophs, and some heterotrophs to generate nitrous oxide by both nitrification and aerobic denitrification, it is imperative to determine the genetic and physiological diversity behind these pathways in ecologically relevant species if we ever hope to control nitrous oxide sources to the atmosphere. The literature is particularly sparse on the genetics and enzymology of nitrifying heterotrophs, particularly the nature of ammonia- and hydroxylamine-oxidizing enzymes. Furthermore, differences in genomic inventories and gene environments strongly indicate that microorganisms, even closely related species, metabolize inorganic N compounds differently. As observed in comparative physiological studies of *Nitrosomonas* versus *Nitrosospira* species of the AOB, even minor differences in metabolism can have strong effects on relative contributions of nitrogen oxide intermediates to the environment (Dundee and Hopkins, 2001; Wrage et al., 2004; Shaw et al., 2005). Likewise, some methanotrophic isolates are incapable of ammonia cometabolism, whereas others thrive under relatively high concentrations of both ammonium and nitrite (Nyerges and Stein, 2009). Because of these physiological differences, only some methanotrophic isolates are even capable of producing nitrous oxide (Nyerges, 2008). Thus, much remains to be characterized to gain a complete understanding of the diversity of nitrogen oxide-producing pathways.

In addition to basic metabolic and genetic studies, there is a continuing need to develop and refine methodologies for surveying and monitoring microbial communities in situ. Isotopic fractionation techniques have revolutionized the way we measure relative contributions and rates of nitrous oxide production in the environment and are particularly useful in combination with other methods like physicochemical analysis and gene diversity surveys.

The global community is starting to recognize that nitrogen pollution is a major threat to human health and the environment (Galloway et al., 2008). Unfortunately, some of our solutions to global issues, such as replacing fossil fuels with agriculturally derived biofuels, essentially ignore effects on the nitrogen cycle. This chapter has described microbial populations and processes that make nitrous oxide in response to increased fertilizer use, nitrogen deposition, and hypoxia. An integrated understanding of biogenic feedbacks through pathways like heterotrophic nitrification, nitrifier denitrification, and aerobic denitrification are essential if we are to manage, and perhaps mitigate, detrimental effects of anthropogenic nitrogen saturation.

REFERENCES

Anshuman, A., A. Khardenavis, A. Kapley, and H. J. Purohit. 2007. Simultaneous nitrification and denitrification by diverse *Diaphorobacter* sp. *Appl. Microbiol. Biotechnol.* **77:**403–409.

Arp, D. J., and L. Y. Stein. 2003. Metabolism of inorganic N compounds by ammonia-oxidizing bacteria. *Crit. Rev. Biochem. Mol. Biol.* **38:**471–495.

Barraclough, D., and G. Puri. 1995. The use of ^{15}N pool dilution and enrichment to separate the heterotrophic and autotrophic pathways of nitrification. *Soil Biol. Biochem.* **27:**17–22.

Beaumont, H. J. E., N. G. Hommes, L. A. Sayavedra-Soto, D. J. Arp, D. M. Arciero, A. B. Hooper, H. V. Westerhoff, and R. J. M. van Spanning. 2002. Nitrite reductase of *Nitrosomonas europaea* is not essential for production of gaseous nitrogen oxides and confers tolerance to nitrite. *J. Bacteriol.* **184:**2557–2560.

Beaumont, H. J. E., S. I. Lens, W. N. M. Reijnders, H. V. Westerhoff, and R. J. M. van Spanning. 2004a. Expression of nitrite reductase in *Nitrosomonas europaea* involves NsrR, a novel nitrite sensitive transcription repressor. *Mol. Microbiol.* **54:**148–158.

Beaumont, H. J. E., B. van Schooten, S. I. Lens, H. V. Westerhoff, and R. J. M. van Spanning. 2004b. *Nitrosomonas europaea* expresses a nitric oxide reductase during nitrification. *J. Bacteriol.* **186:**4417–4421.

Beaumont, H. J. E., S. I. Lens, H. V. Westerhoff, and R. J. M. van Spanning. 2005. Novel *nirK* cluster genes in *Nitrosomonas europaea* are required for NirK-dependent tolerance to nitrite. *J. Bacteriol.* **187:**6849–6851.

Belser, L. W., and E. L. Mays. 1980. The specific inhibition of nitrite oxidation by chlorate and its use in assessing nitrification in soils and sediments. *Appl. Environ. Microbiol.* **39:**505–510.

Bergmann, D. J., J. A. Zahn, A. B. Hooper, and A. A. DiSpirito. 1998. Cytochrome P460 genes from the methanotroph *Methylococcus capsulatus* Bath. *J. Bacteriol.* **180:**6440–6445.

Beyer, S., S. Gilch, O. Meyer, and I. Schmidt. 2009. Transcription of genes coding for metabolic key functions in *Nitrosomonas europaea* during aerobic and anaerobic growth. *J. Mol. Microbiol. Biotechnol.* **16:**187–197.

Bock, E. 1995. Nitrogen loss caused by denitrifying *Nitrosomonas* cells using ammonium or hydrogen as electron donors and nitrite as electron acceptor.

Arch. Microbiol. **163:**16–20.

Bodelier, P. L. E., and H. J. Laanbroek. 2004. Nitrogen as a regulatory factor of methane oxidation in soils and sediments. *FEMS Microbiol. Ecol.* **47:**265–277.

Cantera, J. J. L., and L. Y. Stein. 2007a. Interrelationship between nitrite reductase and ammonia-oxidizing metabolism in *Nitrosomonas europaea. Arch. Microbiol.* **188:**349–354.

Cantera, J. J. L., and L. Y. Stein. 2007b. Molecular diversity of nitrite reductase (*nirK*) genes in nitrifying bacteria. *Environ. Microbiol.* **9:**765–776.

Casciotti, K., and B. B. Ward. 2005. Phylogenetic analysis of nitric oxide reductase gene homologues from aerobic ammonia-oxidizing bacteria. *FEMS Microbiol. Ecol.* **52:**197–205.

Casciotti, K., D. Sigman, and B. Ward. 2003. Linking diversity and stable isotope fractionation in ammonia-oxidizing bacteria. *Geomicrobiol. J.* **20:**335–353.

Casciotti, K. L., and B. B. Ward. 2001. Dissimilatory nitrite reductase genes from autotrophic ammonia-oxidizing bacteria. *Appl. Environ. Microbiol.* **67:**2213–2221.

Castignetti, D., R. Yanong, and R. Gramzinski. 1984. Heterotrophic nitrification among denitrifiers. *Appl. Environ. Microbiol.* **47:**620–623.

Charpentier, J., L. Farias, N. Yoshida, N. Boontanon, and P. Raimbault. 2007. Nitrous oxide distribution and its origin in the central and eastern South Pacific Subtropical Gyre. *Biogeosciences* **4:**729–741.

Colliver, B. B., and T. Stephenson. 2000. Production of nitrogen oxide and dinitrogen oxide by autotrophic nitrifiers. *Biotech. Adv.* **18:**219–232.

Conrad, R. 1996. Soil microorganisms as controllers of atmospheric trace gases (H_2, CO, CH_4, OCS, N_2O, and NO). *Microbiol. Rev.* **60:**609–640.

Crossman, L. C., J. W. B. Moir, J. J. Enticknap, D. J. Richardson, and S. Spiro. 1997. Heterologous expression of heterotrophic nitrification genes. *Microbiology* **143:**3775–3783.

Daum, M., W. Zimmer, H. Papen, K. Kloos, K. Nawrath, and H. Bothe. 1998. Physiological and molecular biological characterization of ammonia oxidation of the heterotrophic nitrifier *Pseudomonas putida. Curr. Microbiol.* **37:**281–288.

De Boer, W., and G. A. Kowalchuk. 2001. Nitrification in acid soils: micro-organisms and mechanisms. *Soil Biol. Biochem.* **33:**853–866.

DiSpirito, A.A., L. R. Taaffe, J. D. Lipscomb, and A. B. Hooper. 1985. A 'blue' copper oxidase from *Nitrosomonas europaea. Biochim. Biophys. Acta* **827:**320–326.

Dundee, L., and D. W. Hopkins. 2001. Different sensitivities to oxygen of nitrous oxide production by *Nitrosomonas europaea* and *Nitrosolobus multiformis. Soil Biol. Biochem.* **33:**1563–1565.

Erickson, R. H., and A. B. Hooper. 1972. Preliminary characterization of a variant CO-binding heme protein from *Nitrosomonas. Biochim. Biophys. Acta* **275:**231–244.

Focht, D. D., and W. Verstraete. 1977. Biochemical ecology of nitrification and denitrification, p. 135–214. *In* M. Alexander (ed.), *Advances in Microbial Ecology*. Plenum Press, New York, NY.

Friedrich, C. G., D. Rother, F. Bardischewsky, A. Quentmeier, and J. Fischer. 2001. Oxidation of reduced inorganic sulfur compounds by bacteria: emergence of a common mechanism? *Appl. Environ. Microbiol.* **67:**2873–2882.

Galloway, J. N., A. R. Townsend, J. W. Erisman, M. Bekunda, Z. Cai, J. R. Freney, L. A. Martinelli, S. P. Seitzinger, and M. A. Sutton. 2008. Transformation of the nitrogen cycle: recent trends, questions, and potential solutions. *Science* **320:**889–892.

Garbeva, P., E. M. Baggs, and J. L. Prosser. 2007. Phylogeny of nitrite reductase (*nirK*) and nitric oxide reductase (*norB*) genes from *Nitrosospira* species isolated from soil. *FEMS Microbiol. Lett.* **266:**83–89.

Goreau, T. J., W. A. Kaplan, S. C. Wofsy, M. B. McElroy, F. W. Valois, and S. W. Watson. 1980. Production of NO_2^- and N_2O by nitrifying bacteria at reduced concentrations of oxygen. *Appl. Environ. Microbiol.* **40:**526–532.

Goring, C. A. I. 1962. Control of nitrification by 2-chloro-6-(trichloromethyl)-pyridine. *Soil Sci.* **93:**211–218.

Hakemian, A. S., and A. C. Rosenzweig. 2007. The biochemistry of methane oxidation. *Ann. Rev. Biochem.* **76:**223–241.

Hooper, A. B. 1968. A nitrite-reducing enzyme from *Nitrosomonas europaea.* Preliminary characterization with hydroxylamine as electron donor. *Biochim. Biophys. Acta* **162:**49–65.

Hooper, A. B., P. C. Maxwell, and K. R. Terry. 1978. Hydroxylamine oxidoreductase from *Nitrosomonas*: absorption spectra and content of heme and metal. *Biochemistry* **17:**2984–2989.

Hynes, R. K., and R. Knowles. 1982. Effect of acetylene on autotrophic and heterotrophic nitrification. *Can. J. Microbiol.* **28:**334–340.

IPCC. 2006. IPCC Guidelines for National Greenhouse Gas Inventories, prepared by the National Greenhouse Gas Inventories Programme, Chapter 11. *In* H. S. Eggleston, L. Buendia, K. Miwa, T. Ngara, and K. Tanabe (ed.), *N_2O Emissions from Managed Soils, and CO_2 Emissions from Lime and Urea Application.* IGES, Hayama, Japan.

Ishaque, M., and A. H. Cornfield. 1976. Evidence for heterotrophic nitrification in an acid Bangladesh soil lacking autotrophic nitrifying organisms.

Trop. Agric. **53:**157–160.

Jetten, M. S. M., P. de Bruijn, and J. G. Kuenen. 1997a. Hydroxylamine metabolism in *Pseudomonas* PB16: involvement of a novel hydroxylamine oxidoreductase. *Antonie van Leeuwenhoek* **71:**69–74.

Jetten, M. S. M., S. Logemann, G. Muyzer, L. A. Robertson, S. d.Vries, M. C. M. v.Loosdrecht, and J. G. Kuenen. 1997b. Novel principles in the microbial conversion of nitrogen compounds. *Antonie van Leeuwenhoek* **71:**75–93.

Joo, H.-S., M. Hirai, and M. Shoda. 2005. Nitrification and denitrification in high-strength ammonium by *Alcaligenes faecalis*. *Biotech. Lett.* **27:**773–778.

Jordan, F. L., J. J. L. Cantera, M. E. Fenn, and L.Y. Stein. 2005. Autotrophic ammonia-oxidizers contribute minimally to nitrification in a nitrogen-saturated forest soil. *Appl. Environ. Microbiol.* **71:**197–206.

Killham, K. 1986. Heterotrophic nitrification, p. 117–126. *In* J. I. Prosser (ed.), *Nitrification*. IRL Press, Oxford, United Kingdom.

Kim, S. W., S. Fushinobu, S. M. Zhou, T. Wakagi, and H. Shoun. 2009. Eukaryotic *nirK* genes encoding copper-containing nitrite reductase: originating from the protomitochondrion? *Appl. Environ. Microbiol.* **75:**2652–2658.

Klotz, M. G., and J. M. Norton. 1998. Multiple copies of ammonia monooxygenase (*amo*) operons have evolved under biased AT/GC mutational pressure in ammonia-oxidizing autotrophic bacteria. *FEMS Microbiol. Lett.* **168:**303–311.

Klotz, M. G., and L. Y. Stein. 2008. Nitrifier genomics and evolution of the nitrogen cycle. *FEMS Microbiol. Lett.* **278:**146–456.

Klotz, M. G., D. J. Arp, P. S. G. Chain, A. F. El-Sheikh, L. Hauser, N. G. Hommes, F. W. Larimer, S. Malfatti, J. M. Norton, A. T. Poret-Peterson, L. Vergez, and B. B. Ward. 2006. The complete genome sequence of the marine, nitrifying purple sulfur bacterium, *Nitrosococcus oceani* ATCC 19707. *Appl. Environ. Microbiol.* **72:**6299–6315.

Klotz, M. G., M. C. Schmid, M. Strous, H. J. M. op den Camp, M. S. M. Jetten, and A. B. Hooper. 2008. Evolution of an octahaem cytochrome *c* protein family that is key to aerobic and anaerobic ammonia oxidation by bacteria. *Environ. Microbiol.* **10:**3150–3163.

Kool, D. M., C. Müller, N. Wrage, O. Oenema, and J. W. Van Groenigen. 2009. Oxygen exchange between nitrogen oxides and H_2O can occur during nitrifier pathways. *Soil Biol. Biochem.* **41:**1632–1641.

Kurokawa, M., Y. Fukumori, and T. Yamanaka. 1985. A hydroxylamine-cytochrome *c* reductase occurs in the heterotrophic nitrifier *Arthrobacter globiformis*. *Plant Cell Physiol.* **26:**1439–1442.

Lawton, T. J., L. A. Sayavedra-Soto, D. J. Arp, and A. C. Rosenzweig. 2009. Crystal structure of a two-domain multicopper oxidase: implications for the evolution of multicopper blue proteins. *J. Biol. Chem.* **284:**10174–10180.

Lieberman, R. L., and A. C. Rosenzweig. 2004. Biological methane oxidation: regulation, biochemistry, and active site structure of particulate methane monooxygenase. *Crit. Rev. Biochem. Mol. Biol.* **39:**147–164.

Lin, Y., H. Kong, Y. He, L. Kuai, and Y. Inamori. 2004. Simultaneous nitrification and denitrification in a membrane bioreactor and isolation of heterotrophic nitrifying bacteria. *Jpn. J. Water Treat. Biol.* **40:**105–114.

Lipschultz, F., O. C. Zafiriou, S. C. Wofsy, M. B. McElroy, F. W. Valois, and S. W. Watson. 1981. Production of NO and N_2O by soil nitrifying bacteria. *Nature* **294:**641–643.

Ma, W. K., R. E. Farrell, and S. D. Siciliano. 2008. Soil formate regulates the fungal nitrous oxide emission pathway. *Appl. Environ. Microbiol.* **74:**6690–6696.

Mandernack, K. W., C. A. Kinney, D. Coleman, Y.-S. Huang, K. H. Freeman, and J. Bogner. 2000. The biogeochemical controls of N_2O production and emission in landfill cover soils: the role of methanotrophs in the nitrogen cycle. *Environ. Microbiol.* **2:**298–309.

Matsuzaka, E., N. Nomura, H. Maseda, H. Otagaki, T. Nakajima-Kambe, T. Nakahara, and H. Uchiyama. 2003. Participation of nitrite reductase in conversion of NO_2^- to NO_3^- in a heterotrophic nitrifier, *Burkholderia cepacia* NH-17, with denitrification activity. *Microb. Environ.* **18:**203–209.

Mével, G., and D. Prieur. 2000. Heterotrophic nitrification by a thermophilic *Bacillus* species as influenced by different culture conditions. *Can. J. Microbiol.* **46:**465–473.

Miller, D. J., and D. J. D. Nicholas. 1985. Characterization of a soluble cytochrome oxidase/nitrite reductase from *Nitrosomonas europaea*. *J. Gen. Microbiol.* **131:**2851–2854.

Mobarry, B. K., M. Wagner, V. Urbain, B. E. Rittman, and D. A. Stahl. 1996. Phylogenetic probes for analyzing abundance and spatial organization of nitrifying bacteria. *Appl. Environ. Microbiol.* **62:**2156–2162.

Moir, J. W. B., J.-M. Wehrfritz, S. Spiro, and D. J. Richardson. 1996a. The biochemical characterization of a novel non-haem-iron hydroxylamine oxidase from *Paracoccus denitrificans* GB17. *Biochem. J.* **319:**823–827.

Moir, J. W. B., L. C. Crossman, S. Sprio, and D. J. Richardson. 1996b. The purification of ammonia monooxygenase from *Paracoccus denitrificans*. *FEBS*

Lett. **387:**71–74.

Molina, J. A. E., and M. Alexander. 1972. Oxidation of nitrite and hydroxylamine by *Aspergillus flavus,* peroxidase and catalase. *Antonie van Leeuwenhoek* **38:**505–512.

Murphy, L. M., F. E. Dodd, F. K. Yousafzai, R. R. Eady, and S. S. Hasnain. 2002. Electron donation between copper containing nitrite reductases and cupredoxins: the nature of protein-protein interaction in complex formation. *J. Mol. Biol.* **315:**859–871.

Nemergut, D. R., and S. K. Schmidt. 2002. Disruption of *narH, narJ,* and *moaE* inhibits heterotrophic nitrification in *Pseudomonas* strain M19. *Appl. Environ. Microbiol.* **68:**6462–6465.

Nicol, G. W., S. Leininger, C. Schleper, and J. L. Prosser. 2008. The influence of soil pH on the diversity, abundance and transcriptional activity of ammonia oxidizing archaea and bacteria. *Environ. Microbiol.* **10:**2966–2978.

Nojiri, M., H. Koteishi, T. Nakagami, K. Kobayashi, T. Inoue, K. Yamaguchi, and S. Suzuki. 2009. Structural basis of inter-protein electron transfer for nitrite reduction in denitrification. *Nature* **462:**U117–U132.

Norton, J. M., J. J. Alzerreca, Y. Suwa, and M. G. Klotz. 2002. Diversity of ammonia monooxygenase operon in autotrophic ammonia-oxidizing bacteria. *Arch. Microbiol.* **177:**139–149.

Norton, J. M., M. G. Klotz, L. Y. Stein, D. J. Arp, P. J. Bottomley, P. S. G. Chain, L. J. Hauser, M. L. Land, F. W. Larimer, M. W. Shin, and S. R. Starkenburg. 2008. Complete genome sequence of *Nitrosospira multiformis*, an ammonia-oxidizing bacterium from the soil environment. *Appl. Environ. Microbiol.* **74:**3559–3572.

Numata, M., T. Saito, T. Yamazaki, Y. Fukumori, and T. Yamanaka. 1990. Cytochrome P-460 of *Nitrosomonas europaea*: further purification and further characterization. *J. Biochem.* **108:**1016–1021.

Nyerges, G. 2008. New insights into ammonia and nitrite metabolism by methanotrophic bacteria. Ph.D. thesis. University of California, Riverside, CA.

Nyerges, G., and L. Y. Stein. 2009. Ammonia cometabolism and product inhibition vary considerably among species of methanotrophic bacteria. *FEMS Microbiol. Lett.* **297:**131–136.

Ono, Y., A. Enokiya, D. Masuko, K. Shoji, and T. Yamanaka. 1999. Pyruvic oxime dioxygenase from the heterotrophic nitrifier *Alcaligenes faecalis*: purification, and molecular and enzymatic properties. *Plant Cell Physiol.* **401:**47–52.

Otani, Y., K. Hasegawa, and K. Hanaki. 2004. Comparison of aerobic denitrifying activity among three cultural species with various carbon sources. *Water Sci. Tech.* **50:**15–22.

Otte, S., J. Schalk, J. G. Kuenen, and M. S. M. Jetten. 1999. Hydroxylamine oxidation and subsequent nitrous oxide production by the heterotrophic ammonia oxidizer *Alcaligenes faecalis*. *Appl. Microbiol. Biotechnol.* **51:**255–261.

Papen, H., and R. von Berg. 1998. A most probably number method (MPN) for the estimation of cell numbers of heterotrophic nitrifying bacteria in soil. *Plant Soil* **199:**123–130.

Perez, T., D. Garcia-Montiel, S. Trumbore, S. Tyler, P. De Camargo, M. Moreira, M. Piccolo, and C. Cerri. 2006. Nitrous oxide nitrification and denitrification ^{15}N enrichment factors from Amazon forest soils. *Ecol. Applic.* **16:**2153–2167.

Poret-Peterson, A. T., J. E. Graham, J. Gulledge, and M. G. Klotz. 2008. Transcription of nitrification genes by the methane-oxidizing bacterium, *Methylococcus capsulatus* strain Bath. *ISME J.* **2:**1213–1220.

Poth, M. 1986. Dinitrogen production from nitrite by a *Nitrosomonas* isolate. *Appl. Environ. Microbiol.* **52:**957–959.

Ralt, D., R. F. Gomez, and S. R. Tannerbaum. 1981. Conversion of acetohydroxamate and hydroxylamine to nitrite by intestinal microorganisms. *Eur. J. Appl. Microbiol. Biotechnol.* **12:**226–230.

Ritchie, G. A. F., and D. J. D. Nicholas. 1974. The partial characterization of purified nitrite reductase and hydroxylamine oxidase from *Nitrosomonas europaea*. *Biochem. J.* **138:**471–480.

Robertson, L. A., and J. G. Kuenen. 1990. Combined heterotrophic nitrification and aerobic denitrification in *Thiosphaera pantotropha* and other bacteria. *Antonie van Leeuwenhoek* **57:**139–152.

Robertson, L. A., E. W. J. Van Niel, R. A. M. Torremans, and J. G. Kuenen. 1988. Simultaneous nitrification and denitrification in aerobic chemostat cultures of *Thiosphaera pantotropha*. *Appl. Environ. Microbiol.* **54:**2812–2818.

Robertson, L. A., R. Cornelisse, P. De Vos, R. Hadioetomo, and J. G. Kuenen. 1989. Aerobic denitrification in various heterotrophic nitrifiers. *Antonie van Leeuwenhoek* **56:**289–300.

Röckstrom, J., W. Steffen, K. Noone, A. Persson, F. S. Chapin, E. F. Lambin, T. M. Lenton, M. Scheffer, C. Folke, H. J. Schellnhuber, B. Nykvist, C. A. de Wit, T. Hughes, S. van der Leeuw, H. Rodhe, S. Sorlin, P. K. Snyder, R. Costanza, U. Svedin, M. Falkenmark, L. Karlberg, R. W. Corell, V. J. Fabry, J. Hansen, B. Walker, D. Liverman, K. Richardson, P. Crutzen, and J. A. Foley. 2009. A safe operating space for humanity. *Nature* **461:**472–475.

Sakai, K., K. Takano, T. Tachiki, and T. Tochikura. 1988. Purification and properties of an enzyme oxidizing nitrite to nitrate from *Candida rugosa*. *Agric. Biol. Chem.* **52:**2783–2789.

Sakai, K., Y. Ikehata, Y. Ikenaga, M. Wakayama, and M. Moriguchi. 1996. Nitrite oxidation by heterotrophic bacteria under various nutritional and aerobic conditions. *J. Ferment. Bioeng.* **82:**613–617.

Sakai, K., H. Nisijima, Y. Ikenaga, M. Wakayama, and M. Moriguchi. 2000. Purification and characterization of nitrite-oxidizing enzyme from heterotrophic *Bacillus badius* I-73, with special concern to catalase. *Biosci. Biotechnol. Biochem.* **64:**2727–2730.

Schalk, J., S. de Vries, J. G. Kuenen, and M. S. M. Jetten. 2000. Involvement of a novel hydroxylamine oxidoreductase in anaerobic ammonium oxidation. *Biochemistry* **39:**5405–5412.

Schimel, J. P., M. K. Firestone, and K. S. Killham. 1984. Identification of heterotrophic nitrification in a Sierran forest soil. *Appl. Environ. Microbiol.* **48:**802–806.

Schmidt, E. L. 1973. Nitrate formation by *Aspergillus flavus* in ure and mixed culture in natural environments. *Trans. 7th Int. Congr. Soil Sci.* **2:**600–605.

Schmidt, H. L., R. A. Werner, N. Yoshida, and R. Well. 2004. Is the isotopic composition of nitrous oxide an indicator for its origin from nitrification or denitrification? A theoretical approach from referred data and microbiological and enzyme kinetic aspects. *Rap. Comm. Mass Spectrom.* **18:**2036–2040.

Schmidt, I. 2009. Chemoorganoheterotrophic growth of *Nitrosomonas europaea* and *Nitrosomonas eutropha*. *Curr. Microbiol.* **59:**130–138.

Schmidt, I., and E. Bock. 1997. Anaerobic ammonia oxidation with nitrogen dioxide by *Nitrosomonas eutropha*. *Arch. Microbiol.* **167:**106–111.

Schmidt, I., O. Sliekers, M. Schmid, E. Bock, J. Fuerst, J. G. Kuenen, M. S. M. Jetten, and M. Strous. 2003. New concepts of microbial treatment processes for the nitrogen removal in wastewater. *FEMS Microbiol. Rev.* **27:**481–492.

Schmidt, I., R. J. M. van Spanning, and M. S. M. Jetten. 2004. Denitrification and ammonia oxidation by *Nitrosomonas europaea* wild-type, and NirK- and NorB-deficient mutants. *Microbiology UK* **150:**4107–4114.

Shaw, L. J., G. W. Nicol, Z. Smith, J. Fear, J. I. Prosser, and E. M. Baggs. 2005. *Nitrosospira* spp. can produce nitrous oxide via a nitrifier denitrification pathway. *Environ. Microbiol.* **8:**214–222.

Shrestha, N. K., S. Hadano, T. Kamachi, and I. Okura. 2002. Dinitrogen production from ammonia by *Nitrosomonas europaea*. *Appl. Catal.* **237:**33–39.

Spiller, H., E. Dietsch, and E. Kessler. 1976. Intracellular appearance of nitrite and nitrate in nitrogen-starved cells of *Ankistrodesmus braunii*. *Planta* **129:**175–181.

Starkenburg, S. R., D. J. Arp, and P. J. Bottomley. 2008. Expression of a putative nitrite reductase and the reversible inhibition of nitrite-dependent respiration by nitric oxide in *Nitrobacter winogradskyi* NB-255. *Environ. Microbiol.* **10:**3036–3042.

Stein, L. Y., and Y. L. Yung. 2003. Production, isotopic composition, and atmospheric fate of biologically produced nitrous oxide. *Ann. Rev. Earth Planet. Sci.* **31:**329–356.

Stein, L. Y., D. J. Arp, P. M. Berube, P. S. G. Chain, L. Hauser, M. S. M. Jetten, M. G. Klotz, F. W. Larimer, J. M. Norton, H. J. M. op den Camp, M. Shin, and X. Wei. 2007. Whole-genome analysis of the ammonia-oxidizing bacterium, *Nitrosomonas eutropha* C91: implications for niche adaptation. *Environ. Microbiol.* **9:**2993–3007.

Stouthammer, A. H., A. P. N. de Boer, J. van der Oost, and R. J. M. van Spanning. 1997. Emerging principles of inorganic nitrogen metabolism in *Paracoccus denitrificans* and related bacteria. *Antonie van Leeuwenhoek* **71:**33–41.

Stroo, H. F., T. M. Klein, and M. Alexander. 1986. Heterotrophic nitrification in an acid forest soil and by an acid-tolerant fungus. *Appl. Environ. Microbiol.* **52:**1107–1111.

Sutka, R. L., N. E. Ostrom, P. H. Ostrom, H. Gandhi, and J. A. Breznak. 2003. Nitrogen isotopomer site preference of N_2O produced by *Nitrosomonas europaea* and *Methylococcus capsulatus* Bath. *Rap. Comm. Mass Spectrom.* **17:**738–745.

Sutka, R. L., N. E. Ostrom, P. H. Ostrom, J. A. Breznak, H. Gandhi, A. J. Pitt, and F. Li. 2006. Distinguishing nitrous oxide production from nitrification and denitrification on the basis of isotopomer abundances. *Appl. Environ. Microbiol.* **72:**638–644.

Tachiki, T., K. Sakai, K. Yamamoto, M. Hatanaka, and T. Tochikura. 1988. Conversion of nitrite to nitrate by nitrite-resistant yeasts. *Agric. Biol. Chem.* **52:**1999–2005.

Takaya, N., S. Kuwazaki, Y. Adachi, S. Suzuki, T. Kikuchi, H. Nakamura, Y. Shiro, and H. Shoun. 2003. Hybrid respiration in the denitrifying mitochondria of *Fusarium oxysporum*. *J. Biochem.* **133:**461–465.

Van Gool, A. P., and E. L. Schmidt. 1973. Nitrification in relation to growth in *Aspergillus flavus*. *Soil Biol. Biochem.* **5:**259–265.

van Niel, E. W. J., P. A. M. Arts, B. J. Wesselink, L. A. Robertson, and J. G. Kuenen. 1993. Competition between heterotrophic and autotrophic nitrifiers for ammonia in chemostat cultures. *FEMS Microbiol. Ecol.* **102:**109–118.

Verstraete, W. 1975. Heterotrophic nitrification in soils and aqueous media—a review. *Bull. Acad. Sci. USSR Biol. Ser.* **4:**515–530.

Verstraete, W., and M. Alexander. 1972. Heterotrophic nitrification by *Arthrobacter* sp. *J. Bacteriol.* **110:**955–961.

Wehrfritz, J.-M., A. Reilly, S. Spiro, and D. J. Richardson. 1993. Purification of hydroxylamine oxidase from *Thiosphaera pantotropha*: identification of electron acceptors that couple heterotrophic nitrification to aerobic denitrification. *FEBS Lett.* **335:**246–250.

Wehrfritz, J.-M., J. P. Carter, S. Spiro, and D. J. Richardson. 1997. Hydroxylamine oxidation in heterotrophic nitrate-reducing soil bacteria and purification of a hydroxylamine-cytochrome *c* oxidoreductase from a *Pseudomonas* species. *Arch. Microbiol.* **166:**421–424.

Well, R., I. Kurganova, V. L. de Gerenyu, and H. Flessa. 2006. Isotopomer signatures of soil-emitted N_2O under different moisture conditions—a microcosm study with arable loess soil. *Soil Biol. Biochem.* **38:**2923–2933.

Well, R., H. Flessa, L. Xing, X. T. Ju, and V. Romheld. 2008. Isotopologue ratios of N_2O emitted from microcosms with NH_4^+-fertilized arable soils under conditions favoring nitrification. *Soil Biol. Biochem.* **40:**2416–2426.

White, J. P., and G. T. Johnson. 1982. Aflatoxin production correlated with nitrification in *Aspergillus flavus* group species. *Mycologia* **74:**718–723.

Whittaker, M., D. Bergmann, D. Arciero, and A. B. Hooper. 2000. Electron transfer during the oxidation of ammonia by the chemolithotrophic bacterium *Nitrosomonas europaea*. *Biochim. Biophys. Acta* **1459:**346–355.

Wrage, N., G. L. Velthof, M. L. van Beusichem, and O. Oenema. 2001. Role of nitrifier denitrification in the production of nitrous oxide. *Soil Biol. Biochem.* **33:**1723–1732.

Wrage, N., G. L. Velthof, O. Oenema, and H. J. Laanbroek. 2004. Acetylene and oxygen as inhibitors of nitrous oxide production in *Nitrosomonas europaea* and *Nitrosospira briensis*: a cautionary tale. *FEMS Microbiol. Ecol.* **47:**13–18.

Yamagishi, H., M. B. Westley, B. N. Popp, S. Toyoda, N. Yoshida, S. Watanabe, K. Koba, and Y. Yamanaka. 2007. Role of nitrification and denitrification on the nitrous oxide cycle in the eastern tropical North Pacific and Gulf of California. *J. Geophys. Res.-Biogeosci.* **112:**G02015.

Yoshinari, T. 1984. Nitrite and nitrous oxide production by *Methylosinus trichosporium*. *Can. J. Microbiol.* **31:**139–144.

Zahn, J. A., C. Duncan, and A. A. DiSpirito. 1994. Oxidation of hydroxylamine by cytochrome P-460 of the obligate methylotroph *Methylococcus capsulatus* Bath. *J. Bacteriol.* **176:**5879–5887.

AMMONIA-OXIDIZING ARCHAEA

III

PHYSIOLOGY AND GENOMICS OF AMMONIA-OXIDIZING *ARCHAEA*

Hidetoshi Urakawa, Willm Martens-Habbena, David A. Stahl

6

INTRODUCTION

The oxidation of ammonia to nitrite by autotrophic microorganisms was long thought mediated by a few restricted groups of *Bacteria*. The recent discovery of ammonia-oxidizing *Archaea* (AOA) of apparent great abundance in both marine and terrestrial environments necessitates a reevaluation of microbial control of the nitrogen cycle, which will require a deeper understanding of the ecophysiology of the AOA and their relationship to the better characterized ammonia-oxidizing bacteria (AOB) (Schleper et al., 2005; Leininger et al., 2006; Prosser and Nicol, 2008). The recent description of a first nitrifying archaeon, *Nitrosopumilus maritimus* strain SCM1 (Könneke et al., 2005), opened up the possibility for a much more detailed characterization of a representative of these widespread microorganisms. The genome sequence of SCM1 shares highly similar gene content and gene organization with marine planktonic *Crenarchaeota* identified through 16S ribosomal RNA (DeLong et al., 1992, 1994; Fuhrman et al., 1992, 1993) and metagenomic analyses (Béjà et al., 2002; Venter et al., 2004). Furthermore, initial physiological characterization of SCM1 has shown that it will grow under extreme ammonia limitation characteristic of most marine systems, growing exponentially until total ammonia is depleted to below 10 nM (Martens-Habbena et al., 2009). This exceptionally high-ammonia affinity is fully consistent with the observed high abundance of *Crenarchaeota* in nutrient-depleted ocean waters (Massana et al., 1997; Karner et al., 2001; Mincer et al., 2007; Varela et al., 2008). Thus, the now recognized genomic and physiological features of SCM1 make this strain an excellent model organism to study principles of the ecology, physiology, biochemistry, and genetics of these ubiquitous and newly recognized participants in the nitrogen cycle. Insights into the genetics and physiology of mesophilic *Crenarchaeota* by investigation of strain SCM1 and environmental genomic studies should also further facilitate the isolation of additional strains, as needed to illuminate in more detail the physiological diversity of these organisms.

Strain SCM1 is affiliated with the Group I *Crenarchaeota*, one of the most abundant clades of marine bacterioplankton (Fig. 1) (Karner et al., 2001; Giovannoni and Stingl, 2005; Mincer et al., 2007; Varela et al., 2008). Metagenomic surveys first hinted at the presence of archaeal ammonia oxidizers (Schleper et al., 2005; Treusch et al., 2005), and extensive

Hidetoshi Urakawa, Willm Martens-Habbena, and David A. Stahl, Department of Civil and Environmental Engineering, University of Washington, Seattle, WA 98195-5014.

Nitrification, Edited by Bess B. Ward, Daniel J. Arp, and Martin G. Klotz

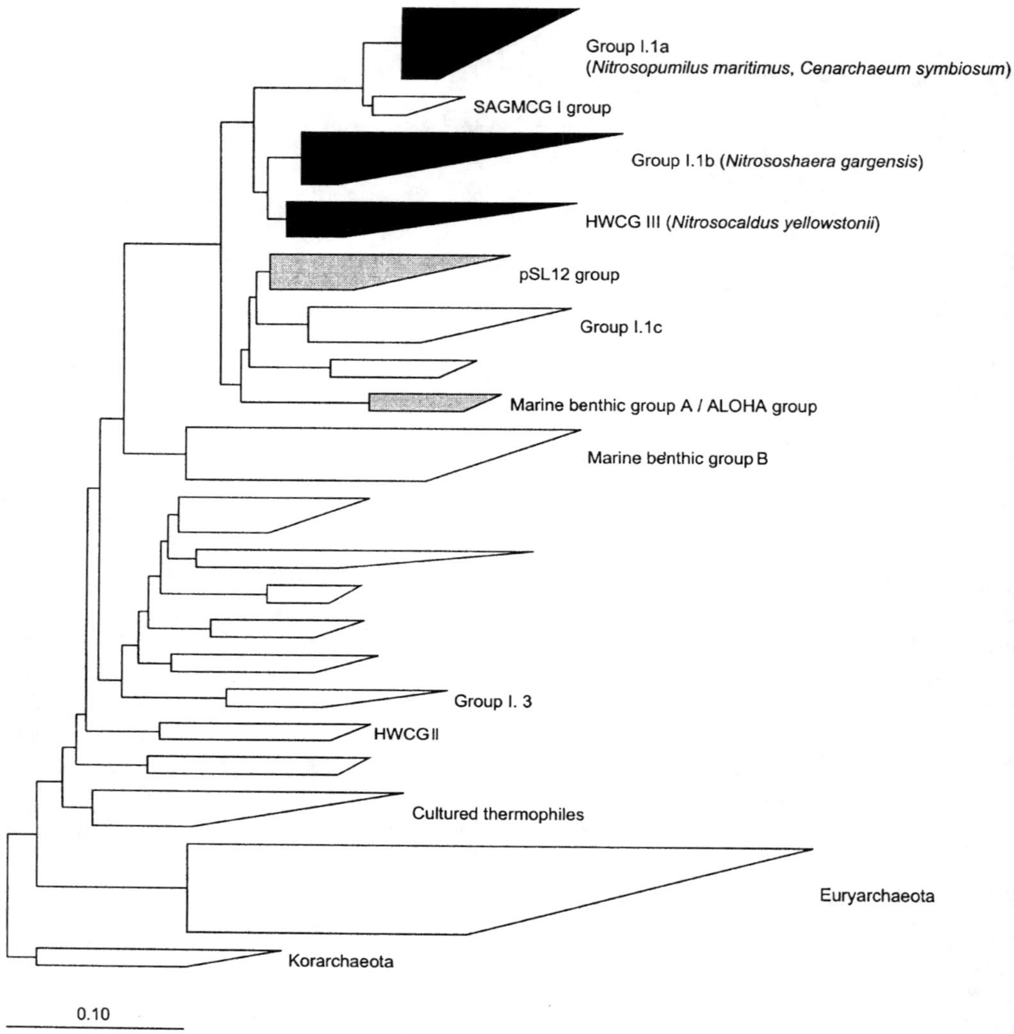

FIGURE 1 Phylogenetic tree of *Crenarchaeota* based on 16S rRNA sequences (more than 800 bp, ≈800 sequences). Clades containing recognized candidate species and enrichment cultures are shown in black. Clades containing environmental clones potentially linking to ammonia oxidation are shown in gray. SAGMCG I, South African Gold Mine Crenarchaeotic Group I; HWCG, Hot Water Crenarchaeotic Group. *Crenarchaeota* Group I.1a is known as marine group I *Crenarchaeota*, and Group I.1b is known as soil *Crenarchaeota*.

environmental surveys using PCR to quantify and compare genes coding for one subunit of a putative archaeal ammonia monooxygenase (*amoA*) suggested the capacity for ammonia oxidation is widely distributed within the Group I clade and other crenarchaeal clades identified in both terrestrial and aquatic habitats (Francis et al., 2005). It was speculated that *Archaea* may often be more significant contributors to nitrification than the better described bacterial ammonia oxidizers (Leininger et al., 2006; Prosser and Nicol, 2008). However, indications of a role in nitrification are as yet almost entirely based on molecular profiling, and their quantitative contribution to ammonia oxidation still remains unclear (Prosser and Nicol, 2008; Jia and Conrad, 2009).

Following the isolation of strain SCM1 from a tropical marine aquarium (Könneke et al., 2005), additional enrichment cultures have been reported from geothermal environments (de la Torre et al., 2008; Hatzenpichler et al., 2008). Cultivated ammonia-oxidizing *Crenarchaeota* have now been described for three major clades, marine group (Group I.1a), soil group (Group I.1b), and thermophilic group (Hot Water Crenarchaeotic Group III [HWCG III]) (Fig. 1), spanning tremendous phylogenetic breadth and extending the upper temperature limit of nitrification to more than 70°C (Könneke et al., 2005; de la Torre et al., 2008; Hatzenpichler et al., 2008; Prosser and Nicol, 2008). Enrichment cultures obtained from Garga hot spring (Buryat Republic, Russia) were first tentatively identified based on immunofluorescence microscopy as members of the genus *Nitrosomonas* (Lebedeva et al., 2005). Later, more detailed molecular analysis revealed that this ammonia-oxidizing enrichment culture was dominated by the member of the Group I.1b *Crenarchaeota* with a growth optimum of 45°C and ammonia oxidation activity up to 55°C (Lebedeva et al., 2005; Hatzenpichler et al. 2008). Ammonia-oxidizing enrichment cultures active at even higher temperatures were simultaneously reported from hot springs in Yellowstone National Park (de la Torre et al., 2008). One of the Yellowstone enrichment cultures belonging to the HWCG III group of thermophilic *Crenarchaeota*, "*Candidatus* Nitrosocaldus yellowstonii" (strain HL72), has a growth optimum of 65°C and grows up to 74°C, thus extending the upper temperature limit of ammonia oxidation by ≈20°C. The first full genome sequence of a mesophilic crenarchaeon was reconstructed from the sponge-associated symbiont, "*Candidatus* Cenarchaeum symbiosum" (Hallam et al., 2006a, 2006b). However, the lack of culturability has greatly limited the physiological characterization of "*Candidatus* C. symbiosum" (Preston et al., 1996). For example, although the genome sequence indicated a capacity for ammonia oxidation, this inference has yet to be substantiated. Thus, the availability of AOA pure cultures provides the first direct associations between gene presence, gene expression, and activity. In addition to affording material for biochemical, genetic, and physiological characterization, it offers a framework for developing and testing hypotheses about AOA ecology and that of their uncultivated relatives. It would be most surprising if all members of crenarchaeal clades having ammonia-oxidizing affiliates were restricted to an autotrophic lifestyle, using ammonia as the sole source of energy (Agogué et al., 2008). For example, the now available genome sequence and preliminary experiments for SCM1 suggest some capacity for utilization of heterotrophic carbon sources (W. Martens-Habbena, personal communication). Such predictions can now be tested and related to the ecology and distribution of environmental populations.

This chapter is divided into two major sections. We begin with a brief comparative description of the physiology of AOA in relation to the better-characterized AOB. It is by no means an attempt to summarize all characteristics of the tremendous diversity of putative AOA detected in marine and terrestrial environments. Rather, it serves primarily to point out the pressing need for the isolation and detailed investigation of other novel lineages of AOA and AOB. The second section is a discussion of features that have been gleaned from the genome sequence and its relevance to environmental genomic studies. Although most currently recognized archaeal ammonia oxidizers are as yet *Candidatus*, we will generally refer to them by their suggested species names in the following sections.

PHYSIOLOGY AND ULTRASTRUCTURE

The recent recognition that ammonia oxidation occurs within widespread lineages of Group I *Crenarchaeota* has raised questions regarding their significance to the nitrogen cycle in natural environments. In natural environments, microorganisms occupy varying nutrient-related ecological niches and, accordingly, exhibit different lifestyles. The physiology and ecological role of AOB has been investigated in some detail for more than a century. Among the hallmarks of these organisms is their strong resistance to ammonium starvation, making these bacteria well adapted to survive in nutrient-depleted natural environments. However, active growth requires high concentrations of ammonium not present in most pristine natural marine and terrestrial environments (Ward, 1986; Prosser, 1989). Although, to date, only a single pure culture of an ammonia-oxidizing archaeon, *N. maritimus* strain SCM1, has been reported (Könneke et al., 2005), initial characterization of this strain has revealed striking differences to known AOB and suggest a paradigm change in our understanding of microbially catalyzed ammonia oxidation. In particular, a now-documented capacity to grow at ammonia concentrations far below that required to sustain the growth of AOB suggests that *Archaea* play a very significant role in the global nitrogen cycle (Martens-Habbena et al., 2009).

Cellular Architecture

N. maritimus is among the smallest of free-living microorganisms. Cells are regular rods 0.5 to 0.9 μm in length and 0.25 μm in width (Fig. 2). Actively growing cells each contain approximately 10 fg of protein (≈16 to 20 fg [dry weight] cell^{-1}) and a thousand ribosomes. A single copy of its 1.645 Mbp genome comprises 8 to 10% of cellular dry mass. Electron micrographs show no evidence of intracellular structures; carboxysomes, glycogen, and polyphosphate particles are absent. However, a subpopulation of actively growing cells contains a small amorphous region of greater electron density, possibly of high phosphorous content and functioning in phosphorous storage (Z. Yu and G. Jensen, personal communication). Similar to previously characterized thermophilic *Crenarchaeota*, strain SCM1 does not contain a peptidoglycan-containing cell wall structure or an outer membrane-surrounded periplasm. Instead, the cytoplasmic membrane of SCM1 is surrounded by an S-layer, made up of a dense symmetrically arrayed surface protein (Fig. 2). The cells do not contain discernable intracellular membrane structures or invaginations of the cytoplasmic membrane found in AOB. These differences between AOA and AOB cell wall architecture is of particular interest since the outer periplasmic region of AOB hosts a number of key enzymes in the ammonia oxidation pathway (e.g., hydroxylamine oxidoreductase) and thus separates toxic intermediates from the cytoplasm.

Analysis of the cytoplasmic membrane of strain SCM1 unequivocally confirmed the synthesis of glycerol dialkyl glycerol tetraether (GDGT) membrane lipids by members of the mesophilic *Crenarchaeota* (Schouten et al., 2008). The core membrane lipids of strain SCM1 consist of glycerol-linked biphytanyl hydrocarbon chains with zero to four cyclopentane rings and phosphohexose headgroups. The most abundant core GDGT membrane lipid of strain SCM1, crenarchaeol, contains four cyclopentane and one cyclohexane ring not found previously in thermophilic *Crenarchaeota*. This lipid has often been used as a diagnostic signature for *Crenarchaea* associated with temperate marine and terrestrial environments (Schouten et al., 2000, 2002; Ingalls et al., 2006) and has been suggested to be of functional significance. Due to asymmetric structures, crenarchaeol and other GDGTs with a higher degree of cyclization were initially proposed to enhance membrane fluidity and thereby to have contributed to the adaptive radiation of an ancestral thermophilic lineage into temperate and permanently cold environments (Schouten et al., 2000; Zhang et al., 2006). However, crenarchaeol has more recently been identified in the thermophilic enrichment culture *Nitrosocaldus yellowstonii* and in hot springs harboring abundant populations of Group I *Crenarchaeota* (Zhang et al., 2006; de la Torre et al., 2008; Pitcher et al., 2009). Thus, synthesis of crenarchaeol per se is not diagnostic of organisms restricted to lower-temperature habitats.

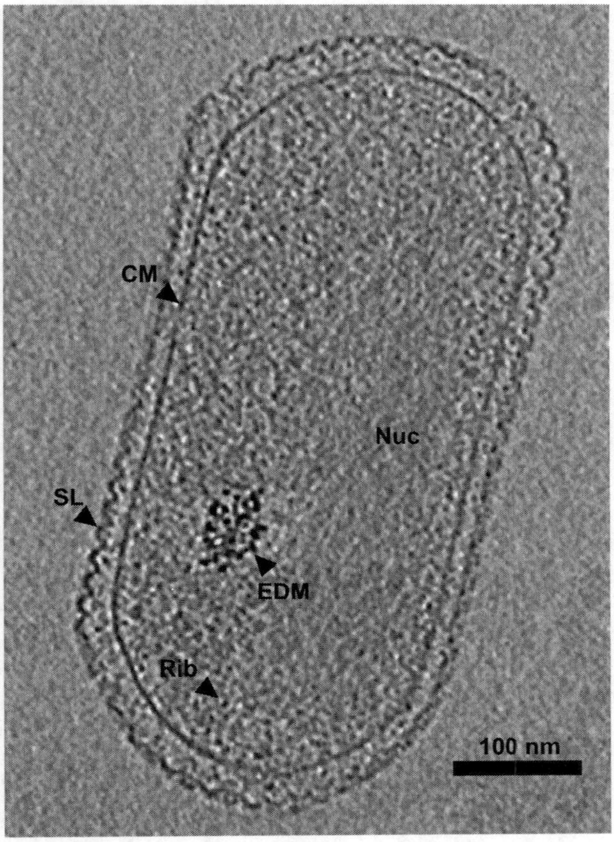

FIGURE 2 Cryo-electron tomographic section of a *N. maritimus* cell. CM, cytoplasmic membrane; SL, S layer; Rib, ribosome; Nuc, nucleoid; EDM, electron-dense matter. (Picture courtesy of Ziheng Yu and Grant Jensen.)

The SCM1 cell volume of approximately 0.023 μm^3 is similar to that of the oligotrophic marine microorganism *Pelagibacter ubique* (Rappé et al., 2002) and between 10- and >100-fold smaller than that of known AOB (Martens-Habbena et al., 2009). Even cells growing in batch culture mark the low end of sizes found in natural seawater (range of 0.026 to 0.4 μm^3) (Lee and Fuhrman, 1987; Simon and Azam, 1989) and may approach the lower limit for free-living organisms (Button, 2000). Reduction of cell size and its associated cytological features have been considered important evolutionary adaptations of oligotrophic microorganisms to life in nutrient-depleted environments (Harder and Dijkhuizen, 1983; Roszak and Colwell, 1987; Button, 2000). Thus, the finding of comparable cell and genome sizes of SCM1 and strains belonging to abundant clades of heterotrophic marine bacteria, like *P. ubique*, may suggest similar adaptive responses to life in nutrient-depleted marine environments, marked by metabolic specialization and a high ratio of surface to volume.

Growth and Activity

Consistent with the hypothesis of narrow catabolic specificity of *Archaea*, the only identified energy metabolism of strain SCM1 thus far is the oxidation of ammonium to nitrite. Growth of SCM1 has not been observed with methane or other organic or inorganic electron donors, suggesting that this strain relies obligately on ammonia oxidation as the source of metabolic energy. SCM1 reaches growth rates comparable to AOB of 0.027 h^{-1} (T_d ~ 26 h) at 30°C. However, in contrast to most AOB, SCM1 grows only at a narrow temperature range between 20 and 30°C at stable pH values between 7.0 and approximately 7.8. No growth occurs below pH 7.0, and ammonia oxidation activity ceases below pH 6.7. In further contrast to many AOB strains, growth is inhibited by ammonium and nitrite concentrations as low as 2 to 3 mM, thus restricting the maximum cell densities in batch cultures to ~5 × 10^7 cells ml^{-1} (equivalent to ~0.5 mg of protein $liter^{-1}$). A similar low tolerance of ammonium was reported for *N. gargensis*, indicating that low ammonium tolerance could be common among AOA. Growth of SCM1 is also very sensitive to perturbations. Even slight temporal changes of temperature or slow shaking increase the doubling time. Optimal growth has so far been achieved only in static batch cultures. Nevertheless, during exponential growth, SCM1 attains ammonia oxidation activities of up to 52 μmol of ammonium mg of $protein^{-1}$ h^{-1}, which is comparable to AOB strains (range of 30 to 80 μmol of ammonium mg of $protein^{-1}$ h^{-1}) (Prosser, 1989; Ward, 1987). However, due to the vastly different cell sizes, the maximum per cell rate of ammonia oxidation of SCM1 thus far observed (0.53 fmol $cell^{-1}$ h^{-1}) is more than 10-fold lower than those of AOB.

In striking contrast to known AOB strains, growth of SCM1 in batch culture continues until ammonium is depleted below 10 nM (Martens-Habbena et al., 2009). This is more than 100-fold lower than the minimum concentration required for growth of cultivated AOB strains (Keen and Prosser, 1987; Prosser, 1989; Bollmann et al., 2002). This substrate threshold, that is the minimum substrate concentration permitting sufficient metabolic activity to meet maintenance energy requirements, is generally far below half-saturation constants and thus often neglected when describing kinetic characteristics. The ammonium concentrations in nutrient-depleted open oceans and unfertilized natural soils are well below the substrate threshold of AOB with known kinetic characteristics. Thus, characterized AOB undergo severe starvation in nutrient-depleted natural marine and terrestrial environments (Jones and Morita, 1985; Prosser, 1989; Bollmann et al., 2002). In contrast, SCM1 will continue to grow at ammonia levels prevailing in nutrient-limited open oceans (Martens-Habbena et al., 2009). It has also been suggested that *Archaea* have a lower maintenance energy than *Bacteria*, which has been attributed to catabolic specialization and a less ion-permeable cytoplasmic membrane (van de Vossenberg et al., 1998; Valentine, 2007). Low maintenance energy would be consistent with the adaptation of *N. maritimus* and related marine *Archaea* to chronically nutrient-limited environments. As discussed later, adaptations to low nutrient are also observed in the genome inventory in *N. maritimus*. The distinct differences between *N. maritimus*-like AOA and AOB, some of which are considered extremely starvation tolerant, suggest that AOA and AOB have evolved distinctly different strategies to cope with nutrient source deprivation.

Ammonia Oxidation in SCM1: Stoichiometry and Kinetics

More detailed insight into the kinetic characteristics and stoichiometry of ammonia oxidation in strain SCM1 was obtained from microrespirometry and ammonium uptake experiments with cells from the late exponential or early stationary growth phase (Fig. 3). Early-stationary-phase cells had an oxygen uptake rate of 0.7 μmol of O_2 mg of $protein^{-1}$ h^{-1}. Oxygen uptake increased more than 50-fold to up to 36 μmol of O_2 mg of $protein^{-1}$ h^{-1} after addition of ammonium to cells. Similar to AOB, ammonium and oxygen were consumed in a 1:1.5 ratio. The maximum oxygen uptake rates were observed at as low as 2 μM ammonium, and the apparent half-saturation constant (K_m) for ammonium was 0.132 μM total ammonium (~3 nM NH_3). In contrast to its high-ammonium affinity and the ability to grow at very low ammonium concentrations, SCM1 has a rather low affinity to

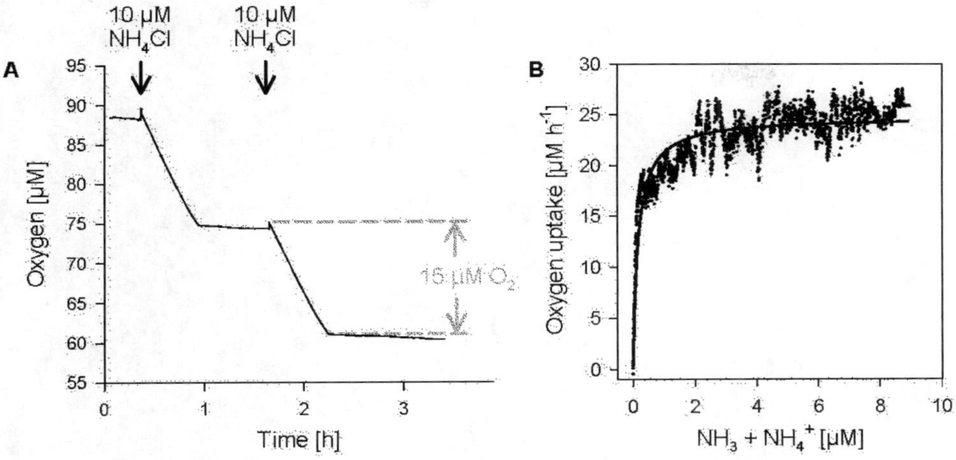

FIGURE 3 Stoichiometry and kinetics of ammonia oxidation by *N. maritimus*. (A) Trace of oxygen uptake by aliquots of early-stationary phase cells (cell density, ~5.0×10^7 cells ml^{-1}, 1 mM nitrite) obtained by microrespirometry. Ammonium added to resting cells was oxidized without significant lag time with a ratio of 1 mol of ammonium to 1.5 mol of O_2. (B) Michaelis-Menten kinetics calculated from oxygen uptake rates (second part) in panel A.

oxygen. Although the apparent K_m for oxygen determined by respirometry was ~4 µM and in a range typical for aerobic microorganisms, SCM1 did not grow under low oxygen tension or completely anoxic conditions. Thus, either other lineages of AOA or low-oxygen-adapted AOB would be more competitive under oxygen-limiting conditions (Laanbroek and Gerards, 1993; Laanbroek et al., 1994). Alternatively, long periods of acclimatization are required to permit growth of SCM1 under low-oxygen conditions.

Multiple lines of evidence now support the view that AOA have a significant role as oligotrophic ammonia oxidizers in various natural environments. However, this has most often been inferred from diagnostic gene or transcript abundance rather than direct comparison of nitrification activities of the two lineages. The kinetic characteristics of strain SCM1 now provide more direct evidence of competitive advantage of AOA under low-nutrient conditions. All investigated AOB have more than a 200-fold higher apparent K_m value than does strain SCM1 (Table 1). In fact, more than 50% of maximum activity of SCM1 could be elicited by single additions of 200 nM ammonium to resting cells. Due to its extremely low apparent K_m and comparable maximum activities, the specific affinity of SCM1 ($V_{max} \times K_m^{-1}$ = 68,700 liters g [wet weight]$^{-1}$ h^{-1}) surpasses those of all characterized AOB by more than 200-fold and is among the highest substrate affinities reported for any microorganism (Button, 1998; Martens-Habbena et al., 2009).

Remarkably, the specific affinity of strain SCM1 for ammonium surpasses the ammonium affinity of even the most oligotrophic ammonium-assimilating organisms, more than 30-fold greater than oligotrophic heterotrophic bacteria and diatoms characterized to date. This margin is sufficient to sustain the ammonia demand of nitrifying *Crenarchaea* in direct competition with heterotrophs and phototrophs for ammonia. In turn, this implies that AOA may be contributing much more substantially to nitrogen transformations than previously anticipated in both marine and terrestrial habitats. For example, Leininger et al. (2006) reported that AOA predominate over AOB in natural unfertilized soils. In con-

TABLE 1 Comparison of kinetic characteristics of ammonia oxidation by *N. maritimus*, AOB, enrichment cultures, and in situ kinetics of nitrification in natural samples, as well as ammonia assimilation by phytoplankton and heterotrophic microorganisms

Parameter	Species/sample type	Strain/ comments	Water		pH	K_m (μM)		Maximum specific affinity, a^0 (liters g of cells^{-1} h^{-1})	Reference(s)
			Type[a]	Temp		Growth	Activity		
Ammonia-oxidizing strains	*Nitrosomonas eutropha*	GH22	FW	25	7.4	890			Suwa et al., 1994
		FH11	FW	25	7.4	3,970			Suwa et al., 1994
		Nm53	FW	30	7.8		750		Stehr et al., 1995
	Nitrosomonas europaea	n.g.	FW	30	7.0	51		278	Belser and Schmidt, 1980; Keen and Prosser, 1989
		Nm 89	FW	30	7.8		420		Stehr et al., 1995
		n.g.	FW	25	7.5		1,200		Suzuki, 1974
	Nitrosomonas communis	Nm58	FW	30	7.8		3,300		Stehr et al., 1995
		Nm85	FW	30	7.8		1,100		Stehr et al., 1995
	Nitrosococcus oceani	ATCC 19707	SW	23	7.5		245	61	Watson, 1965; Ward, 1987
	Nitrosospira briensis (cluster 3)	ATCC 25971 planktonic	FW	25	7.5		159		Bollmann et al., 2005
		Wall growth	FW	25	7.5		98.8		Bollmann et al., 2005
	Nitrosospira cluster 2	B6	FW	22	7.8	275			Jiang and Bakken, 1999
	Nitrosospira cluster 0	40Kl	FW	22	7.8	80			Jiang and Bakken, 1999
	Nitrosospira cluster 3	L115	FW	22	7.8	310			Jiang and Bakken, 1999
	Nitrosospira cluster 3	AF	FW	22	7.8	208			Jiang and Bakken, 1999
	Nitrosomonas oligotropha	AL211	FW	25	7.4	34		315	Suwa et al., 1994
		FL28	FW	25	7.4	80			Suwa et al., 1994
		Nm84	FW	30	7.8		30		Stehr et al., 1995
		Nm86	FW	30	7.8		40		Stehr et al., 1995

		Nm49	FW	30	7.8		75		Stehr et al., 1995
	N. maritimus	SCM1	SW	30	7.4		0.132	68,700	Martens-Habbena et al., 2009
Ammonia-oxidizing enrichments	Soil	Open grassland	FW	23	6.2		819		Stark and Firestone, 1996
	Soil	Canopy covered	FW	23	6.2		46		Stark and Firestone, 1996
	Soil	Mixed conifer forest	FW	23	6.2		27		Stark and Firestone, 1996
In situ nitrification	Soil	Open grassland	FW	23	6.2		40		Stark and Firestone, 1996
	Soil	Canopy covered	FW	23	6.2		1.5		Stark and Firestone, 1996
	Ocean water	Off California coast	SW	15	8.1	<0.1			Olson, 1981
	Ocean water	Upper Cariaco Basin	SW	15	8.1	0			Hashimoto et al., 1983
Ammonia assimilation	*Pseudo-nitzschia delicatissima*	ICMB-F2B2	SW	20	8.1		0.38		Loureiro et al., 2009
	Emiliana huxleyi	BT-6	SW	18	8.1		0.1		Eppley et al., 1969
	Thalassiosira pseudonana	13-1	SW	20	8.1		0.02	1,929	Eppley and Renger, 1974
	Vibrio logei	NCIMB 1143	SW	6	7.2	7		20	Reay et al., 1999
	Hydrogenophaga pseudoflava	NCIMB 13125	FW	5	7.2	4			Reay et al., 1999
	E. coli	NCIMB 09001	FW	15	7.2	9			Reay et al., 1999
	E. coli	NCIMB 09001	FW	35	7.2	74			Reay et al., 1999

[a]FW, fresh water; SW, seawater.

trast, AOB are presumably the more active population of ammonia oxidizers in fertilized soils (Leininger et al., 2006; Jia and Conrad, 2009). Whereas the contribution of AOA to natural greenhouse gas emissions remains to be resolved, the kinetic data of *N. maritimus* strongly suggest that nitrification might have a more profound role in the global nitrogen cycle than previously anticipated.

GENOME ANALYSIS OF AOA

Genome Trends of *N. maritimus*

The genome of *N. maritimus* was approved for sequencing in 2006 by the DOE Microbial Genomics Program, and the complete genome analysis was published in 2010 (Walker et al., 2010). *N. maritimus* SCM1 contains a single chromosome of 1,645,259 bp encoding 1,842 predicted open reading frames (ORFs). No plasmid was found. As often observed for other archaeal genomes, no origin of replication is discernable based on gene content or GC skew (Fig. 4). Its size and GC content (34.2%) are in the lower range of characterized archaeal genomes (Table 2 and Fig. 5), and very distinct from those of the closely related sponge symbiont *Cenarchaeum symbiosum* (ca. 97% 16S rRNA sequence identity) (Hallam et al., 2006a). The symbiont has a larger genome (2.045 Mbp), of average size among characterized *Archaea*, that is of significantly higher average G+C content (57.4%), although it retains a similar GC content of genes coding for the 16S and 23S ribosomal RNAs with *N. maritimus* (50 to 52%). The two genomes share little conservation of synteny, and much of the divergence in gene content is associated with discrete regions (genomic islands), the latter accounting for most of the increased size of the *C. symbiosum* genome (Fig. 6). The smaller genome of *N. maritimus* is associated with a slightly increased protein coding ORF density (1.19 ORF/kb) relative to *C. symbiosum* (0.986 ORF/kb). These distinctive differences in genome architecture presumably reflect changes associated with the shift from a free living to a symbiotic lifestyle.

The identified ORFs code for 1,797 proteins and a full complement of genes for essential RNAs, including one copy each of 5S/16S/23S ribosomal RNAs, RNase P, signal recognition particle RNA, and 44 transfer RNAs (Fig. 4). In addition, the genome contains six candidate genes for C/D box small RNAs. Most or all C/D box small RNAs guide the precise positioning of posttranscriptional 2′-O-methyl group addition to rRNAs or tRNAs, a process also occurring in eukaryotes (but not *Bacteria*). As is typical for *Archaea*, the gene coding for the 5S rRNA is not closely linked to those for the 16S and 23S rRNAs. All other sequenced *Crenarchaeota*, including *C. symbiosum*, contain at least 45 tRNAs (Table 2). The apparent absence of Pro_{CGG} and Arg_{CCG} in *N. maritimus* may be related to the low G+C content of its genome and a preference for codons ending in A/T. One exceptionally large gene (Nmar_1073; 28.8 kbp) of higher G+C content (40.8%), annotated as fibronectin type III domain protein, is conspicuous in the chromosome map (Fig. 4). It is a member of a class of genes coding for secreted cell-surface proteins having a characteristic signature of amino acid usage that has so far been identified in 47 taxa of *Bacteria* and *Euryarchaeota* (Reva and Tümmler, 2008). They are acidic and hydrophilic and lack cysteine. The *N. maritimus* giant gene shares all of these signature features (Fig. 7) and appears to be the first documented instance of occurrence in the *Crenarchaeota*. Our preliminary microarray gene expression analysis showed that this gene was highly expressed under the normal culture condition (D. J. Arp, personal communication).

Approximately 60% of protein coding genes in SCM1 have predicted functions (Table 2). This is slightly lower than for AOB and thermophilic *Crenarchaeota* (64 to 75%), reflecting limited knowledge about a major clade of *Archaea* that has only recently been brought into culture. *N. maritimus* has a lower density of most Clusters of Orthologous Groups of proteins (COGs) when compared with AOB genomes (Table 2) but is relatively enriched in genes for energy conversion (C), coenzyme

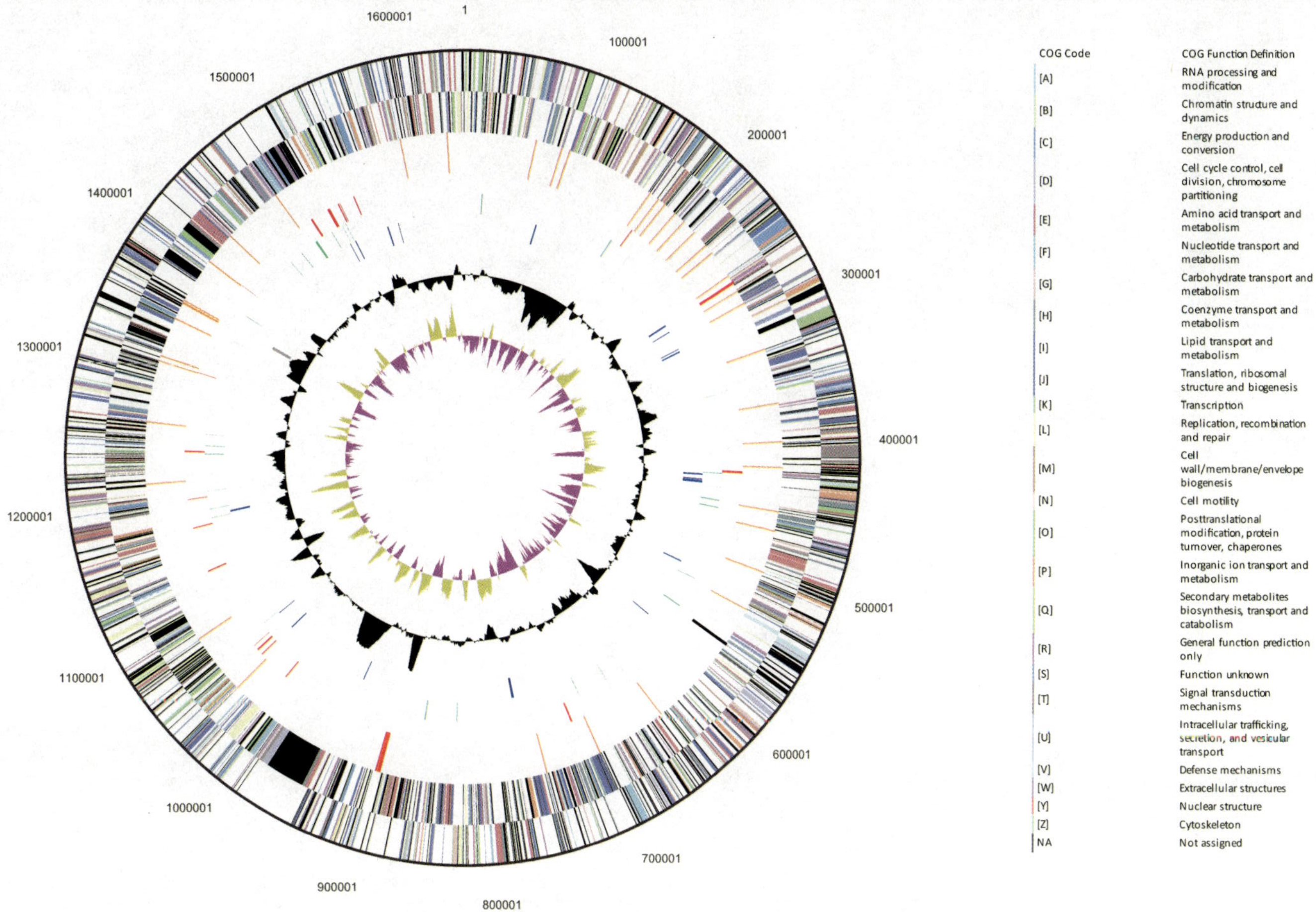

FIGURE 4 Diagram of the *N. maritimus* circular chromosome. Rings, from outside to the center: 1, genes of forward strand (color by COG categories); 2, genes on reverse strand (color by COG categories); 3, RNA genes (tRNAs orange, rRNAs red, other RNAs black); 4, copper-containing protein genes (red); 5, genes involved in transcription and regulation (green); 6, genes annotated as transporters (blue); 7, putative ammonia-monooxygenase genes (silver); 8, G+C content; 9, GC skew.

TABLE 2 Genome features of *N. maritimus* and related *Crenarchaea* and AOB[a]

Parameter	*N. maritimus* SCM1	*C. symbiosum* A	*Sulfolobus acidocaldarius* DSM 639	*Pyrobaculum aerophilum* IM2	*N. oceani* ATCC 19707	*N. europaea* ATCC 19718	*N. eutropha* C71	*N. multiformis* ATCC 25196
DNA, total no. of base pairs	1,645,259	2,045,085	2,225,959	2,222,430	3,522,111	2,812,094	2,781,824	3,234,309
DNA coding no. of base pairs	1,497,096	1,877,407	1,967,324	1,995,981	3,062,653	2,502,800	2,438,266	2,774,849
G+C content	34.2%	57.4%	36.7%	51.4%	50.3%	50.7%	48.5%	53.9%
Gene number total	1,997	2,066	2,344	2,628	3,190	2,631	2,695	2,885
No. of protein coding genes	1,797	2,017	2,285	2,575	3,132	2,572	2,639	2,827
Pseudogenes	6	0	59	91	115	111	89	22
RNA genes	49	49	59	53	58	59	56	58
rRNA genes	4	3	3	3	6	4	3	3
tRNA genes	44	45	49	50	45	41	41	43
Other RNA genes	1	1	7	0	7	14	12	12
Protein coding genes with function prediction	1,083	1,008	1,975	1,344	2,021	1,789	1,972	2,026
Protein coding genes without function prediction	714	1,009	310	1,231	1,111	783	667	801
Protein coding genes connected to KEGG pathways	476	437	636	578	919	795	833	862
Protein coding genes not connected to KEGG pathways	1,321	1,580	1,649	1,997	2,213	1,777	1,806	1,965
Protein coding genes with COGs	1,131	1,008	1,563	1,503	2,290	1,995	1,952	2,102
No. of plasmids	0	0	0	0	1	0	2	3

[a]Most data are adapted from JGI release (version 2.8, April 2009).

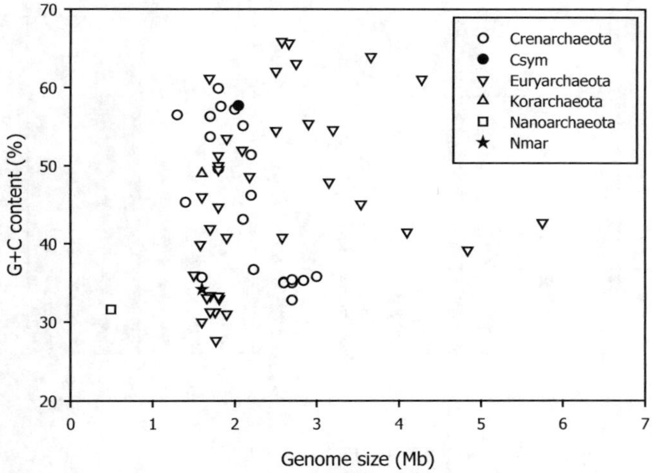

FIGURE 5 Archaeal genome size and G+C content. Csym, *C. symbiosum*; Nmar, *N. maritimus*.

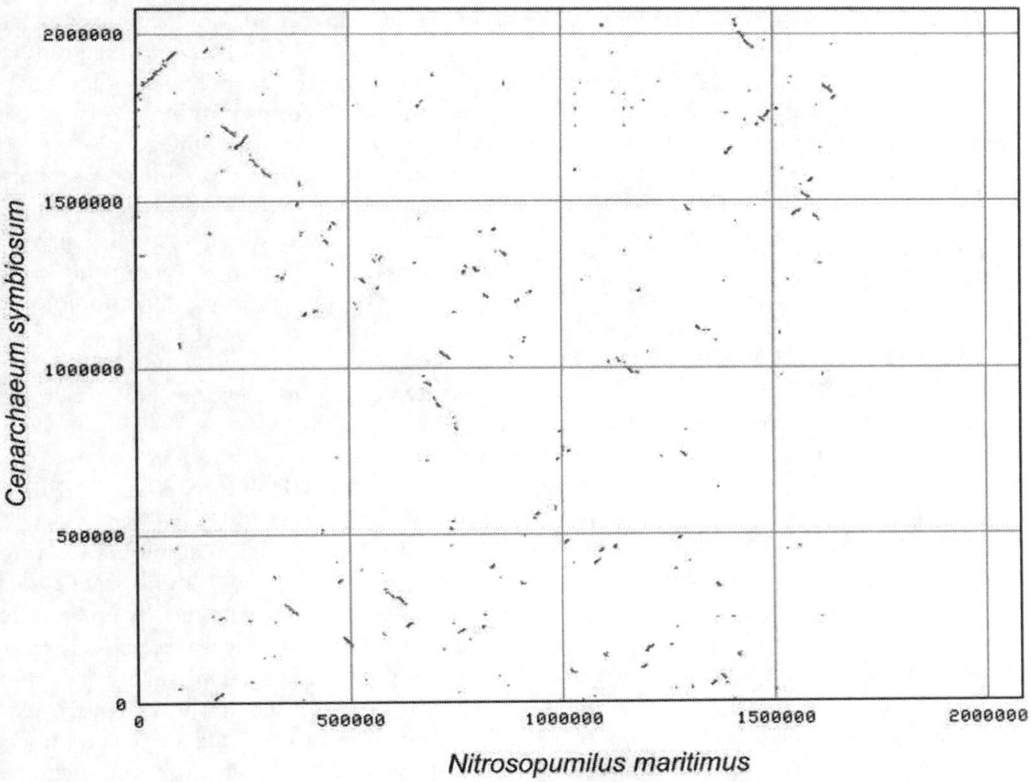

FIGURE 6 Synteny plot comparing *N. maritimus* and *C. symbiosum* genomes. The comparison was done in the protein level with the program Promer (Kurtz et al., 2004). The DNA sequences were translated in all six reading frames and compared.

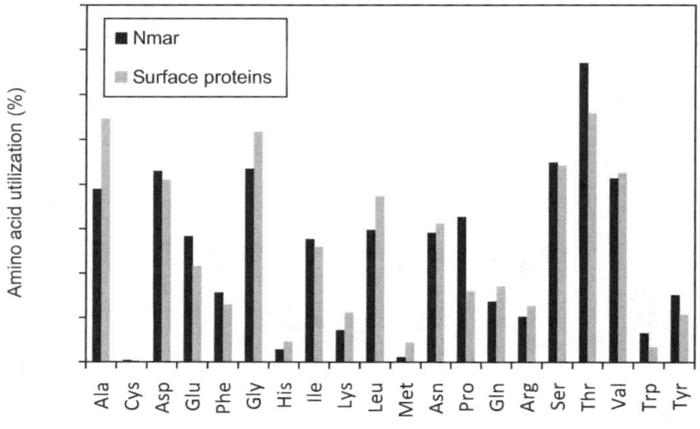

FIGURE 7 Amino acid utilization of the largest gene in *N. maritimus* and surface proteins. The data on surface proteins ($n = 38$) are adapted from Reva and Tümmler (2008).

transport/metabolism (H), translation genes (J), and transcription (K) (Fig. 8). Although relative enrichment of core functions (e.g., energy conversion, translation) is not unexpected for a small genome organism, the greater density of transport/metabolism genes relative to other *Crenarchaea* genomes suggests significant metabolic versatility of *N. maritimus*.

Mechanisms for Ammonia Oxidation and Electron Transfer

DO *ARCHAEA* AND *BACTERIA* SHARE A COMMON BIOCHEMISTRY OF AMMONIA OXIDATION?

Previously isolated and characterized betaproteobacterial and gammaproteobacterial AOB share a common energy metabolic pathway consisting of an ammonia monooxygenase, a hydroxylamine oxidoreductase, and a typical bacterial electron transport system composed of ubiquinones, menaquinones, and *b*- and *c*-type cytochromes (see Chapters 2 and 4). Although the stoichiometry of ammonia oxidation to nitrite for *N. maritimus* and other characterized AOB is identical, the contributing biochemical pathways appear to be different. Other than coding for an evolutionarily divergent ammonia monooxygenase (AMO), it has no central cytochrome network or a homolog of hydroxylamine oxidoreductase. Numerous genes encoding for small blue copper proteins (similar to plastocyanins and sulfocyanins) indicate that *N. maritimus* utilizes primarily copper-dependent mechanisms of electron transfer instead of the iron-based bacterial system. Genes encoding complexes I, III, and IV strongly suggest a similar mechanism for regenerating NADH by reverse electron transfer, although these enzymes presumably interact with the plastocyanin-like redox proteins. The presence of genes homologous with the bacterial copper-containing nitrite reductase indicates *N. maritimus* also has some capacity to use an electron sink other than oxygen.

AMMONIA OXIDATION AND ENERGY TRANSFORMATION

The first step in bacterial ammonia oxidation relies upon a membrane-bound AMO composed of three major subunits (AmoA, AmoB, and AmoC) (Fig. 9). Since *N. maritimus* appears to lack a well-defined periplasmic space typical of AOB (Fig. 2), if the active site of the AMO is oriented away from the cytoplasm it may require a mechanism to retain metabolic intermediates (such as hydroxylamine) or novel intermediate(s) as suggested in an alternative model (discussed below). If the active site faces

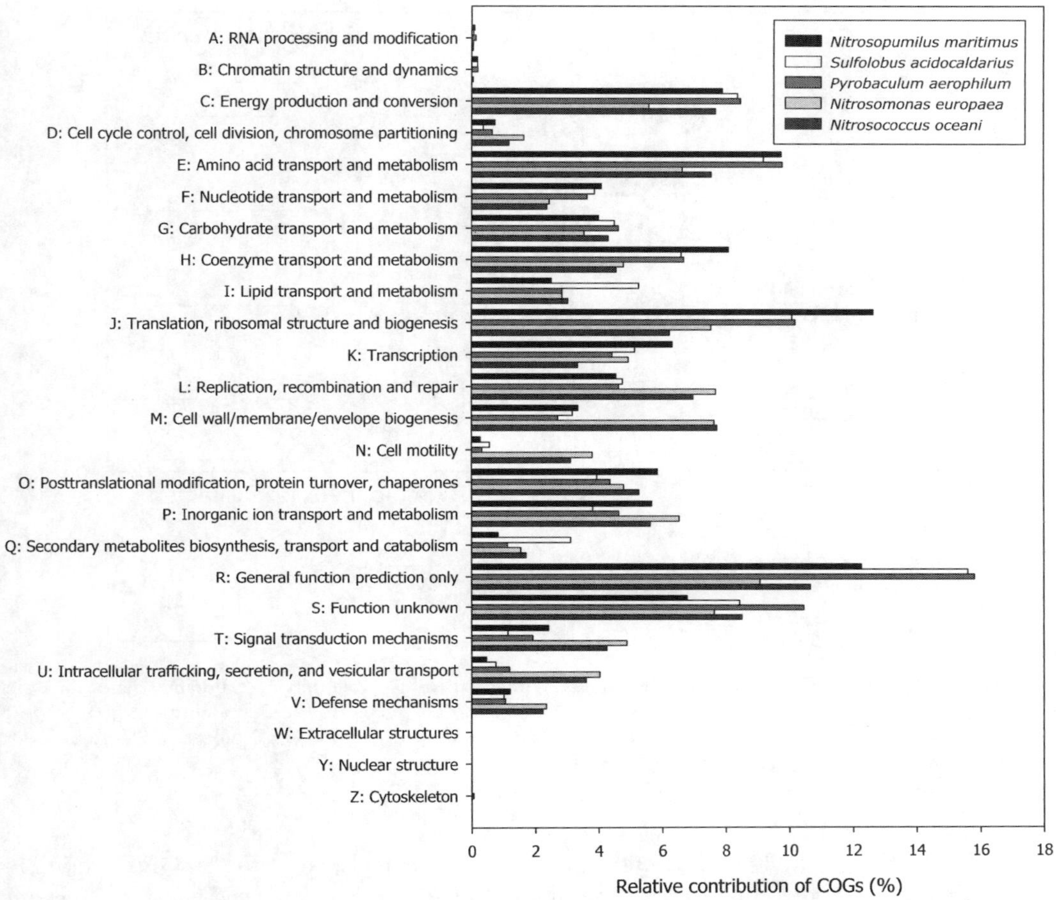

FIGURE 8 Relative contribution in COGs among representative *Crenarchaea* and AOB.

the cytoplasm, the reduced nitrogen substrate (ammonia or ammonium) could derive either from ammonia diffusing freely across the membrane or from transport by two high-affinity ammonium transporters (Nmar_0588 and 1698). However, a role for these transporters in ammonia assimilation versus ammonia oxidation has not yet been resolved.

The genes for the bacterial AMO all share an *amoCAB* organization, generally present as one to three nearly identical copies of operons (Chain et al., 2003; Klotz et al., 2006; Arp et al., 2007). In the case of *Crenarchaeota*, both *N. maritimus* and *C. symbiosum* have a single copy of each gene, but in contrast to bacterial gene organization, *N. maritimus* encodes one of three alternative types of *amo* gene organization so far observed among different lineages of the *amoA*-containing *Crenarchaeota* (Hallam et al., 2006b). The *amoBCA*-like gene organization, with *amoB* in an opposite orientation, found in *N. maritimus* is present in *C. symbiosum* and other marine *Crenarchaeota* (Fig. 9) (Hallam et al., 2006b). The *amoBCA* organization of *N. maritimus* (in contrast to the *amoCAB* order in bacteria) is common among related marine assemblages identified in metagenome libraries or recovered using PCR primers that span the operon but differs from the crenarchaeal soil fosmid (Treusch et al., 2005) and two recently described thermophilic AOA (*N. yellowstonii* and *N. gar-*

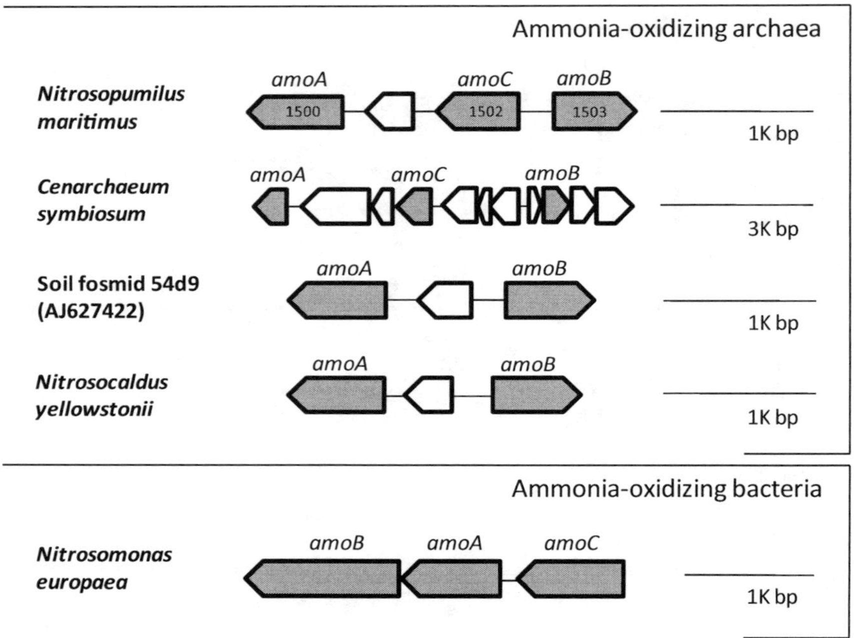

FIGURE 9 Organization of the *amo* gene clusters in archaeal and bacterial nitrifiers. Putative gene names and the *N. maritimus* gene locus numbers are shown within each ORF (gray arrows). The identification and size of proteins encoded in the *N. maritimus* locus are *amoA* (216 amino acids), hypothetical protein (120 amino acids), *amoB* (189 amino acids), and *amoC* (190 amino acids).

gensis) (de la Torre et al., 2008, Hatzenpichler et al., 2008). These representatives of the soil and thermophile clades do not retain a close linkage of all three genes. The *amoA* and *amoB* genes remain associated, but the *amoC* appears to be located elsewhere on the chromosome (Fig. 9). *N. maritimus* also encodes a hypothetical protein of unknown function between the *amoC* and *amoA* genes. So far, this hypothetical gene is present in all *amoBCA* clusters in the genomes of free-living AOA and shares sequence similarity with a segment of a larger gene in *C. symbiosum*. Moreover, *C. symbiosum* contains four additional ORFs located between the *amoB* and *amoC* genes. Thus, in addition to significant differences in overall gene content, *C. symbiosum* also encodes a structurally distinct variant of the putative archaeal AMO gene, suggestive of a different physiological function in this closely related psychrophilic symbiont.

Surprisingly, the three ORFs homologous to the bacterial *amo* genes are the only genomic inventory known to be relevant to the bacterial pathway for ammonia oxidation. Although the bacterial and archaeal AMOs and the bacterial particulate methane monooxygenase all share common ancestry, bacterial AMOs showed higher similarity to bacterial particulate methane monooxygenase than the archaeal AMOs, suggesting significant structural differences between the archaeal and bacterial AMOs (Klotz and Stein, 2008). Structural comparisons show that *N. maritimus* and other AOA lack the conserved cupredoxin fold in the C terminus of the bacterial AmoB subunit thought to be catalytically important (Walker et al., 2010). The archaeal deviation from the highly conserved bacterial AMO structure and hydroxylamine-ubiquinone redox module, composed of HAO and two c-type cytochromes (Klotz and Stein, 2008), is suggestive of a novel biochemistry that may not involve hydroxylamine as an intermediate.

PROPOSED ARCHAEAL AMMONIA-OXIDATION PATHWAY

In consideration of the now-recognized environmental significance of the AOA and a distinctive biochemistry suggested by the genome sequence of SCM1, we feel that some speculation about an alternative biochemistry is justified (Fig. 10). There are two general mechanistic alternatives: either a novel biochemistry exists for the oxidation of hydroxylamine or, alternatively, hydroxylamine is not a product of the divergent AMO. If hydroxylamine is an intermediate, its oxidation might be mediated by one of the periplasmic multicopper oxidases (MCOs) predicted from the genome sequence. Given the lack of cytochrome *c* proteins, the four electrons would then be transferred to a quinone reductase via small blue copper-containing plastocyanin-like electron carriers. Alternatively, hydroxylamine may not be an intermediate in the AOA pathway. Thus, we also consider that possibility that nitroxyl (HNO) is the product of the archaeal AMO. Nitroxyl, also known as nitrosyl hydride, is a highly reactive compound recently recognized as having biological significance in a number of biological systems (Miranda et al. 2003a, 2003b; Fukuto et al., 2005a; 2005b; 2005c). Nitroxyl could be formed by a novel monooxygenase function of archaeal AMO. Alternatively, the archaeal AMO may act as a dioxygenase and insert two oxygen atoms into ammonia (Gibson et al., 1995), producing nitroxyl from the spontaneous decay of HNOHOH. A physiological advantage of using nitroxyl as an intermediate would be obviation of the requirement for investing electrons in the initial monooxygenase reaction. For nitrifiers inhabiting nutrient poor environments, eliminating the requirement for a reductant to initiate respiration would presumably offer significant ecological advantage as well as contribute to the recently reported low K_m and high substrate affinity of the organism for ammonia (above, and Martens-Habbena et al., 2009). As hypothesized for further oxidation of an alternative hydroxylamine intermediate, one of the multicopper oxidase-like proteins could function as a nitroxyl oxidoreductase to oxidize nitroxyl to nitrite with associated extraction of two protons and two electrons in the presence of water (Fig. 10). The proposed NXOR would relay the two extracted electrons into the quinone pool as described above.

Both the AOA and AOB systems result in a net production of two electrons transferred into the quinone pool and useable for the generation of a proton-motive force (ATP) and reductant (NADH) through either complexes III (Nmar_1542-4) and IV (Nmar_0182-5) structures. The production of reductant (NADH) would require complex I (NuoABCDHIJKMLN, Nmar_0276-86) to run in reverse, driven by a proton motive force (Fig. 10). As with all other sequenced crenarchaeal genomes, genes coding for the complex I enzyme lack three subunits responsible for NADH-binding and oxidation, suggesting interaction with alternative electron carriers such as ferredoxin. A functional complex III similar to that found in other *Crenarchaeota* was reconstructed in *N. maritimus* from three genes encoding a transmembrane cytochrome *b* subunit (Nmar_1543), a Rieske-type Fe-S cluster subunit (Nmar_1544) as the core and a variable third plastocyanin-like subunit (Nmar_1542) substituting for a usual heme-containing protein such as cytochrome *c* or *f*. This alternate complex III from *N. maritimus* is the first known example of a copper protein (Nmar_1542) being utilized as the third subunit. Aerobic respiration conferred by genes coding for a terminal heme-copper oxidase (Nmar_0182-5), also containing a plastocyanin-like subunit (rather than a heme containing cytochrome *c*), would provide additional mechanism for proton-motive force generation.

AMMONIA TRANSPORTERS

The genome of *N. maritimus* has two genes coding for Amt proteins (ammonia transporters in the Amt/Mep/Rh family of integral membrane proteins), recognized to serve

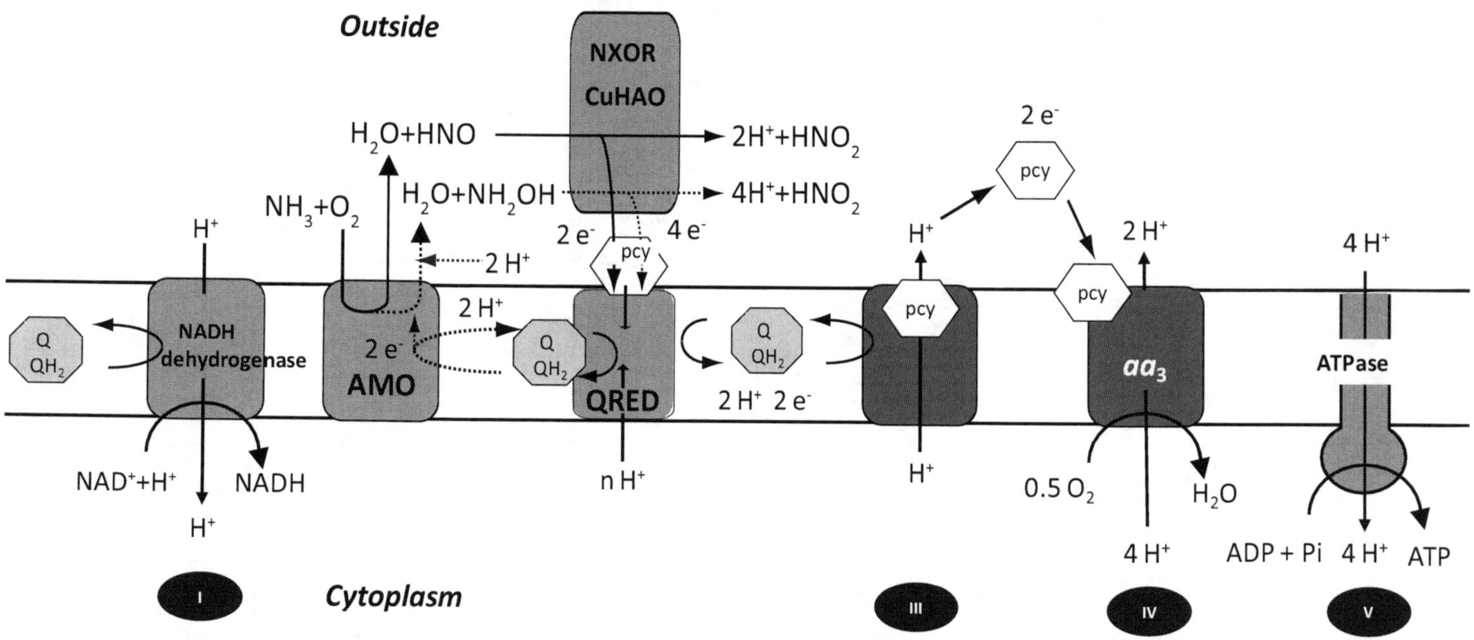

FIGURE 10 Proposed archaeal ammonia-oxidation pathway. NXOR, putative nitroxyl oxidoreductase; CuHAO, copper hydroxylamine oxidoreductase. The dashed line indicates the pathway having hydroxylamine as intermediate. Octagons containing Q and QH_2 represent the oxidized and reduced quinone pool, respectively. Each complex is numbered: complex I, NADH:ubiquinone oxidoreductase; complex III, cytochrome *c*-ubiquinone oxidoreductase; complex IV, terminal oxidase; complex V, ATP synthase. Plastocyanin-like electron carriers are shown as hexagons containing pcy.

for ammonia uptake in all three domains of life. The *N. maritimus* genome also contains four genes coding for nitrogen regulatory PII proteins (GlnK) (Nmar_0586, 0587, 1317 and 1523), two of which flank one of the *amt* genes (Nmar_0588). At a high cellular ratio of ADP/ATP, this regulatory protein associates with Amt blocking ammonia uptake. When bound to its effectors (ATP and 2-oxoglutarate), signaling cellular energy sufficiency and a nitrogen requirement, a conformational change of PII results in its release from the Amt (Khademi et al., 2004). However, as earlier discussed, it is unclear if the archaeal Amt protein mediate uptake of ammonium solely for biosynthetic needs or for both biosynthesis and ammonia oxidation. A study conducted by Schmidt and colleagues (2004) suggested that bacterial ammonia oxidizers have the ability to accumulate significant concentrations of ammonia, implicating their ammonia transporters in both biosynthetic and energy generating pathways. Ammonium assimilation by SCM1 almost certainly uses the Amt transporter (Andrade and Einsle, 2007; Tremblay et al., 2009). However, ammonia oxidation may not be transporter-dependent. The extensive theoretical and experimental studies of oligotrophic bacteria demonstrated that cellular specific affinity for a nutrient substrate (a direct measure of an organism's nutrient collection ability) increases with the number of collection sites on the cell (e.g., transporters [Button, 1994]). Specific affinity is a function of bimolecular collision frequency (encounter between substrate and transporter), transporter size, and transporter density on the cell surface. Since collision frequency is very low in dilute environments, the density of transporters is an important rate-limiting factor. Downstream enzymes in a metabolic pathway can be present at much lower concentrations without being saturated by the product of the initial substrate collection reaction (Button, 1994). The low K_m and high substrate affinity of *N. maritimus* SCM1 could result from a high density of Amt transporters (Martens-Habbena et al., 2009). Alternatively, collection and oxidation of ammonia may be inseparable processes and mediated directly through a high affinity AMO, requiring a high density of this enzyme. Both models would equally account for observations that growth of *N. maritimus* and *N. gargensis* is inhibited by ammonia concentrations greater than 2 to 3 mM (Hatzenpichler et al., 2008; Martens-Habbena et al., 2009). Saturation of the next step in the pathway could result in the buildup of potentially reactive intermediates that inhibit growth.

Evidence for a High Demand for Copper

We suggest that an understanding of copper homeostasis is fundamental to an understanding of ecological success of the AOA. Both copper and iron are thought to be environmentally limiting co-factors for AOB (Loveless and Painter, 1968; Bédard and Knowles, 1989; Ensign et al., 1993) and we have found that provision of chelated trace elements significantly improves cultivation success of AOA (Martens-Habbena, personal communication). In the open ocean, both are present in low concentrations and primarily bound to organic matter. However, cooper can be present at concentrations up to two orders of magnitude higher than iron (Coale and Bruland, 1988). Preferential use of copper for redox chemistry could reduce direct competition with other marine microorganisms for iron. As a related example, it has been shown that diatoms adapted to low iron in the open ocean use plastocyanin instead of the functionally equivalent iron-containing homolog, cytochrome c_6 (Peers and Price, 2006). Future transcriptional analysis of SCM1 response to varying copper and iron availability will hopefully serve to identify systems for copper-specific transport, homeostasis, and stress response.

COPPER-BASED ELECTRON TRANSFER SYSTEMS

In addition to the multiple plastocyanin-like proteins of the respiratory complex, the *N. maritimus* genome codes for a variety of additional copper-containing proteins (Fig. 4). In addition to a copper-containing AMO, it also codes for more than eight multicopper oxidases, two annotated as copper-containing nitrite reductase gene (*nirK*) (Nmar_1259 and 1667) (see discussion of copper-containing nitrite reductases below and Fig. 11), all of which presumably impose a very high copper requirement. However, copper is also toxic, generating superoxide, hydroxyl radicals, and other reactive oxygen species via a reaction analogous to the Fenton reaction of ferrous iron (Huckle et al. 1993; Rensing and Grass, 2003).

As reviewed by Rensing and Grass (2003), mechanisms for copper efflux include a copper-responsive MerR-like transcriptional activator (CueR) that controls the Cu efflux (cue) system. CueR has been shown to regulate both *copA* and *cueO*. CopA is a P-Type ATPase that functions in copper efflux. CueO is a multicopper oxidase, having a laccase activity (p-diphenol: O_2 oxidoreductase), that confers protection to periplasmic proteins from copper-induced damage via an as yet unresolved mechanism (Grass and Rensing, 2001). In *Enterococcus hirae*, copper homeostasis has been attributed to the paired activities of two P-type ATPases, CopA and CopB (Solioz and Odermatt, 1995). One (CopB) functions in copper extrusion, whereas the other (CopA) was suggested to serve for copper uptake during conditions of copper limitation (Solioz and Odermatt, 1995; Rensing and Grass. 2003). A second system (Cus) is controlled by a two-component (CusRS) sensor/regulator pair that activates the adjacent *cusCFBA* operon. The CusCBA system is homologous to proton/cation antiporters involved in export of metal ions, xenobiotics, and drugs in a variety of bacteria. Related CusCBA-like complexes that span the periplasm function in metal export in *Pseudomonas*, *Ralstonia*, *Synechococcus*, *Salmonella*, and *Escherichia coli*. CusF is a periplasmic copper-binding metallochaperone recently shown to participate in the direct, and specific, transfer of copper to CusB (Bagai et al., 2008). A number of bacteria (including *Synechocystis* and *E. hirae*) contain proteins related to the eukaryotic ATX1 metallochaperone, suggested to function in the bacterial cytoplasmic trafficking of copper (Cavet et al., 2003). *Nitrosomonas europaea* contains a CopA-type ATPase, and the genomes of other sequenced AOB code for presumptive orthologs of the *copB*, *copC*, and *copD* genes. In the *N. maritimus* genome, one gene identified as copper resistance D domain protein (Nmar_1652) showed similarity with genes coding copper-resistant protein (CopC and CopD) of some *Archaea* and *Bacteria*. However, no clear homologs to other described systems of copper homeostasis, including CopA-type ATPases, characterized metallochaperones (Solioz and Stoyanov, 2003), or bacterial metallothioneins (Gold et al., 2008), are evident in the SCM1 genome.

COPPER HOMEOSTASIS

The genome sequence points to two types of enzyme systems that may be involved in copper handling or response to oxidative stress: genes encoding multicopper oxidases as described above and DsbA-type proteins. The exceptionally high number of sequences having low but significant similarity to DsbA-type proteins is also of possible relevance. There are 10 copies of genes coding for members of this subfamily of the thioredoxin family in the *N. maritimus* genome, but only a single copy in *N. europaea* and similar low copy numbers in other microbial genomes. The DsbA-type members of this thioredoxin family of proteins catalyze disulfide-bond formation or isomerization during protein folding and have been shown to protect *E. coli* from incorrect disulfide bond formation of periplasmic proteins catalyzed by copper (Hiniker et al., 2005). Alternatively, or in addition, these proteins may serve to protect the cell from reactive nitrogen species generated as intermediates in the ammonia oxidation pathway (as suggested in Fig. 10). For example, one of the more relevant features of nitroxyl chemistry is its ability to react as an electrophile with thiols (Fukuto et al. 2005b). Thus, these genes may be of significance to the suggested novel pathway for ammonia oxidation.

FIGURE 11 Alignment of copper-containing nitrite reductase/multicopper oxidase gene sequences (Nmar_1259 and 1667, annotated as *nirK*) with Global Ocean Sampling data sets and cultured microorganisms. Shading of amino acids: identical (white on black) and similar (black on gray) sequences.

COPPER-CONTAINING NITRITE REDUCTASES

The first evidence of the presence of copper-containing nitrite reductase gene (*nirK*)-like sequence in mesophilic Crenarchaeota was found in a 43 K soil fosmid clone 54d9 with 16S-23S rRNA genes and *amoAB* genes (Table 3) (Treusch et al., 2005). *N. maritimus* has two related genes at 91% amino acid identity to these plausible copper-containing nitrite reductases (Nmar_1259 and 1667) (Fig. 11). Although very similar sequences have been found in Sargasso Sea metagenome data and other open ocean data sets, a protein BLAST search revealed that these genes are more similar to bacterial genes than to those of cultured *Archaea*, suggesting the possibility of lateral gene transfer. One of the *nirK*-like genes (Nmar_1259) shares 29% amino acid identity with the multicopper oxidase of *N. europaea* and 27% amino acid identity to the *Nitrobacter winogradskyi* gene. Although a nitric oxide reductase (*norQ*) gene, catalyzing the reduction of NO to N_2O, was reported in the genome of *C. symbiosum* (Hallam et al., 2006b), a homolog in the *N. maritimus* genome (76% amino acid identity with the *C. symbiosum* gene) appears to be an ATPase associated with other cellular activities (Nmar_1515). Thus, although *N. maritimus* and probably *C. symbiosum* likely do encode a bacterial-type nitrite reductase, a capacity to produce N_2O has not been resolved.

Carbon Fixation and Autotrophic Growth

N. maritimus and all known AOB grow chemolithoautotrophically on ammonia. In contrast to the AOB, which use the Calvin cycle for autotrophic growth, *N. maritimus* appears to use the recently described 3-hydroxypropionate/4-hydroxybutyrate pathway, a variant of the 3-hydroxypropionate/malyl-CoA pathway discovered in *Chloroflexus* species (Holo, 1989; Strauss and Fuchs, 1993). This novel pathway is used by thermophilic *Crenarchaeota,* including *Sulfolobus* and *Metallosphaera sedula* (Berg et al., 2007) (Fig. 12). The presence of genes in the *N. maritimus* genome coding for a biotin-dependent carboxylase (Nmar_0272-74) (Fig. 12, step 8) together with genes for the enzymes methylmalonyl-CoA epimerase, methylmalonyl-CoA mutase (Nmar_0953-4 and 0958) (Fig. 12, step 12), and 4-hydroxybutyryl-CoA dehydratase (Nmar_0207) (Fig. 12, step 4) clearly indicate the usage of a 3-hydroxypropionate/4-hydroxybutyrate-type pathway for autotrophic carbon fixation. Although the formation of succinyl-CoA from acetyl-CoA is common to both the *Chloroflexus* and archaeal pathways, its formation is achieved via different reaction sequences, suggestive of convergent evolution (Berg et al., 2007). From that point, the two pathways diverge. Succinyl-CoA is converted via 4-hydroxybutyrate into acetoacetyl-CoA, which is then cleaved into two molecules of acetyl-CoA (Fig. 12, steps 1 to 7).

TABLE 3 Features of *N. maritimus* genomes and crenarchaeal gene fragments[a]

Parameter	Result for:						
	N. maritimus SCM1	*C. symbiosum* A	Fosmid 4B7	Cosmid DeepAnt-EC39	Fosmid 74A4	45-H-12	Soil fosmid 54d9
Size (bp)	1,645,259	2,045,085	39,297	33,347	43,902	39,411	43,377
% Coding	91.9%	91.2%	89.1%	86.1%	84.0%	70.1%	72.9%
G+C content	34.2%	57.4%	34.4%	34.1%	32.6%	43.0%	36.4%
ORF density (ORF/kb)	1.19	0.986	0.992	1.17	1.12	0.893	0.991
Average ORF length (bp)	757	924	898	737	753	785	736

[a]The detail and environmental background of fosmids and cosmid libraries were described by Beja et al. (2002), Lopez-Garcia et al. (2004), Nunoura et al. (2005), and Treusch et al. (2005).

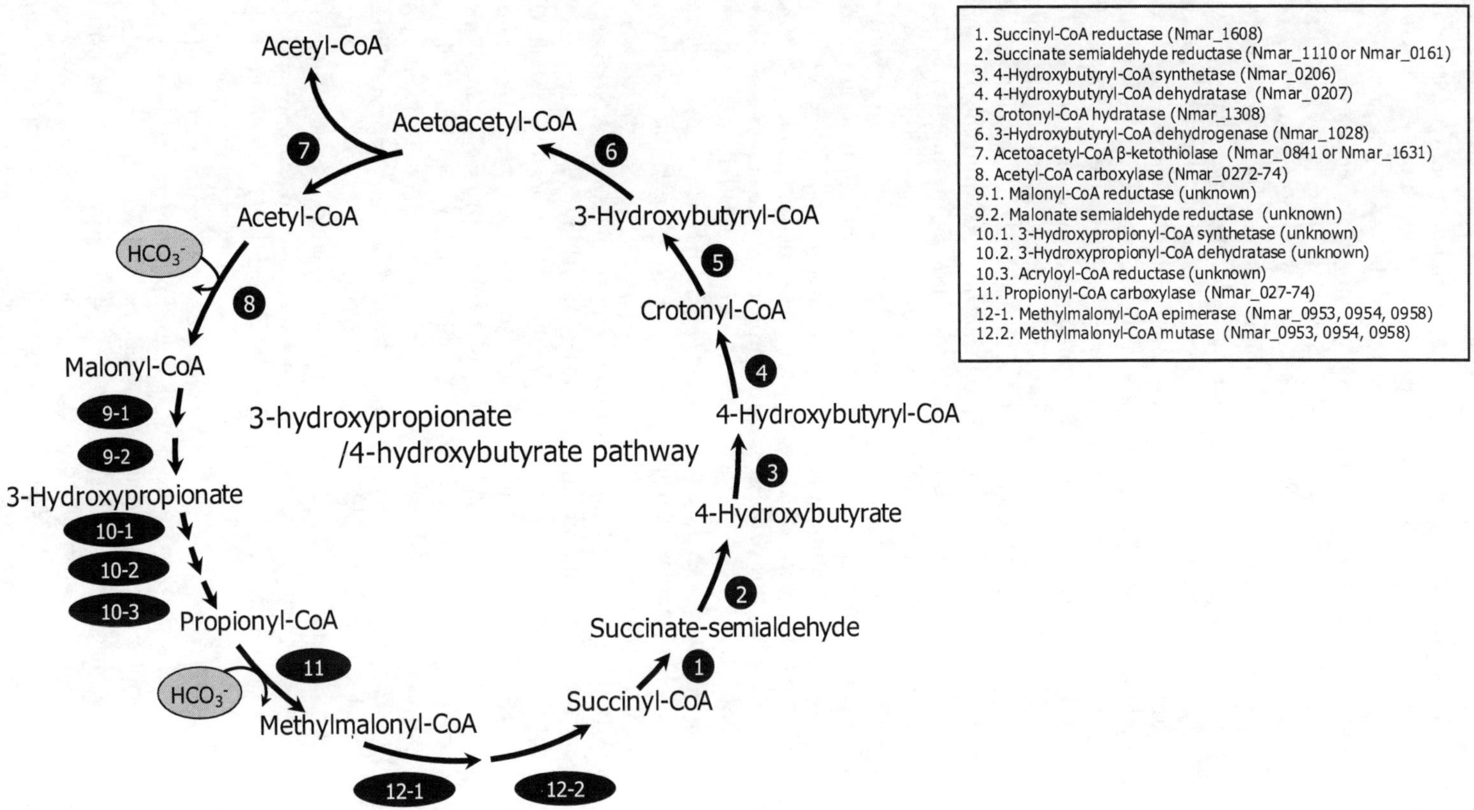

FIGURE 12 Proposed autotrophic 3-hydroxypropionate/4-hydroxybutyrate pathway in *N. maritimus*. Enzymes catalyzing each reaction are numbered. Annotated genes are coded as a locus tag in parentheses.

Pyruvate synthesis seems to occur via the reductive carboxylation of acetyl-CoA through pyruvate:ferredoxin *oxidoreductase*. The *N. maritimus* genome contains two genes coding for subunits of a single 2-oxoacid:ferredoxin oxidoreductase (Nmar_0413-4). Although the broad specificity of these enzymes makes definitive functional assignment difficult, the obligate requirement for an enzyme capable of acetyl-CoA carboxylation (forming pyruvate) strongly suggests that this gene encodes a pyruvate:ferredoxin oxidoreductase. Pyruvate is most likely routed into central metabolism via conversion to phosphoenopyruvate (PEP) by pyruvate:phosphate dikinase (Nmar_0951) and subsequent formation of oxaloacetate by PEP carboxykinase (Nmar_0392). This reaction sequence is also consistent with the absence of a gene coding for pyruvate kinase (EC 2.7.1.40), widely distributed in both *Bacteria* and *Archaea* lineages (Fig. 13).

Although the genome encodes numerous genes of enzymes of the tricarboxylic acid (TCA) cycle, reductive usage of this pathway for autotrophic carbon fixation can be excluded. A complete reductive TCA cycle generally requires a 2-oxoglutarate:ferredoxin oxidoreductase and a citrate cleaving enzyme. Generally, organisms utilizing the reductive TCA cycle for carbon fixation contain unique genes encoding for two oxoacid:ferredoxin oxidoreductases, one specifically active for pyruvate and one for 2-oxoglutarate. Although these appear to be absent in *N. maritimus*, *in vitro* enzyme activity measurements and/or *in vivo*-labeling experiments are needed to verify this hypothesis. Nonetheless, it is highly likely that *N. maritimus* utilizes an incomplete (or horseshoe-type) TCA cycle for strictly biosynthetic purposes, not for carbon fixation.

Biosynthesis of Amino Acids

Pathways for the biosynthesis of all standard amino acids, with the exception of proline, have been identified in the *N. maritimus* genome. This is comparable to inferences made earlier from the *C. symbiosum* genome sequence, in which genes supporting all pathways except proline were reported (Hallam et al., 2006a). Either ornithine or glutamate serves as precursor for proline biosynthesis. Glutamate is directly synthesized by aspartate-2-oxoglutarate transaminase (aminotransferase class I and II; EC 2.6.1.1; Nmar_0546), but there are no clear orthologs for the most common pathway of proline synthesis from glutamate (γ-glutamyl kinase, γ-glutamyl phosphate reductase, and Δ^1-pyroline-5-carboxylate reductase). The urea cycle, from which ornithine is derived, is potentially present if the gene (Nmar_0925) annotated as putative arginase/agmatinase/formiminoglutamase is arginase. However, a gene coding for ornithine cyclodeaminase (EC 4.3.1.12), catalyzing conversion of ornithine to proline, is not evident in the genome. Thus, as for most *Archaea*, the mechanism for proline biosynthesis is not resolved.

In a select overview, the following pathways for amino acid biosynthesis are among those identified. Asparagine is directly synthesized by asparagine synthase (EC 6.3.5.4; Nmar_0935) from asparate. Serine appears to be synthesized via the phosphorylated pathway, catalyzed by D-3-phosphoglycerate dehydrogenase (SerA; (EC 1.1.1.95; Nmar_1258), phosphoserine aminotransferase (SerC; EC 2.6.1.52), and phosphoserine phosphatase (SerB; EC 3.1.3.3) (Nmar_0666). However, the gene for the latter (SerC) has not yet been identified. Glycine is derived from serine by serine hydroxymethyltransferase (GlyA; EC 2.1.2.1; Nmar_1793). The gene coding threonine aldolase (EC 4.1.2.5), which catalyzes glycine production from threonine, is not present. Valine, leucine, and isoleucine are synthesized from different precursors by the same enzyme, branched-chain amino acid aminotransferase (EC 2.6.1.42; Nmar_0192). The threonine pathway for isoleucine biosynthesis, using pyruvate and threonine as precursors, was confirmed. The α-aminoadipic acid route pathway for lysine biosynthesis is likely used by *N. maritimus*, not the diaminopimelic acid pathway. All genes in the two pathways for phenylalanine and tyrosine synthesis from chorismate via prephenate are accounted for in the genome.

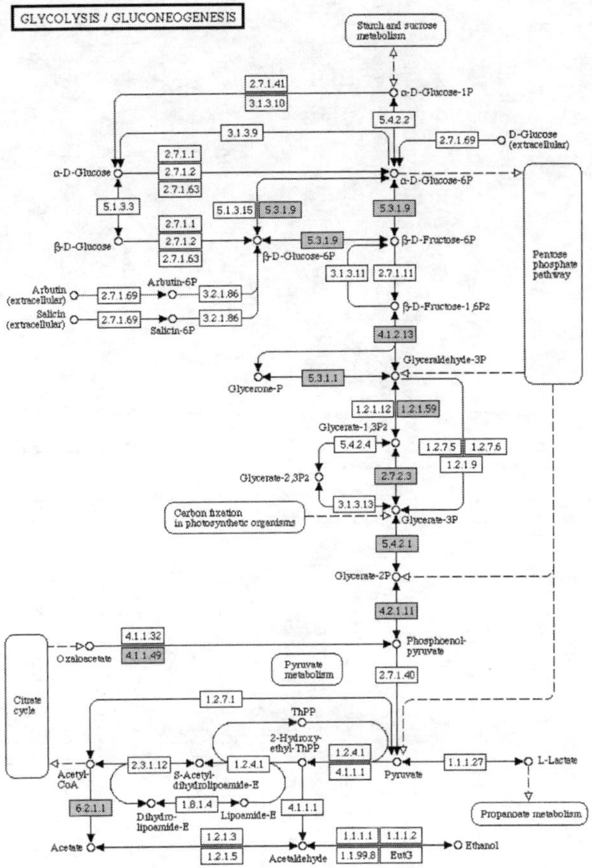

EC Number	*Nitrosopumilus maritimus*	*Sulfolobus tokodaii*	*Pyrococcus furiosus*	*Natranomonas pharaonis*	*Halobacterium salinarium* R1	*Nitrosomonas europaea*	*Nitrosospira multiformis*	*Nitrobacter winogradskyi*	*Escherichia coli* K-12 MG1655	*Prochlorococcus marinus* MT9515	*Synechococcus sp.* WH8102	*Wolinella succinogenes*	*Pelagibacter ubique*	*Wolbachia* sMe1	*Rickettsia typhi*
5.4.2.2		●	●	●		●	●		●	●	●				
2.7.1.1															
2.7.1.2					●	●	●	●	●	●	●				
2.7.1.63								●							
5.3.1.9	●	●	●	●	●	●	●	●	●	●	●	●	●		
2.7.1.11						●	●		●			●			
3.1.3.11				●	●	●		●	●	●	●	●	●	●	
4.1.2.13	●	●	●	●	●	●	●	●	●	●	●	●	●		
5.3.1.1	●	●	●	●	●	●	●	●	●	●	●	●	●	●	
1.2.1.12						●	●	●	●		●	●	●	●	
1.2.1.59	●	●	●	●	●	●				●	●				
2.7.2.3	●	●	●	●	●	●	●	●	●	●	●	●	●	●	
5.4.2.4															
3.1.3.13															
1.2.7.5			●	●								●			
1.2.7.6			●												
1.2.1.9		●	●												
5.4.2.1	●	●	●	●	●	●	●	●	●	●	●	●	●	●	
4.2.1.11	●	●	●	●	●	●	●	●	●	●	●	●	●	●	
2.7.1.40		●	●	●	●	●	●	●	●	●	●				

FIGURE 13 Glycolysis and gluconeogenesis pathway of *N. maritimus* and the distribution of genes of enzymes involved in the process among some *Archaea* and *Bacteria*.

Most of the genes required for synthesis of tryptophan from chorismate and for synthesis of histidine from 5-phosphoribosyl diphosphate via L-histidinol have been identified. Cysteine is likely synthesized by cysteine synthase (EC 2.5.1.47; Nmar_0670). It appears that alanine is synthesized from cysteine via the alanine biosynthesis III pathway and that methionine is synthesized from cysteine via L-homocysteine by methionine synthase (EC 2.1.1.13). However, the methionine synthase (Nmar_1267) is located near a fragmented near-identical gene copy (Nmar_1268), and both contained frame shifts. Thus, a bifunctional gene proximal to these genes may catalyze this reaction (Nmar_1266).

Mixotrophy and Metabolic Versatility

Although *N. maritimus* grows on a completely inorganic medium, the genome sequence suggests significant flexibility in the utilization of organic sources of nitrogen, carbon, and phosphorus. Although it lacks homologs of the putative urea transporter and urease genes identified in *C. symbiosum* (Hallam et al., 2006a), numerous organic transport functions are also evident. These broadly encompass transporters for different amino acids, dipeptides/oligopeptides, sulfonates/taurine, and glycerol. Thus, we anticipate that more detailed physiological characterization will show that *N. maritimus* has some capacity for mixotrophic growth, as previously suggested by isotopic studies of natural populations (Ingalls et al., 2006). Recent culture experiments also support the capacity of SCM1 to grow as a mixotroph (Martens-Habbena, personal communication).

The *N. maritimus* genome contains genes coding for gluconeogenesis via the reverse Embden-Meyerhof pathway (Fig. 13). All genes required for the pentose phosphate pathway and for the synthesis of riboflavin, biotin, vitamin B_{12}, and nicotinate are present. Biosynthesis of thiamine, pantothenate, and folic acid is also supported by the presence of most genes in the corresponding pathways and the more recent observation that growth of SCM1 can be maintained in a completely inorganic medium (Martens-Habbena et al., 2009).

Biosynthesis of Novel Phosphonates

The genome sequence of SCM1 indicated a capacity for synthesis of novel phosphonate(s), compounds structurally similar to phosphate esters but containing a stable carbon-phosphorus bond. Phosphonates serve multiple cellular roles in invertebrates and microorganisms (Kittredge and Roberts, 1969), occurring as phosphonolipids, side groups in exopolysaccharides and glycoproteins, functioning for phosphorus storage (Miceli et al., 1980), and as secondary metabolites of fungi and bacteria (Kononova and Nesmeyanova, 2002). Bioactive phosphonates include the antibiotics fosfomycin and dehydrophos and the herbicide phosphinothricin tripeptide (bialaphos). Characterization of different biosynthetic pathways has shown that biosynthesis generally begins with the same two steps: (i) conversion of PEP to phosphonopyruvate (PnPy) by PEP mutase and (ii) conversion of PnPy to phosphonoacetaldehyde and CO_2 by PnPy decarboxylase. Recently, Shao et al. (2008) have shown that an unexpected intermediate (2-hydroxyethylphosphonate) is also common to the biosynthesis of many of these compounds, produced from phosphonoacetaldehyde by a novel family of Group III metal-dependent alcohol dehydrogenases. All three enzymatic activities appear to be encoded by homologous genes colocalized on the *N. maritimus* genome (Nmar_0158 and 0160-1), unique among available archaeal genome sequences.

Phosphonates comprise a significant fraction of the dissolved organic phosphorus pool of the oceans (most commonly aminoethylphosphonate) and may provide an important resource for organisms inhabiting this often P-limited environment (Clark et al., 1999). For example, the marine diazotroph *Trichodesmium* encodes genes for a C-P lyase that has been suggested to provide this organism adaptive advantage (Dyhrman et al., 2006a, 2006b; Dyhman and Haley, 2006). Since *Archaea* represented by *N. maritimus* are abundant marine bacterioplankton, a capacity to synthesize phosphonates would point to some importance in linking marine C, N, and P cycles. Two systems for phosphorus acquisition may be present. In addition to a high-affinity, high-activity *pstSCAB*-transport system for phosphate (Nmar_479 and 481-3), an ability to use organic phosphates is suggested by the presence of a putative phosphonate transporter (Nmar_0873-5). However, preliminary studies have not shown a capacity to use phosphonates (Martens-Habenna, personal communication), and genes for C-P lyase and known phosphonohydrolases have not been identified in the genome sequence (for review, see Quinn et al., 2007).

Discovery of Ectoine and Hydroxyectoine Biosynthesis in the *Archaea*

Microorganisms synthesize various organic osmolytes (compatible solutes) that accumulate via synthesis and/or uptake with exposure to hyperosmotic conditions and are rapidly expelled under hypoosmotic conditions. These solutes also confer protein stability and are synthesized following temperature shock or entry of cultures into the stationary phase (Bursy et al., 2008). As such, they have also been called "chemical chaperones" (Diamant et al., 2001). Bacterial osmolytes include trehalose, glutamate, proline, glycine betaine, carnitine, and ectoine (for review, see Burg and Ferraris, 2008). Among these, the capacity for ectoine biosynthesis is widely distributed among the *Bacteria*, encoded by a highly conserved gene cluster (*ectABC*), and is particularly common among marine *Bacteria* (indicated in genome sequences of *Marinobacter*, *Oceanicola*, *Oceanobacillus*, *Oceanobacter*, *Oceanospirillum*, and others). However, among characterized *Archaea*, an indication for ectoine biosynthesis is restricted to *N. maritimus*, encoded by an operon-like gene cluster homologous to the bacterial *ectABC* (Fig. 14). Phylogenetic analysis of *ectA* genes suggests the strain SCM1 likely acquired this gene from *Bacteria*. Of particular note, the gene cluster is flanked by two genes encoding transcription factor B (TFB)-type regulatory elements (Nmar_1340-1). These are thought to serve general regulatory functions in *Archaea* and could play a role in the response of *N. maritimus* to temperature and osmotic shock.

Unusual Richness in General Transcription Factors

The large number of transcriptional regulators is indicative of a very adaptive physiology (Fig. 8). The genome contains at least eight TFB (Nmar_0013, 0020, 0517, 0624, 0979, 0987, 1340, and 1341) and two TATA-box binding protein (TBP) genes (Nmar_0598 and 1519), making *N. maritimus* among the densest and richest archaeal genomes currently sequenced. This suggests that this organism of apparently extremely limited metabolic diversity has a contrasting very high adaptive flexibility that may relate to its lifestyle as an extreme oligotroph (Martens-Habbena et al., 2009). Both TFB and TBP are required for start site-specific transcription initiation. In *Archaea* having multiple TFBs and TPBs, they are thought to serve functions similar to the bacterial sigma factors, modulating cellular function in response to fluctuating environmental conditions (Baliga and DasSarma, 2000), with optimal TFB/TPB partners and some essential for viability (Facciotti et al., 2007). Although many *Archaea* encode multiple copies of these transcription factors, only the haloarchaea are known to have more than five TFB genes (Facciotti et al., 2007). The *N. maritimus* genome also encodes representatives of the two families of chromatin proteins, at least five genes related to archaeal histone (Nmar_0579, 1432, 0683, 0788, and 0503) and two homologs of Alba (Nmar_0255 and 0933). These are widely distributed within *Archaea* and thought to maintain chromosomal material in a state that permits polymerase accessibility (Sandman and Reeve, 2005). Differential expression in *Archaea* encoding multiple variant transcriptional regulators may provide another mechanism to alter global chromatin composition and transcription (Sandman and Reeve, 2005). The functional significance of this exceptionally high

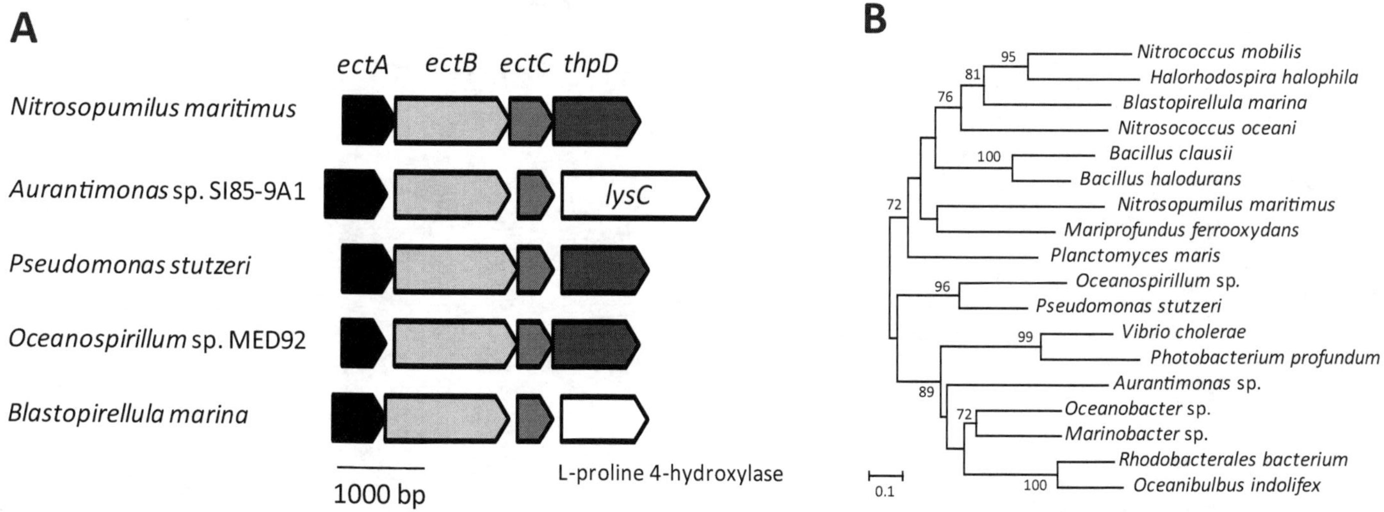

FIGURE 14 The first evidence of ectoine and hydroxyectoine biosynthesis in the *Archaea*. (A) Comparison of ectoine synthesis operon clusters with putative gene names. Microorganisms, except for *Pseudomonas stutzeri*, are of marine origin. Nmar_1346, *ectA*, L-2,4-diaminobutyric acid acetyltransferase; Nmar_1345, *ectB*, diaminobutyrate-2-oxoglutarate aminotransferase; Nmar_1344, *ectC*, ectoine synthase; Nmar_1343, *thpD*, ectoine hydroxylase; *lysC*, aspartate kinase. (B) Phylogenetic relationship of *ectA* gene. The evolutionary history was inferred using the neighbor-joining method. The evolutionary distances were computed using the JTT matrix-based method and are in the units of inferred amino acid substitutions per site. The scale bar shows 0.1 substitutions per site. The bootstrap values greater than 70% (1,000 replicates) are shown next to each node. There were a total of 130 positions in the final data set, and all positions containing gaps and missing data were eliminated from the data set.

number of regulatory factors in an apparently metabolically specialized and small genome organism should be informed by future transcriptional analyses, with the expectation that this will provide important insights into gene ensembles required for growth under different physical/chemical conditions and also advance understanding of gene regulation in mesophilic *Crenarchaeota*.

Novel Hybrid Machinery for Cell Division in *N. maritimus*

NOVEL CELL DIVISION MACHINERY

Unique cell division machinery in the *Crenarchaeota* was recently reported (Lindas et al., 2008; Samson et al., 2008). An operon (*cdvABC*) induced at the onset of genome segregation and cell division was shown to code for cell division machinery related to the eukaryotic endosomal sorting complex. The CdvA, CdvB, and CdvC proteins polymerize between segregating nucleoids, forming successively smaller structures during constriction and cell division (Lindas et al., 2008). With the exception of the *Thermoproteales* and *Nitrosopumilus/Cenarchaeum*, all available archaeal genomes have either the FtsZ or Cdv cell division machinery, not both (*Thermoproteales* division machinery has not been characterized). *Nitrosopumilus* and *Cenarchaeum* are unique in encoding both systems of cell division (*ftsZ*, Nmar_1262; *cdvA*, Nmar_0700; *cdvB*, Nmar_0816; *cdvC*, Nmar_1088) (Lindas et al., 2008). Thus, the functional and evolutionary significance of a unique and possibly hybrid machinery for cell division in *N. maritimus* is of particular interest in future studies, and possibly offers additional support for the hypothesis that representatives of this lineage are at kingdom-level depth within the *Archaea* (Fig. 15) (Brochier-Armanet et al., 2008).

High Similarity to the Marine Metagenome Sequences

Early molecular surveys and comparative genomic analyses revealed the global distribution of planktonic *Crenarchaeota* in marine environments (DeLong, 1992; Fuhrman et al., 1992, 1993; Béjà et al., 2002) but did not provide insight into central energy metabolism or features relevant to their open ocean habitat. Early metagenomics studies (Venter et al., 2004) and the genome sequence of the sponge symbiont *C. symbiosum* provided an indication of a capacity for ammonia oxidation (Hallam et al., 2006a and 2006b). However, the G+C content of *Cenarchaeum* genome deviated significantly from planktonic populations and, in the absence of physiological data and other signature genes in the bacterial pathway for ammonia oxidation, there was no direct support for this inference. In contrast to the symbiont, the genome of *N. maritimus* SCM1 shares remarkable conservation of gene content and gene order with environmental crenarchaeal sequences previously recovered in fosmid libraries and more recent oceanic surveys (Table 3). The Antarctic genome fragments DeepAnt-EC39 (taken from 500 m depth) and fosmid 74A4 (taken from surface water) both share very high gene order with the *N. maritimus* genome.

Recruitment of currently available metagenomic data sets resulted in nearly complete coverage of the *N. maritimus* genome (Fig. 16A). In contrast, coverage of available AOB genomes, such as *N. europaea*, was poor (Fig. 16F). Interestingly, many high identity matches against *N. maritimus* were recruited from both pelagic and coastal sampling sites, with coastal sites having greater similarity (>85% nucleotide identity) (Fig. 16A and B). Notably, the Block Island, New York, coastal site (GS009) yielded the highest number of reads of greater than the 75% nucleotide identity and the largest fractional abundance of total read recruited to the SCM1 genome (Fig. 17). Similarly, significant differences were apparent in two Galapagos Island sites (GS031 and 032). These data suggest that *N. maritimus* and genetically similar marine nitrifying *Crenarchaea* may be adapted to near coastal environments. For example, SCM1 cannot grow at or below 15°C, and this

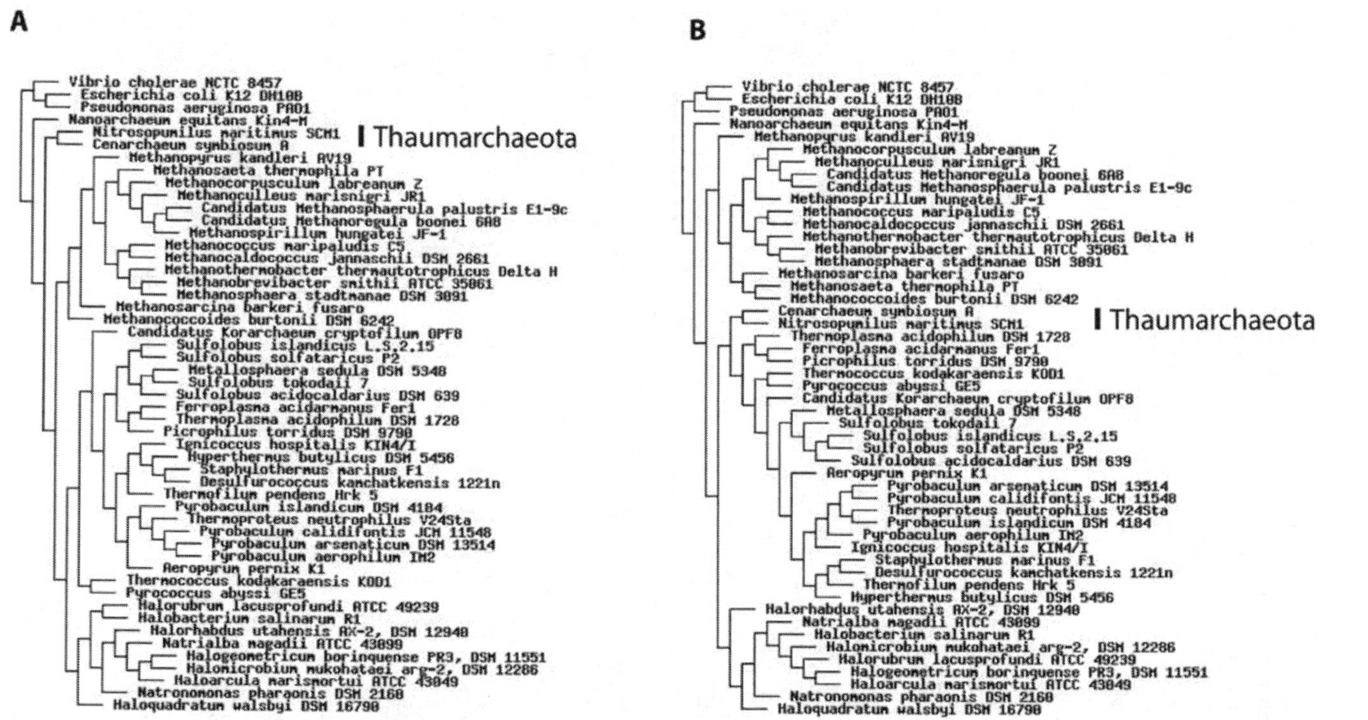

FIGURE 15 Hierarchical clustering of *N. maritimus* and other archaeal genomes based on enzyme (A) and COG (B) clustering methods for the type of protein/functional families. The figures were prepared by using the genome-clustering tool in the integrated microbial genomes system (Markowitz et al., 2006, 2008). The placement in the tree reflects the distance between genomes, whereby the computed distance is based on the similarity of the functional characterization of genomes in terms of a specific protein/functional family. The enzyme-based clustering supports an affiliation of *Nitrosopumilus* and *Cenarchaeum* with the *Crenarchaeota*, whereas the COG-based clustering indicates that these two genera are of independent origin and possibly represent a novel kingdom (Thaumarchaeota).

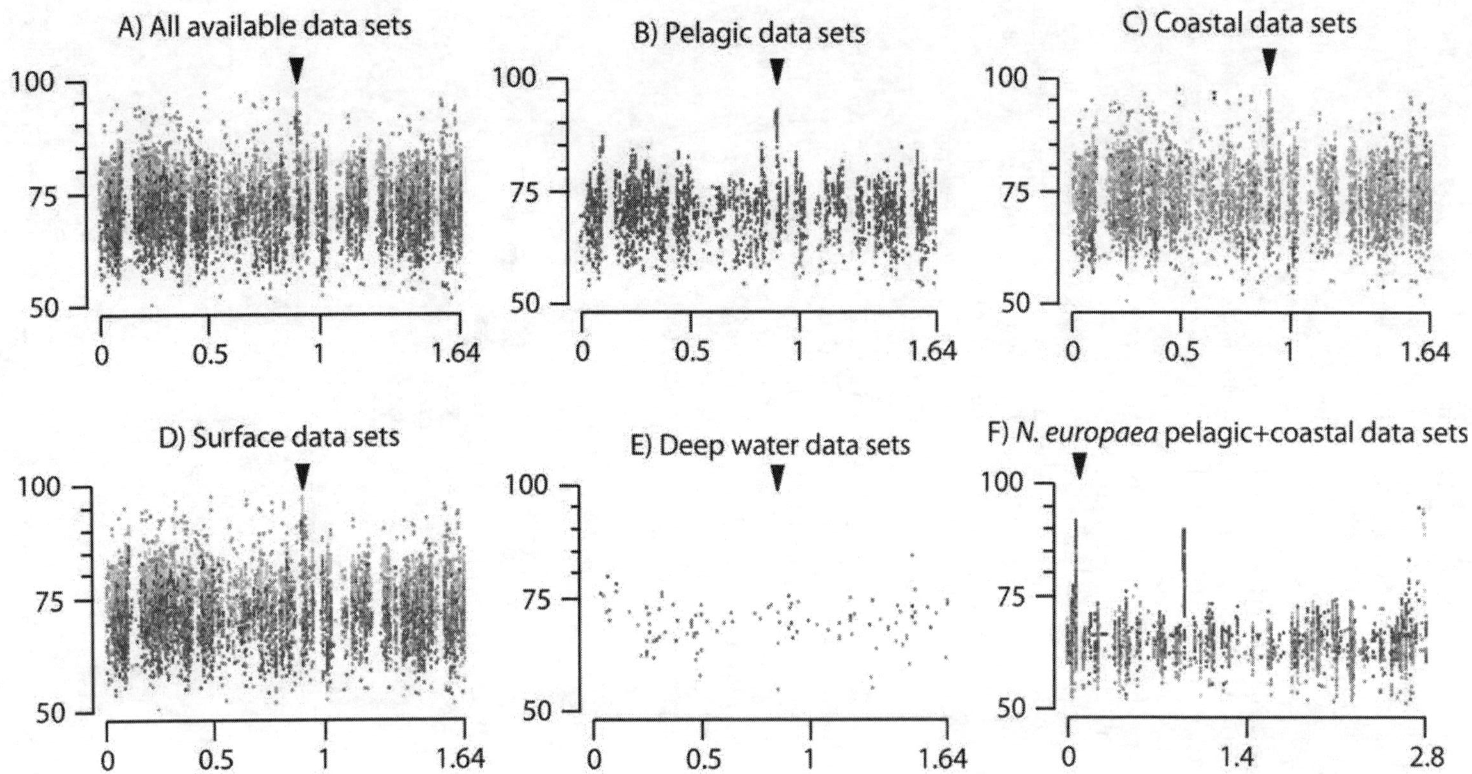

FIGURE 16 Recruitment of GOS and HOT sequences to the *N. maritimus* genome. (A) All available data sets. (B) Pelagic data sets (sampling sites are deeper than 200 m). (C) Coastal data sets (sampling sites are shallower than 200 m). (D) Surface data sets (samples are collected shallower than 200 m). (E) Deep water data sets (samples are collected deeper than 200 m). (F) Pelagic and coastal data sets against *N. europaea* genome. Arrows indicate the position of 16S/23S rRNA genes.

feature would restrict its habitat to coastal, tropical, and near-surface pelagic waters (Fig. 17D and E). This is consistent with variation in gene organization of marine *Crenarchaeota* genomic fragments recovered from different sampling depths and is suggestive of depth-related habitat types (Lopez-Garcia, 2004; Hallam et al., 2006b). Surprisingly, a metagenomic library from Lake Gatun, Panama (GS020), contained many crenarchaeal sequences (4.9% of the total reads examined) with nucleotide identities to SCM1 comparable to the pelagic data sets. Thus, although the data from freshwater sites are as yet very limited, in addition to their well-documented abundance in marine and pristine soil environments (Prosser and Nicol, 2008, and references therein), lakes may be another important habitat for nitrifying *Crenarchaeota*.

EVOLUTION

A Third Kingdom of *Archaea*?

A phylogenetic position of *C. symbiosum* (and *N. maritimus*) basal to the two formally described archaeal kingdoms has been suggested by analysis of ribosomal proteins and patterns of gene presence/absence among the three domains (Brochier-Armanet et al., 2008). This has been used to propose a third kingdom, the Thaumarchaeota, within the *Archaea*. However, the tree topology is sensitive to the method of phylogeny inference. Also, hierarchical clustering of available archaeal genomes based on enzyme or COG distribution does not yield a consistent placement. The COG-based clustering places *N. maritimus* and *Cenarchaeum* basal to the *Euryarchaeota* and *Crenarchaeota*, whereas the enzyme-based clustering suggests that they are deeply diverging members of the *Crenarchaeota* (Fig. 15). Thus, resolution of phylogenetic position will likely only be confirmed by inclusion of additional genome data from more divergent members of the ammonia-oxidizing *Crenarchaeota*, as should soon come available from the recently cultivated *N. yellowstonii* and *N. gargensis*.

Did Ammonia Oxidation Originate within the *Bacteria* or *Archaea*?

The discovery of AOA raised a number of fundamental questions concerning the origin and biochemistry of ammonia oxidation. Both AOA and AOB share a related oxygenase, the AMO, and it is possible that an early hori-

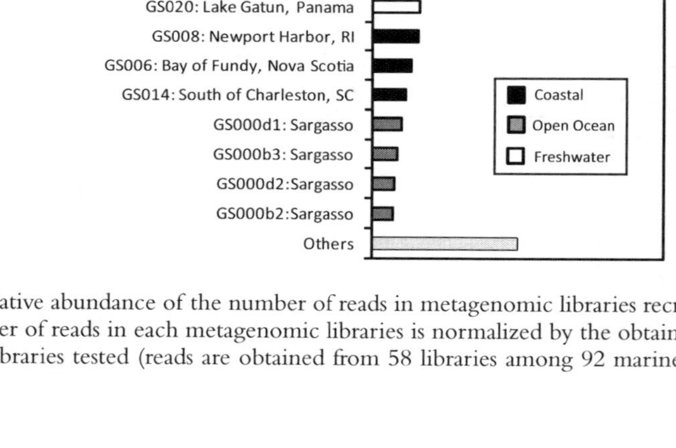

FIGURE 17 Relative abundance of the number of reads in metagenomic libraries recruited to the *N. maritimus* genome. The number of reads in each metagenomic libraries is normalized by the obtained total number of reads (n = 5,773) in all libraries tested (reads are obtained from 58 libraries among 92 marine and freshwater libraries examined).

zontal gene transfer accounts for its presence in both domains. However, apart from this one enzyme, their biochemistry of ammonia oxidation is likely distinct. The AMO is also evolutionarily entwined with the methane monooxygenase, and it is not clear which aerobe (methanotroph versus ammonia oxidizer) first emerged as Earth's atmosphere became increasingly oxidizing. A geochemical dilemma relating to an early emergence of a copper-based aerobic metabolism is oxygenation of the atmosphere in the Proterozoic Eon. Modeling and geochemical data suggest that soluble copper was scarce during the "ferro-sulfidic" archaean eon, became even less available during the primarily "sulfidic" Proterozoic period, and increased significantly only during the last billion years of Earth's history (Canfield, 1998; Anbar and Knoll, 2002; Dupont et al., 2006). The discovery of an early-branching ammonia-oxidizing thermophile (*N. yellowstonii*) points to a possible early thermophilic origin within the *Archaea* (de la Torre et al., 2008). A thermophilic ancestry is also suggested by oceanic representatives of an archaeal lineage (pSL12 and ALOHA groups) (Fig. 1) (Mincer et al., 2007). However, an early evolution of a copper-based metabolism must be reconciled with scarcity of soluble copper during most of Earth's history. These questions will likely only be resolved through more intensive characterization, both through cultivation and molecular surveys of microorganisms that have specialized to grow on simple, and primordial, growth substrates.

CONCLUSIONS AND FUTURE PERSPECTIVES

The discovery of ammonia oxidation within the domain of the *Archaea* has challenged a century-old paradigm of nitrification limited to a few proteobacterial genera. The existence of oligotrophic archaeal ammonia oxidizers among the *Archaea*, their widespread occurrence in nature, and their prevalence in nutrient-poor environments indicates a significant role of AOA in the global nitrogen cycle. Genomic and metagenomic studies have now shed initial light on the gene inventory of mesophilic AOA. These studies indicate not only that a tremendous diversity of *Crenarchaeota* may thrive by ammonia oxidation but also that a distinct biochemistry of ammonia oxidation serves energy conservation in these organisms. AOA are now recognized to inhabit a broad spectrum of environments, from permanently cold marine to geothermal environments and thus have a significantly broader temperature range than known bacterial counterparts. Broad temperature range, as well as oligotrophy among AOA, shows that apparently unfavorable thermodynamics of ammonia oxidation and the requirement of reverse electron transport for biosynthetic needs do not pose significant limitations on the competitiveness of archaeal nitrifiers for scarce energy resources such as ammonia. Detailed physiological and biochemical studies will be required to decipher this novel biochemistry and its significance for the environmental adaptation of AOA.

Both genomic and physiological characterization of SCM1 point to a greater importance of chemoautotrophy in the ocean water column than previously recognized. Strain SCM1 encodes all major biosynthetic pathways for autotrophic growth in its genome and has the ability to grow in completely inorganic media. Its genome shares significant gene content and synteny with planktonic *Crenarchaea*. Its affinity for ammonia likely renders it competitive with both heterotrophs and phototrophs for this scarce resource. Despite their apparent metabolic specialization, genomic and biogeochemical studies indicate some capacity for assimilation of simple organic molecules by Group I *Crenarchaeota*. Direct evidence for an entirely heterotrophic lifestyle is still lacking, and it remains to be shown how significant heterotrophy or even alternative energy sources may be for these organisms. We anticipate that further comparative genomic studies will provide more complete insight into the metabolic adaptations of mesophilic and thermophilic *Crenarchaeota*, facilitate isolation of novel (possibly heterotrophic strains), and help to constrain their evolutionary relationships.

Two decades following the discovery of Group I *Archaea* in temperate environments, evidence for oligotrophic ammonia oxidation via a novel biochemistry and apparently limited metabolic versatility of SCM1 now raise a number of ecological and biogeochemical questions. The now likely existence of three independent pathways for aerobic and anaerobic ammonia oxidation by three phylogenetically distinct groups of organisms makes for a complex interplay of different pathways and microbial groups. Mechanistic insights into the physiology and biochemistry of these pathways (e.g. identification of metabolites, isotope fractionation patterns) and identification of selective metabolic inhibitors would foster future biogeochemical studies designed to inform the contribution of these groups to nitrification, nitrogen removal, and associated greenhouse gas emissions. The high substrate affinity among AOA may suggest a niche for nitrification even under high competition for reduced nitrogen, altering the nitrogen and carbon cycles in nitrogen-limited marine and terrestrial environments in unexpected ways. The existence of distinct biochemistries in phylogenetically vastly divergent organisms further suggest that AOA and AOB may respond significantly differently even to simple environmental cues (e.g. nutrient limitation, change of pH, or temperature). Thus, it is apparent that disentangling the complex interplay of nitrifiers will require a more complete understanding of their physiological diversity. This understanding, essential for designing appropriately constrained biogeochemical studies, will be advanced primarily through the characterization of additional isolates.

ACKNOWLEDGMENTS

We thank all colleagues included in the *N. maritimus* SCM1 genome annotation. This work was supported by the Department of Energy Microbial Genome Program, National Science Foundation Microbial Interactions and Processes Grant MCB-0604448 (to D.A.S.), National Science Foundation Molecular and Cellular Biosciences Grant MCB-0920741 (to D.A.S.), and National Science Foundation Biological Oceanography Grant OCE-0623174 (to D.A.S.).

REFERENCES

Agogué, H., M. Brink, J. Dinasquet, and G. J. Herndl. 2008. Major gradients in putatively nitrifying and non-nitrifying Archaea in the deep North Atlantic. *Nature* **456:**788–791.

Anbar, A. D., and A. H. Knoll. 2002. Proterozoic ocean chemistry and evolution: a bioinorganic bridge? *Science* **297:**1137–1142.

Andrade, S. L., and O. Einsle. 2007. The Amt/Mep/Rh family of ammonium transport proteins. *Mol. Membr. Biol.* **24:**357–365.

Arp, D. J., P. S. Chain, and M. G. Klotz. 2007. The impact of genome analyses on our understanding of ammonia-oxidizing bacteria. *Annu. Rev. Microbiol.* **61:**503–528.

Bagai, I., C. Rensing, N. J. Blackburn, and M. M. McEvoy. 2008. Direct metal transfer between periplasmic proteins identifies a bacterial copper chaperone. *Biochemistry* **47:**11408–11414.

Baliga, N. S., and S. DasSarma. 2000. Saturation mutagenesis of the haloarchaeal bop gene promoter: identification of DNA supercoiling sensitivity sites and absence of TFB recognition element and UAS enhancer activity. *Mol. Microbiol.* **36:**1175–1183.

Bédard, C., and R. Knowles. 1989. Physiology, biochemistry, and specific inhibitors of CH_4, NH_4^+, and CO oxidation by methanotrophs and nitrifiers. *Microbiol. Rev.* **53:**68–84.

Béjà, O., E. V. Koonin, L. Aravind, L. T. Taylor, H. Seitz, J. L. Stein, D. C. Bensen, R. A. Feldman, R. V. Swanson, and E. F. DeLong. 2002. Comparative genomic analysis of archaeal genotypic variants in a single population and in two different oceanic provinces. *Appl. Environ. Microbiol.* **68:**335–345.

Berg, I. A., D. Kockelkorn, W. Buckel, and G. Fuchs. 2007. A 3-hydroxypropionate/4-hydroxybutyrate autotrophic carbon dioxide assimilation pathway in archaea. *Science* **318:**1782–1786.

Bollmann, A., M.-J. Bär-Gilissen, and H. J. Laanbroek. 2002. Growth at low ammonium concentrations and starvation response as potential factors involved in niche differentiation among ammonia-oxidizing bacteria. *Appl. Environ. Microbiol.* **68:**4751–4757.

Bollmann, A., I. Schmidt, A. M. Saunders, and M. H. Nicolaisen. 2005. Influence of starvation on potential ammonia-oxidizing activity and *amoA* mRNA levels of *Nitrosospira briensis*. *Appl. Environ. Microbiol.* **71:**1276–1282.

Brochier-Armanet, C., B. Boussau, S. Gribaldo, and P. Forterre. 2008. Mesophilic crenarchaeota: proposal for a third archaeal phylum, the Thaumarchaeota. *Nat. Rev. Microbiol.* **6:**245–252.

Burg, M. B., and J. D. Ferraris. 2008. Intracellular organic osmolytes: function and regulation. *J. Biol. Chem.* **283:**7309–7313.

Bursy, J., A. U. Kuhlmann, M. Pittelkow, H. Hartmann, M. Jebbar, A. J. Pierik, and E. Bremer. 2008. Synthesis and uptake of the compatible solutes ectoine and 5-hydroxyectoine by *Streptomyces coelicolor* A3(2) in response to salt and heat stresses. *Appl. Environ. Microbiol.* **74:**7286–7296.

Button, D. K. 1994. The physical base of marine bacterial ecology. *Microbial. Ecol.* **28:**273–285.

Button, D. K. 1998. Nutrient uptake by microorganisms according to kinetic parameters from theory as related to cytoarchitecture. *Microbiol. Mol. Biol. Rev.* **62:**636–645.

Button, D. K. 2000. Effect of nutrient kinetics and cytoarchitecture on bacterioplankter size. *Limnol. Oceanogr.* **45:**499–505.

Canfield, D. E. 1998. A new model for Proterozoic ocean chemistry. *Nature* **396:**450–453.

Cavet, J. S., G. P. Borrelly, and N. J. Robinson. 2003. Zn, Cu and Co in cyanobacteria: selective control of metal availability. *FEMS Microbiol. Rev.* **27:**165–181.

Chain, P., J. Lamerdin, F. Larimer, W. Regala, V. Lao, M. Land, L. Hauser, A. Hooper, M. Klotz, J. Norton, L. Sayavedra-Soto, D. Arciero, N. Hommes, M. Whittaker, and D. Arp. 2003. Complete genome sequence of the ammonia-oxidizing bacterium and obligate chemolithoautotroph *Nitrosomonas europaea*. *J. Bacteriol.* **185:**2759–2773.

Clark, L. L., E. D. Ingall, and R. Benner. 1999. Marine organic phosphorus cycling: novel insights from nuclear magnetic resonance. *Am. J. Sci.* **299:**724–737.

Coale, K. H., and K. W. Bruland. 1988. Copper complexation in the northeast Pacific. *Limnol. Oceanogr.* **33:**1084–1101.

Coale, K. H., and K. W. Bruland. 1990. Spatial and temporal variability in copper complexation in the north Pacific. *Deep Sea Res. Part I* **37:**317–336.

de la Torre, J. R., C. B. Walker, A. E. Ingalls, M. Könneke, and D. A. Stahl. 2008. Cultivation of a thermophilic ammonia oxidizing archaeon synthesizing crenarchaeol. *Environ. Microbiol.* **10:**810–818.

DeLong, E. F. 1992. Archaea in coastal marine environments. *Proc. Natl. Acad. Sci. USA* **89:**5685–5689.

DeLong, E. F., K. Y. Wu, B. B. Prezelin, and R. V. Jovine. 1994. High abundance of archaea in Antarctic marine picoplankton. *Nature* **371:**695–697.

Diamant, S., N. Eliahu, D. Rosenthal, and P. Goloubinoff. 2001. Chemical chaperones regulate molecular chaperones in vitro and in cells under combined salt and heat stresses. *J. Biol. Chem.* **276:**39586–39591.

Dupont, C. L., S. Yang, B. Palenik, and P. E. Bourne. 2006. Modern proteomes contain putative imprints of ancient shifts in trace metal geochemistry. *Proc. Natl. Acad. Sci. USA* **103:**17822–17827.

Dyhrman, S. T., and S. T. Haley. 2006. Phosphorus scavenging in the unicellular marine diazotroph *Crocosphaera watsonii*. *Appl. Environ. Microbiol.* **72:**1452–1458.

Dyhrman, S. T., P. D. Chappell, S. T. Haley, J. W. Moffett, E. D. Orchard, J. B. Waterbury, and E. A. Webb. 2006a. Phosphonate utilization by the globally important marine diazotroph *Trichodesmium*. *Nature* **439:**68–71.

Dyhrman, S. T., S. T. Haley, S. R. Birkeland, L. L. Wurch, M. J. Cipriano, and A. G. McArthur. 2006b. Long serial analysis of gene expression for gene discovery and transcriptome profiling in the widespread marine coccolithophore *Emiliania huxleyi*. *Appl. Environ. Microbiol.* **72:**252–260.

Ensign, S. A., M. R. Hyman, and D. J. Arp. 1993. In vitro activation of ammonia monooxygenase from *Nitrosomonas europaea* by copper. *J. Bacteriol.* **175:**1971–1980.

Eppley, R. W., and E. H. Renger. 1974. Nitrogen assimilation of an oceanic diatom in nitrogen-limited continuous culture. *J. Phycol.* **10:**15–23.

Eppley, R. W., J. N. Rogers, and J. J. McCarthy. 1969. Half-saturation constants for uptake of nitrate and ammonium by marine phytoplankton. *Limnol. Oceanogr.* **14:**912–920.

Facciotti, M. T., D. J. Reiss, M. Pan, A. Kaur, M. Vuthoori, R. Bonneau, P. Shannon, A. Srivastava, S. M. Donohoe, L. E. Hood, and N. S. Baliga. 2007. General transcription factor specified global gene regulation in Archaea. *Proc. Natl. Acad. Sci. USA* **104:**4630–4935.

Francis, C. A., K. J. Roberts, J. M. Beman, A. E. Santoro, and B. B. Oakley. 2005. Ubiquity and diversity of ammonia-oxidizing archaea in water columns and sediments of the ocean. *Proc. Natl. Acad. Sci. USA* **102:**14683–14688.

Fuhrman, J. A., K. McCallum, and A. A. Davis. 1992. Novel major archaebacterial group from marine plankton. *Nature* **356:**148–149.

Fuhrman, J. A., K. McCallum, and A. A. Davis. 1993. Phylogenetic diversity of subsurface marine microbial communities from the Atlantic and Pacific Oceans. *Appl. Environ. Microbiol.* **59:**1294–1302.

Fukuto, J. M., M. D. Bartberger, A. S. Dutton, N. Paolocci, D. A. Wink, and K. N. Houk. 2005a. The physiological chemistry and biological activity of nitroxyl (HNO): the neglected, misunderstood, and enigmatic nitrogen oxide. *Chem. Res. Toxicol.* **18:**790–801.

Fukuto, J. M., A. S. Dutton, and K. N. Houk. 2005b. The chemistry and biology of nitroxyl (HNO): a chemically unique species with novel and important biological activity. *Chembiochemistry* **6:**612–619.

Fukuto, J. M., C. H. Switzer, K. M. Miranda, and

D. A. Wink. 2005c. Nitroxyl (HNO): chemistry, biochemistry, and pharmacology. *Annu. Rev. Pharmacol. Toxicol.* **45:**335–355.

Gibson, D. T., S. M. Resnick, K. Lee, J. M. Brand, D. S. Torok, L. P. Wackett, M. J. Schocken, and B. E. Haigler. 1995. Desaturation, dioxygenation, and monooxygenation reactions catalyzed by naphthalene dioxygenase from *Pseudomonas* sp. strain 9816-4. *J. Bacteriol.* **177:**2615–2621.

Giovannoni, S. J., and U. Stingl. 2005. Molecular diversity and ecology of microbial plankton. *Nature* **437:**343–348.

Gold, B., H. Deng, R. Bryk, D. Vargas, D. Eliezer, J. Roberts, X. Jiang, and C. Nathan. 2008. Identification of a copper-binding metallothionein in pathogenic mycobacteria. *Nat. Chem. Biol.* **4:**609–616.

Grass, G., and C. Rensing. 2001. CueO is a multicopper oxidase that confers copper tolerance in *Escherichia coli*. *Biochem. Biophys. Res. Commun.* **286:**902–908.

Hallam, S. J., K. T. Konstantinidis, N. Putnam, C. Schleper, Y. Watanabe, J. Sugahara, C. Preston, J. de la Torre, P. M. Richardson, and E. F. DeLong. 2006a. Genomic analysis of the uncultivated marine crenarchaeote *Cenarchaeum symbiosum*. *Proc. Natl. Acad. Sci. USA* **103:**18296–18301.

Hallam, S. J., T. J. Mincer, C. Schleper, C. M. Preston, K. Roberts, P. M. Richardson, and E. F. DeLong. 2006b. Pathways of carbon assimilation and ammonia oxidation suggested by environmental genomic analyses of marine *Crenarchaeota*. *PLoS Biol.* **4:**520–536.

Harder, W., and L. Dijkhuizen. 1983. Physiological responses to nutrient limitation. *Annu. Rev. Microbiol.* **37:**1–23.

Hashimoto, L. K., W. A. Kaplan, S. C. Wofsy, and M. B. McElroy. 1983. Transformations of fixed nitrogen and N_2O in the Cariaco Trench. *Deep Sea Res.* **30:**575–590.

Hatzenpichler, R., E. V. Lebecleva, E. Spieck, K. Stoecker, A. Richter, H. Daims, and M. Wagner. 2008. A moderately thermophilic ammonia-oxidizing crenarchaeote from a hot spring. *Proc. Natl. Acad. Sci. USA* **105:**2134–2139.

Hiniker, A., J. F. Collet, and J. C. Bardwell. 2005. Copper stress causes an in vivo requirement for the *Escherichia coli* disulfide isomerase DsbC. *J. Biol. Chem.* **280:**33785–33791.

Hjort, K., and R. Bernander. 1999. Changes in cell size and DNA content in *Sulfolobus* cultures during dilution and temperature shift experiments. *J. Bacteriol.* **181:**5669–5575.

Holo, H. 1989. *Chloroflexus aurantiacus* secretes 3-hydroxypropionate, a possible intermediate in the assimilation of CO_2 and acetate. *Arch. Microbiol.* **151:**252–256.

Holo, H., and I. F. Nes. 1989. High-frequency transformation, by electroporation, of *Lactococcus lactis* subsp. *cremoris* grown with glycine in osmotically stabilized media. *Appl. Environ. Microbiol.* **55:**3119–3123.

Huckle, J. W., A. P. Morby, J. S. Turner, and N. J. Robinson. 1993. Isolation of a prokaryotic metallothionein locus and analysis of transcriptional control by trace metal ions. *Mol. Microbiol.* **7:**177–187.

Ingalls, A. E., S. R. Shah, R. L. Hansman, L. I. Aluwihare, G. M. Santos, E. R. M. Druffel, and A. Pearson. 2006. Quantifying archaeal community autotrophy in the mesopelagic ocean using natural radiocarbon. *Proc. Natl. Acad. Sci. USA* **103:**6442–6447.

Jia, Z., and R. Conrad. 2009. *Bacteria* rather than *Archaea* dominate microbial ammonia oxidation in an agricultural soil. *Environ. Microbiol.* **11:**1658–1671.

Jiang, Q. Q., and L. R. Bakken. 1999. Comparison of *Nitrosospira* strains isolated from terrestrial environments. *FEMS Microbiol. Ecol.* **30:**171–186.

Jones, R. D., and R. Y. Morita. 1985. Survival of a marine ammonium oxidizer under energy-source deprivation. *Mar. Ecol. Prog. Ser.* **26:**175–179.

Karner, M. B., E. F. DeLong, and D. M. Karl. 2001. Archaeal dominance in the mesopelagic zone of the Pacific Ocean. *Nature* **409:**507–510.

Keen, G. A., and J. I. Prosser. 1987. Steady state and transient growth of autotrophic nitrifying bacteria. *Arch. Microbiol.* **147:**73–79.

Khademi, S., J. O'Connell 3rd, J. Remis, Y. Robles-Colmenares, L. J. Miercke, and R. M. Stroud. 2004. Mechanism of ammonia transport by Amt/MEP/Rh: structure of AmtB at 1.35 A. *Science* **305:**1587–1594.

Kittredge, J. S., and E. Roberts. 1969. A carbon-phosphorus bond in nature. *Science* **164:**37–42.

Klotz, M. G., and L. Y. Stein. 2008. Nitrifier genomics and evolution of the nitrogen cycle. *FEMS Microbiol. Lett.* **278:**146–156.

Klotz, M. G., D. J. Arp, P. S. G. Chain, A. F. El-Sheikh, L. J. Hauser, N. G. Hommes, F. W. Larimer, S. A. Malfatti, J. M. Norton, A. T. Poret-Peterson, L. M. Vergez, and B. B. Ward. 2006. Complete genome sequence of the marine, chemolithoautotrophic, ammonia-oxidizing bacterium *Nitrosococcus oceani* ATCC 19707. *Appl. Environ. Microbiol.* **72:**6299–6315.

Könneke, M., A. E. Bernhard, J. R. de la Torre, C. B. Walker, J. B. Waterbury, and D. A. Stahl. 2005. Isolation of an autotrophic ammonia-oxidizing marine archaeon. *Nature* **437:**543–546.

Kononova, S. V., and M. A. Nesmeyanova. 2002. Phosphonates and their degradation by microorganisms. *Biochemistry (Moscow)* **67:**184–195.

Kurtz, S., A. Phillippy, A. L. Delcher, M. Smoot, M. Shumway, C. Antonescu, and S. L. Salzberg. 2004. Versatile and open software for comparing large genomes. *Genome Biol.* **5:**R12.

Laanbroek, H. J., and S. Gerards. 1993. Competition for limiting amounts of oxygen between *Nitrosomonas europaea* and *Nitrobacter winogradskyi* grown in mixed continuous cultures. *Arch. Microbiol.* **159:**453–459.

Laanbroek, H. J., P. L. E. Bodelier, and S. Gerards. 1994. Oxygen consumption kinetics of *Nitrosomonas europaea* and *Nitrobacter hamburgensis* grown in mixed continuous cultures at different oxygen concentrations. *Arch. Microbiol.* **161:**156–162.

Lebedeva, E. V., M. Alawi, C. Fiencke, B. Namsaraev, E. Bock, and E. Spieck. 2005. Moderately thermophilic nitrifying bacteria from a hot spring of the Baikal rift zone. *FEMS Microbiol. Ecol.* **54:**297–306.

Lee, S., and J. A. Fuhrman. 1987. Relationships between biovolume and biomass of naturally derived marine bacterioplankton. *Appl. Environ. Microbiol.* **53:**1298–1303.

Leininger, S., T. Urich, M. Schloter, L. Schwark, J. Qi, G. W. Nicol, J. I. Prosser, S. C. Schuster, and C. Schleper. 2006. Archaea predominate among ammonia-oxidizing prokaryotes in soils. *Nature* **442:**806–809.

Lindas, A. C., E. A. Karlsson, M. T. Lindgren, T. J. Ettema, and R. Bernander. 2008. A unique cell division machinery in the Archaea. *Proc. Natl. Acad. Sci. USA* **105:**18942–18946.

Lopez-Garcia, P., C. Brochier, D. Moreira, and F. Rodriguez-Valera. 2004. Comparative analysis of a genome fragment of an uncultivated mesopelagic crenarchaeote reveals multiple horizontal gene transfers. *Environ. Microbiol.* **6:**19–34.

Loureiro, S., C. Jauzein, E. Garcés, Y. Collos, J. Camp, and D. Vaqué. 2009. The significance of organic nutrients in the nutrition of *Pseudo-nitzschia delicatissima* (Bacillariophyceae). *J. Plankt. Res.* **31:**399–410.

Loveless, J. E., and H. A. Painter. 1968. Influence of metal ion concentrations and pH value on growth of a *Nitrosomonas* strain isolates from activated sludge. *J. Gen. Microbiol.* **52:**1–14.

Lundgren, M., and R. Bernander. 2005. Archaeal cell cycle progress. *Curr. Opin. Microbiol.* **8:**662–668.

Lundgren, M., L. Malandrin, S. Eriksson, H. Huber, and R. Bernander. 2008. Cell cycle characteristics of *Crenarchaeota*: unity among diversity. *J. Bacteriol.* **190:**5362–5367.

Markowitz, V. M., F. Korzeniewski, K. Palaniappan, E. Szeto, G. Werner, A. Padki, X. L. Zhao, I. Dubchak, P. Hugenholtz, I. Anderson, A. Lykidis, K. Mavromatis, N. Ivanova, and N. C. Kyrpides. 2006. The integrated microbial genomes (IMG) system. *Nucleic Acids Res.* **34:**D344–D348.

Markowitz, V. M., E. Szeto, K. Palaniappan, Y. Grechkin, K. Chu, I. M. A. Chen, I. Dubchak, I. Anderson, A. Lykidis, K. Mavromatis, N. N. Ivanova, and N. C. Kyrpides. 2008. The integrated microbial genomes (IMG) system in 2007: data content and analysis tool extensions. *Nucleic Acids Res.* **36:**D528–D533.

Martens-Habbena, W., P. M. Berube, H. Urakawa, J. de la Torre, and D. A. Stahl. 2009. Ammonia oxidation kinetics determines niche separation of nitrifying Archaea and Bacteria. *Nature* **461:**976–979.

Massana, R., A. E. Murray, C. M. Preston, and E. F. DeLong. 1997. Vertical distribution and phylogenetic characterization of marine planktonic *Archaea* in the Santa Barbara Channel. *Appl. Environ. Microbiol.* **63:**50–56.

Miceli, M. V., T. O. Henderson, and T. C. Myers. 1980. 2-Aminoethylphosphonic acid metabolism during embryonic development of the planorbid snail *Helisoma. Science* **209:**1245–1247.

Mincer, T. J., M. J. Church, L. T. Taylor, C. Preston, D. M. Karl, and E. F. DeLong. 2007. Quantitative distribution of presumptive archaeal and bacterial nitrifiers in Monterey Bay and the North Pacific Subtropical Gyre. *Environ. Microbiol.* **9:**1162–1175.

Miranda, K. M., R. W. Nims, D. D. Thomas, M. G. Espey, D. Citrin, M. D. Bartberger, N. Paolocci, J. M. Fukuto, M. Feelisch, and D. A. Wink. 2003a. Comparison of the reactivity of nitric oxide and nitroxyl with heme proteins. A chemical discussion of the differential biological effects of these redox related products of NOS. *J. Inorg. Biochem.* **93:**52–60.

Miranda, K. M., N. Paolocci, T. Katori, D. D. Thomas, E. Ford, M. D. Bartberger, M. G. Espey, D. A. Kass, M. Feelisch, J. M. Fukuto, and D. A. Wink. 2003b. A biochemical rationale for the discrete behavior of nitroxyl and nitric oxide in the cardiovascular system. *Proc. Natl. Acad. Sci. USA* **100:**9196–9201.

Nunoura, T., H. Hirayama, H. Takami, H. Oida, S. Nishi, S. Shimamura, Y. Suzuki, F. Inagaki, K. Takai, K. H. Nealson, and K. Horikoshi. 2005. Genetic and functional properties of uncultivated thermophilic crenarchaeotes from a subsurface gold mine as revealed by analysis of genome fragments. *Environ. Microbiol.* **7:**1967–1984.

Olson, R. J. 1981. N-15 tracer studies of the primary nitrite maximum. *J. Mar. Res.* **39:**203–226.

Peers, G., and N. M. Price. 2006. Copper-containing plastocyanin used for electron transport by an oceanic diatom. *Nature* **441:**341–344.

Pitcher, A., S. Schouten, and J. S. Sinninghe

Damsté. 2009. In situ production of crenarchaeol in two California hot springs. *Appl. Environ. Microbiol.* **75:**4443–4451.

Preston, C. M., K. Y. Wu, T. F. Molinski, and E. F. DeLong. 1996. A psychrophilic crenarchaeon inhabits a marine sponge: *Cenarchaeum symbiosum* gen. nov., sp. nov. *Proc. Natl. Acad. Sci. USA* **93:**6241–6246.

Prosser, J. I. 1989. Autotrophic nitrification in bacteria, p. 125–181. *In* A. H. Rose and D. W. Tempest (ed.), *Advances in Microbial Physiology*, vol. 30. Academic Press, San Diego, CA.

Prosser, J. I., and G. W. Nicol. 2008. Relative contributions of archaea and bacteria to aerobic ammonia oxidation in the environment. *Environ. Microbiol.* **10:**2931–2941.

Quinn, J. P., A. N. Kulakova, N. A. Cooley, and J. W. McGrath. 2007. New ways to break an old bond: the bacterial carbon-phosphorus hydrolases and their role in biogeochemical phosphorus cycling. *Environ. Microbiol.* **9:**2392–2400.

Rappé, M. S., S. A. Connon, K. L. Vergin, and S. J. Giovannoni. 2002. Cultivation of the ubiquitous SAR11 marine bacterioplankton clade. *Nature* **418:**630–633.

Reay, D. S., D. B. Nedwell, J. Priddle, and J. C. Ellis-Evans. 1999. Temperature dependence of inorganic nitrogen uptake: reduced affinity for nitrate at suboptimal temperatures in both algae and bacteria. *Appl. Environ. Microbiol.* **65:**2577–2584.

Rensing, C., and G. Grass. 2003. *Escherichia coli* mechanisms of copper homeostasis in a changing environment. *FEMS Microbiol. Rev.* **27:**197–213.

Reva, O., and B. Tümmler. 2008. Think big-giant genes in bacteria. *Environ. Microbiol.* **10:**768–777.

Roszak, D. B., and R. R. Colwell. 1987. Survival strategies of bacteria in the natural environment. *Microbiol. Rev.* **51:**365–379.

Samson, R. Y., T. Obita, S. M. Freund, R. L. Williams, and S. D. Bell. 2008. A role for the ESCRT system in cell division in archaea. *Science* **322:**1710–1713.

Sandman, K., and J. N. Reeve. 2005. Archaeal chromatin proteins: different structures but common function? *Curr. Opin. Microbiol.* **8:**656–661.

Schleper, C., G. Jurgens, and M. Jonuscheit. 2005. Genomic studies of uncultivated archaea. *Nat. Rev. Microbiol.* **3:**479–488.

Schmidt, I., C. Look, E. Bock, and M. S. M. Jetten. 2004. Ammonium and hydroxylamine uptake and accumulation in *Nitrosomonas*. *Microbiology* **150:**1405–1412.

Schouten, S., E. C. Hopmans, R. D. Pancost, and J. S. Damste. 2000. Widespread occurrence of structurally diverse tetraether membrane lipids: evidence for the ubiquitous presence of low-temperature relatives of hyperthermophiles. *Proc. Natl. Acad. Sci. USA* **97:**14421–14426.

Schouten, S., E. C. Hopmans, E. Schefuß, and J. S. Sinninghe Damsté. 2002. Distributional variations in marine crenarchaeotal membrane lipids: a new tool for reconstructing ancient sea water temperatures? *Earth Planet. Sci. Lett.* **204:**265–274.

Schouten, S., E. C. Hopmans, M. Baas, H. Boumann, S. Standfest, M. Konneke, D. A. Stahl, and J. S. Sinninghe Damsté. 2008. Intact membrane lipids of "*Candidatus* Nitrosopumilus maritimus," a cultivated representative of the Cosmopolitan Mesophilic Group I Crenarchaeota. *Appl. Environ. Microbiol.* **74:**2433–2440.

Shao, Z., J. A. Blodgett, B. T. Circello, A. C. Eliot, R. Woodyer, G. Li, W. A. van der Donk, W. W. Metcalf, and H. Zhao. 2008. Biosynthesis of 2-hydroxyethylphosphonate, an unexpected intermediate common to multiple phosphonate biosynthetic pathways. *J. Biol. Chem.* **283:**23161–23168.

Simon, M., and F. Azam. 1989. Protein content and protein synthesis rates of planktonic marine bacteria. *Mar. Ecol. Prog. Ser.* **51:**201–213.

Solioz, M., and A. Odermatt. 1995. Copper and silver transport by CopB-ATPase in membrane vesicles of *Enterococcus hirae*. *J. Biol. Chem.* **270:**9217–9221.

Solioz, M., and J. V. Stoyanov. 2003. Copper homeostasis in *Enterococcus hirae*. *FEMS Microbiol. Rev.* **27:**183–195.

Stark, J. M., and M. K. Firestone. 1996. Kinetic characteristics of ammonium-oxidizer communities in a California oak woodland-annual grassland. *Soil Biol. Biochem.* **28:**1307–1317.

Stehr, G., B. Böttcher, P. Dittberner, G. Rath, and H. P. Koops. 1995. The ammonia-oxidizing nitrifying population of the River Elbe estuary. *FEMS Microbiol. Ecol.* **17:**177–186.

Strauss, G., and G. Fuchs. 1993. Enzymes of a novel autotrophic CO2 fixation pathway in the phototrophic bacterium *Chloroflexus aurantiacus*, the 3-hydroxypropionate cycle. *Eur. J. Biochem.* **215:**633–643.

Suwa, Y., Y. Imamura, T. Suzuki, T. Tashiro, and Y. Urushigawa. 1994. Ammonia-oxidizing bacteria with different sensitivities to $(NH_4)_2SO_4$ in activated sludges. *Water Res.* **28:**1523–1532.

Suzuki, I., U. Dular, and S. C. Kwok. 1974. Ammonia or ammonium ion as substrate for oxidation by *Nitrosomonas europaea* cells and extracts. *J. Bacteriol.* **120:**556–558.

Tremblay, P. L., and P. C. Hallenbeck. 2009. Of blood, brains and bacteria, the Amt/Rh transporter family: emerging role of Amt as a unique microbial sensor. *Mol. Microbiol.* **71:**12–22.

Treusch, A. H., S. Leininger, A. Kletzin, S. C. Schuster, H. P. Klenk, and C. Schleper. 2005. Novel genes for nitrite reductase and Amo-related proteins indicate a role of uncultivated mesophilic crenarchaeota in nitrogen cycling. *Environ. Microbiol.* **7:**1985–1995.

Valentine, D. L. 2007. Adaptations to energy stress dictate the ecology and evolution of the Archaea. *Nat. Rev. Microbiol.* **5:**316–323.

van de Vossenberg, J. L., A. J. Driessen, and W. N. Konings. 1998. The essence of being extremophilic: the role of the unique archaeal membrane lipids. *Extremophiles* **2:**163–170.

Varela, M. M., H. M. van Aken, E. Sintes, and G. J. Herndl. 2008. Latitudinal trends of Crenarchaeota and Bacteria in the meso- and bathypelagic water masses of the Eastern North Atlantic. *Environ. Microbiol.* **10:**110–124.

Venter, J. C., K. Remington, J. F. Heidelberg, A. L. Halpern, D. Rusch, J. A. Eisen, D. Wu, I. Paulsen, K. E. Nelson, W. Nelson, D. E. Fouts, S. Levy, A. H. Knap, M. W. Lomas, K. Nealson, O. White, J. Peterson, J. Hoffman, R. Parsons, H. Baden-Tillson, C. Pfannkoch, Y. H. Rogers, and H. O. Smith. 2004. Environmental genome shotgun sequencing of the Sargasso Sea. *Science* **304:**66–74.

Walker, C. B., J. R. de la Torre, M. G. Klotz, H. Urakawa, N. Pinel, D. J. Arp, C. Brochier-Armanet, P. S. Chain, P. P. Chan, A. Gollabgir, J. Hemp, M. Hügler, E. A. Karr, M. Könneke, M. Shin, T. J. Lawton, T. Lowe, W. Martens-Habbena, L. A. Sayavedra-Soto, D. Lang, S. M. Sievert, A. C. Rosenzweig, G. Manning, and D. A. Stahl. 2010. *Nitrosopumilus maritimus* genome reveals unique mechanisms for nitrification and autotrophy in globally distributed marine crenarchaea. *Proc. Natl. Acad. Sci. USA* **107:**8818–8823.

Ward, B. B. 1986. Nitrification in marine environments, p. 157–184. *In* J. I. Prosser (ed.), *Nitrification*. IRL Press, Oxford, United Kingdom.

Ward, B. B. 1987. Kinetic studies on ammonia and methane oxidation by *Nitrosococcus oceanus*. *Arch. Microbiol.* **147:**126–133.

Watson, S. W. 1965. Characteristics of a marine nitrifying bacterium, *Nitrosocystis oceanus* sp. N. *Limnol. Oceanogr.* **10:**R274–R289.

Zhang, C. L., A. Pearson, Y.-L. Li, G. Mills, and J. Wiegel. 2006. Thermophilic temperature optimum for crenarchaeol synthesis and its implication for archaeal evolution. *Appl. Environ. Microbiol.* **72:**4419–4422.

DISTRIBUTION AND ACTIVITY OF AMMONIA-OXIDIZING ARCHAEA IN NATURAL ENVIRONMENTS

Graeme W. Nicol, Sven Leininger, and Christa Schleper

7

ARCHAEA: IMPORTANT GLOBAL PLAYERS IN BIOGEOCHEMICAL CYCLES?

Discovery of Archaeal Ubiquity in the Environment

Before the discovery of large abundances of archaea in "nonextreme" terrestrial and aquatic environments, it was believed that archaea play only a marginal role in most global element cycles, due to their restriction to extreme habitats such as salt-saturated lakes (halophiles) and high-temperature terrestrial springs and deep-sea vents (thermoacidophiles, hyperthermophiles). Perhaps the only recognized exception was the distribution of methanogens (i.e., archaea that are ubiquitously distributed in anaerobic environments and represent the sole biological source of the greenhouse gas methane) (Garcia et al., 2000). However, in the early 1990s, the use of cultivation-independent molecular techniques led to the discovery of crenarchaeal and euryarchaeal lineages in most moderate and aerobic environments (and often in a vast number), indicating that archaea might contribute more significantly to biogeochemical cycles than previously thought (for review, see Delong, 1998). However, it took several years before any aspect of their physiology or ecological role could be determined.

Initial 16S rRNA gene surveys of moderate archaea in marine environments indicated the presence of three previously undetected lineages that were termed Group I, affiliated with the kingdom of crenarchaeota, and Groups II and III within the kingdom euryarchaeota (Delong, 1992). In particular, organisms within Group I (a lineage distinct from, but specifically associated with, cultured hyperthermophilic organisms) were found to be abundant in many moderate habitats. These 16S rRNA gene sequences from marine and soil samples were quickly recognized to be separated into two distinct clades, referred to as Group 1.1a and 1.1b lineages, respectively (Fig. 1). Despite an abundant and seemingly ubiquitous distribution, aspects of the physiology and energy metabolism of these organisms remained unknown for over a decade. Initial indirect insights into their physiology were from studies of archaea in marine habitats using stable isotope, microautoradiography, or natural radiocarbon analyses which indicated that both modes of carbon assimilation occurred within marine archaea, i.e., autotrophic mode

Graeme W. Nicol, Institute of Biological & Environmental Sciences, University of Aberdeen, Aberdeen AB24 3UU, United Kingdom. *Sven Leininger*, Sars International Centre for Marine Molecular Biology, N-5008 Bergen, Norway. *Christa Schleper*, Department of Genetics in Ecology, University of Vienna, A-1090 Vienna, Austria.

Nitrification, Edited by Bess B. Ward, Daniel J. Arp, and Martin G. Klotz

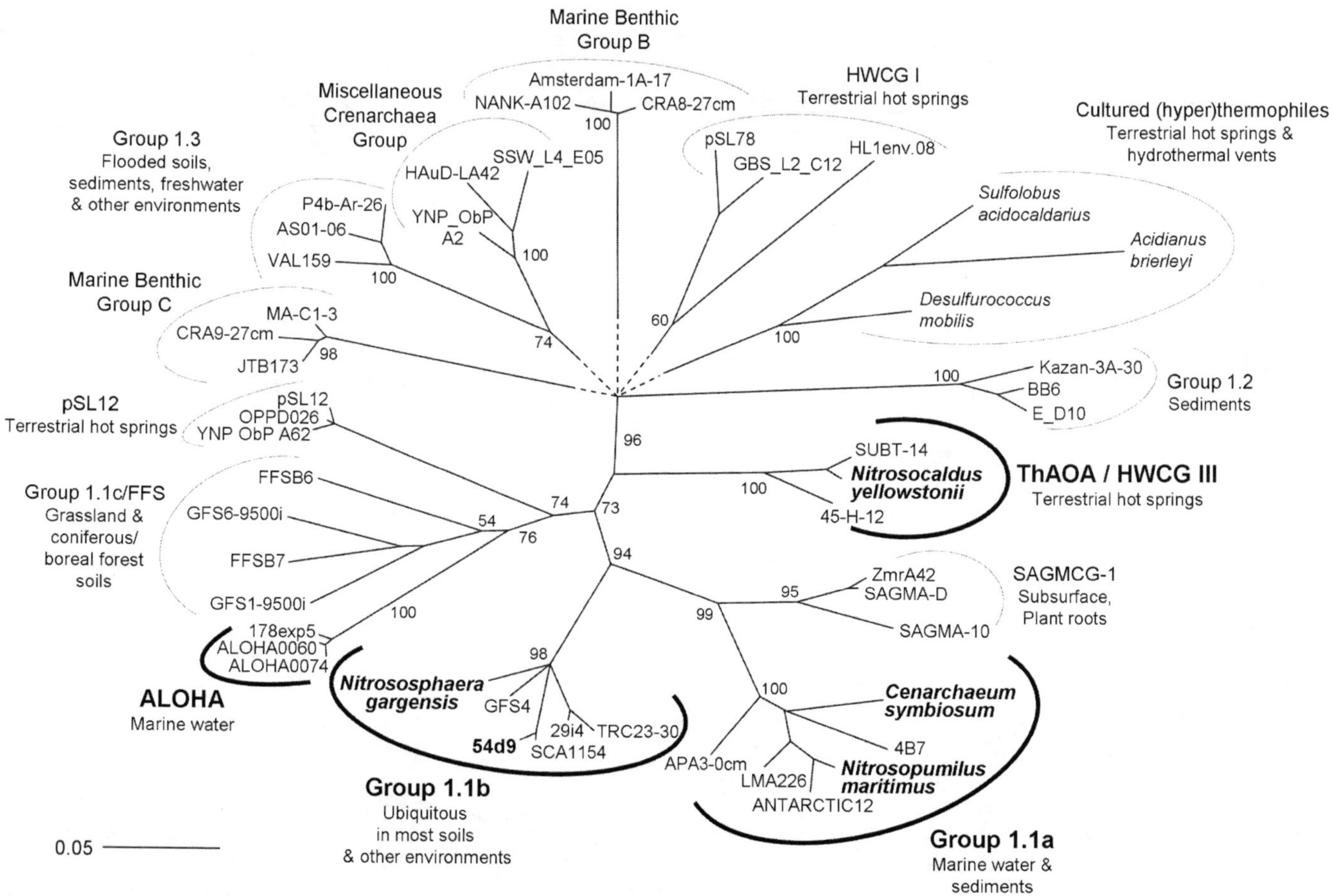

FIGURE 1 Phylogenetic analysis of crenarchaeal 16S rRNA gene sequences recovered from marine and terrestrial environments together with cultivated AOA and hyperthermophiles. Sequence names in bold represent cultivated organisms or genomic fragments containing both 16S rRNA and AMO subunit genes. Lineages with a bold arc represent those known to have associated AMO subunit genes. Dashed lines at multifurcating nodes indicate manual adjustment reflecting low bootstrap support for any relative branching order. The scale bar represents an estimated 0.05 changes per nucleotide position, and numbers at nodes indicate the most conservative value of bootstrap support from three treeing methods.

(using inorganic carbon as a nutrient source) (e.g. Kuypers et al., 2001; Pearson et al., 2001; Wuchter et al., 2003) and heterotrophic mode (using organic carbon compounds as nutrients) (Ouverney and Fuhrman, 2000; Herndl et al., 2005; Teira et al., 2006a; Ingalls et al., 2006). However, identification of specific energy metabolisms remained elusive.

INSIGHTS FROM COMMUNITY GENOMICS (METAGENOMICS)

One of the major challenges in studying microorganisms from natural habitats is the inability to cultivate many of them in the laboratory. One approach that has managed to provide valuable hypotheses and insights into the physiology and ecology of uncultured microorganisms has been the direct cloning and sequencing of large contiguous fragments of environmental DNA (Treusch et al., 2004). These fragments represent discrete portions of genomes from a complex and diverse natural microbial consortium (often referred to as the "metagenome"), rather than partial fragments of individual genes such as 16S rRNA. These genome fragments are then cloned into bacterial artificial chromosomes, or more commonly, bacterial artificial chromosome-derived fosmid vectors, which are archived in *Escherichia coli* clone libraries. The libraries can then be screened for genes of interest by PCR, by hybridization methods with oligonucleotide probes, or by end-sequencing of plasmid inserts (Handelsman, 2004; Treusch and Schleper, 2005).

The first fragments of a genome from an uncultured archaeon was obtained from a marine metagenomic library produced by Stein et al. (1996), with clone "4B7" containing an insert size of 38.5 kb, and identified as being affiliated with Group I based on the presence of a 16S rRNA gene. Other metagenomic libraries from microorganisms associated with a marine sponge, marine plankton, as well as soil-contained Group I genomic fragments (Schleper et al., 1998; Beja et al., 2000, 2002; Quaiser et al., 2002 ; Lopez-Garcia et al., 2004; Treusch et al., 2004). Although these analyses revealed many genomic features of mesophilic archaea, they lacked evidence indicating the fundamental energy metabolism.

The first insight into a specific energy metabolism came from a fosmid clone derived from a soil-derived metagenomic library. Based on 16S and 23S rRNA genes, the fosmid clone "54d9" was identified as belonging to the ubiquitous "soil group" Group 1.1b (Fig. 1). In addition, it contained two open reading frames (ORFs) that coded for putative alpha and beta subunits (AmoA and AmoB, respectively) of an ammonia monooxygenase (AMO) as well as a gene whose product was highly similar to copper-dependent nitrite reductases (NirK) (Treusch et al., 2005). An in silico comparison to environmental sequences deposited in public databases showed that the soil-derived archaeal *amoA* and *amoB* genes were highly similar to archaea-associated scaffolds from the whole genome shotgun sequencing project of the Sargasso Sea (Venter et al., 2004; Schleper et al., 2005; Treusch et al., 2005). In that project, short sequences had been generated from small insert metagenomic libraries, and single reads had, subsequently, been assembled in silico (Venter et al., 2004). Additionally, the genomic fragments from marine archaea assembled in the whole genome shotgun project contained genes coding for the C-subunit of an AMO, apparently organized in a cluster together with *amoA* and *amoB* in a *BCA* gene order and contrasted with the *CAB* arrangement observed in ammonia-oxidizing bacteria (AOB) (Nicol and Schleper, 2006). The similarity of the soil and marine-derived AmoA sequences to the alpha subunits of the bacterial AMO and the particulate methane monooxygenase (pMMO) from bacterial methane oxidizers was only about 40% (~25% identity) on the amino acid level. In contrast, the similarity between the two related proteins AMO and pMMO in bacteria is much higher with up to 74% (~50% identity). Furthermore, the putative *amo/pmo* genes of archaea were considerably shorter than those of their bacterial homologues. Further evidence supporting the hypothesis that the respective archaeal

ORFs indeed coded for subunits of an AMO/pMMO-related protein was provided through the comparison with structural data obtained of the pMMO of *Methylococcus capsulatus* (Lieberman and Rosenzweig, 2005). Many amino acid residues potentially involved in copper-binding metal centers were also found to be highly conserved in the archaeal variants (Treusch et al., 2005). Moreover, microcosm experiments were conducted to study transcription of the archaeal *amoA* genes. A significant increase in transcriptional activity of the putative *amoA* gene was observed upon incubation with NH_4^+, suggesting that the *amo*-like genes indeed coded for a monooxygenase involved in the oxidation of ammonia (Treusch et al., 2005).

Nitrosopumilus maritimus: a Cultivated Archaeal Ammonia Oxidizer

The definitive proof for the ammonia-oxidizing capability of moderate archaea was provided through the cultivation of *N. maritimus* (Könneke et al., 2005). Based on its 16S rRNA gene sequence, the isolate SCM1 could be assigned to Marine Group I (or Group 1.1a). It also contained genes for the A, B, and C subunits of the archaeal AMO, which were highly similar to the *amoA*, *amoB*, and *amoC* sequences of planktonic archaea (Venter et al., 2004) and to *amoA* and *amoB* of soil archaea (Treusch et al., 2005) with 93 to 98% and 80 to 90% identity on the amino acid and nucleotide level, respectively. Cells of *N. maritimus* are rod shaped and particularly small with a maximal diameter of 0.22 and 0.9 μm in length. These cells also resembled marine archaea investigated by fluorescent in situ hybridization, in particular the sponge symbiont *Cenarchaeum symbiosum* (Preston et al., 1996) and other planktonic cells (DeLong et al., 1999). SCM1 grew at rates of 0.78 day^{-1} and was shown to convert ammonia into nitrite stoichiometrically with an approximate rate of ~4 to 14 fmol of NO_2^- $cell^{-1}$ day^{-1} in the absence of organic compounds, thus indicating an autotrophic metabolism. Moreover, growth was inhibited when organic substances were added. More recently, Martens-Habbena et al. (2009) demonstrated that SCM1 has the lowest half-saturation constant and highest specific affinity for ammonium ever reported for an ammonia-oxidizing prokaryote, indicating that this organism is adapted to thriving in environments with extremely low nutrient levels, such as the open ocean.

Archaeal ammonia oxidizers of Group 1.1a were also enriched by cultivation from coastal water of the North Sea, again showing conversion of ammonia to nitrite with correlating numbers of archaea and accumulation of nitrite (Wuchter et al., 2006). The enrichment contained a single phylotype with 99% 16S rRNA gene sequence identity to *N. maritimus*. The single *amoA* gene identified in this enrichment was also highly similar to that of *N. maritimus* (91% on nucleotide and 98% on amino acid level) and to Sargasso Sea *amoA* sequences (90% and 95%, respectively). Quantitative measurements of 16S rRNA and *amoA* genes in this enrichment suggested a 1:1 ratio of these genes per crenarchaeal genome. This is in contrast to ratios within AOB that may have up to three *amoA* genes and one 16S rRNA gene per cell. This enrichment had an estimated nitrification rate of 2 to 4 fmol of converted NH_3 $cell^{-1}$ day^{-1}, which was in agreement with that of *N. maritimus*.

Genome Analysis of the AMO-Containing Archaeon *C. symbiosum*

The symbiotic archaeon *C. symbiosum* that resides in the tissue of the sponge *Axinella mexicana* (Preston et al., 1996) served as an early model to study molecular aspects of the nonthermophilic marine crenarchaeota (Schleper et al., 1997). It exists as two distinct but closely related populations (strains A and B) with a nucleotide sequence divergence of 0.7% in the 16S rRNA gene and microheterogeneity within protein-coding genes and spacer regions (Schleper et al., 1998). Recently, the complete genome sequence of *C. symbiosum* was determined from a metagenomic library providing further insights into the potential physiological properties of uncultured ammonia-oxidizing archaea (AOA) (Hallam et al., 2006a, 2006b). The genome contained ORFs for a putative AMO (containing all three subunits) highly similar to that of *N. maritimus* (Könneke et al., 2005) and the soil crenarchaeote (Treusch et al., 2005). Interestingly, the presence of a gene encoding the enzyme hydroxylamine oxidoreductase, required for the subsequent step in nitrification (oxidation of hydroxylamine to nitrite), has not been detected. Currently, there is no known homologue for hydroxylamine oxidoreductase present in archaea, and it seems that archaeal ammonia oxidizers may use another enzyme or alternative pathway to produce nitrite. Analysis of the *C. symbiosum* genome and additional environmental archaeal sequences indicate that AOA might use the 3-hydroxypropionate cycle for CO_2 fixation (Hallam et al., 2006a) with a complete 3-hydroxypropionate/4-hydroxybutyrate pathway present in *N. maritimus* and *C. symbiosum* as well as in the data set of the Sargasso Sea metagenome (Berg et al., 2007). Interestingly, a gene whose product is highly similar to copper-dependent nitrite reductases (NirK) was detected along with *amo*-like genes on fosmid 54d9 (Treusch et al., 2005). Recent genomic and metagenomic studies indicate that homologues of this gene are widely distributed in AOB (Cantera and Stein, 2007) and also archaea, the only exception being *C. symbiosum* (Bartossek et al., 2010).

Thaumarchaeota and Not Crenarchaeota?

Phylogenetic analysis of 16S rRNA genes from marine and terrestrial habitats indicated, in most studies, that the lineage comprising AOA were distinct from, but specifically associated with, the hyperthermophilic crenarchaeota lineage. One surprising finding from the recent genomic data from both cultivated and uncultivated mesophilic archaea is that this phylogenetic placement based on 16S rRNA genes alone may be misleading, and that AOA may actually belong to a different phylum that is as distinct from crenarchaeota as crenarchaeota are from the euryarchaeota. Brochier-Armanet et al. (2008) analyzed the genome of *C. symbiosum* and, based on phylogenetic analyses of ribosomal protein encoding genes and genome content, proposed that *C. symbiosum* and its (AOA) relatives be placed in a lineage called "thaumarchaeota" (from the Greek *Thaum*, meaning wonder). The increasing amount of genomic data from other studies, including the complete genomes of *N. maritimus* and *N. gargensis*, and metagenomic data appear to strongly support this hypothesis (Bartossek et al., 2010; Spang et al., 2010).

DIVERSITY AND DISTRIBUTION OF AOA

So far, the capacity for archaea to oxidize ammonia seems to be restricted to the Group 1 organisms and maybe a few associated lineages. The *amo* genes of AOB and AOA are distantly related, and molecular approaches to study the distribution and diversity of ammonia oxidizers rely on specific PCR primers that amplify the archaeal *amoA* or the bacterial *amoA* variant (Francis et al., 2005; Treusch et al., 2005). After the identification of *amo* genes on archaeal genome fragments (Schleper et al., 2005; Treusch et al., 2005) and the cultivation of the ammonia-oxidizing archaeon *N. maritimus* (Könneke et al., 2005), it became clear that AOA are globally distributed like mesophilic crenarchaeaota (thaumarchaeota). This was first observed by a molecular study amplifying the archaeal *amo*A gene from diverse marine as well as terrestrial ecosystems (Francis et al., 2005). Phylogenetic analyses revealed two primary evolutionary groups that reflected largely the habitat from which they were retrieved: (i) the soil and sediment group, containing most of the terrestrial sequences, and (ii) the marine water column and sediment group, consisting mostly of sequences recovered from marine-associated habitats (Fig. 2). It was also quite clear that this mirrored, to a large extent, earlier 16S rRNA gene-based phylogenies where the two dominant lineages could be associated largely with soil and marine habitats (Fig. 1). Although others exist, these two major groups still persist even when considering the many

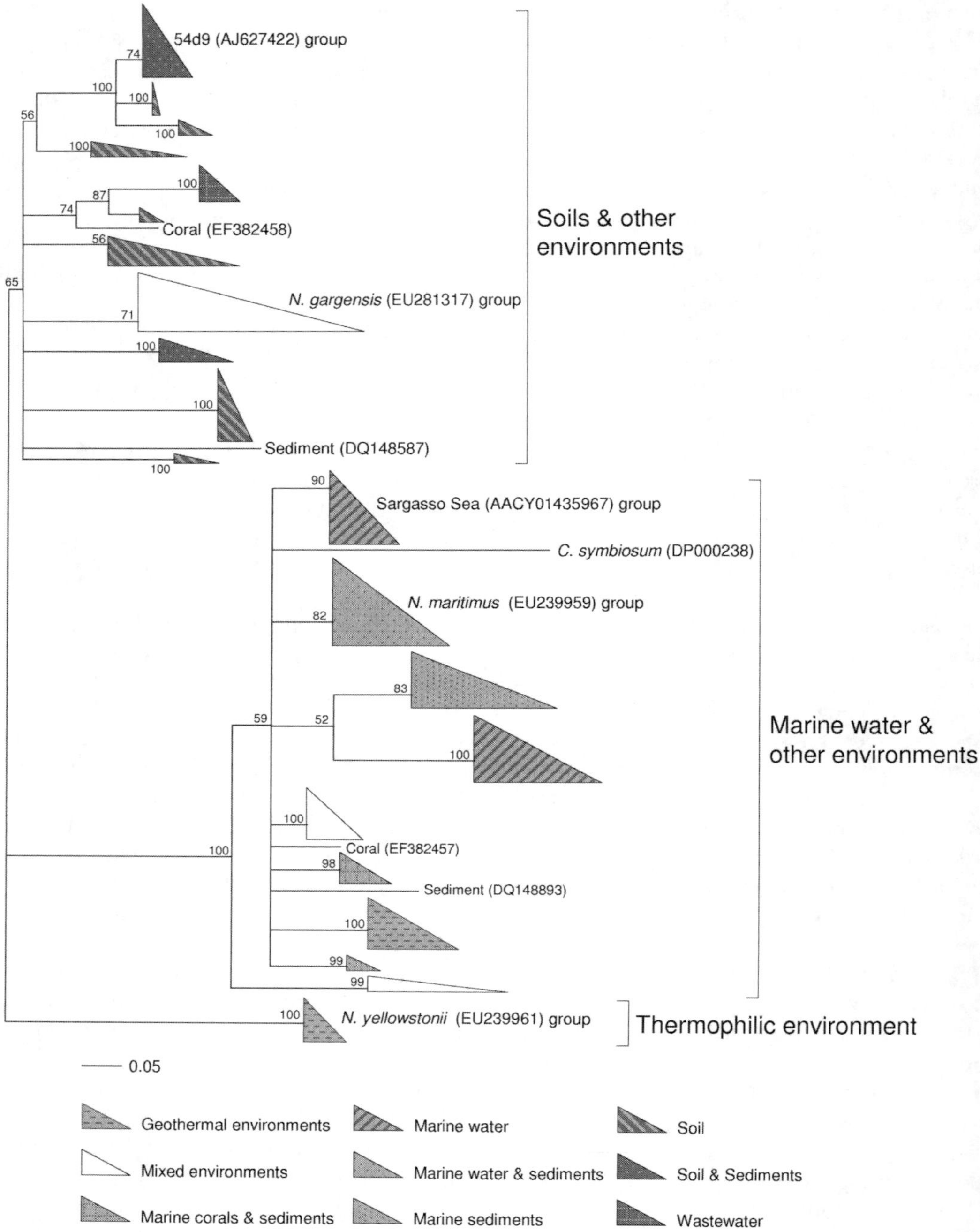

FIGURE 2 Phylogenetic analysis showing the diversity of AMO subunit A genes associated with archaea and the environments from which they were obtained. The height and length of each triangle is proportional to the number of taxa included in this analysis and maximum individual branch length, respectively. The scale bars represent an estimated 0.05 changes per nucleotide position, and numbers at nodes indicate the most conservative value of bootstrap support from three treeing methods. (From Prosser and Nicol, 2008, with permission.)

thousands of archaeal *amoA* sequences deposited in the public databases.

AOA in Soils

Crenarchaeota represent a stable fraction of up to 5% of the total microbial community in many soils (Ochsenreiter et al., 2003) with the predominant lineage in both pristine and agricultural soils being Group 1.1b. The first comprehensive study on the presence and abundance of AOA in soils included 12 pristine and differently managed agricultural soil samples of contrasting physicochemical characteristics spanning a geographical transect from northern to southern Europe (Leininger et al., 2006). Using quantitative PCR (qPCR), AOA *amoA* gene abundance ranged between 7×10^6 and 7×10^8 copies per gram of dry soil. Ratios of AOA to AOB *amo*A genes ranged from 1.5 to over 230 in surface soil layers, indicating a large numerical dominance of the AOA genes. The relative dominance of AOA over AOB also increases with soil depth (Fig. 3) in a situation that is curiously analogous to that seen in open ocean waters (i.e., archaeal numbers remain relatively constant, whereas bacterial numbers decrease) (Leininger et al., 2006; Jia and Conrad, 2009). However, despite the relatively constant numbers of AOA through a soil profile, distinct phylotypes are found associated with specific depths (Fig. 3), indicating specific populations are adapted to different conditions such as organic matter content or oxygen availability. These observations have been repeated in a number of studies, and the apparent numerical dominance of AOA appears to be a general pattern in soils globally (e.g., He et al., 2007; Boyle-Yarwood et al., 2008; Nicol et al., 2008). The quantities of AOA do not appear to be dependant on any physicochemical parameters in soil since they are constantly high in most soil types. However, there is a large amount of evidence that different populations of AOA are selected for soils with contrasting physicochemical conditions. Assuming the distribution of Group 1 16S rRNA phylotypes also indicates, to some extent, the distribution of AOA, a large number of studies have shown that archaeal communities vary across ecological gradients such as grassland management and levels of soil pollution (e.g., Sandaa et al., 1999; Nicol et al., 2003, 2005). Subsequent studies using the AOA marker *amoA* itself show similar trends with the selection of distinct populations under different conditions such as contrasting fertilizer regimens or long-term changes in soil pH (He et al., 2007; Nicol et al., 2008). In fact, a significant effect of long-term mineral fertilizer application (in contrast to manure fertilizers) is the decrease in pH that, in turn, may decrease potential nitrification rates. This has led to the hypothesis that pH rather than nutrient status is a major driver for the presence and abundance of ammonia oxidizers (He et al., 2007; Nicol et al., 2008; Hansel et al., 2008). Another parameter influencing diversity and distribution of archaeal ammonia oxidizers may be temperature. It has been indicated that there is a discrepancy between numbers of AOB and nitrification measurements in soils incubated at temperatures of 30°C or above (Avrahami and Bohannan, 2007). In addition, changes in transcriptionally active AOA and growth of specific populations has been observed at 30°C (Tourna et al., 2008; Offre et al., 2009), where there is no evidence of associated AOB growth, thereby indicating that AOA may be the predominant ammonia oxidizers in soils over particular temperature ranges.

There is increasing evidence that both natural and agricultural soils are habitats for AOA, and they are numerically dominating the ammonia-oxidizing microbial community and occupy a broad range of niches with varying physicochemical parameters. Based on their *amoA* gene phylogeny, most soil-derived AOA cluster within the soil/sediment group, but sequences affiliated with the marine water column/sediment cluster are also detected (He et al., 2007; Hansel et al., 2008). This indicates a large diversity of AOA communities that reflects the complexity and heterogeneity of the soil matrix. However, a major challenge will be to determine whether they truly are the predominantly active ammonia oxidizers

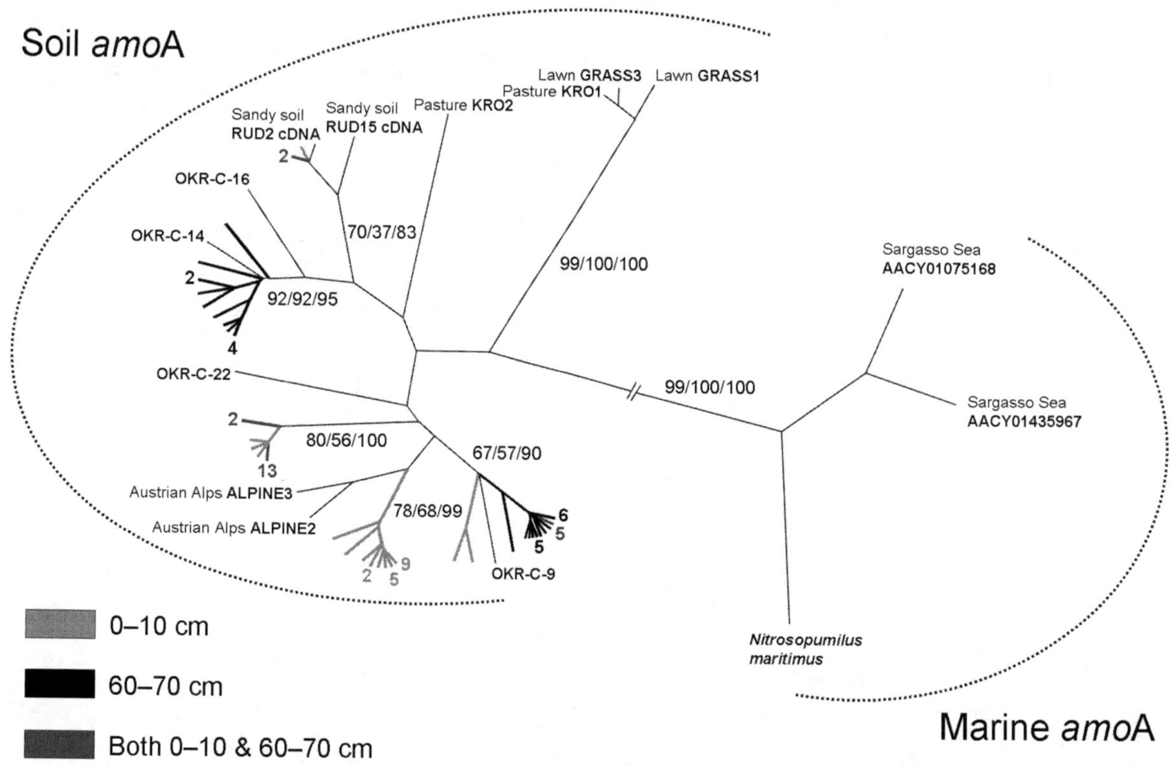

FIGURE 3 Analysis of a soil depth profile of a sandy ecosystem (Rotböll, Darmstadt, Germany). Absolute numbers of *amoA* genes of archaea and bacteria quantified by qPCR and phylogenetic analysis illustrating the relatedness of AOA *amoA* sequences retrieved from two depths (0 to 10 and 60 to 70 cm) are shown. The clustering of sequences from the same depth demonstrates the presence of populations adapted to specific conditions within the soil profile. (Adapted from data obtained by Leininger et al., *Nature* **442:**806–809, 2006, with permission from Macmillan Publishers, Ltd.)

in soil and whether they are major contributors to fertilizer loss in agricultural systems (see below).

AOA in the Ocean

Nearly all recovered AOA *amoA* sequences from the water column cluster within the major water column/sediment clade (Fig. 2). The *amoA* gene sequences of planktonic AOA in surface waters have a low divergence on the nucleotide level and almost complete identity on the amino acid level. A clear characteristic of marine AOA populations is their depth-related phylogenetic partitioning with distinct AOA communities at different water depths and few overlaps in their clustering behavior (Francis et al., 2005; Hallam et al., 2006a; Mincer et al., 2007; Nakagawa et al., 2007; Beman et al., 2008). By developing specific primer sets for the shallow and deep clusters, Beman et al. (2008) showed that AOA bearing the "shallow" *amoA* gene occur also in the deeper waters, but AOA detected in deep regions are not found in shallow waters. The phenomenon of AOA *amoA* community composition dependent on depth might be explained by a possible resistance to photoinhibition acquired by those AOA that thrive in the photic zones (Mincer et al., 2007). However, this might also reflect possible adaptations and phenotypes restricted to certain environmental characteristics. AOA *amoA* abundance has been found to correlate with the two nitrite maxima in shallow and deeper waters (Coolen et al., 2007; Herfort et al., 2007; Beman et al., 2008) and higher levels of nitrate (Mincer et al., 2007), suggesting an involvement in nitrification.

Similar to soil ecosystems, AOA seem to outnumber AOB in marine habitats. Abundances of AOA *amoA* and crenarchaeal 16S rRNA genes show a high correlation with slightly more *amoA* than 16S rRNA gene copies, indicating that most crenarchaeota might be AOA (Wuchter et al., 2006; Herfort et al., 2007). The high numbers of putatively AOA with 10^4 to 10^5 copies ml^{-1} in zones of nitrification activity contrast with the abundances of their bacterial counterparts, which are frequently detected in low numbers or even undetectable (Ward, 2000; Wuchter et al., 2006; Mincer et al., 2007). Furthermore, the quantitative co-occurrence of planktonic archaeal ammonia oxidizers and *Nitrospina*-like nitrite oxidizers in the lower part of the euphotic zone with considerable nitrification rates suggests a metabolic coupling between these two groups sustaining nitrification (Mincer et al., 2007). The abundance of AOA is in line with the general perception that in the open oceans, water column nitrification activity is thought to be less in surface waters and highest at the bottom of the euphotic zone. This may be due to competition for the substrate ammonia between nitrifiers with phytoplanktonic organisms and/or to light inhibition of the AMO enzyme (Ward, 2005) and fits also the seasonal variation of crenarchaeal and euryarchaeal abundances (Murray et al., 19981999; Wuchter et al., 2006; Herfort et al., 2007). It has been shown that there is an inverse correlation between crenarchaeota and chlorophyll *a* (Murray et al., 1998), supporting the idea of the negative impact of phytoplankton on nitrifier communities (Ward, 2005; Herfort et al., 2007). Previous studies could not find a correlation between bacterial ammonia-oxidizing community structures and nitrification rates in ocean waters of the Monterey Bay in California (O'Mullan and Ward, 2005). However, the abundance of AOA in different oceanic provinces has been demonstrated to correlate with regions where nitrification is important (Massana et al., 1997; DeLong et al., 1999; Karner et al., 2001; Teira et al., 2006b)

Interestingly, there is evidence that Group 1.1a archaea might not be the only AOA in the oceans (Mincer et al., 2007). Based on qPCR data, AOA *amoA* genes at 200 m depth in the northern Pacific were far more abundant than archaeal Group 1.1a 16S rRNA genes but revealed a strong correlation to quantities of 16S rRNA genes related to the deep branching (cren)archaeal pSL12 clade, first detected in hot springs (Barns et al., 1996) (Fig. 1). pSL12-like crenarchaeotes were only rarely detected in planktonic samples and had therefore been considered as atypical for pelagic environments. An AOA *amoA* clone library from the same depth, where pSL12 crenarchaeotes were more abundant than others, did not reveal specific *amoA* sequence types potentially characteristic for pSL12-like archaea, indicating that their *amoA* genes were indistinguishable from those of Group 1.1a.

AOA in Sediments

AOA *amoA* sequences recovered from marine estuarine sediments are not only distributed within the marine water/sediment cluster where they are found in specific groups, but some are also affiliated with the major soil/sediment clade (Fig. 2). The trend of the numerical dominance of AOA over AOB is also present in some sediments. AOA have been found to be up to 30 to 80 times more abundant than AOB in different estuarine sediments, and a correlation with environmental parameters indicates that AOA in sediments may prefer low salinity and lower oxygen concentrations (Mosier and Francis, 2008; Santoro et al., 2008). Previous studies on dynamics of AOB regarding community composition, abundance, and nitrification rates from different estuarine sediments revealed a strong effect of salinity gradients, ammonium, and oxygen concentrations as well (de Bie et al., 2001; Cebron et al., 2003; Francis et al., 2003; Bernhard et al., 2005, 2007). Nitrification is an important process in sedimentary biogeochemistry and particularly in estuarine sediments, which can be exposed to high loads of nutrients from agricultural runoff (Beman et al., 2006). In sediments with a low-oxygen concentration, nitrification is directly linked to N loss of fixed nitrogen through denitrification or anammox (Seitzinger, 1988; Galloway et al., 2004; Seitzinger et al., 2006). Based on their abundance, AOA may have a crucial role in this hot spot of global nitrification.

AOA Associated with Marine Invertebrates

A close relative to marine-free living, AOA is the sponge symbiont *C. symbiosum* (Preston et al., 1996). The genome of this organism was the first of this group to be sequenced and investigated in detail (see above) (Hallam et al., 2006a, 2006b). Some genes present in the genome of *C. symbiosum* are absent in planktonic relatives, indicating a sponge-archaeal specificity (Hallam et al., 2006b). Global studies including more AOA *amoA* sequences from different sponges and corals revealed a large sponge- and coral-specific cluster (Beman et al., 2007; Steger et al., 2008) and are congruent with 16S rRNA gene-based studies of crenarchaea (Taylor et al., 2007). In particular, archaea associated with the sponge genus *Axinellida* seem to be host specific (Holmes and Blanch, 2007), and generally, only a few 16S rRNA-defined phylotypes are found associated with an individual sponge. This contrasts, however, with the large AOA *amoA* diversity that has been found associated with a marine sponge (Steger et al., 2008) or perhaps indicates a large amount of microheterogeneity of *amoA* genes within an archaeal species. AOA appear to be stable members of the microbial community associated with many sponge species with a permanent form of symbiosis indicated by the transmission of AOA from adults to larvae observed in different sponge species by PCR and fluorescent in situ hybridization studies (Steger et al., 2008). Nitrification has been shown to occur in several sponges (Corredor et al., 1988; Diaz and Ward, 1997; Diaz et al., 2004; Jimenez and Ribes, 2007), and they have a complex microbial community associated with N cycling including AOA and nitrite oxidizers, but also anammox planctomycetes and denitrifiers (Hoffmann et al., 2009). Archaea may thus be required for detoxification and sustainment of the hosts' health by removal of nitrogenous waste products (Hoffmann et al., 2009).

AOA in Geothermal Environments

16S rRNA gene sequences affiliated with "moderate" crenarchaeota (or thaumarchaeota) have been recovered from environments of high temperature alongside known thermophilic crenarchaeota and euryarchaeota (Kvist et al., 2005, 2007). There is now unambiguous evidence for the occurrence of AOA in environments of elevated temperature. AOA *amoA* sequences have been detected in "moderately hot" terrestrial environments (~45 to 50°C), including speleotherm structures sampled from a geothermal mine adit in Colorado (Spear et al., 2007), water and associated biofilms of thermal springs located in caves in the Austrian Alps (Weidler et al., 2008), and a Siberian hot spring of 46°C (Lebedeva et al., 2005; Hatzenpichler et al., 2008). From this spring, an enrichment of a single Group 1.1b ("soil") AOA, termed "*Nitrososphaera gargensis*" has been maintained for more than 6 years (Hatzenpichler et al., 2008). Furthermore, the suspicion that AOA are present in the same environments as their (hyper)thermophilic relatives was confirmed by detection of archaeal *amoA* genes from numerous hot springs of diverse temperatures and broad pH ranges located on the Russian Kamchatka peninsula and in Iceland (Reigstad et al., 2008) as well as in Yellowstone National Park (de la Torre et al., 2008), and in further terrestrial hot springs in the United States, China, and Russia (Zhang et al., 2009). Nitrification was measured in an acidic hot muddy pool of 80°C under in situ conditions, demonstrating that this process is indeed found at considerable levels in terrestrial hot springs (Reigstad et al., 2008). Additionally, a thermophilic ammonia-oxidizing archaeon, *Nitrosocaldus yellowstonii*, was obtained in enrichment cultures from a hot spring located in the Yellowstone National Park (de la Torre et al., 2008) with an optimal growth temperature between 65 and 72°C, growing with a stoichiometric conversion of ammonia to nitrite in the absence of an organic carbon source. *amoA* genes of archaea have also been amplified from hydrothermal vent chimneys of the Juan de Fuca Ridge, indicating that this group of organisms can also be found in the hot vents of the deep ocean (Wang et al., 2009).

AOA ACTIVITY IN THE ENVIRONMENT

As outlined previously, based largely on the quantification of AMO genes, AOA generally outnumber their bacterial counterparts in most aquatic and terrestrial environments. However, the dominance of the gene copy number is only an indication that certain populations may be important with regard to an ecosystem process and does not provide any definitive proof of activity. A number of studies have therefore attempted to look at indicators of actual activity in the environment using a range of assays including quantifying growth in response to nitrogen amendments and perturbations, abundance of mRNA transcripts, and incorporation of stable isotopes. In addition, there are now a number of AOA cultivated from the environment (be they individual isolates or highly enriched cultures) that allow some extrapolations to environmental situations. Only the combination of studies that involve exploration of physiological activity and metabolic diversity of archaea grown in the laboratory as well as those in the environments will lead to a comprehensive understanding of the ecological versatility and role of AOA in natural habitats under changing environmental conditions.

AOA Activity in the Marine Environment

Most attempts to correlate activities of AOA to nitrification have been made in the marine environment. Crenarchaeota (thaumarchaeota) are among the most abundant group of prokaryotes in the ocean where they have been estimated to represent up to 20% and more of all cells (Karner et al., 2001).

A number of studies have shown excellent correlation between the abundance of AOA *amoA* genes and nitrification activity. Wuchter et al. (2006) examined a time series in the North Sea where the abundance of archaea and both bacterial and archaeal *amoA* genes were quantified together with measurements of inorganic nitrogen concentrations. In these waters, AOA were up to 100 times more abundant than AOB. Ammonia concentrations were observed to be highest in autumn and winter months and then decreased into spring. Over this period, an increase in abundance of archaeal 16S rRNA genes, archaeal cell numbers, and AOA *amoA* genes was observed that showed good correlation with decreases in ammonia concentrations and increases in nitrate concentrations. Further evidence of AOA dominating ammonia oxidizing activity in the water column has come from direct measurements of ammonia oxidation rates and the abundance of bacterial and archaeal *amoA* genes. Beman et al. (2008) measured the oxidation of ^{15}N-labeled ammonium pools added to marine water sampled from over a depth range of 0 to 100 m in the Gulf of California (Fig. 4). In these samples, AOB were present in either very low or undetectable amounts whereas, again, there was excellent correlation with archaeal 16S rRNA and *amoA* genes. Lam et al. (2007) demonstrated that in samples taken through the water column in the Black Sea, the maximum number of archaeal *amoA* mRNA transcripts detected occurred in the lower oxic zone at approximately 75 m depth and was at the same position as the nitrate maximum. In addition, however, significant gammaproteobacterial expression was also observed in both the lower oxic and suboxic waters, indicating that AOB were also important in nitrification. Statistical analysis indicated that AOA were responsible for most of the ammonia oxidation in oxic water and gammaproteobacteria were mainly responsible in the suboxic zone, despite the presence of relatively high numbers of archaea in this area.

The upper 1,000 m of the open ocean represent the major source of ammonia production and oxidation activity (Wuchter et al., 2006). However, an interesting feature of the distribution of crenarchaeota in the marine environment is that, compared with bacteria, their abundance decreases moderately with increasing depth and therefore represents an increasing proportion of the total prokaryotic community. Analysis of AOA *amoA* genes from a number of studies has revealed that there are two major clades of AOA, described as shallow and deep marine groups (Francis et al., 2005; Hallam et al., 2006). Analysis of the ratio of crenarchaeal 16S rRNA and AOA *amoA* genes has indicated that while all shallow crenarchaeal populations may be capable of ammonia

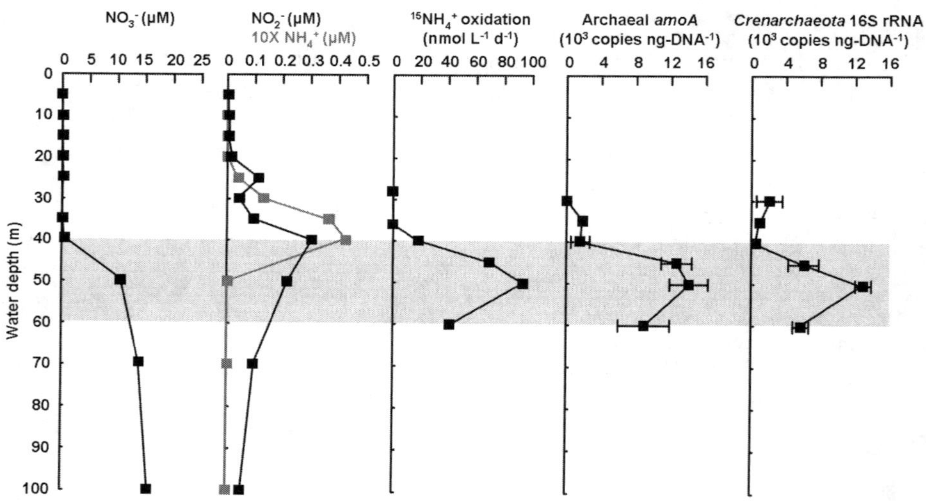

FIGURE 4 Distribution of inorganic nitrogen, ammonia oxidation rates, and archaeal *amoA* and 16S rRNA genes in a 100 m vertical profile in Guaymas Basin, Gulf of California. Areas of active nitrification and correlation with crenarchaeal/AOA numbers and $^{15}NH_4^+$ ammonia oxidation rates are highlighted between dotted lines. (From Beman et al., *ISME Journal* **2:**429–441, 2008, with permission from Macmillan Publishing, Ltd.)

oxidation, ratios of greater than 100:1 for 16S rRNA:*amoA* genes at depths greater than 1,000 m has indicated that archaea in the deep ocean are not autotrophic ammonia oxidizers but may be heterotrophs (Agogue et al., 2008). However, it has also been revealed from analysis of radiocarbon data in archaeal lipids that autotrophy is the dominant metabolism at depth (Ingalls et al., 2006), and a discrepancy between 16S rRNA and *amoA* gene numbers may be due to a lack of specificity in the PCR primers used (Konstantinidis et al., 2009).

AOA Activity in Soil

Although there is increasing evidence from isolates, enrichments, and in situ measurements of ammonia oxidizing activity correlating with the abundance of AOA in marine environments, it is perhaps less clear how important AOA are with respect to their relative contribution to ammonia oxidizing activity in soil. Quantitative analyses of *amoA* genes from several studies generally show the same trend with archaeal *amoA* copies outnumbering their bacterial counterparts by up to two orders of magnitude (e.g., Leininger et al., 2006; He et al., 2007; Nicol et al., 2008). However, attempts to correlate this abundance with activity have provided contrasting results.

Relative Contribution of AOA and AOB to Ammonia Oxidation in Soil

In agricultural soil microcosms with ammonia concentrations below 0.07 mM (being replenished by continual ammonification of organic N), Tourna et al. (2008) demonstrated that with varying levels of nitrifying activity (as a result of varying temperature), changes in transcript profiles were associated with archaeal rather than bacterial populations. In the same soil, Offre et al. (2009) also demonstrated that this observed AOA transcriptional activity resulted in selection and substantial growth of AOA populations but no AOB-associated growth (Fig. 5). The archaeal populations that grew were sensitive to low concentrations of acetylene, and their growth was completely inhibited at low concentrations of 0.01%, thus demonstrating that AOA growth only occurred when ammonia oxidation activity occurred. The surprising result, however, was that this growth was not associated with the abundant soil lineage Group 1.1b, but with populations related to the "marine" lineage Group 1.1a (Fig. 1).

These results contrast with the findings of Jia and Conrad (2009), who reported that it was the growth of AOB (and not AOA) populations that correlated with ammonia oxidation in soil microcosms supplemented with an inorganic nitrogen fertilizer (resulting in a substantially greater concentration of 7 mM). In this study, although the growth of Group 1.1b AOA populations was observed, this growth occurred even when ammonia oxidation was completely inhibited by acetylene. In addition, after establishing soil microcosms with a 5% $^{13}CO_2$ headspace, DNA stable-isotope probing revealed incorporation of inorganic carbon occurred only into the genomic DNA of AOB, indicating that they were the only major contributors to autotrophic ammonia oxidation. Together, these results indicate that the Group 1.1b archaea were not autotrophically oxidizing ammonia in that soil.

Initially, these results may seem somewhat contradictory. However, differences in the experimental designs perhaps indicate fundamental differences in AOA and AOB physiology. In particular, ammonia concentration may be a major driver for relative activity of AOA and AOB in soil. In pasture soil supplemented with concentrations of ammonium typical of mineral fertilization strategies or animal excretions, specific populations of AOB have been shown to grow at different concentrations of ammonium (ranging from 7 to 70 mM), whereas the addition of similarly high concentrations has no selective effect (and presumably no stimulation of growth) of archaea in the same soils (Nicol et al., 2004; Mahmood et al., 2006). In the study by Offre et al. (2009), AOA (but not AOB) growth was observed in actively nitrifying microcosms, but with particularly low ammonia concentrations. In contrast, the growth of only AOB populations

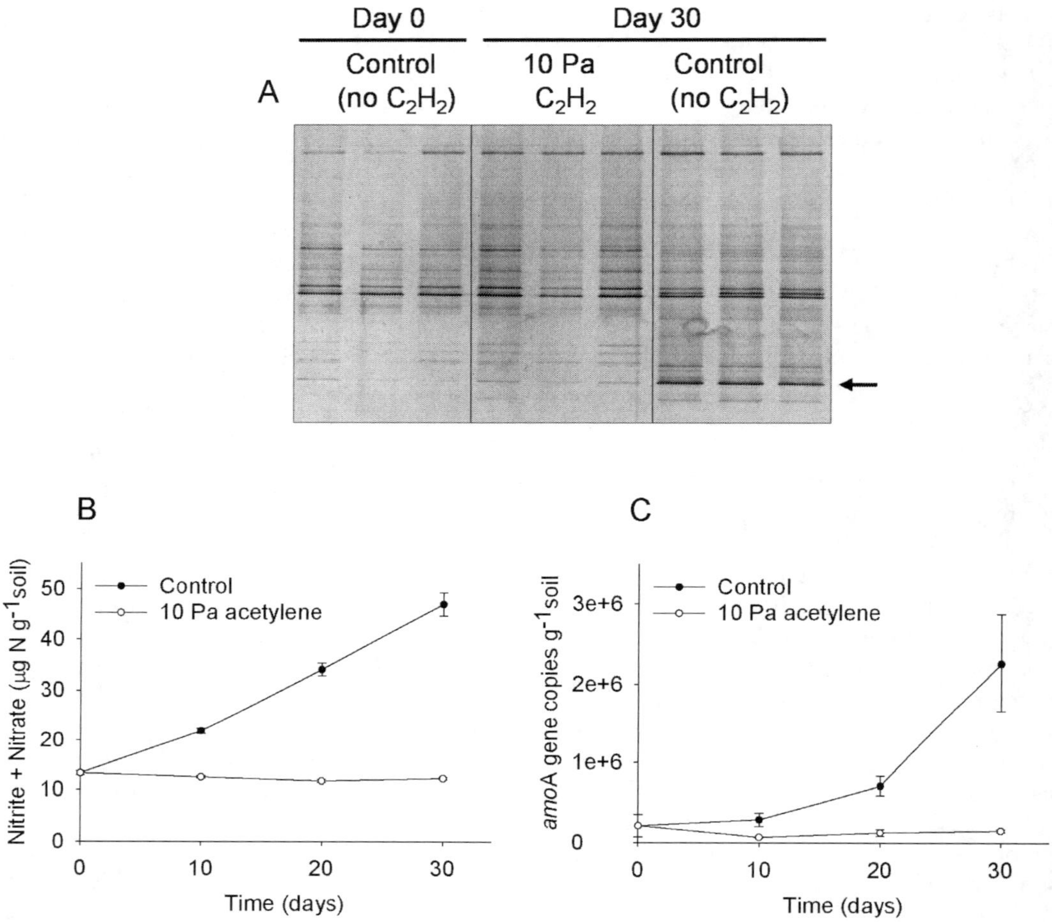

FIGURE 5 Growth of acetylene-sensitive AOA in nitrifying soil microcosms. (A) Denaturing gradient gel electrophoresis (DGGE) analysis of AOA communities after PCR amplification of *amoA* genes. Each lane represents an individual microcosm. The arrow indicates the growth of a specific population of AOA in control microcosms only (no acetylene). (B) Demonstration of complete inhibition of ammonia-oxidizing activity in microcosms with a 10 Pa acetylene headspace partial pressure. (C) qPCR assay specific for the AOA population highlighted in the DGGE profile. Growth of the AOA occurred only in those microcosms with active nitrification. (Adapted from data obtained by Offre et al., *FEMS Microbiol. Ecol.* **70:**99–108, 2009, with permission.)

linked to ammonia oxidizing activity occurred in microcosms supplemented with higher concentrations of mineral fertilizer.

Further evidence of contrasting physiologies within archaeal populations in soil was demonstrated by Schauss et al. (2009). In two different soils amended with manure, AOB populations were shown to increase in size in both a neutral pH silty loam soil and a moderately acidic loamy sand soil. Interestingly, AOA populations only increased in size in the silty loam. While this was the first demonstration of growth stimulation of AOA in soil upon fertilization, it remains unclear why no effect was seen in the other soil. In these experiments, a direct link between nitrification activity and the growth of, specifically, AOA populations was also demonstrated. In microcosms amended with the antibiotic sulfadiazine, growth of AOB was inhibited while nitrification still occurred. Model calculations revealed that in such microcosms, a substantial contribution of ammonia oxidation must be assigned to AOA (Fig. 6).

Metabolic Diversity in Ammonia-Oxidizing Archaea?

These recent studies indicate that there may be some metabolic diversity within the dominating soil lineage Group 1.1b, indicating capability of mixotrophic or heterotrophic growth potentially independent of ammonia oxidizing activity. Distinct clades within Group 1.1b are known to have contrasting genomic features, including substantial variations in gene density (Quaiser et al., 2002; Treusch et al., 2004, 2005) and the length and level of conservation of intertranscribed spacer regions (Nicol et al., 2006). The presence of genes for AMO subunits on the soil fosmid clone 54d9 first gave an indication of an autotrophic metabolism, and the cultivation of the moderately thermophilic *N. gargensis* (affiliated to the soil group) (Fig. 1) demonstrated autotrophic growth with ammonia oxidation. However, the growth of soil AOA populations without ammonia oxidation activity or the incorporation of labeled carbon dioxide has suggested heterotrophic growth (Jia and Conrad, 2009). It may therefore also be the case that the presence of AMO genes in Group 1.1b organisms is not indicative of an exclusively autotrophic lifestyle, but perhaps analogous to heterotrophic nitrite oxidizers, some Group 1.1b populations might possess a broader metabolic repertoire with both heterotrophic and autotrophic capabilities.

Influence of pH Variations in Soil

pH is a major factor in determining the diversity and abundance of different phylotypes, and there is also strong evidence that soil pH is an important determinant of bacterial diversity and community structure on a global scale (Fierer and Jackson, 2006). The mechanisms by which soil pH influences the growth and activity of many microbial functional groups have been determined through a combination of physiological and soil microcosm studies. For example, rates of nitrification and, in particular, ammonia oxidation are significantly reduced in acidic soils (De Boer and Kowalchuk, 2001), and batch growth of pure cultures of AOB in liquid medium does not occur below pH 6.5 (Allison and Prosser, 2001). In AOB, the AMO substrate is the nonionized form of ammonia, and reduced growth and activity of ammonia oxidizers under acidic conditions is therefore believed to result from ionization of NH_3 to NH_4^+, reducing availability of NH_3 for diffusion and increasing the energy demand for NH_4^+ transport. Although inhibition studies have indicated that acidophilic heterotrophs may contribute to ammonia oxidation in some acidic soils, autotrophic oxidation of inorganic ammonia is significant in many acid soils. Nicol et al. (2008) quantified the abundance of AOA and AOB *amoA* gene, and transcript copies across a pH gradient that had been maintained over a range from 4.5 to 7.5 and represents 1,200-fold difference in available ammonia. They demonstrated that AOA and AOB contrasted in the relative amounts of transcript to gene copy, with AOA relatively more transcriptionally active in acidic soil and AOB more so nearer neutral pH. In addition, distinct phylotypes were found to be associated with specific pH ranges, revealing that certain lineages are probably adapted to acidic soil conditions with their associated low levels of un-ionized ammonia, providing a niche that preferentially excludes activities of AOB.

Extrapolation from the Activity of Laboratory Cultures: Adaptation to Low Ammonia?

The growth of AOA at relatively low concentrations of ammonia has been demonstrated for all three organisms brought into culture or enriched to date: *N. maritimus*, *N. gargensis*, and *N. yellowstonii*. All three organisms are grown in media containing between 0.5 and 1.5 mM ammonium, whereas AOB are typically grown in media in the range of 1.5 to 25 mM and have a maximum ammonia tolerance ranging from 50 to 1,000 mM (Koops et al., 2003). Using radiolabeled substrates, Hatzenpichler et

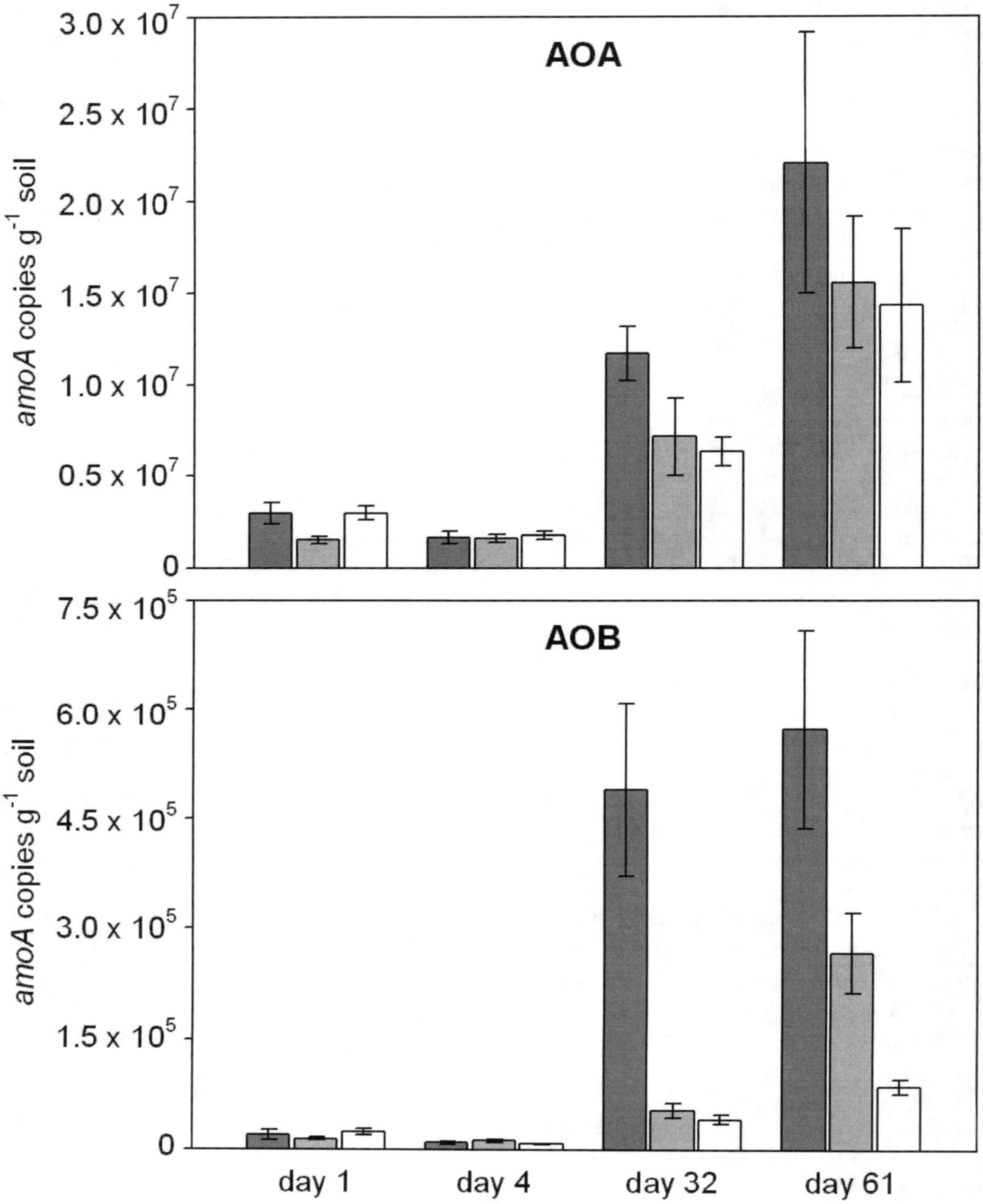

FIGURE 6 Abundances of AOB and AOA in nitrifying soil mesocosms amended with fertilizer and various amounts of the antibiotic sulfadiazine. *amoA* genes were quantified by qPCR. Dark gray bars, 0 mg of sulfadiazine (kg of soil)$^{-1}$ added; light gray bars, 10 mg kg^{-1} added; open bars, 100 mg kg^{-1} added. While both AOA and AOB communities increased in size upon fertilization, the AOA population was less sensitive to sulfadiazine and thus was probably responsible for most of the ammonia oxidation measured after antibiotic treatment. Note the large numbers of archaea compared with AOB (different scales in the figure). (From Schauss et al., *Environ. Microbiol.* **11:**446–456, 2009, with permission.)

al. (2008) not only demonstrated the uptake of inorganic carbon (bicarbonate) at very low ammonium concentrations (<1 mM) but also inhibition of substrate uptake at the relatively low concentration of 3.08 mM, which is lower than any reported inhibitory concentration for AOB. It remains to be shown whether these few laboratory cultures allow general conclusions, but it seems possible that AOA may not be as relevant to nitrification processes in high-ammonia environments such as fertilized soils (Di et al., 2009) or wastewater treatment plants (Wells et al., 2009). However, it might as well be possible that other AOA with adaptation to high-ammonia concentrations exist in the environment.

AOA have a particularly small cell size and appear to be smaller than AOB. *N. maritimus* cells (curved rods) are between 0.17 to 0.20 and 0.5 to 0.9 μm (Könneke et al., 2005), and *N. gargensis* (spherical) are around 0.9 μm in diameter (Hatzenpichler et al., 2008), whereas AOB have been reported to be in the range of 1 to 2 μm in size (Koops et al., 2003). Martens-Habbena et al. (2009) recently reported that, compared with characterized AOB, *N. maritimus* has similar levels of ammonia oxidation activity per unit biomass. However, levels of activity per cell were one order of magnitude less than AOB. Therefore, due to their comparatively smaller size, larger numbers of AOA in the environment may not correlate with a proportional contribution to activity.

CONCLUSIONS

Over the first 4 years, great progress has been made in understanding the ecological role of archaea in most ecosystems. With their involvement in ammonia oxidation, the majority of mesophilic crenarchaeota (or thaumarchaeota) are most likely major players in biogeochemical cycling. In some ecosystems that lack AOB and where nitrification takes place, AOA might be even the sole organisms performing this process, as observed, for example, at certain depths in the Gulf of California, Austrian geothermal caves, and terrestrial hot springs (Weidler et al., 2007; Beman et al., 2008; de la Torre et al., 2008; Hatzenpichler et al., 2008; Reigstad et al., 2008). However, in environments such as soils, oceans, and sediments, AOA and AOB cooccur and probably rely on the same substrate, ammonia, for energy production. It remains unclear to what extent these two groups exhibit functional redundancy in these complex ecosystems and whether they manage to reduce competition for resources by residing in different ecological niches. However, evidence accumulates that AOA might be particularly adapted to low ammonia concentrations (Hatzenpichler et al., 2008; Martens-Habbena et al., 2009). Valentine (2007) proposed that a unifying feature of archaeal physiology is their adaption to energy stress (e.g., in toleration to extremes in temperature, salinity, or acidity). Most "typical" aquatic and terrestrial habitats are not those that would be initially thought of as "extreme." However, most studies that attempted to evaluate the actual activity of AOA (and bacteria) seem to reveal that there could, in fact, be an ecological niche differentiation between AOA and AOB. In any circumstance, the oxidation of ammonia could not be considered a rich source of energy generation. Furthermore, it appears that archaea may be more capable of making a living at the more extreme ends of the ammonia-oxidizing spectrum, being most active in environments with particularly low levels of free ammonia such as the open ocean or acidic soils and hot springs.

ACKNOWLEDGMENT

The discovery of AOA has triggered a large number of excellent environmental studies, which could not all be discussed in this chapter. We apologize to all authors whose papers have not been cited here due to space limitations.

REFERENCES

Agogue, H., M. Brink, J. Dinasquet, and G. J. Herndl. 2008. Major gradients in putatively nitrifying and non-nitrifying Archaea in the deep North Atlantic. *Nature.* **456:**788–791.

Allison, S. M., and J. I. Prosser. 2001. Urease activity in neutrophilic autotrophic ammonia-oxidizing bacteria isolated from acid soils. *Soil Biol. Biochem.* **23:**45–51.

Avrahami, S., and B. J. Bohannan. 2007. Response of *Nitrosospira* sp. strain AF-like ammonia oxidizers to changes in temperature, soil moisture content, and fertilizer concentration. *Appl. Environ. Microbiol.* **73:**1166–1173.

Barns, S. M., C. F. Delwiche, J. D. Palmer, and N. R. Pace. 1996. Perspectives on archaeal diversity, thermophily and monophyly from environmental rRNA sequences. *Proc. Natl. Acad. Sci. USA* **93:**9188–9193.

Bartossek, R., G. W. Nicol, A. Lanzen, H.-P. Klenk, and C. Schleper. 2010. Homologues of nitrite reductases in ammonia-oxidizing archaea: diversity and genomic context. *Environ. Microbiol.* **12:1075–1088**.

Beja, O., M. T. Suzuki, E. V. Koonin, L. Aravind, A. Hadd, L. P. Nguyen, R. Villacorta, M. Amjadi, C. Garrigues, S. B. Jovanovich, R. A. Feldman, and E. F. DeLong. 2000. Construction and analysis of bacterial artificial chromosome libraries from a marine microbial assemblage. *Environ. Microbiol.* **2:**516–529.

Beja, O., E. V. Koonin, L. Aravind, L. T. Taylor, H. Seitz, J. L. Stein, D. C. Bensen, R. A. Feldman, R. V. Swanson, and E. F. DeLong. 2002. Comparative genomic analysis of archaeal genotypic variants in a single population and in two different oceanic provinces. *Appl. Environ. Microbiol.* **68:**335–345.

Beman J. M., and C. A. Francis. 2006. Diversity of ammonia oxidizing Archaea and bacteria in the sediments of a hypernutrified subtropical estuary: Bahía del Tóbari, Mexico. *Appl. Environ. Microbiol.* **72:**7677–7777.

Beman, J. M., K. J. Roberts, L. Wegley, F. Rohwer, and C. A. Francis. 2007. Distribution and diversity of archaeal ammonia monooxygenase genes associated with corals. *Appl. Environ. Microbiol.* **73:**5642–5647.

Beman, J. M., B. N. Popp, and C. A. Francis. 2008. Molecular and biogeochemical evidence for ammonia oxidation by marine Crenarchaeota in the Gulf of California. *ISME J.* **2:**429–441.

Berg, I. A., D. Kockelkorn, W. Buckel, and G. Fuchs. 2007. A 3-hydroxypropionate/4-hydroxybutyrate autotrophic carbon dioxide assimilation pathway in Archaea. *Science* **318:**1782–1786.

Bernhard, A. E., T. Donn, A. E. Giblin, and D. A. Stahl. 2005. Loss of diversity of ammonia-oxidizing bacteria correlates with increasing salinity in an estuary system. *Environ. Microbiol.* **7:**1289–1297.

Bernhard, A. E., J. Tucker, A. E. Giblin, and D. A. Stahl. 2007. Functionally distinct communities of ammonia-oxidizing bacteria along an estuarine salinity gradient. *Environ. Microbiol.* **9:**1439–1447.

Boyle-Yarwood, S. A., P. J. Bottomley, and D. D. Myrold. 2008. Community composition of ammonia-oxidizing bacteria and archaea in soils under stands of red alder and Douglas fir in Oregon. *Environ. Microbiol.* **10:**2956–2965.

Brochier-Armanet, C., B. Boussau, S. Gribaldo, and P. Forterre. 2008. Mesophilic crenarchaeota: proposal for a third archaeal phylum, the Thaumarchaeota. *Nat. Rev. Microbiol.* **6:**245–252.

Buckley, D. H., J. R. Graber, and T. M. Schmidt. 1998. Phylogenetic analysis of nonthermophilic members of the kingdom crenarchaeota and their diversity and abundance in soils. *Appl. Environ. Microbiol.* **64:**4333–4339.

Cantera, J.J., and L.Y. Stein, 2007. Molecular diversity of nitrite reductase genes (nirK) in nitrifying bacteria. *Environ. Microbiol.* **9:**765–776.

Cebron, A., T. Berthe, and J. Garnier. 2003. Nitrification and nitrifying bacteria in the lower Seine River and estuary (France). *Appl. Environ. Microbiol.* **69:**7091–7100.

Coolen, M. J., B. Abbas, J. van Bleijswijk, E. C. Hopmans, M. M. Kuypers, S. G. Wakeham, and J. S. Sinninghe Damsté. 2007. Putative ammonia-oxidizing Crenarchaeota in suboxic waters of the Black Sea: a basin-wide ecological study using 16S ribosomal and functional genes and membrane lipids. *Environ. Microbiol.* **9:**1001–1016.

Corredor, J. E., C. R. Wilkinson, V. P. Vicente, J. M. Morell, and E. Otero. 1988. Nitrate release by Caribbean reef sponges. *Limnol. Oceanogr.* **33:**114–120.

de Bie, M. J. M., A. Speksnijder, G. A. Kowalchuk, T. Schuurman, G. Zwart, J. R. Stephen, O. E. Diekmann, and H. J. Laanbroek. 2001. Shifts in the dominant populations of ammonia-oxidizing beta-subclass Proteobacteria along the eutrophic Schelde estuary. *Aquatic Microbiol. Ecol.* **23:**225–236.

De Boer, W., and G. A. Kowalchuk. 2001. Nitrification in acid soils: micro-organisms and mechanisms. *Soil Biol. Biochem.* **33:**853–866.

de la Torre, J. R., C. B. Walker, A. E. Ingalls, M. Konneke, and D. A. Stahl. 2008. Cultivation of a thermophilic ammonia oxidizing archaeon synthesizing crenarchaeol. *Environ. Microbiol.* **10:**810–818.

DeLong, E. F. 1992. Archaea in coastal marine environments. *Proc. Natl. Acad. Sci. USA* **89:**5685–5689.

DeLong, E. F. 1998. Everything in moderation: archaea as "non-extremophiles." *Curr. Opin. Genet. Dev.* **8:**649–654.

DeLong, E. F., L. T. Taylor, T. L. Marsh, and C. M. Preston. 1999. Visualization and enumeration of marine planktonic archaea and bacteria by using polyribonucleotide probes and fluorescent in situ hybridization. *Appl. Environ. Microbiol.* **65:**5554–5563.

Di, H. J., K. C. Cameron, J. P. Shen, C. S. Winefield, M. O'Callaghan, S. Bowatte, and J. Z.

He. 2009. Nitrification driven by bacteria and not archaea in nitrogen-rich grassland soils. *Nat. Geosci.* 2:621–624.

Diaz, M. C., and B. B. Ward. 1997. Sponge-mediated nitrification in tropical benthic communities. *Mar. Ecol. Prog. Ser.* **156:**97–107.

Diaz, M. C., D. Akob, and C. S. Cary. 2004. Denaturing gradient gel electrophoresis of nitrifying microbes associated with tropical sponges. *Boll. Mus. Ist. Biol. Univ. Genova* **68:**279–289.

Galloway, J. N., F. J. Dentener, D. G. Capone, E. W. Boyer, R. W. Howarth, S. P. Seitzinger, G. P. Asner, C. C. Cleveland, P. A. Green, E. A. Holland, D. M. Karl, A. F. Michaels, J. H. Porter, A. R. Townsend, and C. J. Vorosmarty. 2004. Nitrogen cycles: past, present, and future. *Biogeochemistry* **70:**153–226.

Garcia, J.-L., B. K. C. Patel, and B. Ollivier. 2000. Taxonomic, phylogenetic, and ecological diversity of methanogenic Archaea. *Anaerobe* **6:**205–226.

Fierer, N., and R. B. Jackson. 2006. The diversity and biogeography of soil bacterial communities. *Proc. Natl. Acad. Sci. USA* **103:**626–631.

Francis, C. A., G. D. O'Mullan, and B. B. Ward. 2003. Diversity of ammonia monooxygenase (*amoA*) genes across environmental gradients in Chesapeake Bay sediments. *Geobiology* **1:**129–140.

Francis, C. A., K. J. Roberts, J. M. Beman, A. E. Santoro, and B. B. Oakley. 2005. Ubiquity and diversity of ammonia-oxidizing archaea in water columns and sediments of the ocean. *Proc. Natl. Acad. Sci. USA* **102:**14683–14688.

Hallam, S. J., K. T. Konstantinidis, N. Putnam, C. Schleper, Y. Watanabe, J. Sugahara, C. Preston, J. de la Torre, P. M. Richardson, and E. F. DeLong. 2006a. Genomic analysis of the uncultivated marine crenarchaeote Cenarchaeum symbiosum. *Proc. Natl. Acad. Sci. USA* **103:**18296–18301.

Hallam, S. J., T. J. Mincer, C. Schleper, C. M. Preston, K. Roberts, P. M. Richardson, and E. F. DeLong. 2006b. Pathways of carbon assimilation and ammonia oxidation suggested by environmental genomic analyses of marine Crenarchaeota. *PLoS Biol.* **4:**e95.

Handelsman, J. 2004. Metagenomics: application of genomics to uncultured microorganisms. *Microbiol. Mol. Biol. Rev.* **68:**669–685.

Hansel, C. M., S. Fendorf, P. M. Jardine, and C. A. Francis. 2008. Changes in bacterial and archaeal community structure and functional diversity along a geochemically variable soil profile. *Appl. Environ. Microbiol.* **74:**1620–1633.

Hatzenpichler, R., E. V. Lebedeva, E. Spieck, K. Stoecker, A. Richter, H. Daims, and M. Wagner. 2008. A moderately thermophilic ammonia-oxidizing crenarchaeote from a hot spring. *Proc. Natl. Acad. Sci. USA* **105:**2134–2139.

He, J., J. Shen, L. Zhang, Y. Zhu, Y. Zheng, M. Xu, and H. J. Di. 2007. Quantitative analyses of the abundance and composition of ammonia-oxidizing bacteria and ammonia-oxidizing archaea of a Chinese upland red soil under long-term fertilization practices. *Environ. Microbiol.* **9:**2364–2374.

Herfort, L., S. Schouten, B. Abbas, M. J. Veldhuis, M. J. Coolen, C. Wuchter, J. P. Boon, G. J. Herndl, and J. S. Sinninghe Damsté. 2007. Variations in spatial and temporal distribution of Archaea in the North Sea in relation to environmental variables. *FEMS Microbiol. Ecol.* **62:**242–257.

Herndl, G. J., T. Reinthaler, E. Teira, H. van Aken, C. Veth, A. Pernthaler, and J. Pernthaler. 2005. Contribution of archaea to total prokaryotic production in the deep Atlantic Ocean. *Appl. Environ. Microbiol.* **71:**2303–2309.

Hoffmann, F., R. Radax, D. Woebken, M. Holtappels, G. Lavik, H. T. Rapp, M.-L. Schläppy, C. Schleper, and M. M. Kuypers. 2010. **Complex nitrogen cycling in the sponge** *Geodia barretti*. *Environ. Microbiol.* **11:**2228–2243.

Holmes, B., and H. Blanch. 2007. Genus-specific associations of marine sponges with group I crenarchaeotes. *Mar. Biol.* **150:**759–772.

Ingalls, A. E., S. R. Shah, R. L. Hansman, L. I. Aluwihare, G. M. Santos, E. R. Druffel, and A. Pearson. 2006. Quantifying archaeal community autotrophy in the mesopelagic ocean using natural radiocarbon. *Proc. Natl. Acad. Sci. USA* **103:**6442–6447.

Jia, Z., and R. Conrad. 2009. Bacteria rather than Archaea dominate microbial ammonia oxidation in an agricultural soil. *Environ. Microbiol.* **11:**1658–1671.

Jimenez, E., and M. Ribes. 2007. Sponges as a source of dissolved inorganic nitrogen: Nitrification mediated by temperate sponges. *Limnol. Oceanogr.* **52:**948–958.

Karner, M. B., E. F. DeLong, and D. M. Karl. 2001. Archaeal dominance in the mesopelagic zone of the Pacific Ocean. *Nature.* **409:**507–510.

Koops, H.-P., U. Purkhold, A. Pommerening-Rösner, G. Timmermann, and M. Wagner. 2003. The lithoautotrophic ammonia-oxidizing bacteria, p. 778–811. *In* M. Dworkin, S. Falkow, E. Rosenberg, K.-H. Schleifer, and E. Stackebrandt (ed.), *The Prokaryotes: an Evolving Electronic Resource for the Microbiological Community*, 3rd ed., vol. 5. Springer-Verlag, New York, NY.

Konstantinidis, K. T., J. Braff, D. M. Karl, and E. F. DeLong. 2009. Comparative metagenomic analysis of a microbial community residing at a depth of 4,000 meters at station ALOHA in the North Pacific Subtropical Gyre. *Appl. Environ. Microbiol.* **75:**5345–5355.

Kvist, T., A. Mengewein, S. Manzel, B. K. Ahring,

and P. Westermann. 2005. Diversity of thermophilic and non-thermophilic crenarchaeota at 80 degrees C. *FEMS Microbiol. Lett.* **244:**61–68.

Kvist, T., B. K. Ahring, and P. Westermann. 2007. Archaeal diversity in Icelandic hot springs. *FEMS Microbiol. Ecol.* **59:**71–80.

Könneke, M., A. E. Bernhard, J. R. de la Torre, C. B. Walker, J. B. Waterbury, and D. A. Stahl. 2005. Isolation of an autotrophic ammonia-oxidizing marine archaeon. *Nature* **437:**543–546.

Kuypers, M. M., P. Blokker, J. Erbacher, H. Kinkel, R. D. Pancost, S. Schouten, and J. S. Sinninghe Damsté. 2001. Massive expansion of marine archaea during a mid-Cretaceous oceanic anoxic event. *Science* **293:**92–95.

Lam, P., M. M. Jensen, G. Lavik, D. F. McGinnis, B. Muller, C. J. Schubert, R. Amann, B. Thamdrup, and M. M. M. Kuypers. 2007. Linking crenarchaeal and bacterial nitrification to anammox in the Black Sea. *Proc. Natl. Acad. Sci. USA* **104:**7104–7109.

Lebedeva, E. V., M. Alawi, C. Fiencke, B. Namsaraev, E. Bock, and E. Spieck. 2005. Moderately thermophilic nitrifying bacteria from a hot spring of the Baikal rift zone. *FEMS Microbiol. Ecol.* **54:**297–306.

Leininger, S., T. Urich, M. Schloter, L. Schwark, J. Qi, G. W. Nicol, J. I. Prosser, S. C. Schuster, and C. Schleper. 2006. Archaea predominate among ammonia-oxidizing prokaryotes in soils. *Nature* **442:**806–809.

Lieberman, R. L., and A. C. Rosenzweig. 2005. Crystal structure of a membrane-bound metalloenzyme that catalyses the biological oxidation of methane. *Nature* **434:**177–182.

Lopez-Garcia, P., C. Brochier, D. Moreira, and F. Rodriguez-Valera. 2004. Comparative analysis of a genome fragment of an uncultivated mesopelagic crenarchaeote reveals multiple horizontal gene transfers. *Environ. Microbiol.* **6:**19–34.

Mahmood, S., and J. I. Prosser. 2006. The influence of synthetic sheep urine on ammonia oxidizing bacterial communities in grassland soil. *FEMS Microbiol. Ecol.* **56:**444–454.

Martens-Habbena, W., P. M. Berube, H. Urakawa, J. R. de la Torre, and D. A. Stahl. 2009. Ammonia oxidation kinetics determine niche separation of nitrifying Archaea and Bacteria. *Nature* **461:**976–979.

Massana, R., A. E. Murray, C. M. Preston, and E. F. DeLong. 1997. Vertical distribution and phylogenetic characterization of marine planktonic Archaea in the Santa Barbara Channel. *Appl. Environ. Microbiol.* **63:**50–56.

Mincer, T. J., M. J. Church, L. T. Taylor, C. Preston, D. M. Karl, and E. F. DeLong. 2007. Quantitative distribution of presumptive archaeal and bacterial nitrifiers in Monterey Bay and the North Pacific Subtropical Gyre. *Environ. Microbiol.* **9:**1162–1175.

Mosier, A. C., and C. A. Francis. 2008. Relative abundance of ammonia-oxidizing archaea and bacteria in the San Francisco Bay estuary. *Environ. Microbiol.* **10**:3002–3016.

Murray, A. E., C. M. Preston, R. Massana, L. T. Taylor, A. Blakis, K. Wu, and E. F. DeLong. 1998. Seasonal and spatial variability of bacterial and archaeal assemblages in the coastal waters near Anvers Island, Antarctica. *Appl. Environ. Microbiol.* **64:**2585–2595.

Murray, A. E., A. Blakis, R. Massana, S. Strawzewski, U. Passow, A. Alldredge, and E. F. DeLong. 1999. A time series assessment of planktonic archaeal variability in the Santa Barbara Channel. *Aquat. Microb. Ecol.* **20:**129–145.

Nakagawa, T., K. Mori, C. Kato, R. Takahashi, and T. Tokuyama. 2007. Distribution of cold-adapted ammonia-oxidizing microorganisms in the deep-ocean of the northeastern Japan Sea. *Microbes Environ.* **22:**365–372.

Nicol, G. W., and C. Schleper. 2006. Ammonia-oxidising Crenarchaeota: important players in the nitrogen cycle? *Trends Microbiol.* **14:**207–212.

Nicol, G. W., L. A. Glover, and J. I. Prosser. 2003. The impact of grassland management on archaeal community structure in upland pasture rhizosphere soil. *Environ. Microbiol.* **5:**152–162.

Nicol, G. W., G. Webster, L. A. Glover, and J. I. Prosser. 2004. Differential response of archaeal and bacterial communities to nitrogen inputs and pH changes in upland pasture rhizosphere soil. *Environ. Microbiol.* **6:**861–867.

Nicol, G. W., D. Tscherko, T. M. Embley, and J. I. Prosser. 2005. Primary succession of soil Crenarchaeota across a receding glacier foreland. *Environ. Microbiol.* **7:**337–347.

Nicol, G. W., D. Tscherko, L. Chang, U. Hammesfahr, and J. I. Prosser. 2006. Crenarchaeal community assembly and microdiversity in developing soils at two sites associated with deglaciation. *Environ. Microbiol.* **8:**1382–1393.

Nicol, G. W., S. Leininger, C. Schleper, and J. I. Prosser. 2008. The influence of soil pH on the diversity, abundance and transcriptional activity of ammonia oxidizing archaea and bacteria. *Environ. Microbiol.* **10:**2966–2978.

Ochsenreiter, T., D. Selezi, A. Quaiser, L. Bonch-Osmolovskaya, and C. Schleper. 2003. Diversity and abundance of Crenarchaeota in terrestrial habitats studied by 16S RNA surveys and real time PCR. *Environ. Microbiol.* **5:**787–797.

Offre, P., J. I. Prosser, and G. W. Nicol. 2009 Growth of ammonia-oxidizing archaea in soil microcosms is inhibited by acetylene. *FEMS Micro-*

biol. Ecol. **70:**99–108.

O'Mullan, G. D., and B. B. Ward. 2005. Relationship of temporal and spatial variabilities of ammonia-oxidizing bacteria to nitrification rates in Monterey Bay, California. *Appl. Environ. Microbiol.* **71:**697–705.

Ouverney, C. C., and J. A. Fuhrman. 2000. Marine planktonic archaea take up amino acids. *Appl. Environ. Microbiol.* **66:**4829–4833.

Pearson, A., A. P. McNichol, B. C. Benitez-Nelson, J. M. Hayes, and T. I. Eglinton. 2001. Origins of lipid biomarkers in Santa Monica Basin surface sediment: A case study using compound-specific Delta C-14 analysis. *Geochim. Cosmochim. Acta.* **65:**3123–3137.

Preston, C. M., K. Y. Wu, T. F. Molinski, and E. F. DeLong. 1996. A psychrophilic crenarchaeon inhabits a marine sponge: *Cenarchaeum symbiosum* gen. nov., sp. nov. *Proc. Natl. Acad. Sci. USA* **93:**6241–6246.

Prosser, J. I., and G. W. Nichol. 2008. Relative contributions of archaea and bacteria to aerobic ammonia oxidation in the environment. *Environ. Microbiol.* **10:**2931–2941.

Quaiser, A., T. Ochsenreiter, H. P. Klenk, A. Kletzin, A. H. Treusch, G. Meurer, J. Eck, C. W. Sensen, and C. Schleper. 2002. First insight into the genome of an uncultivated crenarchaeote from soil. *Environ. Microbiol.* **4:**603–611.

Reigstad, L. J., A. Richter, H. Daims, T. Urich, L. Schwark, and C. Schleper. 2008. Nitrification in terrestrial hot springs of Iceland and Kamchatka. *FEMS Microbiol. Ecol.* **64:**167–174.

Sandaa, R. A., O. Enger, and V. Torsvik. 1999. Abundance and diversity of Archaea in heavy-metal-contaminated soils. *Appl. Environ. Microbiol.* **65:**3293–2937.

Santoro, A. E., C. A. Francis, N. R. de Sieyes, and A. B. Boehm. 2008. Shifts in the relative abundance of ammonia-oxidizing bacteria and archaea across physicochemical gradients in a subterranean estuary. *Environ. Microbiol.* **10:**1068–1079.

Schauss, K., A. Focks, S. Leininger, A. Kotzerke, H. Heuer, S. Thiele-Bruhn, S. Sharma, B.-M. Wilke, M. Matthies, K. Smalla, J. C. Munch, W. Amelung, M. Kaupenjohann, M. Schloter, and C. Schleper. 2009. Dynamics and functional relevance of ammonia-oxidizing archaea in two agricultural soils. *Environ. Microbiol.* **11:**446–456.

Schleper, C., R. V. Swanson, E. J. Mathur, and E. F. DeLong. 1997. Characterization of a DNA polymerase from the uncultivated psychrophilic archaeon *Cenarchaeum symbiosum. J. Bacteriol.* **179:**7803–7811.

Schleper, C., E. F. DeLong, C. M. Preston, R. A. Feldman, K. Y. Wu, and R. V. Swanson. 1998. Genomic analysis reveals chromosomal variation in natural populations of the uncultured psychrophilic archaeon *Cenarchaeum symbiosum. J. Bacteriol.* **180:**5003–5009.

Schleper, C., G. Jurgens, and M. Jonuscheit. 2005. Genomic studies of uncultivated archaea. *Nat. Rev. Microbiol.* **3:**479–488.

Seitzinger, S. P. 1988. Denitrification in freshwater and coastal marine ecosystems: ecological and geochemical significance. *Limnol. Oceanogr.* **33:**702–724.

Seitzinger, S., J. A. Harrison, J. K. hlke, A. F. Bouwman, R. Lowrance, B. Peterson, C. Tobias, and G. V. Drecht. 2006. Denitrification across landscapes and waterscapes: a synthesis. *Ecol. Appl.* **16:**2064–2090.

Spang, A., R. Hatzenpichler, C. Brochier-Armanet, T. Rattei, P. Tischler, E. Spieck, W. Streit, D. A. Stahl, M. Wagner, and C. Schleper. 2010. Distinct gene set in two different lineages of ammonia-oxidizing archaea supports the phylum Thaumarchaeota. *Trends Microbiol.* **18:**331–340.

Spear, J. R., H. A. Barton, C. E. Robertson, C. A. Francis, and N. R. Pace. 2007. Microbial community biofabrics in a geothermal mine adit. *Appl. Environ. Microbiol.* **73:**6172–6180.

Steger, D., P. Ettinger-Epstein, S. Whalan, U. Hentschel, R. de Nys, M. Wagner, and M. W. Taylor. 2008. Diversity and mode of transmission of ammonia-oxidizing archaea in marine sponges. *Environ. Microbiol.* **10:**1087–1094.

Stein, J. L., T. L. Marsh, K. Y. Wu, H. Shizuya, and E. F. DeLong. 1996. Characterization of uncultivated prokaryotes: isolation and analysis of a 40-kilobase-pair genome fragment from a planktonic marine archaeon. *J. Bacteriol.* **178:**591–599.

Taylor, M. W., R. Radax, D. Steger, and M. Wagner. 2007. Sponge-associated microorganisms: evolution, ecology, and biotechnological potential. *Microbiol. Mol. Biol. Rev.* **71:** 295–347.

Teira, E., H. van Aken, C. Veth, and G. J. Herndl. 2006a. Archaeal uptake of enantiomeric amino acids in the meso- and bathypelagic waters of the North Atlantic. *Limnol. Oceanogr.* **51:**60–69.

Teira, E., P. Lebaron, H. van Aken, and G. J. Herndl. 2006b. Distribution and activity of Bacteria and Archaea in the deep water masses of the North Atlantic. *Limnol. Oceanogr.* **51:**2131–2144.

Treusch, A., and C. Schleper. 2005. Environmental genomics: a novel tool to study uncultivated microorganisms, p. 45–58. *In* C. W. Sensen (ed.), *Handbook of Genome Research.* Wiley-VCH, Weinheim, Germany.

Treusch, A. H., A. Kletzin, G. Raddatz, T. Ochsenreiter, A. Quaiser, G. Meurer, S. C. Schuster, and C. Schleper. 2004. Characterization of large-insert DNA libraries from soil for environmental genomic studies of Archaea. *Environ. Micro-*

biol. **6:**970–980.

Treusch, A. H., S. Leininger, A. Kletzin, S. C. Schuster, H. P. Klenk, and C. Schleper. 2005. Novel genes for nitrite reductase and Amo-related proteins indicate a role of uncultivated mesophilic crenarchaeota in nitrogen cycling. *Environ. Microbiol.* **7:**1985–1995.

Tourna, M., T. E. Freitag, G. W. Nicol, and J. I. Prosser. 2008. Growth, activity and temperature responses of ammonia oxidising archaea and bacteria in soil microcosms. *Environ. Microbiol.* **10:**1357–1364.

Valentine, D. L. 2007. Adaptations to energy stress dictate the ecology and evolution of the Archaea. *Nat. Rev. Microbiol.* **5:**316–323.

Venter, J. C., K. Remington, J. F. Heidelberg, A. L. Halpern, D. Rusch, J. A. Eisen, D. Wu, I. Paulsen, K. E. Nelson, W. Nelson, D. E. Fouts, S. Levy, A. H. Knap, M. W. Lomas, K. Nealson, O. White, J. Peterson, J. Hoffman, R. Parsons, H. Baden-Tillson, C. Pfannkoch, Y. H. Rogers, and H. O. Smith. 2004. Environmental genome shotgun sequencing of the Sargasso Sea. *Science* **304:**66–74.

Wang, S., X. Xiao, L. Jiang, X. Peng, H. Zhou, J. Meng, and F. Wang. 2009. Diversity and abundance of ammonia-oxidizing Archaea in hydrothermal vent chimneys of the Juan de Fuca Ridge. *Appl. Environ. Microbiol.* **75:**4216–4220.

Ward, B. B. 2000. Nitrification and the marine nitrogen cycle, p. 427–453. *In* D. L. Kirchmann (ed.), *Microbial Ecology of the Oceans.* Wiley Series, New York, NY.

Ward, B. B. 2005. Temporal variability in nitrification rates and related biogeochemical factors in Monterey Bay, California, USA. *Mar. Ecol. Prog. Ser.* **292:**97–109.

Weidler, G. W., F. W. Gerbl, and H. Stan-Lotter. 2008. Crenarchaeota and their role in the nitrogen cycle in a subsurface radioactive thermal spring in the Austrian central Alps. *Appl. Environ. Microbiol.* **74:**5934–5942.

Wells, G. F., H.-D. Park, C.-H. Yeung, B. Eggleston, C. A. Francis, and C. S. Criddle. 2009. Ammonia-oxidizing communities in a highly aerated full-scale activated sludge bioreactor: betaproteobacterial dynamics and low relative abundance of Crenarchaea. *Environ. Microbiol.* **11**:2310–2328.

Wuchter, C., S. Schouten, H. T. Boschker, and J. S. Sinninghe Damste. 2003. Bicarbonate uptake by marine Crenarchaeota. *FEMS Microbiol. Lett.* **219:**203–207.

Wuchter, C., B. Abbas, M. J. L. Coolen, L. Herfort, J. van Bleijswijk, P. Timmers, M. Strous, E. Teira, G. J. Herndl, J. J. Middelburg, S. Schouten, and J. S. S. Damste. 2006. Archaeal nitrification in the ocean. *Proc. Natl. Acad. Sci. USA* **103:**12317–12322.

Zhang, C. L., Q. Ye, Z. Huang, W. Li, J. Chen, Z. Song, W. Zhao, C. Bagwell, W. P. Inskeep, C. Ross, L. Gao, J. Wiegel, C. S. Romanek, E. L. Shock, and B. P. Hedlund. 2008. Global occurrence of archaeal *amoA* genes in terrestrial hot springs. *Appl. Environ. Microbiol.* **74:**6417–6426.

ANAEROBIC AMMONIA OXIDATION (ANAMMOX)

IV

METABOLISM AND GENOMICS OF ANAMMOX BACTERIA

Boran Kartal, Jan T. Keltjens, and Mike S. M. Jetten

8

INTRODUCTION

By the end of the 19th century, the biogeochemical nitrogen cycle seemed to be complete, essentially consisting of aerobic nitrification, anaerobic denitrification, and nitrogen fixation. For more than a century, one possibility was overlooked: oxidation of ammonium under anaerobic conditions (anammox). Based on thermodynamic calculations, the Austrian physicist Broda (1977) proposed the existence of lithotrophic organisms that could derive their energy for growth from the oxidation of ammonium coupled to the reduction of nitrate or nitrite and producing dinitrogen gas as the end product. The common notion that the inert ammonium molecule strictly required activation by oxygen as a first step in its oxidation apparently impeded an active search for these organisms. Likewise, occasional attempts to isolate this type of organism may have given negative, unpublished results, due to the lack of patience, suitable culture approaches, or both. Nevertheless, in the mid-1960s, the analysis of nitrogen balance in a highly stratified anoxic fjord already pointed to an unexplainable loss of ammonium under anoxic conditions (Richards, 1965). Three decades later in Delft, The Netherlands, a similar observation in a denitrifying bioreactor (Mulder et al., 1995) opened the possibility that anammox bacteria might exist. By use of enrichment culture techniques with long biomass retention times and batch tests with inhibitors, it was demonstrated that the anammox process was indeed microbiological (Van de Graaf et al., 1995). It was predicted that, because of the reduction of costs of ammonium removal and the reduction of CO_2 emission, the anammox process would provide a very attractive alternative to the current wastewater treatment technology to remove nitrogen (Jetten et al., 1997). Thus, the initial research was focused on the potential of the anammox process as a wastewater treatment technology as much as the fundamental understanding of the anammox bacteria.

In their seminal study, Strous et al. (1999a) were able to separate a bacterial species (>99.6% pure) from the enrichment cultures by density gradient centrifugation (Strous et al., 1999a). Cells specifically produced dinitrogen gas from ammonium and nitrite and were capable of CO_2 fixation. Since the cell suspensions were not pure by classical microbiological standards, the organism, Brocadia anammoxidans, was given the status of *Candidatus* (Strous et al., 1999a). On the basis of 16S rRNA gene phylogeny, this species belonged to

Boran Kartal, Jan T. Keltjens, and Mike S. M. Jetten, Department of Microbiology, Faculty of Science, Radboud University of Nijmegen, Nijmegen, The Netherlands.

Nitrification, Edited by Bess B. Ward, Daniel J. Arp, and Martin G. Klotz

the planctomycetes (Fig. 1). Electron microscopy analysis of these cells revealed a complex, compartmentalized cell architecture, similar to members of the order *Planctomycetales* (Lindsay et al., 2001). The intracellular compartment of the anammox cells, the anammoxosome, constituted 50 to 70% of the cell volume (Lindsay et al., 2001; Van Niftrik et al., 2008b). Ongoing research efforts resulted in seven new described species, albeit none of these in pure culture, divided over five genera: "*Candidatus* Kuenenia," "*Candidatus* Brocadia," "*Candidatus* Scalindua," "*Candidatus* Jettenia," and "*Candidatus* Anammoxoglobus" (Fig. 1) (Strous et al., 1999a; Schmid et al., 2000, 2003; Kartal et al., 2007b, 2008; Quan et al., 2008; Van de Vossenberg et al., 2008). Quite remarkably, three of five genera could be enriched from the same activated sludge inoculum. The genome of the species "*Candidatus* Kuenenia stuttgartiensis" has been assembled from an environmental metagenome (Strous et al., 2006), and the genomes of three other species are underway. Several studies showed that anammox bacteria are not lithotrophic specialists but are metabolically versatile, capable of using a range of organic electron donors and inorganic electron acceptors (see below) (Strous et al., 2006; Kartal et al., 2007a, 2007b, 2008; Van de Vossenberg et al., 2008). In the presence of nitrate and a suitable organic substrate, they reduce nitrate to dinitrogen gas via the combination of dissimilatory nitrate reduction to ammonium (DNRA) and the anammox reaction (Kartal et al., 2007a). From an applied point of view, it is interesting to note that presently (2010) more than 20 full-scale wastewater treatment plants are successfully running using the anammox process, whereas several others are underway (see Chapter 10).

Despite the scientific novelties, the work on anammox was welcomed with initial skepticism. The organisms were thought to be insignificant in natural environments because of their extremely low growth rates, which would potentially hamper their full-scale applications (Zehr and Ward, 2002). At the moment, the worldwide presence and activity of anammox bacteria have been demonstrated in many anoxic or suboxic zones of marine and freshwater ecosystems (Ward, 2003; Arrigo, 2005; Brandes et al., 2007; and many others) (see also Chapter 9). The organisms play a dominant role in the removal of fixed nitrogen both in the Black Sea, the world's largest anoxic basin, and in Benguela and Peru upwelling systems, two of the world's most important primary production sites (Kuypers

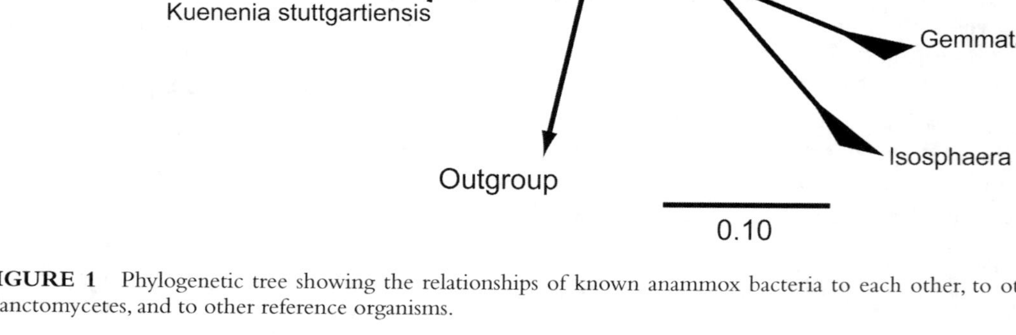

FIGURE 1 Phylogenetic tree showing the relationships of known anammox bacteria to each other, to other Planctomycetes, and to other reference organisms.

et al., 2003, 2005; Hamersley et al., 2007). In the Benguela upwelling system, they may even be the only sink for fixed nitrogen. In all of the studied natural environments, close relatives of the "*Candidatus* Scalindua spp." were found to be the sole anammox species detected (Kuypers et al., 2003, 2005; Penton et al., 2006; Schubert et al., 2006; Schmid et al., 2007) (see Chapter 9). It has been estimated that, in the marine environment, 50% of the dinitrogen gas formed is derived from anammox activity, making the microorganisms important players in the global nitrogen cycle (Ward, 2003; Arrigo, 2005; Brandes et al., 2007; Francis et al., 2007).

In this chapter, we present an overview of the progress that has been made during the last decade with respect to our understanding of the anammox metabolism, focusing on the physiology, cell biology, and information derived from genome sequencing projects. Thereafter, we will discuss the current concepts on the biochemistry and bioenergetics of the anammox bacteria and conclude with the perspectives and urgent issues in this field of research that need to be addressed.

PHYSIOLOGY OF ANAMMOX BACTERIA

Growth of Anammox Bacteria

A challenge in culturing of anammox bacteria was their long doubling time (Van de Graaf et al., 1996). This required a culture technique with high biomass retention capacity operated at constant low-substrate concentrations, reflecting the natural habitat. An approach that fulfilled these demands was the sequencing batch reactor (SBR) technology (Strous et al., 1998). The SBR technique selects on the settling properties of the biomass in the enrichment culture. The reactor is operated in continuous cycles of filling, settling, and withdrawing of the supernatant. By introducing time for settling, cells are accumulated as well-settling biofilm aggregates, which essentially can be kept in the bioreactor for an indefinite amount of time. This procedure is now successfully applied in more than 50 laboratories (Op den Camp et al., 2006).

In the case of anammox, reactors inoculated with activated sludge or an environmental sample were fed with ammonium, nitrite, bicarbonate, and nitrate (to avoid low redox potentials). Anammox bacteria are obligate anaerobes, and the metabolism is (reversibly) inhibited above 2 μM oxygen (Strous et al., 1999b). Reactors are operated under anoxic conditions by sparging with argon, helium, or dinitrogen gas. Growth is possible between 4 and 43°C, the optimal temperature depending on the species under investigation, and at pH 6.7 to 8.3 (with an optimum at pH 8 [Strous et al., 1999b; Kartal et al., 2006; Van de Vossenberg et al., 2008]). In most cases, after a period of 180 to 280 days, at least 70% of the population may consist of anammox bacteria, and the reactor has typically turned red. From the increase in population size and mass balances, it was inferred initially that the organisms double only every 11 to 20 days (Strous et al., 1999b). Recently, however, faster growth rates have been reported (Tsushima et al., 2007; Van der Star et al., 2008).

The Anammox Process

In laboratory-scale reactors operating under steady-state conditions, ammonium, nitrite and bicarbonate were converted according to the following overall equation (equation 1, below) (Strous et al., 1998):

$$1\ NH_4^+ + 1.32\ NO_2^- + 0.066\ HCO_3^- + 0.13\ H^+ \rightarrow 1.02\ N_2 + 0.26\ NO_3^- + 0.066\ CH_2O_{0.5}N_{0.15} + 2.03\ H_2O \quad \textbf{(1)}$$

Under these conditions, bicarbonate serves as the sole carbon source for the synthesis of cell biomass ($CH_2O_{0.5}N_{0.15}$), classifying the organisms as autotrophs. In the process, nitrite plays a dual role: it acts as the electron acceptor in the energy-generating, ammonium-oxidizing reaction (equation 2, below) and as electron donor for the CO_2 reduction to biomass (equation 3, below). In the latter case, nitrite is anaerobically oxidized to nitrate, and, as a consequence, growth is always associated with nitrate production. From the above reaction stoichiometries (equations 1 to 3), it was inferred that 1 mol of carbon is bound per 15 catabolic cycles; likewise, about 4 mol of nitrite was oxidized per mol fixed carbon.

$$NH_4^+ + NO_2^- \rightarrow N_2 + 2\,H_2O\ (\Delta G^{o\prime} = -357\ kJ/mol) \qquad (2)$$

$$0.26\ NO_2^- + 0.066\ HCO_3^- \rightarrow 0.26\ NO_3^- + 0.066\ CH_2O_{0.5}N_{0.15} \qquad (3)$$

$$NH_4^+ + 1.5\ O_2 \rightarrow NO_2^- + 2\ H^+ + H_2O\ (\Delta G^{o\prime} = -275\ kJ/mol) \qquad (4)$$

Anammox bacteria derive a fair amount of energy from the conversion of their substrates into N_2 (equation 2). The free energy change exceeds even the one of the aerobic ammonium oxidation (equation 4, above). However, growth yields, expressed as mol fixed C per mole ammonium converted, are quite comparable between the two groups (Table 1). Ammonium is used by anammox bacteria with high affinity ($K_s < 5$ μM), whereas aerobic nitrifiers are generally less efficient (K_s = 5 to 2,600 μM). Also, nitrite can be metabolized by anammox bacteria down to very low levels ($K_s < 5$ μM). At concentrations above 10 mM, nitrite impairs the metabolism, whereas growth is reversibly inhibited above 20 mM (Strous et al., 1999b). In the literature, different concentrations are reported with respect to the nitrite toxicity (Egli et al., 2001; Strous et al., 1999b); possibly the sensitivity depends on the exposure time.

Considering the quite favorable energy yields and substrate affinities, the metabolic activity of anammox seems low; the rate-limiting step is not known at the moment. Ammonium is oxidized with specific activities ranging between 15 and 80 nmol/mg of protein per min, which is 10-fold less than the activity of aerobic ammonium-oxidizing bacteria (Table 1). The low activity may explain, to some extent, the low growth rates and long doubling times. In nature, ammonium and nitrite concentrations are usually extremely low. Apparently, anammox bacteria have geared their metabolism toward high affinities for their substrates (or more properly, low K_s values) at the expense of the activity.

Anammox Metabolism

The intriguing and only partially answered question is how anammox bacteria are able to activate and oxidize ammonium under anaerobic conditions. The issue has been addressed in batch studies with suspended cells using ^{15}N-labeled substrates (ammonium, nitrite, nitrate) and analyzing the isotope composition of dinitrogen gas formed ($^{14}N^{14}N$, $^{14}N^{15}N$, $^{15}N^{15}N$) by quadrupole mass spectroscopy (Van de Graaf et al., 1997). The studies unequivocally established that the organisms couple ammonia oxidation to nitrite reduction in agreement with the above reaction (equation 2). The labeling experiments with batch cultures gave rise to some new surprises. Inactive cells (i.e., due to nitrite toxicity) can be reactivated by the addition of catalytic amounts

TABLE 1 Comparison between anaerobic (anammox) and aerobic ammonia-oxidizing (nitrification) bacteria[a]

Parameter	Anammox	Nitrification	Unit
Biomass yield	0.08	0.07–0.09	mol (mol of C)$^{-1}$
Aerobic rate	0	200–600	μmol min^{-1} (g of protein)$^{-1}$
Anaerobic rate	15–80	2	μmol min^{-1} (g of protein)$^{-1}$
Growth rate	0.003	0.04	h^{-1}
Doubling time	10.6	0.73	days
K_s NH_4^+	<5	5–2600	μM
K_s NO_2^-	<5	NA	μM
K_s O_2	NA[b]	10–50	μM

[a]Adapted from Jetten et al. (2001).
[b]NA, not applicable.

of hydroxylamine (NH_2OH), which is subsequently converted. This would suggest that the compound could be important in anammox metabolism. Curiously, when amended with hydroxylamine, the intermediary formation of another nitrogen compound was observed. The compound could be identified as hydrazine (N_2H_4), one of the most powerful reductants known in nature. Hydrazine accumulated approximately to the point where hydroxylamine was depleted, and it was metabolized hereafter (Van de Graaf et al., 1997; Strous et al., 1999a; Kartal et al., 2008). Like hydroxylamine, hydrazine was also able to "boost" inactive anammox cells (Strous et al., 1999b). The synthesis of hydrazine is so far unique for anammox bacteria.

The observations described above enabled the proposal of a model for the anammox metabolism consisting of only three subsequent steps (Van de Graaf et al., 1997): (i) the four-electron reduction of nitrite to hydroxylamine; (ii) the condensation of the latter with ammonium to hydrazine; and (iii) the oxidation of hydrazine to nitrogen gas. Herewith, the four electrons are released to drive the first step, nitrite reduction. This model has not been firmly established, especially with respect to the intermediary roles of hydroxylamine and hydrazine. Neither hydrazine nor NH_2OH have ever been demonstrated in anammox cells under physiologically relevant conditions. Concentrations, however, could be too low to detect with current methods. In addition, genome analysis favored a modified scheme that will be discussed later (Strous et al., 2006).

Cell Carbon Fixation

As outlined above, anammox bacteria are autotrophic organisms when growing on ammonium, nitrite, and bicarbonate. Several independent observations as outlined below suggested that anammox bacteria might use the reductive acetyl-CoA (Wood-Ljungdahl) route as the main pathway for CO_2 fixation (Strous et al., 1999a, 2006; Schouten et al., 2004). There are presently no indications consistent with the use of the Calvin-Bassham-Benton cycle, which is widely used by phototrophs and aerobic chemolithotrophs, including nitrifiers, or recently described 3-hydroxypropionate and 3-hydroxypropionate/4-hydroxybutyrate cycles (Berg et al., 2007). The reductive citric acid cycle may also contribute to CO_2 fixation, but the crucial ATP citrate lyase gene seems to be missing from the present genome assembly (Strous et al., 2006). The overall stoichiometries of the Calvin cycle and reductive acetyl-CoA route, calculated for hexose-6-phosphate (hexose-P) formation, are represented by the equations 5 and 6, respectively. At first impression, the latter appears to be less energy demanding for ATP use. However, the acetyl-CoA pathway requires the input of low-redox-potential reducing equivalents ([H]) in certain steps, which might cost ATP-driven reverse electron transport.

$$6\ CO_2 + 12\ NADPH + 18\ ATP \rightarrow \text{hexose-P} + 12\ NADP^+ + 18\ ADP + 17\ Pi \tag{5}$$

$$6\ CO_2 + 24\ [H] + 8\ ATP \rightarrow \text{hexose-P} + 7\ ADP + 7\ Pi \tag{6}$$

Genomic analysis and subsequent activity measurement of the key enzyme, acetyl-CoA synthase/carbon monoxide dehydrogenase, indicated the presence of an operational Wood-Ljundahl pathway in anammox bacteria (Strous et al., 2006). Furthermore, $^{14}CO_2$ labeling experiments and carbon mass balances confirmed the autotrophic lifestyle (Strous et al., 1999a). In addition, isotope ratio mass spectroscopic analysis of anammox biomass and lipid biomarkers showed that these were highly ^{13}C depleted (~47‰ versus CO_2) as expected for CO_2 fixation according to the acetyl-CoA pathway (Schouten et al., 2004).

Source of Nitrite

During growth under autotrophic conditions, anammox bacteria rely on nitrite. In nature, however, the compound is not abundantly present, which raised the question how anammox bacteria obtain nitrite. Several possibilities may be envisaged: nitrification, denitrification, or DNRA. The nitrite supply by (partial) denitrification or DNRA would require an imbalanced induction and expression of nitrate and nitrite reductase, which has been reported to occur in denitrifying and DNRA microorganisms (Baumann et al., 1996; Cole, 1996; Otte et al., 1996; Van de Pas-Schoonen et al., 2005). Denitrifiers mainly use organic electron donors. In habitats that are rich with such carbon compounds, the denitrifying organisms are expected to display high activities and growth rates and compete with DNRA bacteria for the limiting nitrate or nitrite (Strohm et al., 2007). Hence, one may predict that anammox bacteria will be out-competed in such conditions. In ecosystems with limited organic electron donors, anammox bacteria may be able to produce their own nitrite and ammonia (Kartal et al., 2007). The alternative possibility is nitrite formation by aerobic ammonium-oxidizing microbes, although the aerobic lifestyle of these organisms seems to be mutually exclusive to the anaerobic lifestyle of anammox bacteria. Besides that, both guilds are competitors for ammonium. Moreover, under oxic conditions, nitrite is converted to nitrate by nitrite-oxidizing bacteria. However, under oxygen-limited conditions, a different scenario might be possible (Third et al., 2001; Lam et al., 2007). Oxygen respiration by the aerobic ammonium oxidizers may establish the anoxic conditions required for anammox. Anammox bacteria, in turn, can remove the toxic nitrite and the remaining ammonium into nitrogen gas. Such cooperation would benefit both partners.

When oxygen (~5 μM) was carefully introduced to an SBR that contained 80% anammox bacteria, a steady-state culture developed after several weeks (Sliekers et al., 2002). Substrates were converted according to the following stoichiometry (equation 7):

$$1\ NH_4^+ + 0.85\ O_2 \rightarrow 0.11\ NO_3^- + 0.44\ N_2 + 1.14\ H^+ + 1.43\ H_2O \qquad \textbf{(7)}$$

The stoichiometry is expected if aerobic ammonium oxidizers (equation 4) and anammox (equations 1 to 3) contribute to the process in a 56:44 ratio. Aerobic nitrite-oxidizing activity could not be demonstrated, and nitrite oxidizers, such as *Nitrosospira* or *Nitrobacter*, remained beyond detection. Consequently, nitrate formation should be the result of the activity of anammox (equations 1 and 3). Investigation by fluorescence in situ hybridization (FISH) of the microbiota showed that approximately 45% and 40% of the population consisted of aerobic (*Nitrosomonas*-related) ammonium oxidizers and anammox bacteria, respectively. This is remarkable since the former were virtually absent (undetectable with FISH) at the onset of the test. The activity of the aggregates comprising the biomass in the SBR was investigated using ^{15}N-labeled substrates as well as nitrite and O_2 microsensors (Nielsen et al., 2005). FISH analysis of the community revealed that aerobic ammonium oxidation was restricted to the outer shell (<100 μm). Here, oxygen concentrations decreased to undetectable levels. Anammox activity was restricted to the central anoxic parts. In good agreement, larger aggregates (<500 μm) accounted for 68% of the anammox potential, whereas 65% of the aerobic ammonium activity could be attributed to the smaller aggregates (<500 μm). As mentioned, nitrite oxidizers were not found during oxygen limitation. They, however, started growing when the system was operated under ammonium limitation for a prolonged (>1 month) period of time (Third et al., 2001; Kindaichi et al., 2007).

The cooperation between aerobic and anaerobic ammonium oxidizers is also of relevance both in natural habitats and from an applied point of view. Indeed, the partnership of aerobic and anaerobic ammonium oxidizers was recently shown to occur both in man-made and natural ecosystems (Third et al., 2001; Lam et al., 2007). In the Black Sea, anammox bacteria derived the nitrite from ammonium-oxidizing bacteria and crenarchaea occupying different suboxic zones (Lam et al., 2007). Half of the nitrite was supplied by gammaproteobacterial nitrifiers living in the zone where virtually no oxygen could be measured anymore, and the other half of the nitrite was supplied by crenarchaea living in the zone with somewhat higher oxygen concentrations (Lam et al., 2007).

Different types of applications have been developed on the basis of the cooperation between aerobic ammonium oxidizers and anammox. Although different names are used for the processes (Van der Star et al., 2007), the CANON (completely autotrophic nitrogen removal over nitrite) process resembles the direct interaction between the two groups of microorganisms as outlined above (Sliekers et al., 2002) (see Chapter 10). It is very well conceivable that a cooperation of crenarchaea and anammox bacteria, like in the Black Sea, can be used to remove nitrogen from high strength wastewater as well (Kartal, 2008).

Metabolic Versatility of Anammox Bacteria

For some time, anammox bacteria have been considered as specialists that evolved only to respire ammonium and nitrite with high affinities. However, recent research (Güven et al., 2005; Strous et al., 2006; Kartal et al., 2007a, 2007b, 2008) and genome analysis of *K. stuttgartiensis* (see below) revealed that the organisms are much more versatile.

To study the effect of organic compounds on the performance and microbial activities of the anammox bacteria, two SBRs were inoculated with the same activated sludge from which previously "*Candidatus* Brocadia anammoxidans" and "*Candidatus* K. stuttgartiensis" had been enriched (Kartal et al., 2007b, 2008). The two SBRs were supplied with propionate (0.8 mM) and acetate (1 mM), respectively, in the presence of surplus ammonium and nitrate and limiting amounts of nitrite. Over time, when anammox bacteria proliferated, influent concentrations of nitrite and ammonia could be increased from 2.5 mM to 45 mM. The nitrite effluent concentration was below detection at all times. This was also the case for propionate and acetate (both <1 μM), where the inlet concentrations were increased in parallel with ammonium and nitrite up to 15 and 30 mM, respectively. Competition model calculations predicted that if the heterotrophic denitrifiers consumed all the organic acids, a coculture would evolve in which anammox cells would comprise about 30 to 40% of the total biomass. A higher percentage was only expected if anammox bacteria were able to utilize the organic substrates with superior affinity. After 4 months, stable populations had developed in the two reactors. In both cases, anammox bacteria made up approximately 80% of the biomass, but with another unexpected result: the anammox bacteria represented two different species: "*Candidatus* Anammoxoglobus propionicus" in the propionate reactor and "*Candidatus* Brocadia fulgida" in the acetate reactor (Kartal et al., 2007b, 2008). Cells from both reactors were able to couple the oxidation of organic substrates to CO_2 with the reduction of nitrate and nitrite to nitrogen gas (Table 2). Specific rates were 4 to 6% compared to the rate of ammonium oxidation by the cell cultures (15 μmol min^{-1} [g of $protein^{-1}$]). Percoll gradient ultracentrifugation demonstrated that the activity was exclusively associated with the anammox fractions. When cells from other anammox cultures were examined, they showed the immediate capability to convert the organic substrates without induction, although with lower specific activities. The specific activities of propionate and acetate oxidation were highest in *A. propionicus* and *B. fulgida*, respectively, in agreement with the carbon source during enrichment. Apart from the organic acids listed in Table 2, monomethylamine and dimethylamine also served as electron donors for anammox bacteria.

The organic compounds could serve as electron donors in catabolism and/or as a cell carbon source for biosynthesis. Obviously, their role in nitrate reduction supports the first possibility. In addition, ^{13}C isotopic fractionation analysis performed on a series of anammox lipid biomarkers showed no significant differences between CO_2-grown cells and cells grown on propionate (*A. propionicus*) or acetate (*B. fulgida*) (Rattray et al., 2008). Apparently, the organisms use CO_2 as the main building block for lipid biosynthesis. This leaves us with the following puzzle: the oxidation of organic

TABLE 2 Oxidation of organic acids and nitrate reduction by anammox bacteria[a]

Acid	Sp act[b] for:				
	B. anammoxidans	*B. fulgida*	*A. propionicus*	*K. stuttgartiensis*	*Scalindua* sp.
Formate	6.5	7.6	6.7(2.8)	5.8 (3.0)	7
Acetate	0.57	0.95	0.79 (0.7)	0.31 (1.5)	0.7
Propionate	0.12	0.31	0.64 (1.0)	0.12 (0.88)	0.3

[a]Data were taken form Kartal et al. (2007b, 2008b) and Van de Vossenberg et al. (2008).
[b]Specific activities (μmol min^{-1} [g of protein]$^{-1}$) of organic acid oxidation and nitrate reduction (in parentheses).

acids will, at least partially, follow the oxidative route of the acetyl-CoA pathway, while the CO_2 fixation must proceed via the reductive route at the same time. In the genome assembly, only one acetyl-CoA synthase/carbon monoxide dehydrogenase gene cluster is present, while versatile methanogens that can operate the route in two directions have at least two such gene clusters, probably with different functions (Ferry, 1999).

The experiments described above indicate that anammox bacteria can appear as denitrifiers. Common denitrifiers reduce nitrate to nitrogen gas via nitrite, nitric oxide (NO), and N_2O using nitrate, nitrite, NO, and nitrous oxide reductase, respectively. ^{15}N labeling studies performed on physically purified *K. stuttgartiensis* established a different mechanism (Fig. 2) (Kartal et al., 2007a). First, nitrate is reduced to nitrite, which is subsequently reduced in a single step to ammonia, resembling the DNRA mechanism. Hereafter, nitrite and ammonia serve as the substrates for the anammox reaction producing dinitrogen gas as the end product (equation 2). The presence of a dissimilatory nitrite reductase (NrfA) capable of performing the six-electron reduction to ammonia is a widely distributed property, most notably of enteric bacteria (Cole, 1996; Simon, 2002). A calcium-dependent and oxygen-labile nitrite-reducing activity, properties that are typical for NrfA, could be highly enriched from anammox extracts, and a candidate gene cluster is present in the genome (Kartal et al., 2007a). In *K. stuttgartiensis*, the apparent nrf activity is rate limiting, resulting in the frequent, intermediary accumulation of nitrite. No such nitrite formation is found in *A. propionicus* and *B. fulgida* cell suspensions. The presence of the alternative route apparently supplies the organisms with a means to derive ammonia and nitrite from the more general available substrate nitrate.

Apart from nitrite and nitrate, anammox bacteria also can use Fe^{3+} and manganese oxides as electron acceptors in their metabolism (Strous et al., 2006). Next to the organic compounds mentioned, Fe^{2+} can serve as the electron donor for nitrate reduction. The metabolic versatility lends the organisms the opportunity to thrive under conditions where the key substrates, ammonium or nitrite, are limiting. Moreover, it may provide different species with their specific ecological niche.

CELL BIOLOGY OF ANAMMOX

Under the microscope, anammox bacteria show up as simple small coccoid cells with a diameter of approximately 1 μm (Strous et al., 1999a; Van de Graaf et al., 1996). Examination with transmission electron microscopy reveals a much more complicated compartmentalized cell plan, reminiscent of other members of the order *Planctomycetales*, to which anammox is phylogenetically related (Lindsay et al., 2001) (Fig. 3). Cells are surrounded by a thin cell wall in direct connection with an outer (cytoplasmic) membrane. Certain anammox species contain pilus-like appendages (Van de Vossenberg et al., 2008). It is not known whether the cell wall is composed of proteins, like in other Planctomycetes, or peptidoglycan. In the genome of *K. stuttgartiensis*, a near-complete peptidoglycan synthesis operon is present, only lacking the trans peptides cross-linking enzymes (Strous et al., 2006). Next, cells contain a second (intracy-

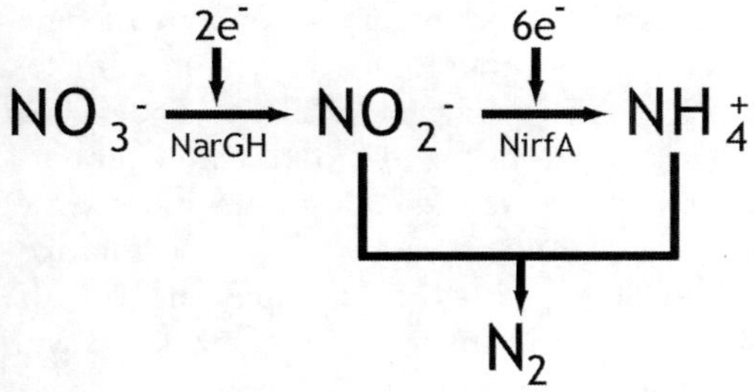

FIGURE 2 Pathway of dinitrogen gas formation from nitrate by anammox bacteria.

toplasmic) and a third membrane, the latter surrounding a central vacuole tentatively termed the anammoxosome (Lindsay et al., 2001). Herewith, three compartments are present, the outermost "paryphoplasm," the riboplasm, and the anammoxosome. The paryphoplasm is a distinctive feature of Planctomycetes in which the cytoplasmic membrane is the actual cell boundary. Its location is the same as the Gram-negative periplasm. The periplasm, however, is in direct contact with the environment by the presence of porin-like channels in the outer membrane (Lindsay et al., 2001). The riboplasm harbors the cell DNA and the ribosomes. Here, storage materials accumulate as glycogen granules (Van Niftrik et al., 2008a). The inner compartment, the anammoxosome, constitutes 50 to 70% of the total cell volume. The membrane surrounding the compartment is often highly curved, possibly to increase the surface-to-volume ratio (Van Niftrik et al., 2008b). Detailed three-dimensional electron tomographic studies established that the anammoxosome represents a true bacterial organelle. It is a closed system (i.e., its membrane was never observed to connect to the intracytoplasmic membrane), and it is vertically inherited to the daughter cell upon cell division (Van Niftrik et al., 2008b).

As in all other living organisms, the anammox membranes are composed of glycerolipid bilayers. The lipids contain a combination of ester-linked (typical of the Bacteria and Eukarya) and ether-linked (typical of the Archaea) fatty acids. Lipids are taxonomic markers and determine the membrane structure. Curiously, anammox bacteria contain a variety of unconventional lipid structures,

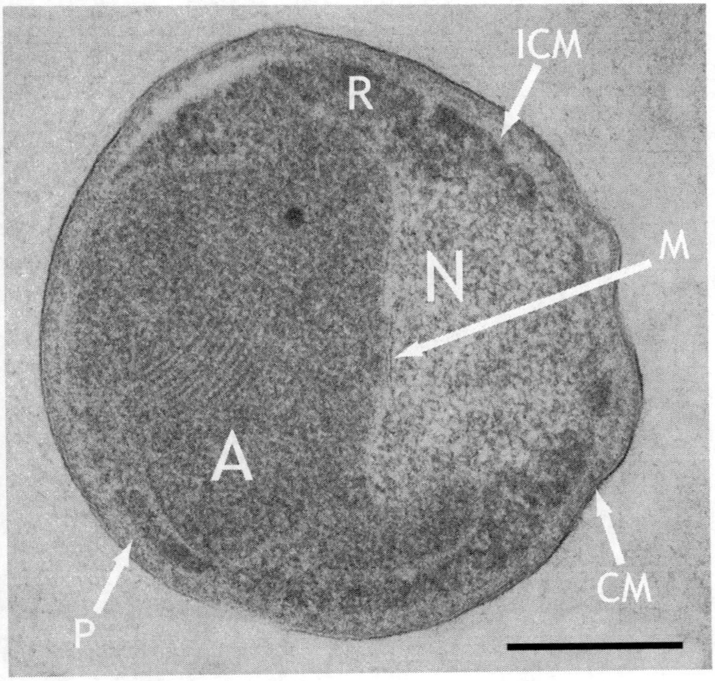

FIGURE 3 Transmission electron microscopy of "*Candidatus* A. propionicus." An anammoxosome (A) containing tubule-like structures, riboplasm (R) containing the nucloid (N) opposed to the anammoxosome membrane (M), paryphoplasm (P) separated from the riboplasm by an intracytoplasmic membrane (ICM), and the cytoplasmic membrane (CM) are shown. Scale bar, 200 nm.

including ring systems composed of five linearly concatenated cyclobutane moieties as well as of six- and four-membered ring systems (Fig. 4) (Sinninghe Damsté et al., 2002, 2005; Kuypers et al., 2003; Schmid et al., 2003; Boumann et al., 2006; Kartal et al., 2007b; Rattray et al., 2008). The concatenated cyclobutane ring systems, termed ladderanes, are unique in nature. Chemical methods to synthesize pentacyclic anammoxic acid require complex chemistry (Mascitti and Corey, 2004). Molecular modeling of the ladderanes showed that these lipids appear to be tightly packed (Sinninghe Damsté et al., 2002). The unusual density makes them impermeable for fluorophores that readily pass through common membranes. Ladderanes constitute 34% of the total lipids. Upon cell fractionation, 53% appeared to be present in a fraction that was enriched with anammoxosomes, indicating that at least a major part of the organelle's membrane is composed of the ladderane lipids (Sinninghe Damsté et al., 2002).

The function of the anammoxosome and the role of the ladderane lipids need further experimentation. Anammox bacteria are chemolithotrophic organisms. This implies that the only means to synthesize ATP is by a chemiosmotic mechanism. Hereby, protons are translocated across a closed semipermeable membrane system in connection with electron transfer processes in the central catabolism, thus establishing a proton motive force (pmf) across the cell membrane. The pmf is composed of a chemical gradient of protons (ΔpH) and of an electrical charge gradient (ΔΨ). The pmf drives ATP synthesis by the action of a membrane-bound proton-translocating enzyme complex, the F_1F_0 ATP synthetase (ATPase). The current view is that, analogous to the mitochondria in eukaryotes, the chemiosmotic processes of anammox bacteria reside in the anammoxosome. Some lines of evidence support the hypothesis. Anammox bacteria are packed with cytochrome *c*-type proteins, lending the organisms a characteristic red color. The cytochromes play a central role in electron transport associated with the anammox metabolism (see below). The staining of the c type cytochromes with diaminobenzidine showed that the proteins are located in the anammoxosome at which the most intense staining occurred within close proximity to the membrane (Van Niftrik et al., 2008a). In addition, one of the most abundant cytochrome *c*-type proteins, a hydroxylamine/hydrazine oxidoreductase (HAO), has been purified (Schalk et al., 2000). Immunogold localization of the antiserum directed against the HAO localized the enzyme specifically to the anammoxosome compartment (Lindsay et al., 2001).

Membranes constitute a barrier for the passage of charged molecules, albeit not a perfect one. Protons can passively slip through the membrane. The slippage occurs at a certain rate, independent of the metabolic activity; thus, dissipating the pmf. It has been calculated that in the highly active mitochondria, passive diffusion accounts for a 10% energy loss (Haines, 2001). In the metabolically inert anammox bacteria, such proton slippage would

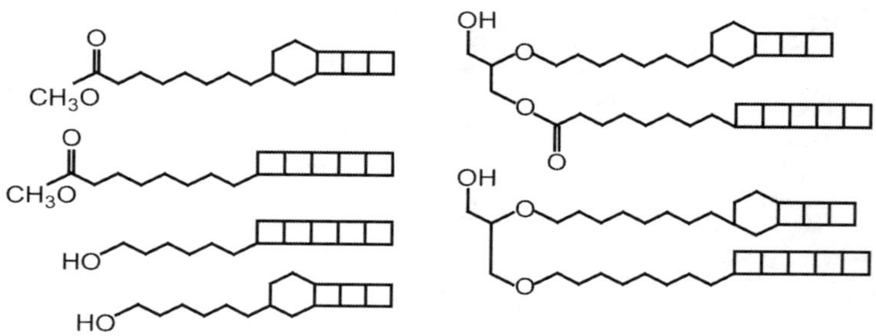

FIGURE 4 General structure of ladderane lipids from anammox bacteria.

be detrimental. Here, the ladderane lipids could come into play. By their dense nature, the membranes could limit proton leakage to a minimum as, for instance, has been shown for the caldarchaeol lipids in thermophilic archaea (Van de Vossenberg et al., 1999). Another possibility could be to minimize the leakage of anammox intermediates out of the cells. A hydrazine drain, for instance, implies a loss of reducing equivalents (four electrons per hydrazine). Replenishment of the electrons requires the oxidation of nitrite by (ATP-driven) reverse electron transport (see below), the oxidation of carbon storage materials (glycogen), or the presence of external electron donors. The two former processes are energy demanding. (Note from equation 6 the energy costs associated with carbon fixation into glycogen.) Obviously, good sealing would prevent the energy loss. However, hydrazine can be detected in the head space and liquid medium of metabolizing cells, albeit under somewhat artificial conditions after the addition of hydroxylamine, which implies that membranes are not completely impermeable for this compound. Still, the ladderane lipids might be important to impede the leakage.

As mentioned, the proposed role of the anammoxosome in the energy metabolism needs quite some experimental verification. Future studies should give an answer to the localization and orientation with respect to the membrane site of the catabolic key enzymes, including ATPase. Experiments with isolated anammoxosomes should establish their role in chemiosmotic processes, which requires the development of solid protocols to isolate the organelles. Such studies might address a potential role of the anammoxosome that has not been given attention, as a reservoir for the storage of ammonium and/or nitrite.

GENOMICS OF ANAMMOX BACTERIA

Elucidation of the "*Candidatus* K. stuttgartiensis" Genome

At the moment, the genome of one anammox species, "*Candidatus* K. stuttgartiensis," has been sequenced to near completion (Strous et al., 2006). The genome could be assembled by a metagenomics approach from a laboratory reactor fed with synthetic wastewater. At the time of sampling, *K. stuttgartiensis* made up 74% of the microbial population; it was the only anammox species present. Apart from *Kuenenia*, 28 different operational taxonomic units were detected, divided over at least six bacterial phyla and two lineages of uncultured bacteria. The DNA sequence was derived from the combination of shotgun and fosmid analyses as well as of bacterial artificial chromosome libraries. The combined information allowed the genome to be assembled into five contigs; the mean read coverage over the contigs was 22-fold. It is estimated that the genome would be more than 98.5% complete, suggesting that about 60 genes are missing.

The almost complete *K. stuttgartiensis* genome (4.2 megabases) codes for 4,663 ORFs. A total of 3,279 genes (70.3%) show significant similarity with genes in other databases, but only 1,385 genes (29.7%) have been annotated a function as yet. The size of the genome and the number of encoded proteins is astonishing, considering the initially conceived character of anammox bacteria as lithotrophic specialists. Ecophysiological studies described above now define the organisms as more versatile. Information from the genome supports the view and predicts new functions with respect to the energy metabolism and anabolism (Table 3).

What Did We Learn from the Genome?

Physiological studies indicated that the anammox process, viz. ammonium oxidation coupled to the reduction of nitrite to produce N_2, would proceed through the intermediary formation of hydroxylamine and hydrazine. During CO_2 fixation, part of the nitrite is oxidized to nitrate. For the uptake of the substrates, five ammonium transporters (Amt), five nitrite/formate transporters (FocA), and two nitrite/nitrate antiporters (NarK) were found in the *K. stuttgartiensis* genome. With respect to the nitrogen metabolism, no less than eight

TABLE 3 Selected genes from genome of *K. stuttgartiesis* coding for enzymes in the central catabolic and anabolic pathways (Strous et al., 2006)

Enzyme/metabolic pathway	Gene/operon	orf/gene cluster[a]
Nitrogen metabolism and substrate uptake systems		
Cytochrome cd_1 nitrite reductase + assemblage	*nirS*	kuste4136–4140
Hydrazine dehydrogenase/oxidase	*hzo*	kustc0694; kustd1340
HH (proposed)	*hh*	kuste2854–2861, kuste2469–2483
Hydroxylamine oxidoreductase	*hao*	kuste2435, kuste2457, kustd2021, kusta0043, kustc1061, kustc0458
Nitrate reductase + accessory proteins	*nar*	kusd1699–1713
Dissimilatory nitrite reductase (putative)		kustc0392–kustc0395
Ammonium transporter	*amt*	kustc0381, kustc1009, kustc1012, kustc1015, kuste3690
Nitrite/formate transporter	*focA*	kusta0004, kusta0009, kustd1720, kustd1721, kuste4324
Nitrite/nitrate antiporters	*narK*	kuste2308, kuste2335
NO reduction/detoxification	*norVW*	kuste2935, kuste3160
Bacterial hemoglobin		kustd1957
Intermediary electron transport		
Cytochrome bc_1 (complex III) + accessory proteins	*bc1*	kuste4569–4574
		kustd1480–1485
		kuste3096–3097
NADH:quinone oxidoreductase (complex I)	*nuoA–N*	kuste2660–2672
Na^+-translocating NADH:quinone oxidoreductase	*nqrA–E*	kuste3325–3329
Cytochromes *c*	Multiple	61 total[b]
Ferredoxins/iron-sulfur proteins	Multiple	17 total[b]
ATP synthesis		
F1F0 ATP synthase	*atpA–G*	kuste3789–3796
	atpA–G	kuste4592–4600
	atpA–I	kustc0572–0579
Archaeal/vacuolar ATPase (proton pumping)	*vatpA–K*	kuste3864–3871
Alternative external electron donors/acceptors		
Ni-Fe hydrogenase, hydrogenase 4	*hydC – I*	kustd1773–1779
Formate:quinone oxidoreductase		kustc0822–0838
Cbb3 terminal oxidase + accessory proteins	*cbb3*	kustc0425–0430
Cell carbon synthesis/intermediary anabolism		
Reductive acetyl-CoA pathway	Complete	kustd1538–1547
Acetyl-CoA synthetase/CO dehydrogenase		kustb0169, kuste4247
H_4folate-dependent steps		kuste2296, kustc0552
Gluconeogenesis	Complete	Various[b]
TCA cycle	Not complete	Various[b], absent: citrate synthase
Other anabolic pathways		
Peptidoglycan synthesis + cell division		kuste2373–2387, others[b]
Fatty acid synthesis		kuste3335–3352
		kuste3603–3608
		kuste2802–2805
		kustd1386–1391

[a]The gene coding is as in the Pedant database (http://pedant.gsf.de).
[b]For a detailed listing, see Strous et al. (2006).

octaheme cytochrome *c* proteins are present. This type of enzymes is known to catalyze hydroxylamine and hydrazine oxidation. Considering the role of hydrazine in anammox metabolism, one or more of these proteins could represent the physiological hydrazine dehydrogenase (see below). Analysis of the genome permitted Strous et al. (2006) to propose two candidate operons (*kuste2469-2483*; *kuste2854-2861*) coding for an enzyme system involved in the synthesis of hydrazine (hydrazine hydrolase [HH]). Moreover, three enzymes active in other nitrogen-converting processes could be identified: nitrite:nitrate oxidoreductase (NarGH), a dissimilatory nitrite reductase (NrfA) (see above), and cd_1 nitrite:nitric oxide oxidoreductase (NirS). The presence of NirS, and the apparent lack of a nitrite:hydroxlamine reductase suggested that NO, rather than hydroxylamine, should be an intermediate in the formation of dinitrogen gas.

The *K. stuttgartiensis* genome supports the view that anammox synthesizes ATP by a chemiosmotic mechanism. Four ATPase operons are apparent including one (*kuste3787-3797*) that contains all F_1F_0 genes and one operon (*kuste3864-3871*) potentially encoding a V/A (archaeal)-type ATPase. The two other operons are incomplete. The presence of three operons encoding the components of the respiratory complex III (cytochrome bc_1) is exceptional in nature (Schneider and Schmidt, 2005). Two gene clusters could be identified as highly homologous to complex I (NuoA-N; NADH:ubiquinone oxidoreductase), whereas a third one codes for a Na^+-translocating NADH:ubiquinone oxidoreductase (NqrA-E). The complexes might act in the formation of NADH for biosynthetic purposes. The presence of complexes I and III suggests a role for ubiquinone or menaquinone in electron transport. In agreement, all ubi/menaquinone biosynthetic genes can be identified.

Altogether, more than 200 genes involved in catabolism and respiration could be annotated, many of these representing c-type cytochromes, which are frequently linked to metabolic enzymes. As pointed out by Strous et al. (2006), the degree of redundancy has been found only for *Geobacter sulfurreducens* (Methe et al., 2003) and *Shewanella oneidensis* (Heidelberg et al., 2002), two highly versatile heterotrophs. The wealth of electron transport proteins suggests an intricate network of branched respiratory chains that link to a variety of external electron donors and terminal electron acceptors. Physiological studies described above presumably have identified only part of these. In agreement with the role of formate as a reductant for nitrate reduction, several operons are at present coding for formate dehydrogenase genes, including a large gene cluster (kustc0822-08838) with high similarity to the formate:quinone oxidoreductase complex genes. One of the putative formate hydrogenases, however, should act in formate synthesis during cell carbon fixation. The presence of a hydrogenase operon (kustd1773-1779) is conspicuous, since no role for hydrogen could be experimentally verified (Methe et al., 2003). The same holds for a gene cluster that codes for a cbb3-type terminal oxidase (kustc0425-0430) taking into account the strict anaerobic nature of anammox bacteria. Alternatively, this oxidase might be operative in detoxifying NO, as are putative norVW and bacterial hemoglobin (Van der Oost et al 1994).

In the genome of *K. stuttgartiensis*, the complete reductive acetyl-CoA pathway for CO_2 fixation can be detected. All other CO_2 fixation pathways are either missing or incomplete. Likewise, the tricarboxylic acid cycle and gluconeogenesis/glycolysis route, both serving intermediary anabolic processes, are found to be complete or nearly complete, lacking only citrate synthase/lyase. As was alluded to above, a peptidoglycan gene cluster is present, which covers 17 of 19 genes known to be involved in its biosynthesis. Remarkably, the gene cluster encodes several cell division (divisome) proteins as well. Anammox bacteria are characterized by the presence of ladderane lipids that have to be synthesized by an as yet unknown pathway (Strous et al., 2006). In the *K. stuttgartiensis* genome, four fatty acid biosynthesis gene clusters are found, two of these interspersed with a number of genes coding for radical SAM proteins. This type of enzyme is known to be active in oxidative radical reactions that likely will be needed to perform cyclization steps, such as in the formation of the concatenated cyclobutane ring systems of the ladderanes.

BIOCHEMISTRY AND BIO-ENERGETICS OF THE ANAMMOX PROCESS

Current View on the Biochemistry and Bio-energetics of the Anammox Process

Previously, a biochemical mechanism was proposed for anammox by Van de Graaf et al. (1997) featuring hydrazine and hydroxylamine as its intermediates. The genome sequence of *K. stuttgartiensis* directed Strous et al. (2006) to postulate an alternative pathway (equations 8 to 10, below) (Fig. 5A), comprising a minimum set of three subsequent reactions: (i) the one-electron reduction of nitrite to NO (equation 8); (ii) the condensation of NO and ammonia in concert with the input of three electrons yielding hydrazine (equation 9); and (iii) the four-electron oxidation of the latter to make the end product, nitrogen gas (equation 10).

$$NO_2^- + 2\,H^+ + \mathbf{e} \rightarrow NO + H_2O \; (E_0' = +0.38\,V) \quad \mathbf{(8)}$$

$$NO + NH_4^+ + 2H^+ + 3\mathbf{e} \rightarrow N_2H_4 + H_2O \; (E_0' = +0.34\,V) \quad \mathbf{(9)}$$

$$N_2H_4 \rightarrow N_2 + 4\,H^+ + 4\mathbf{e} \; (E_0' = -0.75\,V) \quad \mathbf{(10)}$$

Neither hypothesis for the biochemical mechanism for anammox reaction has been backed by experimental evidence. However, several lines of evidence support the intermediary role of NO and hydrazine in metabolizing cells. A cytochrome cd1 nitrite reductase (NirS) is present in the *K. stuttgartiensis* genome. An enzyme (hydrazine dehydrogenase/oxidase) that catalyzes oxidation of hydrazine coupled to the reduction of cytochrome *c* with high specific activity (V_{max} = 6.2 μmol min^{-1} [mg of protein]$^{-1}$) and high substrate affinity (K_m = 5.5 μM) has been purified recently from the anammox strain KSU-1 (Shimamura et al., 2007). The enzyme is abundantly present in the cells. Unfortunately, it has not been verified whether the enzyme reaction proceeds according to equation 10. Quite remarkably, hydroxylamine is a powerful inhibitor of hydrazine oxidation (K_i = 2.4 μM). This complies with the accumulation of hydrazine in metabolizing cells upon hydroxylamine addition. The dimeric octaheme protein (subunit size, 62 kDa) shares 88% and 89% sequence identity to kustc0694 and kustd1340, respectively, found on the *K. stuttgartiensis* genome and annotated as putative hydroxylamine oxidoreductases. Hydrazine synthesis (equation 9) represents an unprecedented reaction. The enzyme HH, which would mediate the reaction, has not yet been described. Genome context analysis indicated two possible gene clusters as the most likely candidates to encode HH.

The current view on the anammox metabolism leaves little room for hydroxylamine. However, cell extracts from anammox bacteria display high hydroxylamine (HAO) activity, and, as mentioned, in the genome of *K. stuttgartiensis*, no less than eight octaheme cytochrome *c* proteins have been annotated as HAOs (Schalk et al., 2000; Strous et al., 2006). One of these has been purified both from *B. anammoxidans* (Schalk et al., 2000) and from strain KSU-1 (Shimamura et al., 2008). The enzyme shows 87% identity with a polypeptide encoded by kustc1061 in the genome of *K. stuttgartiensis*. The *B. anammoxidans* octaheme protein catalyzed the oxidation of hydroxylamine (K_m = 33 μM) with a high rate (V_{max} = 9.6 μmol min^{-1} [mg of protein]$^{-1}$) using a variety of artificial electron acceptors, including cytochrome *c*. Hydrazine also served as a substrate, although the specific activity (V_{max} = 0.54 μmol min^{-1} [mg of protein]$^{-1}$) and hydrazine affinity (K_m = 25 μM) were much lower compared to hydroxylamine and the hydrazine dehydrogenase/oxidase described above. Presently, the reac-

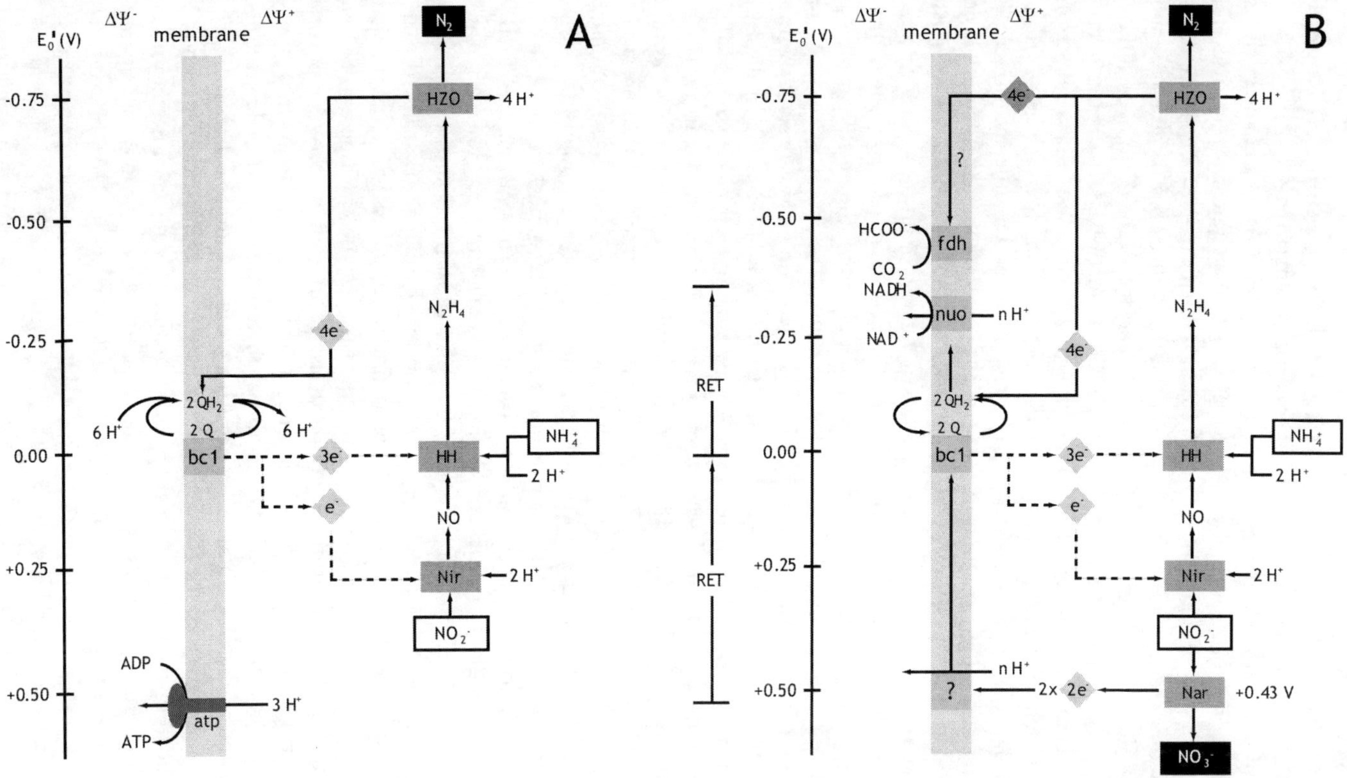

FIGURE 5 Proposed metabolic pathways of *K. stuttgartiensis*. (A) Anammox central catabolism. (B) Combination of central catabolism with nitrate reductase to generate low-redox-potential electrons for the acetyl-CoA pathway. Nir, nitrite reductase; HZO, hydrazine dehydrogenase; Nar, nitrate reductase; Q, ubiquinone; atp, F_1F_0 ATP synthase; fdh, formate dehydrogenase; nuo, NADH:ubiquinone oxidoreductase; RET, reversed electron transport. Light diamonds, cytochromes; dark diamond, ferredoxin; solid arrows, reductions; dashed arrows, oxidations. Note that the localization of the enzymic reactions and the direction of proton translocation is arbitrarily chosen, at which $\Delta\Psi^+$ and $\Delta\Psi^-$ are thought to represent the anammoxosome and riboplasmic compartments, respectively.

tions catalyzed by the various HAOs from anammox bacteria have not been properly defined. Their physiological role thus remains to be established.

Anammox bacteria have to conserve the energy derived from the ammonium oxidation and nitrite reduction (equation 2) by a chemiosmotic principle. The presence of the quinol:cytochrome *c* oxidoreductase complex (bc_1, complex III) would account for a mechanism shown in Fig. 5A. The electrons derived from hydrazine oxidation are transferred via ubiquinone to the cytochrome bc_1 complex. The latter serves a dual role. First, it distributes the electrons toward nitrite reduction (equation 8) and hydrazine synthesis (equation 9). Second, electron transport occurs in concert with the translocation of protons, thus creating a pmf ("proton-motive Q cycle"). Intermediary electron transfer would be accomplished by a set of cytochrome *c*-type proteins. The proton-motive Q cycle allows the translocation of six protons per hydrazine molecule oxidized, which drives ATP synthesis by the F_1F_0 ATPase. Hydrazine is a most powerful reductant ($E_0' = -0.75$ V). Considering a pmf of −0.18 to −0.25 V, as is commonly found in respiring organisms, the redox potential drop ($\Delta E = 0.86$ V) associated with the four-electron transfer from hydrazine to ubiquinone would permit the translocation of 14 to 18 protons. An effective number of only six implies that 60 to 65% of the energy released in the reaction is dissipated as heat.

Many details with respect to the outlined anammox biochemistry and bio-energetics remain to be experimentally established, including determination of the intermediates, the isolation and characterization of the metabolic key enzymes and respiratory complexes involved in the electron transfer processes. Further questions concern the site where the reactions take place and the membrane orientation of proton and electron uptake and release. The very attractive site for the localization would be the anammoxosome, in line with the observation that cytochromes are stained at the inner rim of the organelle.

Cell Carbon Fixation and Reversed Electron Transport

During growth, part of the nitrite is oxidized to nitrate to generate the electrons for CO_2 fixation (equation 3). The Ljungdahl-Wood pathway for cell carbon synthesis requires the input electrons derived from NADH (−0.32 V). CO_2 reduction in formate, acetyl-CoA, and pyruvate syntheses demands electrons with redox potentials as low as −0.42 to −0.5 V. However, thermodynamically, the reactions are very difficult to reconcile: nitrite delivers electrons at +0.43 V. This would only be possible by an extreme case of proton-translocation-driven reverse electron transport, which might be incompatible with the relatively high growth yield, or by the clever use of the reducing power of hydrazine as was previously postulated (Strous et al., 2006) (Fig. 5B). In the latter scenario, hydrazine is recycled by reversed electron transport of electrons from nitrite oxidation up to the level of cytochrome bc_1 (complex III) (Fig. 5B). In this respect, it is interesting to note that the NarGH operon in *K. stuttgartiensis* lacks the NarI, the ubiquinone-binding subunit. Instead six genes are present encoding cytochrome *c*-type proteins, which might facilitate the electron transport. After hydrazine synthesis, part of the electrons derived from its oxidation is channeled toward NAD^+ and CO_2 reduction to sustain carbon fixation. Again, two mechanisms may be envisaged: (i) the presence of an as yet unknown second type of hydrazine dehydrogenase, which uses (low-redox potential) ferredoxin as the electron acceptor; or (ii) a novel type of cytochrome bc_1 complex. With respect to the latter possibility, one may note that the *K. stuttgartiensis* genome encodes three different cytochrome bc_1 operons, two of these linked to an NADH oxidoreductase gene, which is commonly present in complex I (NADH:ubiquinone oxidoreductase) and formate dehydrogenase, but which has never been observed in complex III. If functionally expressed, such new cytochrome bc_1 might directly couple the oxidation of reduced quinone to the reduction of NAD^+ by an alternative Q cycle.

PERSPECTIVES

Studies during the last decade have revealed many unique structural and metabolic properties of the anammox bacteria. The elucidation of the genome of *K. stuttgartiensis* permitted an unprecedented insight into its metabolic potentials. Altogether, the studies are the starting point of our understanding the way the microorganisms perform the formation of nitrogen gas from ammonium and nitrite under anaerobic conditions, as well as in the way the energy released in the reaction is conserved in ATP synthesis and cell carbon fixation. However, many questions still need an answer. The questions relate to the nature and the molecular mechanism of the key enzymes involved in the anammox process and to the nature and organization of the respiratory systems. An especially important aspect herein is the role of the anammoxosome.

At the moment, the genomes of three other anammox bacteria, *B. fulgida*, *A. propionicus*, and the marine species *Scalindua*, are being sequenced (www.jgi.com). The intergenome analysis may define the core set of enzymes that are characteristic for the anammox metabolism and reveal the conservation in the organization of the pertinent operons. In this respect, it is interesting to note that all relevant gene clusters are indeed conserved in the genome of *Scalindua*. Furthermore, the comparisons between the different genomes may give a clue with respect to niche differentiation. Above all, the genome sequencing projects will provide the basic information for future expression studies, both at the gene (genomics) and protein (proteomics) levels. Eventually, such studies will allow an understanding in the way anammox bacteria are able to adapt to changes in the supply of the substrates and other environmental conditions in their highly dynamic habitats.

REFERENCES

Arrigo, K. R. 2005. Marine microorganisms and global nutrient cycles. *Nature* **437:**349–355.

Baumann, B., M. Snozzi, A. J. B. Zehnder, and J. R. vanderMeer. 1996. Dynamics of denitrification activity of Paracoccus denitrificans in continuous culture during aerobic-anaerobic changes. *J. Bacteriol.* **178:**4367–4374.

Berg, I. A., D. Kockelkorn, W. Buckel, and G. Fuchs. 2007. A 3-hydroxypropionate/4-hydroxybutyrate autotrophic carbon dioxide assimilation pathway in archaea. *Science* **318:**1782–1786.

Boumann, H. A., E. C. Hopmans, I. van de Leemput, H. J. M. Op den Camp, J. van de Vossenberg, M. Strous, M. S. M. Jetten, J. S. Sinninghe Damsté, and S. Schouten. 2006. Ladderane phospholipids in anammox bacteria comprise phosphocholine and phosphoethanolamine headgroups. *FEMS Microbiol. Lett.* **258:**297–304.

Brandes, J. A., A. H. Devol, and C. Deutsch. 2007. New developments in the marine nitrogen cycle. *Chem. Rev.* **107:**577–589.

Broda, E. 1977. Two kinds of lithotrophs missing in nature. *Z. Allg. Mikrobiol.* **17:**491–493.

Cole, J. 1996. Nitrate reduction to ammonia by enteric bacteria: redundancy, or a strategy for survival during oxygen starvation? *FEMS Microbiol. Lett.* **136:**1–11.

Egli, K., U. Fanger, P. J. J. Alvarez, H. Siegrist, J. R. van der Meer, and A. J. B. Zehnder. 2001. Enrichment and characterization of an anammox bacterium from a rotating biological contactor treating ammonium-rich leachate. *Arch. Microbiol.* **175:**198–207.

Ferry, J. G. 1999. Enzymology of one-carbon metabolism in methanogenic pathways. *FEMS Microbiol. Rev.* **23:**13–38.

Francis, C. A., J. M. Beman, and M. M. M. Kuypers. 2007. New processes and players in the nitrogen cycle: the microbial ecology of anaerobic and archaeal ammonia oxidation. *ISME J.* **1:**19–27.

Güven, D., A. Dapena, B. Kartal, M. C. Schmid, B. Maas, K. van de Pas-Schoonen, S. Sozen, R. Mendez, H. J. M. Op den Camp, M. S. M. Jetten, M. Strous, and I. Schmidt. 2005. Propionate oxidation by and methanol inhibition of anaerobic ammonium-oxidizing bacteria. *Appl. Environ. Microbiol.* **71:**1066–1071.

Haines, T. H. 2001. Do sterols reduce proton and sodium leaks through lipid bilayers? *Prog. Lipid Res.* **40:**299–324.

Hamersley, M. R., G. Lavik, D. Woebken, J. E. Rattray, P. Lam, E. C. Hopmans, J. S. S. Damste, S. Kruger, M. Graco, D. Gutierrez, and M. M. M. Kuypers. 2007. Anaerobic ammonium oxidation in the Peruvian oxygen minimum zone. *Limnol. Oceanogr.* **52:**923–933.

Heidelberg, J. F., I. T. Paulsen, K. E. Nelson, E. J. Gaidos, W. C. Nelson, T. D. Read, J. A. Eisen, R. Seshadri, N. Ward, B. Methe, R. A. Clayton, T. Meyer, A. Tsapin, J. Scott, M. Beanan, L. Brinkac, S. Daugherty, R. T. DeBoy, R. J. Dodson, A. S. Durkin, D. H. Haft, J. F. Kolonay, R. Madupu, J. D. Peterson, L. A. Umayam, O. White, A. M. Wolf, J. Vamathevan, J. Weidman, M. Impraim, K. Lee, K. Berry, C. Lee, J. Mueller, H. Khouri, J. Gill, T. R. Utterback, L. A. McDonald, T. V. Feldblyum, H. O. Smith, J. C. Venter, K. H. Nealson, and C. M. Fraser. 2002. Genome sequence of the dissimilatory metal ion-reducing bacterium Shewanella oneidensis. *Nat. Biotechnol.* **20:**1118–1123.

Jetten, M. S. M., S. J. Horn, and M. C. M. van Loosdrecht. 1997. Towards a more sustainable municipal wastewater treatment system. *Water Sci. Technol.* **35:**171–180.

Kartal, B. 2008. Ecophysiology of the anammox bacteria. Ph.D. thesis. University of Nijmegen, Nijmegen, The Netherlands.

Kartal, B., M. Koleva, R. Arsov, W. van der Star, M. S. M. Jetten, and M. Strous. 2006. Adaptation of a freshwater anammox population to high salinity wastewater. *J. Biotechnol.* **126:**546–553.

Kartal, B., M. M. M. Kuypers, G. Lavik, J. Schalk, H. J. M. Op den Camp, M. S. M. Jetten, and M. Strous. 2007a. Anammox bacteria disguised as denitrifiers: nitrate reduction to dinitrogen gas via nitrite and ammonium. *Environ. Microbiol.* **9:**635–642.

Kartal, B., J. Rattray, L. van Niftrik, J. van de Vossenberg, M. Schmid, R. I. Webb, S. Schouten, J. A. Fuerst, J. S. Sinninghe Damsté, M. S. M. Jetten, and M. Strous. 2007b. "*Candidatus* Anammoxoglobus propionicus" gen. nov., sp. nov., a new propionate oxidizing species of anaerobic ammonium oxidizing bacteria. *Syst. Appl. Microbiol.* **30:**39–49.

Kartal, B., L. van Niftrik, J. Rattray, J. de Vossenberg, M. C. Schmid, J. S. S. Damste, M. S. M. Jetten, and M. Strous. 2008. "Candidatus Brocadia fulgida": an autofluorescent anaerobic ammonium oxidizing bacterium. *FEMS Microbiol. Ecol.* **63:**46–55.

Kindaichi, T., I. Tsushima, Y. Ogasawara, M. Shimokawa, N. Ozaki, H. Satoh, and S. Okabe. 2007. In situ activity and spatial organization of anaerobic ammonium-oxidizing (anammox) bacteria in biofilms. *Appl. Environ. Microbiol.* **73:**4931–4939.

Kuypers, M. M. M., A. O. Sliekers, G. Lavik, M. Schmid, B. B. Jorgensen, J. G. Kuenen, J. S. Sinninghe Damsté, M. Strous, and M. S. M. Jetten. 2003. Anaerobic ammonium oxidation by anammox bacteria in the Black Sea. *Nature* **422:**608–611.

Kuypers, M. M. M., G. Lavik, D. Woebken, M. Schmid, B. M. Fuchs, R. Amann, B. B. Jorgensen, and M. S. M. Jetten. 2005. Massive nitrogen loss from the Benguela upwelling system through anaerobic ammonium oxidation. *Proc. Natl. Acad. Sci. USA* **102:**6478–6483.

Lam, P., M. M. Jensen, G. Lavik, D. F. McGinnis, B. Muller, C. J. Schubert, R. Amann, B. Thamdrup, and M. M. M. Kuypers. 2007. Linking crenarchaeal and bacterial nitrification to anammox in the Black Sea. *Proc. Natl. Acad. Sci. USA* **104:**7104–7109.

Lindsay, M. R., R. I. Webb, M. Strous, M. S. Jetten, M. K. Butler, R. J. Forde, and J. A. Fuerst. 2001. Cell compartmentalisation in planctomycetes: novel types of structural organisation for the bacterial cell. *Arch. Microbiol.* **175:**413–429.

Mascitti, V., and E. J. Corey. 2004. Total synthesis of (+/-)-pentacycloanammoxic acid. *J. Am. Chem. Soc.* **126:**15664–15665.

Methe, B. A., K. E. Nelson, J. A. Eisen, I. T. Paulsen, W. Nelson, J. F. Heidelberg, D. Wu, M. Wu, N. Ward, M. J. Beanan, R. J. Dodson, R. Madupu, L. M. Brinkac, S. C. Daugherty, R. T. DeBoy, A. S. Durkin, M. Gwinn, J. F. Kolonay, S. A. Sullivan, D. H. Haft, J. Selengut, T. M. Davidsen, N. Zafar, O. White, B. Tran, C. Romero, H. A. Forberger, J. Weidman, H. Khouri, T. V. Feldblyum, T. R. Utterback, S. E. Van Aken, D. R. Lovley, and C. M. Fraser. 2003. Genome of Geobacter sulfurreducens: metal reduction in subsurface environments. *Science* **302:**1967–1969.

Mulder, A., A. A. Vandegraaf, L. A. Robertson, and J. G. Kuenen. 1995. Anaerobic ammonium oxidation discovered in a denitrifying fluidized-bed reactor. *FEMS Microbiol. Ecol.* **16:**177–183.

Nielsen, M., A. Bollmann, O. Sliekers, M. Jetten, M. Schmid, M. Strous, I. Schmidt, L. H. Larsen, L. P. Nielsen, and N. P. Revsbech. 2005. Kinetics, diffusional limitation and microscale distribution of chemistry and organisms in a CANON reactor. *FEMS Microbiol. Ecol.* **51:**247–256.

Op den Camp, H., B. Kartal, D. Guven, L. van Niftrik, S. C. M. Haaijer, W. R. L. van der Star, K. T. van de Pas-Schoonen, A. Cabezas, Z. Ying, M. C. Schmid, M. M. M. Kuypers, J. van de Vossenberg, H. R. Harhangi, C. Picioreanu, M. C. M. van Loosdrecht, J. G. Kuenen, M. Strous, and M. S. M. Jetten. 2006. Global impact and application of the anaerobic ammonium-oxidizing (anammox) bacteria. *Biochem. Soc. Trans.* **34:**174–178.

Otte, S., N. G. Grobben, L. A. Robertson, M. S. M. Jetten, and J. G. Kuenen. 1996. Nitrous oxide production by Alcaligenes faecalis under transient and dynamic aerobic and anaerobic conditions. *Appl. Environ. Microbiol.* **62:**2421–2426.

Penton, C. R., A. H. Devol, and J. M. Tiedje. 2006. Molecular evidence for the broad distribution of anaerobic ammonium-oxidizing bacteria in freshwater and marine sediments. *Appl. Environ. Microbiol.* **72:**6829–6832.

Quan, Z. X., S. K. Rhee, J. E. Zuo, Y. Yang, J. W. Bae, J. R. Park, S. T. Lee, and Y. H. Park. 2008. Diversity of ammonium-oxidizing bacteria in a granular sludge anaerobic ammonium-oxidizing (anammox) reactor. *Environ. Microbiol.* **10:**3130–3139.

Rattray, J. E., J. van de Vossenberg, E. C. Hopmans, B. Kartal, L. van Niftrik, W. I. Rijpstra, M. Strous, M. S. Jetten, S. Schouten, and J. S. Damste. 2008. Ladderane lipid distribution in four genera of anammox bacteria. *Arch. Microbiol.* **190:**51–66.

Richards, F. A. 1965. Anoxic basins and fjords. *In* J. P. Ripley and G. Skirrow (ed.), *Chemical Oceanography*. Academic Press, London, United Kingdom.

Schalk, J., S. de Vries, J. G. Kuenen, and M. S. M. Jetten. 2000. Involvement of a novel hydroxylamine oxidoreductase in anaerobic ammonium oxidation. *Biochemistry* **39:**5405–5412.

Schmid, M., U. Twachtmann, M. Klein, M. Strous, S. Juretschko, M. S. M. Jetten, J. W. Metzger, K. H. Schleifer, and M. Wagner. 2000. Molecular evidence for genus level diversity of bacteria capable of catalyzing anaerobic ammonium oxidation. *Syst. Appl. Microbiol.* **23:**93–106.

Schmid, M., K. Walsh, R. Webb, W. I. C. Rijpstra, K. van de Pas-Schoonen, M. J. Verbruggen, T. Hill, B. Moffett, J. Fuerst, S. Schouten, J. S. Sinninghe Damsté, J. Harris, P. Shaw, M. Jetten, and M. Strous. 2003. "*Candidatus* Scalindua brodae," sp nov., "*Candidatus* Scalindua wagneri," sp nov., two new species of anaerobic ammonium oxidizing bacteria. *Syst. Appl. Microbiol.* **26:**529–538.

Schmid, M. C., N. Risgaard-Petersen, J. van de Vossenberg, M. M. M. Kuypers, G. Lavik, J. Petersen, S. Hulth, B. Thamdrup, D. Canfield, T. Dalsgaard, S. Rysgaard, M. K. Sejr, M. Strous, H. J. M. Op den Camp, and M. S. M. Jetten. 2007. Anaerobic ammonium-oxidizing bacteria in marine environments: widespread occurrence but low diversity. *Environ. Microbiol.* **9:**1476–1484.

Schneider, D., and C. L. Schmidt. 2005. Multiple Rieske proteins in prokaryotes: where and why? *Biochim. Biophys. Acta* **1710:**1–12.

Schouten, S., M. Strous, M. M. M. Kuypers, W. I. C. Rijpstra, M. Baas, C. J. Schubert, M. S. M. Jetten, and J. S. Sinninghe Damsté. 2004. Stable carbon isotopic fractionations associated with inorganic carbon fixation by anaerobic ammonium-oxidizing bacteria. *Appl. Environ. Microbiol.* **70:**3785–3788.

Schubert, C. J., E. Durisch-Kaiser, B. Wehrli, B. Thamdrup, P. Lam, and M. M. M. Kuypers. 2006. Anaerobic ammonium oxidation in a tropical freshwater system (Lake Tanganyika). *Environ. Microbiol.* **8:**1857–1863.

Shimamura, M., T. Nishiyama, H. Shigetomo, T. Toyomoto, Y. Kawahara, K. Furukawa, and T. Fujii. 2007. Isolation of a multiheme protein from an anaerobic ammonium-oxidizing enrichment culture with features of a hydrazine-oxidizing enzyme. *Appl. Environ. Microbiol.* **73:**1065–1072.

Shimamura, M., T. Nishiyama, K. Shinya, Y. Kawahara, K. Furukawa, and T. Fujii. 2008. Another multiheme protein, hydroxylamine oxidoreductase, abundantly produced in an anammox bacterium besides the hydrazine-oxidizing enzyme. *J. Biosci. Bioeng.* **105:**243–248.

Simon, J. 2002. Enzymology and bioenergetics of respiratory nitrite ammonification. *FEMS Microbiol. Rev.* **26:**285–309.

Sinninghe Damsté, J. S., W. I. C. Rijpstra, J. A. J. Geenevasen, M. Strous, and M. S. M. Jetten. 2005. Structural identification of ladderane and other membrane lipids of planctomycetes capable of anaerobic ammonium oxidation (anammox). *FEBS J.* **272:**4270–4283.

Sinninghe Damsté, J. S., M. Strous, W. I. C. Rijpstra, E. C. Hopmans, J. A. J. Geenevasen, A. C. T. van Duin, L. A. van Niftrik, and M. S. M. Jetten. 2002. Linearly concatenated cyclobutane lipids form a dense bacterial membrane. *Nature* **419:**708–712.

Sliekers, A. O., N. Derwort, J. L. C. Gomez, M. Strous, J. G. Kuenen, and M. S. M. Jetten. 2002. Completely autotrophic nitrogen removal over nitrite in one single reactor. *Water Res.* **36:**2475–2482.

Strohm, T. O., B. Griffin, W. G. Zumft, and B. Schink. 2007. Growth yields in bacterial denitrification and nitrate ammonification. *Appl. Environ. Microbiol.* **73:**1420–1424.

**Strous, M., E. Pelletier, S. Mangenot, T. Rattei, A. Lehner, M. W. Taylor, M. Horn, H. Daims, D. Bartol-Mavel, P. Wincker, V. Barbe, N. Fonknechten, D. Vallenet, B. Segurens, C. Schenowitz-Truong, C. Medigue, A. Collingro, B. Snel, B. E. Dutilh, H. J. M. Op den Camp, C. van der Drift, I. Cirpus, K. T. van de Pas-Schoonen, H. R. Harhangi, L. van Niftrik, M. Schmid, J. Keltjens, J. van de Vossenberg, B. Kartal, H. Meier, D. Frishman, M. A.

Huynen, H. W. Mewes, J. Weissenbach, M. S. M. Jetten, M. Wagner, and D. Le Paslier. 2006. Deciphering the evolution and metabolism of an anammox bacterium from a community genome. *Nature* **440:**790–794.

Strous, M., J. J. Heijnen, J. G. Kuenen, and M. S. M. Jetten. 1998. The sequencing batch reactor as a powerful tool for the study of slowly growing anaerobic ammonium-oxidizing microorganisms. *Appl. Microbiol. Biotechnol.* **50:**589–596.

Strous, M., J. A. Fuerst, E. H. M. Kramer, S. Logemann, G. Muyzer, K. T. van de Pas-Schoonen, R. Webb, J. G. Kuenen, and M. S. M. Jetten. 1999a. Missing lithotroph identified as new planctomycete. *Nature* **400:**446–449.

Strous, M., J. G. Kuenen, and M. S. M. Jetten. 1999b. Key physiology of anaerobic ammonium oxidation. *Appl. Environ. Microbiol.* **65:**3248–3250.

Third, K. A., A. O. Sliekers, J. G. Kuenen, and M. S. M. Jetten. 2001. The CANON system (completely autotrophic nitrogen-removal over nitrite) under ammonium limitation: interaction and competition between three groups of bacteria. *Syst. Appl. Microbiol.* **24:**588–596.

Tsushima, I., Y. Ogasawara, T. Kindaichi, H. Satoh, and S. Okabe. 2007. Development of high-rate anaerobic ammonium-oxidizing (anammox) biofilm reactors. *Water Res.* **41:**1623–1634.

Van de Graaf, A. A., A. Mulder, P. Debruijn, M. S. M. Jetten, L. A. Robertson, and J. G. Kuenen. 1995. Anaerobic oxidation of ammonium is a biologically mediated process. *Appl. Environ. Microbiol.* **61:**1246–1251.

Van de Graaf, A. A., P. deBruijn, L. A. Robertson, M. S. M. Jetten, and J. G. Kuenen. 1996. Autotrophic growth of anaerobic ammonium-oxidizing micro-organisms in a fluidized bed reactor. *Microbiology* **142:**2187–2196.

Van de Graaf, A. A., P. deBruijn, L. A. Robertson, M. S. M. Jetten, and J. G. Kuenen. 1997. Metabolic pathway of anaerobic ammonium oxidation on the basis of N-15 studies in a fluidized bed reactor. *Microbiology* **143:**2415–2421.

Van de Pas-Schoonen, K. T., S. Schalk-Otte, S. Haaijer, M. Schmid, H. O. den Camp, M. Strous, J. G. Kuenen, and M. S. M. Jetten. 2005. Complete conversion of nitrate into dinitrogen gas in co-cultures of denitrifying bacteria. *Biochem. Soc. Trans.* **33:**205–209.

Van de Vossenberg, J., A. J. M. Driessen, W. D. Grant, and W. N. Konings. 1999. Lipid membranes from halophilic and alkali-halophilic archaea have a low H^+ and Na^+ permeability at high salt concentration. *Extremophiles* **3:**253–257.

Van de Vossenberg, J., J. E. Rattray, W. Geerts, B. Kartal, L. van Niftrik, E. G. van Donselaar, J. S. Sinninghe Damste, M. Strous, and M. S. Jetten. 2008. Enrichment and characterization of marine anammox bacteria associated with global nitrogen gas production. *Environ Microbiol.* **10:**3120–3129.

Van der Oost, J., A. P. N. De Boer, J.-W. L. De Gier, W. G. Zumft, A. H. Stouthamer and R. J. M. Van Spanning. 1994. The heme-copper oxidase family consists of three distinct types of terminal oxidases and is related to nitric oxide reductase. *FEMS Microbiol. Lett.* **121:**1–10.

Van der Star, W. R., A. I. Miclea, U. G. van Dongen, G. Muyzer, C. Picioreanu, and M. C. van Loosdrecht. 2008. The membrane bioreactor: a novel tool to grow anammox bacteria as free cells. *Biotechnol. Bioeng.* **101:**286–294.

Van der Star, W. R. L., W. R. Abma, D. Blommers, J. W. Mulder, T. Tokutomi, M. Strous, C. Picioreanu, and M. C. M. Van Loosdrecht. 2007. Startup of reactors for anoxic ammonium oxidation: experiences from the first full-scale anammox reactor in Rotterdam. *Water Res.* **41:**4149–4163.

Van Niftrik, L., W. J. C. Geerts, E. G. van Donselaar, B. M. Humbel, R. I. Webb, J. A. Fuerst, A. J. Verkleij, M. S. M. Jetten, and M. Strous. 2008a. Linking ultrastructure and function in four genera of anaerobic ammonium-oxidizing bacteria: cell plan, glycogen storage, and localization of cytochrome c proteins. *J. Bacteriol.* **190:**708–717.

Van Niftrik, L., W. J. C. Geerts, E. G. van Donselaar, B. M. Humbel, A. Yakushevska, A. J. Verkleij, M. S. M. Jetten, and M. Strous. 2008b. Combined structural and chemical analysis of the anammoxosome: a membrane-bounded intracytoplasmic compartment in anammox bacteria. *J. Struct. Biol.* **161:**401–410.

Ward, B. B. 2003. Significance of anaerobic ammonium oxidation in the ocean. *Trends Microbiol.* **11:**408–410.

Zehr, J. P., and B. B. Ward. 2002. Nitrogen cycling in the ocean: new perspectives on processes and paradigms. *Appl. Environ. Microbiol.* **68:**1015–1024.

DISTRIBUTION, ACTIVITY, AND ECOLOGY OF ANAMMOX BACTERIA IN AQUATIC ENVIRONMENTS

Mark Trimmer and Pia Engström

9

INTRODUCTION

The cycling of nitrogen (N) in aquatic ecosystems has been studied extensively, whether it be to understand fundamental aspects of the key biogeochemical cycles of N and carbon (C) on Earth, or the affects of anthropogenic activity on the balance of these cycles and, in particular, the effects of N in coastal regions (Falkowski, 1997; Diaz and Rosenberg, 2008; Conley et al., 2009). While considerable progress had been made in both the science and our understanding of key aspects of the N cycle in aquatic ecosystems, the comparatively recent (i.e., 2002) discovery of anaerobic ammonium oxidation (anammox) as a key player in the aquatic N cycle has had a profound impact on our knowledge base (Thamdrup and Dalsgaard, 2002). Not only has anammox been shown to provide a logical solution to some long-standing "marine N mysteries," but its mere presence undermines some of the key tools/techniques used to study the flux of N in aquatic ecosystems so far (Devol, 2003; Risgaard-Petersen et al., 2003). By coupling the oxidation of ammonium to the reduction of nitrite to produce N_2 gas, anammox provides a mechanism for removing fixed N from ecosystems, which bypasses the classic reliance on aerobic nitrification and subsequent denitrification. Given that the balance between N fixation and its removal via N_2 production (denitrification and anammox) is key to the assimilation of C via primary production that, in turn, modulates CO_2 in the atmosphere, there is a very pressing need for a more comprehensive understanding of the cycling of N in aquatic ecosystems (Codispoti et al., 2001). As such, anammox has received considerable attention itself in the past 7 years, and reviews of its role in the environment are already in the literature (Dalsgaard et al., 2005; Francis et al., 2007). Therefore, this chapter will not spend too much time revisiting the material covered elsewhere but seeks more to provide a synthesis of the broadscale patterns of anammox across a spectrum of aquatic ecosystems and to put forward some hypotheses as to what regulates anammox and the total flux of N in such systems. Research into anammox falls largely into two distinct aquatic ecosystems (Fig. 1): (i) its role in the anaerobic oxidation of ammonium in the suboxic layers of aquatic sediments, where the respective reactions and ecophysiologies are compressed into fractions of centimeters; and (ii) the same ecosystem function,

Mark Trimmer, School of Biological and Chemical Sciences, Queen Mary University of London, London E1 4NS, United Kingdom. *Pia Engström*, Civil and Environmental Engineering, Chalmers University of Technology, SE-412 96 Göteborg, Sweden.

Nitrification, Edited by Bess B. Ward, Daniel J. Arp, and Martin G. Klotz

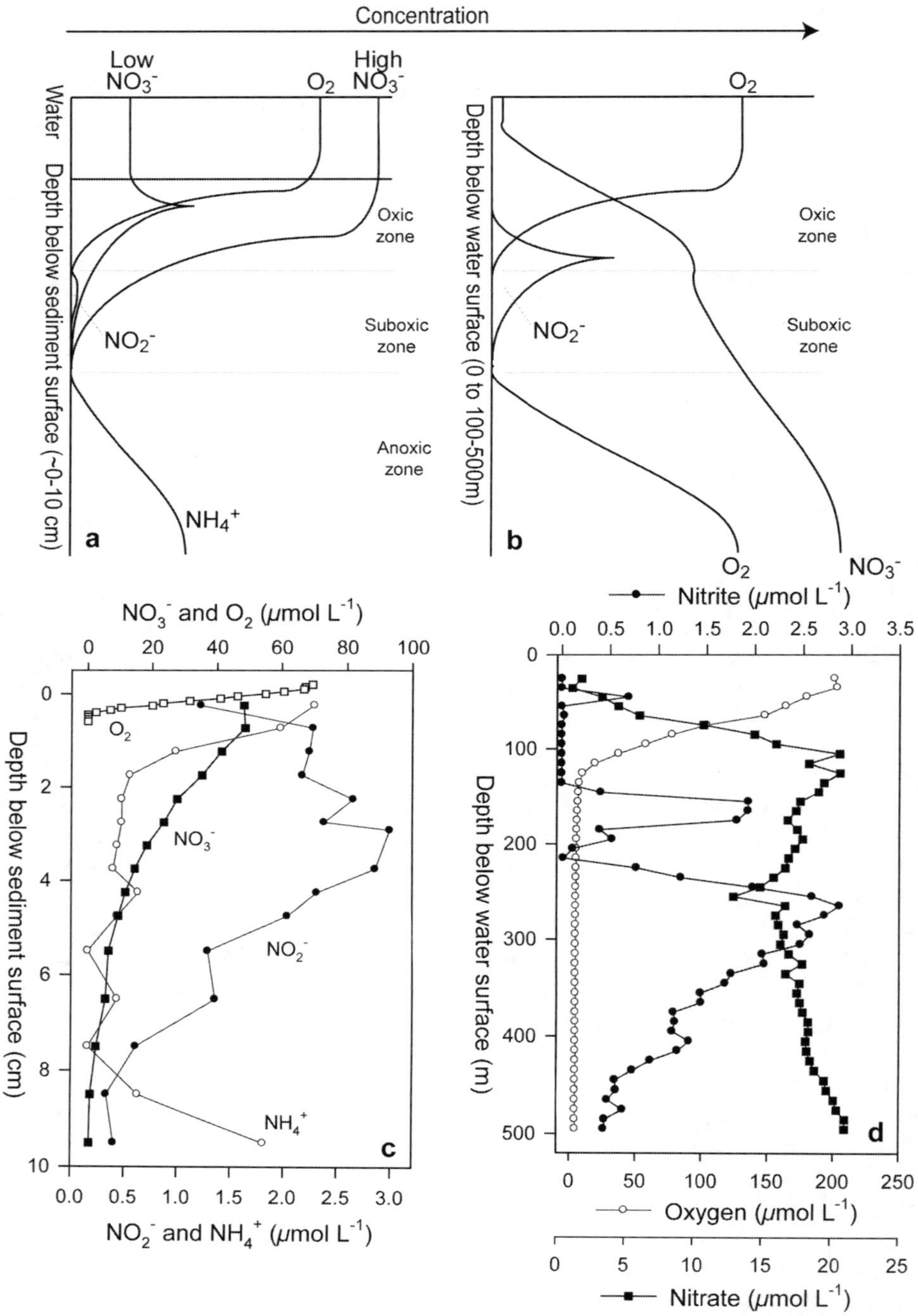

FIGURE 1 Schematic representation of the oxic and suboxic zones in sediments and OMZs to highlight the respective difference in depth scale (a, b) and representative examples for sediment in the Cascadia Basin (c) and the central Arabian Sea (d) (panels c and d are reproduced, respectively, from Engström et al. [2009] and Nicholls et al. [2007], Copyright by the American Society of Limnology and Oceanography, Inc.). Note the high NO_3^- (typical estuarine or deep sea) and low NO_3^- (coastal or shelf sea) depicted in panel a.

but distributed over depths of tens of meters in the oxygen minimum zones (OMZs) of the global ocean, and, as such, we have divided our focus between these two areas.

ECOLOGICAL SIGNIFICANCE OF ANAEROBIC AMMONIUM OXIDATION IN AQUATIC ECOSYSTEMS

In many ecosystems, there is strong regulation of primary production by the availability of N, and denitrification was thought to be the only significant pathway for the removal of N in the suboxic parts of both terrestrial and aquatic ecosystems. The important ecological aspect of a denitrification-only scenario is that any ammonium that accumulated in either the suboxic strata of sediments or the OMZs of the ocean would have to be oxidized to nitrite and/or nitrate by aerobic chemotrophic nitrification bacteria before being "denitrified" to N_2 gas via largely heterotrophic bacterial respiratory pathways. The laboratory-based discovery of anammox, however, suggested that other metabolisms could be present in the environment, and its presence was confirmed in marine sediments in 2002 and two anoxic water basins shortly after (Mulder et al., 1995; Thamdrup and Dalsgaard, 2002; Dalsgaard et al., 2003; Kuypers et al., 2003). Anammox removes fixed N in the readily available form of ammonium from an aquatic ecosystem, without prior oxidation of all of the ammonium to either NO_2^- or NO_3^- via aerobic chemoautotrophic nitrification (Fig. 2). In effect, anammox changes the overall stoichiometry of the mineralization of organic matter by reducing the total requirement for oxygen and increasing the production of N_2 gas, for example, by (i) the oxidation of organic matter with oxygen and subsequent liberation of organically bound N and P (Richards et al., 1965):

$$(CH_2O)_{106}(NH_3)_{16}H_3PO_4 + \underline{106}\ O_2 \rightarrow 106\ CO_2 + 16\ NH_3 + H_3PO_4 + 106\ H_2O \quad \textbf{(1)}$$

(ii) the subsequent complete nitrification of any liberated ammonium:

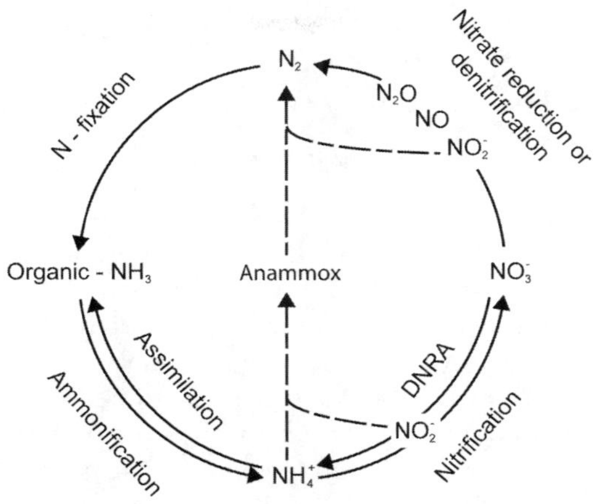

FIGURE 2 Simplified representation of the N cycle highlighting the key characteristic of the anammox reaction. Nitrate reduction refers to the one-step reduction of NO_3^- to NO_2^- and complete denitrification would ultimately generate N_2 gas; assimilatory reduction of NO_3^- to organic N has been omitted. (Reproduced and amended from Trimmer et al. [2003] with permission from The American Society for Microbiology.)

$$16\ NH_3 + \underline{32}\ O_2 \rightarrow 16\ HNO_3 + 16\ H_2O \quad \textbf{(2)}$$

and (iii) the summation of equations 1 and 2:

$$(CH_2O)_{106}(NH_3)_{16}H_3PO_4 + \underline{138}\ O_2 \rightarrow 106\ CO_2 + 16\ HNO_3 + H_3PO_4 + 122\ H_2O \quad \textbf{(3)}$$

Eventually, all of the oxygen will be consumed, and the oxidation of organic matter will proceed via alternative electron acceptors, here, for example, via nitrate (Richards et al., 1965):

$$(CH_2O)_{106}(NH_3)_{16}H_3PO_4 + 84.8\ HNO_3 \rightarrow 106\ CO_2 + 16\ NH_3 + 42.2\ N_2 + 148.4\ H_2O + H_3PO_4 \quad \textbf{(4)}$$

Alternatively, the oxidation of organic matter with partial nitrification of any liberated ammonium to nitrite combined with anammox could proceed according to the following:

$$16\ NH_3 + 24\ O_2 \rightarrow 8\ HNO_2 + 8\ NH_3 + 8\ H_2O \quad \text{partial nitrification} \quad \textbf{(5)}$$

$$8\ NH_3 + 8\ HNO_2 \rightarrow 8\ N_2 + 16\ H_2O$$
anammox (van de Graaf et al., 1995) **(6)**

together gives:

$$(CH_2O)_{106}(NH_3)_{16} + \underline{130}\ O_2 = 106\ CO_2 + 8\ N_2 + 130\ H_2O \quad \textbf{(7)}$$

Note that with a coupled nitrification and denitrification-only scenario (equations 3 and 4), the reduction of nitrate generates further ammonium(ia), which is still "available" to the ecosystem. Now, if we invoke a combined process through a combination of denitrification (strictly this first stage is only nitrate reduction) and anammox through the intermediates ammonium and nitrite, we generate twice as much N_2 gas per mole of nitrate reduced, and any liberated ammonium is "removed" from the ecosystem (Richards et al., 1965; Dalsgaard et al., 2003; Thamdrup et al., 2006):

$$(CH_2O)_{106}(NH_3)_{16}H_3PO_4 + 94.4\ NO_3^- + 94.4\ H^+ \rightarrow 106\ CO_2 + 16\ NH_4^+ + 16\ NO_2^- + \underline{39.2}\ N_2 + 145.2\ H_2O + H_3PO_4 \quad \textbf{(8)}$$

$$16\ NH_4^+ + 16\ NO_2^- \rightarrow \underline{16}\ N_2 + 32\ H_2O$$

to give $\underline{55.2}$ N_2, of which 29 % comes from anammox.

Alternatively, nitrate could be reduced to nitrite through nitrate reduction:

$$(CH_2O)_{106}(NH_3)_{16}H_3PO_4 + 212\ NO_3^- + 16\ H^+ \rightarrow 106\ CO_2 + 212\ NO_2^- + 16\ NH_4^+ + 106\ H_2O + H_3PO_4 \quad \textbf{(9)}$$

or even ammonium through dissimilatory nitrate reduction to ammonium (DNRA):

$$(CH_2O)_{106}(NH_3)_{16}H_3PO_4 + 53\ NO_3^- + 122\ H^+ \rightarrow 106\ CO_2 + 69\ NH_4^+ + 53\ H_2O + H_3PO_4 \quad \textbf{(10)}$$

Overall, with anammox, a fraction of the mineralized ammonium is converted to N_2 gas and lost from the aquatic ecosystem without the consumption of oxygen. Indeed, it was after observing the absence of ammonium in the anoxic water column of a fjord that Richards et al. (1965) reasoned for the presence of "anammox" in the environment, or the simultaneous oxidation of ammonium to N_2 gas coupled to the oxidation of organic matter with nitrate. Note that such stoichiometries apply to the immediate location of each respective metabolism, and the oxidative power supplied by NO_2^-, for example, would still require the consumption of oxygen elsewhere to drive the original oxidation of NH_4^+. Hence, at the scale of the global ocean, the net stoichiometry and mass balance would be the same.

ANAMMOX IN AQUATIC SEDIMENTS

There are two very distinct methods used to measure anammox in sediments, and it is important to appreciate their differences and to bear these in mind when reading this section. The original studies (e.g., Thamdrup and Dalsgaard, 2002) used homogenized slurries of anoxic sediment, and then, subsequently, attempts have been made to quantify anammox and denitrification simultaneously under conditions more representative of the "in situ" conditions (i.e., intact sediment cores) (Risgaard-Petersen et al., 2003, 2004; Trimmer et al., 2006; Minjeaud et al., 2008). Again, we focus first on the results obtained using slurries of homogenized sediment, and then, second, on the data derived with intact sediment cores, and, finally, draw comparisons between the two at the end, without dwelling too long on the respective complexities of each method.

Geographical Distribution and Range of Activity in Homogenized Sediment

To date, the majority of research into anammox in sediments has focused on either marine or estuarine sediments. The anammox metabolism has now been confirmed in numerous deep marine sediments, including the Norwegian Trench, Cascadia Basin, North Atlantic, and Sagami Bay; in shallower coastal shelf sediments (temperate and arctic); and in many estuaries (Table 1). That is not to say that the anammox metabolism is truly ubiquitous, as negative results from anammox screening surveys have been reported, though these are comparatively rare (Risgaard-Petersen et al.,

TABLE 1 Published and unpublished rates of anammox and denitrification measured in anoxic homogenized or slurrified sediment[a]

Source and reference	Water depth (m)	Denitrification (nmol of N cm^{-3} h^{-1})	Anammox (nmol of N cm^{-3} h^{-1})	Total N_2 (nmol of N cm^{-3} h^{-1})	Anammox (%)
Skagerrat, Aarhus Bay; Thamdrup and Dalsgaard, 2002	16–695	0.4–118.6	0.9–2.9	1.3–121.0	2–67
Thames Estuary, United Kingdom; Trimmer et al., 2003	2–4	34.0–154.2	0.2–3.1	34.2–161.3	1–8
Gullmarsfjorden; Engström, 2004	75–116	2.3–3.6	0.9–1.8	2.9–5.3	33–47
Randers Fjord; Risgaard-Petersen et al., 2004	1[b]	31–137	3.8–11.0	42.0–142.2	4–26
Arctic sediments; Rysgaard et al., 2004	36–100	1.0–50.5	0.2–15.1	1.2–65.6	1–34
Skagerrat/Kattegat, Long Island Sound; Engström et al., 2005	16–700	<0.1–45.3	0.3–3.7	0.4–48.8	7–80
Logan and Albert River system, Australia; Meyer et al., 2005	0.5	3.8–85.5	0.0–8.4	3.8–93.3	0–9
Gullmarsfjorden (Alsbäck); P. Engström and S. Hulth (unpublished data)	116[b]	0.7–2.4	0.7–2.5	1.4–4.9	38–52
Hood Canal, Tofino Inlet; Engström et al., 2009	38–147	2.9–8.1	0.7–5.8	3.6–13.9	17–43
Washington Margins; Engström et al., 2009	2,740–3,110	0.2–0.9	0.2–0.9	0.4–1.8	13–51
Cape Fear River Estuary, United States; Dale et al., 2009	1	2.1–4.9	<0.1–0.7	2.1–5.2	1–15
Plum Island Sound Estuary, United States; Koop-Jakobsen and Giblin, 2009	3	4.2–15.6	0.0–0.2	4.2–15.6	0–2

[a]This is a summary illustrating the range at each site; the full data set (n = 98) is available from the corresponding author (M.T.).
[b]Seasonal measurements.

2004; Rich et al., 2008; Koop-Jakobsen and Giblin, 2009). Discounting the upper freshwater tidal limits of estuaries (Trimmer et al., 2003; Meyer et al., 2005; Tal et al., 2005), the presence of anammox bacteria has only been confirmed in one truly freshwater sediment: the Xinyi River in China (Zhang et al., 2007). Our own attempts at screening the sediments of the River Frome (Dorset, United Kingdom) and the floodplain of the River Cole (Oxfordshire, United Kingdom) suggested no potential for anammox, despite very high concentrations of NO_3^- in the overlying water, the presence of organic matter, and appreciable rates of NO_3^- reduction (Sanders and Trimmer, 2006; M. Trimmer, F. Sgouridis, C. M. Heppell, M. Trimmer, and G. Wharton, unpublished). Data on anammox in freshwater sediments are still very scarce, though, and it is impossible to draw any conclusions about its contribution to the production of N_2 in freshwater sediments at this stage.

In their review of anammox in the marine environment, Dalsgaard et al. (2005) presented a clear trend of the potential contribution from anammox to the production of N_2 gas (*ra* %) increasing with water depth, reaching a maximum of 80% at 700 m in the deep Norwegian Trench in the open Skagerrak. Adding to that the data made available in the interim (both published and unpublished between 2005 and 2009) confirms this original trend (Fig. 3a and b). Sediments from water depths of between 1 and 150 m are particularly well represented in the total data set (67 of 96), and there is a clear and simple linear relationship between both increasing *ra* and water depth ($r = 0.89$, $p < 0.001$). Beyond the estuarine and continental shelf sediments, the relationship is not so clear, and, in part, this reflects a comparative lack of data. If anything, there appears to be a plateau at an *ra* of ~50 % (Thamdrup et al., 2006). The potential for anammox to contribute to the majority of N_2 production (*ra* >50%) appears rare and, so far, is restricted to the deep Norwegian Trench and Alsbäck in the Gullmar fjord. To date, 90% of all *ra* values fall below 48%, with a mean for *ra* of 23% (±2 SE, $n = 96$) (Table 1).

It is important to appreciate that the relative increase in the significance of anammox with depth is largely due to a marked decrease in the specific activity of denitrification in deeper waters, whereas anammox decreases to a lesser extent (Fig. 4) (Dalsgaard et al., 2005; Engström et al., 2005). Indeed, the range covered by the denitrification data is orders of magnitude greater than that for anammox across the entire data set (Table 1). This drop in denitrification activity, with the move toward deeper water, has been assigned simply to a concurrent drop in the availability of organic carbon to fuel heterotrophic denitrification, observed as a decrease in total sediment metabolism, sediment chlorophyll content, or rate of ammonification (Engström et al., 2005; Thamdrup and Dalsgaard, 2002). In some deep-water sediments, a higher contribution from anammox to N_2 production has been correlated with an increased sediment content of manganese dioxide, which, in turn, may enable suboxic carbon oxidation to flow via dissimilatory reduction of manganese dioxide, rather than denitrification (Dalsgaard and Thamdrup, 2002; Engström et al., 2005).

Fueling Anammox in Aquatic Sediments

Nitrate is abundant in the estuaries and surrounding coastal seas of the developed world, and, even away from the influence of anthropogenic sources of nitrate, coastal sediments often act as sources of nitrate to the overlying water, suggesting that production (via nitrification) exceeds the requirements of any nitrate-reducing metabolisms in the sediment (Peirels et al., 1991; van Raaphorst et al., 1992; Lohse et al., 1993; Middelburg et al., 1996; Nedwell et al., 2002). In contrast, nitrite is comparatively scarce, and anammox must ultimately be coupled to a source of nitrite in the environment (Dalsgaard and Thamdrup, 2002). Although the first stage of nitrification can produce nitrite in the aerobic layers of sedi-

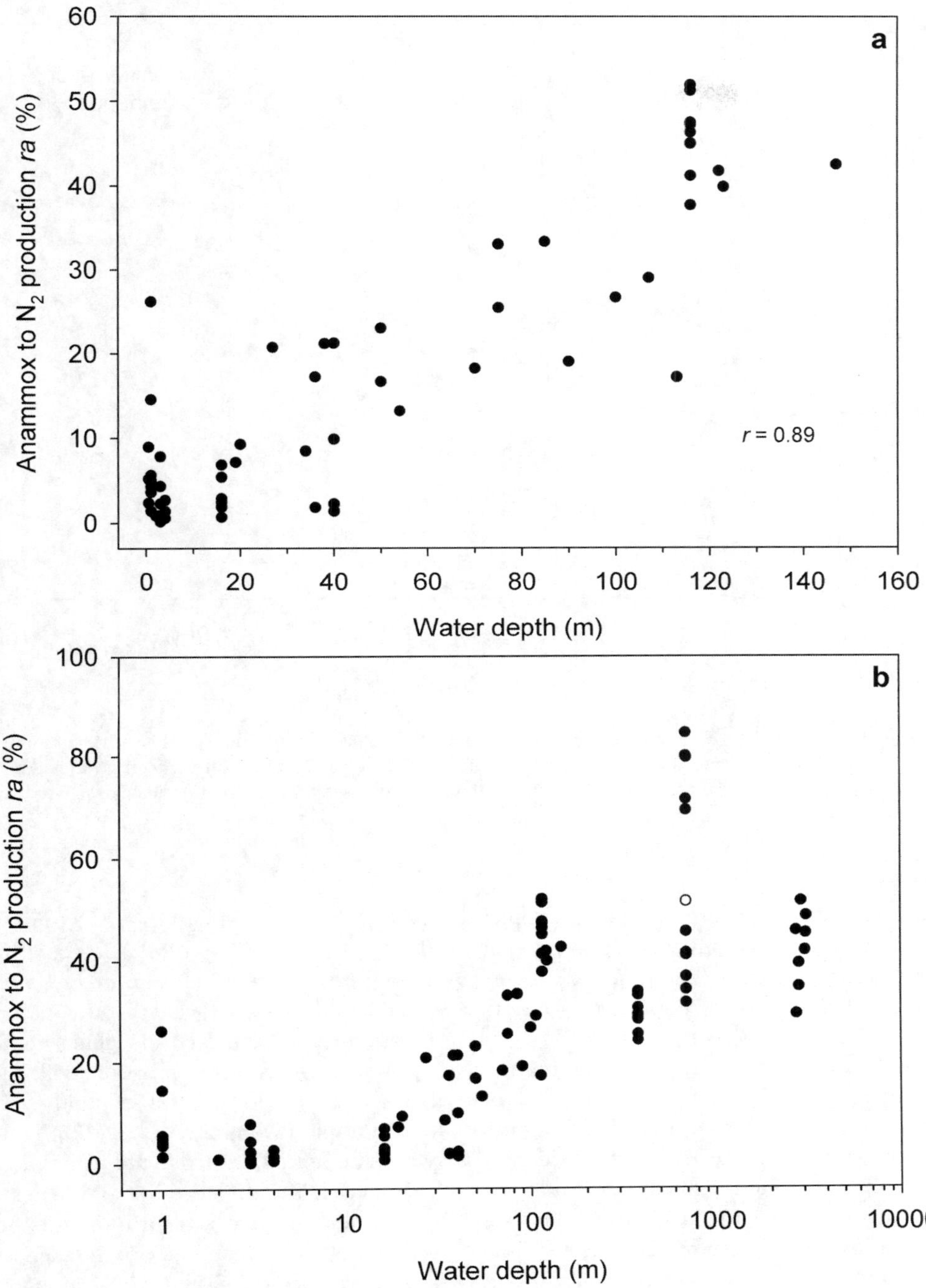

FIGURE 3 Contribution from anammox to the production of N_2 (*ra* %) measured in slurries of anoxic sediment. (a) Relatively simple linear relationship for water depths up to 150 m with the correlation coefficient (*r*). (b) Complete data set against water depth on a common log scale. The open circle is the mean for *ra* at this site (S9 the deep Skagerrak) (see Table 1).

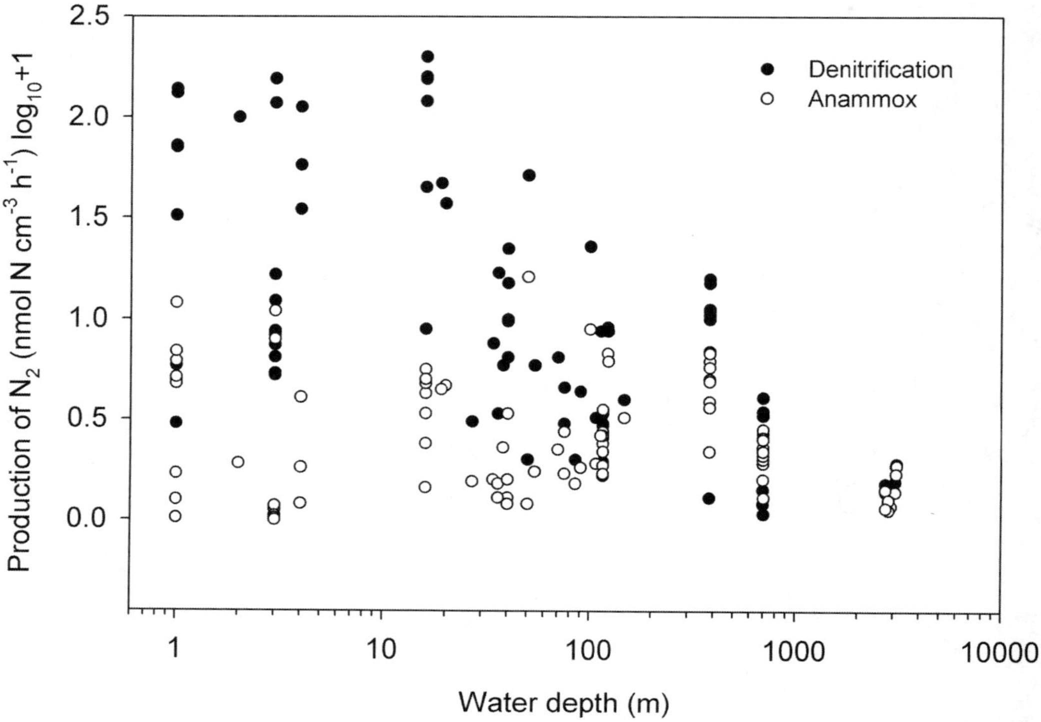

FIGURE 4 Composite of all available anammox- and denitrification-specific activity slurry data (rate) as a function of water depth. Rate data have been log transformed (log_{10}+1) and plotted against water depth on a common log scale. Data are from Table 1; note the difference in both the range and variance associated with the measures of denitrification.

ment, fine-scale micro-profiles and modeling of its distribution in pore water suggest that any nitrite in the underlying suboxic layers results directly from the reduction of nitrate (Meyer et al., 2005).

The total NO_3^--reducing community in sediments probably supports a myriad of metabolisms, some heterotrophic, coupling the reduction of NO_3^- to NO_2^- via the oxidation of organic matter, and others chemoautotrophic, reducing NO_3^- to NO_2^- and NH_4^+ via the oxidation of sulfides, for example (Sayama et al., 2005; Brunet and Garcia-Gil, 1996). Nitrate reduction ($NO_3^- \rightarrow NO_2^-$) results in nitrite as an actual end product of metabolism, which is directly exported from the cell. Nitrite is also an intermediary of the well-described pathways of denitrification ($NO_3^- \rightarrow NO_2^- \rightarrow NO \rightarrow N_2O \rightarrow N_2$) and DNRA ($NO_3^- \rightarrow NO_2^- \rightarrow NH_4^+$). Recent studies with pure cultures of a strain of *Escherichia coli* capable of DNRA showed that the majority of nitrite formed from the intracellular reduction of nitrate was exported from the cell, prior to being reimported and subsequently reduced to ammonia (Wenjing et al., 2008). Hence, it would be difficult to irrefutably link anammox to a specific metabolic source of nitrite.

In their original investigations of the factors regulating anammox in the deep-water sediments of the Skagerrak, Dalsgaard and Thamdrup (2002) measured an almost complete (87%) transient accumulation of NO_2^- from the reduction of NO_3^-. In the OMZ of the central Arabian Sea, the respective ^{15}N labeling of N_2O produced in the presence of either $^{15}NO_3^-$ or $^{15}NO_2^-$ also suggested that all of the nitrite produced from the reduction of nitrate entered the water column before any further reduction to N_2 or N_2O (Nicholls et al., 2007). With pure cultures of *Paracoccus denitrificans*, Blaszczyk (1993) measured up to a 70% accumulation of NO_2^- from the reduction of NO_3^- with growth on minimal medium (ethanol, acetate, or methanol) but no accumulation of NO_2^- with growth on nutrient broth. Hence, it is at least feasible that in low-metabolic/low-carbon systems, both denitrification and/or anammox could start downstream of an initial complete reduction of nitrate to nitrite (Trimmer and Nicholls, 2009).

It is clear that both the denitrifying and anammox bacteria have high affinities for NO_2^-, each with a respective apparent K_m of <3 μmol of NO_2^- liter^{-1} (Dalsgaard and Thamdrup, 2002; Trimmer et al., 2005). Such a high affinity for nitrite is suited to its concentration in the environment, where concentrations of NO_2^- in both the overlying water and sediment are usually two orders of magnitude less than NO_3^- and seldom exceed 5 μmol liter^{-1}, though in comparison to NO_3^-, data for NO_2^- are still scarce (Steif et al., 2002; Meyer et al., 2005). With sediments from the deep Skagerrak, Dalsgaard and Thamdrup (2002) demonstrated that the specific activity of anammox and its contribution to N_2 (*ra*) were both independent of the concentration of nitrite, reflecting its high affinity for the substrate.

Despite such a high affinity for nitrite, anammox can be regulated by the availability of nitrite. For example, in one of the first experiments designed to manipulate anammox in sediment, Risgaard-Petersen et al. (2005) demonstrated that decreasing the concentration NO_3^-, in the overlying water of a sediment from 600 μmol liter^{-1} to 5 to 10 μmol liter^{-1}, in the presence of microphytobenthos (MPB) reduced a sediments capacity for anammox by 85%. With an active layer of microphytobenthos, and scarce nitrate, there was little penetration of nitrate into the suboxic layers to fuel the production of nitrite. Subsequently, Meyer et al. (2005) showed that anammox activity and its contribution to N_2 production were, in fact, correlated with the total amount of nitrite in the nitrate-reducing zone of intact sediment cores from the Albert/Logan estuarine system. This observation helps rationalize the general trend that the significance of anammox increases toward the head of an estuary, where both nitrate and organic carbon tend to be greatest, as both would help maintain heterotrophic denitrification that would, in turn, increase the total supply of nitrite for anammox within the sediment (Trimmer et al., 2003; Meyer et al., 2005; Nicholls and Trimmer, 2009). It also suggests that seasonal changes in the relative abundance of nitrate, nitrite, and carbon may explain some of the scatter in the significance of anammox measured in the most shallow estuarine sites (Fig. 3a, <5 m).

Rysgaard et al. (2004) proposed that denitrification could supply anammox with nitrite, and, as such, the two processes may be positively correlated. Given that the assay routinely used to screen sediments for anammox simultaneously quantifies denitrification (Thamdrup and Dalsgaard, 2002), data are equally abundant for each process, and they are indeed positively correlated, though the scatter suggests bias toward denitrification, especially among the estuarine and shallower coastal sediments (Fig. 5, <20 m water depth). This may, in part, be due to the assay, where anammox and denitrification are routinely quantified after the single addition of either just nitrate or just nitrite (NO_x^-), whereas it has been demonstrated, at least for sediments where the demand for NO_x^- is intense, that the yield of N_2 from anammox is greatly increased if nitrate and nitrite are added together (Fig. 6) (see Trimmer et al., 2005). This would not be an issue for the less reactive sediments from deeper water, as the vast majority of nitrate accumulates transiently as nitrite (Dalsgaard and Thamdrup, 2002) and the relationship between anammox and denitrification is much stronger for the deeper sites (Fig. 5). Even if the two processes are not intrinsically linked through some form of biochemical exchange (e.g., nitrite and ammonium), they broadly occur together.

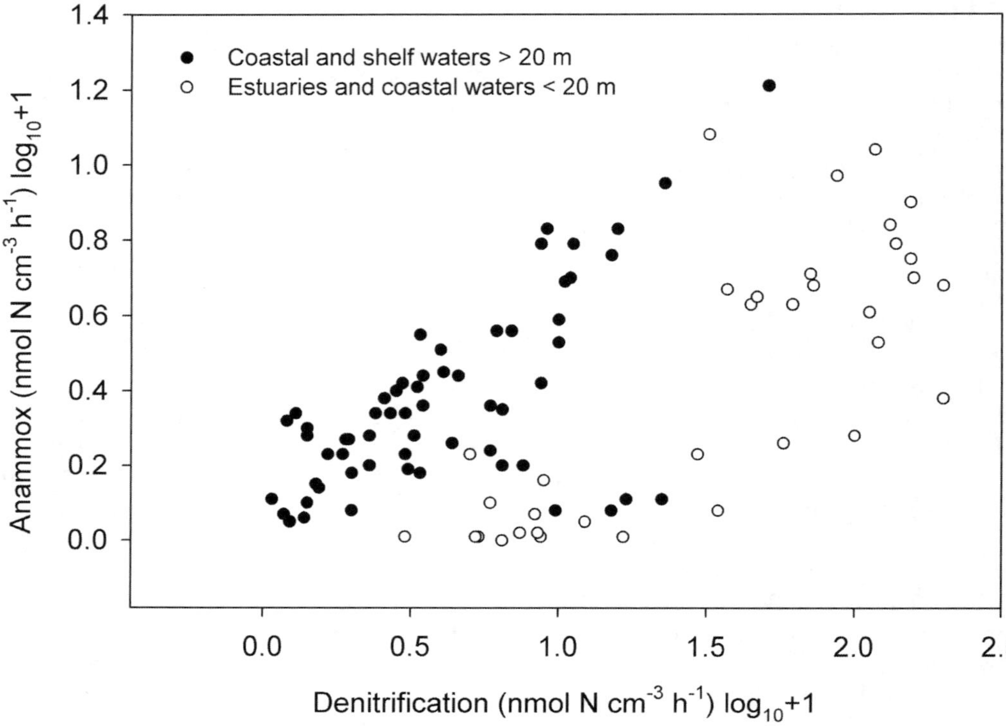

FIGURE 5 A composite of all available anammox-specific activity slurry data (rate) as a function of denitrification. The data are split between coastal and shelf waters deeper than 20 m and estuarine and coastal waters shallower than 20 m. Data have been normalized by common log transformation ($\log_{10}$+1) and are from Table 1.

The DNRA could equally supply anammox with nitrite, but actual simultaneous measurements of the two processes are comparatively rare. Measurements in the less reactive deep water sediments suggest that DNRA is not important, with little of the $^{15}NO_x^-$ (<2%) being recovered as either $^{15}NH_4^+$ or, by mass balance, as that not being recovered as ^{15}N-N_2 gas (~10 %), though assimilation cannot be ruled out of the latter (Dalsgaard and Thamdrup, 2002; Engström et al., 2009; Trimmer and Nicholls, 2009). In the more reactive estuarine sediments, the pattern is more mixed, and the available data only allow the potential for DNRA to be estimated as $^{15}NO_x^-$ not recovered as ^{15}N-N_2 gas (and caution should be applied here [Revsbech, 2006]). Practically all of the added $^{15}NO_x^-$ was recovered as ^{15}N-N_2 gas with sediment from Randers Fjord, an average of 84% with sediment from Chesapeake Bay and between 15% and 69% with sediment from numerous United Kingdom estuaries (Risgaard-Petersen et al., 2004; Rich et al., 2008; Nicholls and Trimmer, 2009). Hence, the potential for DNRA cannot be ruled out in estuarine sediment. Together, these observations are in agreement with the general notion that significant DNRA activity is confined to highly metabolically active and more reduced sediment (Nishio et al., 1982; Christensen et al., 2000).

The confirmatory test routinely used to screen aquatic ecosystems for the anammox metabolism is the sole production of $^{29}N_2$ gas from anoxic samples of either sediment or water incubated in the presence of $^{15}NH_4^+$ and $^{14}NO_2^-$. This is a discrete and sensitive ^{15}N-tool, and its sole use, or that in combination

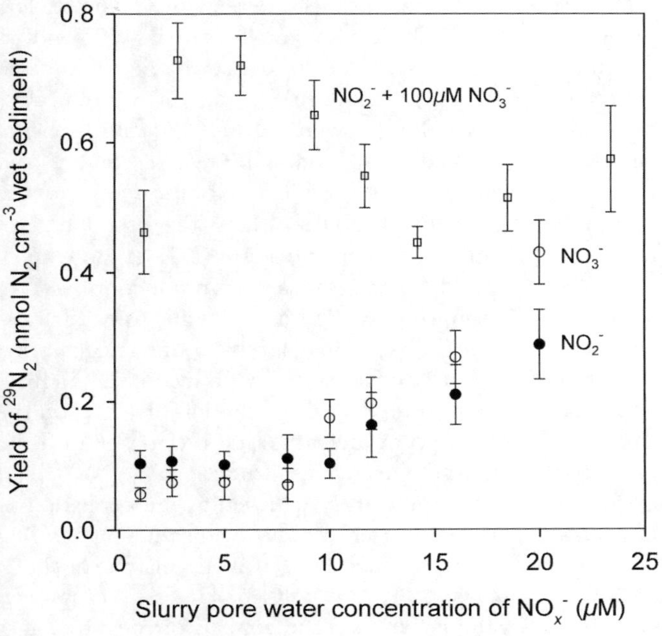

FIGURE 6 The yield of $^{29}N_2$ from the oxidation of $^{15}NH_4^+$ in the presence of either just $^{14}NO_3^-$ or just NO_2^- or a dual labeling experiment with 100 µmol of NO_3^- liter^{-1} and increasing NO_2^-. Clearly, the availability of NO_2^- and NO_3^- affects the significance of anammox. Data are from Trimmer et al. (2005). Sediment is from Grays in the Thames estuary.

with molecular and/or biomarker assays, does indeed suggest that the anammox metabolism in the environment is metabolically and physiologically similar to that characterized during the initial work in laboratory bioreactors (Mulder et al.. 1995; van de Graaf et al.. 1995; Jaeschke et al., 2009) (see Chapters 8 and 10). In addition, however, to this sole reliance on nitrite, anammox bacteria (at least in physically purified suspensions) are capable of reducing nitrate to nitrite and ammonium in the presence of simple organic acids (Kartal et al., 2007). Such a mixotrophic capability would enable anammox to exploit nitrate directly, and this potentially changes their ecological niche from being solely reliant on a source of nitrite to being in direct competition for nitrate with the diverse nitrate-reducing community (Güven et al., 2005; Kartal et al., 2007). Anammox via this alternative nitrate reduction pathway is, however, equivalent to only 10% of that measured via the direct nitrite pathway (Kartal et al., 2007).

If, in sediments, there were an entirely intracellular metabolism of nitrate to ammonium and N_2 gas, then we may expect to see this as an increase in "false" denitrification after the addition of simple organic compounds and/or an imbalance in the estimate of anammox measured with either the $^{15}NH_4^+$ or $^{15}NO_x^-$ assays, but this does not appear to be the case (Engström et al., 2009; Nicholls and Trimmer, 2009; Trimmer and Nicholls, 2009). Finally, it is clear, at least with the slurries of highly reactive estuarine sediment, that anammox cannot compete with the total NO_x^--reducing community for either nitrite or nitrate when they are supplied alone. Anammox is far more significant when some of the demand for NO_x^- is met by the addition of nitrate, which leaves more of the nitrite available for anammox (Trimmer et al., 2005). While there is little evidence to suggest that anammox accesses nitrite to any significant extent directly via the intracellular reduction of nitrate in sediment, this may not always be the case in the suboxic water column (see below).

Anammox in Intact Sediment Cores, the Rationale and Approach

The use of anaerobic sediment slurries was essential to the discovery of anammox in the environment, the pioneering exploration of the factors that regulate its activity and the characterization of its broader biogeographical distribution. Indeed, without the production of $^{29}N_2$ from anoxic slurries enriched with $^{15}NH_4^+$ and $^{14}NO_x^-$, it would have been difficult, if not impossible, to convince the broader community that the anammox metabolism was active in the environment at all. Their use, however, obviously disrupts the natural gradients of substrates and redox in sediments and thereby destroys the chemical microenvironment of the bacteria and processes being studied. To the lay reader, it might appear blatantly obvious that to understand the role of any biogeochemical process in the environment, then it is indeed essential to measure that process under conditions as representative of those in situ as possible, but this is not a trivial task.

The application of ^{15}N to trace the flux of N through aquatic ecosystems, especially that due to denitrification, has received a considerable amount of attention. Since its introduction by Nielsen (1992), the ^{15}N isotope pairing technique (IPT) has become one of the most widely used techniques for measuring N_2 production in intact aquatic sediments (Steingruber et al., 2001). The IPT cannot, however, distinguish between anammox and denitrification as sources of N_2 production, and, more seriously, the presence of anammox violates the central assumptions on which the IPT is built and will tend to overestimate N_2 production (Risgaard-Petersen et al., 2003).

The crux is that in the sole presence of denitrification, ^{15}N- labeled N_2 gas will be produced in the suboxic sediment in relation to the respective availability of $^{14}NO_3^-$ and $^{15}NO_3^-$, that is, a binomial distribution of $^{28}N_2$, $^{29}N_2$, and $^{30}N_2$, reflecting the respective proportions of the two analogues of NO_3^- being reduced (Hauck et al., 1958; Nielsen, 1992). With the co-occurrence of anammox, however, there are now two sources of $^{29}N_2$ ($D^{29}N_2$ from $^{14}NO_3^-$ + $^{15}NO_3^-$ and $A^{29}N_2$ from $^{14}NH_4^+$ + $^{15}NO_x^-$), and the logic underpinning the IPT breaks down (Risgaard-Petersen et al. 2003). This problem was first overcome for the application of ^{15}N to the measurement of anammox and denitrification in anoxic slurries (Thamdrup and Dalsgaard, 2002). The approach is based on the principle that as long as the respective availability (ratio) of $^{14}NO_3^-$ and $^{15}NO_3^-$ being reduced in a slurry is known (r_{14} or F_N), and there is not aerobic nitrification due to the anaerobic conditions, then the production of $^{29}N_2$ due to denitrification ($D^{29}N_2$) can be predicted from the measured production of $^{30}N_2$, assuming that $P^{30}N_2$ is solely due to denitrification ($D^{30}N_2$). Then, any $^{29}N_2$ due to anammox ($A^{29}N_2$) can be found by subtracting $D^{29}N_2$ from the measured total production of $P^{29}N_2$ on the mass spectrometer, and everything else can be derived. Determining the ratio of $^{14}NO_3^-$ and $^{15}NO_3^-$ is usually achieved by preincubation of anoxic slurries to remove any ambient $^{14}NO_3^-$, followed by addition of high purity (^{15}N abundance >99.2%) $^{15}NO_3^-$. The same approach cannot, however, be applied directly to intact sediment cores. With sediment cores, the $^{15}NO_3^-$ is added to the overlying water where it mixes with any $^{14}NO_3^-$ present, but the key parameter r_{14} lies literally beneath the surface in the suboxic sediment. Production of $^{14}NO_3^-$ through nitrification in the oxic sediment will increase the ratio of $^{14}NO_3^-$ to $^{15}NO_3^-$, and r_{14} in the sediment will be higher than the ratio of the two in the overlying water (r_{14w}). Numerous approaches have been proposed to quantify the parameter r_{14} either directly or indirectly to enable simultaneous measurement of anammox and denitrification in intact sediment cores under conditions more representative of the in situ conditions (see Risgaard-Petersen et al., 2003; Trimmer et al., 2006).

Sediment Metabolism and N_2 Production in Intact Sediment

Glud (2008) published a comprehensive review of oxygen dynamics in marine sediments and established a robust relationship between decreasing rates of oxygen uptake and increasing water depth. Essentially, oxygen uptake integrates a broad spectrum of aerobic and anaerobic respiratory pathways and provides a good proxy for total sediment metabolic activity, and, in turn, the availability of organic carbon, both of which broadly decrease with increasing water depth. Here we restrict our analysis to data sets where both oxygen uptake and N_2 production, including anammox, have been measured simultaneously (Table 2). The data are rather clumped around the shallow estuarine (>4 m), coastal shelf (31 to 117 m), and deep water (1,000 to 3,000 m) sediments. We have plotted the data on a double common log scale and fitted them with a simple power function in accordance with other data sets in the literature. Our decay constant of −0.46 (r^2 = 0.76) is lower than the average for the compilation by Glud of −0.66 (mean for both profile- and flux-derived rates of oxygen uptake as ours are a mix of both [Glud, 2008]); regardless, though, the overall pattern of decreasing sediment metabolism with increasing water depth still holds (Fig. 7).

In addition, the total production of N_2 (denitrification plus anammox) decreases as a function of water depth (Fig. 7), with a similar decay constant of −0.41, though the fit is not as good (r^2 = 0.36). This is markedly lower than that recently reported from 50 m on the shelf to 3,000 m on the continental slope in the Chukchi Sea of −0.94 (Chang and Devol, 2009). The production of N_2 in the Chukchi Sea correlates with local primary productivity and flux of carbon to the benthos, whereas as our compilation takes no account of potential differences in the carbon flux across a wide range of locations. Noticeable outliers are the data from the deep Sagami Bay (Glud et al., 2009), where a combination of low oxygen saturation (15% air saturation) and up to 40 μmol NO_3^- liter^{-1} in the overlying water could stimulate N_2 production (*ra* 37%), as long as suitable electron donors are available.

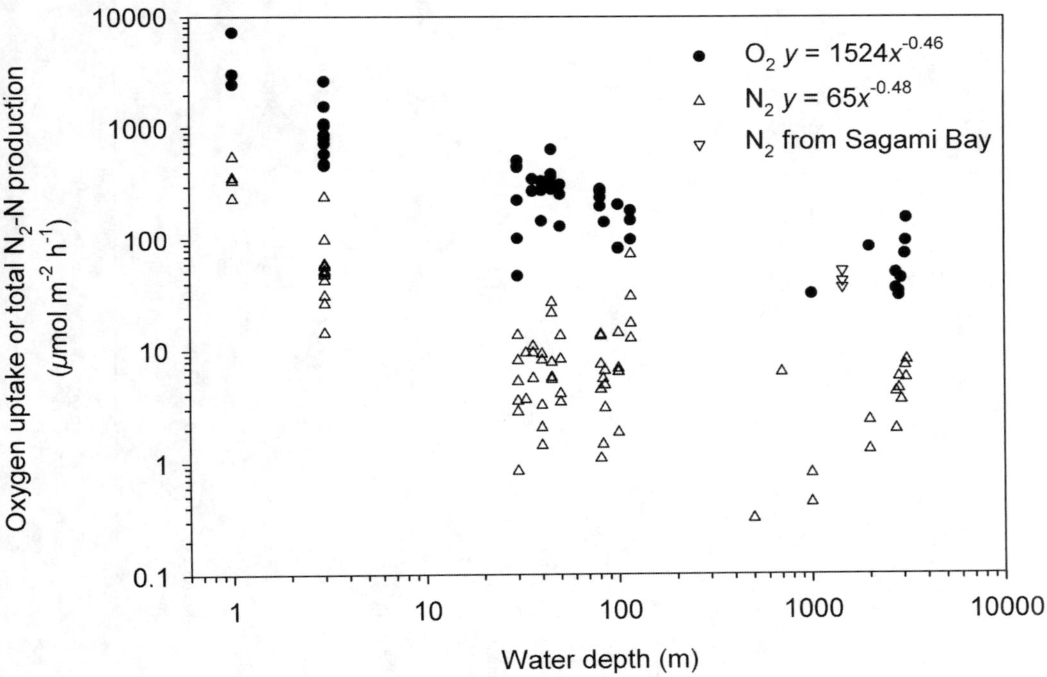

FIGURE 7 Decreasing sediment metabolism as a function of water depth for both oxygen uptake and total N_2 production. Note the inverted triangles for the data from Sagami Bay (Glud et al., 2009), which have been omitted from the nonlinear regression. The data are plotted on a double common log scale, and the coefficients were derived using a simple power function. Data are from Table 2.

TABLE 2 Published and unpublished rates of sedimentary oxygen uptake and anammox and denitrification measured in intact sediment cores[a]

Source and reference(s)	Water depth (m)	$^{14}NO_3^-$ (μmol liter^{-1})	Oxygen uptake (μmol of O_2 m^{-2} h^{-1})	Denitrification (μmol of N m^{-2} h^{-1})	Anammox (μmol of N m^{-2} h^{-1})	Total N_2 (μmol of N m^{-2} h^{-1})
Skagerrat, Kattegat; Risgaard-Petersen et al., 2003	36–700	–	–	1.9–8.3	1.2–4.4	6.3–9.5
Randers Fjord; Risgaard-Petersen et al., 2004	1	120	3,021–7,193[b]	219–335	14–21	233–356
Arctic sediments; Rysgaard et al., 2004	36–100	0.3–15.3	143–345[b]	1.4–10.7	0.0–3.8	1.4–14.3
Washington Margin; Engström et al., 2009	2,740–3,110	36–48	31–155[d]	1.3–4.7	0.7–3.4	1.9–8.1
Gullmarsfjorden (Alsbäck); Trimmer et al., 2006	116	–	–	6.1	6.6	12.7
Medway Estuary, United Kingdom; M. Trimmer et al. (unpublished data)	3	34–69	459–1,569[b]	6.5–63.5	6.5–35.3	14.4–98.8
Gravesend, Thames Estuary, United Kingdom; Trimmer et al., 2006	3	–	2,625[c]	192.9	48.94	241.84
Baltic Sea; Hietanen, 2007; Hietanen and Kuparinen, 2008	33–85	0.3–11.2	–	1.3–8.6	0.1–0.9	1.4–9.5
Colne Estuary, United Kingdom; Dong et al., 2009	1	653	2,470[b]	387.2	157.3	544.5
Sagami Bay; Glud et al., 2009	1,450	40.2	–	23.8–32.9	12.8–18.5	36.5–51.4
North Atlantic; Trimmer and Nicholls, 2009	50–2,000	1.1–20.8	32–131[c]	0.2–5.8	0.1–1.5	0.3–6.8
Gullmarsfjorden; Alsbäck, 2008; Engström and Hulth (unpublished)	116	10.1–17.1	99–181[b]	7.6–57.2	9.8–15.4	17.4–72.5
North Sea; E. Neubacher, R. Parker, and M. Trimmer (unpublished data)	30–81	0.1–9.6	47–632[b]	0.6–21.2	0.2–5.6	0.8–26.9

[a]This is a summary illustrating the range at each site; the full data set (n = 63 for oxygen uptake and n = 78 for N_2 production) is available from the corresponding author (M.T.). –, no data.
[b]Total oxygen uptake (i.e., change in oxygen in the overlying water over time).
[c]M. Trimmer unpublished, measured as part of this work.
[d]Diffusional oxygen uptake modeled using *Profiler* (Peter Berg).

Removing Sagami Bay from our compilation improves the fit (Fig. 7) ($r^2 = 0.36$ and $r^2 = 0.49$ with and without Sagami Bay, respectively) and increases our decay constant to −0.48, which is very similar to that for the decay in oxygen uptake (−0.46). Hence, N_2 metabolism and total sediment metabolism (oxygen uptake) decay at similar rates to each other, all the way from shallow estuarine to deep continental margins, in agreement with the notion that water depth is a very good parameter for describing benthic metabolism (Wenzhöfer and Glud, 2002; Andersson et al., 2004; Glud, 2008). Furthermore, we can make a pairwise comparison using each pair of oxygen uptake and N_2 production measurements ($n = 54$) to estimate an average N_2 mineralization constant (ratio) per mole of oxygen consumed. Doing this suggests that for each mole of oxygen consumed, the sediment releases 0.07 mole of N as N_2 gas (±0.02 95% confidence interval [CI], $n = 54$). If the organic material being mineralized has a Redfield ratio of 6.6C:1N and 1 mol of O_2 consumed by the sediment is equivalent to 1 mol of carbon mineralized to CO_2 (in the simplest sense), then our N_2 mineralization constant of 0.07:1 suggests that 46% of organic N is mineralized and "lost" as N_2 gas, while 54% is potentially available to be returned to the water column, probably as NO_3^- (1/6.6 = 0.15 mol of N mineralized; 0.07/0.15 = 0.46; 1-0.46 = 0.54 remaining). Note that accounting for the oxygen consumed during the nitrification of the originally ammonified N before it is oxidized to N_2 gas makes little difference to this approximate 50:50 "split."

Seitzinger and Giblin (1996) showed that denitrification (as it was solely referred to then) was correlated ($r = 0.8$) with oxygen uptake in sediments across a variety of continental shelf locations. Here we show that this principle holds for total N_2 production and oxygen uptake across a broader range of water depths, with our data bracketing that original range, which is a logical consequence of their similar decay constants (Fig. 8a). (Note the log transformation here compared to that of Seitzinger and Giblin [1996].) What is particularly interesting here is that the two pathways of N_2 production, anammox and denitrification, are strongly correlated with each other, and as total sediment metabolism increases (be it oxygen uptake or N_2 production), both anammox and denitrification also increase (Fig. 8b). Once we move offshore, away from the influence of coastal nitrate, the majority of N_2 production via both denitrification and anammox is going to be coupled to the mineralization of organically bound N to ammonium and its oxidation to nitrate (Seitzinger, 1988). It is perhaps not surprising, therefore, that N_2 production and total sediment metabolism track each other, but what is surprising from this compilation is that the contribution from anammox to the production of N_2 remains far more constant than perhaps would have been previously expected.

Although anammox and denitrification are positively correlated for the slurry data, the relationship is stronger for the intact sediment cores (Fig. 9). The bias toward denitrification is pronounced more clearly in the reactive sediments from shallower waters, where any added NO_x^- may be artificially exposed to H_2S, for example, which would increase the competition for NO_x^-. In addition, slurries will integrate the activity of denitrifiers and anammox bacteria in a volume of sediment and the facultative denitrifiers are, therefore, likely to be over-represented in anoxic slurries prepared from mixed oxic and anoxic strata. In contrast, this distribution will be more accurately captured using intact sediment cores (Trimmer et al., 2006). Overall, though, for the less reactive sediments, either assay gives similar estimates for the significance of anammox to the production of N_2.

Anammox and Denitrification

The mean value for *ra* calculated using all of the sediment core data suggests that anammox contributes 28% (±2 SE, $n = 78$) of total N_2 production, which agrees with the elegant and logical argument for a potential coupling between the two metabolisms put forward by Dalsgaard et al. (2003). For example, if, during

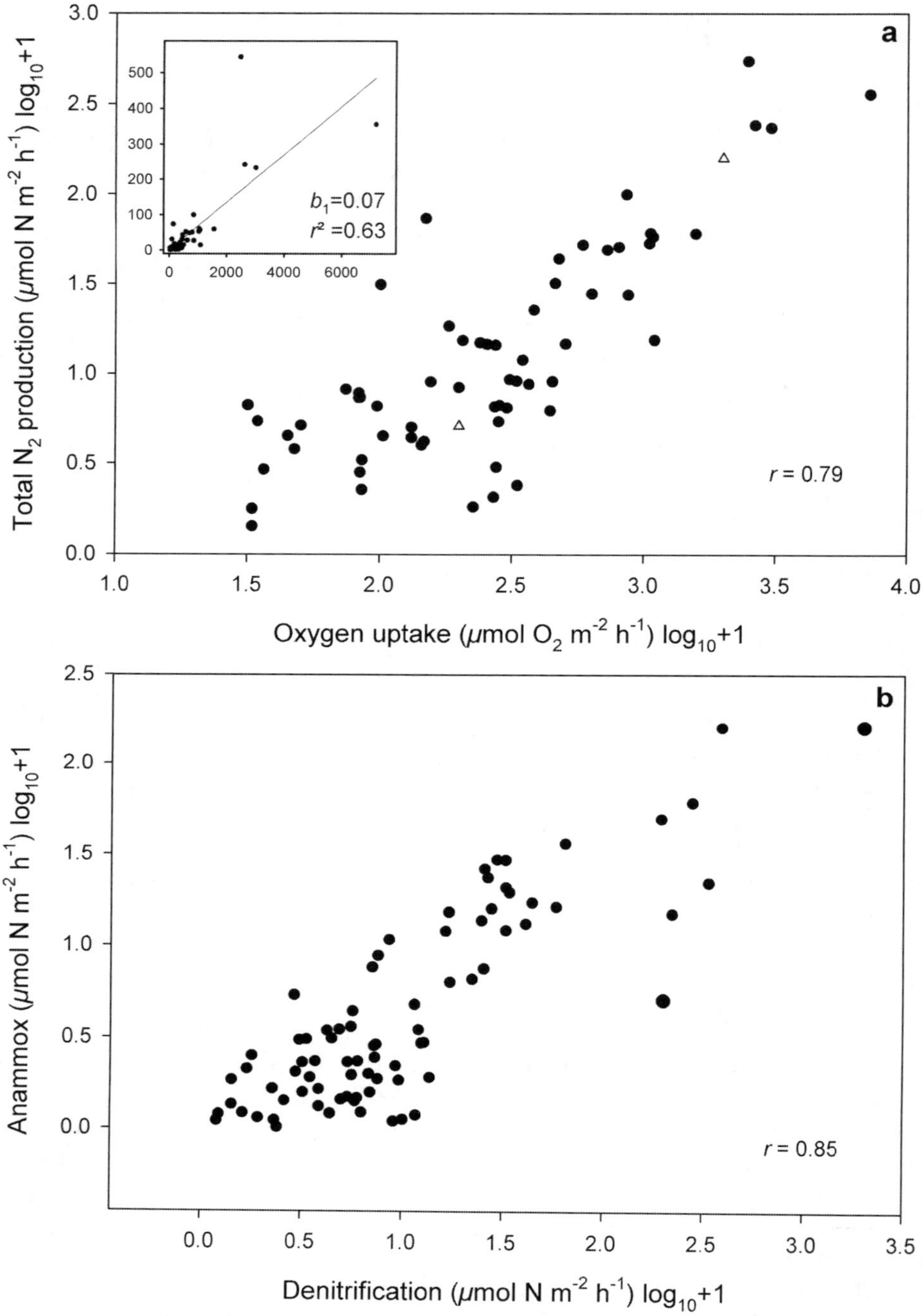

FIGURE 8 Relationships between N_2 metabolisms and total sediment metabolism. (a) Total N_2 production scattered against oxygen uptake. (b) Anammox against denitrification. Data have been normalized by common log transformation ($\log_{10}$+1), and the correlation coefficient (*r*) is given in each panel. In a, the open triangles mark the approximate range of data reported by Seitzinger and Giblin (1996). The inset gives the relationship through the original linear data, where the slope (b_1) is equivalent to the ratio (as $b_0 = 0$) and ratio is equivalent to the pairwise comparison given in the text.

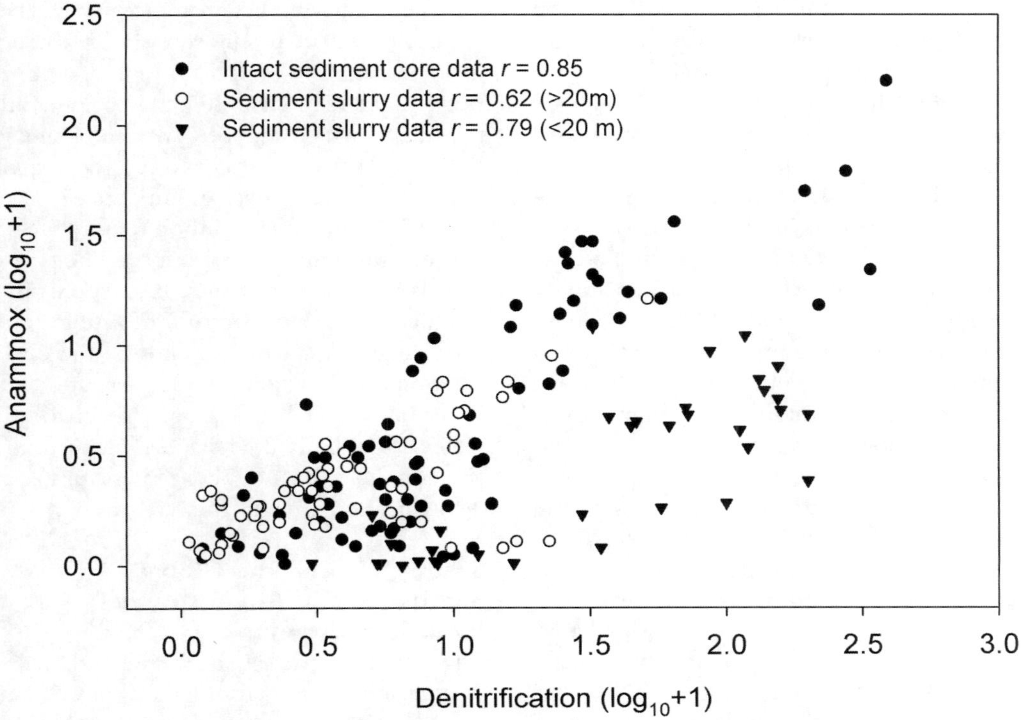

FIGURE 9 Scatter plot of anammox as a function of denitrification to illustrate the bias toward denitrification in the sediment slurries in shallower water. Units for the intact sediment core data are μmol of N m^{-2} h^{-1} (as in Fig. 8b) and for the slurries are nmol of N cm^{-3} h^{-1} (as in Fig. 5). Data have been normalized by common log transformation (log_{10}+1), and the correlation coefficient (*r*) is given in each case.

the oxidation of organic matter (with a Redfield C:N of 6.6:1) via denitrification with NO_3^-, both NO_2^- and NH_4^+ are liberated in equimolar amounts and are, in turn, used by anammox to produce N_2 gas, then they argued that anammox would be responsible for 29% of the N_2 produced (see equation 8) and, indeed, that was close to what they measured. Here we also report a remarkably good agreement, on average, between the predicted significance of anammox to N_2 production and that measured if, as our positive correlation also corroborates, anammox is indeed coupled to denitrification.

Clearly, there are exceptions to this, and it would be contradictory to our own findings not to point these out. Between 2,000 and 500 m in the North Atlantic, we measured a higher contribution from anammox in intact sediment cores of up to 65% and up to 40% in the Washington margin (Engström et al., 2009; Trimmer and Nicholls, 2009). There is considerable scatter in the data (Fig. 8b), and some local differences could, in part, be due to differences in the C:N ratio of the organic matter being mineralized, the total input of organic carbon and local stimulation, or suppression of denitrification (Thamdrup and Dalsgaard, 2002; Glud, 2008). The significance of anammox could increase if the organic matter undergoing mineralization was more enriched with N, relative to Redfield, and the opposite must also hold for organic matter with a higher C:N ratio, when the significance of anammox could be expected to go down (Dalsgaard et al., 2003). As we have already discussed, significant carbon oxidation via manganese oxides or local zones of deoxygenation could alter the balance between denitrification and anammox, but the potential of a link between the two is worthy of further exploration.

The original argument proposed for a coupling between denitrification and anammox was based on the water column study in the Golfo Dulce where, in the anammox zone, ammonium was particularly scarce and a coupling between mineralization and anammox was logical (Dalsgaard et al., 2003). In contrast, in sediments where ammonium is known to accumulate at depth, one might not expect anammox to be ammonium limited. Engström et al. (2009), however, showed that ammonium was absent from the pore water (0.5 cm resolution) within the suboxic nitrate reduction zone for sediment from the Cascadia Basin (2,700 to 3,100 m) and that this pattern was consistent with numerous other deep-water sediments such as the San Clemente Basin (Bender et al., 1989), California margin (Reimers et al., 1992), Panama Basin (Aller et al., 1998), and the western Mexico margin (Hartnett and Devol, 2003).

Whether or not this is also true for the more reactive coastal and estuarine sediments is harder to assess at the moment, as pore water profiles of a sufficient resolution are often either not available or have not been published as part of the research into anammox. Estuarine sediments do, however, largely act as sources of ammonium to the overlying water, and, as such, ammonium appears to be in excess to the sedimentary N requirements (Dollar et al., 1991; Ogilvie et al., 1997). Even if ammonium were not limiting for anammox in muddy estuarine sediments and anammox were only reliant on nitrate reduction for its nitrite, this would not change the degree of coupling between the two.

In some of the original papers on anammox in estuarine sediments, a paradoxical notion was put forward whereby anammox was both reliant on nitrate-reducing bacteria for its nitrite and in competition with these bacteria for this nitrite (Meyer et al., 2005; Trimmer et al., 2005). Our compilation and proceeding argument suggests that anammox may actually depend on nitrite produced as an intermediate in "denitrification" and the notion of competition may be redundant. It could be argued that the nitrite, which supports anammox, must be excess to the requirements of the denitrifying bacteria and may reflect an imbalance between the NO_3^- and organic carbon required to sustain heterotrophic denitrification. The positive correlation between anammox and denitrification across a broad spectrum of activity, however, suggests that this may not be the case, as we would have expected a higher ratio at the lower rates of denitrification, if denitrification were carbon limited. The actual mechanism of any coupling, if at all, between anammox and denitrification remains to be resolved.

Scaling Up and the Global Significance of Anammox in Benthic Sediments

There are many caveats associated with making measurements of benthic metabolism, especially with sediments recovered from the deep sea (see Glud [2008] for an overview). These include leaching of cellular material as a consequence of depressurization, which could affect the accuracy and interpretation of pore water nutrient profiles, and, obviously, cell death, which would impact on the measured rates of sediment metabolism. Hence, absolute true patterns of sedimentary N metabolism may only be uncovered when benthic-landers are equipped with techniques to simultaneously measure anammox and denitrification in situ. That said, the fact that we can see broad-scale patterns in both sedimentary and N_2 and O_2 metabolisms measured using different techniques, across a broad spectrum of water depth, primary production, and season, suggests that the data are robust.

We have shown that N_2 metabolism decays at the same rate as total sediment metabolism (i.e., oxygen uptake). We can now use our N_2 mineralization constant of 0.07:1 (±0.02 95% CI, n = 54) in combination with the compilation by Glud (2008) for global benthic oxygen consumption, to first propose an estimate for global benthic N_2 production and then the significance of anammox to that production. Glud (2008) used his relationship for oxygen uptake and water depth, in combination with global topography data, to estimate a total global sediment consumption for oxygen of 152 Tmol of O_2 year^{-1}. In combination with our own N_2 mineralization constant, this 152 Tmol of O_2 year^{-1} equates to 9 Tmol of N year^{-1} (152 × 0.07) or 126 Tg of N year^{-1} (90 to 162 with a 95% CI) released as N_2 gas from the global benthos. This is toward the upper end of some previous estimates of global benthic denitrification, e.g., 95 Tg of N year^{-1} (±20 [Gruber and Samiento, 1997]) but considerably less than the 230 to 300 Tg of N year^{-1} proposed by others (Middelburg et al., 1996; Codispoti et al., 2001; Codispoti, 2006). Furthermore, of the 126 Tg of N year^{-1}, anammox would, on average, contribute approximately 35 Tg of N year^{-1} (i.e., ~28%) and denitrification would contribute the remaining 91 Tmol of N year^{-1}.

Recent revisions to the global budget for N_2 production have, in part, been deemed necessary to take account of "new" N_2-producing pathways, namely anammox and redox "metal-mediated" denitrification (Codispoti, 2006). Despite hundreds of control incubations used in the routine screening of sediments for anammox, no significant production of ^{15}N-labeled N_2 gas has been measured that could be ascribed to the oxidation of $^{15}NH_4^+$ coupled to either the reduction of metal oxides or sulfate (Hulth et al., 1999; Fernandez-Polanco et al., 2001; Schrum et al., 2009). Only when $^{15}NH_4^+$ and $^{14}NO_x^-$ are incubated together do we get the confirmatory $^{29}N_2$ signature of anammox. In addition, Risgaard-Petersen et al. (2006) demonstrated that benthic foraminifera were capable of complete denitrification, yet their contribution as a novel source of N_2 production in sediments appears limited to date (Risgaard-Petersen et al., 2006; Glud et al., 2009). Anammox is a real and almost ubiquitous component of the estuarine and marine sedimentary N cycle. It is correct to revise estimates of N_2 production based on previously accepted stoichiometric principles (pore water profile and gradient models), but the occasions where it contributes to the majority of N_2 production in sediments appear rare, and it appears more in unison, rather than at odds, with denitrification.

ANAEROBIC AMMONIUM OXIDATION IN OCEANIC OMZs

Global Distribution of OMZs and Suboxic Waters

The vast majority of water that makes up the global ocean (1.34 × 10^9 km^3) is at equilibrium with the atmosphere with respect to oxygen, that is, 100% of air saturation for oxygen. If we assume an average salinity of 35 (psu or 0.035 kg of "salt" [kg of seawater]$^{-1}$) and a representative temperature of 16°C for this water, then it would, at equilibrium with the atmosphere, contain 249 μmol of O_2 liter^{-1}, and it would be termed oxic. If the rate of supply of oxygen to a sediment cannot keep up with the sediment's demand for oxygen (aerobic respiration and reoxidation of reduced chemical species), then the sediment will become suboxic and, eventually, fully anoxic. The same is also true for a column of water, but the local physics of water movement can add compounding complexity, specific to a particular location; a sill at the mouth of a fjord, an isolated water body, strong upwelling, and gyre structures can all govern mixing and reaeration rates. In addition, whereas in a sediment we can cross over from oxic to suboxic strata in a few hundred micrometers to a couple of centimeters (depending on reactivity and permeability of the sediment), the structure of an oceanic OMZ is much larger, with oxygen decreasing over tens to hundreds of meters (Fig. 1). We follow the convention that between the layers of either oxic sediment or oxic water and the deeper suboxic layers, there is an oxycline of decreasing oxygen and, hence, increasing hypoxia and that hypoxia is physiologically stressful for higher organisms at 90 μmol of O_2 liter^{-1} (Diaz and Rosenberg, 2008).

A distinctive characteristic of OMZs is that once oxygen has fallen to 1 to 3 μmol of O_2 liter^{-1} (or between 0.4% and 1.2 % of saturation on the scale above), nitrite starts to accumulate (typically peaks of 2 to 5 μmol of NO_2^- liter^{-1}) in the water column (Codispoti and Christensen, 1985; Naqvi et al., 1992; Morrison et al., 1999; Thamdrup et al., 2006). Hence, oxygen appears physiologically limiting for aerobic respiration, and electrons begin to flow via the reduction of nitrate to nitrite, and we term these water layers, simply, as suboxic. If, however, oxygen depletion in an OMZ, or parts thereof, is particularly intense, the oxidized species exhausted, and redox sufficiently negative, then free sulfide can start to accumulate, and the water would be truly anoxic (e.g., the deeper parts of the Black Sea, Baltic Sea, Golfo Dulce, and coastal regions of the western Arabian Sea) (Naqvi et al., 2000; Hannig et al., 2007; Jensen et al., 2008).

The known OMZs in the modern ocean comprise only about 0.1% (1.34×10^6 km^3) of the total oceanic volume, according to Codispoti et al. (2001). Recently, however, Paulmier and Ruiz-Pino (2009) reformulated this estimate to include waters where the concentration of oxygen falls below 20 μmol of O_2 liter^{-1}; that is, their definition includes waters that are also hypoxic. With the latter formulation, the volume of the ocean's OMZ increases to 10.3×10^6 km^3 or 0.77% of the total oceanic volume. The large OMZs found in the global ocean are as follows: one in the eastern tropical North Pacific (ETNP) off the west coast of Mexico and Guatemala; one in the eastern tropical South Pacific off Peru and Chile (ETSP); and in the northern Arabian Sea and Bay of Bengal in the Indian Ocean (Codispoti et al. 2001; Paulmier and Ruiz-Pino 2009). A lesser known permanent deep OMZ in the east subtropical North Pacific off the west coast of the United States is also accounted for by Paulmier and Ruiz-Pino (2009). Permanent suboxic water columns can also be found in some fjords and basins, such as the Black Sea, and overproductive shelf areas as the water column off the south west coast of Africa.

Such regions of the ocean may be comparatively small, but they play a vital role in the global N cycle. The oceanic OMZ support both the anammox and denitrification metabolisms and are suggested to account for one-third of total marine N_2 production, even though they make up less than 0.1% of the ocean volume. This means, as Codispoti et al. (2001) pointed out, that a small change in volume of these suboxic zones can potentially have a large impact on global N_2 production. Seasonal variability in the five large OMZs is minor on a global scale, except for the Arabian Sea, which thickens by 20% (640 to 790 m) during the summer season (Paulmier and Ruiz-Pino, 2009). The important point to note, however, is that the OMZs associated with the tropics are known to have expanded in the last 50 years and that cases of coastal hypoxia are also increasing sharply (Diaz and Rosenberg, 2008; Stramma et al., 2008). How this will affect the global balance of N and, in turn, C sequestration via primary production is unknown.

Distribution of Anammox in Oxygen Minimum Zones and the Effect of Sulfide

Direct evidence for anammox in an OMZ was first presented from an enclosed bay, The Golfo Dulce, in Costa Rica and from the Black Sea (Dalsgaard et al., 2003; Kuypers et al., 2003). Since then, the anammox metabolism has been confirmed in other suboxic basins and most of the ocean's other OMZs, including the Namibia shelf waters, the ETSP, and the Arabian Sea (Table 3). Anammox has also been found in the water column of a tropical lake, where it contributed about 10% of the total N_2 production (Schubert et al., 2006), but this is the only freshwater site where anammox has been reported, and as with the sediments, it remains very much understudied in freshwater ecosystems. In a Swiss meromictic lake and the brackish Mariager Fjord in Denmark, two sites characterized by narrow suboxic and

TABLE 3 Published anammox and denitrification rates measured with ^{15}N-stable isotopes in water column OMZs[a]

Source and reference	Water depth (m)	$^{15}NH_4^+$ anammox (nmol of N liter^{-1} day^{-1})	$^{15}NH_4^+$ + $^{14}NO_2^-$ anammox (nmol of N liter^{-1} day^{-1})	$^{15}NO_2^-$ anammox (nmol of N liter^{-1} day^{-1})	$^{15}NO_3^-$ anammox (nmol of N liter^{-1} day^{-1})	$^{15}NO_3^-$ denitrification (nmol of N liter^{-1} day^{-1})	Anammox (%)	Anammox cells (×10^4 ml^{-1})
Golfo Dulce; Dalsgaard et al., 2003	120–180	NA[b]	NA	NA	24–408	12–2,568	19–35[c] 7–67	NA
Benguela upwelling; Kuypers et al., 2005	40–130	10–170	NA	NA	27–47[d]	ND[b]	100	0.4–2[e] (0.89 ± 0.15)
ETSP, Chile; Thamdrup et al., 2006	60–150	4–18	NA	0–27	NA	5.8 One depth	100 (one site 76%)	NA
Lake Tanganyika; Schubert et al., 2006	90–110	NA	NA	NA	0–240	467–2,322	0–13	0.1–1.3[e] (0.6 ± 0.36)
Black Sea; Lam et al., 2007	85–110	1–7	NA	3–14[d]	NA	ND	100	0–0.28[f] (0.15 ± 0.057)
ETSP, Peru; Hamersley et al., 2007	25–400	1.5–105	1.2–384	4–48[d]	1–27[d]	ND	100	0.09–13[e] (3.6 ±0.97) 0.10–15[f] (4.2 ± 1.1)
Black Sea; Jensen et al., 2008	85–110	0.7–11	7–10	0.1–14	0–2.8[d]	ND	100	NA
ETSP, Chile; Galan et al., 2009	50	2–17	NA	NA	NA	ND	100	0.3
ETSP, Peru; Lam et al., 2009	–[g]	–[g]	–[g]	–[g]	–[g]	–[g]	–[g]	0.04–0.2[h] (0.075 ± 0.014)
Arabian Sea; Ward et al., 2009	120–200	0.12–4.3	NA	NA	NA	0.24–25	1–13 S2 150 m 95%	1–8[f] (3.8 ± 1.0)
ETSP; Ward et al., 2009	80–250	0.63–8.8	NA	NA	NA	0	100	3–12[f] (8.6 ± 1.5)
Mariager Fjord; Jensen et al., 2009	14–21	ND	NA	ND	ND	4.1–19	0	NA

[a]Only sites where either anammox or denitrification could be measured are shown. This is a summary illustrating the range at each site; the full data set (n = 76 for N_2 production) is available from the corresponding author (M.T.); mean ± SE.

[b]NA, not analyzed; ND, not detected.

[c]Integrated over the whole suboxic zone.

[d]$^{29}N_2$ production (no $^{30}N_2$ detected).

[e]Anammox cell abundance quantified by FISH.

[f]Anammox cell abundance quantified by quantitative PCR.

[g]Same rates as Hamersley et al. (2007).

[h]Scalindua mRNA.

anoxic interfaces affected by sulfide, there was no evidence of anammox in the suboxic water column (Halm et al., 2009; Jensen et al., 2009). Chemolithotrophic denitrification, where the reduction of nitrate is coupled to the oxidation of sulfide, was proposed to be the pathway responsible for N_2 production at both of these sites. A similar pattern of minor anammox activity was reported in the bottom waters of the Golfo Dulce (180 m), where marked sulfide oxidation was occurring at the interface of nitrate reduction and nitrite production (Dalsgaard et al., 2003). In the Baltic Sea, a pronounced stratification of the water column with steep gradients of NO_3^- and H_2S in the redox cline was shown to support chemolithotrophic denitrification coupled to the oxidation of sulfide, but no anammox activity was detected (Brettar and Rheinheimer, 1991; Hannig et al., 2007). However, Hannig et al. (2007) did measure anammox activity in the Baltic after a deepwater renewal resulting in a disappearance of the sulfidic redoxcline and the presence of anammox bacteria was further confirmed in the water column using fluorescent in situ hybridization (FISH) (Hannig et al. 2007).

Jensen et al. (2008) showed that low concentrations of sulfide had a clear inhibitory effect on anammox activity measured in water samples from the Black Sea. In incubations with 4 μmol of H_2S liter^{-1}, anammox rates decreased by up to ~98% compared to the controls (5 to 17 nmol of N_2 liter^{-1} day^{-1} in the controls down to the detection limit of 0.36 nmol of N_2 liter^{-1} day^{-1} in the presence of sulfide). Nitrifying bacteria and heterotrophic denitrification are also inhibited by sulfide (Sørensen et al., 1987; Joye and Hollibaugh, 1995); however, it seems that it is the toxicity of sulfide itself and not an indirect effect, such as a shortage of substrate, that is inhibitory, since anammox is not active in sulfidic zones rich in both NH_4^+ and NO_3^-. The increasing incidence of coastal hypoxia, which is often driven by efflux of sulfide from underlying anoxic sediments, may result in chemolithotrophic denitrification becoming a more significant player in future N cycling scenarios (Naqvi et al., 2000; Diaz and Rosenberg, 2008; Lavik et al., 2009).

Anammox: a Marine N Mystery Solved?

The concentration profiles of O_2, NO_2^-, NO_3^-, and NH_4^+, representative of suboxic water columns and deep-sea sediments shown in Fig. 1, all suggest suboxic conversion of ammonium to N_2 gas. Following Richards's prediction of an "anammox"-like reaction (equation 8, above) where anammox and denitrification act in unison, 29% of the N_2 produced would be due to anammox (Richards et al., 1965). The original Costa Rica study by Dalsgaard et al. (2003) measured an average anammox contribution of 27%, with a positive correlation between the two processes; hence, their case for a Richards style of anammox was strong, and a long-standing marine N mystery appeared to have been resolved (Devol, 2003). Since then, however, the accounts of anammox in a variety of OMZs have not been so straightforward.

Subsequent studies in the shelf waters off Namibia, the ETSP, and the Black Sea all suggested a total dominance of anammox, with no measurable production of N_2 by denitrification, and it has been argued that anammox is the only pathway to form N_2 in the major OMZs (Table 3). The point to bear in mind here, then, is if there is no heterotrophic denitrification, then where does the ammonium and nitrite come from to fuel anammox? The OMZ of the Arabian Sea is commonly regarded as the world's largest, and it is believed to be responsible for 50% of total oceanic water column N_2 production (Devol et al., 2006). Recently, and in contrast to that just outlined, denitrification was reported to be, by far, the dominant pathway for N_2 production in the Arabian Sea (Ward et al., 2009). Previous accounts had argued for "multiple pathways of N_2 production," which could not be ascribed categorically to either complete anammox or denitrification; the ^{15}N labeling of produced N_2O could most easily be explained by a simple reduction of NO_2^-, and, in effect, part of the classic denitrification pathway was known to be present (Nicholls et al., 2007). As Ward et al. (2009) argued, there is plenty of molecular evidence for the apparatus of denitrification in the Arabian Sea, and the ^{15}N data support this. If, however, denitrification dominates the production of N_2 in the Arabian Sea, and assuming "Redfield" for the organic material being mineralized, then there must be an unknown sink for ammonium. With such a minor role for anammox in the Arabian Sea and a predominance of organic matter mineralization coupled to denitrification, then there should be more ammonium present in the water column than can actually be measured (Nicholls et al., 2007). Despite this apparent confusion, some explanations may lie in the respective availability of organic carbon.

Organic Carbon and the Balance between Anammox and Denitrification

The difference in the significance of anammox and denitrification between the ETSP and the Arabian Sea may be explained by denitrification being governed by the availability of organic carbon. This hypothesis is based on observations from a previous study that showed nitrate reduction rates in the ETSP OMZ to be limited by the availability of organic carbon, while that was not the case in the Arabian Sea (Ward et al., 2008, 2009). Several studies have reported a total dominance of anammox in the ETSP OMZ but, at the same time, documented the potential for denitrification through the presence and abundance of the denitrifier *nirS* gene (Hamersley et al., 2007; Lam et al., 2009; Ward et al., 2009). The split between anammox and denitrification could follow seasonal changes in productivity or advected import of organic matter, where denitrification activity can be significant in the ETSP following an extensive phytoplankton bloom (i.e., a pulse of organic carbon).

In parallel to the observations made with the sediments (see above), the range of rates reported for denitrification across OMZs is an order of magnitude greater than that for anammox, namely 0 to 270 nmol of N_2 liter^{-1} day^{-1} (±6 SE, $n = 76$) and 0 to 2,568 nmol of N_2 liter^{-1} day^{-1} (±87 SE, $n = 76$) for anammox and denitrification, respectively. In the Arabian Sea, the highest rate of anammox was 4.3 nmol of N_2 liter^{-1} day^{-1} compared to 8.8 nmol of N_2 liter^{-1} day^{-1} in the ETSP (Ward et al., 2009), while the difference in denitrification was a measured maximum of 25 nmol of N_2 liter^{-1} day^{-1} in the Arabian Sea compared to 0 nmol of N_2 liter^{-1} day^{-1} in the ETSP.

Together, the ETNP and the east subtropical North Pacific make up 68% of the total area of OMZs across the globe, covering 41% and 27%, respectively (Paulmiere and Ruiz-Pino, 2009). To the best of our knowledge, there are no published anammox data for these regions. We can speculate about their potential significance using available data for rates of nitrate reduction with and without the addition of organic carbon, as outlined above (Ward et al., 2008). Experiments were performed with water collected in the three large OMZs of the ETSP, ETNP, and Arabian Sea. Samples collected in the ETNP responded in a very similar way to water collected from the OMZ off Peru (ETSP), with a clear stimulation of nitrate reduction by addition of organic carbon, whereas no significant changes in concentration of inorganic N species (NO_3^-, NO_2^-, NH_4^+) could be measured in the controls. Given that the samples from the Arabian Sea were not carbon limited and until more data from the ETNP becomes available, we assume this region to have an anammox contribution and N_2 production rates more similar to the ETSP than the Arabian Sea.

Sensitivity of Anammox to Oxygen

The sensitivity to oxygen among the anammox bacteria is not fully clear, and studies from bioreactors reported anammox to be reversibly inhibited by oxygen concentrations as low as 1 μmol of O_2 liter^{-1} (Strous et al., 1999). In contrast, anammox bacteria have been shown to be active in low-oxygen and suboxic environments. Hamersley et al. (2007) detected anammox bacteria off the coast of Chile in water with up to 20 μmol of O_2 liter^{-1} and showed that these anammox bacteria could start their metabolism immediately upon establishment of suboxic conditions (Hamersley et al., 2007). An experiment to study anammox activity over a range of different oxygen concentrations could only show anammox activity at oxygen concentrations below 14 μmol of O_2 liter^{-1}, and between 0 and 14 μmol of O_2 liter^{-1}, anammox activity decreased linearly with increasing oxygen concentration (Jensen et al., 2008). It is, however, difficult to compare measured rates of anammox with environmental or ambient concentrations of oxygen, since the water used in the vast majority of ^{15}N incubations is degassed prior to the start of the experiment (Table 3).

Anammox bacteria appear active in both low-oxygen and suboxic waters, and such conditions are often considered as prerequisites for denitrification, since oxygen represses synthesis and activity of denitrifying enzymes (Zumft, 1997), though the effect may be more subtle. Körner and Zumft (1989) showed that each of the denitrifying enzymes in the denitrifying sequence had variable sensitivities to oxygen, whereas as the NO_3^- and NO_2^- reductases are expressed under moderate hypoxia (~30 to 40% of air saturation), N_2O reductase requires lower oxygen (<15%). In addition, N_2O production through denitrification has been measured in water with up to 50 μmol of O_2 liter^{-1} off Northern Chile (Farias et al., 2009).

The sensitivity to oxygen for both anammox and denitrification affects the distribution of their activity. When estimating the boundaries for the extent of oceanic OMZs, Paulmier and Ruiz-Pino (2009) set the maximum concentration for oxygen at 20 μmol liter^{-1}, though this is far in excess of the 2 to 3 μmol of O_2 liter^{-1} associated with the classic signature of nitrite maxima and it has a massive impact on the global estimate of N_2 production in the ocean (see global N budget et the end of this section). Their maximum concentration was based on the highest concentration for oxygen at which water column denitrification had been observed in situ (i.e., 20 μmol of O_2 liter^{-1}) (Smethie et al., 1987). Present-day studies of N_2 production in OMZs suggest that the 20 μmol of O_2 liter^{-1} may be an overestimate. For example, if we compare oxygen concentrations at the depth where the peak in nitrate is also measured (i.e., before it starts to be reduced), they coincide at an oxygen concentration of 15 μmol of O_2 liter^{-1} in the Black Sea, 0 μmol of O_2 liter^{-1} in numerous other studies, and between 4 and ~20 μmol of O_2 liter^{-1} in the ETSP (Thamdrup et al., 2006; Hamersley et al., 2007). The average of these latter values is ~10 μmol of O_2 liter^{-1}, which agrees well with the concentration of oxygen, in the data set below, where nitrite starts to accumulate markedly (Fig. 10a). As indicated above, the accumulation of nitrite is often associated with oxygen below 4 μmol of O_2 liter^{-1}. Our data probably reflect differences in the accuracy of determination, especially for oxygen, and the reader is referred to the more comprehensive data set of Morrison et al. (1999), among others.

Distribution and Supply of Nitrite and Ammonium

A classic feature of the ocean's OMZs is the suboxic peaks of the secondary nitrite maxima, which coincide with low oxygen and a drop in nitrate from ~35 μmol liter^{-1} to 15 μmol liter^{-1} (for example, see Morrison et al. [1999]). The anammox data set reported here (Table 3) agrees with the larger oceanographic data set as a whole, with maximal nitrite associated with both low oxygen and the initial reduction of nitrate from ~35 μmol liter^{-1} to 15 μmol liter^{-1} (Fig. 10a and c). After that and with further nitrate reduction, nitrite is subsequently consumed. Ammonium is usually present at very low concentrations or close to the detection limit in oceanic samples, but the data are not nearly as abundant as they are for oxygen, nitrite, and nitrate. The scarcity of ammonium may reflect some of the inherent problems with measuring low concentrations of ammonium with the traditional indo-phenol blue assay (Holmes et al., 1999).

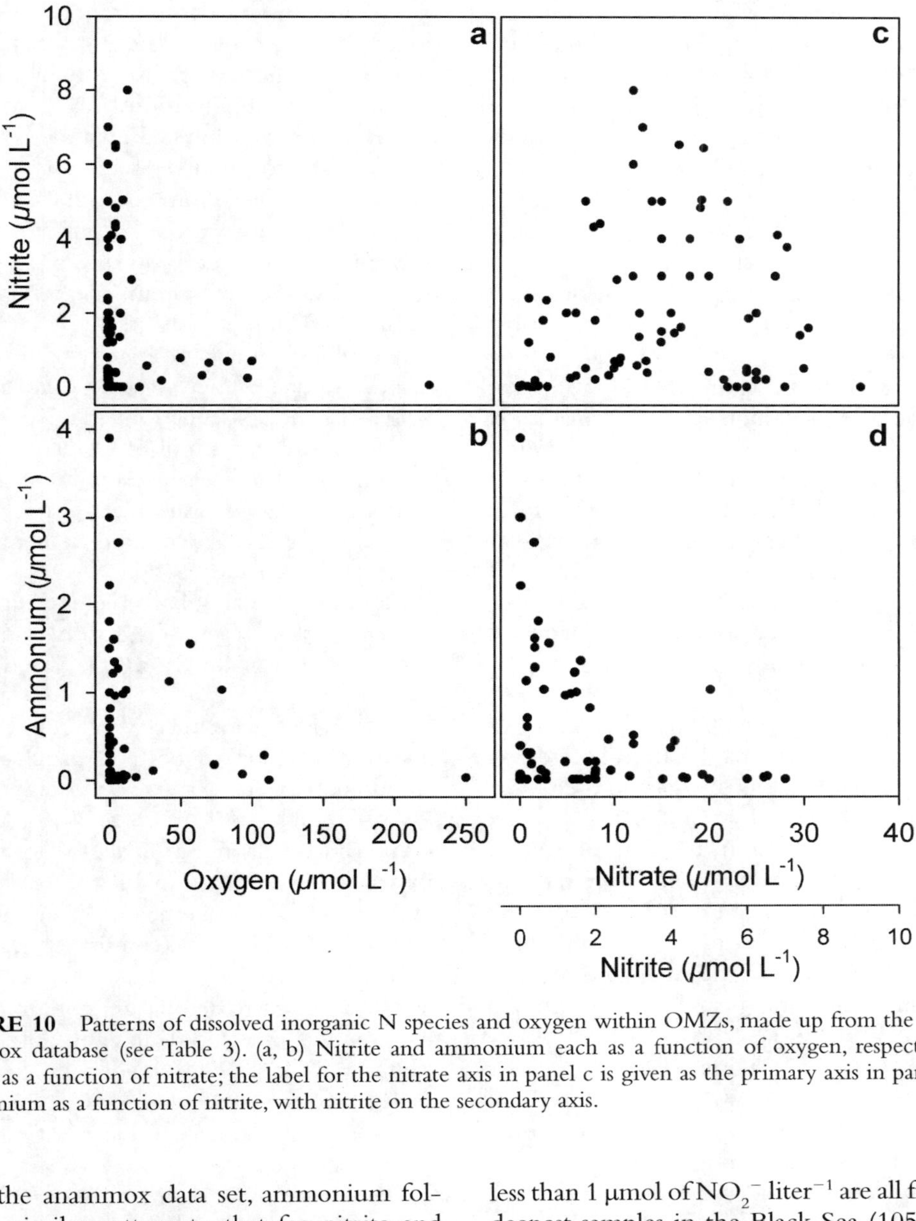

FIGURE 10 Patterns of dissolved inorganic N species and oxygen within OMZs, made up from the available anammox database (see Table 3). (a, b) Nitrite and ammonium each as a function of oxygen, respectively. (c) Nitrite as a function of nitrate; the label for the nitrate axis in panel c is given as the primary axis in panel d. (d) Ammonium as a function of nitrite, with nitrite on the secondary axis.

In the anammox data set, ammonium follows a similar pattern to that for nitrite and oxygen, whereby maximum concentrations of ammonium (~4 μmol of NH_4^+ liter^{-1}) are found close to the limit of detection for oxygen (Fig. 10b). Another interesting feature here is that ammonium and nitrite are present along opposing gradients (Fig 10d). Here concentrations greater than 2 μmol of NH_4^+ liter^{-1} and less than 1 μmol of NO_2^- liter^{-1} are all from the deepest samples in the Black Sea (105 to 110 m), at the interface between the suboxic and truly anoxic waters where ammonium accumulates. Given that anammox requires ammonium and nitrite in a 1:1 ratio, the majority of sites in the suboxic waters have an excess of nitrite. This is, therefore, the characteristic signature for anammox activity: nitrite coming from the reduction of nitrate, that reduction mineralizing ammonium, but that ammonium being absent from the water column.

The majority of water column anammox studies (76% of cases) report anammox activity within these zones, at ambient nitrite concentrations of 3 μmol of NO_2^- liter^{-1} or less (Fig. 11). As discussed in the previous sediment section, anammox has a high affinity for nitrite, with saturation kinetics occurring at or below 3 μmol of NO_2^- liter^{-1} (Dalsgaard and Thamdrup, 2002; Trimmer et al., 2003, 2005). If the scatter shows anything in Fig. 11, it is that nitrite actually accumulates where anammox activity is minimal, perhaps suggesting limitation by ammonium. Nitrite might, therefore, not be expected to limit anammox activity in the OMZs, but there are a few ^{15}N experiments that indicate nitrite limitation.

Nitrite limitation for the anammox reaction can be examined by comparing rates of anammox measured with $^{15}NH_4^+$ and parallel incubations with $^{15}NO_2^-$ or $^{15}NH_4^+$ + $^{14}NO_2^-$. Thamdrup et al. (2006), Hamersley et al. (2007), and Jensen et al. (2008) all presented such data for parallel incubations in the OMZ off Chile and Peru (ETSP) and the suboxic zone in the Black Sea. In the Chilean OMZ, 39% of experiments exhibited faster rates of anammox with the addition of NO_2^-, compared to controls with $^{15}NH_4^+$ only, although, overall, the two sets of measurements are not significantly different (paired *t* test, d.f. = 38, $P = 0.105$). Interestingly, in the Chilean cases that suggested NO_2^- limitation, the in situ concentration of NO_2^- was between 0 and 1.6 μmol liter^{-1} and below the apparent saturation or optimum concentration for anammox (Dalsgaard and Thamdrup, 2002; Trimmer et al., 2005).

Furthermore, in the Black Sea, Jensen et al. (2008) concluded that anammox activity in the suboxic zone was NO_2^- limited, and in a joint study, the rate of NO_2^- consumption was modeled from NO_2^-, NO_3^-, and NH_4^+ concentrations profiles with a reaction-diffusion model (Lam et al., 2007). In the model, nitrite consumption was only about half the rate of ammonium consumption, which suggested that if anammox is the major pathway removing ammonium in the suboxic zone of the Black Sea with a ratio of NO_2^- to NH_4^+ of 1:1, nitrite was supplied by another source other than just through the reduction of nitrate (Jensen et al., 2008). Lam et al. (2007) invoked another potential route of supply for nitrite by verifying nitrification in the suboxic zone of the Black Sea by careful measurement of the accumulation of $^{15}NO_2^-$ and $^{30}N_2$ production in closed incubations to which $^{15}NH_4^+$ had been added. There was no "measurable" oxygen in these incubations (detection limit 2% saturation; ca 5 μmol of O_2 liter^{-1}); nitrification was therefore suggested to be "microaerobic." This novel aspect was corroborated by measuring the active expression of *amoA* genes for Crenarchaea and γAOB (gamma ammonium-oxidizing bacteria) and suggested that γAOB were responsible for this "suboxic" nitrite production in the Black Sea (Lam et al., 2007).

Later, evidence for microaerobic ammonium oxidation coupled to both Crenarchaea and ammonium-oxidizing bacteria was reported for the OMZ off Peru (Lam et al., 2009). Ammonium oxidation could be detected in incubations with less than 2 μmol of O_2 liter^{-1}, and in the upper OMZ off Peru, aerobic ammonium oxidation was estimated to produce at least 65% of the NO_2^- required for anammox, but in the lower part, no microaerobic ammonium oxidation could be detected (Lam et al., 2009). Also, Molina and Farías (2009) suggested that a large part of the ammonium removal in the ETSP OMZ is due to microaerobic nitrification as well as anammox.

Microaerobic ammonium oxidation is a very interesting dynamic. For example, whereas heterotrophic nitrate-reducing bacteria are physiologically limited by oxygen at <4 μmol of O_2 liter^{-1} and start respiring nitrate to nitrite, Crenarchaea and AOB presented by Lam et al. (2009) may be able to operate at concentrations (<2 μmol of O_2 liter^{-1}) that would be considered at the limit for heterotrophic nitrate-reducing bacteria.

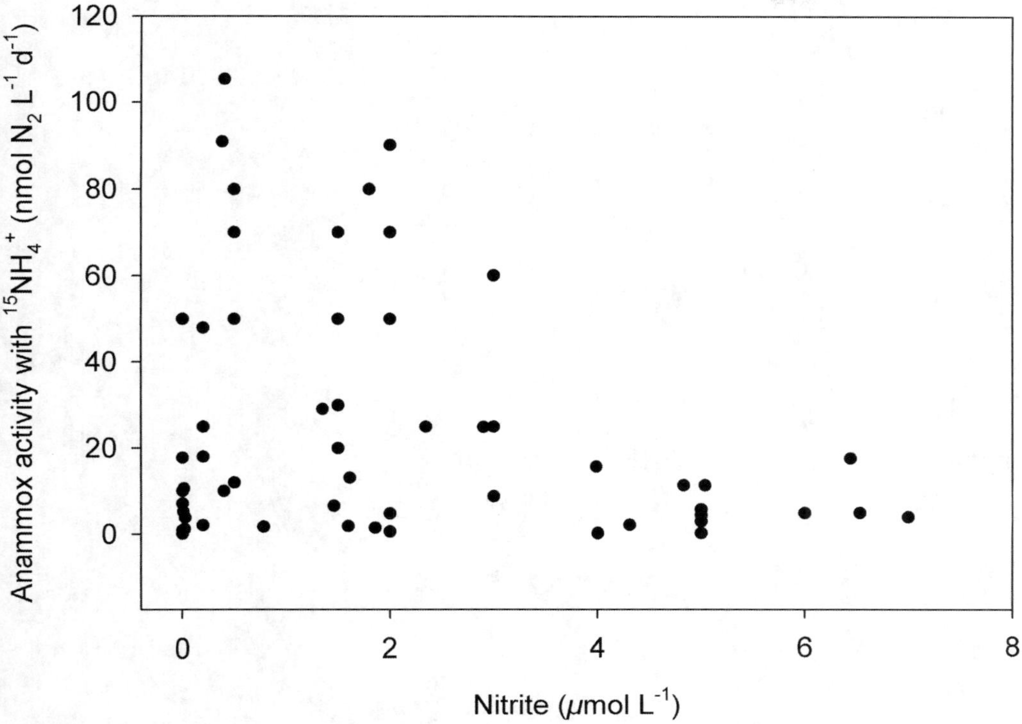

FIGURE 11 Anammox activity measured by enrichment with $^{15}NH_4^+$ as a function of the concentration of ambient nitrite, made up from the available anammox database (see Table 3.). To illustrate the overall trend, two outliers have been removed that had activity of 170 and 270 nmol of N_2 liter^{-1} day^{-1}.

In the studies with direct measurements of anammox (Table 3), the mean in situ concentration for ammonium at the depths where anammox activity was detected was 0.52 μmol of NH_4^+ liter^{-1} (±0.095 SE, n = 74), and 93% of the incubations were collected from water with an ammonium concentration of below 2 μmol liter^{-1} (Fig. 12). Even though all of the studies report a low concentration of ammonium, actual ammonium limitation is suggested only for the ETSP and in Golfo Dulce. In the latter, incubations with 10 μmol of NH_4^+ liter^{-1} increased anammox activity twofold to fourfold compared to incubations at ambient concentrations (0.3 μmol of NH_4^+ liter^{-1} [Dalsgaard et al., 2003]). In the Golfo Dulce, denitrification contributed to over 60% of total N_2 production and was purported to supply both ammonium and nitrite for anammox. Considering ammonium limitation in the ETSP, as mentioned earlier, a comparison between anammox measured with $^{15}NO_2^-$ against that measured with $^{15}NH_4^+$ in the ETSP and the Black Sea did not show any significant effect of ammonium.

Galan et al. (2009) suggested that the availability of ammonium is the main candidate for controlling the abundance and activity of anammox bacteria in the OMZ off northern Chile. Three studies from the ETSP suggest a depth distribution of anammox activity consistent with ammonium limitation, with higher activity close to the upper border of the OMZ where mineralization of organic matter is high and a subsurface ammonium peak could be seen (Thamdrup et al., 2006; Hamersley et al., 2007; Galan et al., 2009). The same pattern where high anammox rates coincide with

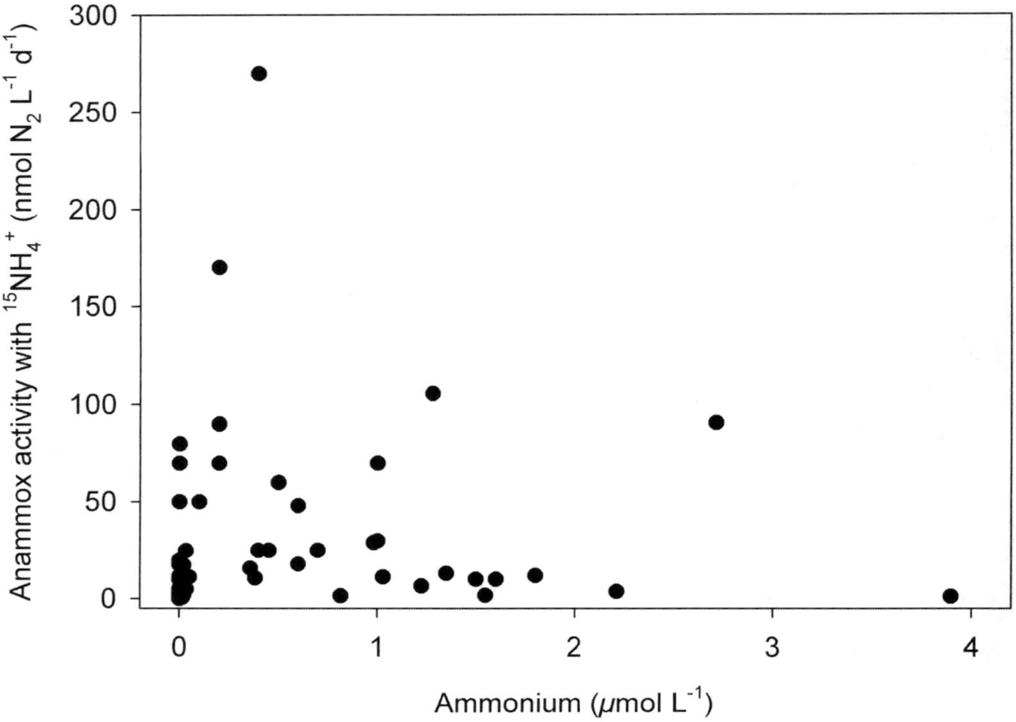

FIGURE 12 Anammox activity measured by enrichment with $^{15}NH_4^+$ as a function of the concentration of ambient ammonium, made up from the available anammox database (see Table 3.).

a subsurface ammonium peak in the oxycline has also been reported from the Benguela upwelling (Kuypers et al., 2005), suggesting that anammox is driven, in part, by aerobically regenerated ammonium diffusing into the suboxic zone.

Studies of the anammox metabolism using cell suspensions of *Kuenenia stuttgartiensis* (>99%) showed that subsequent to additions of $^{15}NO_3^-$, a significant amount of $^{15}NH_4^+$ was recovered along with $^{15}NO_2^-$ as an intermediate (Kartal et al., 2007). Furthermore, 10% of added $^{15}NO_3^-$ was recovered as $^{15}NH_4^+$ in incubations with water from the suboxic zone of the Benguela upwelling, where a high abundance of anammox bacteria was reported. Experiments with both cell suspensions and marine samples could only detect $^{29}N_2$ after addition of $^{15}NO_3^-$, so there is no direct evidence that anammox bacteria use the ammonium they produce for N_2 formation, though this effect may have been dampened by the respective labeling of the ammonium pools. Even though anammox bacteria have the ability to form ammonium via reduction of nitrate in marine environments, all studies from OMZs (Table 3) where additions of $^{15}NH_4^+$ were followed by a linear production of $^{29}N_2$ suggest that anammox bacteria generally use NH_4^+ from an outside source, in concurrence with the sediment findings.

As discussed previously, the ratio of anammox to denitrification measured in Golfo Dulce indicated a tight coupling between the two, yet subsequent accounts of anammox in the ETSP, the Benguela upwelling and the Black Sea suggested that anammox can exist without denitrification, as the latter was absent (Dalsgaard et al., 2003; Kuypers et al., 2005; Thamdrup et al., 2006; Jensen et al., 2008).

If anammox occurs without denitrification in several suboxic water columns, the question comes up as to what other sources of ammonium besides denitrification could drive anammox? The possibility of an "outside" source (i.e., ammonium from water layers above or below the anammox zone), as suggested for the ETSP, may also be present in the Black Sea. In the Black Sea, anammox is most active at the interface between the suboxic and anoxic water layers, fueled by ammonium diffusing up from below, where it accumulates through degradation of organic matter, probably through sulfate reduction, in the underlying sulfidic anoxic waters (Fuchsman et al., 2008; Kuypers, 2003; Lam et al., 2007). The high "outlier" concentrations for ammonium from these depths (Fig. 10b) support this here, but it may not apply elsewhere. Ammonium flux from underlying nutrient-rich water can also be significant in benthic regions covered with dense mats of sulfur bacteria (Fossing et al., 1995). These bacteria oxidize sulfide with nitrate to form ammonium, and these mats are found at the Peruvian and Chilean coast, where anoxic nitrate-rich water meets the sediment; such a situation could also apply to the coastal zone of the Arabian Sea and off the coast of Namibia (Naqvi et al., 2000; Lavik et al., 2009).

In the Black Sea, further experimental data for the natural abundance of ^{15}N in the NO_2^-, NO_3^-, NH_4^+, and N_2 pools, together with modeling, suggested a system that oscillates between two states: (i) where anammox is driven by the upward flux of ammonium from the deep water; and (ii) where particulate organic nitrogen remineralized in the suboxic zone is suggested as a significant source of ammonium (Fuchsman et al., 2008; Konovalov et al., 2008). Furthermore, to successfully model the concentration profiles of NO_2^-, NO_3^-, NH_4^+, and N_2 at steady state in the Black Sea, it was necessary for anammox to be responsible for ~90% of the N_2 production.

More bacteria are known to be capable of nitrate reduction to nitrite rather than complete denitrification (i.e., the production of gaseous nitrogenous compounds) (Zumft 1997). An OMZ where nitrate reduction is the dominant source of ammonium and, in turn, anammox is the pathway that consumes all of the nitrite would accumulate far too much nitrite (equation 9), since a release of 212 mol of nitrite and only 16 mol of ammonium could not be accounted for by the 1:1 stoichiometry of anammox. Natural distributions of inorganic N in OMZs (Fig. 1) support the removal of NO_2^- and NH_4^+ by anammox and a more even supply of NO_2^- and NH_4^+ than could be provided by nitrate reduction. In addition, such a stoichiometry could not, in the long term, produce sufficient N_2 to account for the nitrate deficit in some OMZs (Thamdrup et al., 2006).

Microaerobic organic matter mineralization, microaerobic heterotrophs regenerating N under low, but non-zero, oxygen conditions, could also provide ammonium for anammox. However, if there is microaerobic organic matter oxidation present, there is also a potential for microaerobic ammonium oxidation, which would also remove ammonium.

DNRA is a source of ammonium that was recently shown to exist in the OMZ off the coast of Peru, where it was estimated to provide between 7% and 134% of the ammonium needed for anammox at the shelf and 7% to 34% at an offshore station (Lam et al., 2009). In the same study, nitrate reduction, together with DNRA, was suggested to produce enough ammonium for anammox in that region. The DNRA metabolism has traditionally been believed to be restricted to fully anoxic, sulfidic environments (see sediment section), and the discovery that it could make a significant contribution to N cycling in OMZs changes this perspective of its role in aquatic ecosystems. If DNRA is significant, however, in OMZs, then the basic assumptions that underpin the application of ^{15}N to the measurement of anammox, denitrification, and DNRA in such systems will need revising. Marked DNRA should lead to imbalances in the production of $^{29}N_2$ measured in incubations with $^{15}NO_2^-$ relative to that measured with $^{15}NH_4^+$ (outlined above, "Fueling anammox in aquatic sediments"). Nicholls et al. (2007) suggested that $^{29}N_2$ is produced by a pairing between ^{15}N from nitrite and another source besides extracellular ammonium, perhaps the dissolved organic N pool.

TOTAL GLOBAL BENTHIC AND PELAGIC N BUDGET

At the end of the sediment section, we proposed a global budget for benthic N_2 production of 126 Tg of N year^{-1} (see "Scaling up and the global significance of anammox in benthic sediments," above). In terms of scaling up, the major difference between sediments and OMZs is that the latter are not as easy to define in terms of size and there is no agreement concerning the threshold of oxygen that defines an OMZ (Codispoti et al., 2001; Paulmier and Ruiz-Pino, 2009). Paulmier and Ruiz-Pino (2009) defined OMZs as water with less than 20 μmol of O_2 liter^{-1}, with a minimum concentration in the core of below 3 μmol of O_2 liter^{-1} and denitrification zones as a water body where the nitrate deficit is greater than 10 μmol of NO_3^- liter^{-1}. Codispoti et al. (2001) argued that the vast majority of nitrite accumulation occurs at less than 4 μmol of O_2 liter^{-1} (which is supported by numerous large data sets) and that this is indicative of denitrification. According to Codispoti et al. (2001), therefore, the OMZs occupy 0.1% of the total ocean volume or 1.35×10^6 km^3. The denitrification zone estimated using the nitrate deficit according to Paulmier and Ruiz-Pino (2009) is 2.5 times bigger, at 3.45×10^6 km^3.

Data from direct measurements of anammox and denitrification ($^{15}NH_4^+$ and $^{15}NO_3^-$ incubations, respectively) (Table 3) were used to estimate average rates for total N_2 production in the Arabian Sea and the ETSP. The average rate for total N_2 production in the Arabian Sea was 7 nmol of N_2 liter^{-1} day^{-1} (±3 standard error [SE], n =8), with an anammox contribution of 18% (± 11 SE). Station 1, however, at 150 m (Ward et al., 2009), had an unusually high rate of denitrification (25 nmol of N_2 liter^{-1} day^{-1}), and if removed from the data set, then the average N_2 production in the Arabian Sea falls to 3.4 nmol of N_2 liter^{-1} day^{-1} (±1 SE, n = 7). The average rate for N_2 production in the ETSP is 16 nmol of N_2 liter^{-1} day^{-1} (±5 SE, n =26), which is much higher than the Arabian Sea, and could be due to the fact that all of the study sites in the ETSP were located relatively close to the coast (within 250 km from land), compared to the two stations much further offshore in the central Arabian Sea (Ward et al. 2009). Hence, we have taken an average value for these two areas of 9.9 nmol of N_2 liter^{-1} day^{-1}, which is assumed to be representative of N_2 production between coastal and oceanic sites. Based on this assumption and depending on which OMZ volume is used (e.g., Codispoti et al. [2001; 1.35×10^6 km^3] or Paulmier and Ruiz-Pino [2009; 3.45×10^6 km^3]), we estimate the global water column production of N_2 to be between 136 and 349 Tg of N year^{-1}. Calculating an overall average contribution from anammox to global N_2 production in this form makes little sense, given the large variation between the mean *ra* % for the ETSP and the Arabian Sea, respectively. If, however, the entire OMZ of the Pacific behaves like the ETSP, then, with its respective total size relative to that for the Arabian Sea, we would argue that anammox dominates global water column N_2 production.

The lower value (136 Tg of N year^{-1}) in this estimate is in line with the earlier estimates of 150 Tg of N year^{-1} proposed by Codispoti et al. (2001), while the larger volume proposed by Paulmier and Ruiz-Pinos (2009) suggested that N_2 production in OMZs (349 Tg of N year^{-1}) is potentially more important than that in the sediments (126 Tg of N year^{-1}). The higher value supports Codispoti's (2006) later argument that "there is no reason to reduce the oceanic denitrification rate below 400 Tg of N year^{-1}, and that this rate may be conservative" but it may need to be redefined in terms of its metabolism, as it is not due solely to denitrification, as anammox makes a significant, but as yet not fully quantified, contribution to the production of N_2 gas in the global ocean.

REFERENCES

Aller, R. C., P. O. J. Hall, P. D. Rude, and J. Y. Aller. 1998. Bigoechemical heterogeneity and suboxic diagenesis in hemipelagic sediments of the Panama basin. *Deep Sea Res. Part I* **45:**133–165.

Andersson, H. J., J. W. M. Wijsman, P. M. J. Herman, J. J. Middelburg, K. Soetaert, and C. Heip. 2004. Respiration patterns in the deep ocean. *Geophys. Res. Lett.* **31:**L03304.

Bender, M., R. Jahnke, R. Weiss, W. Martin, D. T. Heggie, J. Orchardo, and T. Sowers. 1989. Organic carbon oxidation and benthic nitrogen and silica dynamics in San Clemente Basin, a continental borderline site. *Geochim. Cosmochim. Acta* **53:**685–697.

Blaszczyk, M. 1993. Effect of medium composition on the denitrification of nitrate by *Paracoccus denitrificans*. *Appl. Environ. Microbiol.* **59:**3951–3953.

Brettar, I., and G. Rheinheimer. 1991. Denitrification in the Central Baltic: evidence for H_2S-oxidation as a motor for denitrification at the oxic-anoxic interface. *Mar. Ecol. Prog. Ser.* **77:**157–169.

Brunet, R. C., and L. J. Garcia-Gil. 1996. Sulfide-induced dissimilatory nitrate reduction to ammonia in anaerobic freshwater sediments. *FEMS Microbiol. Ecol.* **21:**131–138.

Chang, B. X., and A. H. Devol. 2009. Seasonal and spatial patterns of sedimentary denitrification rates in the Chukchi sea. *Deep Sea Res. Part II* **56:**1339–1350.

Christensen, P. B., S. Rysgaard, N. P. Sloth, T. Dalsgaard, and S. Schwæter. 2000. Sediment mineralization, nutrient fluxes, denitrification and dissimilatory nitrate reduction to ammonium in an estuarine fjord with sea cage trout farms. *Aquat. Microb. Ecol.* **21:**73–84.

Codispoti, L. A. 2006. An oceanic fixed nitrogen sink exceeding 400 Tg Na^{-1} vs the concept of homeostasis in the fixed-nitrogen inventory. *Biogeosci. Disc.* **3:**1203–1246.

Codispoti, L. A., J. A. Brandes, J. P. Christensen, A. H. Devol, S. W. A. Naqvi, H. W. Paerl, and T. Yoshinari. 2001. The oceanic fixed nitrogen and nitrous oxide budgets: moving targets as we enter the anthropocene? *Sci. Mar.* **65:**85–105.

Codispoti, L. A., and J. P. Christensen. 1985. Nitrification, denitrification and nitrous oxide cycling in the eastern tropical south Pacific Ocean. *Mar. Chem.* **16:**277–300.

Conley, D. J., H. W. Paerl, R. W. Howarth, D. F. Boesch, S. P. Seitzinger, K. E. Havens, C. Lancelot, and G. E. Likens. 2009. Ecology. Controlling eutrophication: nitrogen and phosphorus. *Science* **323:**1014–1015.

Dale, O. R., C. R. Tobias, and B. K. Song. 2009. Biogeographical distribution of diverse anaerobic ammonium oxidizing (anammox) bacteria in Cape Fear River Estuary. *Environ. Microbiol.* **11:**1194–1207.

Dalsgaard, T., and B. Thamdrup. 2002. Factors controlling anaerobic ammonium oxidation with nitrite in marine sediments. *Appl. Environ. Microbiol.* **68:**3802–3808.

Dalsgaard, T., D. E. Canfield, J. Petersen, B. Thamdrup, and J. Acuña-González. 2003. N_2 production by the anammox reaction in the anoxic water column of Golfo Dulce, Costa Rica. *Nature* **422:**606–608.

Dalsgaard, T., B. Thamdrup, and D. E. Canfield. 2005. Anaerobic ammonium oxidation (anammox) in the marine environment. *Res. Microbial.* **156:**457–464.

Devol, A. H. 2003. Nitrogen cycle—solution to a marine mystery. *Nature* **422:**575–576.

Devol, A. H., A. G. Uhlenhopp, S. W. A. Naqvi, J. A. Brandes, D. A. Jayakumar, H. Naik, S. Gaurin, L. A. Codispoti, and T. Yoshinari. 2006. Denitrification rates and excess nitrogen gas concentrations in the Arabian Sea oxygen deficient zone. *Deep Sea Res. Part II* **53:**1533–1547.

Diaz, R. J., and R. Rosenberg. 2008. Spreading dead zones and consequences for marine ecosystems. *Science* **321:**926–929.

Dollar, S. J., S. V. Smith, S. M. Vink, S. Brebski, and J. T. Hollibaugh. 1991. Annual cycle of benthic nutrient fluxes in Tomales Bay, California, and contribution of the benthos to total ecosystem metabolism. *Mar. Ecol. Prog. Ser.* **79:**115–125.

Dong, L. F., C. J. Smith, S. Papaspyrou, A. Stott, M. A. Osborn, and D. B. Nedwell. 2009. Changes in benthic denitrification, nitrate ammonification, and anammox process rates and nitrate and nitrite reductase gene abundances along an estuarine nutrient gradient (the Colne Estuary, United Kingdom). *Appl. Environ. Microbiol.* **75:**3171–3179.

Engström, P. 2004. The importance of anaerobic ammonium oxidation (anammox) and anoxic nitrification for N removal in coastal marine sediments. Ph.D. thesis. University of Gothenburg, Gothenburg, Sweden.

Engström, P., T. Dalsgaard, S. Hulth, and R. C. Aller. 2005. Anaerobic ammonium oxidation by nitrite (anammox): implications for N_2 production in coastal marine sediments. *Geochim. Cosmochim. Acta* **69:**2057–2065.

Engström, P., C. R. Penton, and A. H. Devol. 2009. Anaerobic ammonium oxidation in deep-sea sediments off the Washington margin. *Limnol. Oceanogr.* **54:**1643–1652.

Falkowski, P. G. 1997. Evolution of the nitrogen cycle and its influence on the biological sequestration of CO_2 in the ocean. *Nature* **387:**272–275.

Farias, L., M. Castro-Gonzalez, M. Cornejo, J. Charpentier, J. Faundez, N. Boontanon, and N. Yoshida. 2009. Denitrification and nitrous oxide cycling within the upper oxycline of the eastern tropical South Pacific oxygen minimum zone. Limnol. Oceanogr. **54:**132–144.

Fernandez-Polanco, F., M. Fernandez-Polanco, N. Fernandez, M. A. Uruena, P. A. Garcia, and S. Villaverde. 2001. New process for simultaneous removal of nitrogen and sulphur under anaerobic conditions. *Water Res.* **35:**1111–1114.

Fossing, H., V. A. Gallardo, B. B. Jørgensen, M. Huttel, L. P. Nielsen, and H. Schulz. 1995. Concentration and transport of nitrate by the mat-forming sulfur bacterium Thioploca. *Nature* **374:**713–715.

Francis, C. A., J. M. Beman, and M. M. M. Kuypers. 2007. New processes and players in the nitrogen cycle: the microbial ecology of anaerobic and archaeal ammonia oxidation. *ISME J.* **1:**19–27.

Fuchsman, C. A., J. W. Murray, and S. K. Konovalov. 2008. Concentration and natural stable isotope profiles of nitrogen species in the Black Sea. *Mar. Chem.* **111:**90–105.

Galan, A., V. Molina, B. Thamdrup, D. Woebken, G. Lavik, M. M. M. Kuypers, and O. Ulloa. 2009. Anammox bacteria and the anaerobic oxidation of ammonium in the oxygen minimum zone off northern Chile. *Deep Sea Res. Part II* **56:**1021–1031.

Glud, R. N. 2008. Oxygen dynamics of marine sediments. *Mar. Biol. Res.* **4:**165–179.

Glud, R. N., B. Thamdrup, H. Stahl, F. Wenzhoefer, A. Glud, H. Nomaki, K. Oguri, N. P. Revsbech, and H. Kitazato. 2009. Nitrogen cycling in a deep ocean margin sediment (Sagami Bay, Japan). *Limnol. Oceanogr.* **54:**723–734.

Gruber, N., and J. L. Sarmiento. 1997. Global patterns of marine nitrogen fixation and denitrification. *Glob. Biogeochem. Cycles* **11:**235–266.

Güven, D., A. Dapena, B. Kartal, M. C. Schmind, B. Maas, K. van de Pas-Schoonen, S. Sozen, R. Mendez, H. J. M. Op den Camp, M. S. M. Jetten, M. Strous, and I. Schmidt. 2005. Propionate oxidation by and methanol inhibition of anaerobic ammonium-oxidizing bacteria. *Appl. Environ. Microbiol.* **71:**1066–1071.

Halm, H., N. Musat, P. Lam, R. Langlois, F. Musat, S. Peduzzi, G. Lavik, C. J. Schubert, B. Singha, J. LaRoche, and M. M. M. Kuypers. 2009. Cooccurrence of denitrification and nitrogen fixation in a meromictic lake, Lake Cadagno (Switzerland). *Environ. Microbiol.* **11:**1945–1958.

Hamersley, M. R., G. Lavik, D. Woebken, J. E. Rattray, P. Lam, E. C. Hopmans, J. S. Sinninghe Damste, S. Kruger, M. Graco, D. Gutierrez, and M. M. M. Kuypers. 2007. Anaerobic ammonium oxidation in the Peruvian oxygen minimum zone. *Limnol. Oceanogr.* **52:**923–933.

Hannig, M., G. Lavik, M. M. M. Kuypers, D. Woebken, W. Martens-Habbena, and K. Juergens. 2007. Shift from denitrification to anammox after inflow events in the central Baltic Sea. *Limnol. Oceanogr.* **52:**1336–1345.

Hartnett, H. E., and A. H. Devol. 2003. Role of a strong oxygen-deficient zone in the preservation and degradation of organic matter: a carbon budget for the continental margins of northwest Mexico and Washington State. *Geochim. Cosmochim. Acta* **67:**247–264.

Hauck, R. D., S. W. Melsted, and P. E. Yankwich. 1958. Use of N-isotope distribution in nitrogen gas in the study of denitrification. *Soil Sci.* **86:**287–296.

Hietanen, S. 2007. Anaerobic ammonium oxidation (anammox) in sediments of the Gulf of Finland. *Aquat. Microb. Ecol.* **48:**197–205.

Hietanen, S., and J. Kuparinen. 2008. Seasonal and short-term variation in denitrification and anammox at a coastal station on the Gulf of Finland, Baltic Sea. *Hydrobiologia* **596:**67–77.

Holmes, R. M., A. Aminot, R. Kerouel, B. A. Hooker, and B. J. Peterson. 1999. A simple and precise method for measuring ammonium in marine and freshwater ecosystems. *Can. J. Fish Aquat. Sci.* **56:**1801–1808.

Hulth, S., R. C. Aller, and F. Gilbert. 1999. Coupled anoxic nitrification manganese reduction in marine sediments. *Geochim. Cosmochim. Acta* **63:**49–66.

Jaeschke, A., C. Rooks, M. Trimmer, J. C. Nicholls, E. C. Hopmans, S. Schouten, and J. S. Sinninghe Damsté. 2009. Comparison of ladderane phospholipid and core lipids as indicators for anaerobic ammonium oxidation (anammox) in marine sediments. *Geochim. Cosmochim. Acta* **73:**2077–2088.

Jensen, M. M., M. M. M. Kuypers, G. Lavik, and B. Thamdrup. 2008. Rates and regulation of anaerobic ammonium oxidation and denitrification in the Black Sea. *Limnol. Oceanogr.* **53:**23–36.

Jensen, M., J. Petersen, T. Dalsgaard, and B. Thamdrup. 2009. Pathways, rates, and regulation of N_2 production in the chemocline of an anoxic basin, Mariager Fjord, Denmark. *Mar. Chem.* **113:**102–113.

Joye, S. B., and J. T. Hollibaugh. 1995. Influence of sulfide inhibition of nitrification on nitrogen regeneration in sediments. *Science* **270:**623–625.

Kartal, B., M. M. M. Kuypers, G. Lavik, J. Schalk, H. J. M. Op den Camp, M. S. M. Jetten, and M. Strous. 2007. Anammox bacteria disguised as denitrifiers: nitrate reduction to dinitrogen gas via nitrite and ammonium. *Environ. Microbiol.* **9:**635–642.

Konovalov, S. K., C. A. Fuchsman, V. Belokopitov, and J. W. Murray. 2008. Modeling the distribution of nitrogen species and isotopes in the water column of the Black Sea. *Mar. Chem.* **111:**106–124.

Koop-Jakobsen, K., and A. E. Giblin. 2009. Anammox in tidal marsh sediments: the role of salinity, nitrogen loading, and marsh vegetation.

Estuar. Coasts **32:**238–245.

Körner, H., and W. G. Zumft. 1989. Expression of denitrification enzymes in response to the dissolved oxygen level and respiratory substrate in continuous culture of *Pseudomonas stutzeri. Appl. Environ. Microbiol.* **55:**1670–1676.

Kuypers, M. M. M., A. O. Sliekers, G. Lavik, M. Schmid, B. Barker Jørgensen, J. G. Kuenen, J. S. Sinninghe Damsté, M. Strous, and M. S. M. Jetten. 2003. Anaerobic ammonium oxidation by anammox bacteria in the Black Sea. *Nature* **422:**608–611.

Kuypers, M. M. M., G. Lavik, D. Woebken, M. Schmid, B. M. Fuchs, R. Amann, B. Barker Jørgensen, and M. S. M. Jetten. 2005. Massive nitrogen loss from the Benguela upwelling system through anaerobic ammonium oxidation. *Proc. Natl. Acad. Sci. USA* **102:**6478–6483.

Lam, P., M. M. Jensen, G. Lavik, D. F. McGinnis, B. Muller, C. J. Schubert, R. Amann, B. Thamdrup, and M. M. M. Kuypers. 2007. Linking crenarchaeal and bacterial nitrification to anammox in the Black Sea. *Proc. Natl. Acad. Sci. USA* **104:**7104–7109.

Lam, P., G. Lavik, M. M. Jensen, J. van de Vossenberg, M. Schmid, D. Woebken, D. Gutierrez, R. Amann, M. S. M. Jetten, and M. M. M. Kuypers. 2009. Revising the nitrogen cycle in the Peruvian oxygen minimum zone. *Proc. Natl. Acad. Sci. USA* **106:**4752–4757.

Lavik, G., T. Stuhrmann, V. Bruchert, A. Van der Plas, V. Mohrholz, P. Lam, M. Mußmann, B. M. Fuchs, R. Amann, U. Lass, and M. M. M. Kuypers. 2009. Detoxification of sulphidic African shelf waters by blooming chemolithotrophs. *Nature* **457:**581–584.

Lohse, L., H. F. P. Malschaert, C. P. Slomp, W. Helder, and W. van Raaphorst. 1993. Nitrogen cycling in North Sea sediments: interaction of denitrification and nitrification in offshore and coastal areas. *Mar. Ecol. Prog. Ser.* **101:**283–296.

Meyer, R. L., N. Risgaard-Petersen, and D. E. Allen. 2005. Correlation between anammox activity and microscale distribution of nitrite in a subtropical mangrove sediment. *Appl. Environ. Microbiol.* **71:**6142–6149.

Middelburg, J. J., K. Soetart, P. M. J. Herman, and C. H. R. Heip. 1996. Denitrification in marine sediments: a model study. *Glob. Biogeochem. Cycles* **10:**661–673.

Minjeaud, L., P. C. Bonin, and V. D. Michotey. 2008. Nitrogen fluxes from marine sediments: quantification of the associated co-occurring bacterial processes. *Biogeochemistry* **90:**141–157.

Molina, V., and L. Farías. 2009. Aerobic ammonium oxidation in the oxycline and oxygen minimum zone of the eastern tropical South Pacific off northern Chile (~20°S). *Deep Sea Res. Part II* **56:**1032–1041.

Morrison, J. M., L. A. Codispoti, S. L. Smith, K. Wishner, C. Flagg, W. D. Gardner, S. Gaurin, S. W. A. Naqvi, V. Manghnani, L. Prosperie, and J. S. Gunderson. 1999. The oxygen minimum zone in the Arabian Sea during 1995. *Deep Sea Res. Part II* **46:**1903–1931.

Mulder, A., A. A. van de Graaf, L. A. Robertson, and J. G. Kuenen. 1995. Anaerobic ammonium oxidation discovered in a denitrifying fluidized bed reactor. *FEMS Microbiol. Ecol.* **16:**177–184.

Naqvi, S. W. A., R. J. Noronha, M. S. Shailaja, K. Somasunda, and R. S. Gupta. 1992. Some aspects of the nitrogen cycling in the Arabian Sea, p. 285–311. *In* B. N. Desai (ed.), *Oceanography of the Indian Ocean.* Oxford and IBH Publishers, New Delhi, India.

Naqvi, S. W. A., D. A. Jayakumar, P. V. Narvekar, H. Naik, V. V. S. S. Sarma, W. D. D'Souza, and J. M. D. George. 2000. Increased marine production of N_2O due to intensifying anoxia on the Indian continental shelf. *Nature* **408:**346–349.

Nedwell, D. B., L. F. Dong, A. Sage, and G. J. C. Underwood. 2002. Variations of the nutrients loads to the mainland U.K. estuaries: correlation with catchment areas, urbanization and coastal eutrophication. *Estuar. Coast. Shelf Sci.* **54:**951–970.

Nicholls, J. C., C. A. Davies, and M. Trimmer. 2007. High resolution profiles and nitrogen isotope tracing reveal and dominant source of nitrous oxide and multiple pathways of nitrogen gas production in the central Arabian Sea. *Limnol. Oceanogr.* **52:**156–168.

Nicholls, J. C., and M. Trimmer. 2009. Widespread relationship between the anammox reaction and organic carbon in estuarine sediments. *Aquat. Microb. Ecol.* **55:**105–113.

Nielsen, L. P. 1992. Denitrification in sediments determined from nitrogen isotope pairing. *FEMS Microbiol. Ecol.* **86:**357–362.

Nishio, T., I. Koike, and A. Hattori. 1982. Denitrification, nitrate reduction, and oxygen consumption in coastal and estuarine sediments. *Appl. Environ. Microbiol.* **43:**648–653.

Ogilvie, B., D. B. Nedwell, R. M. Harrison, A. Robinson, and A. Sage. 1997. High nitrate, muddy estuaries as nitrogen sinks: the nitrogen budget of the River Colne estuary (United Kingdom). *Mar. Ecol. Prog. Ser.* **150:**217–228.

Paulmier, A., and D. Ruiz-Pino. 2009. Oxygen minimum zones (OMZs) in the modern ocean. *Prog. Oceanogr.* **80:**113–128.

Peirels, B., N. Caraco, M. Pace, and J. Cole. 1991. Human influence on river nitrogen. *Nature* **350:**386–387.

Reimers, C. E., R. A. Jahnke, and D. C. McCorkel.

1992. Carbon fluxes and burial rates over the continental slope of California with implications for the global carbon cycle. *Glob. Biogeochem. Cycles* **6:**199–224.

Revsbech, N. P., N. Risgaard-Petersen, A. Schramm, and L. P. Nielsen. 2006. Nitrogen transformations in stratified aquatic microbial ecosystems. *Antonie Leeuwenhoek Int. J. G.* **90:**361–375.

Rich, J. J., O. R. Dale., B. Song., and B. B. Ward. 2008. Anaerobic ammonium oxidation in Chesapeake Bay sediments. *Microbial. Ecol.* **55:**311–320.

Richards, F. A., J. D. Cline, W. W. Broenkow, and L. P. Atkinson. 1965. Some consequences of the decomposition of organic matter in Lake Nitinat, an anoxic fjord. *Limnol. Oceanogr.* **10:**R185–R201.

Risgaard-Petersen, N., L. P. Nielsen, S. Rysgaard, T. Dalsgaard, and R. L. Meyer. 2003. Application of the isotope pairing technique in sediments where anammox and denitrification coexist. *Limnol. Oceanogr. Methods* **1:**63–73.

Risgaard-Peterson, N., R. L. Meyer, M. Schmid, M. S. M. Jetten, A. Enrich-Prast, S. Rysgaard, and N. P. Revsbech. 2004. Anaerobic ammonium oxidation in an estuarine sediment. *Aquat. Microb. Ecol.* **36:**293–304.

Risgaard-Peterson, N., R. L. Meyer, M. and N. P. Revsbech. 2005. Denitrification and anaerobic ammonium oxidation in sediments: effects of microphytobenthos and NO_3^-. *Aquat. Microbiol. Ecol.* **40:**67–76.

Risgaard-Petersen, N., A. M. Langezaal, S. Ingvardsen, M. C. Schmid, M. S. M. Jetten, H. J. M. Op den Camp, J. W. M. Derksen, E. Piña-Ocho, S. P. Eriksson, L. P. Nielsen, N. P. Revsbech, T. Cedhagen, and G. J. van der Zwaan. 2006. Evidence for complete denitrification in a benthic foraminifer. *Nature* **443:**93–96.

Rysgaard, S., R. N. Glud, N. Risgaard-Petersen, and T. Dalsgaard. 2004. Denitrification and anammox activity in Arctic marine sediments. *Limnol. Oceanogr.* **49:**1493–1502.

Sanders, I. A., and M. Trimmer. 2006. In-situ application of $^{15}NO_3^-$ isotope pairing technique to measure denitrification in sediments at the surface water-groundwater interface. *Limnol. Oceanog. Methods* **4:**142–152.

Sayama, M., N. Risgaard-Petersen, L. P. Nielsen, H. Fossing, and P. B. Christensen. 2005. Impact of bacterial NO_3^- transport on sediment biogeochemistry. *Appl. Environ. Microbiol.* **71:**7575–7577.

Schrum, H. N., A. J. Spivack, M. Kastner, and S. D'Holt. 2009. Sulfate-reducing ammonium oxidation: a thermodynamically feasible metabolic pathway in subseafloor sediment. *Geology* **37:**939–942.

Schubert, C. J., E. Durisch-Kaiser, B. Wehrli, B. Thamdrup, P. Lam, and M. M. M. Kuypers. 2006. Anaerobic ammonium oxidation in a tropical fresh water system (Lake Tanganyika). *Environ. Microbiol.* **10:**1857–1863.

Seitzinger, S. P. 1988. Denitrification in freshwater and coastal marine ecosystems: ecological and geochemical significance. *Limnol. Oceanogr.* **33:**702–724.

Seitzinger, S. P., and A. E. Giblin. 1996. Estimating denitrification in North Atlantic continental shelf sediments. *Biogeochemistry* **35:**235–260.

Smethie, W. M. 1987. Nutrient regeneration and denitrification in low oxygen fjords. *Deep Sea Res.* **34:**983–1006.

Steif, P., D. De Beer, and D. Neumann. 2002. Small-scale distribution of interstitial nitrite in freshwater sediment microcosms: the role of nitrate and oxygen availability, and sediment permeability. *Microbiol. Ecol.* **43:**367–378.

Steingruber, S. M., J. Freidrich, R. Gächter, and B. Wehrli. 2001. Measurements of denitrification in sediments with the ^{15}N isotope pairing technique. *Appl. Environ. Microbiol.* **67:**3771–3778.

Stramma, L., C. G. C. Johnson, J. Sprintall, and V. Mohrholz. 2008. Expanding oxygen-minimum zones in the tropical oceans. *Science* **320:**655–658.

Strous, M., J. G. Kuenen, and M. S. M. Jetten. 1999. Key physiology of anaerobic ammonium oxidation. *Appl. Environ. Microbiol.* **65:** 3248–3250.

Sørensen, J., L. K. Rasmussen, and I. Koike. 1987. Micromolar sulfide concentrations alleviate acetylene blockage of nitrous oxide reduction by denitrifying Pseudomonas fluorescens. *Can. J. Microbiol.* **33:**1001–1005.

Tal, Y., J. E. M. Watts, and H. J. Schreier. 2005. Anaerobic ammonia-oxidizing bacteria and related activity in Baltimore Inner Harbor Sediment. *Appl. Environ. Microbiol.* **71:**1816–1821.

Thamdrup, B., and T. Dalsgaard. 2002. Production of N_2 through anaerobic ammonium oxidation coupled to nitrate reduction in marine sediments. *Appl. Environ. Microbiol.* **68:**1312–1318.

Thamdrup, B., T. Dalsgaard, M. M. Jensen, O. Ulloa, L. Farías, and R. Escribano. 2006. Anaerobic ammonium oxidation in the oxygen-deficient waters off northern Chile. *Limnol. Oceanogr.* **51:**2145–2156.

Trimmer, M., and J. C. Nicholls. 2009. Production of nitrogen gas via anammox and denitrification in intact sediment cores along a continental shelf to slope transect in the North Atlantic. *Limnol. Oceanog.* **54:**577–589.

Trimmer, M., J. C. Nicholls, and B. Deflandre. 2003. Anaerobic ammonium oxidation measured in sediments along the Thames Estuary, United Kingdom. *Appl. Environ. Microbiol.* **69:**6447–6454.

Trimmer, M., J. C. Nicholls, N. Morley, C. A.

Davies, and J. Aldridge. 2005. Biphasic behavior of anammox regulated by nitrate and nitrite in an estuarine sediment. *Appl. Environ. Microbiol.* **71:**1923–1930.

Trimmer, M., N. Risgaard-Petersen, J. C. Nicholls, and P. Engström. 2006. Direct measurement of anaerobic ammonium oxidation (anammox) and denitrification in intact sediment cores. *Mar. Ecol. Prog. Ser.* **326:**37–47.

van de Graaf, A. A., A. Mulder, P. De Bruijn, M. S. M. Jetten, L. A. Robertson, and J. G. Kuenen. 1995. Anaerobic oxidation of ammonium is a biologically mediated process. *Appl. Environ. Microbiol.* **61:**1246–1251.

van Raaphorst, W., H. T. Kloosterhuis, E. M. Berghuis, A. J. M. Gieles, J. F. P. Malschaert, and G. J. van Noort. 1992. Nitrogen cycling in two types of sediments of the southern North Sea (Frisian Front, Broad Fourteens): field data and mesocosm results. *Neth. J. Sea. Res.* **28:**293–316.

Ward, B. B., C. B. Tuit, A. Jayakumar, J. J. Rich, J. Moffett, S. Wajih, and A. Naqvi. 2008. Organic carbon, and not copper, controls denitrification in oxygen minimum zones of the ocean. *Deep Sea Res., Part I* **55:**1672–1683.

Ward, B. B., A. H. Devol, J. J. Rich, B. X. Chang, S. E. Bulow, H. Naik, A. Pratihary, and A. Jayakumar. 2009. Denitrification as the dominant nitrogen loss process in the Arabian Sea. *Nature* **461:**78–82.

Wenjing, J., N. Tovell, S. Clegg, M. Trimmer, and J. A. Cole. 2008. A single channel for nitrate uptake, nitrite export and nitrite uptake by Escherichia coli NarU and a role for NirC in nitrite export and uptake. *Biochem. J.* **417:**295–304.

Wenzhöfer, F., and R. N. Glud. 2002. Benthic carbon mineralization in the Atlantic: a synthesis based on in situ data from the last decade. *Deep Sea Res. Part I* **49:**1255–1279.

Zhang, Y., X.-H. Ruan, H. J. M. Op den Camp, T. J. M. Smits, M. S. M. Jetten, and M. C. Schmid. 2007. Diversity and abundance of aerobic and anaerobic ammonium-oxidizing bacteria in freshwater sediments of the Xinyi River (China). *Environ. Microbiol.* **9:**2375–2382.

Zumft, W. G. 1997. Cell biology and molecular basis of denitrification. *Microbiol. Mol. Biol. Rev.* **61:**533–616.

APPLICATION OF THE ANAMMOX PROCESS

Wouter R. L. van der Star, Wiebe R. Abma, Boran Kartal, and Mark C. M. van Loosdrecht

10

INTRODUCTION

The removal of ammonium from wastewater generally takes a long way around: first, complete oxidation takes place to nitrate in nitrification, followed by a reduction to dinitrogen gas in denitrification (for design guidelines, see Metcalf & Eddy et al., 2003). A direct three-electron oxidation to dinitrogen gas was, although thermodynamically feasible, generally considered biochemically impossible:

$$2\ NH_4^+ + 1.5\ O_2 \rightarrow N_2 + 3\ H_2O + 2\ H^+$$
$$\Delta G_R^{0'} = -330\ kJ/mol\ NH_4^+$$

This ammonium oxidation has, indeed, never been observed to be the main catabolic reaction of a single microorganism, but this shortcut nitrogen removal reaction does constitute the overall reaction of the nitritation-anammox process. In this process, nitrification to nitrite (nitritation) performed by aerobic ammonia-oxidizing bacteria (AOB) (see Section II)

$$NH_4^+ + 1.5\ O_2 \rightarrow NO_2^- + H_2O + 2\ H^+$$
$$\Delta G_R^{0'} = -300\ kJ/mol\ NH_4^+$$

is combined with the anammox process (performed by anammox bacteria) (see Chapter 6).

$$NH_4^+ + NO_2^- \rightarrow N_2 + 2\ H_2O$$
$$\Delta G_R^{0'} = -360\ kJ/mol\ NH_4^+$$

The treatment of ammonium with the combination of these two processes is termed the nitritation-anammox process (as well as other names; see Table 1) and constitutes the only full-scale use and most researched configuration of the anammox process in wastewater treatment. In comparison with classical nitrogen removal (via nitrification-denitrification), 60% less aeration (energy) is needed (1.71 g of O_2/g of N instead of 2.86 g of O_2/g of N), and in the absence of denitrification, organic substrate is not necessary at all. In Fig. 1, the differences between the three possible nitrogen removal strategies are shown: nitrification-denitrification, nitritation-anammox process, and nitritation-denitrification (nitrification-denitrification via nitrite). Treatment of ammonium nitrate waste streams, however, is also possible by coupling partial denitrification (from nitrate to nitrite) to the anammox process. The status of this treatment will be treated below (see "One-reactor denitrification-anammox process," below). However, unless specifically mentioned, the processes in this chapter relate to the nitritation-anammox process only.

Wouter R. L. van der Star, Department of Geo-Engineering, Deltares, Delft 2600MH-177, The Netherlands. *Wiebe R. Abma*, Paques B.V., Balk 8560AB-52, The Netherlands. *Boran Kartal*, Department of Microbiology, Radboud University Nijmegen, Nijmegen 6500GL-9010, The Netherlands. *Mark C. M. van Loosdrecht*, Department of Biotechnology, Delft University of Technology, Delft 2628BC-67, The Netherlands.

Nitrification, Edited by Bess B. Ward, Daniel J. Arp, and Martin G. Klotz

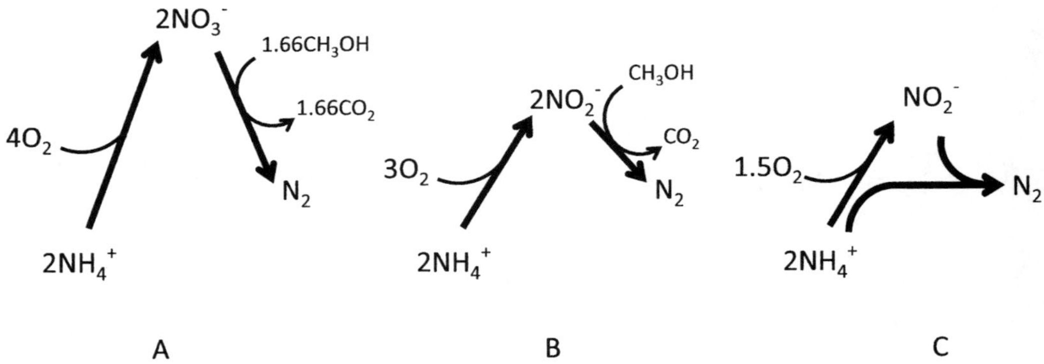

FIGURE 1 The costs of nitrogen removal mainly consist of aeration (electricity) and electron donor and are compared for classical nitrification-denitrification (A), nitritation-denitrification (B), and the nitritation-anammox process (C). Methanol is used as the carbon source for calculatory purposes; only catabolic processes are taken into account.

The nitritation-anammox process was developed at the end of the 20th century in several groups in Europe separately (for an overview, see Van der Star et al., 2007) and has been implemented at full scale in several locations in the past years. In this chapter, an overview of physiological parameters relevant for design and operation of the nitritation-anammox is discussed, based on which different treatment and start-up strategies as well as environmental impact can be evaluated. The chapter concludes with an overview on the state of the art of full-scale implementation of the nitritation-anammox process.

PHYSIOLOGY

Fifteen years of research on the anammox process has resulted in a sufficient overall view of physiological parameters needed for application of the anammox process in wastewater treatment. Surprisingly, however, and probably as a result of the experimental complications of slow growth and the absence of pure cultures, many of the parameters that were identified as critical are not known quantitatively or their mode of action is largely unknown. Although this paragraph is not a comprehensive discussion on physiology, we discuss here the important parameters of anammox bacteria for use in process and reactor design and the present knowledge about their value ranges. For a proper evaluation of the one-reactor nitritation-anammox process also, such parameters for AOB and nitrite-oxidizing bacteria (NOB) are of importance. Ammonia (AOB >10 to 150 mg of N/liter, NOB 0.1 to 1 mg of N/liter) and nitrous acid (NOB >0.2 to 2.8 mg of N/liter) are the main causes for toxicity during nitrification. As the values have been described comprehensively elsewhere (Anthonisen et al., 1976; Wiesmann, 1994), they are not explicitly discussed here.

Stoichiometry and Growth Rate

Although anammox bacteria are autotrophs with a similar energy gain from their catabolic reaction as AOB, their maximum specific growth rate is considerably lower: instead of 1 to 1.2 day^{-1} (typical for AOB [Anthonisen et al., 1976; Sin et al., 2008]), the growth rate of anammox bacteria is to 0.05 to 0.2 day^{-1} (Strous et al., 1999; Tsushima et al., 2007; Van der Star et al., 2008b). Due to the increased maintenance requirement associated with growth at this rate, the maximum biomass yield for both bacteria therefore varies considerably as well: 0.12 Cmol biomass/mol NH_4^+ for AOB, but 0.07 Cmol biomass/mol NH_4^+ for anammox bacteria, resulting in the following overall reaction (Strous et al., 1998):

$$1\ NH_4^+ + 1.32\ NO_2^- + 0.066\ HCO_3^- + 0.13\ H^+ \rightarrow 1.02\ N_2 + 0.066\ CH_{1.8}O_{0.5}N_{0.2} + 0.26\ NO_3^- + 2.03H_2O$$

When the anammox yield is compared to the yield of AOB at a growth rate of 0.05 day^{-1}, both yields are strongly comparable. The nitrate production in the anammox process stems from nitrite oxidation, which functions as the electron-donating redox reaction for the CO_2 fixation:

$$HCO_3^- + 2.3\ NO_2^- \rightarrow CH_{1.8}O_{0.5}N_{0.2} + 2.1\ NO_3^- + 0.2\ H^+ + 0.8\ H_2O$$

Nitrate production is an inevitable part of the overall anammox reaction and can be used to measure the growth of anammox bacteria. The stoichiometric conversion ratio of $NO_2^-:NH_4^+:NO_3^-$ of (circa) (1.1 to 1.3) : (1) : (−0.1 to −0.25) is another characteristic for the anammox process. A higher ratio between nitrite and ammonium and/or the reduced nitrate production generally is an indication of the (co)occurrence of denitrification.

Besides conversion of ammonium and nitrite, anammox bacteria are also capable of reducing nitrate to nitrite and nitrite to ammonium with fatty acids as electron donor. The nitrite or ammonium thus produced serves then as a substrate for the normal anammox catabolism (Kartal et al., 2007) but completely changes the stoichiometry described above. The overall catabolic reaction in this case is the same as in denitrification, but since anammox organisms have never been shown to use the fatty acids directly as C-source for growth and still use the energy-expensive CO_2 fixation for growth, the biomass yield (i.e., sludge production) is expected to be extremely low compared to "classical" denitrification. This low yield, in turn, leads to lower sludge production.

The decay rate (b_{AN}) of anammox bacteria in slow-growing organisms is not easy to assess because, like growth, decay also is slow. Recently, it was estimated at 0.0048 d^{-1} at 35°C (Scaglione et al., 2009) under anaerobic conditions, which is equivalent to an "anammox biomass half-life" of 145 days. The decay is thus about 10-fold lower than the maximum specific growth rate, which is in line with observations for faster-growing organisms.

Anammox bacteria have a very high affinity for their substrates nitrite and ammonium with half-saturation constants estimated at <100 μg of N/liter (Strous et al., 1999) and 3 to 50 μg of N/liter (Van der Star et al., 2008b) (evaluated for nitrite only), respectively. These values are lower than those commonly found for AOB, NOB, or denitrification on nitrite and are thus a competitive advantage.

Toxicity/Inhibition

Knowledge of adverse affects of compounds possibly present in anammox reactors can be important for feasibility studies, reactor design, but also for the development of the start-up strategy. Of the relevant factors that are known (nitrite, phosphate, sulfide), their mode of action (reversibility, effect of exposure time, combination with other substances, is the compound only toxic when cells are active?) and the concentrations at which it occurs is often unknown.

NITRITE INHIBITION/TOXICITY

The most striking inhibitor to the anammox process is its own substrate: nitrite. Unlike inhibition of AOB and NOB, there are indications that it is not the nitrous acid (HNO_2, the undissociated form that most easily is transported over the cell wall by passive transport), but the ion itself (NO_2^-), that is toxic to the organisms (Strous, 2000). The level at which toxicity occurs and its reversibility remains unclear and seems strongly dependent on exposure time. When only short-term effects on nitrite removal rate are evaluated, relatively high values are found (50% reduction only at 400 mg of N/liter [Strous et al., 1999], 630 mg of N/liter [Dapena-Mora et al., 2007], or 37% reduction at 430 mg of N/liter [Kimura et al., 2010], respectively). However, an immediate deviation occurs at these nitrite levels in the nitrite:ammonium ratio, possibly indicating ammonification. Toxicity evaluated during longer periods is observed already at much lower concentrations: prolonged experiments (40 h) showed this stoichiometry change started already above 70 mg of N/liter (Strous et al., 1999), probably indicating a negative response of the culture. In contrast to inhibition at these levels, operational conditions in a full-scale anammox reactor at 10 kg of N/m^3/day took place at typical concentrations of 40 to 80 mg of N/liter of nitrite (Van der Star et al., 2007), indicating that growth can take place at these values.

Much lower inhibition concentrations were reported in an intermittently aerated one-reactor nitritation-anammox process where irreversible toxicity occurred at 50 mg of N/liter and where values as low as 5 mg of N/liter (Wett et al., 2007) were reported to already have a detrimental effect on the process. The finding of these lower toxicity values predominantly in aerated systems suggests an effect of operation on loss of conversion capacity.

Also, the extent to which nitrite inhibition is reversible is subject to much discussion. Anammox on gel-carriers recovered fully in 3 days from a 7-day exposure to 700 mg of N/liter (resulting in a temporary reduction of conversion of 90%) (Kimura et al., 2010). In view of the low growth rate of anammox bacteria, such an increase in conversion can be only due to recovery rather than growth. However, high nitrite levels are often cited as the cause for failure of a reactor (e.g., Fux et al., 2004). This is often, however, a "chicken and egg" discussion. As any cause of reactor failure leads to a stop in the conversion of nitrite, the discovery of a nonactive reactor is generally associated with high nitrite, which in that case are the result rather than the cause of the failure.

SALT TOLERANCE AND ADAPTATION

Anammox bacteria can grow in freshwater and marine conditions. By gradually increasing the salt stress in an enrichment of freshwater bacterium *Kuenenia*, growth at concentrations of up to 30 g/liter was possible (Kartal et al., 2006; Liu et al., 2009). Kartal et al. (2006) also showed that enrichments adapted to growth at 30 g/liter showed conversions of ca. 50% of the normal conversion even at concentrations of 60 to 80 g/liter (well above seawater concentrations), whereas a nonadapted environment already experienced inhibition at 30 g/liter. Short-term manometric batch experiments with several different salts (NaCl, KCl) indicated similar levels of inhibition (Dapena-Mora et al., 2007).

TEMPERATURE

The anammox bacterium *Kuenenia* had its maximum activity between 30 and 35°C with an optimum at 43°C (Strous et al., 1999; Dosta et al., 2008) and an activation energy of 63 to 70 kJ/mol. For marine anammox bacteria, activation energies of 51 kJ/mol (Rysgaard et al., 2004) and 61 kJ/mol (Dalsgaard and Thamdrup, 2002) were determined. As anammox enrichment reactors both on full scale and on lab scale generally consist of *Kuenenia* or *Brocadia*, these higher values are probably the most reliable for use in engineering.

In applications, growth at 15 to 18°C (Dosta et al., 2008; Van de Vossenberg et al., 2008) has been achieved several times. In view of the extremely low temperatures at which anammox bacteria are found in the environment (anammox activity in arctic ice [Rysgaard and Glud, 2004]), the possibility of enriching anammox bacteria at lower temperatures seems likely as well. Such an enrichment, however, will only be feasible with extremely efficient biomass retention systems.

OXYGEN, SULFIDE, AND PHOSPHATE

Under truly anaerobic conditions (in the absence of nitrate/nitrite), even an extremely low rate of sulfate reduction on endogenous substrates is capable of producing sulfide at toxic levels. This toxicity seems not to be reversible. Fifty percent inhibition at 0.3 mM were reported in batch tests (Dapena-Mora et al., 2007). Inhibition data on phosphate are conflicting: phosphate was found (i) to be totally inhibiting at values of 5 mM for an anammox enrichment (Van de Graaf et al., 1997), but (ii) batch tests with 20 mM phosphate did not result in adverse effects for a *Kuenenia* enrichment (Egli et al., 2001), whereas (iii) at 21 mM a 50% activity reduction was recorded in *Kuenenia* batch tests (Dapena-Mora et al., 2007).

The (reversible) inhibition by oxygen is only noted in enrichments where insufficient nitrification or endogenous respiration occurs to successfully remove it, which is generally the case at very high enrichment levels. If this is the case, toxicity occurs already at the lowest measureable oxygen concentrations (0.5% of oxygen saturation [Strous et al., 1997; Van der Star et al., 2008b]). In systems with aerobic respiration present (e.g., AOB) in the granular sludge or biofilm systems, inhibition levels of oxygen are much higher.

ORGANIC COMPOUNDS

Anammox enrichments are strongly and irreversibly inhibited by methanol, which is already toxic at levels up to as low as 0.5 mM (Güven et al., 2005). As the toxic action only occurs in actively metabolizing enrichments, it has been suggested that another compound (possibly formaldehyde) is produced from methanol and thus the actual inhibitor (Isaka et al. 2008).

Short chain fatty acids (formate, acetate, propionate) can be metabolized by anammox bacteria and serve as a mode to produce ammonium and nitrite from nitrate (Kartal et al., 2007). Anammox bacteria can successfully compete with denitrifying microorganisms on these electron donors. However, since the use of these fatty acids could not significantly increase the growth rate of anammox bacteria, anammox activity will occur only in reactor systems with a sufficiently long solid retention time (SRT).

NITROGEN REMOVAL PROCESSES

Ammonium removal with the anammox process always consists of partial nitritation followed by the anammox process. Both processes can take place in one reactor or in two reactors placed in series. This paragraph first introduces the different processes for ammonium removal and their requirements separately, based on which the suitability of different reactor concepts is evaluated. The paragraph is concluded with a brief overview of the denitrification-anammox process.

One-Reactor Processes

When both the nitritation and anammox process take place in the same reactor, oxygen is both a substrate (for AOB) and a toxin (for anammox bacteria). As highly enriched anammox bacteria are (reversibly) inhibited even at very low oxygen levels, truly anoxic conditions should be present in the reactor, in addition to the aerobic conditions required for the growth of NOB. In addition, the SRT should be sufficiently high (several days) to allow growth of anammox bacteria.

REACTOR CONFIGURATIONS

To obtain both oxic and anoxic conditions in the same reactor, three different approaches can be distinguished:

1. Continuous operation, in which the oxygen levels are governed by gradients in biofilm systems (Hippen et al., 1997; Kuai and Verstraete, 1998; Sliekers et al., 2002). In such systems, oxygen is consumed in the outer layer of the biofilm and thus does not penetrate the biofilm completely. The anammox process can thus be performed in the anoxic inner layers making use of the produced nitrite that diffuses further into the biofilm. In an experimental stage, systems have also been evaluated with an inside-out configuration in which oxygen is supplied via hollow membranes (Gong et al., 2008; Syron and Casey, 2008).

2. Time-dependent aeration, in which oxygen levels vary in time (Third et al., 2005; Wett, 2006). In such a system, nitritation takes place during the aerated periods, and the anammox process during the nonaerated periods. However, in view of the low penetration depth of oxygen in biofilm systems, also during aerated periods, the anammox process will likely play a role according to approach 1.

3. Physical transportation of the biomass between the oxic and the anoxic zone, by either (i) alternating between full inundation and presence above the water level (Kuai and Verstraete, 1998) or (ii) transporting the biomass between aerated and nonaerated zones of a reactor (Beier et al., 2008).

OPERATION AT LOW OXYGEN LEVEL

Besides conditions that favor growth of AOB and anammox bacteria, successful operation of nitritation-anammox reactors requires also that NOB are outcompeted. Competition between these three groups can potentially occur on oxygen, ammonium, as well as nitrite (Fig. 2). Maintaining regions (or periods) within the reactor in which the oxygen and nitrite levels are low, while ammonium remains not limiting, would be the ideal competitive environment, as under those conditions the not-favored NOB have to compete for both their electron donor (nitrite) and electron acceptor (oxygen). Both experimentally (Third et al., 2001) and by mathematical modeling (Hao et al., 2005), it was shown that such a system is indeed possible at low bulk oxygen concentrations and that at lower ammonium concentrations (which is disadvantageous for both AOB and anammox bacteria), the system becomes instable.

The oxygen level in the reactor has a strong effect on the morphology of the biofilm (both for biofilms formed on carrier and granular/floccular systems). Operation at low-oxygen concentrations in granular/floccular systems (where no external biomass carrier is used) can lead to the formation of fluffy biofilms or flocs, as in such a configuration AOB have the most efficient access to oxygen (Nielsen et al., 2005). The result in reactors without a biomass carrier is poor settling, which is indeed a general characteristic of this type of reactors (Vlaeminck et al., 2008). On a full scale, this could be solved at the expense of installation of a cyclone capable of keeping the biomass in the reactor system. Furthermore, it is likely that operation at a sufficiently high shear stress (which is also favorable for more rounded and thus better settling biomass) will lead to better biomass retention.

OPERATION AT HIGH OXYGEN LEVEL

Operation at higher bulk oxygen levels is possible as well (Gaul et al., 2005; Abma et al., 2009). Due to the higher penetration of oxygen in the biofilm, the selective pressure toward growth in fluffy structures is therefore much lower. In such systems, however, the bulk liquid conditions (with significant nitrite and oxygen levels) favor growth of NOB in flocs or even in suspension that destabilizes the reactor (De Clippeleir et al., 2009). Due to the fluffy nature, however, the settling rates of this undesired sludge are low. Growth can thus be prevented by only keeping well-settling biomass in the reactor through an internal settler design. In addition to these settling-related measures, care should be taken to keep the hydraulic retention time (HRT) sufficiently high (>1 day, depending on conditions) to prevent the possibility for NOB to grow fully suspended. The latter point is of special concern during start-up strategies in which flow rates are low.

REACTOR CHOICE

Several reactor types can be used to performing the one-reactor nitritation-anammox process in several of its operating strategies. These include granular sludge systems (sequencing batch reactors [SBRs], air lifts, bubble columns) as well as systems with biomass on carrier (moving bed reactors or with carrier materials fixed in the reactor). The main common feature is the ability to achieve high SRTs and substantial mixing. The maximum attainable conversion rate (and thus the design volume) in nitritation-anammox reactors is determined by the limiting factor in the conversion of the substrates of anammox bacteria or AOB. The most likely limiting factors that can be encountered are as follows:

- Oxygen transfer: The transfer of oxygen from the gas phase to the liquid phase; typically dependent on aeration design, aeration flow rate, and reactor height.
- Oxygen penetration: The flux of oxygen into the biofilm. The penetration depth is

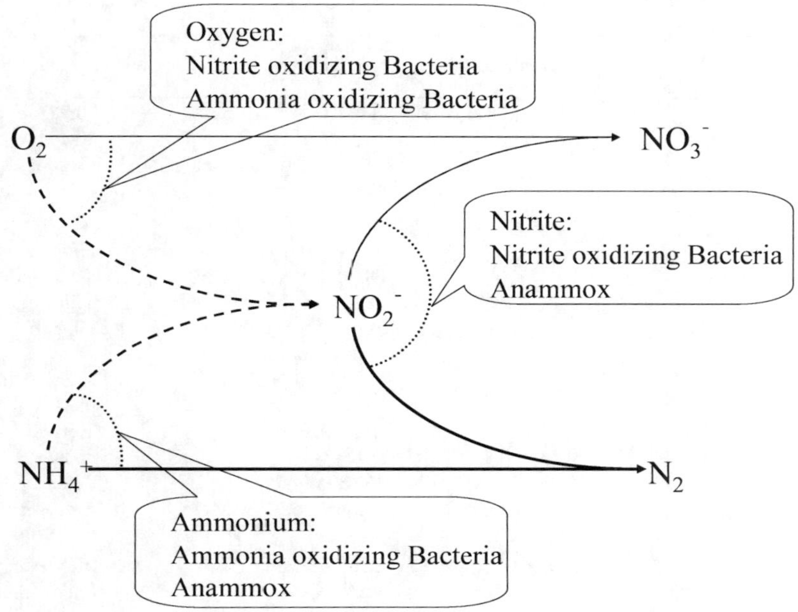

FIGURE 2 Competition between AOB (dashed lines), anammox bacteria (thick solid lines), and NOB (solid lines) for the substrates oxygen, ammonium/ammonia, and nitrite. For a one-reactor nitritation-anammox process, oxygen limitation under sufficient ammonium levels is the most favorable condition.

½th order proportional to the oxygen level according to the relationship below:

$$\Phi_{ox} = \sqrt{2q_{ox,AOB}C_{x,AOB}D_{ox}C_{ox,bulk}}$$

- (adapted from Arvin and Harremoës, 1990), where Φ_{ox} is oxygen flux, $q_{ox,AOB}$ is specific oxygen conversion rate of AOB, D_{ox} is the oxygen diffusion coefficient, and $C_{ox,bulk}$ is bulk oxygen concentration.
- Nitrite penetration: The flux of nitrite into the biofilm (or from the outsides of the biofilm in which it is produced to the inside).
- Biomass retention: The efficiency of biomass retention is lower at higher reactor loading rates due to increased shear within the reactor and/or turbulence in settlers.

In Table 3, these limiting factors are evaluated theoretically for several characteristic reactors at typical operating values (for details, see Van der Star et al., 2007). When biofilm areas are large enough, oxygen transfer turns out to be the main limiting factor, whereas in reactors with less biofilm surface, oxygen penetration becomes limiting. As the latter is strongly dependent on bulk oxygen level (see equation above), operation at higher bulk oxygen levels will, in many reactor configurations, enable higher volumetric conversion levels.

Two-Reactor Processes

In two-reactor configurations, the nitritation and the anammox process are taking place in an aerated and a nonaerated reactor, respectively, which have characteristics and requirements completely different from the one-reactor operation. Both reactors are treated separately below.

NITRITATION REACTOR

In the nitritation reactor, about 55% of the ammonium needs to be converted to nitrite to produce the desired reaction mixture for the anammox process. The challenge in this type

TABLE 1 Process options and names for nitrogen removal involving the anammox process (Van der Star et al., 2007)

Process name proposed in this chapter	No. of reactors	Source of nitrite	Alternative process names	First reference
Two-reactor nitritation-anammox process (Fux et al., 2001)	2	NH_4^+ nitritation	SHARON[a,b]-anammox	Van Dongen et al., 2001
			Two-stage OLAND	Wyffels et al., 2004
			Two-stage deammonification	Trela et al., 2004
One-reactor nitritation-anammox process	1		Aerobic deammonification	Hippen et al., 1997
			OLAND	Kuai and Verstraete, 1998
			CANON	Third et al., 2001
			Aerobic/anoxic deammonification	Hippen et al., 2001
			Deammonification	Seyfried et al., 2001
			SNAP[c]	Lieu et al., 2005
			DEMON[d]	Wett, 2006
			DIB[d]	Ladiges et al., 2006
			PANDA+ [e]	Beier et al., 2008
			One-step ANAMMOX	Abma et al., 2009
One-reactor denitrification-anammox process		NO_3^- denitrification	Anammox[f]	Mulder et al., 1995
			DEAMOX[g]	Kalyuzhnyi et al., 2006
			Denammox[h]	Pathak and Kazama, 2007

[a]SHARON, sustainable high rate ammonium removal over nitrite; OLAND, oxygen-limited autotrophic nitrification-denitrification; CANON, completely autotrophic nitrogen removal over nitrite; SNAP, single-stage nitrogen removal using the anammox process and partial nitritation; DIB, deammonification in interval-aerated biofilm systems; PANDA, partially augmented nitritation denitritation; DEAMOX, denitrifying ammonium oxidation; Denammox, denitrification-anammox process.

[b]The name only refers to nitritation where nitrite oxidation is avoided by choice of residence time and operation at elevated temperature. Sometimes the nitrification-denitrification over nitrite is addressed by with this term.

[c]Name only refers to the process on a biofilm surface layer.

[d]Name only refers to the process in a sequencing batch reactor (SBR) under pH control.

[e]Original name refers to nitritation-denitritation in a single sludge system with anoxic and oxic zones; "+" designates conversion to the nitritation-anammox process.

[f]System where anammox was found originally. Whole process was originally designated as "anammox."

[g]This name only refers to denitrification with sulfide as electron donor.

[h]This name only refers to denitrication with organic matter as electron donor.

of reactor is to (i) prevent in-growth of NOB, which leads to production of nitrate rather than nitrite, and (ii) ensure that only 55% of the ammonium is converted. Many routes for production of nitrite, rather than nitrate, are thinkable based on oxygen limitation or addition of specific inhibitors. Although resulting in nitrite as a final product in the short run (even during several months), these processes often fail to be stable during longer periods due to adaption. However, at particularly high nitrite loads (several hundreds of mg of N/liter) and sufficiently low pH, the toxicity of nitrous acid (HNO_2, >2.8 mg of N/liter) seems to be strong enough to prevent nitrite oxidation in such systems (Wyffels et al., 2003).

An alternative and nitrite-level-independent way of producing nitrite uses the difference in maximum specific growth rate between AOB and NOB (Hellinga et al., 1998). At temperatures above 25°C, this growth rate is higher for AOB than for NOB. By choosing a biomass retention time, which enables growth of AOB while preventing growth of NOB (typically 1 day), only AOB are enriched in this type of operation. In the calculation of this retention time in intermittently aerated reactors, only the time in which the reactor is actually aerated should be taken into account, as this is the only period that the AOB are growing.

How to ensure that only 55% of the ammonium is converted depends on the counter ion of the ammonium in the waste stream. If this is bicarbonate (as is the case in most waste streams), nitritation is pH limited, rather than ammonium limited, as only 50% of the produced protons can be balanced by stripping of CO_2:

$$NH_4HCO_3 + 1.5\ O_2 \rightarrow H_2CO_3\ (= H_2O + CO_2) + NH_4NO_2$$

Consequently, the reaction will stop automatically at 50 to 60% conversion, and an equilibrium pH will be reached (6.3 to 6.6) at which 50 to 60% conversion will occur (Van Dongen et al., 2001). It should be noted that all available alkalinity (1 mol HCO_3^- per mol of produced NH_4NO_2) is necessary to obtain the desired 1:1 ratio. Designing based on full conversion to nitrite of 50% of the stream, and bypassing the other 50% of the waste stream directly to the anammox reactor, also bypasses 50% of the alkalinity and therefore will result in a 50% lower nitrite load to the reactor.

ANAMMOX REACTOR

Critical for the stable operation of anammox reactors are stable and sufficiently long biomass retention and good mixing. The latter is mainly important at the location where the influent enters the reactor, as the concentrations of nitrite in the influent are generally high enough to be toxic and thus mixing should be sufficiently fast that no zones in the reactor exist in which high-nitrite concentrations occur. Premixing with available return streams (for example, with the liquid in a "biomass-free" downcomer of a gas lift reactor) might therefore be advantageous. Biomass retention is important in view of the slow growth of anammox bacteria. It should be noted that the required sludge age in typical cases is not extreme, due to the higher temperatures at which most reactors are operated (coming from warm sludge digesters, or being small thus enabling to be heated economically, partly also because of anammox-reaction-associated heat production). In principle, an SRT of 30 days is sufficient.

Especially in discontinuously operated systems with low biomass densities, flotation is a possible concern (Dapena-Mora et al., 2004), as are sudden (i.e., changing within several days rather than weeks/months) changes in or a too high exposure to shear stress (Arrojo et al., 2008).

The requirements for anammox reactors (good mixing and high biomass retention) are on an industrial scale met in granular sludge reactors where a selective pressure is used for the formation of granules and which consist of separate mixing and settler zones. In the latter zone, a stable upflow velocity (>1 m/h) strongly selects for stable granules. The advantage of this type of reactor is the very high volumetric loading rates possible due to existence of specific biomass areas of up to 3,000 m^2/m^3 and, when internal circulation reactors are used, the use of the produced gas as a free/cheap mixing agent (Van der Star et al., 2007). Other reactors that have been used for the anammox process (like low-weight biofilm carrier materials with a 1-cm diameter [Cema et al., 2006], or biofilm sheets [Fujii et al., 2002]) have a specific biofilm surface area that is much lower, and thus much lower volumetric conversions are reached. An overview of typical maximum specific volumetric conversion rates for different reactor types, and the limitation that these maxima are based on, are shown in Table 3.

One-Reactor Denitrification-Anammox Process

In the denitrification-anammox process, nitrite does not stem from partial oxidation of ammonium but from partial denitrification of nitrate. The electron donor for the denitrification is either sulfide or organic matter. The nitrite produced by partial denitrification of nitrate is, subsequently, combined with ammonium to form dinitrogen gas in the anammox process. The overall catabolic reaction (here with methanol as electron donor) is shown below:

$$NH_4^+ + NO_3^- + 0.33\ CH_3OH \rightarrow N_2 + 0.33\ HCO_3^- + 0.33\ H^+ + 2.33\ H_2O\ \ \Delta G_R^{0\prime} = -560\ kJ/mol\ NH_4^+$$

This is the process that took place in a pilot scale wastewater treatment plant of a baker's yeast factory in Delft, The Netherlands; and was the original process for which the acronym anammox was used (Mulder et al., 1995). When sulfide is the electron donor for denitrification of nitrate to nitrite, care should be taken to keep sulfide levels low enough not to be toxic. Operation under simultaneous conversion of sulfide, however, has been shown experimentally (Kalyuzhnyi et al., 2006), and this conversion is similar to the combination of sulfide oxidation and (also sulfide inhibited) nitrification (Heijnen et al., 1993).

Should both nitrate- and ammonium-containing wastestreams need to be treated on a single site, then the denitrification-anammox process might be an interesting option. Depending on the carbon source that is available, also part of the nitrate reduction to nitrite can be performed by anammox bacteria (Kartal et al., 2007). The reactor requirements for the denitrification-anammox process are likely to be very similar to the anammox reactor in the two-reactor nitritation-anammox process (aee above) as mixing, long sludge age and anoxic conditions are critical as well. During operation, care should be taken that sulfate reduction cannot occur by keeping (low) levels of nitrate present as the produced of sulfide is toxic for anammox bacteria.

Besides evaluations for treatment of ammonium nitrate wastestreams, the process has also been tested in ammonium-fed bioreactors to remove the excess 15% nitrate produced by anammox bacteria when only the anammox process takes place (Pathak et al., 2007). Such a removal step might be desirable if the anammox reactor produces effluent that is not discharged or returned to another part of the treatment system (as is the case in reject water systems). However, the lower nitrate levels again constitute an increased risk of sulfide toxicity due to the presence of anaerobic conditions. The denitrification-anammox process can also be used as an alternative ammonium-removal pathway consisting of full nitrification of 50% of a digester effluent to nitrate, followed by a sulfide-driven denitrification-anammox process of the produced nitrate with the ammonium in the remaining and untreated 50% of the waste stream (Kalyuzhnyi et al., 2006). In such a system, however, extra alkalinity should be added during the nitrification to achieve the required full conversion.

MEASUREMENTS AND CONTROL

Following and controlling of anammox-based processes is, in principle, not different from monitoring normal activated sludge operation. However, the high concentrations in the influent and within the reactors and (resulting) substrate toxicity require some adjustments to "standard" methods and modes of control.

Measuring Physical Parameters

Concentrations of ammonium, nitrite, and nitrate can be routinely assessed using standardized laboratory procedures. However, in case of (expected) problems, it is vital that also fast indicative tests for nitrite, ammonium, and nitrate are available. It is also possible to measure these parameters on line with ion-selective electrodes or periodic (one to four times per hour) automated spectrophotometric tests. Although the quality of available ion-selective electrodes for ammonium, nitrite, and nitrate has strongly improved over the past years, they are notorious for their high drift and short lifetime (order of month), requiring at least weekly recalibration. However, under dedicated care, reasonable results have been obtained also for these sensors (Joss et al., 2009).

Conductivity sensors have proven to be reliable indicators of conversion, as both nitritation and the anammox process reduce the conductivity level significantly in high-strength wastewaters (Cema et al., 2006). Being an indirect measure for conversion, the interpretation should be periodically adjusted to the wastewater characteristics.

Microbial Population

Manometric batch tests have proven to be good and fast indicators of activity of the anammox process (Dapena-Mora et al., 2007). When performed in the presence and absence of oxygen, they can also be used reliably in one-reactor processes for following the start-up, troubleshooting, or testing process changes. However, these gas measurements should be verified by evaluation of pH change and nitrite:ammonium:nitrate conversion ratios that are determined from measurements of conditions at the beginning and at the end of the experiment. Furthermore, structural differences between the batch tests and reactor performance can exist, and therefore the tests are mainly suitable for determining the changes in performance and evaluation of adverse effects of new waste streams.

In addition, fluorescence in situ hybridization (FISH) is a very powerful tool to visualize the distribution of AOB, anammox bacteria, and (undesired) NOB in flocs and granules (see Chapter 8 for an overview). This can be particularly useful to evaluate whether in-growth of NOB is taking place or whether they are proliferating in the bulk of the reactor. However, during start-up of reactors without a suitable inoculum, overall conversion measurements, batch measurements, and FISH are not successful (as concentrations of anammox bacteria are too low), and the only information on growth can be obtained from quantitative PCR (Q-PCR). This method was successfully used during startup on lab scale (Tsushima et al., 2007) and on full scale (Van der Star et al., 2007).

Control in Two-Reactor Processes

In the nitritation reactor treating ammonium bicarbonate, a 1:1 mixture of ammonium and nitrite is automatically established, as the reactor is basically limited by alkalinity. As long as aeration is sufficient to (i) transfer oxygen from the gas phase to the liquid phase (limiting at reactor heights smaller than 4 m) or (ii) transfer CO_2 from the liquid phase to the gas phase (limiting in higher reactors), a 1:1 ratio will develop automatically. This does not mean that it is not desirable to control the dissolved oxygen (DO) concentration, as it is a useful means to limit aeration costs. Furthermore, preliminary findings indicate that, above a certain threshold, aeration is directly proportional to nitric oxide (NO) emission (Kampschreur et al., 2008) and thus NO emission can be reduced by reducing the DO level.

The outcompetition of NOB can be controlled by the sludge age. Note that this can be done by removing sludge, but also by reducing the period during which the reactor is aerated. As growth can take place only during periods of aeration, only the aerated period "counts" as SRT. However, the use of anoxic periods strongly enhances the greenhouse gas emissions (Kampschreur et al., 2007, 2008).

A well-running anammox reactor is nitrite limited and thus need no specific nitrite control. This goes particularly for reject water treatment, where (unlike the main stream) variation is limited and dampened out by the long residence time (typically 30 days) in the sludge digester. The introduction of a stop (or strong reduction) of the influent flow rate to the anammox reactor during a sudden increase in nitrite level (20 to 100 mg of N/liter, dependent on the biomass characteristics), however, is a useful security measure. Instead of nitrite measurement, electrical conductivity also can be used as an effective (and more reliable) indicator of conversion. Too low ammonium levels in the reactor could be an indication of an incorrect nitrite:ammonium influent ratio and require mitigative action (e.g., slight pH adjustment in the nitritation reactor or bypassing a small part of the ammonium-containing influent directly to the anammox reactor).

Control in One-Reactor Processes

In one-reactor processes, oxygen transfer and oxygen level are key to keeping favorable conditions for both anammox bacteria and AOB. In a one-reactor system, aeration flow is controlled by DO: the DO level is increased when nitritation is hampering and decreased during problems with the anammox process. Controlling aeration rate with the ammonium concentration (and conductivity as an analogue), nitrite concentration (Kampschreur et al., 2009), or the ratio between nitrite and ammonium are alternative possibilities. Although aeration control with the intermediate nitrite as an actuator might seem peculiar, it has a similar effect on AOB and anammox bacteria. Higher aeration leads to more nitritation and less anammox activity due to inhibition (or an increased penetration of oxygen), and the opposite is achieved with lower aeration. In a system in which nitrite level limits the anammox process, increased aeration will lead to a higher AOB activity, and thus also to a higher rate of anaerobic ammonium oxidation.

In systems with discontinuous aeration, both aeration rate and aeration time can be used for control. Although the aeration flow can be controlled by all parameters discussed in the continuous system, in all used systems, aeration flow was governed by DO. For the length of the aerated phase, pH was used on full scale, which is possible because of the strongly acidifying action of AOB (Wett, 2006). An alternative option that was successfully used on full scale is conductivity (Joss et al., 2009).

START-UP TIMES AND STRATEGIES

Two-Reactor Processes

The startup of the nitritation process is relatively fast and not complicated. In SHARON (sustainable high rate ammonium removal over nitrite)-type nitritation (in which the residence time is controlled, and in which no biomass retention takes place), 2 weeks after inoculation with nitrifying activated sludge, no significant NOB activity could be found anymore (J .W. Mulder, personal communication) in a full-scale reject water treatment. The conversion from a full-scale nitritation-denitritation reactor (in which denitrification took place in anoxic periods with methanol addition) to a process suitable for nitritation only, was achieved within 4 days by simply stopping methanol addition (Van der Star et al., 2007).

For supplying a suitable feed to the anammox reactor, the nitrite loads during startup can initially not be too high. Only a fraction of the design load can therefore be supplied to the reactor. It is, furthermore, advantageous to dilute this influent with effluent or return streams prior to introduction into the reactor. After considerable conversion is established, the dilution can be reduced, provided that the influent is sufficiently mixed with the reactor liquid. However, (i) at too long hydraulic retention times, there is a risk of sulfate reduction when all nitrite/nitrate is removed by endogenous respiration and (ii) the designed hydraulic system (e.g., airlift recirculation) might work less at low conversion (equal to gas production) rates leading to reduced mixing.

One-Reactor Processes

Contrary to two-reactor processes, it is possible for one-reactor processes to receive (but not treat) a considerable part of the full design load immediately. In general, care should be taken to have during the startup (i) enough residual ammonium and (ii) no toxic nitrite levels, and flotation should be prevented. Several strategies can be followed.

- During startup, first focus can be to achieve nitrification followed by a gradual reduction in aeration to produce favorable conditions for anammox bacteria and to develop competition on nitrite as well as oxygen to washout NOB.
- An alternative strategy is to first operate the reactor as nitritation-denitritation (with methanol/acetate addition for denitrification and use of an SRT of 1 to 4 days to wash out nitrite oxidizers or reducing oxygen levels), followed by a gradual reduction of the added carbon source. This strategy was successfully used for startup in Niederglatt, Switzerland (Joss et al., 2009), and conversions to nitritation-anammox reactors in Breitenberg and Gelsenkirchen, Germany (Walter et al., 2007).
- When considerable biomass is available from previous inoculations, a start-up strategy can be chosen with reactor control that is close to the eventually chosen control system. For example, by operating under pH control (Wett, 2006) or conductivity/ammonium control (Joss et al., 2009), the length of the aerated phase can be controlled in the same way as in fully operational reactors.

However, the load to the reactor should be maintained manually during the first days to avoid ammonia toxicity.

Care should further be taken that NOB, while being outcompeted in granules, are not remaining in the reactor by nitrite oxidation within flocs (with easier access oxygen) (Gaul et al., 2005; De Clippeleir et al., 2009).

Direct Scale-Up or Stepwise Scale-Up

Startup of anammox reactors requires patience, or experience and sufficient amounts of inoculum. When inoculum is not available in sufficient amounts (for the first reactor of a supplier, or for the first reactor in a certain region), full enrichment from a nonenriched inoculum (e.g., nitrifying activated sludge) will take several months under favorable conditions. Unfortunately, evaluation whether such favorable conditions are actually existing in the reactor is complicated because in the first weeks/months, conversions are too low (and variability in influent flow rate or concentration is too high) to be detected from evaluation of ammonium, nitrite, and nitrate levels by nitrogen mass balancing. If teething problems occur in the reactor, they are therefore hard to detect. In the startup of a full-scale anammox reactor in Rotterdam (part of a two-reactor process), Q-PCR was used successfully to identify whether favorable conditions were occurring and to follow the growth of the community even when this was not detectable from changes in nitrite, nitrate, and ammonium levels (Van der Star et al., 2007). However, the information from Q-PCR is only useful if results are available on a regular (weekly, preferably more often) basis and thus can directly be compared with changes in reactor operation.

With the strategy outlined above, it was possible to directly scale up the two-reactor anammox process from a 10-liter scale to a 70-m^3 scale. The alternative approach is classical scale-up by starting up the process in a reactor that is typically 10 times larger than the previous one. The inoculum, but also the experience that is gained in the previous step, can thus be used for startup at a larger scale. With amounts of inoculum that constitute already 10% of the required nitrogen conversion, design or operational errors are detected faster, as the activity of the inoculum is well measurable and will disappear fast (usually within 1 week) when exposed to unfavorable conditions. Figure 3 shows a typical learning curve for a reactor operator. Whereas it took more than 3 years to achieve full conversion in the first reactor, startup of the third one took only a few months. Surprisingly, whether a direct scale-up procedure (The Netherlands [Van der Star et al., 2007]; Germany [Walter et al., 2007]), or step-wise scale-up (Switzerland [Wett, 2006; Joss et al., 2009]), is used, the typical time for reaching full scale is 2 to 3 years. With the proper conditions and experience, startup (also without anammox seed sludge) should be possible within 6 months as has been shown on lab scale.

TREATABLE WASTEWATERS

The application of the anammox process is mainly associated with wastewaters with a high nitrogen content (>200 mg of N/liter) and a low C/N ratio. Although this requirement holds true for the influent of nitritation-anammox systems, it does not necessarily hold true for the type of wastewater for which the treatment can be used. Since no organic material is a requirement in the nitritation-anammox process, wastewaters with a higher C/N ratio can be treated advantageously by combining the nitritation-anammox process with anaerobic treatment. In contrast to nitrification-denitrification where an external electron donor is required, the organic carbon in a wastestream can now be completely converted into biogas. By maximizing biogas production and minimizing power consumption, this combination provides maximal sustainability.

Reject Water of Municipal Wastewater Treatment Plants (WWTP)

The anammox process has been successfully tested and used mainly in the treatment of reject water (sludge digestor liquids). This

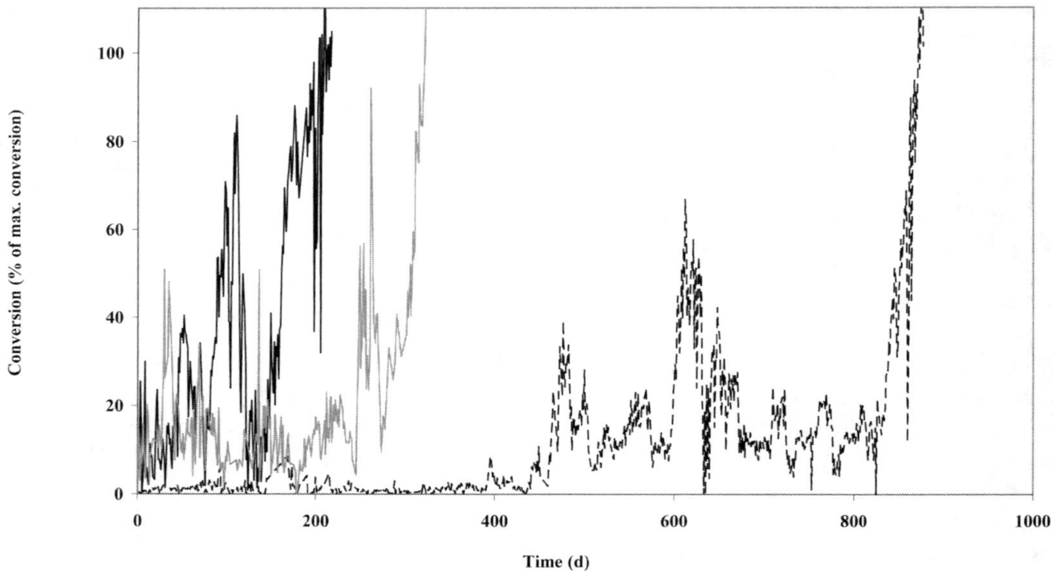

FIGURE 3 Startup of three full-scale nitritation-anammox reactors (two-reactor processes [in Rotterdam and Lichtenvoorde], one-reactor process [in Olburgen]) started up consecutively by the same company. The start-up time decreased in later startups, as a result of availability of biomass for inoculation and application of knowledge.

water stems from anaerobic digestion of the produced sludge of municipal wastewater treatment plants and typical contains 500 to 1,500 mg of N/liter of ammonium (as ammonium bicarbonate). Although reject water flows are low (typically 0.5 to 2% of the main influent flow rate), they contain 5 to 20% of the available nitrogen load. Reject water treatment can therefore significantly contribute to the overall performance with relatively small reactors (for an overview of available technologies, see Van Loosdrecht, 2008). Since no electron donor is available anymore in the wastewater, nitritation-anammox is particularly suitable for this type of wastewater. On real wastewater, both the one-reactor (Hippen et al., 1997) and two-reactor (Van Dongen et al., 2001) process was tested and is now applied on full scale (see Table 2). The nitritation-anammox process can be integrated in the design of a completely new plant (focusing on optimization of sludge production and thus biogas production) or used for the revamp of existing plants that need to handle a larger influent or need to meet stricter effluent regulations.

Digested Food Industry Effluents and Manures

Wastewaters from food industries are generally protein rich (and thus nitrogen rich) and can be efficiently digested. The remaining wastestream can be used in the nitritation-anammox process. A digested potato wastewater is currently treated on full scale in a one-reactor nitritation-anammox process (Abma et al., 2009), as is the digested wastewater of a tannery (in a two-reactor configuration [Abma et al., 2007]). Digested wastewater from a baker's yeast factory and monosodium glutamate (MSG) was also treated.

MSG production produces a wastewater with high nitrogen levels. In view of the low COD/N ratio after digestion, the nitritation-anammox process is a promising alternative. Indeed conversion of wastewater containing 500 mg of N/liter was shown experimentally (Chen et al., 2007) and has been implemented at full scale (Table 2).

Tests on wastewaters (ca. 1 g of N/liter) from treatment of digested seafood and fish canning effluents (two-reactor processes [Dapena-Mora et al., 2006; Lamsam et al., 2008]) were possible without dilution (and thus at the full-salt load of ca. 1 g of NaCl/liter).

Although several reports exist on treatment of manure with the anammox process, only for digested pig manure have full experimental nitritation-anammox systems been operated. In these treatments, anaerobically digested manure (containing 1 g of N/liter of ammonium) was successfully treated in a one-reactor configuration (Hwang et al., 2005) and a two-reactor configuration (Qiao et al., 2009).

Source-Separated Treatment

Several AOB are, in contrast to all known anammox bacteria, capable of urea hydrolysis (see Chapter 2). Although it seems unlikely that energy is generated from this hydrolysis, it produces the ammonium that is required in nitritation. Sliekers et al. (2004) have showed that a urea-based nitritation-anammox process (in which AOB perform urea hydrolysis as well as nitritation) was possible in a one-reactor configuration. A urine analogue (in which urea was replaced by ammonium/ammonia) was, furthermore, used in a two-reactor and a one-reactor nitritation-anammox system (Wilsenach et al., 2006), thus indicating the feasibility of treatment of source-separated urine.

Black water (toilet water) is another source-separated high-N-containing waste stream. After anaerobic digestion of black water, a 1 to 1.5 g of N/liter of ammonium waste stream remains, which was recently shown to be converted by the one-reactor (Vlaeminck et al., 2009) or two-reactor (De Graaff et al., 2011) nitritation-anammox processeses.

Landfill Leachates

In landfill leachates, ammonium concentrations up to 5 g of N/liter occur (although typical concentrations are around 1 g of N/liter), which are presently treated by nitrification-denitrification or nitrification only. Both in well-studied lab-scale systems (two-reactor system [Ruscalleda et al., 2008]) and in converted full-scale nitrification-denitrification reactors (Walter et al. 2007), stable conversions could be reached. The feasibility of the nitritation-anammox process for this type of wastewater already had been shown much earlier, with the discovery of the nitritation-anammox process in reactors that were actually designed for nitrification-denitrification (Table 2).

DESCRIPTIVE TERMINOLOGY

Although the nature of a process does not change by the name that might be given to it, a stable, consistent, and widely accepted terminology is of importance to focus discussions and facilitate the understanding of the field. However, the independent finding of anammox-based processes in different geographical locations, and the hypothesis that AOB are responsible for both the nitritation and the anammox process, has resulted in a multitude of names (often based on anammox, deammonification, and <u>o</u>xygen-<u>l</u>imited <u>a</u>utotrophic <u>n</u>itrification-<u>d</u>enitrification [OLAND]). The number of names has recently been expanded by use of specific names for specific reactor combinations or brands.

This situation can be strongly improved by introducing a descriptive terminology based on the processes taking place and whether this takes place in one or two reactors (Van der Star et al., 2007), using the following:

- Anammox process for the anoxic combination of ammonium and nitrite to form dinitrogen gas.
- One-reactor nitritation-anammox process as the occurrence of the nitrite production and the anammox process in one reactor.
- Two-reactor nitritation-anammox process for the partial oxidation of ammonium to nitrite in an aerated reactor, followed by an anoxic reactor, where only the anammox process takes place.
- One-reactor denitrification-anammox process for the anoxic processes of denitrification from nitrate to nitrite, combined with the anammox process. This was the original process configuration in which the anammox process was discovered (Mulder et al., 1995).

TABLE 2 Overview of full-scale anammox reactors (>50 m^3) using the one-reactor or the two-reactor anammox process (Van der Star et al., 2007)

Process	Location	Reactor type	Wastewater	Volume (m^3)	Area-specific conversion (g of N/m^2/day)	Conversion (kg of N/day)	Maximum volumetric conversion (kg of N/m^3/day)	Limitation	First year of full operation[a]	Organism	Source
Two reactors	Rotterdam NL[b]	Granular sludge	Reject water	70	ND	700	10	Feed (NO_2^-)	2006	*Brocadia*	Van der Star et al., 2007
	Lichtenvoorde NL	Granular sludge	Tannery	100	ND	250	2.5	Feed (NO_2^-)	2006	*Kuenenia*	Van der Star et al.. 2007
	Mie prefecture JP	Granular sludge	Semiconductor	58	ND	220	4	Feed (NH_4^+)	2006	ND	Tokutomi et al., 2007
	Hattingen DE	Moving bed	Reject water	67	5	67	1	Na[c]	(2003)[c]	ND	Thöle et al., 2005
One reactor	China		Yeast production plant	500	ND	1,500	2.5		2010	ND	W. R. Abma, unpublished data
	Zürich CH	SBR	Reject water	2x1400	ND	1400	0.5	Feed	2007	ND	Joss et al., 2009
	Olburgen NL	Air lift	Potato processing plant	600	ND	1200	2.0	Feed	2006	*Brocadia*[d]	Abma et al., 2010
	Himmerflärden SE	Moving bed	Reject water	1400	1.9	420	0.3	Not reported	2007	ND	Ling, 2009
	Heidelberg DE	SBR	Reject water		ND	300		Not reported	2008	ND	Scheider[e]
	St.Gallen CH	SBR	Reject water	2x300	ND	240	0.4	Not reported	2008	ND	Joss et al., 2009
	Gelsenkirchen DE	MBR	Landfill leachate	660	ND	264	0.4	Feed	2005	*Brocadia*/ *Kuenenia*	Denecke et al., 2007
	Strass AT	SBR	Reject water	500	ND	350	0.7	Feed	2006	*Brocadia*[e]	Wett, 2007
	Glarnerland CH	SBR	Reject water	400	ND	240	0.6	Feed	2006	ND	Wett, 2007
	Hattingen DE	Moving bed	Reject water	102	6	102	1	Not reported	2003	ND	Thöle et al., 2005
	Niederglatt CH	SBR	Reject water	160	ND	48	0.3	Feed	2008	ND	Joss et al., 2009

Breitenberg DE	MBR	Landfill leachate			30			2007	ND	M. Denecke, personal communication
Mechernich DE	RDC	Landfill leachate	8010	2	5126	0.64	Aeration	Unknown[g]	ND	Hippen, 1997
Pitsea GB	RDC	Reject water/ leachate	240	7	408	1.7	Feed	Unknown[g]	*Scalindua*	Schmid et al., 2003
Kölliken CH	RDC	Landfill leachate	33	2	13	0.4	Feed	Unknown[g]	ND	Siegrist et al., 1998
Tongliao CN		MSG wastewater	(6700)	ND	(11,000)	(1.65)[h]		(2010)	ND	W. R. Abma, unpublished data
Apeldoorn NL	SBR	Reject water	(2500)	ND	(1,600)	(0.67)		(2010)		ONRI/Technisch Weekblad[h]

[a]In parentheses if not in operation anymore, not in operation yet, or operation not yet published.
[b]NL, The Netherlands; JP, Japan; DE, Germany; CH, Switzerland; SE, Sweden; AT, Austria; GB, United Kingdom; CN, China; ND, not determined.
[c]Reactor was converted to one reactor operation.
[d]Confirmed by FISH; updated (conversion rates) in 2009.
[e]http://www.schneider-electric.de/documents/events-fairs/thementage/downloadbereiche/wasser-seligenstadt-2010/06_Seligenstadt10_Wett_Energie_dt_29-06-2010.pdf
[f]From Innerebner et al. 2007.
[g]System developed automatically. Present status is unknown.
[h]In startup; last reported conversion 1 kg of N/m^3/day.
[i]http://www.onri.nl/projecten/demon (ONRI, December 1, 2009); http://www.technischweekblad.nl/energiezuinige-stikstofverwijdering-in-apeldoorn.39444.lynkx (Technisch Weekblad; December 1, 2009).

- Anammox reactor for the (nonaerated) reactor in which only the anammox process takes place.

As an aid to interpret existing literature, an overview of names used in literature and their suggested generic name is shown in Table 1.

ENVIRONMENTAL IMPACT

The environmental impact of nitrogen removal from wastewater lies mainly in the CO_2 emission associated with aeration energy, CO_2 production during denitrification, and N_2O emission associated with nitrification and denitrification. In the nitritation-anammox process, the reduction of aeration is roughly 60%, and the only direct CO_2 emission stems from the bicarbonate already present as counter ion in the wastewater.

Besides these direct effects, the use of the nitritation-anammox process can be used for more sustainable process design: introducing nitritation-anammox on a wastewater treatment plant after sludge digestion enables a higher loading of the primary settler and thus more energy generation from biogas production. This is a relatively simple revamp of an existing wastewater treatment plant that results in a reduction of net energy consumption of the full plant of 50% (from 2 to 1 W/p [Siegrist et al., 2008]) and thus also from energy-derived CO_2 emissions (from 9 to 5 kg of CO_2/p/year). Should the nitritation-anammox process also become feasible for treatment of low concentrations of ammonium, no COD is necessary anymore for nitrogen removal, and energy production can be increased to match the plants demands completely, thus resulting in complete "energy autarky" (Van Loosdrecht et al., 2001; Siegrist et al., 2008; Kartal et al., 2010).

Besides the impact of CO_2 emission, emission of NO and N_2O also is of importance in the overall evaluation of environmental impact. Both are intermediates in denitrification and are released during nitrification. They contribute directly (N_2O, 296 times the strength of CO_2) or indirectly (NO) to the greenhouse effect. The emission of N_2O from wastewater treatments plants is presently not well known, and only few studies relate N_2O emission to nitrogen load or conversion giving rise to very large variations (e.g., Wicht and Beier, 1995). The NO and N_2O emission of well-running lab-scale anammox reactors is virtually zero (<0.01 % of converted ammonium) (e.g., Strous et al., 1998; Van der Star et al., 2008b). However, in full-scale anammox reactors, emissions of 0.1 to 0.5% were found (Kampschreur et al., 2008; T. Lotti, personal communication), which were suggested to originate from inflowing AOB from the previous nitritation reactor. In the only evaluated nitritation reactor, emissions were estimated to be 2.3% but could be mainly attributed to N_2O production during nonaerated periods. Operation of a continuously aerated nitritation reactor is therefore expected to have much lower emissions. In the mentioned reactor, it was also noted that NO emission (above a certain DO threshold) seemed to be proportional to aeration rate. A too high aeration rate therefore also contributes to higher NO emissions (Kampschreur et al., 2008).

In one-reactor nitritation-anammox systems, emissions of 1.2% (continuously aerated) (Kampschreur et al., 2009), 1.3% (Weissenbacher et al., 2010), 0.6% (intermittent aerated), and 0.4% (continuously aerated) (Joss et al. 2009) were found for full-scale reactors. The large variation in these values (which are slightly higher than in normal wastewater treatment plants) and their variation in time and mode of operation show a lack of understanding/predictability of N_2O emission factors and the importance to evaluate the emission characteristics in lab-scale systems.

The sustainability of the anammox process is emphasized by a reduction of energy-based CO_2 production of 4.2 kg/person/year in an optimized municipal WWTP. Assuming an extra N_2O-derived emission of 0.25% of the nitrogen load (not unrealistic, as also during nitrification-denitrification N_2O is emitted, and taking into account that the high values in the full-scale nitritation reactor were mainly due to the presence of nonaerated periods), the

N_2O derived increase in greenhouse gas emissions is 1.3 kg of CO_2 equivalents/person/year, and the overall system thus still holds a net reduction of greenhouse gases of 2.9 kg of CO_2 equivalents/person/year.

MATHEMATICAL MODELING

Modeling has constituted an important activity in the evaluation and design of anammox reactors directly from the beginning. The extremely slow growth rate did not only make experiments lengthy, but in the laboratory practice, stable operation for several solid retention times has turned out to be a challenge. Biofilm models, however, showed, at the same time it was actually shown in the laboratory, that the one-reactor nitritation-anammox process was a possibility at low DO (Hao et al., 2002; Picioreanu et al., 2004) and under certain conditions in the presence of organic matter (Hao and van Loosdrecht, 2004). It should be noted, however, that until now the absence of reliable parameters of substrate affinity, decay, growth rate, and nitrite toxicity has seriously hampered the predictability of the earlier models.

Modeling of nitrogen conversions in the waste streams that are typical for the anammox process, however, cannot be performed with the models typical for wastewater treatment (like ASM 1,2,3). The high conversion rate of ammonium makes its acidifying effect significant. pH effects due to acidifying nitrification and CO_2 stripping therefore have to be taken into account in both one- and two-stage nitritation-anammox reactors. Besides the absence of pH/chemical speciation in present wastewater models, nitrite also is often not a model component. Inclusion of nitrite requires not only splitting the two microbial processes of nitrification in two model steps, but also splitting denitrification in (at least) two steps. The introduction of the affinity for nitrite in denitrification is problematic, as it should (at least partially) be regarded an intracellular intermediate. A detailed discussion of different modeling concepts for nitrite in recent wastewater treatment models including nitritation and anammox is provided by Sin et al. (2008).

Optimization of a sludge treatment can never be viewed upon on its own: only a detailed analysis of the sludge treatment in combination with the main line of the wastewater treatment plant can result in identification of the most economic operation. To avoid introducing complete (and unnecessary) introduction of pH and nitrite in existing models for the main line of wastewater treatment plants, Volcke et al. (2006) have devised a series of conversion matrices, so that the reject water models and normal models can be combined, thus finally enabling calculation of their interaction and process (design) optimization.

STATE OF THE ART: FULL-SCALE REACTORS

The occurrence of the anammox process in wastewater treatment is not always the result of deliberate action, but the process could develop spontaneously in wastewater treatment systems—with essentially underdesigned aeration—in Pitsea, United Kingdom; Mechernich, Germany; and Kölliken, Switzerland. Although it is likely that these spontaneous one-reactor nitritation-anammox systems could be maintained under dedicated care, it is well thinkable that the anammox activity has changed in time due to changes in operation or reactor loading. In Pitsea, the anammox activity had disappeared virtually completely 2 years after the initial measurements (Schmid et al., 2003). Regardless of the stability of "accidental" nitritation-anammox processes, the amount of spontaneous nitritation processes in the world is probably significant (especially in nitrified landfill leachates).

In the past years, lab-scale results with anammox-based processes constituted a solid basis for the start-up dedicated reactors. Startup of reactors in Hattingen, Germany (tested as a two-reactor system, later converted and fully operational as a one-reactor system), Rotterdam (two-reactor process), and Strass (one-reactor process with intermitted pH controlled aeration) were critical for further expansion in the past 2 years (Fig. 4). Both the availability of inoculum as the increase in experience has resulted in the

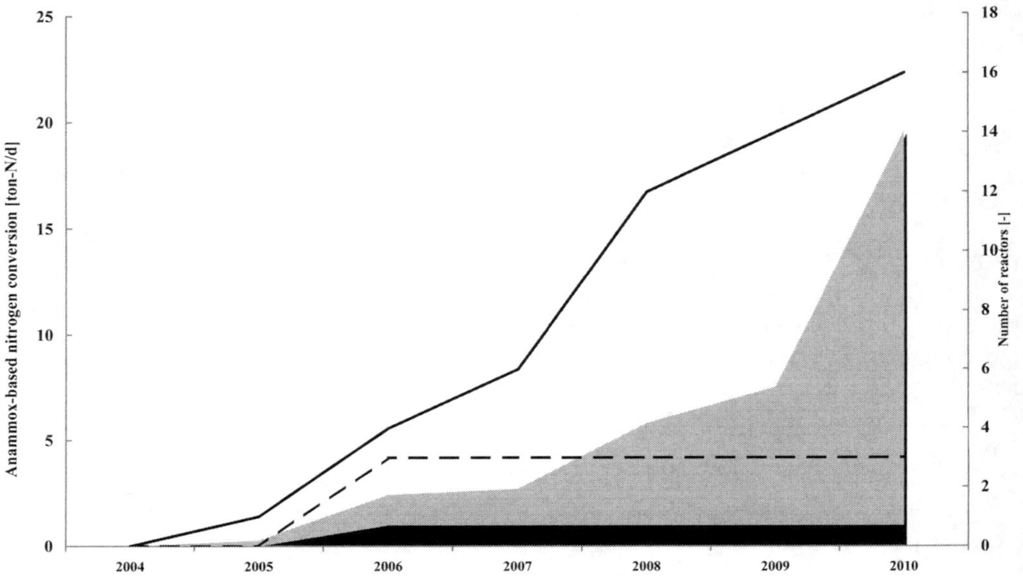

FIGURE 4 Number of fully operational one-reactor (solid line) and two-reactor (dashed line) anammox processes and amount of nitrogen removed (tons of N/day; black, one-reactor process; gray, two-reactor process).

construction of several installations in the past years. With the construction of four large reactors in 2010, the available nitritation-anammox capacity has doubled, and reactors treating up to 10,000 kg of N/day have been constructed. An overview of full-scale nitritation-anammox installations is shown in Table 2.

Characteristics of Full-Scale and Lab-Scale Evaluations

The stoichiometric ratio of the conversion in full-scale anammox reactors is not significantly different from lab-scale reactors as can be seen from the conversions in a full-scale reactor (Van der Star et al., 2007), which were calculated (from a 193 day average) as being 1.31 ± 0.032:1:−0.25 ± 0.006. Also for other parameters, no principle deviations between lab scale and full scale have occurred. The exception may be the N_2O emission in anammox reactors, which is much higher on full scale than on lab scale. The most plausible explanations for these differences are nitrite level and the type of feed: synthetic for lab-scale tests, but AOB-containing partially nitritated wastewater at full scale.

Foaming is a phenomenon that is particularly hard to assess on lab scale and should therefore mainly be evaluated directly on full scale. Although foaming was reported from one-reactor intermittently aerated nitritation-anammox systems (Wett et al., 2007), a significant reduction in foaming was reported when an existing nitritation-denitrification process (with acetic acid addition for denitrification) was converted to operation as a one-reactor nitritation-anammox process (Rekers et al., 2008).

Reactor and Process Choice

Only three of the reactors presently in use are two-reactor processes. Although initially expected to be easier to operate and (thus) cheaper in use for larger plants, the robustness of one-reactor systems and more straightforward startup has led to use of mainly that type of reactor in the past 2 years, and it is expected that this trend will continue.

A multitude of reactor types has, meanwhile, been used on full scale. Of these, granular sludge processes (whether in SBR or in

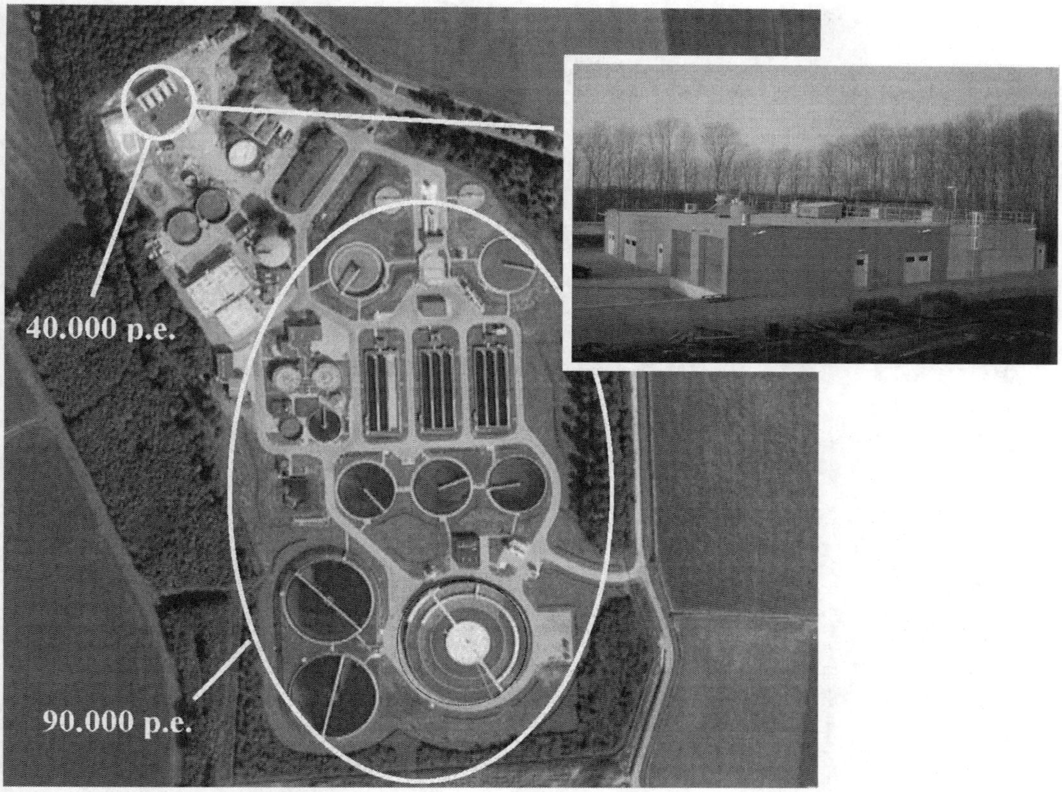

FIGURE 5 Reduction of space requirement by separate treatment of industrial wastewater and reject water (top circle) with a capacity of 40,000 p.e. compared to the sewage treatment plant (bottom circle) with a capacity of 90,000 p.e.; picture of the separate treatment in frame.

airlift-like configurations) have constituted the majority. These reactor types also seem to comply best with the main requirements: good mixing and availability of a high biofilm surface.

The highest volumetric conversion rates were (expectedly) achieved with granular sludge reactors, both in one-reactor (Olburgen, The Netherlands; 2.0 kg of N/m^3/day) and two-reactor (Rotterdam, The Netherlands; 10 kg/m^3/day) systems. The indicated small size of reject water treatment is emphasized with an aereal picture of Olburgen. The combination of sludge digestion, phosphate removal, and nititation-anammox process handles more than 30% of the wastewater on the site, with a much smaller footprint (Fig. 5).

An evaluation of possibly limiting factors in different (nitritation-)anammox reactors showed that even the highest volumetric conversions reached now can be increased in this type of reactor (Table 3), which promises even more efficient reactors in the future.

OUTLOOK

In the past years, information on physiology of anammox bacteria and lab-scale evaluations of possible treatment concepts have resulted in the successful industrial application of the anammox process on more than 10 locations, treating mainly reject water. Because of the cost reductions that were achieved, the stability of the installations, and the ease of their control, in combination with more stringent nitrogen

TABLE 3 Estimated volumetric conversion limitations and fluxes (shown in brackets) in different types of reactors for the anammox process and the one-reactor nitritation-anammox process[a]

Process	Reactor type	Particle diam (m)	Surface area (m^2/m^3)	Maximum volumetric conversion rate (kg of N/m^3/day)[f] (conversion flux [g of N/m^2/day])			
				Limiting process			
				Nitrite penetration[b]	Oxygen penetration[b,d]	Oxygen transfer[c]	Hydrodynamics
Anammox process	Granular sludge	0.001	3,000	**90 (30)**	-	-	**12**
	Biofilm moving bed	0.01	250	**7 (30)**	-	-	ND[e]
	Biofilm packed bed		250	**7 (30)**	-	-	ND
	Biofilm sheets reactor		250	**7 (30)**	-	-	ND
One-reactor nitritation-anammox process	Airlift/bubble column	0.001	3,000	89 (30)	15 (5)	**8**	ND
	Rotating disk contactor		250	7 (30)	**2.5 (10)**	ND	ND
	Moving bed	0.01	250	7 (30)	**1.2 (5)**	8	ND
	Sequencing batch reactor	0.001	3,000	89 (30)	15 (5)	**8**	ND

[a]The strongest limitation for each reactor is shown in boldface (adapted from Van der Star et al. [2008a]; see notes there for motivations of the chosen numbers).
[b]Penetration refers to penetration into the biofilm.
[c]Oxygen transfer is the transfer of oxygen from the gas phase to the liquid phase.
[d]Bulk oxygen concentration is 1 mg/liter. For the rotating disk contactor an average concentration of 4 mg/liter was assumed.
[e]ND, not determined.
[f]–, not applicable.

removal requirements being implemented all over the western world, the nitritation-anammox process is likely to be implemented on a much larger scale in the next years. Besides the application in reject water and treatment of landfill leachates, more suitable wastewaters are likely to be used.

The main research questions for application are the range of application of the nitritation-anammox process and the effect of several toxic or inhibiting compounds. Although identified, their mode of action is often mostly unknown, which hampers the predictability of reactor behavior (especially as a result of hiccups) and also renders present models much less predictive than they could be.

Finally, the recently elaborated possibility of removal of nitrogen at very low ammonium concentrations would be a true game-changer: should this become feasible, then the anammox process can be applied in the main line of municipal wastewater treatment plants and treatment of organic matter in WWTP can be completely focused on digestion. Such treatment is, however, still largely uncharted territory.

ACKNOWLEDGMENTS

We thank Rolf Poldermans for assistance in evaluation of start-up times and Siegfried Vlaeminck for confirming the presence of "Brocadia" in the Rotterdam full-scale anammox reactor by FISH. Tables 1 to 3 were reprinted from Van der Star et al. (*Water Res.* **41:**4149, 2007) with permission from Elsevier; Fig. 5 was reprinted from Abma et al. (*Water Sci. Technol.* **61:**1715–1722, 2010) with permission from IWA.

REFERENCES

Abma, W., C. E. Schultz, J. W. Mulder, M. C. M. van Loosdrecht, W. R. L. van der Star, M. Strous, and T. Tokutomi. 2007. The advance of Anammox. *Water21* **36:**36–37.

Abma, W. R., W. Driessen, R. Haarhuis, and M. C. M. van Loosdrecht. 2010. Upgrading of sewage treatment plant by sustainable and cost-effective separate treatment of industrial wastewater. *Water Sci. Technol.* **61:**1715–1722.

Anthonisen, A. C., R. C. Loehr, T. B. S. Prakasam, and E. G. Srinath. 1976. Inhibition of nitrification by ammonia and nitrous acid. *J. Water Pollut. Control Fed.* **48:**835–852.

Arrojo, B., M. Figueroa, A. Mosquera-Corral, J. L. Campos, and R. Mendez. 2008. Influence of gas flow-induced shear stress on the operation of the Anammox process in a SBR. *Chemosphere* **72:**1687–1693.

Arvin, E., and P. Harremoës. 1990. Concepts and models for biofilm reactor performance. *Water Sci. Technol.* **22**(1–2)**:**171–192.

Beier, M., M. Sander, and K.-H. Rosenwinkel. 2008. Kombination anaerober Vorbehandlung mit dem Verfahren der Deammonifikation zur energieeffizienten Behandlung organisch hoch belasteter Industrieabwässer. [Combination of anaerobic pretreatment and deammonification for efficient treatment of high loaded organic industrial wastewater]. *GWF, Wasser/Abwasser* **149:**80–87.

Cema, G., B. Szatkowska, E. Plaza, J. Trela, and J. Surmacz-Gorska. 2006. Nitrogen removal rates at a technical-scale pilot plant with the one-stage partial nitritation/Anammox process. *Water Sci. Technol.* **54:**209–217.

Chen, X., P. Zheng, R. Jin, B. Hu, S. Zhou, and G. Ding. 2007. Biological nitrogen removal from monosodium glutamate-containing industrial wastewater with the Anaerobic Ammonium Oxidation (ANAMMOX) process. *Huanjing Kexue Xuebao* **27:**747–752.

Dalsgaard, T., and B. Thamdrup. 2002. Factors controlling anaerobic ammonium oxidation with nitrite in marine sediments. *Appl. Environ. Microbiol.* **68:**3802–3808.

Dapena-Mora, A., J. L. Campos, A. Mosquera-Corral, M. S. M. Jetten, and R. Méndez. 2004. Stability of the ANAMMOX process in a gas-lift reactor and a SBR. *J. Biotechnol.* **110:**159–170.

Dapena-Mora, A., J. L. Campos, A. Mosquera-Corral, and R. Méndez. 2006. Anammox process for nitrogen removal from anaerobically digested fish canning effluents. *Water Sci. Technol.* **53:**265–274.

Dapena-Mora, A., I. Fernández, J. L. Campos, A. Mosquera-Corral, R. Méndez, and M. S. M. Jetten. 2007. Evaluation of activity and inhibition effects on Anammox process by batch tests based on the nitrogen gas production. *Enzyme Microb. Technol.* **40:**859–865.

De Clippeleir, H., S. E. Vlaeminck, M. Carballa, and W. Verstraete. 2009. A low volumetric exchange ratio allows high autotrophic nitrogen removal in a sequencing batch reactor. *Bioresour. Technol.* **100:**5010–5015.

de Graaff, M. S., H. Temmink, G. Zeeman, M.C. M. van Loosdrecht, and C. J. M. Buisman. 2011. Autotrophic nitrogen removal from black water: calcium addition as a requirement for settleability. *Water Res.* **45:**63–74.

Denecke, M., V. Rekers, and U. Walter. 2007. Einsparpotentiale bei der biologischen Reinigung von Deponiesickerwasser. [Cost savings potentials in the biological treatment of landfill leachates.] *Muell Abfall* **39:**4–7.

Dosta, J., I. Fernandez, J. R. Vazquez-Padin, A. Mosquera-Corral, J. L. Campos, J. Mata-Alvarez, and R. Mendez. 2008. Short- and long-term effects of temperature on the Anammox process. *J. Hazard. Mater.* **154:**688–693.

Egli, K., U. Fanger, P. J. J. Alvarez, H. Siegrist, J. R. Van der Meer, and A. J. B. Zehnder. 2001. Enrichment and characterization of an anammox bacterium from a rotating biological contactor treating ammonium-rich leachate. *Arch. Microbiol.* **175:**198–207.

Fujii, T., H. Sugino, J. D. Rouse, and K. Furukawa. 2002. Characterization of the microbial community in an anaerobic ammonium-oxidizing biofilm cultured on a nonwoven biomass carrier. *J. Biosci. Bioeng.* **94:**412–418.

Fux, C., M. Bohler, P. Huber, and H. Siegrist. 2001. Stickstoffelimination durch anaerobe Ammoniumoxidation (Anammox) [Nitrogen elimination during anaerobic ammonium oxidation (Anammox).] *Stuttgarter Berichte zur Siedlungswasserwirtschaft* **166:**35–49.

Fux, C., V. Marchesi, I. Brunner, and H. Siegrist. 2004. Anaerobic ammonium oxidation of ammonium-rich waste streams in fixed-bed reactors. *Water Sci. Technol.* **49:**77–82.

Gaul, T., S. Maerker, and S. Kunst. 2005. Start-up of moving bed biofilm reactors for deammonification: the role of hydraulic retention time, alkalinity and oxygen supply. *Water Sci. Technol.* **52:**127–133.

Gong, Z., S. Liu, F. Yang, H. Bao, and K. Furukawa. 2008. Characterization of functional microbial community in a membrane-aerated biofilm reactor operated for completely autotrophic nitrogen removal. *Bioresour. Technol.* **99:**2749–2756.

Güven, D., A. Dapena-Mora, B. Kartal, M. C. Schmid, B. Maas, K. Van de Pas-Schoonen, S. Sozen, R. Méndez, H. J. M. Op den Camp, M. S. M. Jetten, M. Strous, and I. Schmidt. 2005. Propionate oxidation by and methanol inhibition of anaerobic ammonium-oxidizing bacteria. *Appl. Environ. Microbiol.* **71:**1066–1071.

Hao, X., J. J. Heijnen, and M. C. M. Van Loosdrecht. 2002. Sensitivity analysis of a biofilm model describing a one-stage completely autotrophic nitrogen removal (CANON) process. *Biotechnol. Bioeng.* **77:**266–277.

Hao, X. D., and M. C. M. van Loosdrecht. 2004. Model-based evaluation of COD influence on a partial nitrification-Anammox biofilm (CANON) process. *Water Sci. Technol.* **49:**83–90.

Hao, X. D., X. O. Cao, C. Picioreanu, and M. C. M. van Loosdrecht. 2005. Model-based evaluation of oxygen consumption in a partial nitrification-Anammox biofilm process. *Water Sci. Technol.* **52:**155–160.

Heijnen, J. J., M. C. M. van Loosdrecht, R. Mulder, R. Weltevrede, and A. Mulder. 1993. Development and scale-up of an aerobic biofilm air-lift suspension reactor. *Water Sci. Technol.* **27:**253–261.

Hellinga, C., A. A. J. C. Schellen, J. W. Mulder, M. C. M. Van Loosdrecht, and J. J. Heijnen. 1998. The SHARON process: an innovative method for nitrogen removal from ammonium-rich waste water. *Water Sci. Technol.* **37:**135–142.

Hippen, A., K.-H. Rosenwinkel, G. Baumgarten, and C. F. Seyfried. 1997. Aerobic de-ammonification: a new experience in the treatment of wastewaters. *Water Sci. Technol.* **35:**111–120.

Hippen, A., C. Helmer, S. Kunst, K. H. Rosenwinkel, and C. F. Seyfried. 2001. Six years' practical experience with aerobic/anoxic deammonification in biofilm systems. *Water Sci. Technol.* **44:**39–48.

Hwang, I. S., K. S. Min, E. Choi, and Z. Yun. 2005. Nitrogen removal from piggery waste using the combined SHARON and ANAMMOX process. *Water Sci. Technol.* **52:**487–494.

Isaka, K., Y. Suwa, Y. Kimura, T. Yamagishi, T. Sumino, and S. Tsuneda. 2008. Anaerobic ammonium oxidation (anammox) irreversibly inhibited by methanol. *Appl. Microbiol. Biotechnol.* **81:**379–385.

Joss, A., D. Salzgeber, J. Eugster, R. Konig, K. Rottermann, S. Burger, P. Fabijan, S. Leumann, J. Mohn, and H. Siegrist. 2009. Full-scale nitrogen removal from digester liquid with partial nitritation and anammox in one SBR. *Environ. Sci. Technol.* **43:**5301–5306.

Kalyuzhnyi, S., M. Gladchenko, A. Mulder, and B. Versprille. 2006. DEAMOX - New biological nitrogen removal process based on anaerobic ammonia oxidation coupled to sulphide-driven conversion of nitrate into nitrite. *Water Res.* **40:**3637–3645.

Kampschreur, M. J., C. Picioreanu, N. Tan, R. Kleerebezem, M. S. M. Jetten, and M. C. M. Van Loosdrecht. 2007. Unraveling the source of nitric oxide emission during nitrification. *Water Environ. Res.* **79:**2499–2509.

Kampschreur, M. J., W. R. L. Van der Star, H. A. Wielders, J. W. Mulder, M. S. M. Jetten, and M. C. M. Van Loosdrecht. 2008. Dynamics of nitric oxide and nitrous oxide emission during full-scale reject water treatment. *Water Res.* **42:**812–826.

Kampschreur, M. J., R. Poldermans, R. Kleerebezem, W. R. L. van der Star, R. Haarhuis, W. R. Abma, M. S. M. Jetten, and M. C. M. van Loosdrecht. 2009. Emission of nitrous oxide and nitric oxide from a full-scale single-stage nitritation-anammox reactor. *Water Sci. Technol.* **60:**3211–3217.

Kartal, B., M. Koleva, R. Arsov, W. van der Star, M. S. M. Jetten, and M. Strous. 2006. Adaptation of a freshwater anammox population to high salinity wastewater. *J. Biotechnol.* **126:**546–553.

Kartal, B., J. G. Kuenen, and M.C. M. van Loosdrecht. 2010. Sewage treatment with Anammox. *Science* **328:**702–703.

Kartal, B., M. M. M. Kuypers, G. Lavik, J. Schalk, H. J. M. Op den Camp, M. S. M. Jetten, and M. Strous. 2007. Anammox bacteria disguised as denitrifiers: nitrate reduction to dinitrogen gas via nitrite and ammonium. *Environ. Microbiol.* **9:**635–642.

Kimura, Y., K. Isaka, F. Kazama, and T. Sumino. 2010. Effects of nitrite inhibition on anaerobic ammonium oxidation. *Appl. Microbiol. Biotechnol.* **86:**359–365.

Kuai, L., and W. Verstraete. 1998. Ammonium removal by the oxygen-limited autotrophic nitrification-denitrification system. *Appl. Environ. Microbiol.* **64:**4500–4506.

Ladiges, G., R. D. Thierbach, M. Beier, and Focken. 2006. Versuche zur zweistufigen Deammonifikation im Hamburger Klärwerksverbund. (Attempts at two-stage deammonification in the wastewater treament union of Hamburg.) Presented at the 6. Aachener Tagung mit Informationsforum: Sticksoffrückbelastung Stand der Technik 2006, Aachen, Germany.

Lamsam, A., S. Laohaprapanon, and A. P. Annachhatre. 2008. Combined activated sludge with partial nitrification (AS/PN) and anammox processes for treatment of seafood processing wastewater. *J. Environ. Sci. Health, Part A* **43:**1198–1208.

Lieu, P. K., R. Hatozaki, H. Homan, and K. Furukawa. 2005. Single-stage nitrogen removal using Anammox and partial nitritation (SNAP) for treatment of synthetic landfill leachate. *Jpn. J. Water Treat. Biol.* **41:**103.

Ling, D. 2009. Experience from commissioning of full-scale DeAmmon™ plant at Himmerfjärden (Sweden), p. 403–410. *In* Lemtech Konsulting (ed.), Second IWA specialized conference: Nutrient Management in Wastewater Treatment Systems, Krakow, Poland.

Liu, C., T. Yamamoto, T. Nishiyama, T. Fujii, and K. Furukawa. 2009. Effect of salt concentration in anammox treatment using non woven biomass carrier. *J. Biosci. Bioeng.* **107:**519–523.

Metcalf & Eddy, Inc., G. Tchobanoglous, F. R. Burton, and H. D. Stensel. 2003. Wastewater Engineering: Treatment and Reuse. McGraw-Hill, New York, NY.

Mulder, A., A. A. Van de Graaf, L. A. Robertson, and J. G. Kuenen. 1995. Anaerobic ammonium oxidation discovered in a denitrifying fluidized bed reactor. *FEMS Microbiol. Ecol.* **16:**177–184.

Nielsen, M., A. Bollmann, O. Sliekers, M. Jetten, M. Schmid, M. Strous, I. Schmidt, L. H. Larsen, L. P. Nielsen, and N. P. Revsbech. 2005. Kinetics, diffusional limitation and microscale distribution of chemistry and organisms in a CANON reactor. *FEMS Microbiol. Ecol.* **51:**247–256.

Pathak, B. K., and F. Kazama. 2007. Influence of temperature and organic carbon on denammox process, p. 402–413. Proceedings CD of Nutrient Removal 2007, Baltimore, MD.

Pathak, B. K., F. Kazama, Y. Saiki, and T. Sumino. 2007. Presence and activity of anammox and denitrification process in low ammonium-fed bioreactors. *Bioresour. Technol.* **98:**2201–2206.

Picioreanu, C., J.-U. Kreft, and M. C. M. Van Loosdrecht. 2004. Particle-based multidimensional multispecies biofilm model. *Appl. Environ. Microbiol.* **70:**3024–3040.

Qiao, S., T. Yamamoto, M. Misaka, K. Isaka, T. Sumino, Z. Bhatti, and K. Furukawa. 2009. High-rate nitrogen removal from livestock manure digester liquor by combined partial nitritation-anammox process. *Biodegradation* **21:**11–20.

Rekers, V., M. Denecke, and U. Walter. 2008. Betriebserfahrungen mit der anaeroben Deammonifikation von Deponiesickerwasser. DepoTech, Leoben, Austria.

Ruscalleda, M., H. Lopez, R. Ganigue, S. Puig, M. D. Balaguer, and J. Colprim. 2008. Heterotrophic denitrification on granular anammox SBR treating urban landfill leachate. *Water Sci. Technol.* **58:**1749–1755.

Rysgaard, S., and R. N. Glud. 2004. Anaerobic N_2 production in Arctic sea ice. *Limnol. Oceanogr.* **49:**86–94.

Rysgaard, S., R. N. Glud, N. Risgaard-Petersen, and T. Dalsgaard. 2004. Denitrification and anammox activity in Arctic marine sediments. *Limnol. Oceanogr.* **49:**1493–1502.

Scaglione, D., S. Caffaz, E. Bettazzi, and C. Lubello. 2009. Experimental determination of Anammox decay coefficient. *J. Chem. Technol. Biotechnol.* **84:**1250–1254.

Schmid, M., K. Walsh, R. Webb, W. Irene, C. Rijpstra, K. Van de Pas-Schoonen, M. J. Verbruggen, T. Hill, B. Moffett, J. Fuerst, S. Schouten, J. S. Sinninghe Damsté, J. Harris, P. Shaw, M. Jetten, and M. Strous. 2003. Candidatus "Scalindua brodae," sp. nov., Candidatus "Scalindua wagneri," sp. nov., two new species of anaerobic ammonium oxidizing bacteria. *Syst. Appl. Microbiol.* **26:**529–538.

Seyfried, C. F., A. Hippen, C. Helmer, S. Kunst,

and K. H. Rosenwinkel. 2001. One-stage deammonification: nitrogen elimination at low costs. *Water Sci. Technol,* **1:**71–80.

Siegrist, H., S. Reithaar, G. Koch, and P. Lais. 1998. Nitrogen loss in a nitrifying rotating contactor treating ammonium-rich wastewater without organic carbon. *Water Sci. Technol.* **38:**241–248.

Siegrist, H., D. Salzgeber, J. Eugster, and A. Joss. 2008. Anammox brings WWTP closer to energy autarky due to increased biogas production and reduced aeration energy for N-removal. *Water Sci. Technol.* **57:**383–388.

Sin, G., D. Kaelin, M. J. Kampschreur, I. Takacs, B. Wett, K. V. Gernaey, L. Rieger, H. Siegrist, and M. C. M. van Loosdrecht. 2008. Modelling nitrite in wastewater treatment systems: a discussion of different modelling concepts. *Water Sci. Technol.* **58:**1155–1171.

Sliekers, A. O., N. Derwort, J. L. C. Gomez, M. Strous, J. G. Kuenen, and M. S. M. Jetten. 2002. Completely autotrophic nitrogen removal over nitrite in one single reactor. *Water Res.* **36:**2475–2482.

Sliekers, A. O., S. Haaijer, M. Schmid, H. Harhangi, K. Verwegen, J. G. Kuenen, and M. S. M. Jetten. 2004. Nitrification and Anammox with urea as the energy source. *Syst. Appl. Microbiol.* **27:**271–278.

Strous, M. 2000. Microbiology of anaerobic ammonium oxidation. PhD thesis. Delft University of Technology, Delft, The Netherlands.

Strous, M., E. Van Gerven, J. G. Kuenen, and M. Jetten. 1997. Effects of aerobic and microaerobic conditions on anaerobic ammonium-oxidizing (Anammox) sludge. *Appl. Environ. Microbiol.* **63:**2446–2448.

Strous, M., J. J. Heijnen, J. G. Kuenen, and M. S. M. Jetten. 1998. The sequencing batch reactor as a powerful tool for the study of slowly growing anaerobic ammonium-oxidizing microorganisms. *Appl. Microbiol. Biotechnol.* **50:**589–596.

Strous, M., J. G. Kuenen, and M. S. M. Jetten. 1999. Key physiology of anaerobic ammonium oxidation. *Appl. Environ. Microbiol.* **65:**3248–3250.

Syron, E., and E. Casey. 2008. Membrane-aerated biofilms for high rate biotreatment: performance appraisal, engineering principles, scale-up, and development requirements. *Environ. Sci. Technol.* **42:**1833–1844.

Third, K. A., A. O. Sliekers, J. G. Kuenen, and M. S. M. Jetten. 2001. The CANON system (completely autotrophic nitrogen-removal over nitrite) under ammonium limitation: interaction and competition between three groups of bacteria. *Syst. Appl. Microbiol.* **24:**588–596.

Third, K. A., J. Paxman, M. Schmid, M. Strous, M. S. M. Jetten, and R. Cord-Ruwisch. 2005. Enrichment of Anammox from activated sludge and its application in the CANON process. *Microb. Ecol.* **49:**236–244.

Thöle, D., A. Cornelius, and K.-H. Rosenwinkel. 2005. Großtechnische Erfahrungen zur Deammonifikation von Schlammwasser auf der Kläranlage Hattingen. [Full scale experiences with deammonification of sludge liquor at Hattingen wastewater treatment plant.] *GWF, Wasser/Abwasser* **146:**104–109.

Tokutomi, T., H. Yamauchi, S. Nishihara, M. Yoda, and W. R. Abma. 2007. Demonstration of full-scale Anammox process in Japan. 4th IWA Leading Edge Conference & Exhibition on Water and Wastewater Technologies, Singapore. (Proceedings CD.)

Trela, J., E. Płaza, B. Szatkowska, B. Hultman, J. Bosander, and A.-G. Dahlberg. 2004. Deammonifakation som en ny process för behandling av avloppsströmmar med hög kvävehalt. [Deammonification as a new process for treatment of wastewater with a high nitrogen content.] *Vatten* **60:**119–127.

Tsushima, I., T. Kindaichi, and S. Okabe. 2007. Quantification of anaerobic ammonium-oxidizing bacteria in enrichment cultures by real-time PCR. *Water Res.* **41:**785–794.

Van de Graaf, A. A., P. De Bruijn, L. A. Robertson, M. S. M. Jetten, and J. G. Kuenen. 1997. Metabolic pathway of anaerobic ammonium oxidation on the basis of ^{15}N studies in a fluidized bed reactor. *Microbiology* **143:**2415–2421.

Van de Vossenberg, J., J. E. Rattray, W. Geerts, B. Kartal, L. Van Niftrik, E. G. Van Donselaar, J. S. S. Damsté, M. Strous, and M. S. M. Jetten. 2008. Enrichment and characterization of marine anammox bacteria associated with global nitrogen gas production. *Environ. Microbiol.* **10:**3120–3129.

Van der Star, W. R. L., W. R. Abma, D. Blommers, J. W. Mulder, T. Tokutomi, M. Strous, C. Picioreanu, and M. C. M. Van Loosdrecht. 2007. Startup of reactors for anoxic ammonium oxidation: experiences from the first full-scale anammox reactor in Rotterdam. *Water Res.* **41:**4149–4163.

Van der Star, W. R. L., W. R. Abma, D. Blommers, J. W. Mulder, T. Tokutomi, M. Strous, C. Picioreanu, and M. C. M. Van Loosdrecht. 2008a. Startup of reactors for anoxic ammonium oxidation: experiences from the first full-scale anammox reactor in Rotterdam. *Water Res.* **42:**1825–1826. [Erratum, *Water Res.* **41:**18, 2007.]

Van der Star, W. R. L., A. I. Miclea, L. G. J. M. Van Dongen, G. Muyzer, C. Picioreanu, and M. C. M. Van Loosdrecht. 2008b. The membrane bioreactor: a novel tool to grow anammox bacteria as free cells. *Biotechnol. Bioeng.* **101:**286–294.

Van Dongen, U., M. S. M. Jetten, and M. C.

M. Van Loosdrecht. 2001. The SHARON®-Anammox® process for treatment of ammonium rich wastewater. *Water Sci. Technol.* **44**(1)**:**153–160.

van Loosdrecht, M. C. M. 2008. Innovative nitrogen removal, p. 139–153. *In* M. Henze, M. C. M. van Loosdrecht, G. Ekama, and D. Brdjanovic (ed.), *Biological Wastewater Treatment, Principles, Modelling and Design.* IWA Publishing, London, United Kingdom.

Van Loosdrecht, M. C. M., M. S. M. Jetten, and W. Abma. 2001. Improving the sustainability of ammonium removal. *Water21* **30:**50–52.

Vlaeminck, S. E., L. F. F. Cloetens, M. Carballa, N. Boon, and W. Verstraete. 2008. Granular biomass capable of partial nitritation and anammox. *Water Sci. Technol.* **58:**1113–1120.

Vlaeminck, S. E., A. Terada, B. F. Smets, D. Van der Linden, N. Boon, W. Verstraete, and M. Carballa. 2009. Nitrogen removal from digested black water by one-stage partial nitritation and anammox. *Environ. Sci. Technol.* **43:**5035–5041.

Volcke, E. I. P., M. C. M. van Loosdrecht, and P. A. Vanrolleghem. 2006. Continuity-based model interfacing for plant-wide simulation: a general approach. *Water Res.* **40:**2817–2828.

Walter, U., V. Rekers, and M. Denecke. 2007. Betriebserfahrungen mit der Deammonifikation bei der biologischen Behandlung von Deponiesickerwasser. *Umweltmagazin* **37:**14–16.

Weissenbacher, N., I. Tatacs, S. Murthy, M. Fuerhacker, and B. Wett. 2010. Gaseous nitrogen and carbon emissions from a full-scale deammonification plant. *Water Environ. Res.* **82:**169–175.

Wett, B. 2006. Solved upscaling problems for implementing deammonification of rejection water. *Water Sci. Technol.* **53:**121–128.

Wett, B., S. Murthy, I. Takács, M. Hell, G. Bowden, A. Deur, and M. O'Shaughnessy. 2007. Key parameters for control of demon deammonification process, p. 424–236. Proceedings CD of "Nutrient Removal 2007," Baltimore, MD.

Wicht, H., and M. Beier. 1995. N_2O emission aus nitrifizierenden und denitrificierenden Kläranlagen. [N_2O emission from nitrifying wastewater treatment plants.] *Korrespondenz Abwasser* **42:**404–406, 411–413.

Wiesmann, U. 1994. Biological nitrogen removal from wastewater, p. 113–154. *In* A. Fiechter (ed.), *Advances in Biochemical Engineering/Biotechnology*, vol. 51. Springer-Verlag, Berlin, Germany.

Wilsenach, J., M. C. Schmid, and M. C. M. van Loosdrecht. 2006. Biological nitrogen removal from urine, p. 139–161. *In* J. Wilsenach (ed.), *Treatment of Source Separated Urine and Its Effects on Wastewater Systems.* Delft University of Technology, Delft, The Netherlands.

Wyffels, S., P. Boeckx, K. Pynaert, W. Verstraete, and O. Van Cleemput. 2003. Sustained nitrite accumulation in a membrane-assisted bioreactor (MBR) for the treatment of ammonium-rich wastewater. *J. Chem. Technol. Biotechnol.* **78:**412–419.

Wyffels, S., P. Boeckx, K. Pynaert, D. Zhang, O. Van Cleemput, G. Chen, and W. Verstraete. 2004. Nitrogen removal from sludge reject water by a two-stage oxygen-limited autotrophic nitrification denitrification process. *Water Sci. Technol.* **49:**57–64.

NITRITE-OXIDIZING BACTERIA

V

METABOLISM AND GENOMICS OF NITRITE-OXIDIZING BACTERIA: EMPHASIS ON STUDIES OF PURE CULTURES AND OF *NITROBACTER* SPECIES

Shawn R. Starkenburg, Eva Spieck, and Peter J. Bottomley

11

INTRODUCTION

Nitrite-oxidizing bacteria (NOB) play a key role in nitrification by oxidizing nitrite (NO_2^-) to nitrate (NO_3^-). Despite NO_2^- being an energy-poor substrate that is generated ubiquitously by oxidation of ammonia (NH_3) under aerobic conditions, it rarely accumulates in natural oxic environments. This is a testimony to the versatility of NOB to effectively couple the process of nitrification and consume NO_2^- over a wide range of environmental conditions. We might anticipate, therefore, that NO_2^- oxidation should be widely distributed among the prokaryotes and that several different strains would have emerged as model organisms for detailed study. Yet, whereas the phenotype of NO_2^- oxidation is found in several genera distributed among different phylogenetic lineages of Bacteria (*Nitrobacter*, *Nitrococcus*, *Nitrospina*, *Nitrospira*, *Nitrotoga*), virtually all knowledge of the physiology and biochemistry of NO_2^- oxidation has been derived from studies of a limited number of strains of *Nitrobacter* species. Furthermore, a disproportionately large fraction of the literature on the metabolism and biochemistry of NOB was published >20 years ago during the "golden era" of laboratory-based research on pure cultures of chemolithoautotrophs (Aleem and Sewell, 1984; Bock et al., 1991; Hooper and DiSpirito, 1985; Wood, 1986; Yamanaka and Fukumori, 1988). There are several examples of both novel and controversial findings that were made during that era that remain unconfirmed or unresolved even in 2011. Furthermore, in contrast to many other examples of environmentally significant, microbially mediated processes, a genetically manipulatable strain of NOB has not been developed that can be used to provide unequivocal genetic evidence to discriminate between models formulated from physiological or biochemical evidence alone.

In recent years, non-cultivation-dependent molecular techniques have clearly shown that the genus *Nitrospira* is often the numerically dominant NOB in many habitats including soils and waste water treatment plants (Juretschko et al., 1998; Schramm et al., 1999; Daims et al., 2001; Bartosch et al., 2002). Unfortunately, very few representatives of *Nitrospira* have been obtained in pure culture, and details of their physiology and biochemistry are lacking. As a consequence, this chapter will remain biased toward research

Shawn R. Starkenburg, Los Alamos National Laboratory, Bioscience Division MS888, P.O. Box 1663, Los Alamos, NM 87545. *Eva Spieck*, Universität Hamburg, Biozentrum Klein Flottbek, Mikrobiologie & Biotechnologie, D-22609 Hamburg, Germany. *Peter J. Bottomley*, Departments of Microbiology and Crop and Soil Science, Oregon State University, Corvallis, OR 97331-3804.

Nitrification, Edited by Bess B. Ward, Daniel J. Arp, and Martin G. Klotz

findings generated from *Nitrobacter*. Recently, the genomes of three *Nitrobacter* species/strains were sequenced, and details of the annotations have been published (Starkenburg et al., 2006, 2008c). In this chapter, we will attempt to place the physiology and biochemistry of NOB into context with information derived from the annotated *Nitrobacter* genomes. Furthermore, since *Nitrobacter* is an Alphaproteobacterium found in the family *Bradyrhizobiaceae*, and is closely related to the genera *Bradyrhizobium* and *Rhodopseudomonas* by 16S rDNA sequence similarity (>96%) (Seewaldt et al., 1982; Teske et al., 1994), genomic comparisons have been made in an attempt to identify the core gene "complement" that defines *Nitrobacter* and to uncover what distinguishes the chemolithoautotrophic *Nitrobacter* from its metabolically versatile, phototrophic/organotrophic relatives (Starkenburg et al., 2008c).

TAXONOMY/SYSTEMATICS

The process of NO_2^- oxidation is found in morphologically and phylogenetically diverse lineages of Bacteria (Koops and Pommerening-Röser, 2001; Spieck and Bock, 2005; Alawi et al., 2007). Although evidence exists for some Crenarchaea possessing and expressing the gene that encodes for subunit A of ammonia monooxygenase in marine (Konneke et al., 2005; Wuchter et al., 2006), hydrothermal (de la Torre et al., 2008; Hatzenpichler et al., 2008), and some soil (Leininger et al., 2006; Nicol et al., 2008; Prosser and Nicol, 2008) environments, NO_2^--oxidizing Archaea have not been identified or isolated. Whereas *Nitrobacter* is located in the Alphaproteobacteria lineage, *Nitrococcus* is found in the Gammaproteobacteria, *Nitrospina* is provisionally placed in the Deltaproteobacteria, and *Nitrospira* occupies its own deep-branching lineage (Spieck and Bock, 2005). Strains of *Nitrospina* and *Nitrococcus* have been recovered exclusively from marine environments and are characterized as obligately halophilic. Excellent descriptions of the different genera of NOB that include their history, details of growth media, growth conditions, cell morphologies, and basic growth characteristics can be found in previous review articles (Watson et al., 1989; Bock et al., 1991; Spieck and Bock, 2005).

Recently, Spieck and colleagues described successful isolation and culturing of novel isolates and enrichments of *Nitrospira*, namely "*Candidatus* Nitrospira bockiana" (Lebedeva et al., 2008) and "*Candidatus* Nitrospira defluvii" (Spieck et al., 2006). A novel Betaproteobacterium with NO_2^--oxidizing capability ("*Candidatus* Nitrotoga arctica") was recently highly enriched from a Siberian Arctic permafrost-affected soil (Alawi et al., 2007). Additionally, an NO_2^--oxidizing anoxygenic phototroph closely related to *Thiocapsa* of the Gammaproteobacteria was also isolated (Griffin et al., 2007). Because *Nitrospira* spp. usually require low NO_2^- concentrations for growth and grow slowly with low cell yield (Watson et al., 1986; Ehrich et al., 1995), it is not surprising that little inroad has been made into their metabolism and biochemistry. Nonetheless, Daims et al. (see Chapter 12) provide an excellent review of insights gained into the physiology and ecology of *Nitrospira* that were obtained primarily by using genomics tools and other cultivation-independent methods. Table 1 summarizes the basic properties of the cultured species of NOB, including the most recently isolated strains.

NITROBACTER GENOMICS

General Characteristics

To date, five different genomes within three genera of NOB (*Nitrobacter*, *Nitrococcus*, and *Nitrospira*) have been sequenced (Table 2). Comprehensive genomic annotations and analysis have been published on three sequenced *Nitrobacter* genomes, *N. hamburgensis* X14, *Nitrobacter winogradskyi* Nb255, and *Nitrobacter* sp. strain NB311A. An unfinished, draft sequence from *Nitrococcus mobilis* NB231 (isolated from equatorial surface waters in the Pacific Ocean) was recently completed, although a comprehensive analysis has not been published (https://moore.jcvi.org/moore/SingleOrganism.do?speciesTag=NB231&pageAttr=pageMain). Most

TABLE 1 Differentiating properties among the genera of NOB

Parameter	Result for:				
	Nitrobacter	*Nitrococcus*	*Nitrospina*	*Nitrospira*	*Nitrotoga*
Phylogeny	Alphaproteobacteria	Gammaproteobacteria	Deltaproteobacteria	Nitrospirae	Betaproteobacteria
G+C (%)	59–62	61	58	50–54	ND[a]
Morphology	Pleomorphic rod	Coccus	Slender rod	Helical	Coccoid/short rod
ICMs	Polar	Random	None	None	None
Typical fatty acids[b]	18:1*cis*11 16:0	18:1cis11 16:1*cis*9 16:0	16:1*cis*9 14:0 16:0	16:1*cis*7[c] 16:1*cis*11 16:0,11me[d] 16:0	16:1*cis*9 16:0 10,12,&14:0 OH
Carboxysomes	Yes	Yes	None	None	None
β-Subunit NXR (kDa)	65	65	48	46	ND

[a]ND, not determined.
[b]Data are from Alawi et al. (2007), Lipski et al. (2001), and Spieck et al. (2006).
[c]Composition varies with species.
[d]Present in some moderately thermophilic species of *Nitrospira*.

recently, the genome of "*Candidatus* Nitrospira defluvii" was sequenced; highlights from the initial annotation have been described by Daims et al. (see Chapter 12).

The *Nitrobacter* genomes range in size from 3.4 to 5 Mbp and encode approximately 3,117 to 4,716 total genes, respectively. On average, the *Nitrobacter* genomes (~4.1 Mbp) are much smaller than their phototrophic/organotrophic alphaproteobacterial relatives, *Bradyrhizobium japonicum* (~8.3 Mbp) and *Rhodopseudomonas palustris* (~5.4 Mbp). Approximately 2,179 genes were found to be conserved among the three *Nitrobacter* genomes (Fig. 1), which represents the majority (86%) of the genes found in the smallest genome of *N. winogradskyi*. Nevertheless, each genome was found to contain unique genetic material (13 to 29% of the sequence space) that may be relevant to the ecological niche of each bacterium (Table

TABLE 2 General genomic characteristics of NOB

Parameter	Result for:			
	N. hamburgensis X14	*N. winogradskyi* NB255	*Nitrobacter* sp. strain NB311A draft[a]	*N. mobilis* NB231
Origin	Soil	Soil	Marine	Marine
Chromosome bases	4,406,967	3,402,093	~4,105,362	~3,617,638
G+C (%)	61.6	62	62	60
Total genes	4,716	3,118	4,256	3,503
Genes without predicted function	1,848	993	1,461	1,185
Pseudogenes	347	21	ND[b]	ND[b]
Paralogs	634	283	478	ND
Paralog groups	251	74	143	ND
Plasmids	3	0	ND	ND
pPB13	294,829 bp			
pPB12	188,318 bp			
pPB11	121,408 bp			

[a]16S rRNA is 100% identical to *N. winogradskyi* Nb-255.
[b]ND, not determined (an accurate count of the pseudogenes or the presence of plasmids was not possible from the unfinished draft sequence of NB311A).

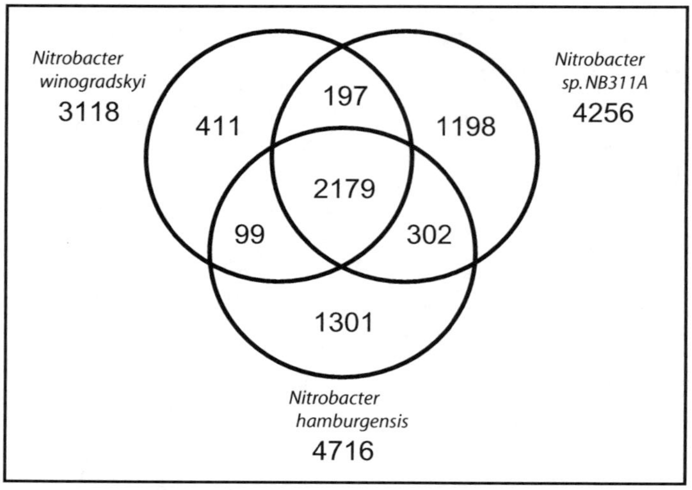

FIGURE 1 Global gene conservation in *Nitrobacter.* Each circle represents the total number of gene types in each genome. Overlapping regions depict the number of gene types shared between the respective genomes. The numbers outside the circles indicate the total number of genes identified in each genome, including paralogs/gene duplications. (Reproduced from *Applied and Environmental Microbiology* [Starkenburg et al., 2008c] with permission.)

3). With regard to *N. winogradskyi*, of the 411 genes not found in either of the other two *Nitrobacter* species, 124 were assigned a putative function, including an alkane-sulfonate monooxygenase, two nitrate/sulfonate/bicarbonate ABC transporters, and synthesis genes for the pyrroloquinoline quinone cofactor. All three genomes were found to encode a putative Na^+/H^+ antiporter (*nhaA*), which supports and extends observations of halotolerance in *Nitrobacter*, yet several genes unique to the NB311A genome (a chloride channel, a Na^+/Ca^{2+} antiporter, many cation-dependent ATPases, and ectoine-like osmoprotectants) may indicate that this strain has additional mechanisms to manage osmotic stress and to survive in marine environments.

The genome of *N. hamburgensis* is the largest of the *Nitrobacter* genomes and appears to have maintained a greater level of metabolic flexibility and adaptability than the other sequenced representatives. Among its unique genetic material, *N. hamburgensis* contains putative genes that code for unique terminal oxidases and cytochromes, nitric oxide reductase (NOR), formate dehydrogenase, sulfur oxidation, unique copies of carbon monoxide dehydrogenase-like genes, lactate dehydrogenase, and other anapleurotic enzymes. Many paralogous and nonparalogous duplications of genes involved in key metabolic functions (nitrite oxidoreductase [NXR], terminal oxidases, and ribulose-bisphosphate carboxylase [RuBisCO]) also infer an increase in metabolic capacity and/or the ability to differentially express these gene clusters based on different environmental conditions. Despite its seemingly broader base of metabolic potential, the *N. hamburgensis* genome is less organized and more fragmented than the other NOB. Approximately 8% of the genome encodes pseudogenes (n = 347), and it contains a higher number of mobile genetic elements and phage remnants. In contrast, the *N. winogradskyi* genome contains only 21 pseudogenes and half the number of paralogs.

N. hamburgensis Plasmids

Before sequencing of its genome, little was known about the metabolic role of the plasmids in *N. hamburgensis.* Approximately 494 genes are encoded on the plasmids, although only half of these could be assigned a function (Starkenburg et al., 2008c). The genes on the largest plasmid, pPB13, appear to be biased toward carbon/energy metabolism. The small plasmid, pPB11, is dominated by conjugation/pilus formation genes. pPB12 appears to be a functional hybrid of the other two plasmids, containing gene clusters for conjugation, energy, and carbon metabolism, plus a suite of genes for heavy metal resistance. Notably, the only copy in the genome of an ATP-dependent glucokinase is located on pPB12. An interesting feature of pPB13 is the presence of a large "autotrophic island" (~28 kb gene cluster) that encodes the large and small subunits of a Type I RuBisCO, and the only set of genes that encode for carboxysome formation. Most of the plasmid-born genes are unique to *N. hamburgensis*, yet ~21 kb of the ~28 kb autotrophic island are conserved in the chromosomes of *N. winogradskyi* and NB311A. Several other Calvin-Benson-Bassham cycle enzymes are also located on pPB13, including a second, nonparalogous copy of a Type I RuBisCO and

TABLE 3 Unique genes and putative functional biases in the genus *Nitrobacter*

Putative category	Unique genes in:		
	N. winogradskyi	*N. hamburgensis*	*Nitrobacter* strain Nb311A
Transport	NO_3^-/sulfonate/CO_3^{2-}	Ammonia permease	TonB systems
	Iron uptake systems	K^+ transport	Ca^{2+}/Na^{2+} antiporter
	Fe/Ni/Co	Uncharacterized ABC transport components	Cl^- channel
	PO_4^{3-} porin		Chromate
	Uncharacterized ABC transporter components		Uncharacterized ABC transporter components
			Mg/cobalt sulfate permeases
			Co/Zn/Cd efflux
Carbon metabolism		Formate dehydrogenase	
		Carbon monoxide DH[a]-like	
		D- or L-lactate DH-like	
		Malate DH, pyruvate-formate lyase	
		Homogentisate/phenylacetate degradation	
Energetics		Cyt c oxidase	
		Cyt bd ubiquinol oxidase	
		Cytochome b_{561}	
		Cytochrome P_{460}	
		Flavoredoxin reductase	
		Nitric oxide reductase (*sNOR*)	
Replication			DNA replication/repair
			DNA polymerase IV, III
			minCDE septum formation
Miscellaneous	Histidine biosynthesis	Type II secretion	Polysaccharide synthesis, export
	Multiple FecIR genes	Serine proteases	UspA stress genes
	Pyrroloquinoline quinone biosynthesis	Conjugal transfer	Ectoine synthase
		Heavy metal resistance	Cation ATPases
		Arsenite oxidase	Pyoverdine synthesis
			Hydroxymate siderophore syn. (IucC family)

[a]DH, dehydrogenase.

single copies of fructose-1,6-bisphosphatase, phosphoribulokinase, and ketose-bisphosphate aldolase. This second RuBisCO gene cluster is duplicated on the chromosome (99% similarity) although a neighboring LysR-type regulator has diverged significantly, suggesting that each RuBisCO may be differentially regulated.

Core *Nitrobacter* Genome Analysis

To further assess the genomic basis of NO_2^- oxidation, Starkenburg et al. (2008c) constructed a composite or "core" genome, which was composed of genes conserved in all three *Nitrobacter* genomes. Using the core genome as a query database, all core genes with high sequence identity to a gene(s) in any strain of *R. palustris* or *B. japonicum* were removed. Approximately 116 gene types were found to be uniquely conserved in each *Nitrobacter* genome. Within this 116 gene "subcore," 46 genes had no match in the Kyoto Encyclopedia of Genes and Genomes (KEGG) genome database, and it is not clear what role they play in these NOB. Among the functionally annotated subcore gene set (41 genes), two gene clusters appear to encode polysaccharide synthesis proteins, some of which have little sequence similarity to any Alphaproteobacterial proteins. Many of the subcore genes with functional annotations were found to be associated with NO_2^- metabolism, transport, and regulation, including the gene cluster encoding the subunits of NXR and the *c*-type cytochromes, and a putative NsrR-like regulatory protein adjacent to *nirK*. Of note, the *R. palustris* and *B. japonicum* genomes contain homologs of several genes in the *Nitrobacter* subcore inventory, yet the latter were found to be more orthologous to genes outside the alphaproteobacterial lineage.

In summary, this analysis suggested that the collection of genes that is responsible for the key metabolic machinery required for NO_2^- oxidation in *Nitrobacter* may reflect the ecological niche of these bacteria (rather than its phylogeny) possibly achieved through assimilation, modification, and expression of genes acquired from more distant bacterial lineages. It will be of considerable interest to compare the core genes of *Nitrobacter* with those of other NOB from phylogenetically distinct lineages if and when they become available.

ULTRASTRUCTURAL FEATURES OF NOB

Transmission electron microscopic images (TEMs) of NOB provide evidence for intracellular morphological variations that still await detailed investigation of their physiological significance. TEMs of *Nitrobacter* and *Nitrococcus* reveal a complex network of intracytoplasmic membranes (ICMs) that penetrates into the cytoplasm (Fig. 2). In the case of *Nitrobacter*, ICMs are often located at one pole of the cell, and electron-dense particles extensively cover the cytoplasmic side of the membrane (Spieck et al., 1996a, 1996b). Immunolabeling with antibodies targeted at the subunits of NXR indicates that these membrane-associated particles are the subcellular location of NXR (Spieck et al., 1996a, 1998). Examination of ultrathin sections of *Nitrobacter* led researchers to suggest that the cell wall structure might be atypical of a gram-negative Proteobacterium because the peptidoglycan layer seems to be absent (Watson et al., 1989; Bock et al., 1991). TEMs of *Nitrospina*, *Nitrospira*, and *Nitrotoga* are notable for the absence of ICMs (Fig. 3 and 4) and, in the case of *Nitrospira* and *Nitrotoga* species, for the presence of an unusually large periplasmic space (Watson et al., 1989; Ehrich et al., 1995; Spieck and Bock, 2005; Alawi et al., 2007). *Nitrobacter* and *Nitrospira* cells located in mixed-culture-nitrifying bioreactors are often embedded in a capsule and exist in aggregates in close proximity to ammonia-oxidizing bacteria (AOB). The relative contributions of AOB- and NOB-encoded exocellular products to the capsular matrix of these aggregates remain unknown. In that context, it is interesting to reiterate that one of the sets of genes unique to the subcore of *Nitrobacter* is implicated in capsule biosynthesis. TEMs of NOB often show that a substantial portion of the cytoplasm is occupied by cell inclusions such as carboxysomes and poly β hydroxybutyrate, glycogen, and polyphosphate granules (Watson et al., 1989; Spieck and Bock, 2005). Whereas carboxysomes are routinely detected in cells of *Nitrobacter* and *Nitrococcus*, they have not been observed in *Nitrospira*, *Nitrospina*, or *Nitrotoga*. A discussion of the current status of knowledge about carboxysomes is presented later in this chapter, and genomic evidence for a potentially different mechanism of CO_2 fixation by "*Candidatus* Nitrospira defluvii" is presented by Daims et al. (see Chapter 12).

GROWTH CHARACTERISTICS

NO_2^- Level, pH, and Temperature

The well-studied model strains of *N. winogradskyi*, *N. hamburgensis*, and *N. vulgaris* grow optimally at 25 to 30°C, pH 7.5 to 8.0, and tolerate relatively high concentrations of NO_2^- (10 to 45 mM) in the growth medium. Evidence has also surfaced that NOB can grow under more diverse conditions. For example, an acidophilic *Nitrobacter* (IOacid) recovered from an acidic deciduous forest soil grew optimally at pH 5.5 (Hankinson and Schmidt, 1988), while alkaliphilic *Nitrobacter* strains were described that grew well at pH 9.5 and were subsequently named *N. alkalicus* (Sorokin et al., 1998). Some NOB appear to be cold adapted, as "*Candidatus* Nitrotoga arctica" was observed to grow between 4°C and 17°C (Alawi et al., 2007). *Nitrospira moscoviensis* was enriched from a corroded steam pipe located in a Moscow heating system and grows optimally at 39°C (Ehrich et al., 1995), while "*Candidatus* Nitrospira bockiana," obtained from a similar source, expressed optimum NO_2^- oxidation at 42°C (Lebedeva et al., 2008). Enrichments of NOB from a hot spring source in the Buryat Republic, Russia, were obtained that expressed optimum NO_2^- oxidation at 50°C, and stable enrichments were obtained that grew reproducibly at 42, 48, and 55°C, respectively (Lebedeva et al., 2005). One of the issues found to be critical for successful enrichment and isolation of *N. moscoviensis* was the need to maintain a low NO_2^- level (0.7 to 1.4 mM) in the growth medium. In fact, it was noted that optimum growth of *N. moscoviensis* occurred at 0.35 mM NO_2^- and growth was inhibited at 15 mM NO_2^- (Ehrich et al., 1995). However, it has been shown, subsequently, that there is considerable variation in NO_2^- tolerance among isolates and enrichment cultures of *Nitrospira* (*N. defluvii*, 20 to 25 mM; *N. bockiana*, 18 mM; *N. marina*, 6 mM). Perhaps this is a reflection of the wide range of environments in which *Nitrospira* spp. have been found to be highly successful. In another study, *Nitrospira* was enriched from natural environments if $NaNO_2$ was kept ≤0.2 g liter^{-1} (2.8 mM), whereas *Nitrobacter* was enriched if $NaNO_2$ was used at 2 g liter^{-1} (28 mM) (Bartosch et al., 2002). Subsequently, evidence was obtained for differential NO_2^- sensitivity among *Nitrospira* lineages in natural populations. In activated sludge samples, it was shown that *Nitrospira* sublineages 1 and 2 could be distinguished because sublineage 1 was more tolerant of higher NO_2^- concentrations than sublineage 2 (Maixner et al., 2006). Given the need for metabolic versatility among NOB to ensure that nitrification remains a tightly coupled process in different environments, we predict that NOB from diverse phylogenetic backgrounds and with diverse physiological properties will continue to be discovered.

Heterotrophic Growth

There is still no concrete evidence about what discriminates obligately chemolithoautotrophic strains of NOB from those with chemoorganotrophic or mixotrophic growth potential. *Nitrobacter* strains show varying degrees of chemoorganotrophic growth that is mostly restricted to C_2 to C_3 substrates (acetate, pyruvate, lactate, and glycerol). In contrast, *Nitrospina* and *Nitrococcus* were reported initially to be obligate chemolithoautotrophs (Watson and Waterbury, 1971). Whereas growth of the marine isolate *Nitrospira marina* occurred better under mixotrophic than chemolithoautotrophic conditions (Watson et al., 1986), *N. moscoviensis* was reported initially to be an obligate

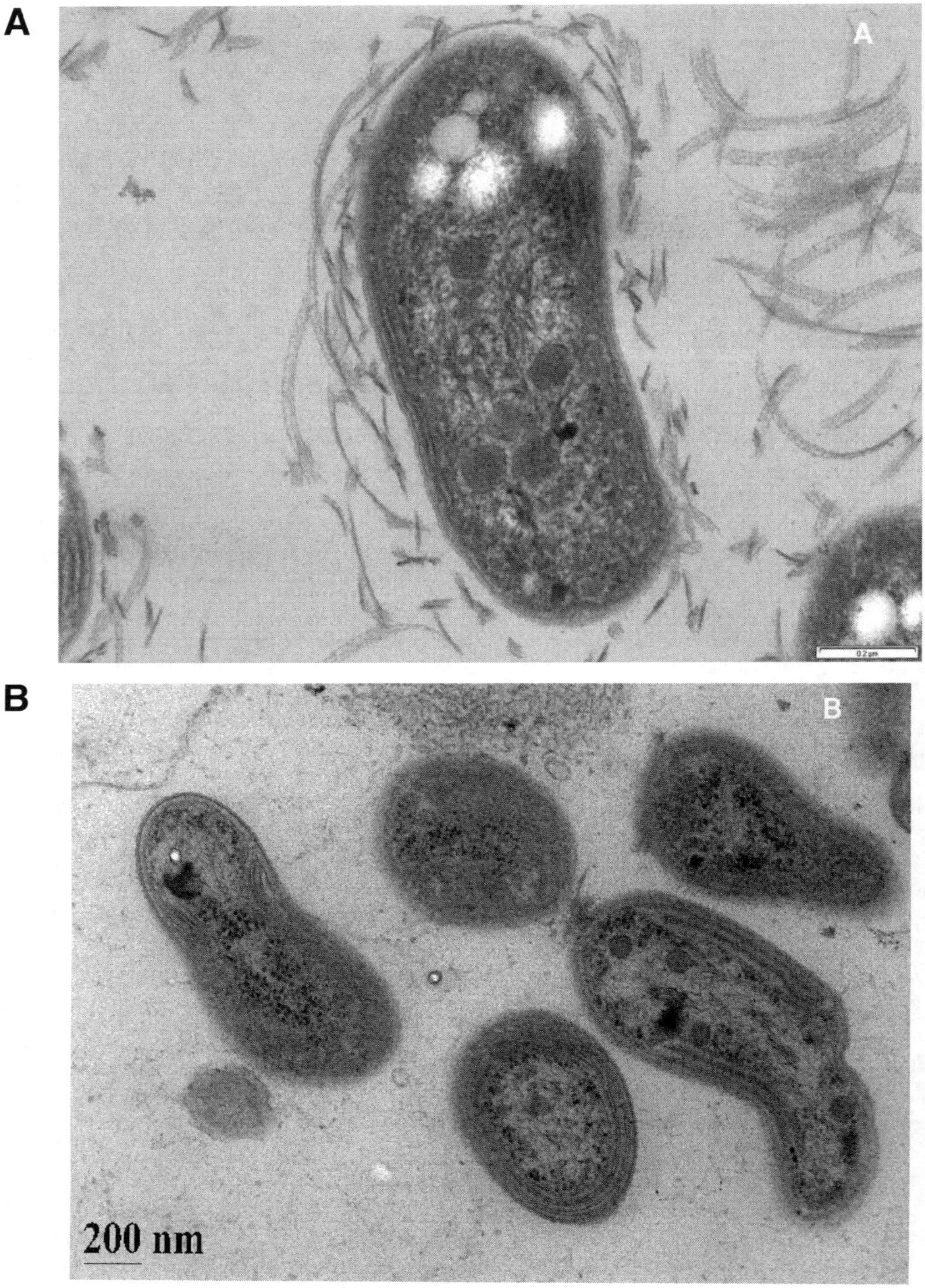

FIGURE 2 Electron micrographs of *Nitrobacter* and *Nitrococcus*. (A) *N. winogradskyi* Nb255 (image by William Hickey). (B) *Nitrobacter* spp. enriched from activated sludge. (C) *N. mobilis* 231 (images in panels B and C by Eva Spieck).

C

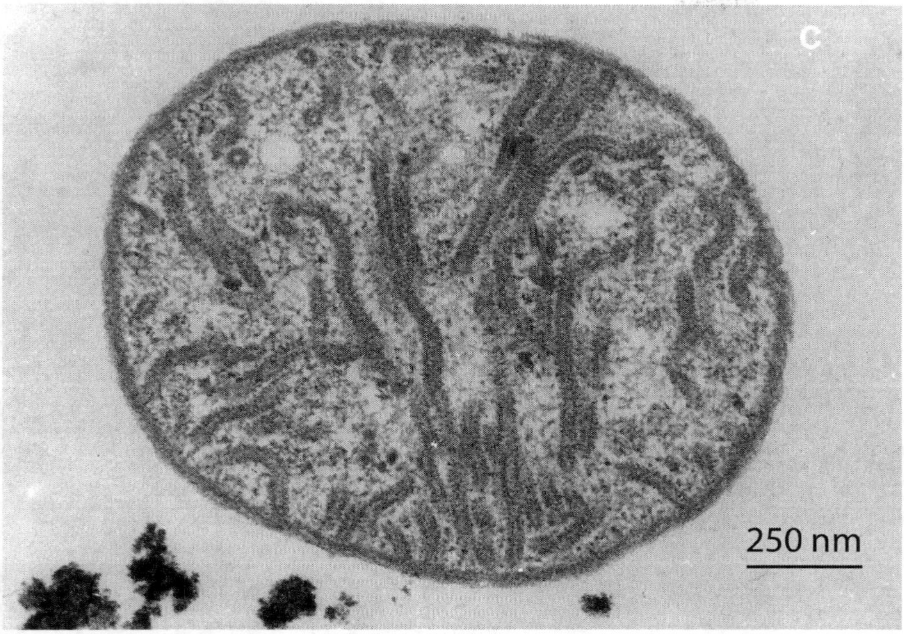

chemolithoautotroph. Reports in the older literature showed that chemoorganotrophic growth of *Nitrobacter* species was improved if they were cultured in a combination of either pyruvate or acetate, a complex organic N source such as peptone, yeast extract, or casamino acids and in the absence of supplemental NO_2^- and NH_4^+ (Smith and Hoare, 1968; Bock, 1976). Although these data provide circumstantial evidence of *Nitrobacter*'s ability to use organic N sources, it remains to be experimentally proven that organic N sources can be directly assimilated.

PHYSIOLOGY AND METABOLISM

Genetic and Biochemical Structure of the NO_2^--Oxidizing System

Clearly, the defining property of NOB is their unique ability to oxidize NO_2^- to NO_3^- and recover the energy and reductant for growth. During the 1980s, several publications described the purification and characterization of the nitrite-oxidizing enzyme (NXR) and its associated hemes and electron carriers (Yamanaka et al., 1981; Yamanaka and Fukumori, 1988; Bock et al., 1991). Unfortunately, the subject has been neglected in recent years, despite the fact that much uncertainty remains about the details of the enzyme structure and its mechanism. NXR is thought to be a membrane-associated heterodimeric protein consisting of one large α subunit, NxrA (130 kDa), and one smaller β subunit, NxrB (65 kDa) (Meincke et al., 1992). It is putatively associated with a molybdopterin cofactor (Kruger et al., 1987; Meincke et al., 1992). The method chosen for NXR purification has influenced both estimates of the number and size of putative polypeptides thought to be associated with it and the specific types of hemes (Bock et al., 1991). The presence or absence of specific hemes is thought to influence whether or not the NXR preparation possesses NO_2^--dependent ferricytochrome *c* reduction, or is only capable of reducing artificial electron acceptors such as ferricyanide and chlorate (Bock et al., 1991). It is equivocal which of the NXR

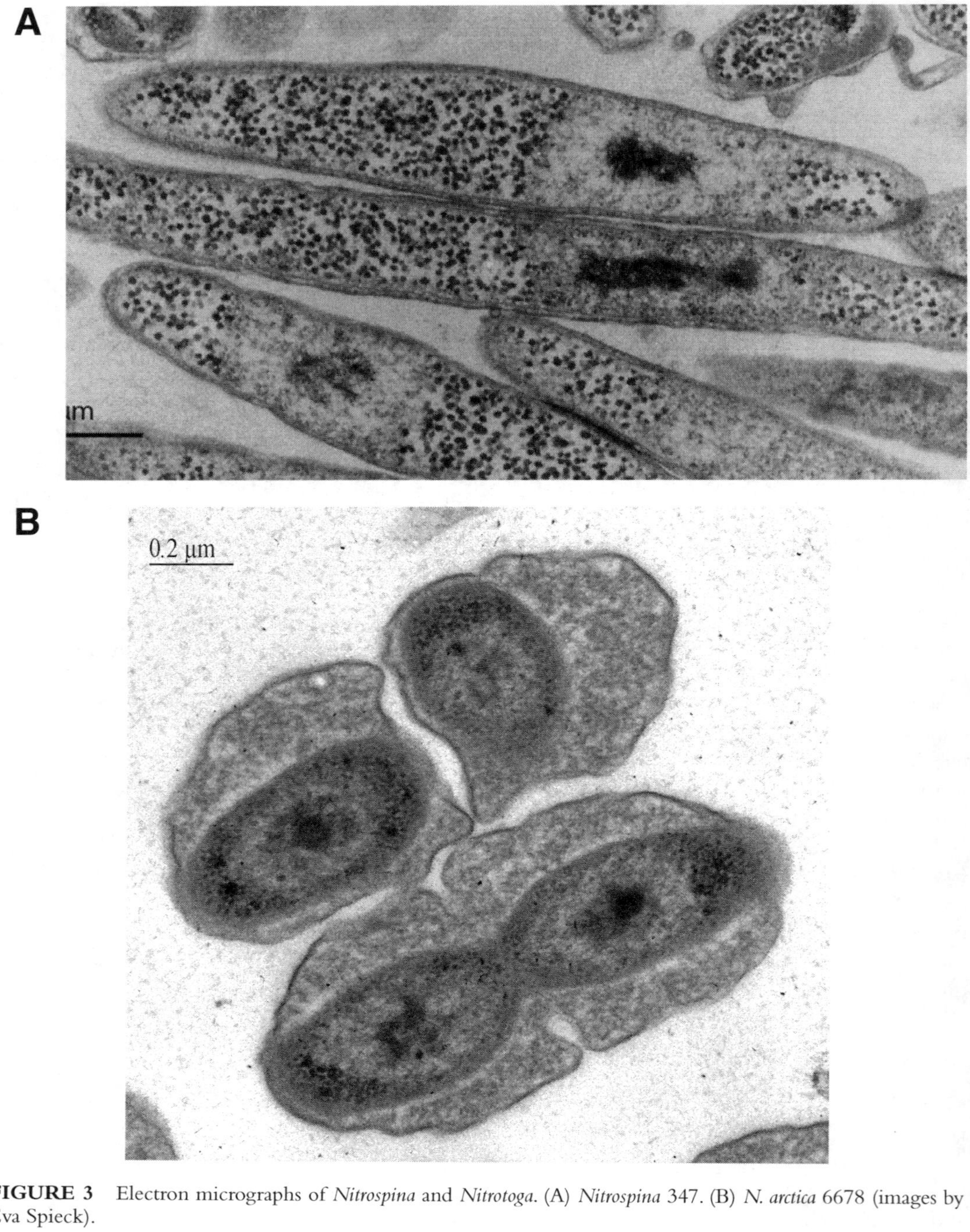

FIGURE 3 Electron micrographs of *Nitrospina* and *Nitrotoga*. (A) *Nitrospina* 347. (B) *N. arctica* 6678 (images by Eva Spieck).

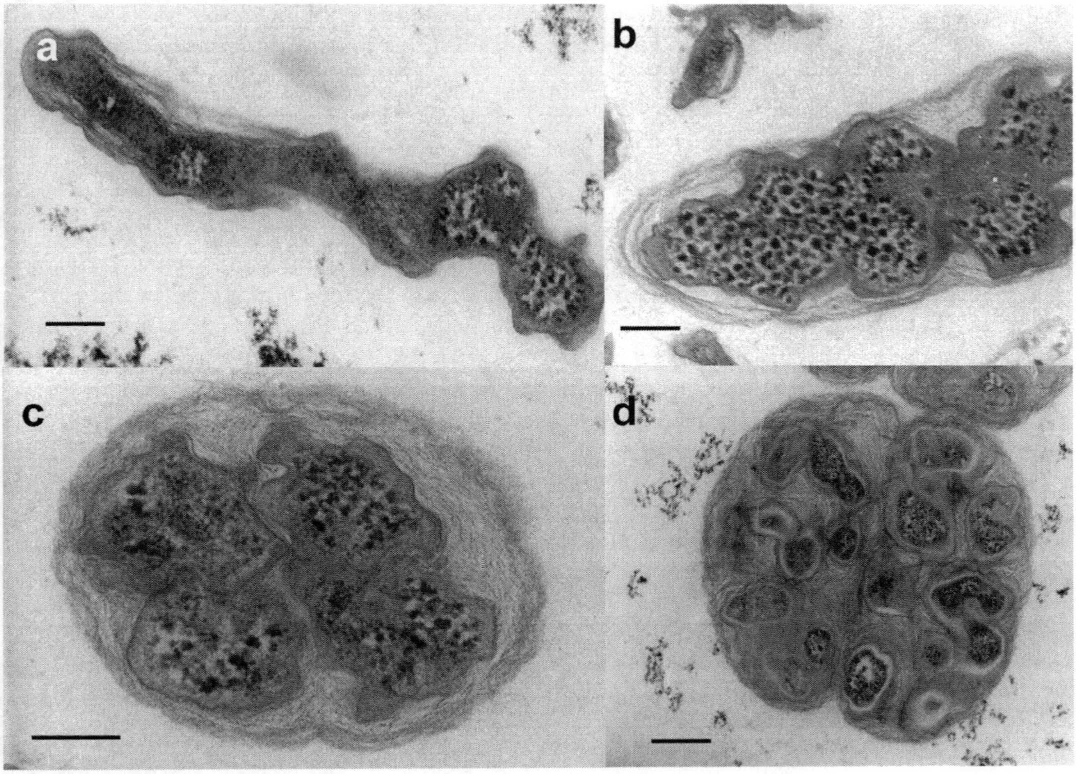

FIGURE 4 Electron micrographs of "*Candidatus* Nitrospira bockiana." (a) Planktonic cells. (b) Planktonic cell surrounded by EPS. (c) Microcolony surrounded in EPS. (d) Pleomorphic microcolony. Bars, 0.25 μm (panels a to c) or 0.5 μm (panel d). (Reproduced from the *International Journal of Systematic and Evolutionary Microbiology* [Lebedeva et al., 2008] with permission from the Society of General Microbiology.)

subunits definitively possesses the catalytic site of the enzyme, yet indirect evidence favors NxrA since it is more similar to the catalytic α subunit of dissimilatory NO_3^- reductase than is NxrB. Although signal peptides were not detected in the *nxr*A and *nxr*B genes, NxrA is predicted to have transmembrane-spanning domains, implying that it might anchor the NXR complex to the cytoplasmic membrane (Starkenburg et al., 2006). Spieck et al. (1996a, 1996b, 1998) had suggested previously that NXR from *Nitrobacter* is membrane associated and localized on the cytoplasmic face of the cell membrane. H. Daims (personal communication) noted that contradictory results were obtained when he and his coworkers searched for transmembrane-spanning domains in *nxrA* with different bioinformatic tools. Therefore, if such features are actually missing from NxrA, it raises the possibility that another protein subunit might be required to anchor NXR to the membrane.

Multiple copies of *nxrA* and *nxrB* are carried by the genomes of *Nitrobacter*, but only one central gene cluster encodes both the structural proteins of NXR and putative accessory proteins (Fig. 5) that include NxrC (a homolog of NarJ, which is thought to insert the Mo cofactor into related dissimilatory nitrate reductases and might do the same for NxrA) and NxrD (NarI homolog), which encodes a *b*-type cytochrome that might serve as an electron acceptor (or donor) to the Mo cofactor.

Upstream of NxrA lies a gene encoding a cytochrome *c*, which might be a key electron carrier associated with the NXR complex and directly involved in the oxidation and reduction of NO_2^-. Indeed, Sundermeyer-Klinger et al. (1984) described a membrane-bound cytochrome *c* (32 kDa), which they speculated might be a third subunit of the NXR. A NarK-type gene is also encoded in this gene cluster, which likely functions as a nitrite and/or nitrate transporter. Effective transport of NO_2^- is critically important in any model of NO_2^- oxidation that predicts the following: (i) NXR is located on the cytoplasmic side of the membrane; (ii) NO_3^- accumulation might interfere with NO_2^- oxidation; and (iii) NO_2^- interferes with NO_3^- reduction under anaerobic conditions (see below a discussion of dissimilatory nitrate reduction by NOB).

Variation in the molecular organization of the NXR operon exists among the different NOB lineages (Fig. 5). All sequenced genomes contain at least one copy of *nxrA* and *nxrB*. The *Nitrobacter nxr* gene clusters contain a small open reading frame, *nxrX*, located between *nxrA* and *nxrB*. This gene appears to be a unique feature of the *Nitrobacter*-encoded NXR and is not found in the *N. mobilis* (Vanparys et al., 2006) or "*Candidatus* Nitrospira defluvii" genomes (see Chapter 12). Although nothing is known about the product of *nxrX*, the gene shares high sequence homology with peptidyl prolyl *cis-trans* isomerases, which suggests that *nxrX* may be required for correct folding/maturation of NXR. Further sequence analysis by Vanparys et al. (2006) indicated that the gene sequences of *nxrA* and *nxrB* of *N. mobilis* are only 69% similar to those of *N. winogradskyi*, and the *nxr* genes of *Nitrospira* are presumed to be even more distantly related. The outcome of the comparative genomic analysis of *nxr* agrees with earlier western blot immunoanalysis, which showed that the molecular weight of NxrB is smaller in *Nitrospira* and *Nitrospina* (46 to 48 kDa) than in *Nitrobacter* and *Nitrococcus* (65 kDa) (Aamand et al., 1996; Bartosch et al., 1999) and that an antibody raised against the α subunit of NXR of the *Nitrobacter* strains (130 kDa) was genus specific, indicating some variation among alpha subunits of the different NOB genera.

Nitrite Oxidation Mechanism and Associated Electron Flow

NXR from *N. winogradskyi* was reported to be an iron-sulfur molybdoenzyme containing about 1 to 2 Mo atoms per molecule of NXR (Yamanaka and Fukumori, 1988). A preparation of NXR originating from *N. winogradskyi*, and derived from a purification procedure that involved a detergent extraction step, contained both heme a_1 and heme c and was able to catalyze the reduction of cytochrome c_{550}. When isolated from heat-treated membranes, NXR consisted of two subunits (115 kDa and 65 kDa) and the enzyme fraction contained molybdenum, iron, zinc, and copper (Meincke et al., 1992). Based upon findings from electron paramagnetic resonance spectroscopy, Meincke et al. (1992) proposed that both molybdenum and iron-sulfur centers are involved in the oxidation of nitrite to nitrate resulting in reduction of Mo(VI) to Mo(IV), which is followed by reoxidation of Mo centers via Mo(V) as the electrons are transferred to other carriers. Electrons originating from NO_2^- oxidation by NXR are postulated to be released from the heme a_1 component of NXR, transferred to cyt c_{550}, and subsequently passed to cytochrome oxidase (Fig. 6) (Yamanaka and Fukumori, 1988). Both soluble and membrane *c*-type cytochrome have been purified from *N. winogradskyi* and shown to serve as electron donors to cytochrome *c* oxidase (Tanaka et al., 1983; Nomoto et al., 1993). Cytochrome *c* oxidase has been purified and shown to be a 67-kDa protein that contains two molecules of heme *a* and two Cu atoms, and it is characterized as a cytochrome aa_3 type (Chaudhry et al., 1980; Yamanaka et al., 1981). Two electrons are released during the oxidation of NO_2^- to NO_3^-, and the third O atom in NO_3^- is derived from H_2O (Aleem, 1965; Aleem et al., 1965).

Genome analysis of *N. winogradskyi* has confirmed the existence of the genes that encode

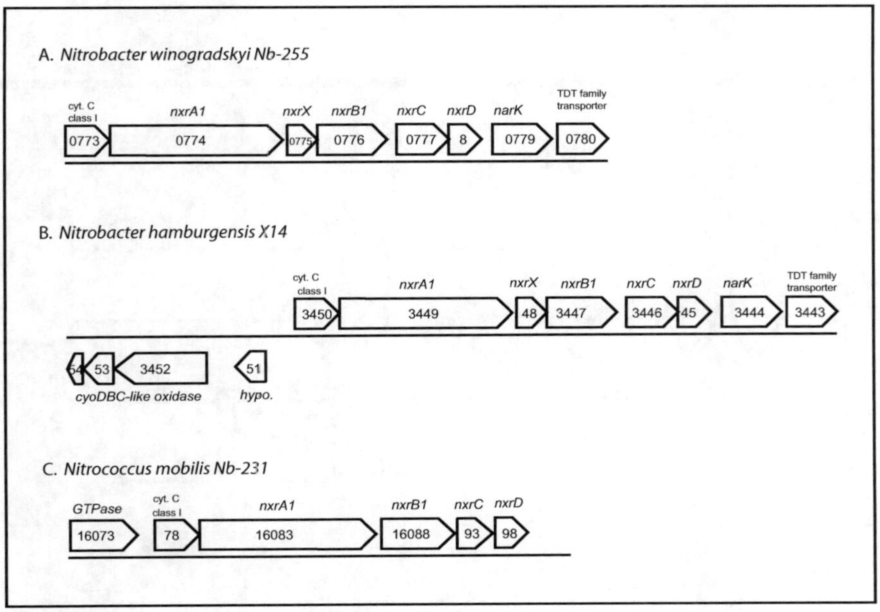

FIGURE 5 Organization of NXR operons of *Nitrobacter* and *Nitrococcus* species. Each arrow represents one gene. Locus numbers are indicated within the arrows, and putative gene names are indicated above each arrow.

for both the soluble and membrane-bound forms of cytochrome c_{550} (Starkenburg et al., 2006). Other genes putatively encoding for additional *c*-type cytochromes were identified, but further work is needed to assign functions to them. Interestingly, a gene encoding for a cytochrome a_1 was not found in the *nxr* operon of *N. winogradskyi*, and heme a could not be found in an active preparation of NXR purified from *N. hamburgensis* (Sundermeyer-Klinger et al., 1984). The latter enzyme preparation was incapable of using ferricytochrome *c* as an electron acceptor for NO_2^- oxidation. Clearly, consensus about the composition and properties of NXR from these two *Nitrobacter* species has not been achieved!

NO_2^- Oxidation and Energy Generation

Many of the details of energy generation from NO_2^- oxidation also remain uncertain, primarily because of limited research and few publications on the topic over the past 20 years or so. Because NO_2^- was oxidized by inverted membrane vesicles, it was suggested that NXR might be located on the cytoplasmic face of the cell membrane (Cobley, 1976a, 1976b). Because uncouplers slowed the reduction of cyt *c* from NO_2^- oxidation, the membrane potential was implicated in the cytochrome reduction mechanism. In contrast, using reconstituted liposomes and purified NXR constituents, Nomoto et al. (1993) obtained no evidence for the membrane potential being required for NO_2^- oxidation.

If the following equations accurately describe the empirical mechanism of NO_2^- oxidation,

$$2\,NO_2^- + 2\,H_2O \rightarrow 2\,NO_3^- + 4\,H^+ + 4e^-$$
$$4\,H^+ + 4e^- + O_2 \rightarrow 2\,H_2O$$

then the oxidation of two molecules of NO_2^- produces $4H^+$ on the cytoplasmic side of the membrane, and concomitant conversion of one molecule O_2 to 2 H_2O consumes $4H^+$. If this combination of redox activities occurs on the same side of the cytoplasmic membrane, it would prevent the formation of a H^+

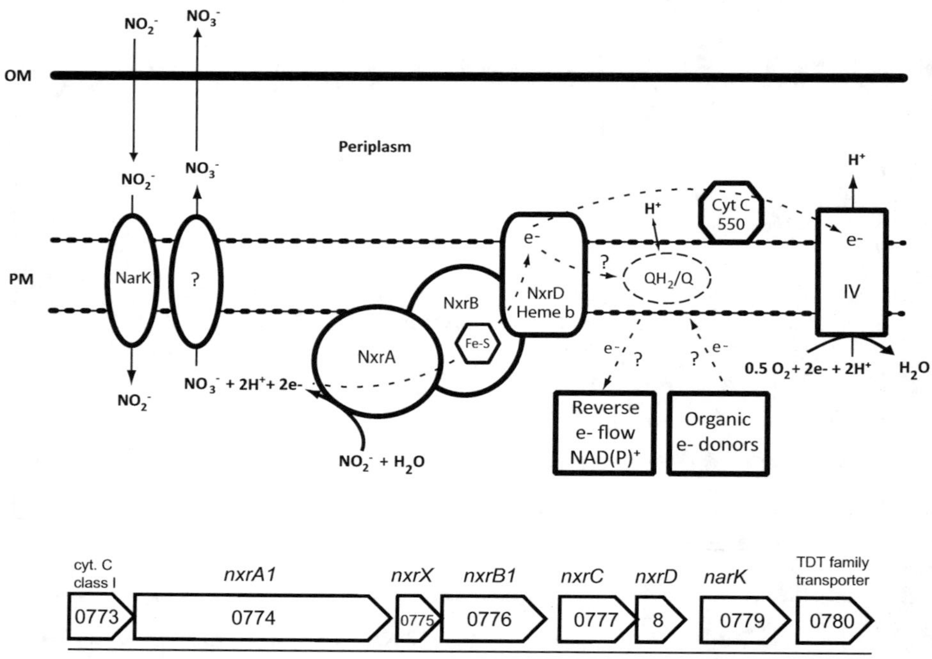

FIGURE 6 Putative organization of NXR and associative electron transport in the cytoplasmic membrane of a *Nitrobacter* sp. Many details remain unknown or uncertain. Although the quinone pool plays a key role in most prokaryotic electron transport and transmembrane H^+ translocation schemes, a complete biosynthetic pathway for menaquionone or ubiquinone biosynthesis could not be annotated in *Nitrobacter*. The details of the NO_3^-/NO_2^- transport mechanism are unknown. The stoichiometry of proton translocation by cytochrome oxidase is unknown. The interrelationships between the members of the electron transport scheme involved with NXR for carrying out NO_2^- oxidation versus dissimilatory NO_3^- reduction are unknown. The specific mechanism of association of NXR with the cytoplasmic membrane is unknown.

gradient, However, since it is now generally accepted that $4H^+$ are transferred per turnover of conventional terminal cytochrome oxidase (Mathews et al., 2000), this should result in a net transfer of $2H^+$ per mole of NO_2^- oxidized, a H^+/O of 2, and generation of 1 ATP per 1.5 NO_2^- oxidized (assuming that $3H^+$ are translocated per ATP formed and that the H^+ gradient is not dissipated during reverse electron flow). Attempts to show H^+ pumping by cytochrome *c* oxidase from *N. winogradskyi* reconstituted into phospholipid vesicles were unsuccessful (Sone et al., 1983; Sone, 1986), as were attempts to show H^+ pumping by spheroplasts of *N. winogradskyi* (Hollocher et al., 1982). In contrast, others were successful in measuring H^+ translocation by whole cells of *N. winogradskyi* (Wetzstein and Ferguson, 1985). Interestingly, there is immunocytochemical evidence that NXR of *N. moscoviensis* is located either on the outside of the cell membrane or in the periplasmic space (Spieck et al., 1996a, 1998). This orientation of NXR would permit development of a conventional proton gradient, assuming that H^+ are generated on the periplasmic side of the membrane and that cytochrome *c* oxidase consumes H^+ on the cytoplasmic side of the membrane (see Chapter 12).

Because the rate of NO_2^--dependent O_2 uptake is quite rapid in *Nitrobacter*, cytochrome *c* oxidase must be a major sink for cytochrome

c_{550}-derived reductant, thereby facilitating the flow of electrons from NO_2^- oxidation by keeping cytochrome c_{550} mostly oxidized, and by consuming the H^+ generated during NO_2^- oxidation by NXR. Nonetheless, some NO_2^--derived reductant must flow against the thermodynamic gradient to produce NAD(P)H and FAD(H) to meet the needs of CO_2, NO_2^-, and SO_4^{2-} reduction, and for other biosynthetic reactions. Reverse electron flow requires energy input and should result either in consumption of ATP and/or dissipation of the proton gradient, or both. Just how NOB control the disproportionation of NO_2^--derived reductant still remains an unsolved mystery. Furthermore, it has been pointed out that reverse electron flow from NO_2^- oxidation accompanied by formation of NAD(P)H has not been experimentally demonstrated in NOB (Spieck and Bock, 2005). A recent model of the anaerobic NH_4^+-oxidizing mechanism (anammox) of "*Candidatus* Kuenenia stuttgartensis" also invokes a disproportionation of NO_2^- to both NO_3^- and nitric oxide (NO) as follows. Electrons obtained from NO_2^- oxidation to NO_3^- are speculated to fuel the reduction of more NO_2^- to NO, which then reacts in a reductant-consuming manner with NH_3 or NH_4^+ to form hydrazine (N_2H_4) (Strous et al., 2006). Perhaps this recent work on the mechanism of anammox will stimulate new efforts to better understand disproportionate electron flow from NO_2^- in *Nitrobacter* and other NOB.

DISSIMILATORY NO_3^- REDUCTION

It has been shown that NXR can behave as a dissimilatory NO_3^- reductase, optimally reducing NO_3^- to NO_2^- with NADH as the electron donor under anaerobic conditions at pH 6 to 7 (Tanaka et al., 1983; Sundermeyer-Klinger et al., 1984; Freitag et al., 1987). This property fits with the fact that *nxrA* and *nxrB* are homologs of *narGH* that comprise the large and small subunits of dissimilatory nitrate reductase found in some denitrifying bacteria (Kirstein and Bock, 1993; Moreno-Vivian and Ferguson, 1998). Because there is no evidence for genes representing other kinds of dissimilatory nitrate reductases in the genomes of *Nitrobacter* (Starkenburg et al., 2006, 2008c), and since *N. winogradskyi* grows anaerobically on pyruvate with NO_3^- as electron acceptor (Freitag et al., 1987; Bock et al., 1988), it is assumed that NXR functions as the dissimilatory nitrate reductase during anaerobic growth.

Evidence for further reduction of NO_2^- under anaerobic conditions is sparse. A copper-containing protein with NO_2^- reductase activity was shown to copurify with NXR from *N. vulgaris* strain Ab1, and some preliminary data suggest that NO is formed as an end product of NO_2^- reduction (Ahlers et al., 1990). A gene cluster that encodes a putative NirK-type nitrite reductase was found to be conserved in all three sequenced *Nitrobacter* genomes (Fig. 4), and *nirK* gene products reduce NO_2^- to NO during denitrification (Berks et al., 1995; Zumft, 1997; Tavares et al., 2006). Interestingly, this *nirK* gene cluster is most similar in sequence and organization to homologs in the AOB, *Nitrosomonas europaea* and *N. eutropha* (Cantera and Stein, 2007b), which suggests a broader function in nitrification reactions. Assuming that NirK is actively involved in NO_2^- reduction in *Nitrobacter*, a question arises about the fate of the product NO.

Although there is a report that N_2O is a significant terminal product of NO_3^- respiration in *N. vulgaris* strain Ab1, quantitative data were not presented (Freitag et al., 1987). Interestingly, the genomes of *N. winogradskyi* and *Nitrobacter* NB311A lack homologs of NOR (Starkenburg et al., 2008c), yet the genome of *N. hamburgensis* possesses a NOR indicating a genomic potential to reduce NO to N_2O. NO-dependent production of NADH in NO_2^--starved cells of *N. winogradskyi* strain Engel was determined by following the increase in absorbance of NADH at 340 nm in a whole-cell assay (Freitag and Bock, 1990). Confirmatory data were not obtained using a more conventional biochemical/enzymatic determination of NADH. NO-dependent NADH production occurred 200 times more rapidly than did NO_2^--dependent NO formation, presumably because of lower thermodynamic constraints. Unfortunately, the NO-consuming mechanism was not identified. Starkenburg et al (2008b) showed that the putative *nirK* in *N. winogradskyi* was upregulated significantly in response to incubation under subambient O_2 levels (≤10%), and upregulation was absolutely dependent upon the presence of NO_2^- (Cantera and Stein, 2007b). However, no evidence was obtained for NO_2^--dependent NO formation/accumulation. Interestingly, only micromolar concentrations of NO were required to completely inhibit NO_2^--dependent O_2 uptake, which recovered completely and immediately after NO was consumed. Both NO- and NO_2^--dependent O_2 consumption were inhibited by 1 mM CN^-, implying that NO consumption might occur via cytochrome oxidase activity.

Although the role of NirK and NO in *Nitrobacter* is unclear, Starkenburg et al. (2008b) speculated that NO might be involved in regulating forward versus reverse electron flux under conditions of low O_2 (Fig. 7). If NirK-dependent NO formation from NO_2^- occurred on the periplasmic side of the cytoplasmic membrane, it would consume H^+ and reductant. It has been calculated that if 25% of the electrons normally transferred to cytochrome oxidase were directed in a reverse direction to generate biosynthetic reductant, the H^+/O ratio would drop to 1.0, less ATP would be formed, and cell yield would decline (Poughon et al., 2001). It is possible that, under low O_2, NO inhibition of cytochrome oxidase activity facilitates electron flow from NO_2^- via NO to NADH and thereby promotes the synthesis of poly-β-hydroxybutyrate (PHB), which is known to occur in other bacteria under O_2-limited, nongrowth conditions. TEMs of *Nitrobacter* grown under O_2-limited conditions clearly show that the cells accumulate PHB to high levels (Freitag et al., 1987). Indeed, *Nitrobacter*'s close relative, *Bradyrhizobium*, is also well known for accumulating large amounts of PHB in the nongrowing symbiotic (bacteroid) state that exists in the O_2-limited environment of the legume root nodule. The following issues remain to be resolved as a legacy of Freitag and Bock's work: (i) the identity and physical location of the NO production and consumption mechanisms, (ii) the details of the mechanism of NO-dependent NADH formation, and (iii) the impact of NO formation and reverse electron transport on the H^+ gradient.

NITROGEN ASSIMILATION FOR BIOSYNTHESIS

Under chemolithoautotrophic growth conditions, *Nitrobacter* is routinely grown with NO_2^- as the sole source of N, which means that a fraction of NO_2^- is reductively assimilated into NH_4^+ and then into biomass. NO_2^- assimilation is likely mediated by an NADPH-dependent nitrite reductase encoded by *nirB* and *nirD* and identified on the *Nitrobacter* genomes. This raises questions about how effectively the assimilatory NO_2^- reductase competes with NXR for NO_2^- and how the reductant that moves by reverse electron flow against the redox gradient is regulated and partitioned among the competing biosynthetic processes.

After surveying a large number of papers on NOB, we concluded that NH_4^+ is not added routinely to growth medium as an N source for growth, despite the fact that in actively nitrifying environments, NH_4^+ will be available for assimilation by NOB. Although *Nitrobacter* genomes do not contain genes encoding for assimilatory NO_3^- reductase, they do contain genes that encode for NH_4^+ assimilatory enzymes such as glutamine synthetase, glutamate synthase, and glutamate dehydrogenase. Several genes associated with the regulation of NH_4^+ assimilation are present in the *Nitrobacter* genome, including *ntrB* (NRII) and *ntrC* (NR1), the uridylyl transferase and removing enzyme (GlnD), PII protein (GlnB), and the GS adenylylation (and deadenylylation) enzyme (GlnE). Interestingly, there is a copy of a gene encoding for a PII-type of regulatory protein adjacent to RuBisCO

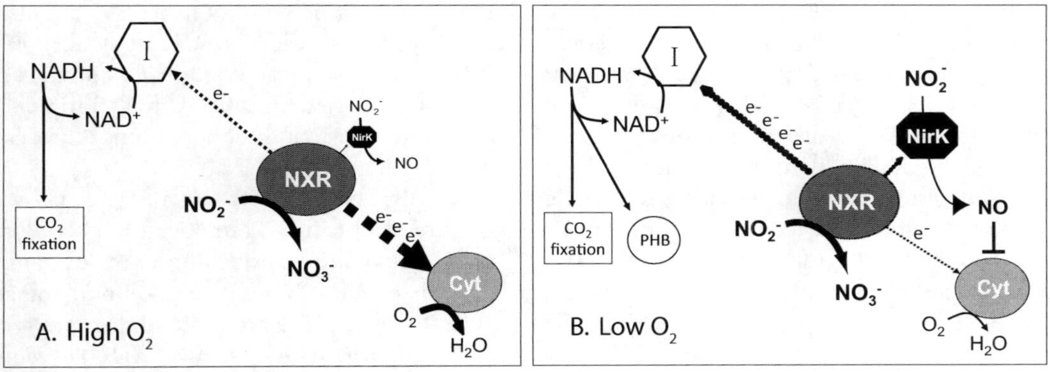

FIGURE 7 Model of NirK function and NO metabolism in *N. winogradskyi*. (A) In the presence of O_2, most electrons are thought to be directed toward respiration. (B) Under low-oxygen conditions, NirK expression increases, promoting the potential for NO_2^--dependent NO production and favoring electron flux toward reductant generation. Excess reductant would be consumed via nitrite reduction and PHB synthesis to maintain a balanced redox state. Cyt, cytochrome oxidase; NirK, nitrite reductase; I, NADH dehydrogenase (Complex I). (Reproduced from *Environmental Microbiology* [Starkenburg et al., 2008b] and the Society for Applied Microbiology/Blackwell Publishing with permission.)

(Fig. 7B). To our knowledge, the only other known example of this is found in the chemolithotrophic bacterium *Thiobacillus denitrificans* ATCC 25259. Although the role of the PII-like gene is unknown in *Nitrobacter*, its presence adjacent to both of the RuBisCO operons could indicate coordinated regulation of N and C metabolism. Indeed, there is precedent for extended roles of PII-like proteins in regulation of ammonia transport in *Azospirillum brasiliense*, high-affinity CO_2 transport in *Synechococcus*, and in a RuBisCO mutant of *Rhodobacter sphaeroides*, where *glnB* expression is no longer repressed by NH_4^+(Arcondeguy et al., 2001).

Genes encoding for the transport of branched and polar amino acids and peptides were identified in the genome of *N. winogradskyi* (Starkenburg et al., 2006). As mentioned previously, it has been common to use complex organic N sources such as casamino acids, peptone, and yeast extract for chemoorganotrophic growth of *Nitrobacter* (Smith and Hoare, 1968; Bock, 1976; Steinmuller and Bock, 1976). Since mineral sources of N were not added to the medium, we must conclude that *Nitrobacter* assimilates organic N, unless the complex N sources were contaminated with sufficient NH_4^+ to support growth. Furthermore, in the instances where chemoorganotrophic growth was reported to be faster than lithoautotrophic growth, we must assume that organic N assimilation is energy efficient (Bock et al., 1983, 1990). Studies are needed to determine whether the expression of genes associated with organic N transport facilitates transport of organic N sources into the cell.

CARBON STORAGE AND METABOLISM

Autotrophy and Carboxysomes

All cultured NOB have autotrophic growth potential, with many being obligate autotrophs using CO_2 as their only source of carbon. Although CO_2 fixation is mediated by a Type 1 ribulose 1,5-bisphosphate carboxylase-oxygenase in many NOB (Bock et al., 1986; Harris et al., 1988), recent genomic evidence has suggested the possibility of a different mechanism of CO_2 fixation by "*Candidatus* Nitrospira defluvii" and is described by Daims et al. (see Chapter 12).

Two copies of *cbbL* and *cbbS*, encoding the large and small subunits of RuBisCO, respectively, were identified in the genome of *N. winogradskyi* Nb255 as well as a complement of enzymes capable of carrying out the reactions of the Calvin-Benson-Bassham cycle. It is of interest that the two RuBisCO copies are not paralogs; one copy is similar to the genes found in the Alphaproteobacteria relatives *B. japonicum* and *R. palustris*, whereas the other RuBisCO genes are most similar to those found in *Thiobacillus* and the AOB, *Nitrosomonas* and *Nitrosospira*. In the latter cases, the RuBisCO genes are associated with carboxysome genes (Fig. 8B) and organized in a manner almost identical to that found in the Gammaproteobacteria, *Acidithiobacillus ferrooxidans*, and *T. denitrificans*.

An important question to ask is whether or not there is differential control of the two forms of RuBisCO and carboxysome structural genes in NOB. In this context, it is well documented that the ratio of soluble-to-particulate RuBisCO activities varies with both CO_2 availability and culture age. As noted in a previous section, *Nitrobacter* and *Nitrococcus* strains produce carboxysomes and contain both soluble and particulate (i.e., carboxysome-associated) forms of RuBisCO activity (Shively et al., 1977; Watson et al., 1989), whereas TEMs of *Nitrospira*, *Nitrospina*, and *Nitrotoga* strains do not reveal carboxysomes, and, in a few test cases, they have been shown to only possess soluble CO_2-fixing activity (Watson and Waterbury, 1971; Watson et al., 1986; Alawi et al., 2007).

Although some of the earliest pioneering work on carboxysomes was performed on *Nitrobacter* (Peters, 1974; Shively et al., 1977; Biedermann and Westphal, 1979), most research on these structures over the past 25 years has involved cyanobacteria or *Thiobacillus* spp. (Codd, 1988; Yeates et al., 2008). Recent data indicate that both RuBisCO and carbonic anhydrase (CA) are located in carboxysomes, with the outer shell impeding diffusion of CO_2 out of the carboxysome and CA-concentrating CO_2 in the carboxysome to maximize RuBisCO activity (Long et al., 2007; Cot et al., 2008; Yeates et al., 2008). Long et al. (2007) showed that cyanobacteria form a complex of RuBisCo with a specific carboxysome shell protein (CcmM) and with carbonic anhydrase (CCaA) (Fig. 8A).

Worthy of mention, all three *Nitrobacter* genomes contained homologs of molybdopterin-containing carbon monoxide dehydrogenase (Mo-CODH) (Starkenburg et al., 2008c). These genes are most similar and syntenous to Mo-CODH genes in *B. japonicum* USDA 110 and *R. palustris* CGA009 genomes. *B. japonicum* USDA 110 is capable of aerobic growth on CO as a sole carbon and energy source (Lorite et al., 2000) and can oxidize CO at the expense of nitrate reduction, but without growth, under anaerobic conditions (King, 2006).

Although consumption and growth on CO by commonly studied species of *Nitrobacter* have not been reported, the fact that the *N. hamburgensis* genome contains more complete copies of Mo-CODH-like genes than it does of NXR suggests that these proteins are (or were) physiologically important in the lifestyle of *Nitrobacter*.

Organic Carbon Metabolism

It has been recognized for many years that *Nitrobacter* species show potential for chemoorganotrophic growth and that the substrate range is restricted to acetate and a few C_3 molecules (pyruvate, glycerol, D-lactate). Neither phosphofructokinase nor phosphogluconate dehydratase was identified in the genome of *N. winogradskyi*, thereby preventing sugar catabolism by either the Embden-Myerhof or the Entner-Doudoroff pathway, respectively. Furthermore, the *N. winogradskyi* genome lacks genes that encode for sugar transporters. The reason for the inability of *N. hamburgensis* to grow on hexoses remains unclear, however, given that a complete Embden-Meyerhof pathway was annotated and that a gene cluster with homology to an ABC-type general sugar transporter was also identified (Starkenburg et al., 2008c). Growth on C_2 and C_3 compounds

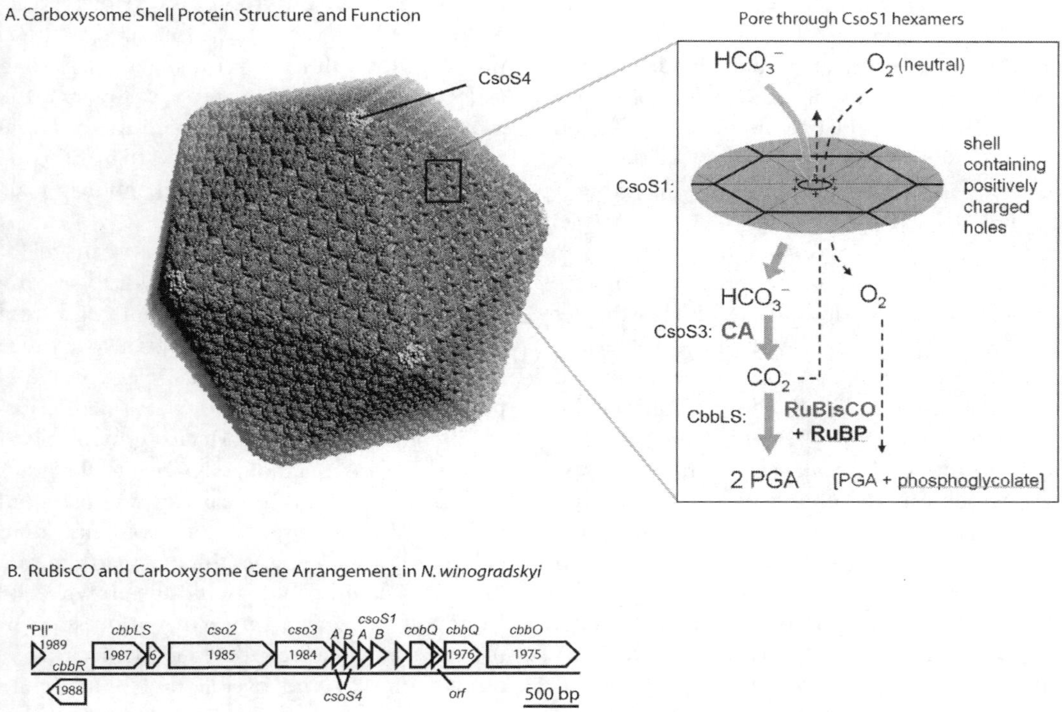

FIGURE 8 (A) Carboxysome shell protein structure and function. (B) RuBisCO and carboxysome gene arrangement in *N. winogradskyi*. CsoS1 forms hexamers that pack into a two-dimensional molecular layer. CsoS4 forms the vertices of the shell. Electrostatic pores (positively charged) through CsoS1 may function to transport bicarbonate (negatively charged) into and out of the carboxysome. Within the CsoS1/CsoS4 protein shell, the carboxysome encapsulates the CO_2-fixing enzymes, ribulose-1,5-bisphosphate carboxylase/oxygenase (RuBisCO; CbbLS) and carbonic anhydrase (CA; CsoS3), to enhance the efficiency of CO_2 fixation and the formation of two molecules of 3-phosphoglycerate (PGA). The function of CsoS2 is unknown. (Assembled from images provided by Todd Yeates.)

by *Nitrobacter* implies possession of glyoxylate bypass and other anapleurotic carboxylating enzymes. The glyoxylate pathway genes were annotated, and several genes encoding for enzymes that facilitate metabolism of pyruvate, acetate, and glycerol were identified.

There is evidence for adaptation of *Nitrobacter* to chemoorganotrophic growth conditions. Using difference spectrophotometric techniques, Kirstein et al. (1986) concluded that two different *b*-type cytochromes were produced during lithoautotrophic and chemoorganotrophic growth conditions, respectively. Difference spectra of CN- treated cells detected one version of cytochrome *b* in lithoautotrophically and mixotrophically grown cells, whereas no CN^--binding effect was detected in the spectrum of chemoorganotrophically cells. In contrast, a CO difference spectrum revealed additional peaks in chemoorganotrophic cells, which was interpreted to mean that an additional cytochrome *b* was produced during chemoorganotrophic growth in a defined medium containing acetate and NH_4^+ as sole C and N sources, respectively. Because a-type cytochromes were undetected in chemoorganotrophically grown cells, the authors speculated that the CO-binding *b*-type cytochrome might function as a cytochrome *o* oxidase.

Presumably, the absence of any *a*-type cytochromes would also fit with the fact that NXR is repressed under chemoorganotrophic growth conditions (Steinmuller and Bock, 1977). The relative abundance of heme *b*:heme *c* was determined to vary with composition of the chemoorganotrophic growth medium (Kirstein et al., 1986). For example, during growth on pyruvate and yeast extract, the ratio of heme *b*:heme *c* was 0.5:1, whereas during growth on acetate and NH_4^+ the *b*:*c* ratio was much higher with the cytochrome c_{550} absorption peak being virtually absent. Starkenburg et al. (2008c) found four copies of genes encoding *b*-type cytochromes on the chromosome of *N. hamburgensis* and one copy of a cytochrome *bd* ubiquinol oxidase located on a plasmid. Conventional models of bacterial respiration place *b*-type cytochromes in respiratory complex III, which functions at a redox potential similar to that of NO_2^- oxidation. Furthermore, in a conventional denitrification pathway, low- and high-potential *b*-type hemes are located on opposite sides of the cytoplasmic membrane to facilitate transmembrane electron transfer (Moreno-Vivian et al., 1999). The possible multiplicity of roles of *b*-type cytochromes in gating electrons produced via NO_2^- or organic C oxidation, and used for either O_2 or NO_3^- reduction and for reverse electron flow to NAD(P)H, is worthy of further biochemical and molecular studies.

Effects of Mixed Carbon and Energy Sources

Despite the ability to utilize and adapt to some organic carbon sources, several questions still remain regarding what controls the rate of chemoorganotrophic versus lithoautotrophic growth in NOB and how *Nitrobacter* utilizes, regulates, and responds to different carbon and energy sources. Some reports have indicated that chemoorganotrophic growth of *Nitrobacter* was much slower than chemolithoautotrophic growth (Smith and Hoare, 1968; Bock, 1976; Starkenburg et al., 2008a), whereas others have shown that *N. hamburgensis* X14 and *N. vulgaris* strain Z grow faster chemoorganotrophically than chemolithoautotrophically (Bock et al., 1983, 1990). Nevertheless, compared to bona fide organotrophs, the growth rates of *Nitrobacter* on organic carbon are very slow, and the growth yields of mixotrophic cultures of *Nitrobacter* are only modestly higher than lithoautotrophically grown cells (Starkenburg et al., 2008a).

The genomes of *Nitrobacter* contain all of the genes required for complexes I to IV of a respiratory electron transport chain, suggesting they have the theoretical potential to oxidize NAD(P)H through a complete electron transport chain to O_2. In this context, Starkenburg et al. (2008b) showed that D-lactate-grown cells of *N. hamburgensis* metabolized D-lactate at a faster rate than lithoautotrophically grown cells and that the rate of lactate-dependent O_2 uptake in D-lactate-grown cells was higher than the rate measured in lithoautotrophically grown cells. Although these data prove that processing of D-lactate is enhanced during chemoorganotrophic growth, the overall rate of D-lactate consumption was still barely sufficient to support the sum of the rate of lactate-dependent O_2 uptake and the rate of lactate-C assimilation. In the former study, the generation time of *N. hamburgensis* X14 grown chemoorganotrophically was 48 h after a 4-day lag period (Starkenburg et al., 2008a). This slow growth rate on D-lactate could be due to (i) rate-limiting transport of lactate into the cell, (ii) low specific activity of D-lactate dehydrogenase for D-lactate and/or that D-lactate is a secondary less favored substrate for another enzyme (e.g., β hydroxybutyrate dehydrogenase, glycolate oxidase), or (iii) a rate-limiting step in the upper electron transport system slows the oxidation of NADH or FADH relative to the oxidation of NO_2^-. A lactate transporter was not identified in any *Nitrobacter* genome (Starkenburg et al., 2006, 2008c), and no evidence was found of a regulatory gene upstream of the putative D-lactate dehydrogenase.

Despite the physiological adaptations to organic carbon observed by several investigators, in general, the results collectively point to a preference of *Nitrobacter* toward a lithoautotrophic lifestyle. As indicated above, organotrophic growth rates in *N. winogradskyi* are slower than lithoautotrophic growth rates, and when CO_2 was stripped from cultures of either *N. winogradskyi* or *N. hamburgensis* containing both organic carbon and NO_2^-, growth was not observed or was suppressed until the NO_2^- was consumed. It has been reported on several occasions that atmospheric levels of CO_2 are minimally required for optimal organotrophic growth of *Nitrobacter* (Delwiche and Feinstein, 1965; Ida and Alexander, 1965; Starkenburg et al., 2008a). Additionally, in *N. hamburgensis*, lactate consumption was shown to be suppressed, and CO_2 fixation continued to occur whenever NO_2^- was present (Starkenburg et al., 2008a). At best, organic C metabolism appears to be a supplemental system for growth under lithoautotrophic conditions. In one report, NO_2^- stimulated acetate assimilation by both lithoautotrophically and chemoorganotrophically grown cells of *N. winogradskyi* (Smith and Hoare, 1968) and in another, NO_2^-, reduced the rate of lactate consumption in mixotrophic and organotrophically grown cells (Starkenburg et al., 2008a).

To summarize, the data suggest that at least some strains of *Nitrobacter* can utilize organic C for growth when it is the sole source of energy, yet if NO_2^- is present, *Nitrobacter*'s heterotrophic potential is hampered by an inefficient mechanism to suppress NO_2^- consumption and CO_2 fixation.

Carbon Storage Compounds

As mentioned previously, TEMs indicate that *Nitrobacter* accumulates large amounts of PHB when grown chemoorganotrophically with NO_3^- under low O_2 conditions (Freitag et al., 1987). PHB synthesis and storage has been studied extensively in *Nitrobacter*'s close relative *Bradyrhizobium*, where it has been suggested that the availability of NH_4^+ is critical for promoting C assimilation into amino acids and N limitation directs reduced C into PHB synthesis (Trainer and Charles, 2006). The relative availabilities of reduced and oxidized inorganic N might play a role in the allocation of reductant in the NOB. As an aside, there is a putative link between PHB catabolism and the work of Starkenburg et al. (2008a), who showed that *N. hamburgensis* would grow on D-lactate but not L-lactate (Starkenburg et al., 2008a). Although this work was prompted by the presence of genes in the *N. hamburgensis* chromosome that were annotated to encode for enzymes with the ability to oxidize D-lactate, a protein was purified from *R. palustris* that was 50% as active in oxidizing D-β-hydroxybutyrate as D-lactate (Horikiri et al., 2004). Further work is needed to determine whether D-β-hydroxybutyrate CoA dehydrogenase and the PHB-degrading pathway are inadvertently involved in catabolism and growth of *N. hamburgensis* on D-lactate.

NOB BEHAVIOR IN COCULTURE WITH AOB

It has often been observed that the properties of nitrification measured in the soil environment, in particular, are often quite different from those measured in pure cultures of AOB and NOB (Prosser, 1989; Stark and Firestone, 1996; De Boer and Kowalchuk, 2001; Booth et al., 2005). While this might be due to different physiologies of archaeal and heterotrophic ammonia oxidizers relative to AOB, it remains a possibility that AOB-NOB associations might express different nitrifying properties than they do when grown independently in pure culture. In this context, the following examples are worthy of discussion.

Nitrite Metabolism

Recent studies indicate that the nitrite reductase (*nirK*) of AOB is upregulated in response to increasing NO_2^- concentration and decreasing pH (Beaumont et al., 2004; Beaumont et al., 2005). The growth yields of *nirK* and *norB* mutants of *N. europaea* were lower than wild type, suggesting that NO_2^- accumulation might interfere with efficient oxidation of hydroxylamine, which subsequently undergoes chemical autooxidation (Schmidt et al., 2004). At low O_2, up to 17% of NH_3 oxidized by *N. europaea* ended up in products other than NO_2^- and N_2O (Cantera and Stein, 2007a). In most natural oxic environments, NOB effectively remove NO_2^-, and AOB do not normally need to deal with the consequences of NO_2^- accumulation. On the other hand, some *Nitrospira* are very sensitive to NO_2^- levels, which might be due to the lack of effective NO_2^- protection. Perhaps *Nitrospira* spp. rely on being some critical distance away from any NO_2^- source, or only thrive in close contact with AOB under NH_4^+-limited conditions when NO_2^- production is limited and *Nitrospira* can handle the flux.

pH Tolerance

As mentioned elsewhere, it is common for nitrification to occur in low-pH environments where cultivated nitrifiers cannot function. Gieseke et al. (2006) obtained evidence for nitrification occurring in biofilms at pH 4 and concluded that the nitrifiers had adapted to low pH rather than avoiding it. This work complimented the older studies where aggregated cells were able to nitrify at low pH and single dispersed cells were not (De Boer et al., 1991). Furthermore, it has been shown that NXR is optimally active as a NO_2^- oxidoreductase at pH 8, whereas its activity as a NO_3^- reductase increases as pH is lowered to pH 6 to 7 (Tanaka et al., 1983). Given the combined sensitivity of NO_2^- and NO_3^- transformations to pH, it does raise the possibility that AOB/NOB aggregations might modify their functions to better tolerate acidic conditions known to deleteriously affect them as individuals.

O_2 Concentration

It has been documented that *N. europaea* has a lower K_s for O_2 than *N. winogradskyi*, which was shown by the accumulation of NO_2^- in cocultures at low O_2 (Laanbroek and Gerards, 1993). There is evidence that *N. hamburgensis* can adapt to low O_2 in a coculture with *N. europaea* by lowering its K_s for O_2 (Laanbroek et al., 1994). Indeed, as mentioned previously, genes encoding respiratory terminal oxidases are more abundant in *N. hamburgensis* than in *N. winogradskyi* (Starkenburg et al., 2008c), and Kirstein et al. (1986) postulated that *N. hamburgensis* expressed a different cytochrome oxidase under chemoorganotrophic growth conditions. However, NOB also possess *nirK*, and *N. hamburgensis* possesses *norB*, indicating that it might adapt to coculture under low O_2 by carrying out a more complete denitrification of NO_2^- to N_2O and prevent NO accumulation. To what extent the physiological activities of AOB and NOB change in response to coculture in biofilms or aggregates, and particularly under low-O_2 and low-pH conditions, awaits further detailed study.

CONCLUSIONS AND IMPLICATIONS

In contrast to AOB, the metabolism and biochemistry of NOB have been seriously neglected for almost 20 years and studied by a mere handful of dedicated research groups. Our knowledge about some of the most fundamental biochemical properties of the NOB remains incomplete and is often still controversial (mechanism of reverse electron transport, mechanism of NADH generation, mechanism of ATP generation, control of NO_2^--oxidizing versus NO_3^--reducing properties of NXR, and the role of NO in reductant generation, to name but a few). The reasons for the limited substrate range that supports chemoorganotrophic growth, as well as for the growth variations of NOB under chemoorganotrophic conditions remain unknown.

The genome analysis of *Nitrobacter* has provided some new and confirmatory insights into the biology of this genus. The picture that has emerged over the past 10 years regarding the wide phylogenetic distribution of the NOB across many different environments leads us to predict that other NOB (and possibly NO_2^--oxidizing archaea) with different growth properties and niches await discovery. The insights gained from the genomic analyses of *Nitrobacter* and *Nitrospira* suggest that the mechanistic details of NO_2^- oxidation and CO_2 fixation might differ among the NOB.

Through a comparative analysis of *Nitrobacter* genomes with the close non-NO_2^--

oxidizing alphaproteobacterial relatives *B. japonicum* and *R. palustris*, the genetic basis of NO_2^- oxidation is beginning to take shape. Further cross-lineage comparisons of all NOB genomes will expand our understanding of the metabolic commonalities and variations that support life via oxidation of NO_2^-.

ACKNOWLEDGMENTS

We are indebted to colleagues Todd Yeates and William Hickey for providing illustrations and figures for the chapter. We express our sincere appreciation to Holger Daims for a constructive appraisal of our chapter and for drawing our attention to insights gained from his extensive experiences with NOB, and to the comments of an anonymous reviewer. We thank the editors for accurate editing of our chapter.

For our own research contributions to this subject (S.R.S. and P.J.B.), we acknowledge financial support from the U.S. National Science Foundation, the U.S. Department of Energy, and the Oregon Agricultural Experiment Station. E.S. acknowledges financial support from the German Research Foundation and the Federal Ministry of Education and Research.

REFERENCES

Aamand, J., T. Ahl, and E. Spieck. 1996. Monoclonal antibodies recognizing nitrite oxidoreductase of *Nitrobacter hamburgensis*, *N. winogradskyi*, and *N. vulgaris*. *Appl. Environ. Microbiol.* **62:**2352–2355.

Ahlers, B., W. Konig, and E. Bock. 1990. Nitrite reductase activity in *Nitrobacter vulgaris*. *FEMS Microbiol. Lett.* **67:**121–126.

Alawi, M., A. Lipski, T. Sanders, E. M. Pfeiffer, and E. Spieck. 2007. Cultivation of a novel cold-adapted nitrite oxidizing betaproteobacterium from the Siberian Arctic. *ISME J.* **1:**256–264.

Aleem, M. I. 1965. Path of carbon and assimilatory power in chemosynthetic bacteria. I. *Nitrobacter agilis*. *Biochim. Biophys. Acta.* **107:**14–28.

Aleem, M. I. H., and D. L. Sewell. 1984. Oxido-reductase systems in *Nitrobacter agilis*, p. 185–210. *In* W. R. Strohl and O. H. Tuovinen (ed.), *Microbial Chemoautotrophy*. Ohio State University Press, Columbus, OH.

Aleem, M. I., G. E. Hoch, and J. E. Varner. 1965. Water as the source of oxidant and reductant in bacterial chemosynthesis. *Proc. Natl. Acad. Sci. USA* **54:**869–873.

Arcondeguy, T., R. Jack, and M. Merrick. 2001. P(II) signal transduction proteins, pivotal players in microbial nitrogen control. *Microbiol. Mol. Biol. Rev.* **65:**80–105.

Bartosch, S., I. Wolgast, E. Spieck, and E. Bock. 1999. Identification of nitrite-oxidizing bacteria with monoclonal antibodies recognizing the nitrite oxidoreductase. *Appl. Environ. Microbiol.* **65:**4126–4133.

Bartosch, S., C. Hartwig, E. Spieck, and E. Bock. 2002. Immunological detection of *Nitrospira*-like bacteria in various soils. *Microb. Ecol.* **43:**26–33.

Beaumont, H. J., S. I. Lens, W. N. Reijnders, H. V. Westerhoff, and R. J. van Spanning. 2004. Expression of nitrite reductase in *Nitrosomonas europaea* involves NsrR, a novel nitrite-sensitive transcription repressor. *Mol. Microbiol.* **54:**148–158.

Beaumont, H. J., S. I. Lens, H. V. Westerhoff, and R. J. van Spanning. 2005. Novel *nirK* cluster genes in *Nitrosomonas europaea* are required for NirK-dependent tolerance to nitrite. *J. Bacteriol.* **187:**6849–6851.

Berks, B. C., S. J. Ferguson, J. W. Moir, and D. J. Richardson. 1995. Enzymes and associated electron transport systems that catalyse the respiratory reduction of nitrogen oxides and oxyanions. *Biochim. Biophys. Acta.* **1232:**97–173.

Biedermann, M., and K. Westphal. 1979. Chemical composition and stability of Nb1-particles from *Nitrobacter agilis*. *Arch. Microbiol.* **121:**187–191.

Bock, E. 1976. Growth of *Nitrobacter* in the presence of organic matter. II. Chemoorganotrophic growth of *Nitrobacter agilis*. *Arch. Microbiol.* **108:**305–312.

Bock, E., H. Sundermeyer-Klinger, and E. Stackebrandt. 1983. New facultative lithoautotrophic nitrite-oxidizing bacteria. *Arch. Microbiol.* **136:**281–284.

Bock, E., H. P. Koops, and H. Harms. 1986. Cell biology of nitrifiers, p. 17–38. *In* J. I. Prosser (ed.), *Nitrification*. IRL, Oxford, United Kingdom.

Bock, E., P. A. Wilderer, and A. Freitag. 1988. Growth of *Nitrobacter* in the absence of dissolved oxygen. *Water Res.* **22:**245–250.

Bock, E., H. P. Koops, U. C. Möller, and M. Rudert. 1990. A new facultatively nitrite oxidizing bacterium, *Nitrobacter vulgaris* sp. nov. *Arch. Microbiol.* **153:**105–110.

Bock, E., H. P. Koops, H. Harms, and B. Ahlers. 1991. The biochemistry of nitrifying organisms, p. 171–200. *In* J. M. Shively and L. L. Barton (ed.), *Variations in Autotrophic Life*. Academic Press, San Diego, CA.

Booth, M. S., J. M. Stark, and E. Rastetter. 2005. Controls on nitrogen cycling in terrestrial ecosystems: a synthetic analysis of literature data. *Ecol. Monogr.* **75:**139–157.

Cantera, J. J., and L. Y. Stein. 2007a. Role of nitrite reductase in the ammonia-oxidizing pathway of *Nitrosomonas europaea*. *Arch. Microbiol.* **188:**349–354.

Cantera, J. J., and L. Y. Stein. 2007b. Molecular diversity of nitrite reductase genes (*nirK*) in nitrifying bacteria. *Environ. Microbiol.* **9:**765–776.

Chaudhry, G. R., I. Suzuki, and H. Lees. 1980.

Cytochrome oxidase of *Nitrobacter agilis*: isolation by hydrophobic interaction chromatography. *Can. J. Microbiol.* **26:**1270–1274.

Cobley, J. G. 1976a. Energy-conserving reactions in phosphorylating electron-transport particles from *Nitrobacter winogradskyi*. Activation of nitrite oxidation by the electrical component of the protonmotive force. *Biochem. J.* **156:**481–491.

Cobley, J. G. 1976b. Reduction of cytochromes by nitrite in electron-transport particles from *Nitrobacter winogradskyi*: proposal of a mechanism for H^+ translocation. *Biochem. J.* **156:**493–498.

Codd, G. A. 1988. Carboxysomes and ribulose bisphosphate carboxylase/oxygenase. *Adv. Microb. Physiol.* **29:**115–164.

Cot, S. S., A. K. So, and G. S. Espie. 2008. A multiprotein bicarbonate dehydration complex essential to carboxysome function in cyanobacteria. *J. Bacteriol.* **190:**936–945.

Daims, H., J. L. Nielsen, P. H. Nielsen, K. H. Schleifer, and M. Wagner. 2001. In situ characterization of *Nitrospira*-like nitrite-oxidizing bacteria active in wastewater treatment plants. *Appl. Environ. Microbiol.* **67:**5273–5284.

De Boer, W., and G. A. Kowalchuk. 2001. Nitrification in acid soils: miroorganisms and mechanisms. *Soil. Biol. Biochem.* **33:**853–866.

De Boer, W., P. J. Gunnewiek, M. Veenhuis, E. Bock, and H. J. Laanbroek. 1991. Nitrification at low pH by aggregated chemolithotrophic bacteria. *Appl. Environ. Microbiol.* **57:**3600–3604.

de la Torre, J. R., C. B. Walker, A. E. Ingalls, M. Konneke, and D. A. Stahl. 2008. Cultivation of a thermophilic ammonia oxidizing archaeon synthesizing crenarchaeol. *Environ. Microbiol.* **10:**810–818.

Delwiche, C. C., and M. S. Feinstein. 1965. Carbon and energy sources for the nitrifying autotroph *Nitrobacter*. *J. Bacteriol.* **60:**102–107.

Ehrich, S., D. Behrens, E. Lebedeva, W. Ludwig, and E. Bock. 1995. A new obligately chemolithoautotrophic, nitrite-oxidizing bacterium, *Nitrospira moscoviensis* sp. nov. and its phylogenetic relationship. *Arch. Microbiol.* **164:**16–23.

Freitag, A., and E. Bock. 1990. Energy conservation in *Nitrobacter*. *FEMS Microbiol. Lett.* **66:**157–162.

Freitag, A., M. Rudert, and E. Bock. 1987. Growth of *Nitrobacter* by dissimilatoric nitrate reduction. *FEMS Microbiol. Lett.* **48:**105–109.

Griffin, B. M., J. Schott, and B. Schink. 2007. Nitrite, an electron donor for anoxygenic photosynthesis. *Science* **316:**1870.

Hankinson, T. R., and E. L. Schmidt. 1988. An acidophilic and aneutrophilic *Nitrobacter* strain Isolated from the numerically predominant nitrite-oxidizing population of an acid forest soil. *Appl. Environ. Microbiol.* **54:**1536–1540.

Harris, S., A. Ebert, E. Schutze, M. Diercks, E. Bock, and J. M. Shively. 1988. Two different genes and gene products for the large subunit of ribulose-1,5-bisphosphate carboxylase/oxygenase (RuBisCOase) in *Nitrobacter hamburgensis*. *FEMS Microbiol. Lett.* **49:**267–271.

Hatzenpichler, R., E. V. Lebedeva, E. Spieck, K. Stoecker, A. Richter, H. Daims, and M. Wagner. 2008. A moderately thermophilic ammonia-oxidizing crenarchaeote from a hot spring. *Proc. Natl. Acad. Sci. USA* **105:**2134–2139.

Hollocher, T. C., S. Kumar, and D. J. Nicholas. 1982. Respiration-dependent proton translocation in *Nitrosomonas europaea* and its apparent absence in *Nitrobacter agilis* during inorganic oxidations. *J. Bacteriol.* **149:**1013–1020.

Hooper, A. B., and A. A. DiSpirito. 1985. In bacteria which grow on simple reductants, generation of a proton gradient involves extracytoplasmic oxidation of substrate. *Micobiol. Rev.* **49:**140–157.

Horikiri, S., Y. Aizawa, T. Kai, S. Amachi, H. Shinoyama, and T. Fujii. 2004. Electron acquisition system constructed from an NAD-independent D-lactate dehydrogenase and cytochrome c2 in *Rhodopseudomonas palustris* No. 7. *Biosci. Biotechnol. Biochem.* **68:**516–522.

Ida, S., and M. Alexander. 1965. Permeability of *Nitrobacter agilis* to organic compounds. *J. Bacteriol.* **90:**151–156.

Juretschko, S., G. Timmermann, M. Schmid, K. H. Schleifer, A. Pommerening-Roser, H. P. Koops, and M. Wagner. 1998. Combined molecular and conventional analyses of nitrifying bacterium diversity in activated sludge: *Nitrosococcus mobilis* and *Nitrospira*-like bacteria as dominant populations. *Appl. Environ. Microbiol.* **64:**3042–3051.

King, G. M. 2006. Nitrate-dependent anaerobic carbon monoxide oxidation by aerobic CO-oxidizing bacteria. *FEMS Microbiol. Ecol.* **56:**1–7.

Kirstein, K., and E. Bock. 1993. Close genetic relationship between *Nitrobacter hamburgensis* nitrite oxidoreductase and *Escherichia coli* nitrate reductases. *Arch. Microbiol.* **160:**447–453.

Kirstein, K. O., E. Bock, D. J. Miller, and D. J. D. Nicholas. 1986. Membrane-bound b-type cytochromes in *Nitrobacter*. *FEMS Microbiol. Lett.* **36:**63–67.

Konneke, M., A. E. Bernhard, J. R. de la Torre, C. B. Walker, J. B. Waterbury, and D. A. Stahl. 2005. Isolation of an autotrophic ammonia-oxidizing marine archaeon. *Nature* **437:**543–546.

Koops, H.-P., and A. Pommerening-Röser. 2001. Distribution and ecophysiology of the nitrifying bacteria emphasizing cultured species. *FEMS Microbiol. Ecol.* **37:**1–9.

Kruger, B., O. Meyer, A. Nagel, J. R. Andreesen, M. Meincke, E. Bock, S. Blamle, and W. G.

Zumft. 1987. Evidence for the presence of bactopterin in the eubacterial molybdoenzymes nicotinic acid dehydrogenase, nitrite oxidoreductase and respiratory nitrate reductase. *FEMS Microbiol. Lett.* **48:**225–227.

Laanbroek, H. J., and S. Gerards. 1993. Competition for limiting amounts of oxygen between *Nitrosomonas europaea* and *Nitrobacter winogradskyi* grown in mixed continuous cultures. *Arch. Microbiol.* **159:**453–459.

Laanbroek, H. J., P. Bodelier, and S. Gerards. 1994. Oxygen consumption kinetics of *Nitrosomonas europaea* and *Nitrobacter hamburgensis* grown in mixed continuous cultures at different oxygen concentrations. *Arch. Microbiol.* **161:**156–162.

Lebedeva, E. V., M. Alawi, C. Fiencke, B. Namsaraev, E. Bock, and E. Spieck. 2005. Moderately thermophilic nitrifying bacteria from a hot spring of the Baikal rift zone. *FEMS Microbiol. Ecol.* **54:**297–306.

Lebedeva, E. V., M. Alawi, F. Maixner, P. G. Jozsa, H. Daims, and E. Spieck. 2008. Physiological and phylogenetic characterization of a novel lithoautotrophic nitrite-oxidizing bacterium, '*Candidatus* Nitrospira bockiana.' *Int. J. Syst. Evol. Microbiol.* **58:**242–250.

Leininger, S., T. Urich, M. Schloter, L. Schwark, J. Qi, G. W. Nicol, J. I. Prosser, S. C. Schuster, and C. Schleper. 2006. Archaea predominate among ammonia-oxidizing prokaryotes in soils. *Nature* **442:**806–809.

Lipski, A., E. Spieck, A. Makolla, and K. Altendorf. 2001. Fatty acid profiles of nitrite-oxidizing bacteria reflect their phylogenetic heterogeneity. *Syst. Appl. Microbiol.* **24:**377–384.

Long, B. M., M. R. Badger, S. M. Whitney, and G. D. Price. 2007. Analysis of carboxysomes from *Synechococcus* PCC7942 reveals multiple Rubisco complexes with carboxysomal proteins CcmM and CcaA. *J. Biol. Chem.* **282:**29323–29335.

Lorite, M. J., J. Tachil, J. Sanjuán, O. Meyer, and E. J. Bedmar. 2000. Carbon monoxide dehydrogenas activity in *Bradyrhizobium japonicum*. *Appl. Environ. Microbiol.* **66:**1871–1876.

Maixner, F., D. R. Noguera, B. Anneser, K. Stoecker, G. Wegl, M. Wagner, and H. Daims. 2006. Nitrite concentration influences the population structure of *Nitrospira*-like bacteria. *Environ. Microbiol.* **8:**1487–1495.

Mathews, C. K., K. E. Van Holde, and K. G. Ahern. 2000. *Biochemistry*, 3rd ed. Benjamin Cummings, San Francisco, CA.

Meincke, M., E. Bock, D. Kastrau, and P. M. H. Kroneck. 1992. Nitrite oxidoreductase from *Nitrobacter hamburgensis*: redox centers and their catalytic role. *Arch. Microbiol.* **158:**127–131.

Moreno-Vivian, C., and S. J. Ferguson. 1998. Definition and distinction between assimilatory, dissimilatory and respiratory pathways. *Mol. Microbiol.* **29:**661–669.

Moreno-Vivian, C., P. Cabello, M. Martinez-Luque, R. Blasco, and F. Castillo. 1999. Prokaryotic nitrate reduction: molecular properties and functional distinction among bacterial nitrate reductases. *J. Bacteriol.* **181:**6573–6584.

Nicol, G. W., S. Leininger, C. Schleper, and J. I. Prosser. 2008. The influence of soil pH on the diversity, abundance and transcriptional activity of ammonia oxidizing archaea and bacteria. *Environ. Microbiol.* **10:**2966–2978.

Nomoto, T., Y. Fukumori, and T. Yamanaka. 1993. Membrane-bound cytochrome c is an alternative electron donor for cytochrome aa3 in *Nitrobacter winogradskyi*. *J. Bacteriol.* **175:**4400–4404.

Peters, K. R. 1974. Reconstruction of capside structures in isometrical viruses with an equidensities rotation method. [Author's translation.] *Mikroskopie* **30:**270–280.

Poughon, L., C. G. Dussap, and J. B. Gros. 2001. Energy model and metabolic flux analysis for autotrophic nitrifiers. *Biotechnol. Bioeng.* **72:**416–433.

Prosser, J. I. 1989. Autotrophic nitrification in bacteria, p. 125–181. *In* A. H. Rose and J. F. Wilkinson (ed.), *Advances in Microbial Physiology*. Academic Press, London, United Kingdom.

Prosser, J. I., and G. W. Nicol. 2008. Relative contributions of archaea and bacteria to aerobic ammonia oxidation in the environment. *Environ. Microbiol.* **10:**2931–2941.

Schmidt, I., R. J. van Spaaning, and M. S. Jetten. 2004. Denitrification and ammonia oxidation by *Nitrosomonas europaea* wild-type, and NirK- and NorB-deficient mutants. *Microbiology* **150:**4107–4114.

Schramm, A., D. de Beer, J. C. van den Heuvel, S. Ottengraf, and R. Amann. 1999. Microscale distribution of populations and activities of *Nitrosospira* and *Nitrospira* spp. along a macroscale gradient in a nitrifying bioreactor: quantification by in situ hybridization and the use of microsensors. *Appl. Environ. Microbiol.* **65:**3690–3696.

Seewaldt, E., K. H. Schleifer, E. Bock, and E. Stackebrandt. 1982. The close phylogenetic relationship of *Nitrobacter* and *Rhodopseudomonas palustris*. *Arch. Microbiol.* **131:**287–290.

Shively, J. M., E. Bock, K. Westphal, and G. C. Cannon. 1977. Icosahedral inclusions (carboxysomes) of *Nitrobacter agilis*. *J. Bacteriol.* **132:**673–675.

Smith, A. J., and D. S. Hoare. 1968. Acetate assimilation by *Nitrobacter agilis* in relation to its "obligate autotrophy." *J. Bacteriol.* **95:**844–855.

Sone, N. 1986. Measurement of proton pump activity of the thermophilic bacterium PS3 and *Nitrobacter agilis* at the cytochrome oxidase level using total membrane and heptyl thioglucoside. *J. Biochem.* (Tokyo) **100:**1465–1470.

Sone, N., Y. Yanagita, K. Hon-nami, Y. Fukumori, and T. Yamanaka. 1983. Proton pump activity of *Nitrobacter agilis* and *Thermus thermophilus* cytochrome c oxidase. *FEBS Lett.* **155:**150–154.

Sorokin, D. Y., G. Muyzer, T. Brinkhoff, J. G. Kuenen, and M. S. Jetten. 1998. Isolation and characterization of a novel facultatively alkaliphilic *Nitrobacter* species, *N. alkalicus* sp. nov. *Arch. Microbiol.* **170:**345–352.

Spieck, E., and E. Bock. 2005. The lithoautotrophic nitrite-oxidizing bacteria, p. 149–153. *In* D. J. Brenner, N. R. Krieg, and J. T. Staley (ed.), *Bergey's Manual of Systematic Bacteriology*, 2nd ed. Springer, New York, NY.

Spieck, E., J. Aamand, S. Bartosch, and E. Bock. 1996a. Immunocytochemical detection and location of the membrane-bound nitrite-oxidoreductase in cells of *Nitrobacter* and *Nitrospira*. *FEMS Microbiol. Lett.* **139:**71–76.

Spieck, E., S. Muller, A. Engel, E. Mandelkow, H. Patel, and E. Bock. 1996b. Two-dimensional structure of membrane-bound nitrite oxidoreductase from *Nitrobacter hamburgensis*. *J. Struct. Biol.* **117:**117–123.

Spieck, E., S. Ehrich, J. Aamand, and E. Bock. 1998. Isolation and immunocytochemical location of the nitrite-oxidizing system in *Nitrospira moscoviensis*. *Arch. Microbiol.* **169:**225–230.

Spieck, E., C. Hartwig, I. McCormack, F. Maixner, M. Wagner, A. Lipski, and H. Daims. 2006. Selective enrichment and molecular characterization of a previously uncultured *Nitrospira*-like bacterium from activated sludge. *Environ. Microbiol.* **8:**405–415.

Stark, J. M., and M. K. Firestone. 1996. Kinetic characteristics of ammonium oxidizer communities in a California oak woodland-annual grassland. *Soil Biol. Biochem.* **28:**1307–1317.

Starkenburg, S. R., P. S. Chain, L. A. Sayavedra-Soto, L. Hauser, M. L. Land, F. W. Larimer, S. A. Malfatti, M. G. Klotz, P. J. Bottomley, D. J. Arp, and W. J. Hickey. 2006. Genome sequence of the chemolithoautotrophic nitrite-oxidizing bacterium *Nitrobacter winogradskyi* Nb-255. *Appl. Environ. Microbiol.* **72:**2050–2063.

Starkenburg, S. R., D. J. Arp, and P. J. Bottomley. 2008a. D-Lactate metabolism and the obligate requirement for CO_2 during growth on nitrite by the facultative lithoautotroph *Nitrobacter hamburgensis*. *Microbiology* **154:**2473–2481.

Starkenburg, S. R., D. J. Arp, and P. J. Bottomley. 2008b. Expression of a putative nitrite reductase and the reversible inhibition of nitrite-dependent respiration by nitric oxide in *Nitrobacter winogradskyi* Nb-255. *Environ. Microbiol.* **10:**3036–3042.

Starkenburg, S. R., F. W. Larimer, L. Y. Stein, M. G. Klotz, P. S. Chain, L. A. Sayavedra-Soto, A. T. Poret-Peterson, M. E. Gentry, D. J. Arp, B. Ward, and P. J. Bottomley. 2008c. Complete genome sequence of *Nitrobacter hamburgensis* X14 and comparative genomic analysis of species within the genus *Nitrobacter*. *Appl. Environ. Microbiol.* **74:**2852–2863.

Steinmuller, W., and E. Bock. 1976. Growth of *Nitrobacter* in the presence of organic matter. I. Mixotrophic growth. *Arch. Microbiol.* **108:**299–304.

Steinmuller, W., and E. Bock. 1977. Enzymatic studies on autotrophically, mixotrophically and heterotrophically grown *Nitrobacter agilis* with special reference to nitrite oxidase. *Arch. Microbiol.* **115:**51–54.

Strous, M., E. Pelletier, S. Mangenot, T. Rattei, A. Lehner, M. W. Taylor, M. Horn, H. Daims, D. Bartol-Mavel, P. Wincker, V. Barbe, N. Fonknechten, D. Vallenet, B. Segurens, C. Schenowitz-Truong, C. Medigue, A. Collingro, B. Snel, B. E. Dutilh, H. J. Op den Camp, C. van der Drift, I. Cirpus, K. T. van de Pas-Schoonen, H. R. Harhangi, L. van Niftrik, M. Schmid, J. Keltjens, J. van de Vossenberg, B. Kartal, H. Meier, D. Frishman, M. A. Huynen, H. W. Mewes, J. Weissenbach, M. S. Jetten, M. Wagner, and D. Le Paslier. 2006. Deciphering the evolution and metabolism of an anammox bacterium from a community genome. *Nature* **440:**790–794.

Sundermeyer-Klinger, H., W. Meyer, B. Warninghoff, and E. Bock. 1984. Membrane-bound nitrite-oxidoreductase of *Nitrobacter*: evidence for a nitrate reductase system. *Arch. Microbiol.* **140:**153–158.

Tanaka, Y., Y. Fukumori, and T. Yamanaka. 1983. Purification of cytochrome a1c1 from *Nitrobacter agilis* and characterization of the nitrite oxidation system of the bacterium. *Arch. Microbiol.* **135:**265–271.

Tavares, P., A. S. Pereira, J. J. Moura, and I. Moura. 2006. Metalloenzymes of the denitrification pathway. *J. Inorg. Biochem.* **100:**2087–2100.

Teske, A., E. Alm, J. M. Regan, S. Toze, B. E. Rittmann, and D. A. Stahl. 1994. Evolutionary relationships among ammonia- and nitrite-oxidizing bacteria. *J. Bacteriol.* **176:**6623–6630.

Trainer, M. A., and T. C. Charles. 2006. The role of PHB metabolism in the symbiosis of rhizobia with legumes. *Appl. Microbiol. Biotechnol.* **74:**377–386.

Vanparys, B., P. Bodelier, and P. De Vos. 2006. Validation of the correct start codon of *norX/nxrX* and universality of the *norAXB/nxrAXB* gene cluster in *Nitrobacter* species. *Curr. Microbiol.* **53:**255–257.

Watson, S. W., and J. B. Waterbury. 1971. Characteristics of two marine nitrite oxidizing bacteria,

Nitrospina gracilis nov. gen. nov. sp. and *Nitrococcus mobilis* nov. gen. nov. sp. *Arch. Microbiol.* **77:**203–230.

Watson, S. W., E. Bock, F. W. Valois, J. B. Waterbury, and U. Schlosser. 1986. *Nitrospira marina*, gen. nov. sp. nov.: a chemolithotrophic nitrite oxidizing bacterium. *Arch. Microbiol.* **144:**1–7.

Watson, S. W., E. Bock, H. Harms, H. P. Koops, and A. B. Hooper. 1989. Nitrifying bacteria, p. 1808–1834. *In* J. T. Staley (ed.), *Bergey's Manual of Systematic Bacteriology*. Williams and Wilkins, Baltimore, MD.

Wetzstein, H. G., and S. J. Ferguson. 1985. Respiration dependent proton translocation and the mechanism of proton motive force generation in *Nitrobacter winogradskyi*. *FEMS Microbiol. Lett.* **30:**87–92.

Wood, P. M. 1986. Nitrification as a bacterial energy source, p. 39–62. *In* J. I. Prosser (ed.), *Nitrification*. IRL Press, Washington, DC.

Wuchter, C., B. Abbas, M. J. Coolen, L. Herfort, J. van Bleijswijk, P. Timmers, M. Strous, E. Teira, G. J. Herndl, J. J. Middelburg, S. Schouten, and J. S. Sinninghe Damsté. 2006. Archaeal nitrification in the ocean. *Proc. Natl. Acad. Sci. USA* **103:**12317–12322.

Yamanaka, T., and Y. Fukumori. 1988. The nitrite oxidizing system of *Nitrobacter winogradskyi*. *FEMS Microbiol. Rev.* **4:**259–270.

Yamanaka, T., Y. Kamita, and Y. Fukumori. 1981. Molecular and enzymatic properties of "cytochrome aa3"-type terminal oxidase derived from *Nitrobacter agilis*. *J. Biochem.* (Tokyo) **89:**265–273.

Yeates, T. O., C. A. Kerfeld, S. Heinhorst, G. C. Cannon, and J. M. Shively. 2008. Protein-based organelles in bacteria: carboxysomes and related microcompartments. *Nat. Rev. Microbiol.* **6:**681–691.

Zumft, W. G. 1997. Cell biology and molecular basis of denitrification. *Microbiol. Mol. Biol. Rev.* **61:**533–616.

DIVERSITY, ENVIRONMENTAL GENOMICS, AND ECOPHYSIOLOGY OF NITRITE-OXIDIZING BACTERIA

Holger Daims, Sebastian Lücker, Denis Le Paslier, and Michael Wagner

12

INTRODUCTION

Nitrite-oxidizing bacteria (NOB) catalyze the oxidation of nitrite to nitrate, which is the second step of nitrification and a key process of the biogeochemical nitrogen cycle. As nitrite is usually scarce in natural habitats, the activity of NOB is tightly linked to that of ammonia oxidizers that convert ammonia to nitrite, thus supplying the NOB with their substrate. In absence of NOB, nitrite would accumulate and finally reach concentrations toxic for other microbes and eukaryotes in the environment. The product of nitrite oxidation, nitrate, not only is a main nitrogen source for other microorganisms and for plants but also serves as electron acceptor in microbial nitrate respiration under oxygen-limited conditions. The ecological key role of NOB implies that these bacteria are widely distributed in nature and have adapted to a great variety of environmental conditions. They inhabit not only all kinds of moderate aquatic and terrestrial ecosystems but also live in extreme settings like permafrost soils (Alawi et al., 2007) and geothermal springs (Lebedeva et al., 2005). This impressive ecophysiological versatility of NOB is paralleled by a considerable phylogenetic diversity within this functional group, whose known representatives belong to five genera from two different phyla of the bacterial domain. That said, it appears surprising that their ubiquity and the extent of their diversity were unknown until recently. After the discovery of the first described nitrite oxidizer by Sergej Winogradsky more than a century ago (Winogradsky, 1892), the methods used to enrich, isolate, and characterize NOB did not significantly change for many decades. These techniques, which were based on the cultivation of NOB in the laboratory, led to the discovery of the type strains of all presently known NOB genera. Without any doubt, this was a great achievement in view of the difficulties to enrich and purify the slowly growing NOB and to maintain their cultures in the laboratory. However, the success of these approaches was limited to those representatives that grew in the artificial media and under the applied incubation conditions. Soon after cultivation-independent molecular methods to identify and characterize bacteria had become available, the diversity and environmental distribution of yet uncultured members of the respective NOB genera were discovered. In particular, within the genera *Nitrobacter* and *Nitrospira*, the molecular tools revealed a great diversity of

Holger Daims, Sebastian Lücker, and Michael Wagner, Department of Microbial Ecology, University of Vienna, 1090 Vienna, Austria. *Denis Le Paslier,* CEA/Genoscope, CNRS UMR8030, Evry, France.

Nitrification, Edited by Bess B. Ward, Daniel J. Arp, and Martin G. Klotz

previously overlooked environmental strains. Today, the respective advantages of cultivation-based and molecular methods are combined to study novel NOB, for example, by adapting the cultivation strategies to conditions prevalent in habitats where yet uncultured NOB have been detected. This integrative approach has already led to the description of new candidate NOB species (Spieck et al., 2006; Alawi et al., 2007), and it has paved the way to a detailed characterization of such NOB by genome reconstruction from nitrite-oxidizing enrichments.

In addition to the roles NOB play in natural ecosystems, these bacteria are of high importance in biotechnology. Nitrification is a key process for nitrogen removal from sewage in biological wastewater treatment. Without functional nitrification in wastewater treatment plants, natural ecosystems would be flooded with ammonia from household sewage and industrial waste. Soon this would lead to excessive eutrophication of lakes and rivers, and this effect, together with the toxicity of ammonia and nitrite, would cause a dramatic decline of aquatic organisms. Thus, it is safe to say that the activity of nitrifiers (including NOB) in wastewater treatment plants is critical for maintaining environmental health, especially in times of an increasing human population and growing urban areas in many regions of the world. Unfortunately, nitrite oxidation in sewage treatment plants is prone to failure due to the slow growth of NOB and their sensitivity to disturbances such as a changing wastewater composition. This applies especially to industrial systems but also to small municipal wastewater treatment plants like those built in rural areas and developing countries. However, NOB are not always beneficial. Nitrification causes nitrogen losses from agricultural soils, because nitrate is leached out of the upper soil layers more quickly than ammonia. A comprehensive biological understanding of NOB will be important to improve the functional stability of wastewater treatment systems and to reduce detrimental effects of nitrification in agriculture.

The first part of this chapter provides an overview of the phylogenetic diversity and distribution of NOB in the environment and in engineered systems. Special emphasis is put on yet uncultured NOB, which have turned out to be the most widely distributed and abundant nitrite oxidizers known to date. Subsequently, we focus on the ecophysiology and ecological niche differentiation of the genera *Nitrobacter* and *Nitrospira*. These lineages not only represent the most versatile groups of NOB but also are of major importance as model organisms for nitrification research (*Nitrobacter*) and as key nitrifiers in biological wastewater treatment (*Nitrospira*). Due to difficulties to culture *Nitrospira* in the laboratory, still very little is known about these NOB, although they seem to be essential for nitrogen cycling in most ecosystems. The third part of this chapter addresses most recent insights into the biology of *Nitrospira*, which are based on the first sequenced genome from this genus.

DIVERSITY AND ENVIRONMENTAL DISTRIBUTION OF NOB

NOB are phylogenetically a relatively heterogeneous functional group. To date, five genera of aerobic chemolithoautotrophic NOB have been described: *Nitrobacter*, *Nitrococcus*, *Nitrospina*, *Nitrospira*, and the candidate genus "Nitrotoga." Recently, Griffin et al. (2007) enriched photoautotrophic NOB from freshwater sediments and from sewage. These enrichments were able to use nitrite as electron donor for anoxygenic photosynthesis and stoichiometrically oxidized nitrite to nitrate during this light-dependent process. One phototrophic NOB, a gammaproteobacterium designated as "strain KS," was then purified from activated sludge. The source of this strain is somewhat intriguing, because microbes are hardly exposed to light in the usually very turbid activated sludge suspensions in wastewater treatment plants. To which extent this newly discovered metabolism contributes to nitrite turnover in sewage and in other ecosystems awaits clarification in future research. Here we focus mainly on the "classical" groups of nitrite oxidizers, because, at present, much more is known about the distribution and ecological impact of these NOB.

Most known genera of NOB belong to one of the major lineages in the phylum *Proteobacteria*. The genus *Nitrobacter* is a member of the *Alphaproteobacteria* (Woese et al., 1984; Stackebrandt et al., 1988), *Nitrococcus* is a member of the *Gammaproteobacteria* (Teske et al., 1994), and candidate genus "Nitrotoga" is a member of the *Betaproteobacteria* (Alawi et al., 2007). The anoxygenic phototrophic strain KS is closely related to the gammaproteobacterial purple sulfur bacterium *Thiocapsa roseopersicina* (Griffin et al., 2007). The genus *Nitrospina* was provisionally assigned to the *Deltaproteobacteria* (Teske et al., 1994), but analyses using larger 16S rRNA sequence databases suggested that *Nitrospina* more likely belongs to a separate bacterial phylum (e.g., Schloss and Handelsman, 2004). Finally, the genus *Nitrospira* is a major lineage of the distinct bacterial phylum *Nitrospirae* (Ehrich et al., 1995) and thus not closely related to the proteobacterial NOB. This phylum also includes the two genera *Leptospirillum* (aerobic chemolithoautotrophic iron oxidizers) and *Thermodesulfovibrio* (anaerobic sulfate reducers) (Ehrich et al., 1995) and, in addition, the magnetotactic organism "*Candidatus* Magnetobacterium bavaricum" (Spring et al., 1993). Figure 1 shows the phylogenetic positions of the main lineages of NOB and their affiliations to related ammonia-oxidizing or non-nitrifying organisms.

The Genus *Nitrobacter*

Traditionally, members of the genus *Nitrobacter* have served as model organisms for studying the physiology of nitrite oxidizers. *Nitrobacter* are easier to culture than the other known NOB, and growing sufficient biomass is no major obstacle in physiological and biochemical studies. Recently, the genomes of three cultured *Nitrobacter* species have been fully sequenced, and detailed comparative genome analyses have been performed (Starkenburg et al., 2006, 2008) (see also Chapter 11). Thus, most current knowledge on the biology of nitrite oxidizers is based on work done with *Nitrobacter* pure cultures.

At present, the genus *Nitrobacter* contains four validly described species. *N. winogradskyi* (Winslow et al., 1917; Watson, 1971) and *N. hamburgensis* (Bock et al., 1983) were originally isolated from soils, *N. alkalicus* (Sorokin et al., 1998) from highly alkaline Siberian soda lake sediments and soda soil, and *N. vulgaris* (Bock et al., 1990) from various habitats including soil, freshwater and brackish water, and sewage. An additional species designation, "Nitrobacter agilis" (Nelson, 1931), is now considered invalid due to insufficient phenotypic difference between the type strains of *N. agilis* and *N. winogradskyi* (e.g., Pan, 1971; Watson, 1971).

The variety of sources for *Nitrobacter* isolates points out that this genus is widely distributed and that the different representatives must be adapted to a broad range of environmental conditions. *Nitrobacter* were isolated from samples as unusual as the weathering crust of ultrabasic rocks (Lebedeva et al., 1978), and they are thought to be involved in the biodeterioration of natural building stone that can be deeply colonized by nitrifying bacteria (Mansch and Bock, 1998). Although acidic environments are generally regarded as being unfavorable to nitrifiers, two *Nitrobacter* isolates were obtained from a forest soil with a pH between 4.3 and 5.2 (Hankinson and Schmidt, 1988). Furthermore, the known *Nitrobacter* species show different growth characteristics in presence of organic matter (e.g., Steinmüller and Bock, 1976). This high ecological flexibility is in sharp contrast to the apparently low phylogenetic diversity within the genus *Nitrobacter*, which is observed when small-subunit ribosomal RNA sequences are used as phylogenetic markers. In trees based on 16S rRNA gene sequences, the isolated *Nitrobacter* strains cluster very closely together, because they share high 16S rRNA sequence similarities above 99% (Orso et al., 1994) (Fig. 1). *Nitrobacter* also show high 16S rRNA similarities to their closest non-nitrifying relatives, *Rhodopseudomonas palustris* and *Bradyrhizobium japonicum* (Seewaldt et al., 1982; Orso et al., 1994), suggesting

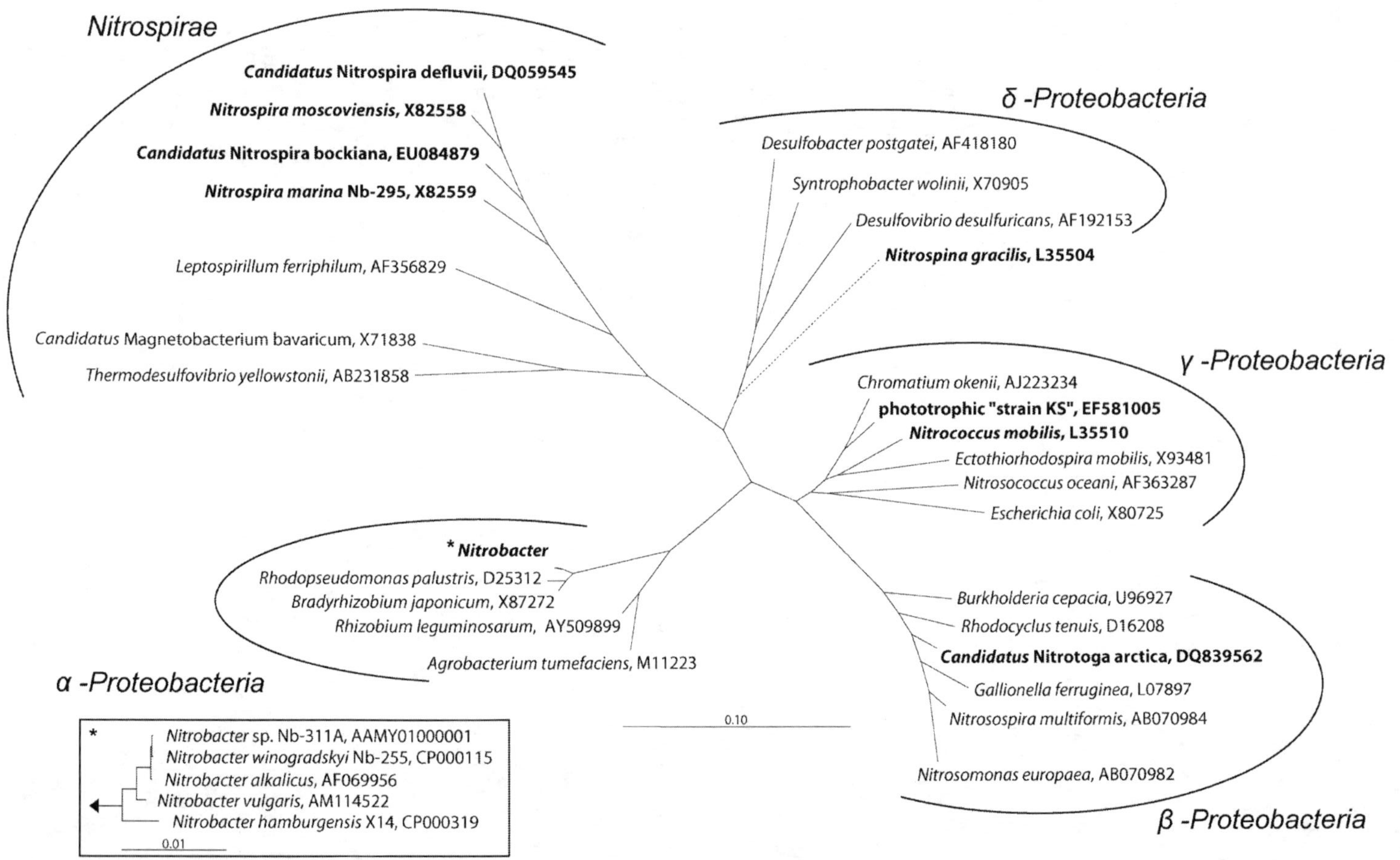

FIGURE 1 Phylogenetic tree, based on 16S rRNA gene sequences, showing the phylogenetic affiliations of the currently known NOB to selected ammonia-oxidizing and nonnitrifying bacteria. Names of nitrite oxidizers are printed bold. Database accession numbers are indicated for all 16S rRNA gene sequences. Arcs delimit bacterial phyla. The inset shows the phylogeny within the genus *Nitrobacter*, which is condensed to a single branch in the large tree. The stippled line at the *Nitrospina* branch indicates the uncertain phylogenetic affiliation of this genus. The tree topology was determined by maximum likelihood analysis of the sequences and by using a 50% sequence conservation filter for the *Bacteria*. The scale bars indicate 0.1 (large tree) or 0.01 (inset) estimated change per nucleotide.

that the genus *Nitrobacter* and its capability to oxidize nitrite evolved only recently (Orso et al., 1994). Therefore, it is often difficult to decide whether environmentally retrieved 16S rRNA sequences with sequence similarity to *Nitrobacter* represent novel members of this genus or non-nitrite-oxidizing relatives (Orso et al., 1994; Freitag et al., 2005). Furthermore, the 16S rRNA is too conserved to unambiguously resolve evolutionary lineages within the genus *Nitrobacter*.

Several strategies have been followed to overcome this limitation. Navarro et al. (1992a) used a combined approach, which included DNA-DNA hybridization, quantification of the genomic GC content, and analysis of rRNA gene restriction patterns, to examine the genetic diversity of 22 cultured *Nitrobacter* strains. Based on the obtained data, they grouped the analyzed strains in three "genomic species": *N. winogradskyi*, *N. hamburgensis*, and one unnamed group. The *N. winogradskyi* group was further split into three distinct "subspecies," one of them representing the reference strain of *N. agilis*. *N. vulgaris* was not analyzed in this study. These results, which have gained additional support from a more recent study using nitrite oxidoreductase as phylogenetic marker (Vanparys et al., 2007; see also below), start to reveal the microdiversity in the genus *Nitrobacter*. In another study, Navarro et al. (1992b) differentiated *Nitrobacter* strains by restriction fragment length polymorphism analysis of the intergenic spacer region (IGS) between the 16S and 23S rRNA genes. Because the IGS shows a higher mutation rate than the ribosomal RNA genes, it offers a better resolution than rRNA to distinguish very closely related bacteria. Several *Nitrobacter* isolates from soil and lake samples were analyzed. Interestingly, this molecular approach revealed diverse *Nitrobacter* populations in the samples and led to the conclusion that not large-scale biogeography, but local ecological niches, may shape the composition of *Nitrobacter* communities (Navarro et al., 1992b). Grundmann et al. (2000) used a modified IGS-based approach that included also a 5′ part of the 23S rRNA gene. Phylogenetic studies of the PCR-amplified, sequenced, and concatenated IGS and partial rRNA genes yielded high-resolution trees that clearly distinguished the analyzed *Nitrobacter* reference strains from each other and from *B. japonicum* and *R. palustris*. When this method was applied to soil samples, the diversity (in terms of restriction patterns) of *Nitrobacter* isolated from small soil clumps was found to be as large as the diversity of the reference strains, which had been isolated in different geographical areas (Grundmann and Normand, 2000). Furthermore, identical restriction patterns were found only among *Nitrobacter* isolated from the same soil clump, but not on a larger scale. A high microdiversity of soil *Nitrobacter* was also supported by serotyping with fluorescently labeled antibodies raised against *Nitrobacter* reference strains. Several different serotypes were detected in the same small pieces of soil (Grundmann and Normand, 2000), but one should consider that serotyping, which has long been used to differentiate *Nitrobacter* isolates (Fliermans et al., 1974; Stanley and Schmidt, 1981), discriminates phenotypic rather than genetic features. Altogether, these results suggest that a significant genetic (and phenotypic) diversity of *Nitrobacter* exists in the environment, which is not resolved when only the highly conserved 16S rRNA genes are used as markers.

A different molecular approach to identify *Nitrobacter* in environmental samples uses genes encoding subunits of the key enzyme of nitrite oxidation, nitrite oxidoreductase (Nxr), as functional and phylogenetic markers. By using PCR primers targeting the gene of the Nxr alpha subunit of *Nitrobacter*, *nxrA*, Poly et al. (2008) obtained partial sequences of this gene from all four *Nitrobacter* species and performed phylogenetic analyses. Consistent with the taxonomic classification, these *nxrA* sequences formed four distinct branches in phylogenetic trees. Paralogous *nxrA* genes, which exist in *Nitrobacter* genomes (Starkenburg et al., 2006, 2008), grouped consistently in these clusters. Additional phylogenetic lineages were formed by *nxrA* sequences retrieved from soil samples, strongly suggesting that yet unknown *Nitrobacter* strains, or other NOB carrying *nxrA* genes similar to those of *Nitrobacter*, existed in these soils (Poly et al., 2008). Interestingly, the diversity and phylogenetic distribution of *nxrA* in the soil samples differed with the mode of land use. Only one sequence type was found in a formerly intensively cultivated fallow soil, but five distinct *nxrA* types were detected in a pasture soil. This *nxrA*-based approach was extended to a denaturing gradient gel electrophoresis protocol, which was then applied to study the NOB communities in different grassland soils exposed to either light or intensive grazing (Wertz et al., 2008). Although no grazing regimen-specific *nxrA* phylotypes were obtained, statistical analyses of the *nxrA* denaturing gradient gel electrophoresis profiles showed that the NOB community composition was influenced by the grazing regime. Phylogeny revealed that the *nxrA* sequences, which had been retrieved from these soils, differed from the *nxrA* genes of cultured NOB (Wertz et al., 2008). Two other genes from the Nxr cluster, *nxrB* and *nxrX*, also are useful molecular markers for the differentiation of *Nitrobacter* isolates (Vanparys et al., 2007) but have not yet been applied as tools to characterize uncultured *Nitrobacter* populations in environmental samples.

In summary, the aforementioned molecular analyses have revealed that soil and freshwater environments harbor a greater diversity of different *Nitrobacter* strains than previously anticipated. Genetic heterogeneity within communities of closely related microorganisms seems not to be uncommon in bacteria and has been demonstrated, by environmental genomics, for example, in cyanobacteria (Coleman et al., 2006) and in chemolithotrophic iron oxidizers (Simmons et al., 2008). It is tempting to speculate that distinct *Nitrobacter* strains, which coexist on very small spatial scales (Grundmann and Normand, 2000), are ecotypes adapted to (slightly) different ecological niches in structurally complex environments such as soil or sediment. Future research should clarify whether these strains indeed are ecophysiologically different and how their diversity influences ecosystem functioning.

The Genera *Nitrococcus* and *Nitrospina*

Only one species has been described in either of the two genera *Nitrococcus* and *Nitrospina*: *Nitrococcus mobilis* and *Nitrospina gracilis*, respectively (Fig. 1). These NOB are of marine origin and were originally isolated from South Pacific (*N. mobilis*) and South Atlantic (*N. gracilis*) ocean water samples (Watson and Waterbury, 1971). An additional Pacific isolate of *N. gracilis* (Teske et al., 1994) and molecular data (Mincer et al., 2007) indicate a presumably global distribution of this organism.

Recently, Mincer et al. (2007) used quantitative PCR of 16S rRNA genes to record depth abundance profiles of ammonia-oxidizing *Archaea* and *Nitrospina*-like bacteria in coastal (Monterey Bay) and open ocean (North Pacific Subtropical Gyre) water samples. In all cases, the abundances of these nitrifiers markedly increased below the euphotic zone at a depth of approximately 100 to 200 m. The higher cell densities of nitrifiers in the subeuphotic zone can be explained by the light sensitivity of these organisms (Mincer et al., 2007, and references cited therein). These results suggest that communities consisting of ammonia-oxidizing *Archaea* and *Nitrospina* are important for nitrification in marine ecosystems. In the same study, fosmid and bacterial artificial chromosome (BAC) clone libraries were established from the coastal and open ocean samples and were screened for 16S rRNS genes related to *N. gracilis*. Among the several positive clones, one 64-kb-long BAC clone insert was selected for full sequencing and analysis. This genomic fragment contained, in addition to a 16S rRNA gene similar to *Nitrospina*, 88 protein-encoding open reading frames (ORFs). Sequence comparisons of these ORFs to the environmental whole genome shotgun database retrieved similar sequences from the Sargasso Sea dataset, the whale fall microbial mat/bone datasets, and the farm surface soil dataset (Mincer et al., 2007). The presence of sequences similar to *Nitrospina* ORFs in the marine metagenomic datasets supports the view that *Nitrospina* are widely distributed in the oceans. The search hits obtained from the farm soil dataset might indicate the existence of yet unidentified terrestrial *Nitrospina*-like NOB. Alternatively, these sequences could originate from organisms other than *Nitrospina*.

As the genome of *N. mobilis* has been sequenced and is publicly available, the genes encoding the nitrite oxidoreductase (Nxr) of this bacterium have been phylogenetically analyzed together with the respective genes of *Nitrobacter*. The *nxrA*, *nxrB, and nxrX* forms of these different NOB were found to be similar, but clearly distinguishable, in phylogenetic trees (Vanparys et al., 2007; Poly et al., 2008). Screenings for *nxrA* in various soils retrieved several different *nxrA* genes of the *Nitrobacter* type, but none related to *Nitrococcus* (Poly et al., 2008; Wertz et al., 2008). To date, neither *Nitrospina* nor *Nitrococcus* has been identified unequivocally, by cultivation or by molecular methods, in a nonmarine habitat.

The Candidate Genus "Nitrotoga"

The majority of known nitrifying microbes are mesophilic organisms with growth optima between 28°C and 39°C (e.g., Ehrich et al., 1995; Könneke et al., 2005), although thermophilic ammonia-oxidizing archaea have recently been described (de la Torre et al., 2008; Hatzenpichler et al., 2008). Still, very little is known about the diversity and distribution of NOB in extreme habitats. In geothermal springs, *Nitrospira*-like bacteria have been found (see below), but whether specialized NOB are active at very low temperatures remained unclear until recently. Major habitats, which would be expected to harbor cold-adapted nitrifiers, are the permafrost-affected soils that cover wide regions of the northern hemisphere. Indeed, Alawi et al. (2007) recently established, from Siberian permafrost-affected soil, an enrichment culture that oxidized nitrite at a temperature as low as 4°C, whereas no nitrite oxidation was observed at 25°C. Subsequent use of Nxr-targeted fluorescent antibodies, 16S rRNA-based phylogeny, and fluorescence in situ hybridization (FISH) with 16S rRNA-targeted probes identified the enriched organism as a novel nitrite-oxidizing betaproteobacterium (Alawi et al., 2007) (Fig. 1). Electron microscopy revealed an interesting ultrastructure of this bacterium with a particularly large periplasmic space. The newly discovered organism was named "*Candidatus* Nitrotoga arctica" (Alawi et al., 2007). The environmental distribution of this novel lineage of NOB has not been investigated in great detail yet. The original report mentions that some previously published environmental 16S rRNA sequences grouped together with "*Candidatus* Nitrotoga arctica" in phylogenetic trees (16S rRNA sequence similarities in this group ranged from 96.3 to 99%). These rRNA sequences were retrieved from sewage, polluted river biofilm, subglacial environments, and lake sediments. Moreover, Alawi et al. (2007) mentioned that *Nitrotoga*-like NOB were selectively enriched from municipal activated sludge by using the same cultivation conditions as for "*Candidatus* Nitrotoga arctica." High 16S rRNA similarities are not necessarily indicative of similar physiological traits. However, the aforementioned molecular data and the enrichment from activated sludge strongly suggest that NOB related to "Nitrotoga" are not restricted to permafrost soil and may be more widely distributed, but hitherto overlooked, nitrifiers in cold and moderate environments.

The Genus *Nitrospira*

The first described *Nitrospira* species, *Nitrospira marina*, was isolated from seawater sampled in the Gulf of Maine (Watson et al., 1986). Nine years later, Ehrich et al. (1995) purified the second species, *Nitrospira moscoviensis*, from a completely different habitat: the urban heating system of Moscow. It took almost another decade until the third pure culture was obtained, again from a corroded steel pipe of the Moscow heating system, and described as "*Candidatus* Nitrospira bockiana" (Lebedeva et al., 2008). A fourth culture was highly enriched from nitrifying activated sludge but has not been purified yet (Spieck et al., 2006). According to its origin from sewage, this nitrite oxidizer has been named "*Candi-*

datus Nitrospira defluvii." The longtime intervals between the descriptions of new *Nitrospira* species or high enrichments are caused by the slow growth of these bacteria and by the tediousness and difficulties to separate *Nitrospira* from other NOB and from heterotrophic contaminants. For example, the isolation of "*Candidatus* Nitrospira bockiana" took as long as 12 years (Lebedeva et al., 2008), and the enrichment of "*Candidatus* Nitrospira defluvii" was a labor-intensive process with several iterations of dilution series, incubation steps in mineral nitrite medium for several months, and purifications by density gradient centrifugation (Spieck et al., 2006). The complicated cultivation was mainly responsible for a long-lasting lack of knowledge on the diversity and environmental distribution of *Nitrospira*. This situation changed only when cultivation-independent molecular methods became available. A breakthrough was the discovery that uncultured *Nitrospira* were abundant in technical systems such as nitrifying aquarium filters (Hovanec et al., 1998), lab-scale reactors (Burrell et al., 1998), and, most importantly from the applied perspective, full-scale wastewater treatment plants (Juretschko et al., 1998). As these results were obtained by using the "rRNA approach" (Amann et al., 1995) and FISH with rRNA-targeted probes, possible biases of cultivation-based approaches could be excluded. With an increasing number of 16S rRNA gene libraries established from different kinds of environmental samples, it soon became clear that *Nitrospira* are not limited to the oceans and man-made habitats such as heating systems and sewage treatment facilities. Ribosomal RNA sequences related to this genus were retrieved from various soils, freshwater and sediment samples, the Nullarbor caves in Australia, marine sponge tissue, and additional activated sludge and biofilm samples (Daims et al., 2001, and references cited therein). Analyses based on these sequence data showed that the genus *Nitrospira* could be subdivided into four groups ("sublineages") if the following criteria were applied. In phylogenetic trees, the members of each sublineage had to form a monophyletic branch, which was supported by all applied treeing methods and bootstrap values of at least 90%. All 16S rRNA sequences, which were grouped in the same sublineage, also had to share a similarity of at least 94.9%. The sequence similarities among members of different sublineages were found to be always below 94% (Daims et al., 2001). Meanwhile, more 16S rRNA sequences related to *Nitrospira* have become available, and application of these grouping criteria to current *Nitrospira* phylogenies has now revealed six sublineages within this genus (Fig. 2). Interestingly, these sublineages seem not to be equally distributed in nature but show pronounced habitat specificity.

Sublineage I, which contains the enriched "*Candidatus* Nitrospira defluvii" (Fig. 2), consists of *Nitrospira* sequences obtained from many kinds of nitrifying sewage treatment systems. These organisms have been found in small lab-scale reactors as well as in pilot-scale systems and full-scale wastewater treatment plants. They occur in chemostats, sequencing batch (biofilm) reactors, trickling filters, and activated sludge tanks. In numerous studies, FISH with rRNA-targeted probes or quantitative PCR techniques showed that sublineage I *Nitrospira* are a major component of the nitrifying communities in engineered systems with abundances of usually 1 to 20% of all bacteria (e.g., Juretschko et al., 1998; Okabe et al., 1999; Daims et al., 2001). Until now, no members of sublineage I have been detected in pristine natural habitats, but *Nitrospira* similar to those in sewage were found in the effluent of a wastewater treatment plant and also in the receiving river downstream of the plant (Cebron and Garnier, 2005). Thus, *Nitrospira* sublineage I must be particularly well adapted to the conditions in nitrifying bioreactors, and their niche in natural ecosystems has not been identified yet. In contrast, sublineage II has the broadest distribution of all known *Nitrospira* lineages. These organisms have been found, by the rRNA approach and by FISH, in wastewater treatment plants and a laboratory-scale reactor (Schramm et al., 1998; Maixner et al., 2006),

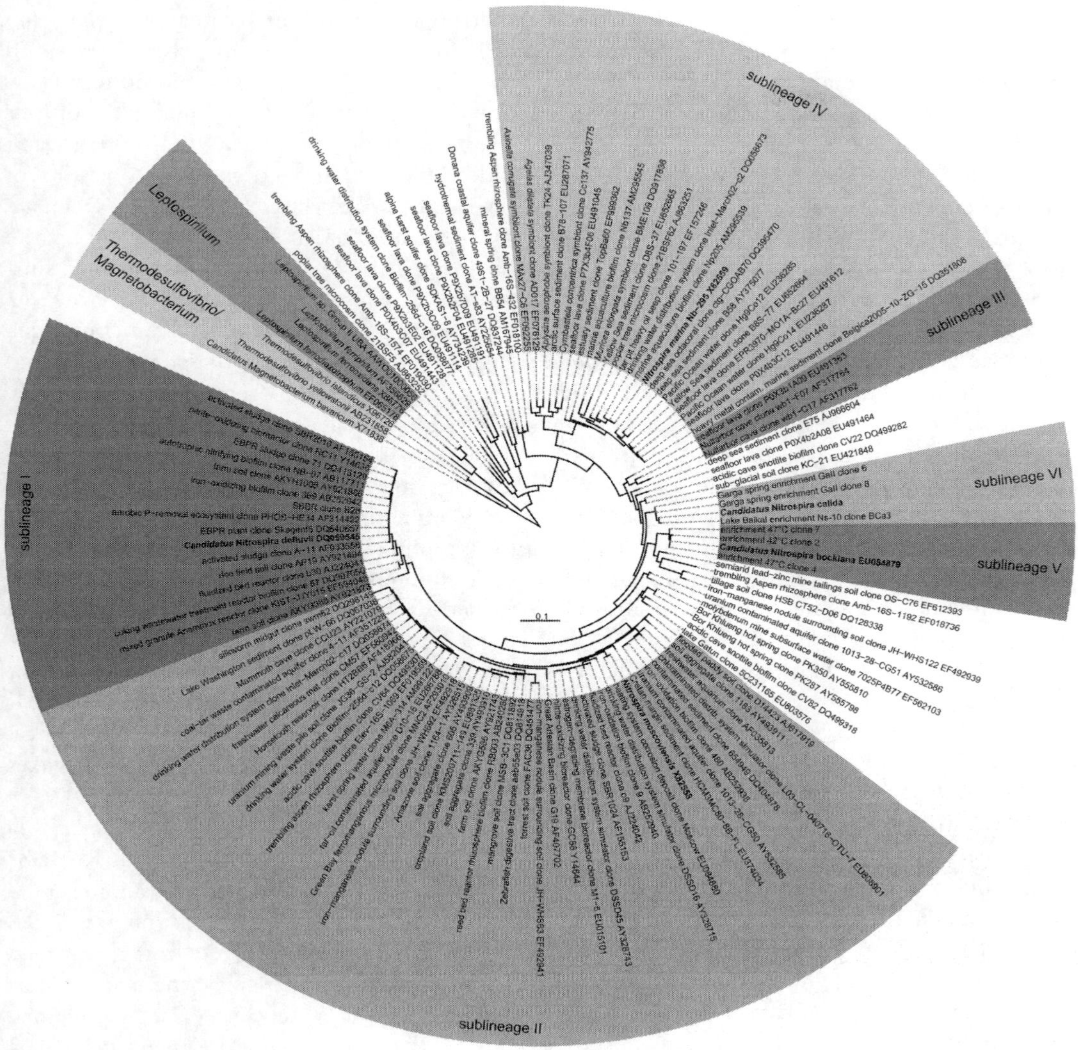

FIGURE 2 Phylogenetic tree, based on 16S rRNA gene sequences, showing the phylum *Nitrospirae*. Shaded regions delimit the nitrite-oxidizing *Nitrospira* sublineages I to VI and the non-nitrifying *Leptospirillum* and *Thermodesulfovibrio-Magnetobacterium* groups. Unshaded sequences between *Nitrospira* sublineage IV and *Leptospirillum* cannot yet be assigned to any genus, and the physiology of the respective organisms is unknown. Unshaded sequences within the genus *Nitrospira* cannot be assigned to one of the sublineages by using the criteria proposed by Daims et al. (2001). Database accession numbers are indicated for all 16S rRNA gene sequences. The tree topology was determined by maximum likelihood analysis of the sequences and by using a 50% sequence conservation filter for the genus *Nitrospira*. The scale bar indicates 0.1 estimated change per nucleotide.

but also in different soils, rhizosphere samples, freshwater habitats, drinking water distribution systems, and groundwater. The cultured *Nitrospira moscoviensis* from the Moscow heating system belongs to sublineage II, too (Fig. 2).

Sublineage III consists of only a few 16S rRNA clones from the Nullarbor cave system in Australia (Holmes et al., 2001). These sequences were obtained from microbial mantles growing on the roofs and walls of submerged parts of the Nullarbor caves. Interestingly, relatively high levels of nitrite were found in the water column of these caves, which lacked macroorganisms and organic material except of the microbial biomass. Therefore, a key role as primary producers in a nitrite-dependent microbial cave ecosystem was proposed for the nitrite-oxidizing, autotrophic *Nitrospira* (Holmes et al., 2001).

Sublineage IV, which contains the species *N. marina*, comprises the halophilic and marine *Nitrospira*. It harbors planktonic marine *Nitrospira* and 16S rRNA sequences retrieved from marine sediments including deep sea sampling sites. Remarkably, it also contains *Nitrospira* found to be microbial symbionts of marine sponges (Hentschel et al., 2002). Nitrification seems to play an important role in sponge tissues, because ammonia, which is a metabolic waste product of sponges, must be removed to avoid its accumulation and toxic effects (Taylor et al., 2007). The first step of nitrification in sponges is performed by ammonia-oxidizing bacteria (AOB) and archaea (Taylor et al., 2007; Steger et al., 2008). *Nitrospira* of cluster IV seem to be the major nitrite oxidizers in this symbiosis of multiple partners.

Our perception of the environmental distribution of nitrite oxidizers was significantly expanded when Lebedeva et al. (2005) obtained nitrite-oxidizing enrichments from the Garga hot spring, which is located in the northeastern part of Baikal rift zone. These enrichments showed activity at temperatures up to 60°C with an optimum of 50°C. Molecular analyses of 16S rRNA genes identified the enriched thermophilic NOB as members of the genus *Nitrospira* (Lebedeva et al., 2005). Subsequent phylogenetic analyses confirmed this finding and added a novel *Nitrospira* sublineage, number VI (Fig. 2), which contains the Garga strains and *Nitrospira* from a few other hot springs. The discovery of *Nitrospira* in hot springs is noteworthy as it adds to our picture of biogeochemical nitrogen cycling in extreme environments. Furthermore, the existence of thermophilic nitrite oxidizers is consistent with the recent identification of thermophilic ammonia-oxidizing *Archaea* (de la Torre et al., 2008; Zhang et al., 2008) and the hypothesis that archaeal ammonia oxidation evolved under thermophilic conditions (Hatzenpichler et al., 2008). If nitrification indeed evolved in extreme environments of the ancient earth, it is tempting to speculate that thermophilic *Nitrospira* were among the first bacteria to exploit nitrite as the main substrate of their energy metabolism. However, sublineage VI is not a deep-branching lineage within the genus *Nitrospira* (Fig. 2), and its members possibly represent a secondary adaptation to life at high temperatures of organisms with mesophilic ancestors. Future research may clarify whether additional, and deep-branching, thermophilic or even hyperthermophilic *Nitrospira* lineages occur in hot environments. Microbially catalyzed ammonia oxidation, and subsequent accumulation of nitrate, were recently demonstrated at temperatures above 80°C and at pH 3 in hot springs on Iceland (Reigstad et al., 2008). It remains to be determined whether biological or chemical nitrite oxidation, or both, takes place under such extreme conditions.

The most recently obtained *Nitrospira* isolate, "*Candidatus* Nitrospira bockiana" (Lebedeva et al., 2008), belongs to sublineage V (Fig. 2). Like *N. moscoviensis*, this organism was enriched and finally isolated from the urban heating system of Moscow where it inhabits internal corrosion deposits of steel pipelines. Despite the similar habitat, "*Candidatus* Nitrospira bockiana" and *N. moscoviensis* differ with respect to their cell morphology, temperature optima for growth, tolerance against nitrite, and dominant lipids (Table 1) (Lebedeva et al., 2008). Interesting observations were made with enrichments of "*Candidatus* Nitrospira bockiana" that still contained contaminants. These enrichments, which mainly contained a *Nocardioides*-like bacterium in addition to *Nitrospira*, showed a broader temperature range and were more tolerant against higher nitrite concentrations than the finally obtained pure culture (Lebedeva et al., 2008). Whether these differences resulted from the isolation pro-

TABLE 1 Selected characteristics of known *Nitrospira* sublineages

Parameter	Result for sublineage[a]: I	II	III	IV	V	VI
Cell morphology	Short, slightly curved cells or spiral-shaped rods	Irregularly shaped cells or spiral-shaped rods	Putatively spirals	"Comma-shaped" cells or spiral-shaped rods	Spirals, curved and straight rods, or coccoid cells	Spirals
Size (μm)	0.2–0.4 × 0.7–1.7	0.2–0.4 × 0.9–2.2	ND	0.3–0.4 × 0.8–1.0	0.3–0.6 × 1.0–2.5 or 0.9 × 0.9	0.2–0.4 × 1.0–1.7
Turns	1–4	1–3	ND	1–12	1–4	ND
Tendency to aggregate	Strong	Present	ND, growth in biofilm	Weak	Present	Present
Growth temp (°C)	28–32	39	18.9[b]	28	42	40–60
Isolate or enrichment	"*Ca.* Nitrospira defluvii"[c]	*N. moscoviensis*[d]	16S rRNA gene clones only[e]	*N. marina*[f]	"*Ca.* Nitrospira bockiana"[g]	GaII[h]
Dominant membrane lipids	16:1cis11, 16:0	16:1cis11, 16:0, 16:0 11 methyl	ND	16:1cis7, 16:1cis11, 16:0	16:1cis7, 16:0, 16:0 11 methyl	ND
Nitrite concn (mM)[i]	3 (20–25)	0.35 (15)	0.2[g]	1 (6)	0.3–3 (18)	1
Anaerobic metabolism (as determined in expts)	ND	Respiration with nitrate	ND	Strictly aerobic	ND	ND
Use of organic substrates (as determined in expts)	Pyruvate	No organotrophic growth observed	ND	Glycerole and pyruvate	No organotrophic growth observed	ND
Intracellular storage compounds	Glycogen and polyphosphate	Poly-β-hydroxybutyrate and polyphosphate	ND	Glycogen and polyphosphate	Glycogen and polyphosphate	ND

[a]ND, not determined; *Ca.*, *Candidatus*.
[b]Measured in cave water surrounding the biofilm.
[c]Spieck et al. (2006).
[d]Ehrich et al. (1995).
[e]Holmes et al. (2001).
[f]Watson et al. (1986).
[g]Lebedeva et al. (2008).
[h]Lebedeva et al. (2005).
[i]Nitrite concentrations used in growth media or measured in natural habitat; values in parentheses indicate maximal tolerated concentrations.

cess or from interactions between "*Candidatus* Nitrospira bockiana" and the accompanying microorganisms has not been clarified.

Table 1 summarizes key properties of the six known *Nitrospira* sublineages. Based on the currently available molecular and cultivation-based data, it appears that the genus *Nitrospira* is the most diverse and widely distributed group of NOB.

Nitrite Oxidizers and Wastewater Treatment

For decades, mainly *Nitrobacter* were regarded to be responsible for nitrite oxidation in wastewater treatment plants. This "textbook opinion" emerged because traditional cultivation techniques retrieved *Nitrobacter* isolates from almost every nitrifying activated sludge sample tested. As nitrification is a key process of biological wastewater treatment, the growth characteristics of *Nitrobacter* cultures were used for the design of sewage treatment plants and for the modeling of nitrification in engineered systems (Bever et al., 1995). However, this view changed rapidly when cultivation-independent techniques, in particular FISH with rRNA-targeted oligonucleotide probes, exposed the apparent importance of *Nitrobacter* in wastewater treatment as an artefact of the cultivation-based approach. In the majority of the examined activated sludge samples, no *Nitrobacter* cells were detectable by FISH, indicating that their abundance was well below the detection limit of this method (10^3 to 10^4 cells per ml) (Wagner et al., 1996). Due to these low cell densities, relevance of *Nitrobacter* for nitrite oxidation in these wastewater treatment plants could be excluded. FISH also revealed that yet uncultured *Nitrospira* are much more abundant NOB in sewage treatment systems (Juretschko et al., 1998). Low numbers of *Nitrobacter* cells, which occur in most wastewater treatment plants, explain why these bacteria have been detected by cultivation- or PCR-based methods in activated sludge. Higher amounts of *Nitrobacter* have been found, by using FISH, in only a few full-scale or pilot sewage treatment systems that received wastewater with high nitrogen loads (e.g., Mobarry et al., 1996; Daims et al., 2001; Gieseke et al., 2003).

Occasional reports of high *Nitrobacter* densities in full-scale systems receiving normal domestic wastewater (Coskuner and Curtis, 2002) await verification by experiments using the well-established, standardized FISH protocols. In this context, however, it should be mentioned that FISH, in general, is not without biases (Wagner et al., 2003). Thus, in theory, larger amounts of *Nitrobacter* in wastewater treatment plants might have escaped detection by FISH and related techniques so far. This appears unlikely considering the large number of different studies that have used FISH to screen activated sludge for *Nitrobacter*. Nevertheless, a new approach that is independent of FISH, PCR, and cultivation biases would be desirable to check whether the density of *Nitrobacter* in full-scale wastewater treatment plants really is minor compared to the abundance of *Nitrospira*. Recently, we have addressed this question in the course of a comprehensive metagenomics-based analysis. First, the metagenome of a nitrifying wastewater treatment plant was cloned in *Escherichia coli* in BAC vectors. The resulting BAC library consisted of more than 500,000 clones. Subsequently, more than 320,000 BAC paired ends were sequenced, and the obtained sequence reads were compared to all hitherto sequenced bacterial and archaeal genomes using stringent matching criteria (at least 90% nucleotide sequence similarity over at least 80% of the read length). Figure 3 shows the results of this analysis for the nitrifiers. A large number of BAC end sequences (>7,000) were highly similar to corresponding regions in the genome of "*Candidatus* Nitrospira defluvii," a recently sequenced representative of *Nitrospira* sublineage I (see below for more information on this genome). In contrast, not more than 49 sequence reads matched to the genomes of any sequenced *Nitrobacter* strain (Fig. 3), suggesting that sublineage I *Nitrospira* are much more frequent in the metagenome and in the wastewater treatment plant than *Nitrobacter*. To confirm that the large number of *Nitrospira* hits was not due to sequence similarities

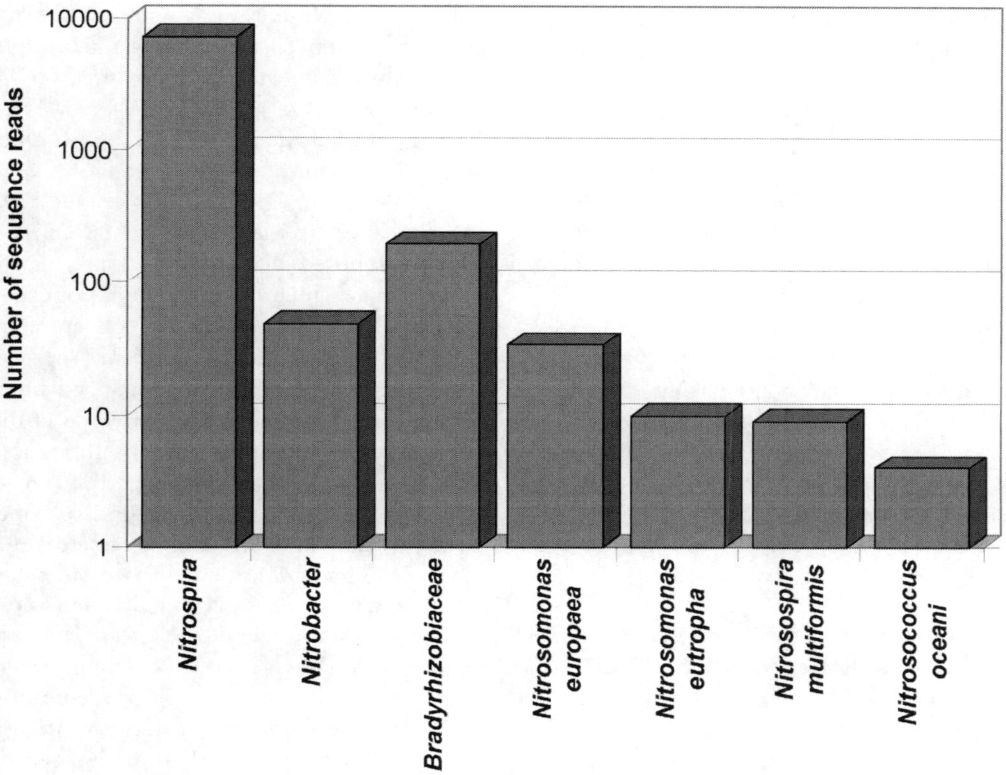

FIGURE 3 Numbers of sequence reads that were obtained from the metagenome of a nitrifying activated sludge and had a high sequence similarity to genomes of nitrifying bacteria. The number of reads indicated for the *Bradyrhizobiaceae* includes all reads obtained for *Nitrobacter* plus the reads obtained for other members of the *Bradyrhizobiaceae* (refer to the main text for details). Note that the scaling of the *y* axis is logarithmic.

between "*Candidatus* Nitrospira defluvii" and bacteria other than *Nitrospira* represented in the metagenome, the complete genome of "*Candidatus* Nitrospira defluvii" was compared to all other sequenced prokaryotic genomes by using the same matching criteria. This separate test yielded only one hit between *Nitrospira* and another organism (data not shown). Thus, based on currently available genome data and on the applied matching criteria, the likelihood of numerous false-positive *Nitrospira* hits in the metagenome analysis is very low, and the number of *Nitrospira* hits indeed reflects a high abundance of *Nitrospira* in the activated sludge. However, not only the number of *Nitrobacter* hits but also the metagenome hits obtained for AOB are surprisingly low (Fig. 3), suggesting that ammonia oxidizers not closely related to those that have been sequenced on the genome level are dominant in the analyzed wastewater treatment plant. To exclude that *Nitrobacter* strains not represented by the available genome sequences are important in this plant, all hits obtained for members of the *Bradyrhizobiaceae*, including *Nitrobacter*, were determined. As *Nitrobacter* are very closely related to other members of this family—for example, *B. japonicum* (Fig. 1)—any sequence reads related to the *Bradyrhizobiaceae* might actually belong to *Nitrobacter* strains whose gene content differs from the fully sequenced *Nitrobacter* representatives. Nevertheless, even the total number of all *Bradyrhizobiaceae* hits was far below the number of hits for *Nitrospira* (Fig. 3). This outcome of a comprehensive, PCR-independent community analysis by environmental genomics strongly corroborates the results of previous work based on FISH (Wagner et al., 1996; Juretschko et al., 1998). Furthermore, FISH-microautoradiography experiments with nitrifying activated sludges usually show that only *Nitrospira* are strongly labeled with radioactive bicarbonate in the presence of nitrite, excluding another abundant nitrite-oxidizing population in the analyzed samples (our unpublished data). In conclusion, a battery of cultivation-independent molecular approaches has demonstrated that *Nitrobacter* cannot be relevant for nitrite oxidation in most wastewater treatment plants. Notable exceptions are systems treating high strength sewage, which contains higher nitrogen levels than are usually found in municipal and industrial wastewaters.

ECOPHYSIOLOGY AND NICHE PARTITIONING OF *NITROBACTER* AND *NITROSPIRA*

Since the discovery of a high diversity of yet uncultured NOB, especially in the genus *Nitrospira*, the ecophysiology of these bacteria has received some attention in nitrification research. This applies, in particular, to the NOB living in wastewater treatment plants, because their metabolic activity and growth characteristics are directly linked to the operational performance and stability of nitrifying bioreactors. One major question is why members of the *Nitrospira* and not of *Nitrobacter* are the predominant NOB in most sewage treatment plants. Experience with the available isolates and enrichments has shown that members of *Nitrobacter* are generally easier to culture and can grow faster than those of *Nitrospira* under laboratory conditions. Nevertheless, the mostly uncultured *Nitrospira* spp. are more competitive in situ in activated sludge systems. To explain this phenomenon, Schramm et al. (1999) performed experiments to estimate the K_s value of uncultured *Nitrospira* spp. for nitrite. This was achieved by combining quantitative FISH with microelectrode measurements in a nitrifying biofilm. Interestingly, their results suggested that *Nitrospira* spp. have a much higher affinity for nitrite than do *Nitrobacter* spp. The estimated K_s (NO_2^-) of *Nitrospira* was as low as 10 μM, whereas the measured values of *Nitrobacter* pure cultures are in the range of 60 to 600 μM (Prosser, 1989; Hunik et al., 1993). On the basis of these data, Schramm et al. (1999) proposed that *Nitrospira* spp. may be K-strategists, which can reach high population densities even when nitrite concentrations are very low. In contrast, *Nitrobacter* spp. were suggested to be r-strategists that need higher nitrite concentrations but can grow faster than *Nitrospira* spp. if nitrite is not a limiting factor. This would indicate a selective advantage for *Nitrospira* spp. under the nitrite-limited conditions prevalent in domestic wastewater treatment plants and in most natural habitats, where usually nitrite does not accumulate but is oxidized to nitrate or denitrified. *Nitrobacter* spp. would then depend on microenvironments with locally higher nitrite concentrations, which may be found, for example, in the rhizosphere (Freitag et al., 2005). If this hypothesis indeed reflects the ecological strategies of these NOB, intermediate nitrite concentrations should enable a (temporary) coexistence of *Nitrobacter* and *Nitrospira* spp. Indeed, Bartosch et al. (2002) enriched both *Nitrospira* and *Nitrobacter* spp. from soil samples if media containing 0.2 g of nitrite per liter was used but obtained only *Nitrobacter* enrichments in media containing 2 g of nitrite per liter. Interestingly, a pilot-scale sequencing batch biofilm reactor that contained large numbers of *Nitrobacter* also contained *Nitrospira* (Daims et al., 2001; Gieseke et al., 2003). The operational mode of this reactor was a cyclic sequence of filling with new wastewater, an aeration period for nitrification, and removal of the supernatant. The reactor received reject water from activated sludge dewatering, which contained very high ammonia concentrations. During each operational cycle, a transient but pronounced increase of the nitrite concentration occurred. These temporal nitrite gradients probably created ecological niches for both groups of NOB and were the basis of their stable coexistence. Because the biomass grew in a sessile biofilm on carrier material, both populations were held back in the reactor. A selective loss of biomass of the weaker competitor, which takes place in continuously operated activated sludge tanks, was not observed.

The "K/r-hypothesis" for *Nitrospira* and *Nitrobacter* has been tested and supported in several follow-up studies using laboratory-scale nitrifying bioreactors (e.g., Nogueira and Melo, 2006). This leads to the speculation that selective effects of nitrite may also extend to members of the same genus. For example, the aforementioned habitat specificity of the six *Nitrospira* sublineages might be influenced by different nitrite optima of these organisms. Maixner et al. (2006) analyzed a nitrifying biofilm from a wastewater treatment plant, which contained both sublineage I and II *Nitrospira* in addition to AOB. FISH with probes targeting AOB or one of the two *Nitrospira* types showed that all three populations were abundant in the biofilm. However, sublineage I *Nitrospira* seemingly were located closer to the cell clusters of AOB than were sublineage II *Nitrospira*. This difference in spatial arrangement was confirmed by applying digital image analysis and spatial statistics on the biofilm. Based on these data, the hypothesis was raised that small-scale local nitrite concentration gradients may exist in the biofilm, with highest nitrite concentrations close to the AOB microcolonies that convert ammonia to nitrite. Such gradients could determine the differential distribution of the two *Nitrospira* populations if sublineage I *Nitrospira* prefer higher nitrite concentrations than sublineage II *Nitrospira*. Indeed, a mathematical model of nitrite production, diffusion, and consumption in the biofilm suggested the existence of local nitrite gradients. In a separate experiment, activated sludge containing both sublineage I and II *Nitrospira* was incubated with different nitrite concentrations, and the two *Nitrospira* populations were quantified by FISH with specific probes during this experiment. Consistent with the observations made in the biofilm, this experiment showed that higher nitrite concentrations favored the growth of sublineage I and selected against sublineage II *Nitrospira* (Maixner et al., 2006). Hence, nitrite concentrations affect the composition of *Nitrospira* communities and most likely also their local distribution in spatially complex habitats. On an imaginary scale with pure K- and pure r-strategists on either end, the *Nitrospira* sublineages may take different positions close to the K-end, whereas *Nitrobacter* species may take positions closer to the r-end.

Present data indicate that the K/r-hypothesis might explain the predominance of *Nitrospira* spp. in wastewater treatment plants and in many kinds of pristine ecosystems, where ambient nitrite concentrations are very low. However, this hypothesis may apply not only to nitrite. Past research showed that *Nitrospira* can outcompete *Nitrobacter* in biofilm regions with a low dissolved oxygen (DO) concentration (e.g., Schramm et al., 2000; Downing and Nerenberg, 2008), suggesting that *Nitrospira* spp. also have a lower K_s for O_2 than do *Nitrobacter* spp. A high-oxygen affinity would also help *Nitrospira* spp. to compete for O_2 with ammonia oxidizers and heterotrophic microorganisms. Moreover, oxygen availability may have selective effects even within the genus *Nitrospira*. Park and Noguera (2008) operated two laboratory-scale chemostats in parallel, which contained the same nitrifying activated sludge as inoculum but differed in the DO concentrations. The low DO reactor was dominated by sublineage I *Nitrospira* for the whole duration of the experiment (271 days), whereas a population shift occurred in the high DO reactor that led to coexistence of sublineage I and II *Nitrospira*. Whether these results indicate a higher O_2 affinity or a lower O_2 tolerance of sublineage I *Nitrospira* was not determined in the study. Both the nitrite and DO concentrations seem to be important environmental factors influencing the competition and environmental distribution of different NOB, but they probably are not the only key parameters. Past research showed that cultured *Nitrobacter* spp. are not obligately autotrophic nitrifiers but can use simple organic compounds such as pyruvate, acetate, and formate for mixotrophic and even for chemoorganotrophic growth (Bock, 1976). Higher growth yields in the presence of pyruvate than without organic carbon were also reported for nitrite-oxidizing *N. marina* (Watson et al., 1986), whereas no use of organic substrates was observed for *N. moscoviensis* pure cultures (Ehrich et al., 1995). A non-cultivation-dependent tool, FISH combined with microautoradiography, revealed that uncultured *Nitrospira* in activated sludge were able to assimilate carbon from bicarbonate or pyruvate, whereas use of acetate was not observed by this method (Daims et al., 2001). Thus, the quality and concentrations of available organic compounds may select for different NOB if these bacteria differ in their substrate usage spectra and their capabilities for mixotrophic or even chemoorganotrophic growth.

Especially in wastewater and at industrially contaminated sites, inhibitory substances hampering the activity and growth of NOB could have strong selective effects, too. For example, nitrite oxidation by *Nitrobacter* pure cultures is inhibited in presence of chlorate (ClO_3^-) or chlorite (ClO_2^-) (Lees and Simpson, 1957; Hynes and Knowles, 1983). This effect is caused by chlorite, which destroys essential cytochromes of *Nitrobacter* spp. (Lees and Simpson, 1957). The toxic chlorite is formed from chlorate when *Nitrobacter* spp. use chlorate instead of oxygen as a terminal electron acceptor for nitrite oxidation. (Per)chlorate and chlorite are contaminants that originate, for example, from the rocket fuel and paper-bleaching industries. Based on the observations made with *Nitrobacter* pure cultures, one would expect high concentrations of these compounds to inhibit nitrite oxidation in polluted soil or water. However, an environmental genomics approach in combination with heterologous gene expression has revealed that sublineage I *Nitrospira*, "*Candidatus* Nitrospira defluvii," living in activated sludge possess a gene coding for a highly active form of chlorite dismutase (CLD) (Maixner et al., 2008). This remarkable enzyme converts chlorite to chloride (Cl^-) and oxygen (O_2) by a not yet fully elucidated mechanism (van Ginkel et al., 1996). The CLD of "*Candidatus* Nitrospira defluvii" is expressed when the organism is incubated in nitrite-containing medium, suggesting a protective role of the enzyme in case that chlorate and/or chlorite are also present (Maixner et al., 2008). Previously, CLD was known to exist only in chlorite-resistant heterotrophic *Proteobacteria* that use (per)chlorate as an alternative electron acceptor for anaerobic respiration. The presence of CLD in sublineage I *Nitrospira* indicates that these NOB might also be resistant to chlorite and may link the bioremediation of (per)chlorate with the nitrogen cycle. Moreover, chlorite can occur in wastewater as a byproduct of the chlorination of activated sludge. Chlorination has been common practice to fight excess growth of filamentous bacteria in wastewater treatment plants. Thus, resistance to chlorite might be an additional reason for the high abundance of *Nitrospira* spp. in activated sludge. However, alternative functions of CLD in *Nitrospira* must be considered (Maixner et al., 2008), and the main biological role of this enzyme in "*Candidatus* Nitrospira defluvii" may not have been determined yet. *Nitrospira* spp. occur also in municipal drinking water distribution systems where microbial growth is controlled by adding chloramines (Regan et al., 2003). Although chloramines are effective as disinfectants, these compounds can support nitrification in drinking water utilities by introducing ammonia as substrate for ammonia-oxidizing microbes. The nitrifiers, which are primary producers, can fuel the heterotrophic growth of other organisms, leading to biological instability of drinking water and to violations of hygienic standards. Interestingly, the uncultured AOB and NOB in drinking water systems seem to be more resistant to chloramines than pure cultures of nitrifiers (Regan et al., 2003). The mechanisms underlying this resistance are not known.

ENVIRONMENTAL GENOMICS AND FULL-GENOME ANALYSIS OF "*CANDIDATUS* NITROSPIRA DEFLUVII"

Nitrospira spp. are difficult to culture, and larger amounts of biomass are hard to obtain from the few available isolates. Therefore, this genus is dramatically undersampled with respect to genomics despite its high ecological importance as the most diverse and widely distributed group of NOB and as the key NOB in wastewater treatment. Only recently, the first complete *Nitrospira* genome sequence has been determined from an enrichment of "*Candidatus* Nitrospira defluvii." This sublineage I *Nitrospira* strain was enriched from nitrifying activated sludge (Spieck et al., 2006) and has not yet been purified from accompanying microbes. Nevertheless, its genome could be sequenced by using an environmental genomics approach that had been developed for sequencing an uncultured anaerobic ammonium-oxidizing bacterium (Strous et al., 2006). Briefly, this approach relied on BAC and plasmid shotgun cloning and paired-end sequencing to construct contigs and scaffolds from the extracted community genomic DNA. A key step was the identification of an anchor point for assembly, which in the case of "*Candidatus* Nitrospira defluvii" was one contig containing the complete ribosomal RNA gene operon of this organism. Using this long contig (Maixner et al., 2008) as a starting point, the remaining parts of the *Nitrospira* genome were obtained in iterative sequencing, binning, and assembly steps. Following final gap closure, the complete genome of "*Candidatus* Nitrospira defluvii" with a size of approximately 4.3 million base pairs was available for automated and manual downstream analyses. Table 2 summarizes key features of the genome.

The following sections provide an overview of selected metabolic characteristics of "*Candidatus* Nitrospira defluvii" as derived from the genomic data.

Nitrite Oxidation and Nitrite Oxidoreductase

The key enzyme of chemolithotrophic nitrite oxidation is nitrite oxidoreductase (Nxr). It has been studied most intensively in *Nitrobacter*, where Nxr is a membrane-bound enzyme containing one large (alpha) and one small (beta) subunit that are encoded by the genes *nxrA* and *nxrB*, respectively (Spieck et al., 1996b; Starkenburg et al., 2006). Depending on the method used to isolate the Nxr holoenzyme from *Nitrobacter* cell membranes, a putative third subunit was detected in some studies (Tanaka et al., 1983; Sundermeyer-Klinger et al., 1984). Consistently, candidate genes for putative subunits other than NxrA and NxrB have been identified in sequenced *Nitrobacter* genomes (Starkenburg et al., 2006). Nxr belongs to the dimethyl sulfoxide (DMSO) reductase family of molybdopterin-containing enzymes. Other prominent members of this structurally and functionally diverse enzyme family are, for example, dissimilatory nitrate reductases as well as chlorate, selenate, and arsenate reductases (McDevitt et al., 2002; Thorell et al., 2003). Most representatives of this group consist of at least two subunits. The alpha subunit contains the active center with the substrate-binding site and the molybdopterin cofactor. The beta subunit, which contains multiple iron-sulfur clusters, functions in the transport of electrons between the alpha subunit and additional subunits or the membrane-bound electron transport chain, respectively (Kisker et al., 1998). This functional modularity is conserved within this enzyme family (Rothery et al., 2008), but direct experimental evidence for the location of the substrate-binding site in the alpha subunit of Nxr is still lacking. The Nxr of *Nitrobacter* oxidizes nitrite to nitrate, but the reaction is reversible so that Nxr plays an additional role in denitrification by this organism (Sundermeyer-Klinger et al., 1984).

Spieck et al. (1998) purified, by heat treatment of cytoplasmic membrane preparations, the nitrite-oxidizing system of *N. moscoviensis*. Antibodies, which primarily targeted the Nxr beta (NxrB) subunit of *Nitrobacter*, were used to

TABLE 2 Selected features of the "*Candidatus* Nitrospira defluvii" genome

Feature	Result
Genome size	4,317,083 bp
GC content	59.03%
No. of genes	4,321
No. of rRNA genes	3
No. of tRNA genes	46
No. of ORFs with predicted functions	2,147
Coding density	89.45%
Repeated regions	2.29%
No. of transposon-related genes (including fragments)	46

identify a putative NxrB protein in the extracted mixture of prominent *Nitrospira* membrane proteins. The same protein fraction also contained a putative Nxr alpha (NxrA) subunit, whose apparent molecular mass (130 kDa) was similar to the mass of NxrA of *Nitrobacter*, plus two proteins of unknown function. The presence of the two unknown proteins in the extract suggested that the nitrite-oxidizing system of *Nitrospira* differs from that of *Nitrobacter* in its composition and possibly in functional properties (Spieck et al., 1998).

On the basis of comparisons of the sequence to that of the *nxr* genes of *Nitrobacter* and *Nitrococcus* and the known molecular masses of the NxrA and NxrB subunits of *N. moscoviensis* (Spieck et al., 1998), putative *nxr* genes have been identified in the sequenced genome of "*Candidatus* Nitrospira defluvii." The genome contains two copies each of *nxrA* and *nxrB*, which are colocalized in two clusters (*nxrA1B1* and *nxrA2B2*) that are separated by approximately 24 kb from each other. The two NxrA copies share an amino acid identity of 86.6%, whereas the two NxrB copies are identical. Whether the differences in the primary structures of the two NxrA versions have functional effects, for example, on the affinity for nitrite or on substrate range and specificity, must be verified in future analyses. The gene upstream of *nxrA2* encodes a sigma-54-dependent transcriptional regulator with a CheY-like response regulator receiver region, which might be involved in the regulation of *nxrA2B2* expression based upon a two-component signaling system. A different sigma-54 dependent transcriptional regulator is located upstream of the *nxrA1B1* cluster, indicating that the two gene clusters may be regulated differently. The next gene downstream of *nxrB2* is remotely similar to a small [2Fe-2S] ferredoxin (Bfd) of *E. coli*. Bfd is thought to be involved in the release of iron from the iron storage protein bacterioferritin or from siderophores, in the insertion of iron into heme, or in iron-dependent gene regulation (Quail et al., 1996). Indeed, a gene encoding bacterioferritin is present at another location in the "*Candidatus* Nitrospira defluvii" genome, suggesting that this compound is used for iron storage in *Nitrospira*. Thus, it is tempting to speculate that the product of the *bfd*-like gene may supply iron for the Nxr holoenzyme. Alternative roles of Bfd as an iron-dependent expression regulator for *nxrAB* or as part of the Nxr protein complex cannot be ruled out either. The genome also encodes three proteins similar to the gamma subunits of DMSO reductase family members such as nitrate, chlorate, and selenate reductases. The predicted molecular masses of these proteins resemble one of the two unknown proteins in the membrane extract from *N. moscoviensis* (Spieck et al., 1998), indicating that these are candidates for a gamma subunit (NxrC) of the *Nitrospira* Nxr. The gamma subunits of the related reductases are heme-containing proteins and shuttle electrons from the electron transport chain to the beta subunits of these enzymes (e.g., Berks et al., 1995; Thorell et al., 2003). During nitrite oxidation, electrons must be shuttled in the opposite direction (i.e., from the Nxr alpha subunit to the beta and gamma subunits and further to the electron transport chain). Each of the three NxrC candidates contains one predicted transmembrane helix, suggesting an additional function as membrane anchor of the Nxr protein complex. However, none of the three putative *nxrC* genes is located very close to the *nxrA1B1* and *nxrA2B2* gene clusters. Moreover, the pairwise amino acid identities of the three NxrC candidates are as low as 27 to 33%, indicating that they do not share the same functions and might be parts of different, yet uncharacterized, protein complexes involved in oxidation/reduction processes and electron transport. Consistent with a putative role in nitrite oxidation and energy conservation, one of the candidate *nxrC* genes is located in close neighborhood to several genes encoding components of electron transport chains such as c-type cytochromes and a putative ubiquinol-cytochrome *c* oxidoreductase. The genome of "*Candidatus* Nitrospira defluvii" also encodes a protein whose molecular mass is similar to the fourth major protein, which was detected in the membrane extract from *N. moscoviensis* but was not further analyzed (Spieck et al., 1998). This protein is similar to cytochrome *c* and to the gamma subunits of selenate and chlorate reductase, contains two binding sites for c-type hemes, and has predicted transmembrane helices. Based on these properties, it is another candidate for an NxrC subunit in *Nitrospira*. Future experimental work may resolve whether at least one of the four NxrC candidates is part of the Nxr holoenzyme in "*Candidatus* Nitrospira defluvii."

Interestingly, immunogold labeling revealed that the membrane-associated Nxr of *Nitrospira* is arranged in hexagonal patterns on the outer side of the cell membrane and thus faces the periplasmic space. In contrast, the Nxr of *Nitrobacter* is located on the cytoplasmic side of the membrane (Spieck et al., 1996a). A periplasmic localization of the *Nitrospira* Nxr is also supported by sequence analysis of the *nxr* genes. Like other, structurally similar, periplasmic enzymes of the DMSO reductase family (McDevitt et al., 2002; Thorell et al., 2003), both *nxrA* and *nxrC* encode N-terminal signal peptides for the export of the gene products into the periplasmic space via the twin-arginine protein translocation (Tat) pathway (NxrA) or the Sec-pathway (all NxrC candidates). NxrB has no signal peptide but is assumed to fold together with NxrA in the cytoplasm and to be translocated with the alpha subunit via the Tat pathway, which transports proteins in their folded state. Such a "hitchhiker" mechanism (Rodrigue et al., 1999) is most likely used also by the beta subunits of chlorate reductase and dimethyl sulfide dehydrogenase (McDevitt et al., 2002; Thorell et al., 2003) lacking a signal peptide, too. The location of Nxr in the periplasm bears some advantages for *Nitrospira*. For nitrite oxidation, no transport of nitrite into the cell and no excretion of nitrate are needed if this reaction takes place in the periplasm. In contrast, *Nitrobacter* with a cytoplasmic Nxr depends on nitrate/nitrite transport across the cytoplasmic membrane (Starkenburg et al., 2006). Nitrite oxidation (with H_2O as the source of the additional O-atom in nitrate) on the periplasmic side is coupled to electron translocation through the cytoplasmic membrane and to the consumption of protons in the cytoplasm during the reduction of O_2. Most likely, this leads to a proton gradient for ATP production by a membrane-bound ATP synthase (complex V) in *Nitrospira*. The mechanism of energy conservation in *Nitrobacter*, where both nitrite oxidation and O_2 reduction take place on the cytoplasmic side of the inner membrane, is not understood (Bock and Wagner, 2001).

Comparative sequence analyses showed that the two NxrA copies of "*Candidatus* Nitrospira defluvii" are substantially different from the NxrA of *Nitrobacter* and *Nitrococcus*. The NxrA proteins of the latter two NOB are closely affiliated with each other and with the alpha subunit of respiratory membrane-bound nitrate reductase (NarG) (Fig. 4). In contrast, both *Nitrospira* NxrA copies group together with a putative nitrite-oxidizing/nitrate-reducing enzyme of the anaerobic ammonium oxidizer (anammox organism) "*Candidatus* Kuenenia stuttgartiensis" (Strous et al., 2006) (Fig. 4). This distinct phylogenetic affiliation may reflect ecologically significant differences in structural and catalytic properties. For example, a different substrate affinity of NxrA could be the molecular basis of the aforementioned K/r-hypothesis for *Nitrospira* and *Nitrobacter*. Furthermore, the phylogeny of NxrA raises questions about the evolution of NOB. The tree (Fig. 4) suggests a common ancestor for the NxrA of *Nitrospira* and the protein of "*Candidatus* Kuenenia stuttgartiensis." Which organism possessed this ancestral enzyme and where did it live, under oxic conditions like *Nitrospira* or in anoxic habitats like the anammox bacterium *Kuenenia*? Although the phylogeny shown in Fig. 4 might be blurred by past lateral gene transfer events, it is tempting to speculate that the aerobic nitrite-oxidizing system of *Nitrospira* was derived from an anaerobic protein complex and that the ancestor of *Nitrospira* was adapted to life under hypoxic conditions. This hypothesis will also be addressed in the context of autotrophic carbon fixation (see below). Finally, the distant relationship between the NxrA forms of *Nitrospira* and those of *Nitrobacter* and *Nitrococcus* suggests that the use of nitrite as an energy source was invented more than once in the course of bacterial evolution.

Autotrophic Carbon Fixation

Like the other NOB, *Nitrospira* spp. are able to use CO_2 or bicarbonate as the sole carbon source for autotrophic growth. The genome of "*Candidatus* Nitrospira defluvii" encodes three carbonic anhydrases, one of the eukaryotic (alpha) type and two of the prokaryotic type (gamma). The alpha-type carbonic anhydrase most likely is located in the periplasm and might play a crucial role for bicarbonate uptake by converting bicarbonate to CO_2, which can freely diffuse through the cytoplasmic membrane. Consistent with this assumption, the genome lacks a known bicarbonate transporter.

All previously genome-sequenced bacterial nitrifiers use the Calvin cycle for autotrophic carbon fixation (Chain et al., 2003; Klotz et al., 2006; Starkenburg et al., 2006, 2008; Stein, 2007). The genome of "*Candidatus* Nitrospira defluvii" contains a gene similar to the large subunit of ribulose-1,5-bisphosphate carboxylase (RubisCO), which at first glance suggests that *Nitrospira* rely on the same pathway as the other nitrifying bacteria. However, a closer look revealed that this *Nitrospira* protein is phylogenetically affiliated to Form IV RubisCO-like proteins (Fig. 5). In other bacteria, Form IV RubisCO-like proteins lack a robust carboxylase activity (Hanson and Tabita, 2001) but function as 2,3-diketo-5-methylthiopentyl-1-phosphate enolase in the methionine salvage pathway (Ashida et al., 2003). At the substrate binding and catalytic sites, the primary structure of the *Nitrospira* RubisCO-like protein differs from highly conserved amino acid residue patterns found in the carboxylating RubisCO Forms I to III of plants, autotrophic bacteria, and archaea (Hanson and Tabita, 2001) (data not shown). Therefore, a role of this Form IV RubisCO-like protein in autotrophic carbon fixation seems to be unlikely in *Nitrospira*. This conclusion is further supported by the absence of carboxysomes in *Nitrospira* cells (Watson et al., 1986; Ehrich et al., 1995; Lebedeva et al., 2008). The genome of "*Candidatus* Nitrospira defluvii" also lacks any homolog of another key enzyme of the Calvin cycle, phosphoribulokinase (or ribulose-5-phosphate kinase), strongly suggesting that *Nitrospira* uses a different pathway for carbon fixation. Indeed, the genome encodes all proteins needed for carbon fixation via the reverse tricarboxylic acid (rTCA) cycle, in particular fumarate reductase, 2-oxoglutarate:ferredoxin oxidoreductase (OGOR), and ATP-citrate lyase. These are considered to be the key enzymes of the rTCA cycle. The other necessary enzymes are shared between the rTCA and the oxidative TCA cycles (Hügler et al., 2005). The CO_2 fixation product acetyl-CoA is carboxylated to pyruvate by pyruvate:ferredoxin oxidoreductase (POR), which also is encoded in the *Nitrospira*

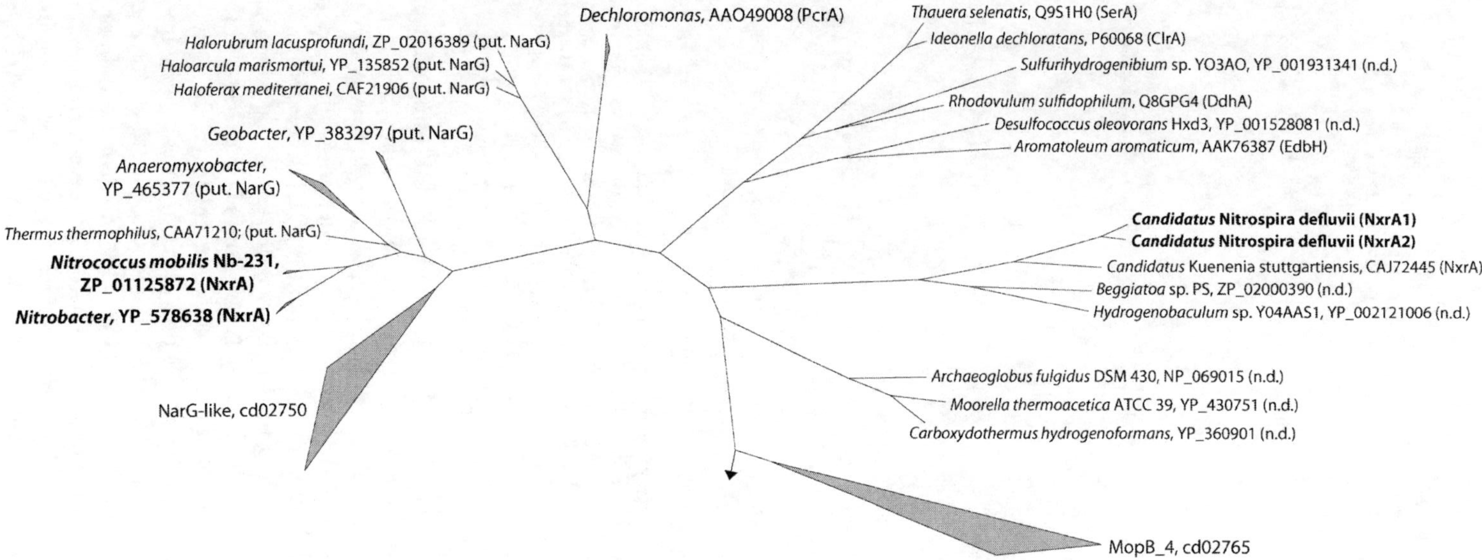

FIGURE 4 Phylogenetic tree, based on protein sequences, showing selected alpha subunits from the DMSO reductase family of molybdopterin-containing enzymes. Sequences of nitrite oxidoreductase alpha subunits (NxrA) of NOB are printed in boldface. Database accession numbers are indicated for all protein sequences. The tree topology was determined by maximum likelihood analysis of the sequences of the molybdopterin- and [Fe-S]-binding domains and by excluding N-terminal signal peptides. Protein subunit names indicate the functions of the respective enzymes: NarG, dissimilatory membrane-bound nitrate reductase; NxrA, nitrite oxidoreductase; PcrA, perchlorate reductase; SerA, selenate reductase; ClrA, chlorate reductase; DdhA, dimethyl sulfide dehydrogenase; EdbH, ethylbenzene dehydrogenase; put., putative; n.d., function not determined. (Figure courtesy of Frank Maixner.)

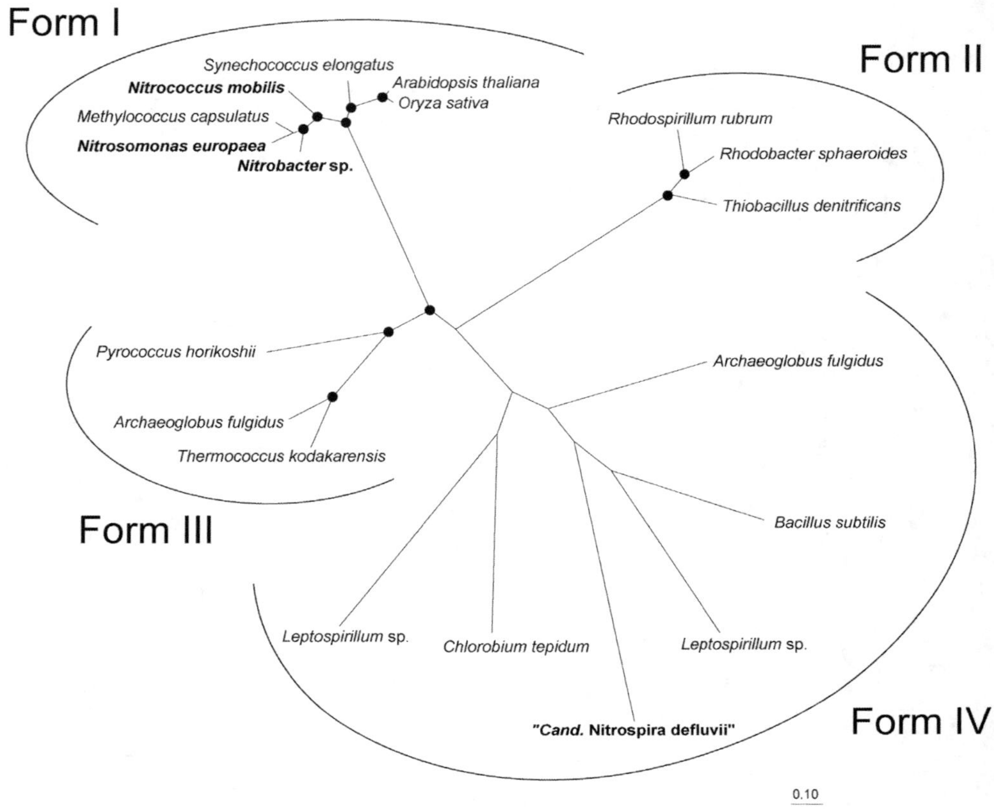

FIGURE 5 Phylogenetic tree, based on protein sequences, showing ribulose-1,5-bisphosphate carboxylase (RubisCO) of selected organisms. Arcs delimit RubisCO forms I to IV. Black dots indicate a high parsimony bootstrap support (>90%, 100 iterations) for the respective tree branches. Names of nitrifying bacteria are printed in boldface. The tree topology was determined by Fitch-Margoliash analysis of the sequences (with global rearrangement and randomized input order [seven jumbles]). The scale bar indicates 0.1 estimated changes per residue.

genome. In contrast to the membrane-bound dehydrogenases of the oxidative TCA cycle, the ferredoxin-containing oxidoreductases of the reverse pathway are soluble proteins (Evans et al., 1966). Other known CO_2 fixation pathways were not identified, and the finding that "*Candidatus* Nitrospira defluvii" most likely uses the rTCA cycle was a surprise. The dependence of this pathway on reduced ferredoxin poses a requirement for reverse electron transport from the electron donor nitrite over a relatively large difference in reduction potential (E_0') from +0.43 V for the NO_3^-/NO_2^- redox pair to about −0.39 V for ferredoxin. In contrast, NOB using the Calvin cycle must overcome the slightly smaller reduction potential difference between nitrite and NAD^+ by reverse electron transport (from +0.43 to −0.32 V). Thus, reverse electron transport costs more energy in *Nitrospira* than in other NOB. On the other hand, the synthesis of one triose phosphate via the Calvin cycle requires nine ATP and reductants from six NAD(P)H, whereas the rTCA cycle requires only five ATP and reductants from three NAD(P)H, two ferredoxins, and one unknown donor (Lengeler et al., 1999). Because of the lower ATP demand, the rTCA cycle is not necessarily a competitive disadvantage for *Nitrospira* compared to NOB using the Calvin cycle. *Nitrospira* spp. might also use reductants that have a lower potential than nitrite (e.g., H_2) to reduce ferredoxin if such reductants are available in their environmental microenvironments. The ability of *N. moscoviensis* to use H_2 as electron donor was documented by Ehrich et al. (1995), who observed this organism to reduce nitrate with H_2 under anoxic incubation conditions. Also puzzling is the paradox that *Nitrospira* oxidize nitrite under aerobic conditions, whereas the rTCA cycle is most common in anaerobic and microaerophilic microbes, because the ferredoxin-containing key enzymes POR and OGOR are sensitive to oxygen (Campbell et al., 2006). However, the POR and OGOR of "*Candidatus* Nitrospira defluvii" are highly similar to homologous enzymes in *Hydrogenobacter thermophilus*, an organism known to use the rTCA cycle for autotrophy under oxic conditions (Yamamoto et al., 2006). The similar *Nitrospira* versions of POR and OGOR might be relatively oxygen resistant, too. The presence of the rTCA cycle in *Nitrospira* might also explain why high densities of these NOB were observed in deeper and rather oxygen-depleted regions of biofilms (Okabe et al., 1999; Gieseke et al., 2003), where oxygen-sensitive enzymes would benefit from low DO concentrations. Consistent with this assumption, Okabe et al. (1999) observed a low abundance of *Nitrospira* close to the surface of a nitrifying biofilm and therefore suggested that *Nitrospira* may be inhibited by high oxygen levels. This is contrasted by other reports of high *Nitrospira* densities close to the surface of nitrifying aggregates, where DO concentrations were higher than in the inner parts of the aggregates (e.g., Schramm et al., 1999). However, even in the presence of high bulk DO concentrations, the oxygen levels may be lower in local microniches, for example, because O_2 is consumed by adjacent ammonia oxidizers or heterotrophs. If *Nitrospira* spp. indeed are confined to niches with a low-oxygen tension, their affinity to oxygen must be high enough to enable nitrite oxidation under oxygen-limited conditions.

The rTCA cycle is widely distributed among the *Bacteria* and *Archaea* and has been proposed to be the most ancient pathway for autotrophic carbon fixation (e.g., Wächtershäuser, 1990). All known NOB that use the Calvin cycle are *Proteobacteria*, and, in particular, *Nitrobacter* is considered to be a relatively recently evolved genus (Orso et al., 1994). In contrast, the genus *Nitrospira* is a deep-branching lineage within the phylum *Nitrospirae* that constitutes a separate, main line of descent among the *Bacteria*. Their distinct phylogenetic position and impressive diversity, and the use of an early evolved CO_2 fixation pathway, indicate that *Nitrospira* occupied its niche early in evolution. This process likely occurred by a transition from an anaerobic to an aerobic or microaerophilic lifestyle as reflected by the similarity of the carbon fixation pathway and the nitrite oxidoreductase to the respective enzymes in anaerobic bacteria. Aided by the now available genomic data, specific physiological experiments should be performed to determine whether modern *Nitrospira* have retained other metabolic features of anaerobic or microaerophilic microbes and to which extent such features may enable *Nitrospira* to survive without their apparent key substrates nitrite and oxygen.

Use of Organic Carbon Compounds

As mentioned earlier in this chapter, cultivation-independent methods showed that *Nitrospira* in a nitrifying biofilm were able to assimilate pyruvate (Daims et al., 2001). The capacity to assimilate organic carbon from pyruvate and some other simple compounds is reflected by the genome of "*Candidatus* Nitrospira defluvii." Interestingly, this includes acetate although no acetate uptake was observed in situ previously (Daims et al., 2001). Moreover, based on the genomic data, "*Candidatus* Nitrospira defluvii" may be able to gain energy from the oxidation of organic compounds. The Embden-Meyerhof-Parnas glycolytic pathway is complete, and, in addition to the reductive version (see above), the genome encodes the complete oxidative TCA cycle. The genes for complexes I to V of a conventional electron transport chain, which is needed for oxidative phosphorylation with reductants derived from organic carbon, are also present. Furthermore, the genome contains a putative soluble, NAD-dependent formate dehydrogenase and a formate transporter. Further experiments are needed to verify that the chemoorganoheterotrophic pathways are expressed and functional in "*Candidatus* Nitrospira defluvii." Meanwhile, it is tempting to speculate that this *Nitrospira* representative may not be an obligate autotroph but could take some advantage of the organic carbon supply in wastewater treatment plants. Its substrate spectrum would also be determined by the presence or absence of suitable transport systems. The genome contains various putative transporters for organic compounds including sugars, but based solely on sequence data, the substrate ranges of these transporters cannot be determined yet.

In conclusion, the last years have witnessed a steep increase of our knowledge on NOB, which was made possible by a combination of traditional and molecular approaches. These new insights have radically changed our perception of the biodiversity, physiological versatility, and ecological importance of NOB. The recent discovery of previously overlooked NOB suggests that our picture of the biology of nitrite oxidation as a key step in the global nitrogen cycle is still far from complete.

ACKNOWLEDGMENTS

We gratefully acknowledge the constructive comments and suggestions on various parts of this chapter provided by colleague Peter Bottomley. The ideas expressed in this chapter also benefited from discussions with colleagues Jim Prosser, Andreas Schramm, Eva Spieck, and Frank Maixner. We are indebted to Thomas Rattei for performing various computational analyses on genomic data from *Nitrospira*.

REFERENCES

Alawi, M., A. Lipski, T. Sanders, E.-M. Pfeiffer, and E. Spieck. 2007. Cultivation of a novel cold-adapted nitrite oxidizing betaproteobacterium from the Siberian Arctic. *ISME Journal* **1:**256–264.

Amann, R. I., W. Ludwig, and K.-H. Schleifer. 1995. Phylogenetic identification and in situ detection of individual microbial cells without cultivation. *Microbiol. Rev.* **59:**143–169.

Ashida, H., Y. Saito, C. Kojima, K. Kobayashi, N. Ogasawara, and A. Yokota. 2003. A functional link between RuBisCO-like protein of *Bacillus* and photosynthetic RuBisCO. *Science* **302:**286–290.

Bartosch, S., C. Hartwig, E. Spieck, and E. Bock. 2002. Immunological detection of Nitrospira-like bacteria in various soils. *Microb. Ecol.* **43:**26–33.

Berks, B. C., M. D. Page, D. J. Richardson, A. Reilly, A. Cavill, F. Outen, and S. J. Ferguson. 1995. Sequence analysis of subunits of the membrane-bound nitrate reductase from a denitrifying bacterium: the integral membrane subunit provides a prototype for the dihaem electron-carrying arm of a redox loop. *Mol. Microbiol.* **15:**319–331.

Bever, J., A. Stein, and H. Teichmann (ed.). 1995. *Weitergehende Abwasserreinigung.* R. Oldenbourg Verlag, München, Germany.

Bock, E. 1976. Growth of *Nitrobacter* in the presence of organic matter. II. Chemoorganotrophic growth of *Nitrobacter agilis*. *Arch. Microbiol.* **108:**305–312.

Bock, E., and M. Wagner. 2001. Oxidation of inorganic nitrogen compounds as an energy source, p. 457–495. *In* M. Dworkin, S. Falkow, E. Rosenberg, K. H. Schleifer, and E. Stackebrandt (ed.), *The Prokaryotes: a Handbook on the Biology of Bacteria*, 3rd ed. Springer Science+Business Media, New York, NY.

Bock, E., H. Sundermeyer-Klinger, and E. Stackebrandt. 1983. New facultative lithoautotrophic nitrite-oxidizing bacteria. *Arch. Microbiol.* **136:**281–284.

Bock, E., H.-P. Koops, U. C. Möller, and M. Rudert. 1990. A new facultatively nitrite oxidizing bacterium, *Nitrobacter vulgaris* sp. nov. *Arch. Microbiol.* **153:**105–110.

Burrell, P. C., J. Keller, and L. L. Blackall. 1998. Microbiology of a nitrite-oxidizing bioreactor. *Appl. Environ. Microbiol.* **64:**1878–1883.

Campbell, B. J., A. S. Engel, M. L. Porter, and K. Takai. 2006. The versatile epsilon-proteobacteria: key players in sulphidic habitats. *Nat. Rev. Microbiol.* **4:**458–468.

Cebron, A., and J. Garnier. 2005. *Nitrobacter* and *Nitrospira* genera as representatives of nitrite-oxidizing bacteria: detection, quantification and growth along the lower Seine River (France). *Water Res* **39:**4979–4992.

Chain, P., J. Lamerdin, F. Larimer, W. Regala, V. Lao, M. Land, L. Hauser, A. Hooper, M. Klotz, J. Norton, L. Sayavedra-Soto, D. Arciero, N. Hommes, M. Whittaker, and D. Arp. 2003. Complete genome sequence of the ammonia-oxidizing bacterium and obligate chemolithoautotroph *Nitrosomonas europaea*. *J. Bacteriol.* **185:**2759–2773.

Coleman, M. L., M. B. Sullivan, A. C. Martiny, C. Steglich, K. Barry, E. F. Delong, and S. W. Chisholm. 2006. Genomic islands and the ecology and evolution of *Prochlorococcus*. *Science* **311:**1768–1770.

Coskuner, G., and T. P. Curtis. 2002. In situ characterization of nitrifiers in an activated sludge plant: detection of *Nitrobacter* spp. *J. Appl. Microbiol.* **93:**431–437.

Daims, H., J. L. Nielsen, P. H. Nielsen, K. H. Schleifer, and M. Wagner. 2001. In situ characterization of *Nitrospira*-like nitrite-oxidizing bacteria active in wastewater treatment plants. *Appl. Environ. Microbiol.* **67:**5273–5284.

de la Torre, J. R., C. B. Walker, A. E. Ingalls, M. Konneke, and D. A. Stahl. 2008. Cultivation of a thermophilic ammonia oxidizing archaeon synthesizing crenarchaeol. *Environ. Microbiol.* **10:**810–818.

Downing, L. S., and R. Nerenberg. 2008. Effect of oxygen gradients on the activity and microbial community structure of a nitrifying, membrane-aerated biofilm. *Biotechnol. Bioeng.* **101:**1193–1204.

Ehrich, S., D. Behrens, E. Lebedeva, W. Ludwig, and E. Bock. 1995. A new obligately chemolithoautotrophic, nitrite-oxidizing bacterium, *Nitrospira moscoviensis* sp. nov. and its phylogenetic relationship. *Arch. Microbiol.* **164:**16–23.

Evans, M. C., B. B. Buchanan, and D. I. Arnon. 1966. A new ferredoxin-dependent carbon reduction cycle in a photosynthetic bacterium. *Proc. Natl. Acad. Sci. USA* **55:**928–934.

Fliermans, C. B., B. B. Bohlool, and E. L. Schmidt. 1974. Autecological study of the chemoautotroph *Nitrobacter* by immunofluorescence. *Appl. Microbiol.* **27:**124–129.

Freitag, T. E., L. Chang, C. D. Clegg, and J. I. Prosser. 2005. Influence of inorganic nitrogen management regime on the diversity of nitrite-oxidizing bacteria in agricultural grassland soils. *Appl. Environ. Microbiol.* **71:**8323–8334.

Gieseke, A., L. Bjerrum, M. Wagner, and R. Amann. 2003. Structure and activity of multiple nitrifying bacterial populations co-existing in a biofilm. *Environ. Microbiol.* **5:**355–369.

Griffin, B. M., J. Schott, and B. Schink. 2007. Nitrite, an electron donor for anoxygenic photosynthesis. *Science* **316:**1870.

Grundmann, G. L., and P. Normand. 2000. Microscale diversity of the genus *Nitrobacter* in soil on the basis of analysis of genes encoding rRNA. *Appl. Environ. Microbiol.* **66:**4543–4546.

Grundmann, G. L., M. Neyra, and P. Normand. 2000. High-resolution phylogenetic analysis of NO_2^--oxidizing *Nitrobacter* species using the rrs-rrl IGS sequence and rrl genes. *Int. J. Syst. Evol. Microbiol.* **50:**1893–1898.

Hankinson, T. R., and E. L. Schmidt. 1988. An acidophilic and a neutrophilic *Nitrobacter* strain isolated from the numerically predominant nitrite-oxidizing population of an acid forest soil. *Appl. Environ. Microbiol.* **54:**1536–1540.

Hanson, T. E., and F. R. Tabita. 2001. A ribulose-1,5-bisphosphate carboxylase/oxygenase (RubisCO)-like protein from *Chlorobium tepidum* that is involved with sulfur metabolism and the response to oxidative stress. *Proc. Natl. Acad. Sci. USA* **98:**4397–4402.

Hatzenpichler, R., E. V. Lebedeva, E. Spieck, K. Stoecker, A. Richter, H. Daims, and M. Wagner. 2008. A moderately thermophilic ammonia-oxidizing crenarchaeote from a hot spring. *Proc. Natl. Acad. Sci. USA* **105:**2134–2139.

Hentschel, U., J. Hopke, M. Horn, A. B. Friedrich, M. Wagner, J. Hacker, and B. S. Moore. 2002. Molecular evidence for a uniform microbial community in sponges from different oceans. *Appl. Environ. Microbiol.* **68:**4431–4440.

Holmes, A. J., N. A. Tujula, M. Holley, A. Contos, J. M. James, P. Rogers, and M. R. Gillings. 2001. Phylogenetic structure of unusual aquatic microbial formations in Nullarbor caves, Australia. *Environ. Microbiol.* **3:**256–264.

Hovanec, T. A., L. T. Taylor, A. Blakis, and E. F. DeLong. 1998. *Nitrospira*-like bacteria associated with nitrite oxidation in freshwater aquaria. *Appl. Environ. Microbiol.* **64:**258–264.

Hügler, M., C. O. Wirsen, G. Fuchs, C. D. Taylor, and S. M. Sievert. 2005. Evidence for autotrophic CO_2 fixation via the reductive tricarboxylic acid cycle by members of the epsilon subdivision of proteobacteria. *J. Bacteriol.* **187:**3020–3027.

Hunik, J. H., H. J. G. Meijer, and J. Tramper. 1993. Kinetics of *Nitrobacter agilis* at extreme substrate, product and salt concentrations. *Appl. Microbiol. Biotech.* **40:**442–448.

Hynes, R. K., and R. Knowles. 1983. Inhibition of chemoautotrophic nitrification by sodium chlorate and sodium chlorite: a reexamination. *Appl. Environ. Microbiol.* **45:**1178–1182.

Juretschko, S., G. Timmermann, M. Schmid, K.-H. Schleifer, A. Pommerening-Röser, H.-P. Koops, and M. Wagner. 1998. Combined molecular and conventional analyses of nitrifying bacterium diversity in activated sludge: *Nitrosococcus mobilis* and *Nitrospira*-like bacteria as dominant populations. *Appl. Environ. Microbiol.* **64:**3042–3051.

Kisker, C., H. Schindelin, D. Baas, J. Retey, R. U. Meckenstock, and P. M. Kroneck. 1998. A structural comparison of molybdenum cofactor-containing enzymes. *FEMS Microbiol. Rev.*

22:503–521.

Klotz, M. G., D. J. Arp, P. S. Chain, A. F. El-Sheikh, L. J. Hauser, N. G. Hommes, F. W. Larimer, S. A. Malfatti, J. M. Norton, A. T. Poret-Peterson, L. M. Vergez, and B. B. Ward. 2006. Complete genome sequence of the marine, chemolithoautotrophic, ammonia-oxidizing bacterium *Nitrosococcus oceani* ATCC 19707. *Appl. Environ. Microbiol.* **72:**6299–6315.

Könneke, M., A. E. Bernhard, J. R. de la Torre, C. B. Walker, J. B. Waterbury, and D. A. Stahl. 2005. Isolation of an autotrophic ammonia-oxidizing marine archaeon. *Nature* **437:**543–546.

Lebedeva, E. V., N. N. Lialikova, and I. Bugel'skii. 1978. Participation of nitrifying bacteria in the disintegration of serpentinous ultrabasic rock. *Mikrobiologiia* **47:**1101–1107.

Lebedeva, E. V., M. Alawi, C. Fiencke, B. Namsaraev, E. Bock, and E. Spieck. 2005. Moderately thermophilic nitrifying bacteria from a hot spring of the Baikal rift zone. *FEMS Microbiol. Ecol.* **54:**297–306.

Lebedeva, E. V., M. Alawi, F. Maixner, P. G. Jozsa, H. Daims, and E. Spieck. 2008. Physiological and phylogenetical characterization of a new lithoautotrophic nitrite-oxidizing bacterium, "*Candidatus* Nitrospira bockiana" sp. nov. *Int. J. Syst. Evol. Microbiol.* **58:**242–250.

Lees, H., and J. R. Simpson. 1957. The biochemistry of the nitrifying organisms. V. Nitrite oxidation by *Nitrobacter*. *Biochem. J.* **65:**297–305.

Lengeler, J. W., G. Drews, and H. G. Schlegel (ed.). 1999. *Biology of the Prokaryotes*, 1st ed. Georg Thieme Verlag, Stuttgart, Germany.

Maixner, F., D. R. Noguera, B. Anneser, K. Stoecker, G. Wegl, M. Wagner, and H. Daims. 2006. Nitrite concentration influences the population structure of *Nitrospira*-like bacteria. *Environ. Microbiol.* **8:**1487–1495.

Maixner, F., M. Wagner, S. Lücker, E. Pelletier, S. Schmitz-Esser, K. Hace, E. Spieck, R. Konrat, D. Le Paslier, and H. Daims. 2008. Environmental genomics reveals a functional chlorite dismutase in the nitrite-oxidizing bacterium "*Candidatus* Nitrospira defluvii." *Environ. Microbiol.* **10:**3043–3056.

Mansch, R., and E. Bock. 1998. Biodeterioration of natural stone with special reference to nitrifying bacteria. *Biodegradation* **9:**47–64.

McDevitt, C. A., P. Hugenholtz, G. R. Hanson, and A. G. McEwan. 2002. Molecular analysis of dimethyl sulphide dehydrogenase from *Rhodovulum sulfidophilum*: its place in the dimethyl sulphoxide reductase family of microbial molybdopterin-containing enzymes. *Mol. Microbiol.* **44:**1575–1587.

Mincer, T. J., M. J. Church, L. T. Taylor, C. Preston, D. M. Karl, and E. F. DeLong. 2007. Quantitative distribution of presumptive archaeal and bacterial nitrifiers in Monterey Bay and the North Pacific Subtropical Gyre. *Environ. Microbiol.* **9:**1162–1175.

Mobarry, B. K., M. Wagner, V. Urbain, B. E. Rittmann, and D. A. Stahl. 1996. Phylogenetic probes for analyzing abundance and spatial organization of nitrifying bacteria. *Appl. Environ. Microbiol.* **62:**2156–2162.

Navarro, E., M. P. Fernandez, F. Grimont, A. Claysjosserand, and R. Bardin. 1992a. Genomic heterogeneity of the genus *Nitrobacter*. *Int. J. Syst. Bacteriol.* **42:**554–560.

Navarro, E., P. Simonet, P. Normand, and R. Bardin. 1992b. Characterization of natural populations of *Nitrobacter* spp. using PCR/RFLP analysis of the ribosomal intergenic spacer. *Arch. Microbiol.* **157:**107–115.

Nelson, D. H. 1931. Isolation and characterization of *Nitrosomonas* and *Nitrobacter*. *Zbl. Bakt. II. Abt.* **83:**280–311.

Nogueira, R., and L. F. Melo. 2006. Competition between *Nitrospira* spp. and *Nitrobacter* spp. in nitrite-oxidizing bioreactors. *Biotechnol. Bioeng.* **95:**169–175.

Okabe, S., H. Satoh, and Y. Watanabe. 1999. In situ analysis of nitrifying biofilms as determined by in situ hybridization and the use of microelectrodes. *Appl. Environ. Microbiol.* **65:**3182–3191.

Orso, S., M. Gouy, E. Navarro, and P. Normand. 1994. Molecular phylogenetic analysis of *Nitrobacter* spp. *Int. J. Syst. Bacteriol.* **44:**83-86.

Pan, P. H. 1971. Lack of distinction between *Nitrobacter agilis* and *Nitrobacter winogradskyi*. *J. Bacteriol.* **108:**1416–1418.

Park, H. D., and D. R. Noguera. 2008. *Nitrospira* community composition in nitrifying reactors operated with two different dissolved oxygen levels. *J. Microbiol. Biotechnol.* **18:**1470–1474.

Poly, F., S. Wertz, E. Brothier, and V. Degrange. 2008. First exploration of *Nitrobacter* diversity in soils by a PCR cloning-sequencing approach targeting functional gene *nxrA*. *FEMS Microbiol. Ecol.* **63:**132–140.

Prosser, J. I. 1989. Autotrophic nitrification in bacteria. *Adv. Microb. Physiol.* **30:**125–181.

Quail, M. A., P. Jordan, J. M. Grogan, J. N. Butt, M. Lutz, A. J. Thomson, S. C. Andrews, and J. R. Guest. 1996. Spectroscopic and voltammetric characterisation of the bacterioferritin-associated ferredoxin of *Escherichia coli*. *Biochem. Biophys. Res. Commun.* **229:**635–642.

Regan, J. M., G. W. Harrington, H. Baribeau, R. De Leon, and D. R. Noguera. 2003. Diversity of nitrifying bacteria in full-scale chloraminated distribution systems. *Water Res.* **37:**197–205.

Reigstad, L. J., A. Richter, H. Daims, T. Urich, L. Schwark, and C. Schleper. 2008. Nitrification in terrestrial hot springs of Iceland and Kamchatka. *FEMS Microbiol. Ecol.* **64:**167–174.

Rodrigue, A., A. Chanal, K. Beck, M. Muller, and L. F. Wu. 1999. Co-translocation of a periplasmic enzyme complex by a hitchhiker mechanism through the bacterial tat pathway. *J. Biol. Chem.* **274:**13223–13228.

Rothery, R. A., G. J. Workun, and J. H. Weiner. 2008. The prokaryotic complex iron-sulfur molybdoenzyme family. *Biochim. Biophys. Acta* **1778:**1897–1929.

Schloss, P. D., and J. Handelsman. 2004. Status of the microbial census. *Microbiol. Mol. Biol. Rev.* **68:**686–691.

Schramm, A., D. de Beer, M. Wagner, and R. Amann. 1998. Identification and activities in situ of *Nitrosospira* and *Nitrospira* spp. as dominant populations in a nitrifying fluidized bed reactor. *Appl. Environ. Microbiol.* **64:**3480–3485.

Schramm, A., D. de Beer, J. C. van den Heuvel, S. Ottengraf, and R. Amann. 1999. Microscale distribution of populations and activities of *Nitrosospira* and *Nitrospira* spp. along a macroscale gradient in a nitrifying bioreactor: quantification by in situ hybridization and the use of microsensors. *Appl. Environ. Microbiol.* **65:**3690–3696.

Schramm, A., D. De Beer, A. Gieseke, and R. Amann. 2000. Microenvironments and distribution of nitrifying bacteria in a membrane-bound biofilm. *Environ. Microbiol.* **2:**680–686.

Seewaldt, E., K. H. Schleifer, E. Bock, and E. Stackebrandt. 1982. The close phylogenetic relationship of *Nitrobacter* and *Rhodopseudomonas palustris. Arch. Microbiol.* **131:**287–290.

Simmons, S. L., G. Dibartolo, V. J. Denef, D. S. Goltsman, M. P. Thelen, and J. F. Banfield. 2008. Population genomic analysis of strain variation in *Leptospirillum* group II bacteria involved in acid mine drainage formation. *PLoS Biol.* **6:**e177.

Sorokin, D. Y., G. Muyzer, T. Brinkhoff, J. G. Kuenen, and M. S. M. Jetten. 1998. Isolation and characterization of a novel facultatively alkaliphilic *Nitrobacter* species, *N. alkalicus* sp. nov. *Arch. Microbiol.* **170:**345–352.

Spieck, E., J. Aamand, S. Bartosch, and E. Bock. 1996a. Immunocytochemical detection and location of the membrane-bound nitrite oxidoreductase in cells of *Nitrobacter* and *Nitrospira. FEMS Microbiol. Lett.* **139:**71–76.

Spieck, E., S. Müller, A. Engel, E. Mandelkow, H. Patel, and E. Bock. 1996b. Two-dimensional structure of membrane-bound nitrite oxidoreductase from *Nitrobacter hamburgensis. J. Struct. Biol.* **117:**117–123.

Spieck, E., S. Ehrich, J. Aamand, and E. Bock. 1998. Isolation and immunocytochemical location of the nitrite-oxidizing system in *Nitrospira moscoviensis. Arch. Microbiol.* **169:**225–230.

Spieck, E., C. Hartwig, I. McCormack, F. Maixner, M. Wagner, A. Lipski, and H. Daims. 2006. Selective enrichment and molecular characterization of a previously uncultured *Nitrospira*-like bacterium from activated sludge. *Environ. Microbiol.* **8:**405–415.

Spring, S., R. Amann, W. Ludwig, K. H. Schleifer, H. van Gemerden, and N. Petersen. 1993. Dominating role of an unusual magnetotactic bacterium in the microaerobic zone of a freshwater sediment. *Appl. Environ. Microbiol.* **59:**2397–2403.

Stackebrandt, E., R. G. E. Murray, and H. G. Trüper. 1988. *Proteobacteria* classis nov., a name for the phylogenetic taxon that includes the "purple bacteria and their relatives." *Int. J. Syst. Bacteriol.* **38:**321–325.

Stanley, P. M., and E. L. Schmidt. 1981. Serological diversity of *Nitrobacter* spp. from soil and aquatic habitats. *Appl. Environ. Microbiol.* **41:**1069–1071.

Starkenburg, S. R., P. S. Chain, L. A. Sayavedra-Soto, L. Hauser, M. L. Land, F. W. Larimer, S. A. Malfatti, M. G. Klotz, P. J. Bottomley, D. J. Arp, and W. J. Hickey. 2006. Genome sequence of the chemolithoautotrophic nitrite-oxidizing bacterium *Nitrobacter winogradskyi* Nb-255. *Appl. Environ. Microbiol.* **72:**2050–2063.

Starkenburg, S. R., F. W. Larimer, L. Y. Stein, M. G. Klotz, P. S. Chain, L. A. Sayavedra-Soto, A. T. Poret-Peterson, M. E. Gentry, D. J. Arp, B. Ward, and P. J. Bottomley. 2008. Complete genome sequence of *Nitrobacter hamburgensis* X14 and comparative genomic analysis of species within the genus *Nitrobacter. Appl. Environ. Microbiol.* **74:**2852–2863.

Steger, D., P. Ettinger-Epstein, S. Whalan, U. Hentschel, R. de Nys, M. Wagner, and M. W. Taylor. 2008. Diversity and mode of transmission of ammonia-oxidizing archaea in marine sponges. *Environ. Microbiol.* **10:**1087–1094.

Stein, L. Y., D. J. Arp, P. M. Berube, P. S. G. Chain, L. Hauser, M. S. M. Jetten, M. G. Klotz, F. W. Larimer, J. M. Norton, H. J. M. Op den Camp, M. Shin, and X. Wei. 2007. Whole-genome analysis of the ammonia-oxidizing bacterium, *Nitrosomonas eutropha* C91: implications for niche adaptation. *Environ. Microbiol.* **9:**2993–3007.

Steinmüller, W., and E. Bock. 1976. Growth of Nitrobacter in the presence of organic matter. I. Mixotrophic growth. *Arch Microbiol* **108:**299–304.

Strous, M., E. Pelletier, S. Mangenot, T. Rattei, A. Lehner, M. W. Taylor, M. Horn, H. Daims,

D. Bartol-Mavel, P. Wincker, V. Barbe, N. Fonknechten, D. Vallenet, B. Segurens, C. Schenowitz-Truong, C. Medigue, A. Collingro, B. Snel, B. E. Dutilh, H. J. Op den Camp, C. van der Drift, I. Cirpus, K. T. van de Pas-Schoonen, H. R. Harhangi, L. van Niftrik, M. Schmid, J. Keltjens, J. van de Vossenberg, B. Kartal, H. Meier, D. Frishman, M. A. Huynen, H. W. Mewes, J. Weissenbach, M. S. Jetten, M. Wagner, and D. Le Paslier.** 2006. Deciphering the evolution and metabolism of an anammox bacterium from a community genome. *Nature* **440:**790–794.

Sundermeyer-Klinger, H., W. Meyer, B. Warninghoff, and E. Bock. 1984. Membrane-bound nitrite oxidoreductase of *Nitrobacter*: evidence for a nitrate reductase system. *Arch. Microbiol.* **140:**153–158.

Tanaka, Y., Y. Fukumori, and T. Yakamaka. 1983. Purification of cytochrome a_1c_1 from *Nitrobacter agilis* and characterization of nitrite oxidation system of the bacterium. *Arch. Microbiol.* **135:**265–271.

Taylor, M. W., R. Radax, D. Steger, and M. Wagner. 2007. Sponge-associated microorganisms: evolution, ecology, and biotechnological potential. *Microbiol. Mol. Biol. Rev.* **71:**295–347.

Teske, A., E. Alm, J. M. Regan, T. S., B. E. Rittmann, and D. A. Stahl. 1994. Evolutionary relationships among ammonia- and nitrite-oxidizing bacteria. *J. Bacteriol.* **176:**6623–6630.

Thorell, H. D., K. Stenklo, J. Karlsson, and T. Nilsson. 2003. A gene cluster for chlorate metabolism in *Ideonella dechloratans*. *Appl. Environ. Microbiol.* **69:**5585–5592.

van Ginkel, C. G., G. B. Rikken, A. G. Kroon, and S. W. Kengen. 1996. Purification and characterization of chlorite dismutase: a novel oxygen-generating enzyme. *Arch. Microbiol.* **166:**321–326.

Vanparys, B., E. Spieck, K. Heylen, L. Wittebolle, J. Geets, N. Boon, and P. De Vos. 2007. The phylogeny of the genus *Nitrobacter* based on comparative rep-PCR, 16S rRNA and nitrite oxidoreductase gene sequence analysis. *Syst. Appl. Microbiol.* **30:**297–308.

Wächtershäuser, G. 1990. Evolution of the first metabolic cycles. *Proc. Natl. Acad. Sci. USA* **87:**200–204.

Wagner, M., M. Horn, and H. Daims. 2003. Fluorescence in situ hybridization for the identification of prokaryotes. *Curr. Opin. Microbiol.* **6:**302–309.

Wagner, M., G. Rath, H.-P. Koops, J. Flood, and R. Amann. 1996. In situ analysis of nitrifying bacteria in sewage treatment plants. *Water Sci. Tech.* **34:**237–244.

Watson, S. W. 1971. Taxonomic considerations of the family *Nitrobacteraceae* Buchanan. Requests for opinions. *Int. J. Syst. Bacteriol.* **21:**254–270.

Watson, S. W., and J. B. Waterbury. 1971. Characteristics of two marine nitrite oxidizing bacteria, *Nitrospina gracilis* nov. gen. nov. sp. and *Nitrococcus mobilis* nov. gen. nov. sp. *Arch. Mikrobiol.* **77:**203–230.

Watson, S. W., E. Bock, F. W. Valois, J. B. Waterbury, and U. Schlosser. 1986. *Nitrospira marina* gen. nov. sp. nov.: a chemolithotrophic nitrite-oxidizing bacterium. *Arch. Microbiol.* **144:**1–7.

Wertz, S., F. Poly, X. Le Roux, and V. Degrange. 2008. Development and application of a PCR-denaturing gradient gel electrophoresis tool to study the diversity of *Nitrobacter*-like *nxrA* sequences in soil. *FEMS Microbiol. Ecol.* **63:**261–271.

Winogradsky, S. 1892. Contributions a la morphologie des organismes de la nitrification. *Arch. Sci. Biol.* (St. Petersburg) **1:**88–137.

Winslow, C. E. A., J. Broadhurst, R. E. Buchanan, J. C. Krummwiede, L. A. Rogers, and G. H. Smith. 1917. The families and genera of the bacteria. Preliminary report of the Committee of the Society of American Bacteriologists on characterization and classification of bacterial types. *J. Bacteriol.* **2:**505–566.

Woese, C. R., E. Stackebrandt, W. G. Weisburg, B. J. Paster, M. T. Madigan, V. J. Fowler, C. M. Hahn, P. Blanz, R. Gupta, K. H. Nealson, and G. E. Fox. 1984. The phylogeny of the purple bacteria: the alpha subdivision. *Syst. Appl. Microbiol.* **5:**315–326.

Yamamoto, M., H. Arai, M. Ishii, and Y. Igarashi. 2006. Role of two 2-oxoglutarate:ferredoxin oxidoreductases in *Hydrogenobacter thermophilus* under aerobic and anaerobic conditions. *FEMS Microbiol. Lett.* **263:**189–193.

Zhang, C. L., Q. Ye, Z. Huang, W. Li, J. Chen, Z. Song, W. Zhao, C. Bagwell, W. P. Inskeep, C. Ross, L. Gao, J. Wiegel, C. S. Romanek, E. L. Shock, and B. P. Hedlund. 2008. Global occurrence of archaeal *amoA* genes in terrestrial hot springs. *Appl. Environ. Microbiol.* **74:**6417–6426.

PROCESSES, ECOLOGY, AND ECOSYSTEMS

VI

NITRIFICATION IN THE OCEAN

Bess B. Ward

13

INTRODUCTION

Nitrogen is one of the essential nutrients for life, and the one most often limiting for biological production in vast regions of the world's oceans. Thus, the marine nitrogen cycle tends to function efficiently in the ocean, and the net nitrogen inventory is governed by the balance between losses due to denitrification and inputs from nitrogen fixation. Within the cycle of fixed nitrogen transformations, ammonium is produced from the breakdown of organic matter during remineralization by heterotrophs such as zooplankton and microbes in the water column and by benthic invertebrates, worms, and microbes in the sediments. In oxygenated environments, ammonium rarely accumulates, however, as it is rapidly oxidized to nitrite and nitrate by the process of nitrification. In surface waters, nitrate is rapidly consumed by photosynthetic phytoplankton, while it tends to accumulate in the deep ocean where this assimilatory process cannot occur. Nitrate that makes its way to oxygen-depleted environments in the water column and sediments is usually lost as dinitrogen gas by denitrification or anammox. In very low oxygen environments, ammonium can accumulate if suitable oxidants are not available. Thus, nitrification links other critical processes in the nitrogen cycle, performing essential transformations in all ocean environments except those that are extremely reducing.

Nitrification in the ocean has been comprehensively reviewed in the context of the overall marine nitrogen cycle in a recent book (Capone et al., 2008). Therefore, following a brief overview and introduction, the present chapter will focus only on very recent developments and their implications for nitrogen cycling in the marine environment. (i) For oceanographers, in general, the extensive verification that nitrification occurs in the euphotic zone, the sunlit surface layer of the ocean, has important ramifications for estimating primary production and modeling the biological carbon cycle. (ii) The most important new insights for marine nitrification are related to the recent discovery that Archaea are abundant ammonia oxidizers in the ocean. Their abundance far exceeds that of the bacterial nitrifiers, which raises major questions about the mass balance of nitrogen and carbon fluxes in the mesopelagic zone. The dominance of archaeal nitrifiers also has implications for their ecology and physiology with respect to environmental regulation of nitrification and the production of nitrous oxide. (iii) Anaerobic ammonia oxidation (anammox)

Bess B. Ward, Department of Geosciences, Princeton University, Princeton, NJ 08544.

Nitrification, Edited by Bess B. Ward, Daniel J. Arp, and Martin G. Klotz

is now recognized as a major fixed nitrogen loss term in oxygen-depleted zones (ODZs) of the ocean, where denitrification had formerly been assumed to be the primary process.

DISTRIBUTION AND RATES OF NITRIFICATION IN OCEAN

The role of nitrification in marine systems is to link the oxidizing (nutrient regeneration) and reducing (assimilation, respiration) processes of the nitrogen cycle by converting ammonium (NH_4^+) to nitrate (NO_3^-) (Fig. 1). Nitrification does not influence the net N inventory of the ocean directly except by small losses to the gaseous pool of nitrous oxide (N_2O; see later), but it does determine the distribution of N among important dissolved inorganic nitrogen pools. Organic matter, initially produced by primary production and cycled through the marine food web, is degraded by microbes, and organic nitrogen is eventually mineralized to ammonium. Most microbes and phytoplankton can easily assimilate ammonium, and it rarely accumulates to significant concentrations in oxygenated ocean waters. Ammonia-oxidizing bacteria (AOB) and ammonia-oxidizing archaea (AOA) oxidize ammonium to nitrite, and nitrite oxidizers (assumed to be bacteria, nitrite-oxidizing bacteria [NOB]), convert the nitrite to nitrate, which can be a very important N source for many kinds of phytoplankton.

Nitrate concentrations in the surface ocean are usually very low due to utilization by phytoplankton, except in "high-nutrient low-chlorophyll" regions and when supplied by episodic events such as regional upwelling. The deep nitrate reservoir can be made available to phytoplankton by mixing, upwelling, and seasonal overturn. These physical processes bring cold deep nitrate-rich water up to the surface where, in the presence of light, phytoplankton can assimilate the nitrate. Thus, although nitrate is not usually abundant in surface waters, it is a very important nitrogen source for phytoplankton.

The significance of ammonium and nitrate as sources of N for primary production in the ocean is usually understood in the context of the new production paradigm (Dugdale and Goering, 1967; Eppley and Peterson, 1979) (Fig. 2). New production, or export production, is the term necessary to estimate deep carbon burial, the rate at which the biological pump removes carbon, nitrogen, and associated materials from the surface ocean and sequesters them in sediments for long-term burial. The new production paradigm has been a very useful construct for understanding and modeling biogeochemical processes in the ocean.

Nitrogen is often the limiting nutrient for primary production, and it can be supplied either by recycled ammonium within the euphotic zone (regenerated nutrients: supporting regenerated production) or by supply of nitrogen from outside the euphotic zone (new nitrogen: from nitrogen fixation or vertical transport via upwelling or mixing of NO_3^- from the deep ocean reservoir, supporting new production). The new production paradigm explicitly states that nitrification does not occur in the euphotic zone (Fig. 2, Left), so that all nitrate must be supplied from external sources and is equated to new nitrogen. If the system is in steady state, then loss of material from the euphotic zone by grazing or sinking (export) is balanced by the input of new nitrogen. If nitrogen fixation is minimal, then the supply of nitrate from below is approximately equal to the export loss term. If nitrogen is limiting and nitrate therefore does not accumulate when supplied to the euphotic zone, then the rate of nitrate assimilation is equivalent to the rate of export production. The simplicity of this model has great practical attractions because it implies that a simple measurement of the rates of nitrate and ammonium uptake using stable isotope tracer experiments can quantify the rates of new and regenerated production, without the need for ecosystem-wide experiments or sediment traps, etc. Nitrification is implicitly important in the conception and measurement of new production because nitrification is the ultimate source of the nitrate that supports new production, but it is assumed that this nitrification

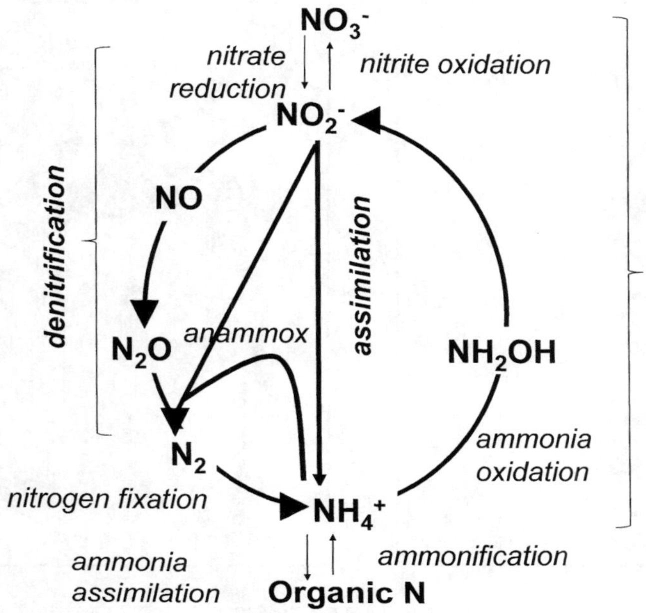

FIGURE 1 The biological nitrogen cycle, showing the role of nitrification in linking the oxidized and reduced components of the dissolved inorganic nitrogen pools.

does not occur in the same layer where that production occurs.

Most measurements of nitrification in the ocean and estuarine environments have been made in the upper layers. Where depth profiles are available, however, it is found that the highest nitrification rates, both ammonium oxidation and nitrite oxidation, occur in a region near the bottom of the euphotic zone. A maximum is often observed in the nitrification rate at a depth in the water column where the light intensity is 5 to 10% of surface light intensity, which also often includes the depth zone of the primary nitrite maximum and the deep chlorophyll maximum (Ward et al., 1984; Ward, 1987b; Lipschultz et al., 1990; Sutka et al., 2004).

Although nitrification rates are usually low in the surface layer (approximately 0 to 50 m), the rate of nitrate production via nitrification that occurs within the photic zone is often on the same order as the rate of nitrate utilization by phytoplankton (Ward et al., 1989; Dore and Karl, 1996; Clark et al., 2008; Fernandez and Raimbault, 2007). For example, in Monterey Bay, California (~1,000 m total depth), the photic zone usually extends to 30 to 50 m, and a primary nitrite maximum typically occurs between 20 and 50 m (Ward, 2005b). Ammonium oxidation rates range between 0 and 80 nM day^{-1}, often with a rate maximum at 30 to 50 m. Using a model to evaluate the distribution of ^{15}N and ^{18}O in nitrate throughout the upper 200 m in Monterey Bay, Wankel et al. (2007) concluded that 15 to 27% of nitrate assimilation was supported by nitrification. Isotope tracer experiments in the same region detected nitrification in the photic zone and showed that a large fraction of ammonium utilization (21 to 33%) supported nitrification rather than phytoplankton assimilation (Ward, 2005b). In the Southern California Bight, a more oligotrophic region to the south of Monterey Bay, nitrification could supply the entire nitrate assimilation demand in the lower depths of the euphotic zone (Ward et al.,

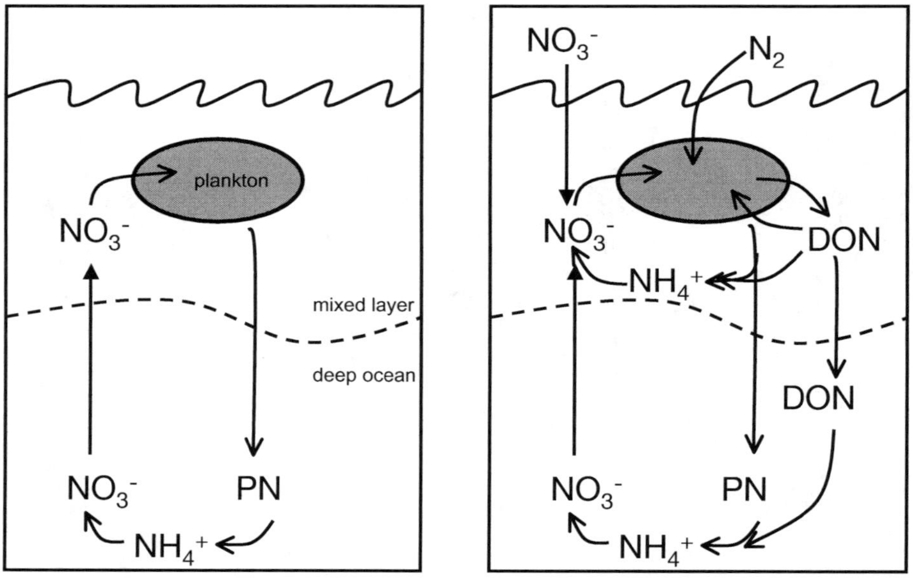

FIGURE 2 Schematic of the role of nitrification in the surface ocean. Plankton, phytoplankton and zooplankton, the grazing food web; PN, particulate nitrogen, living or dead; DON, dissolved organic nitrogen. (Left) Nitrification occurs in the deep ocean, and nitrate is supplied to the euphotic zone by mixing. This physical separation between the processes of nitrate assimilation and regeneration, as described in the New Production Paradigm (Eppley and Peterson, 1979), means that at steady state, the rate of nitrate assimilation is equivalent to the rate of export production (sinking or otherwise removal of PN from the euphotic zone). (Right) Nitrification occurs in the euphotic zone as well as at depth, implying that nitrate assimilation cannot be equated to export production. Other processes that complicate the simple application of the New Production Paradigm are also shown: DON is a much greater flux than previously imagined, and nitrogen fixation can be a significant source of new production is some regions of the ocean.

1989). Thus, it appears that some of the nitrate assimilated by phytoplankton cannot be called new nitrogen, and the uptake rate of nitrate cannot be equated with new production. At least part of the primary production supported by nitrate must be considered regenerated production, and nitrate is a rapidly cycled regenerated nutrient (Fig. 2, Right). The degree to which nitrate assimilation and nitrate regeneration are separated in time and space no doubt varies regionally and seasonally and is a topic sorely in need of additional investigation.

Rates of nitrification reported for the open ocean are in the range of a few to a few hundred nanomolar per day (Ward, 2006, 2008; Yool et al., 2007) and have been detected as deep as 3,000 m (Ward and Zafiriou, 1988). Most of the reports are of ammonium oxidation rates. Far fewer reports of nitrite oxidation rates have been published, and many of the high rates reported may be overestimates. A compilation of published rates of nitrification in various aquatic environments was presented by Ward (2008).

The rate of nitrification in the deep ocean is minimal, because the flux of ammonium from organic matter decomposition decreases with increasing depth. The accumulation of nitrate as the main form of nitrogen in the deep sea results from very low production and the lack of any significant consumption. An exception to this deep sea condition is found in the vicinity of hydrothermal vent plumes, where ammonium concentrations are elevated to hundreds of nanomolar (Lilley et al., 1993) and oxidation rates (up to 91 nM day^{-1}) of the same magnitude as those detected in surface water have been reported (Lam et al., 2004).

AOB AND AOA

The NH_3-oxidizing bacteria (AOB), including the marine AOB, fall into two major phyla in the Proteobacteria (Purkhold et al., 2003; Koops et al., 2003) (see Chapter 2). The β subdivision contains the genera *Nitrosospira* and *Nitrosomonas*, while *Nitrosococcus* is in the γ subdivision. On the basis of 16S rRNA sequence analysis, several main clusters and a large amount of microdiversity within clusters of the β AOB have been detected in marine environments (Stephen et al., 1996, Bano and Hollibaugh, 2000; O'Mullan and Ward, 2005). *Nitrosomonas* sequences were more often associated with enrichment cultures and *Nitrosospira* with clone libraries (Stephen et al., 1996; Smith et al., 2001), implying that even in the well-known AOB, the most important strains in the environment are not represented in the culture collections. *Nitrosospira*-like sequences related only to other environmental sequences are often dominant in AOB sequence clone libraries retrieved from oceanic (Bano and Hollibaugh, 2000; Hollibaugh et al., 2002; O'Mullan and Ward, 2005; Molina et al., 2007), estuarine (Francis et al., 2003; Bernhard et al., 2005), and coastal (Ando et al., 2009) environments.

Marine *Nitrosococcus*, although ubiquitous (Ward and O'Mullan, 2002), are apparently less abundant and have received much less study. Compared to the large amount of data available for the β AOB, there are few reports of *Nitrosococcus*-like sequences obtained from the marine environment and no reports of finding *Nitrosococcus halophilus*-like sequences. This lack of information on *Nitrosococcus* may be a matter of PCR primer specificity and limited research effort at present, but it just as likely implies that *Nitrosococcus* is a minor component of the AOB assemblage.

Until recently, the only cultivated ammonia oxidizers were bacteria. These cultures have provided the basis of physiological inferences about ecological niches and environmental regulation of nitrification in the ocean. The most important development in the study of nitrification in the ocean in the last decade is the discovery of AOA (Konneke et al., 2005) (see Chapters 6 and 7). The implications of this discovery may not result in big changes in our understanding of the rates and distribution of nitrification in the ocean. Some of the new findings, however, do compel us to reconsider our current picture of carbon and nitrogen fluxes in the mesopelagic ocean (see next section).

After their initial identification as a domain separate from Bacteria (Woese et al., 1978), Archaea, including the kingdom *Crenarchaeota*, were assumed to be mostly extremophiles, typical of extremely hot, hypersaline, acidic, or anoxic environments. The first report of ubiquitous archaeal genes in 16S rRNA clone libraries from the ocean (Fuhrman et al., 1992) was met with skepticism. This report was subsequently substantiated (Delong, 1992), and the great numerical abundance of *Crenarchaeota* (Karner et al., 2001) proved them to be ubiquitous in the ocean, by definition due to its vastness, a nonextreme environment. First attempts to determine the metabolic repertoire of this vast but previously unknown component of the microbial ecosystem showed that the ubiquitous *Crenarchaeota* assimilated radiolabeled amino acids (Ouverney and Fuhrman, 2000), suggesting that the dominant metabolism of the group was heterotrophy. This was consistent with the assumption that most microbes in the deep sea are heterotrophic, subsisting on the rain of organic matter that reaches the deep sea from primary production by phytoplankton in the surface waters.

This view began to change when ammonia-oxidizing genes of apparent archaeal origin were discovered in environmental metagenomic libraries from soils (Schleper et al., 2005) and the ocean (Venter et al., 2004). Homologues of the ammonia monooxygenase gene from AOB were definitively linked with 16S rRNA genes from the most common group of *Crenarchaeota* in soils (Treusch et al., 2005). The physiological link was clearly established by cultivation from a seawater aquarium of a strain of Archaea that oxidizes NH_4^+ to NO_2^-, apparently using a pathway with stoichiometry similar to that of the AOB (Konneke et al., 2005). *Nitrosopumilus maritimus* was inhibited by high levels of organic matter, but preliminary information from its genome shows that it possesses the capability to acquire and assimilate a range of organic compounds. If *N. maritimus* is representative of marine crenarchaeal AOA, the metabolic type appears to be mixotrophy.

The 16S rRNA sequence of only one cultivated marine AOA is currently available, and that sequence places *N. maritimus* within the low-temperature marine Crenarchaeota, a group that is abundant in seawater and distinct from low-temperature Crenarchaeota in soils (Konneke et al., 2005). Phylogenetic analysis of several hundred AOA partial *amoA* sequences identified major groups that clustered by environment (Francis et al., 2005). Marine AOA are found in two of the major clades of the mesophilic *Crenarchaeota*, as defined by 16S rRNA phylogeny (Prosser and Nicol, 2008). Based on clone libraries of *amoA* gene sequences, AOA-type *amoA* genes are ubiquitous in aquatic environments, and different clades characterize different environments (Prosser and Nicol, 2008), including different depth horizons (Beman et al., 2008; De Corte et al., 2009; Yakimov et al., 2009). Although both rRNA and *amoA* genes for both AOA and AOB are easily retrieved from marine environments for clone library analysis, it now appears that the AOA are much more abundant than AOB in marine systems (Wuchter et al., 2006; Mincer et al., 2007; De Corte et al., 2009). Prosser and Nicol (2008) concluded that all available evidence now indicates that AOA are not only more abundant than AOB in the marine environment and elsewhere, but that they are likely responsible for most of the ammonium oxidation.

In estuaries, the dominance of AOA over AOB is not ubiquitous. Some estuarine sediments are clearly dominated by AOA (Caffrey et al., 2007) and other by AOB (Magalhães et al., 2009). In San Francisco Bay sediments (Mosier and Francis, 2008) and the Chesapeake Bay water column (Bouskill et al., unpublished data), the ratio of AOA:AOB increased with increasing salinity, such that AOB were most abundant in the oligohaline regions of the estuary and AOA greatly outnumbered AOA in the more oceanic conditions of the lower estuary. Even within the AOB, community composition varies with salinity in estuarine waters, with *Nitrosomonas* types being more prevalent in oligohaline regions and *Nitrosospira* dominated in the higher-salinity regions (Caffrey et al., 2003; Berhnard et al., 2005; Ward et al., 2007). A recent review (Erguder et al., 2009) documented the occurrence of AOA under a wide range of environmental conditions, including very low pH and measureable sulfide, where AOB have not been detected. Clearly the AOA are a very diverse group, although the few clades found in the open ocean are probably very restricted in their physicochemical tolerances.

NOB IN THE OCEAN

The phylogeny of NO_2^--oxidizing bacteria (Bock and Wagner, 2003) (see Chapter 12), based on 16S rRNA sequences, shows that the best-known autotrophic NO_2^- oxidizer, *Nitrobacter*, comprises a coherent genus in the subdivision of the Proteobacteria (Teske et al., 1994). Like *Nitrosococcus oceani*, *Nitrococcus mobilis* (Watson and Waterbury, 1971) belongs to the γ subdivision of the Proteobacteria, the only example of both NH_3- and NO_2^--oxidizing phenotypes occurring in the same subdivision. *Nitrospina gracilis*, the only species in this genus, represented by two isolates (Watson and Waterbury, 1971), is assigned to the δ subdivision of the Proteobacteria. Possibly the most unusual nitrifier is the genus *Nitrospira*, which is represented by only two isolates and does not share a common lineage with the other nitrifiers. A novel *Nitrospira* strain, *Nitrospira moscoviensis*, isolated from a heating system in Moscow, Russia, was assigned to a new genus outside of the Proteobacteria (Ehrich et al., 1995). These authors reanalyzed the *Nitrospira marina* sequence, which Teske et al. (1994) had placed in the δ-Proteobacteria, and concluded that *Nitrospira* belonged outside the Proteobacteria, in a deeply branching cluster related to *Leptospirilla*. The marine isolate (Watson et al., 1986) was obtained from the Gulf of Maine, and the authors reported that similar cells were present in many enrichments, suggesting it is a common member of the marine nitrifier assemblage. Although much less is known about the composition of nitrite oxidizers than ammonia oxidizers in the ocean, recent evidence suggests that most of the nitrite oxidizers are bacteria of the *Nitrospina* lineage (Mincer et al., 2007).

PHYSIOLOGY AND ECOLOGY OF MARINE NITRIFIERS

Despite their somewhat restricted phylogenetic range, the bacterial nitrifiers are polyphyletic, and the phenotype has apparently arisen independently numerous times. The homology of the functional genes (*amo*, *hao*) involved in the ammonium-oxidizing physiology implies gene transfer events, however, rather than independent evolution of these enzymes. The fact that the *amoA* genes from AOB and AOA are homologous raises the question of the ultimate origin of the NH_3-oxidizing phenotype (see Chapter 4). If the ancestral Crenarchaeota were thermophiles, it is possible that NH_3 oxidation originally arose in thermophiles and spread from the archaea to the bacteria.

An understanding of the phylogeny of nitrifying bacteria is relevant to the study of nitrification in aquatic habitats because it has implications for detection and quantification methods. Although the bacterial nitrifiers are polyphyletic, they are not so diverse as to be unmanageable; their affiliation within a small group of lineages makes them amenable to identification and detection using a relatively small suite of molecular probes. This approach forms the basis of much current knowledge on the diversity and distribution of autotrophic nitrifying bacteria and has already made important contributions to the study of nitrifying archaea as well.

Before the discovery of the AOA, the ecophysiology of ammonium oxidation was interpreted in terms of the biochemical and physiological capabilities of the AOB. Thus, characteristics such as light inhibition and high substrate affinity, formerly attributed to AOB, must now be reconsidered in terms of AOA. The sensitivity of AOB and NOB to light in both pure culture and enrichments is well documented (Muller-Neugluck and Engel, 1961; Horrigan et al., 1981; Guerrero and Jones, 1996). The physiological basis of light sensitivity is assumed to be the abundant cytochromes in AOB and NOB, which are reported to confer wavelength-specific inhibition (Guerrero and Jones, 1996). Light inhibition has been interpreted to explain the primary nitrite maximum (Olson, 1981b) and to support the absence of nitrification in the photic zone, in accord with the new production paradigm. The degree to which this physiological characteristic is expressed in the real world is debatable, because other factors such as substrate limitation and bacterivory might easily determine nitrifier distribution and activity, and phytoplankton alone could be responsible for the primary nitrite maximum (Lomas and Lipschultz, 2006). It is not yet known whether cultivated AOA are inhibited by natural light. If the AOA in the marine environment are subject to light inhibition, it must be via a different mechanism, because the archaea are sorely lacking in heme proteins, having instead a preponderance of Cu-containing proteins (see Chapter 6).

Cultivated marine AOB demonstrate a classic response of increased ammonium oxidation rate related to increasing ammonium concentration (Carlucci and Strickland, 1968; Ward, 1987), but attempts to demonstrate such a response with natural assemblages from the ocean have generally failed (Olson, 1981a; Ward and Kilpatrick, 1990). One explanation for this lack of a kinetic response is that the affinity of the natural assemblage for ammonia was so high (their half-saturation constants so low), that the rates of oxidation were saturated at measurable levels of substrate enrichment. This implies that the most important components of the natural assemblage of AOBs have not been cultivated and is consistent with the very high affinity for ammonium demonstrated by the cultivated AOA, *N. maritimus* (Martens-Habbena et al., 2009). Ergruber et al. (2009) documented the retrieval of AOA *amoA* genes from habitats with a very high ammonia concentration, but the AOA of the true oceanic environment rarely encounter ammonium concentrations higher than 1 μM. Substrate affinity might well be an important determinant of the biogeography of AOA clades.

Microaerophily has been attributed to cultivated AOB (Gundersen, 1966; Carlucci and McNally, 1969) and NOB (Laanbroek et al., 1994) in culture and in the environment. The apparent link between nitrification and denitrification in facilitating fixed N loss in stratified environments (e.g., sediments, biofilms) (Jensen et al., 1996; Laursen and Seitzinger, 2002; Revsbech et al., 2006; Krishnan et al., 2008) is consistent with the ability of bacterial nitrifiers to tolerate and grow at very low oxygen tension. AOA have been detected in low O_2 waters (Beman et al., 2008), suggesting that they too may have microaerophilic capabilities. Nitrification at low oxygen concentrations is often linked to N_2O production (Jorgensen et al., 1984; Usui et al., 2001; Meyer et al., 2008) (see below). The anaerobic metabolism of AOB is reported to include production of N_2 gas (Bock et al., 1995; Zart and Bock, 1998), but the pathways and significance of anaerobic metabolism by nitrifiers in the ocean is not well understood. It seems likely, however, that much of the nitrification-related activities in low-oxygen environments that were previously attributed to AOB must now be investigated in relation to AOA metabolism.

CARBON AND NITROGEN FLUXES IN THE OCEAN

The degree to which AOA represent heterotrophic or autotrophic biomass is one of the most intriguing and important unresolved issues. Ingalls et al. (2006) addressed this question by analyzing the radiocarbon content of archaeal lipids in surface waters and waters from 670 m in the subtropical North Pacific Ocean. The isotopic signatures of the lipids showed that Archaea in surface waters incorporate modern carbon into their cell membranes—this does not differentiate between autotrophy and heterotrophy, as most dissolved organic carbon (DOC) and dissolved inorganic carbon in the surface ocean is of modern origin. However, at 670 m, archaeal lipids were isotopically enriched relative to the ambient DOC pool, indicating that incorporation of ambient DOC could not be the main source of carbon for these cells. Ingalls et al. (2006) concluded that incorporation of dissolved inorganic carbon, through some autotrophic CO_2 assimilation pathway, was the source of 83% of the archaeal membrane lipids and, by extension, archaeal biomass, at 670 m. What cannot be distinguished is whether this contribution of autotrophy implies that 83% of the cells are completely autotrophic, or the whole archaeal assemblage is mixotrophic, obtaining 83% of its carbon from CO_2 and 17% from DOC. It is also possible that more of the biomass is heterotrophic than this calculation implies: if heterotrophic biomass is mostly supported by the rain of fresh organic matter, then even organisms that utilize DOC would incorporate modern carbon (Aristegui et al., 2002), rather than the older DOC that makes up most of the standing stock of DOC at depth. Thus, even heterotrophs at depth could have a relatively modern isotopic signature.

If a large fraction of the microbial biomass below the euphotic zone of the ocean (100 to 1,000 m) is autotrophic, depending directly on the assimilation of CO_2 rather than DOC or particulate organic carbon, the current conception of oceanic carbon and nitrogen cycling requires major revision. In the currently accepted scenario, the only significant input of organic carbon (i.e., primary production) occurs via photosynthesis in the surface ocean. Other sources of primary production, such as anoxygenic photosynthetic bacteria in microbial mats or stratified basins, are interesting and geologically important but quantitatively unimportant to total primary production. The same conclusion applies to the chemosynthetic communities of hot vents and cold seeps: they are locally important but contribute only a small fraction of the total ocean carbon budget. Thus, primary production in the surface ocean supports virtually all metabolism in the ocean, either directly through grazing or indirectly by contributing to the vertical flux of material leaving the photic zone as intact cells or waste material. This dependence on surface production is consistent with the typical pattern of exponential decay with increasing depth of many oceanographically interesting quantities: vertical flux of total particulate material, as measured with sediment traps (Suess, 1980; Martin et al., 1987; Berelson, 2001) and vertical distribution and flux of dissolved organic matter (Santinelli et al., 2006) and of various organic constituents (Wakeham et al., 1997; Sheridan et al., 2002); total microbial abundance and abundance of various phylogenetically defined subsets of cells (Karner et al., 2001; Morris et al., 2002). The decrease in concentration and flux of particulate and dissolved material with increasing depth implies loss of this material via consumption and respiration by heterotrophic organisms, including bacteria and zooplankton (Steinberg et al., 2002).

If a large fraction of the microbial biomass below the euphotic zone is supported by autotrophy, then it is not the organic carbon portion of the vertical flux that controls microbial metabolism at depth, but rather the flux of ammonium resulting from the mineralization of the nitrogen component of the organic material. Suess (1980) and Martin et al. (1987) used the vertical flux of organic carbon to compute the rate and distribution of respiration in the water column as a function of overlying primary production or loss just below the euphotic zone. Just as oxygen utilization can be computed on the assumption that organic material is respired with a standard stoichiometry, the rate of N regeneration is also predictable (Martin et al., 1987). N is remineralized initially as ammonium, but the deep nitrogen reservoir is nitrate, not ammonium, implying that nitrification is also coupled directly to organic matter mineralization. The few deep profiles of nitrification rates support such a quantitative relationship (Ward and Zafiriou, 1988).

Vertical flux measurements and relationships derived to describe oxygen consumption and nutrient regeneration imply that half of the organic material removed from the surface layer by sinking is consumed within the upper 300 m of the northeast Pacific, 75% by 500 m and 90% by 1,500 m (Martin et al., 1987). The increase in C:N and C:P ratios of organic material with depth implies that nitrogen and phosphorus are regenerated even more rapidly and at shallower depths than the carbon component. Thus, most of the nitrification supported by the ammonium flux from mineralization of sinking material should occur within the upper 1,000 to 1,500 m of the open ocean. Simulated in situ measurements of nitrification rates support this conclusion (Ward 1987b; Ward and Zafiriou, 1988). Measurement of nitrification rates using ^{15}N tracers yields rate estimates independent of the kind of nitrifying organism responsible for the process. Thus, these arguments based on stoichiometry and vertical distributions provide constraints on the metabolic characteristics of the microbes in the ocean, and on the overall nature and distribution of biogeochemical fluxes in the mesopelagic ocean.

By considering independent estimates of the abundance of *Crenarchaeota*, their contribution to autotrophic biomass, and the implications for nitrite oxidation, we can estimate the total contribution of nitrifiers to microbial biomass in the ocean. Then, using measures of the fluxes of C and N into the mesopelagic realm from sediment trap and other data, we can estimate the rate of biomass turnover (growth rate) of the autotrophic biomass that such flux could sustain. The supply and demand terms that result suggest that if autotrophic biomass constitutes a very large portion of the total biomass, it grows very slowly.

Karner et al. (2001) reported that *Crenarchaeota* constituted on the order of 15 to 40% of total 4′,6-diamidino-2-phenylindole-stained cells (varying with season) in the same depth interval (600 to 1,000 m) at a station very close to the station from which the autotrophy estimate (Ingalls et al., 2006) was obtained. These are the kind of cells whose lipids carry the ^{14}C signatures of autotrophy, implying that up to 40% of the microbial biomass at 670 m is supported mostly by autotrophy. If ammonia oxidation is the basis of this autotrophy, there are additional major implications for C and N fluxes and the abundance of NOB in the deep ocean. Nitrite, the assumed product of archaeal ammonia oxidation (Konneke et al., 2005), does not accumulate in this depth zone. The subtropical North Pacific contains a typical oceanic oxygen minimum zone (OMZ), but oxygen is not depleted to the extent that denitrification occurs and nitrite concentrations never exceed a few nanomolar (http://hahana.soest.hawaii.edu/hot/hot_jgofs.html). Thus, any nitrite produced by ammonia oxidation is likely rapidly oxidized further to nitrate, the most abundant form of nitrogen in the deep ocean. This requires the involvement of a substantial biomass of nitrite-oxidizing microbes. If we assume that ammonia and nitrite oxidation are equally energetically efficient (i.e., the ratio of inorganic N turnover to biomass is the same for ammonium and nitrite oxidation), then to convert the nitrite to nitrate requires a biomass of NOB equivalent to the biomass of AOB plus AOA. If all the *Crenarchaeota* enumerated by Karner et al. (2001) are AOA, this implies that a biomass equal to 15 to 40% of the microbial cells must be nitrite oxidizers. All known NOB are autotrophic. Although the capability for limited mixotrophy has been observed in culture (Smith and Hoare, 1968) and is consistent with published genomes of *Nitrobacter* (Starkenburg et al., 2006, 2008), autotrophy is still assumed to be the dominant metabolism of known nitrite oxidizers. This reasoning implies that 30 to 80% of the total microbial biomass of the deep ocean is supported by a predominantly autotrophic metabolism.

Using sediment trap data from the HOT station off Hawaii, close to the site of the studies by Karner et al. (2001) and Ingalls et al. (2006), we can estimate a flux of organic matter into the mesopelagic zone. This is the material that must provide the autotrophic substrates needed for AOA. Existing data on the abundance and distribution of total microbial and crenarchaeal cells can be used to estimate the amount of C and N required to support the mesopelagic ecosystem.

Cell Numbers and N Demand

We take the total number of crenarchaeal cells estimated from the 150- to 1,000-m interval (Karner et al., 2001) and assume as an extreme to set the boundaries of the argument that all of these cells are AOA. To support this biomass by an ammonia-oxidizing metabolism, we assume that AOA have an average cell C content of 10 fg and a cellular C:N of 5. This means that the microbial biomass represented in crenarchaeal cells between 150 and 1,000 m is 2.38×10^{13} cells m^{-2}. Among bacterial nitrifiers, nitrification is an inefficient process, requiring ~25 mol of N oxidized for each mole of CO_2 fixed via the Calvin cycle. Archaeal nitrifiers are suspected to use the 3-hydroxypropionate pathway to fix CO_2 (Hallam et al., 2006), a less efficient mechanism. The cell and nitrite production data of Konneke et al. (2005) suggest a very similar ratio of N oxidized to CO_2 fixed, however. Therefore, we will assume a C:N ratio of 25 for the Archaeal and bacterial processes and for the combined oxidation of ammonium to nitrite and then to nitrate. Thus, for every mole of C converted to nitrifier biomass, 25 mol of ammonium must be oxidized. Thus, the amount of N required to support autotrophic C fixation at the level of 83% of the biomass in crenarchaeal cells is 0.41 mol of N m^{-2}. Next, we compare the rate of supply to estimate the growth rate of this biomass that could be sustained by the measured vertical flux. For example, if AOA have a generation time of 1 day, then this amount must be supplied daily. An equal number or amount of NOB biomass involved in nitrite oxidation does not, however, double the N demand because the nitrite is derived stoichiometrically from ammonium.

Vertical Flux of Organic Material

The source of the ammonium that is oxidized to support nitrification is regeneration of organic material by heterotrophic microbes.

The organic carbon is supplied from the dissolution or degradation of particulate organic matter flux through the water column. The magnitude of the vertical supply term was estimated from the flux at 150 m (2,300 μmol of C m^{-2} day^{-1} and 280 μmol of N m^{-2} day^{-1}) (Christian et al., 1997). If we assume that the flux at 1,000 m is only 10% of that at 150 m (i.e., 90% is remineralized within that interval) (Martin et al., 1987), then the total supply of organic C and N to the 150- to 1,000-m interval is 2,070 and 252 μmol m^{-2} day^{-1}, respectively.

The turnover of this mineralized nitrogen represents nitrification at an integrated nitrification rate of 252 μmol of N m^{-2} day^{-1}, assuming nitrification is directly coupled to mineralization. The integrated mineralization/nitrification rate estimated by Martin et al. (1987) for a central equatorial Pacific station a few degrees northwest of the ALOHA station (Karner et al., 2001) was 458 μmol of N m^{-2} day^{-1}, twice the rate estimated by Christian et al. (1997). There are few direct measurements of integrated nitrification rates with which to compare these flux-derived estimates or the nitrification rate needed to support the crenarchaeal biomass. Ward and Zafiriou (1988) reported integrated nitrification rates based on ^{15}N tracer incubation experiments on the order of 1,000 μmol m^{-2} day^{-1} for a series of stations (>2,000 m deep) in the Eastern Tropical North Pacific close to Baja California. Estimates of the vertical fluxes of N in this region based on sediment trap measurements range from <100 to 500 μmol N m^{-2} day^{-1} (Martin et al., 1987; Voss et al., 2001). Thus, the range of estimated integrated nitrification rates or vertical flux supply terms are all on the order of 100 to 1,000 μmol m^{-2} day^{-1}. This leads to an estimated turnover time for autotrophic biomass (AOA plus NOB) of 410 to 4,100 days.

To evaluate the contribution of crenarchaea to nitrification in another way, we can compute the per cell ammonia oxidation rate implied by these data. If all the crenarchaeal cells are ammonia oxidizers, then the per cell ammonia oxidation rate is obtained by dividing the total crenarchaeal abundance in the 850-m interval by the integrated nitrification rate. An integrated nitrification rate of 252 μmol m^{-2} day^{-1} implies a per cell rate of 10^{-17} mol $cell^{-1}$ day^{-1} per cell, a rate ~10^3 to 10^4 slower than that documented for cultivated AOB (Ward, 1987a) and cultivated AOA (Konneke, 2005). If cells in culture are actively growing at a generation time of ~1 day^{-1} and nitrification rate scales linearly to growth rate, then these nitrification rates are consistent with the very long generation times of the nitrifier biomass calculated above.

Agogue et al. (2008) measured dark $^{14}CO_2$ uptake directly and found that AOA *amoA* gene numbers explained half of the variability in CO_2 uptake rates. From their regression, a cell number turnover time can be computed. For cell numbers of 6×10^4 ml^{-1} (the abundance reported by Karner et al. [2001] at 100 m) and 5×10^3 ml^{-1} (abundance at 1,000 m), this relationship yields a turnover time of ~1,000 to 3,000 days, respectively. Because this relationship was derived from *amoA* copy numbers, it relates directly to the number of AOA, rather than the total number of Crenarchaeota, and provides a more direct argument for the slow growth rates of AOA in the deep ocean.

One of the major assumptions in these calculations is that all the crenarchaeal cells enumerated by Karner et al. (2001) were AOA, and recent data suggest that this may not be so. By comparing total crenarchaeal abundance and the abundance of *amoA* genes, it appears that not all of the crenarchaeal cells detected in the deep ocean are capable of ammonia oxidation (Wuchter et al., 2006; DeCorte et al., 2009). In the Mediterranean Sea, the copy number of *amoA* genes decreased from ~4×10^3 ml^{-1} between 200 and 500 m to less than 10 copies ml^{-1} below 950 m (DeCorte et al., 2009). The ratio of *amoA* gene copy number to crenarchaeal 16S rRNA genes was always less than 1 and less than 0.05 at depths greater than 750 m. An extensive data set of crenarchaeal and *amoA* AOA abundances in the North Atlantic documented AOA/Crenarchaea ratios of nearly 1 in the upper 150 m, decreasing to 0.1 to less than

0.01 in the bathypelagic (Agogue et al., 2008). In the Arctic and Antarctic Oceans, Kalanetra et al. (2009) reported distinctly different clades of crenarchaeal 16S rRNA genes in different water masses. The ratio of AOA *amoA* genes to crenarchaeal 16S rRNA genes was 2.0 for one water mass and 0.15 for another, suggesting that the latter crenarchaeal group was not predominantly ammonia oxidizing (Kalanetra et al., 2009).

These studies suggest that most of the deep-ocean Crenarchaea may not be ammonia oxidizers and that the abundance of autotrophic AOA decreases with depth in a manner consistent with the ammonium supply term from mineralization. If true for the central Pacific, then not all the cells enumerated by Karner et al. (2001) and identified as *Crenarchaeota* may contribute to the autotrophic biomass. Mincer et al. (2008) reported the opposite trend at station ALOHA, however, finding similar numbers of crenarchaeal 16S rRNA genes and archaeal *amoA* genes at depths between 300 and 1,000 m. This report also included estimates of the NOB *Nitrospina* abundance, based on quantitative PCR of 16S rRNA genes; *Nitrospina* was much less abundant than the AOA, fewer than 1,000 cells ml^{-1} (Mincer et al., 2007). If, however, AOA clades show significant depth differentiation, such that not all clades are equally efficiently quantified by the reported methods, the actual depth distributions and abundances of AOA may still be poorly known.

This effort to reconcile the measured integrated N mineralization/nitrification rates, the per-cell ammonia oxidation rates, and the nitrogen demand for ammonia oxidizer-based autotrophy points out the obvious distinction between biomass and flux. Either the abundance of AOA must be much lower than the total crenarchaeal abundance, or the entire autotrophic assemblage must be growing very slowly. The estimate of autotrophic biomass (Ingalls et al., 2006) does not depend on cell counts and constitutes a compelling argument in favor of predominantly autotrophic metabolism among marine archaea. It now seems important to identify the archaeal cells associated with that autotrophic signature and to reconcile the estimates of archaeal abundance with lipid distributions.

The lipid data of Ingalls et al. (2006) strongly support the predominance of autotrophy in the upper 1,000 m of the central Pacific. Similar data from a subsequent study found that the contribution of autotrophy in archaeal lipids was greater at 670 m than at 915 m (Hansman et al., 2009). This is the same depth range within which independent measurements of organic flux and nitrification rates imply that most nitrification occurs, as both flux and nitrification rates decrease with increasing depth. Thus, the contribution of autotrophy based on nitrification and the rate of nitrification are expected to approach zero below 1,000 to 1,500 m. This suggests that most of the crenarchaeal cells documented by Karner et al. (2001) as a significant portion of the biomass below 1,000 m are probably not nitrifiers.

The ratio of D:L-aspartic acid uptake has been used as an indicator of the relative abundance of *Crenarchaeota* to *Eubacteria* (Teira et al., 2006); this ratio and the relative contribution of crenarchaea to the total prokaryotic abundance were positively correlated. Their distributions were not directly correlated with depth but appeared to be associated with particular water masses in the North Atlantic. The main depth trend that was apparent in the data was generally lower D:L ratios in surface waters and increased relative crenarchaeal abundance with depth, in a manner similar to that shown by Karner et al. (2001). Crenarchaeal contribution to total microbial biomass was even higher than documented in the previous study. These findings are consistent with a heterotrophic metabolism for deep sea Archaea.

These patterns suggest that AOA are most abundant in the near-surface layers, like most biological processes and biomass, and are likely responsible for much of the nitrification, which occurs primarily in this layer. Even above 1,000 m, however, not all crenarchaea are AOA, and even if AOA plus AOB and NOB are equally represented in the biomass, most of the total biomass is not autotrophic. Below 1,000 m, the AOA are not only much less abundant but comprise a decreasing proportion of the total archaeal assemblage. This pattern is consistent with the distribution of overall nitrification rates measured directly using tracer methods and suggested from the mineralization rates computed from the pattern of vertical fluxes of organic matter.

Together, these data point out areas in which future research must reconcile conflicting inferences from the presently available information. The following description of our view of the deep ocean N cycle incorporates the recent information on nitrifier distributions and abundance.

1. In the surface layer (photic zone), most prokaryotes other than cyanobacteria are heterotrophs and are engaged in recycling pathways directly supported by photosynthesis and the grazing food web.

2. A significant fraction of the microbial metabolism in the depth interval between the euphotic zone and 1,000 m (the base of the main thermocline) is autotrophic (including AOA, AOB, and NOB), based on nitrification supported by the mineralization flux of ammonium. The remaining bacterial component of the microbial assemblage in this interval is heterotrophic, supported by decomposition of sinking organic matter or consumption of microbial products produced by autotrophs in situ. It is important to note that both heterotrophy and autotrophy in the mesopelagic zone are supported ultimately by reducing power harvested from the sun by photosynthesis in the overlying waters. Nitrifier autotrophy does not constitute "new" primary production.

3. Below 1,000 m, most prokaryotes are heterotrophic, persisting on ever more recalcitrant organic matter derived from the surface layer and the small input of fresh material. The autotrophic signal from nitrification in the mesopelagic zone does not contribute significantly to the vertical flux of organic matter reaching the deep sea because it is produced in small particles that do not sink directly and do not support a grazing food chain that might produce sinking particles as waste.

4. Due to its dependence on organic matter mineralization as the supply of ammonium, the distribution of nitrification is linked to the vertical flux of organic matter, and is maximal in the vicinity of the primary nitrite maximum at the base of the euphotic zone, and decreases exponentially with depth.

NITROUS OXIDE AND NITROGEN CYCLING IN OXYGEN MINIMUM ZONES

The scenario described above applies to most of the open ocean. An interesting and important exception occurs in the small areas of the ocean that contain oxygen-depleted water. Nitrogen cycling in low-oxygen waters is poorly understood, although the chemistry of these regions has been studied for a long time, and may involve complex interactions among nitrification, denitrification, and anammox (Fig. 3). In most of the ocean, a broad oxygen minimum depth interval (OMZ) occurs where oxygen consumption during mineralization of organic matter proceeds at a rate slightly faster than oxygen can be supplied by the ocean circulation. This interval generally extends from a depth of a few hundred meters to ~1,000 m, where oxygen reaches minimum levels between 50 and 100 μM. In only three regions of the ocean does oxygen depletion become severe enough to induce denitrification: the Eastern Tropical North and South Pacific and the Arabian Sea. In these regions, oxygen concentrations fall below 10 μM and are often as low as 1 μM or undetectable (ODZ). Nitrate deficits in these depth zones imply that bacterial respiration at the expense of nitrate has replaced oxygen respiration. How much of the fixed N loss that occurs in these regions is attributable to denitrification and how much is attributable to anammox (anaerobic ammonia oxidation) are the subjects of another current controversy in the nitrogen cycle.

ODZs are important in the oceanic N cycle because, although they comprise only 0.1 to 0.2% of the total ocean volume, they account for about 30% of the oceanic fixed nitrogen loss (Codispoti et al., 2001) and perhaps 10% of the total global N_2O flux to the atmosphere (Bange et al., 2000). Most of this N_2O is produced in the upper 600 m of the ocean (Suntharalingam and Sarmiento, 2000), and ODZs are hotspots of N_2O emissions.

Nitrification has long been linked to ODZ regions because of the microaerophilic nature of nitrifying bacteria and the strong correlation between oxygen utilization (apparent oxygen utilization [AOU]) and N_2O concentration on an ocean-wide basis (Cohen and Gordon, 1978; Nevison et al., 2003). AOU represents the amount of oxygen consumed during organic carbon mineralization via oxygen respiration, and the resulting oxidation of the mineralized ammonium to nitrate. The AOU/N_2O relationship implies that about 0.1 nM N_2O accumulates for every micromolar of O_2 consumed and holds for seawater containing saturating oxygen concentrations down to ~6 μM (Nevison et al., 2003). The fact that nitrous oxide concentration is positively correlated with AOU implies that N_2O is a byproduct of complete mineralization through nitrification, and only at very low oxygen concentrations where the relationship fails is denitrification an important source or sink.

Both marine and terrestrial AOB produce N_2O in culture, and the fraction of NH_4^+ oxidized to N_2O, rather than to NO_2^-, has been reported to increase with decreasing oxygen concentration (Goreau et al., 1980; Lipschultz et al., 1981). Thus, it has long been debated whether the N_2O maxima that occur in ODZs are derived from nitrification at low oxygen concentrations or from denitrification at oxygen concentrations too low to support aerobic respiration. The AOU relationship described above argues strongly that most of the N_2O in the ocean is derived from nitrification, although in restricted ODZ regions it can be shown that denitrification is a net source (Bange et al., 2005).

The relationship between AOU and N_2O is consistent with the known metabolism of AOB. It is not known, however, whether AOA contribute to the production of N_2O in the ocean, because it is not known whether AOA produce N_2O. The metabolic pathways involved in ammonia oxidation by AOA are currently under investigation, but there appear to be important differences in the pathways of AOB and AOA, at least as represented in the single AOA genome currently available (see Chapter 6). In the *N. maritimus* genome, there are no clear genes encoding hydroxylamine oxidoreductase, so the pathway to nitrite remains unknown. One pathway for nitrous oxide production in AOB involves the reduction of nitrite, via a copper-containing nitrite reductase, to nitric oxide and then to nitrous oxide. Known cultivated AOB possess both genes, nitrite reductase and nitric oxide reductase, which encode the enzymes required for this reductive pathway (Casciotti and Ward, 2001, 2005; Shaw et al., 2006). The *nirK* genes of AOB are apparently of multiple ancestry and have different regulatory motifs (Cantera and Stein, 2007) (see Chapter 4). Archaeal genome fragments from both the Sargasso Sea and soil contained *nirK* homologs that more closely resembled those of known nitrifiers than those of known denitrifiers (Treusch et al., 2005), suggesting that archaeal ammonia oxidizers, like their AOB analogs, are also capable of nitrite reduction to nitric oxide. The nitric oxide reductase step is less obvious in AOA, but a related gene has been detected in the genome of the archaeal sponge symbiont that also possesses *amoA* and *nirK* (Hallam et al., 2006).

The correlation between AOU and N_2O mentioned above suggests that if AOA are the major nitrifiers in the ocean, then they probably also produce N_2O as a side product of ammonia oxidation in the same stoichiometric relationship attributed to AOB. An alternative explanation is that N_2O is not produced by nitrification, but simply as a trace side product of respiration and mineralization in general heterotrophic bacteria. A diverse range of nitric oxide reductase genes has been described recently (Hemp et al., 2008), and they occur in many organisms other than nitrifiers. In their assumed role as detoxifying enzymes, they might be ubiquitous in heterotrophs. Thus, the link between AOU and N_2O would be a result of the heterotrophic component of organic matter remineralization, not a specific product of nitrification. This would be an important change in our understanding of the regulation of N_2O production in the ocean and other environments as well.

LINKS BETWEEN CONVENTIONAL NITRIFICATION AND ANAMMOX

The AOB, nitrite-oxidizing bacteria, and Archaea discussed so far are known or assumed to be obligate aerobes, and in the case of ammonia oxidation, molecular oxygen is involved in the initial ammonium oxidation step as well as in respiration. The anaerobic oxidation of ammonium to dinitrogen is thermodynamically favorable and had long been suspected on the basis of chemical profiles in sediments and stratified water columns (Richards, 1965; Richards and Broenkow, 1971; Bender et al., 1977). Anaerobic ammonia oxidation (anammox) and the organisms responsible for it were finally identified, first in wastewater (Mulder et al., 1995; van de Graaf et al., 1995) and then in the marine environment (Dalsgaard et al., 2003; Kuypers et al., 2003, 2005). The role of anammox in the marine environment is quite different from the role of conventional nitrification; anammox results in the loss of fixed nitrogen from the system and thus is actually a form of denitrification (Devol, 2008). Anammox is reviewed elsewhere in this volume (see Chapters 9 and 10); only the interactions between aerobic nitrifiers and anammox in the marine environment will be discussed here.

Characterized enrichment cultures of anammox derived from wastewater proved to be an obligate consortium containing the anammox organism, a planctomycete, and an aerobic ammonia oxidizer as the primary components (Sliekers et al., 2003). The substrates for anammox are ammonium and nitrite, which combine to form dinitrogen gas (Fig. 1). The wastewater supplied the ammonium to both the anammox organism and to the AOB. The AOB then produced the nitrite that was used by the planctomycetes, while scrubbing the last traces of oxygen to provide the anoxic environment required by the planctomycetes. A similar coupling between nitrification and anammox was suggested (Lam et al., 2007) in the oxic/anoxic interface region of the Black Sea. Both bacterial and archaeal ammonia oxidizers appeared to be involved in providing nitrite for anammox in these very low oxygen waters.

Loss of fixed N in ODZ regions had long been assumed to be due to conventional denitrification, carried out by facultatively anaerobic heterotrophic bacteria that switch to respiration of nitrogen oxides when oxygen disappears. Direct measurement of N_2 production rates using stable isotope tracer methods have now shown that anammox is the dominant process in some ODZs. The nutritional demands of anammox are quite different from those of denitrification; denitrification produces nitrite, but anammox requires a supply of nitrite from elsewhere. Part of the answer might be linkage to aerobic nitrification, analogous to the wastewater situation of the original anammox enrichments or in the narrow OMZ of the Black Sea (Lam et al., 2007). It does not seem possible, however, to support the oxygen flux necessary to sustain the obligately aerobic AOA or AOB in an OMZ of several hundred meters thickness with an almost zero oxygen concentration gradient (Revsbech et al., 2009); the diffusion of oxygen would be extremely slow and unable to support aerobic ammonia-oxidizing metabolism. Coupling between denitrifying bacteria and anammox might be expected to occur in ODZ waters. In the absence of aerobic ammonia oxidation, the ammonium and nitrite required by anammox could be supplied by organic matter breakdown and nitrate respiration by denitrifiers. The solution to this quandary remains to

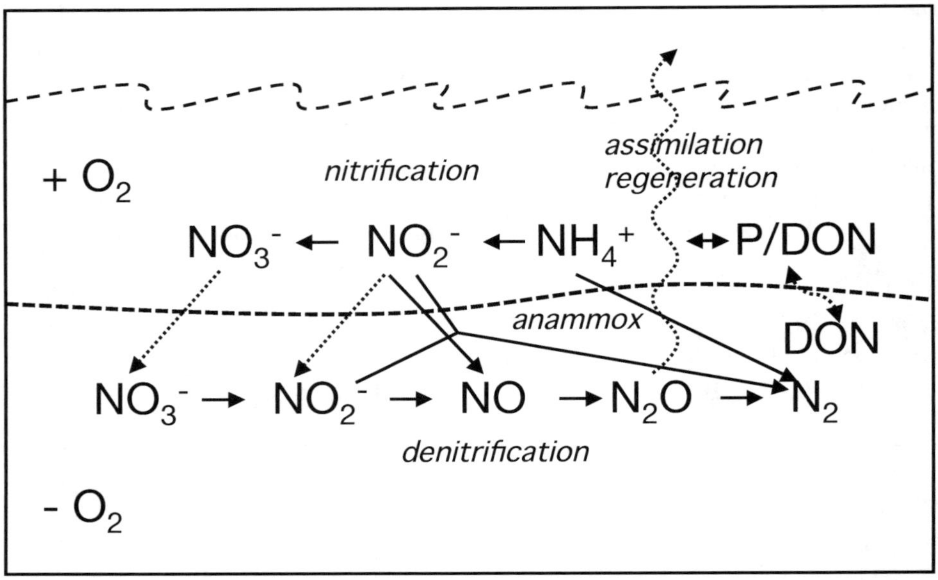

FIGURE 3 Possible linkages between nitrification and denitrification, including anammox, across an oxic/anoxic interface. The interface could be at the sediment/water interface or in the gradient at the upper boundary of an open ocean OMZ. P/DON, particulate/dissolved organic nitrogen, which is supplied to the system by primary production in overlying waters. The dashed lines imply diffusion, while the solid arrows represent microbial transformations of dissolved nitrogen compounds.

be found, but part of the answer may lie in the temporal variability of conditions in the ODZ, or an undiscovered anaerobic metabolism of AOA.

FUTURE DIRECTIONS

Recent discoveries in the marine nitrogen cycle and in nitrification, in particular, point out the important gaps in our understanding, both at the level of microorganisms and at the ecosystem level. This range of scales intersects with the question of what regulates the rates and distributions of nitrification in the ocean. This can be addressed at the system level using manipulation experiments to measure nitrification rates using various methods (Ward, 2005a) and investigating the effect of specific environmental parameters such as light, ammonium, organic carbon, etc. Regardless of which organisms are responsible for the rates, this kind of experiment can elucidate the oceanographic patterns and controls on the process.

At the microbiological level, both cultures and genomic studies will provide insights into the metabolic capabilities and responses of the newly recognized archaeal component. The most important questions to be resolved concern the AOA: the pathway of ammonia oxidation in AOA, their nutritional flexibility, their sensitivity to light, oxygen and ammonium concentrations, and the capability of AOA to produce N_2O. This kind of information is available for cultivated AOB, but the most important AOBs in the ocean, at least of the basis of clone libraries, have not been cultured. Therefore, the same questions remain important subjects of investigation for AOB. Nitrifiers of all kinds are notoriously difficult to cultivate and maintain, so barring some breakthrough in culturing approaches, it seems unlikely that new cultures will provide the necessary material for direct study of physiology of the environmentally important strains. Genomic/transcriptomic information and incubation and isotope approaches hold much promise for dissecting natural assemblages without cultivation.

The vast majority of work on nitrification in the ocean, as elsewhere, has been on ammonia oxidation. The documented great abundance of AOA compels us to investigate the next step in the process as well. Who are the missing NOB? The most important nitrite oxidizers in terrestrial systems are apparently in the *Nitrospira* lineage, while current evidence points to *Nitrospina* for the marine environment. The genome work has so far focused on *Nitrobacter*, so we face the same ignorance for NOB as for the much better studied AOB. Are there NOA out there to be discovered? What is the nutritional repertoire of marine nitrite oxidizers? The questions of autotrophy, abundance, and substrate supply again are important gaps in our knowledge.

REFERENCES

Agogue, H., M. Brink, J. Dinasquet, and G. J. Herndl. 2008. Major gradients in putatively nitrifying and non-nitrifying Archaea in the deep North Atlantic. *Nature* **456:**788–791.

Ando, Y., T. Nakagawa, R. Takahashi, K. Yoshihara, and T. Tokuyama. 2009. Seasonal changes in abundance of ammonia-oxidizing archaea and ammonia-oxidizing bacteria and their nitrification in sand of an eelgrass zone. *Microb. Environ.* **24:**21–27.

Bange, H. W., T. Rixen, A. M. Johansen, R. L. Siefert, R. Ramesh, V. Ittekkot, M. R. Hoffmann, and M. O. Andreae. 2000. A revised nitrogen budget for the Arabian Sea. *Glob. Biogeochem. Cycles* **14:**1283–1297.

Bange, H. W., S. W. A. Naqvi, and L. A. Codispoti. 2005. The nitrogen cycle in the Arabian Sea. *Progr. Oceanogr.* **65:**145–158.

Bano, N., and J. T. Hollibaugh. 2000. Diversity and distribution of DNA sequences with affinity to ammonia-oxidizing bacteria of the beta subdivision of the class Proteobacteria in the Arctic Ocean. *Appl. Environ. Microbiol.* **66:**1960–1969.

Beman, M. J., B. N. Popp, and C. A. Francis. 2008. Molecular and biogeochemical evidence for ammonia oxidation by marine Crenarchaeota in the Gulf of California. *ISME J.* **2:**429–441.

Bender, M. L., K. A. Fanning, P. N. Froelich, and G. R. Heath. 1977. Interstitial nitrate profiles and oxidation of sedimentary organic matter in the eastern Equatorial Atlantic. *Science* **198:**605–609.

Berelson, W. M. 2001. The flux of particulate organic carbon into the ocean interior: a comparison of four U.S. JGOFS regional studies. *Oceanography* **14:**59–67.

Bernhard, A. E., T. Donn, A. E. Giblin, and D. A. Stahl. 2005. Loss of diversity of ammonia-oxidizing bacteria correlates with increasing salinity in an estuary system. *Environ. Microbiol.* **7:**1289–1297.

Bock, E., and M. Wagner. 2006. Oxidation of inorganic nitrogen compounds as an energy source, p. 457–495. *In* M. Dowkin (ed.), *The Prokaryotes: an Evolving Electronic Resource for the Microbiological Community*. Springer Verlag, New York, NY.

Bock, E., I. Schmidt, R. Stuven, and D. Zart. 1995. Nitrogen loss caused by denitrifying *Nitrosomonas* cells using ammonium or hydrogen as electron donors and nitrite as electron acceptor. *Arch. Microbiol.* **163:**16–20.

Caffrey, J. M., N. Bano, K. Kalanetra, and J. T. Hollibaugh. 2007. Ammonia oxidation and ammonia-oxidizing bacteria and archaea from estuaries with differing histories of hypoxia. *ISME J.* **1:**660–662.

Cantera, J. J. L., and L. Y. Stein. 2007. Molecular diversity of nitrite reductase genes (nirK) in nitrifying bacteria. *Environ. Microbiol.* **9:**765–776.

Capone, D. G., D. A. Bronk, M. R. Mulholland, and E. J. Carpenter (ed.). 2008. *Nitrogen in the Marine Environment*, 2nd ed. Academic Press, Burlington, MA.

Carlucci, A. F., and P. M. McNally. 1969. Nitrification by marine bacteria in low concentrations of substrate and oxygen. *Limnol. Oceanogr.* **14:**736.

Casciotti, K. L., and B. B. Ward. 2001. Nitrite reductase genes in ammonia-oxidizing bacteria. *Appl. Environ. Microbiol.* **67:**2213–2221.

Casciotti, K. L., and B. B. Ward. 2005. Phylogenetic analysis of nitric oxide reductase gene homologues from aerobic ammonia-oxidizing bacteria. *FEMS Microbiol. Ecol.* **52**:197–205.

Christian, J. R., M. R. Lewis, and D. M. Karl. 1997. Vertical fluxes of carbon, nitrogen, and phosphorus in the North Pacific Subtropical Gyre near Hawaii. *J. Geophys. Res.* **102:**15667–15677.

Clark, D. R., A. P. Rees, and I. Joint. 2008. Ammonium regeneration and nitrification rates in the oligotrophic Atlantic Ocean: implications for new production estimates. *Limnol. and Oceanogr.* **53:**52–62.

Codispoti, L., J. Brandes, J. Christensen, A. Devol, S. Naqvi, H. Paerl, and T. Yoshinari. 2001. The oceanic fixed nitrogen and nitrous oxide budgets: moving targets as we enter the anthropocene? *Sci. Marina* **65:**85–105.

Cohen, Y., and L. I. Gordon. 1978. Nitrous oxide in the oxygen minimum of the eastern tropical North Pacific: evidence for its consumption during denitrification and possible mechanisms for its production. *Deep Sea Res.* **6:**509–525.

Dalsgaard, T., D. E. Canfield, J. Petersen, B. Thamdrup, and J. Acuna-Gonzalez. 2003. N_2 production by the anammox reaction in the anoxic water column of Golfo Dulce, Costa Rica. *Nature* **422:**606–608.

De Corte, D., T. Yokokawa, M. M. Varela, H. Agogue, and G. J. Herndl. 2009. Spatial distribution of Bacteria and Archaea and *amoA* gene copy numbers throughout the water column of the Eastern Mediterranean Sea. *ISME J.* **3:**147–158.

Delong, E. F. 1992. Archaea in coastal marine environments. *Proc. Natl. Acad. Sci. USA* **89**: 5685–5689.

Devol, A. H. 2008. Denitrification, including Anammox, p. 263–301. *In* D. G. Capone, D. A. Bronk, M. R. Mulholland, and E. J. Carpenter (ed.), *Nitrogen in the Marine Environment*, 2nd ed. Academic Press, Burlington, MA.

Dore, J. E., and D. M. Karl. 1996. Nitrification in the euphotic zone as a source for nitrite, nitrate, and nitrous oxide at station ALOHA. *Limnol. Oceanogr.* **41:**1619–1628.

Dugdale, R. C., and J. J. Goering. 1967. Uptake of new and regenerated forms of nitrogen in marine production. *Limnol. Oceanogr.* **12:**196–206.

Ehrich, S., D. Behrens, E. Lebedeva, W. Ludwig, and E. Bock. 1995. A new obligately chemolithoautotrophic, nitrite-oxidizing bacterium, *Nitrospira moscoviensis* sp. nov., and its phylogenetic relationship. *Arch. Microbiol.* **164:**16–23.

Eppley, R. W., and B. J. Peterson. 1979. Particulate organic matter flux and planktonic new production in the deep ocean. *Nature* **282:**677–680.

Erguder, T. H., N. Boon, L. Wittebolle, M. Marzorati, and W. Verstraete. 2009. Environmental factors shaping the ecological niches of ammonia-oxidizing archaea. *FEMS Microbiol. Rev.* **33:**855–869.

Fernandez, C., and P. Raimbault. 2007. Nitrogen regeneration in the NE Atlantic Ocean and its impact on seasonal new, regenerated and export production. *Mar. Ecol. Prog. Ser.* **337:**79–92.

Francis, C. A., G. D. O'Mullan, and B. B. Ward. 2003. Diversity of ammonia monooxygenase (*amoA*) genes across environmental gradients in Chesapeake Bay sediments. *Geobiology* **1:**129–140.

Francis, C. A., K. J. Roberts, M. J. Beman, A. E. Santoro, and B. B. Oakley. 2005. Ubiquity and diversity of ammonia-oxidizing archaea in water columns and sediments of the ocean. *Proc. Natl. Acad. Sci. USA* **102:**14683–14688.

Fuhrman, J. A., K. McCallum, and A. A. Davis. 1992. Novel major archaebacterial group from marine plankton. *Nature* **356:**148–149.

Goreau, T. J., W. A. Kaplan, S. C. Wofsy, M. B. McElroy, F. W. Valois, and S. W. Watson. 1980. Production of NO_2^- and N_2O by nitrifying bacteria at reduced concentrations of oxygen. *Appl. Environ. Microbiol.* **40:**526–532.

Guerrero, M. A., and R. D. Jones. 1996. Photoinhibition of marine nitrifying bacteria.1. Wavelength-dependent response. *Mar. Ecol. Prog. Ser.* **141:**183–192.

Gundersen, K. 1966. The growth and respiration of *Nitrosocystis oceanus* at different partial pressures of oxygen. *J. Gen. Microbiol.* **42:**387–396.

Hallam, S. J., T. J. Mincer, C. Schleper, C. M. Preston, K. Roberts, P. M. Richardson, and E. F. DeLong. 2006. Pathways of carbon assimilation and ammonia oxidation suggested by environmental genomic analyses of marine Crenarchaeota. *PLoS Biol.* **4:**520–536.

Hansman, R. L., S. Griffin, J. T. Watson, E. R. M. Druffel, A. E. Ingalls, and A. Pearson. 2009. The radiocarbon signature of microorganisms in the mesopelagic ocean. *Proc. Natl. Acad. Sci. USA* **106:**6513–6518.

Hemp, J., L. A. Pace, M. G. Klotz, L. Y. Stein, T. J. Martinez, and R. B. Gennis. 2008. Diversity of the heme-copper nitric oxide reductases: evidence for multiple independent origins. *Abstr. Am. Soc. Microbiol.* **108:**462.

Hollibaugh, J. T., N. Bano, and H. W. Ducklow. 2002. Widespread distribution in polar oceans of a 16S rRNA gene sequence with affinity to *Nitrosospira*-like ammonia-oxidizing bacteria. *Appl. Environ. Microbiol.* **68:**1478–1484.

Horrigan, S. G., A. F. Carlucci, and P. M. Williams. 1981. Light inhibition of nitrification in sea-surface films. *J, Mar. Res.* **39:**557–565.

Ingalls, A. E., S. R. Shah, R. L. Hansman, L. I. Aluwihare, G. M. Santos, E. R. M. Druffel, and A. Pearson. 2006. Quantifying archaeal community autotrophy in the mesopelagic ocean using natural radiocarbon. *Proc. Natl. Acad. Sci. USA* **103:**6442–6447.

Jensen, K. M., M. H. Jensen, and E. Kristensen. 1996. Nitrification and denitrification in Wadden Sea sediments (Konigshafen, Island of Sylt, Germany) as measured by nitrogen isotope pairing and isotope dilution. *Aquat. Microb. Ecol.* **11:**181–191.

Jorgensen, K. S., H. B. Jensen, and J. Sorensen. 1984. Nitrous-oxide production from nitrification and denitrification in marine sediment at low oxygen concentrations. *Can. J. Microbiol.* **30:**1073–1078.

Kalanetra, K. M., N. Bano, and J. T. Hollibaugh. 2009. Ammonia-oxidizing Archaea in the Arctic Ocean and Antarctic coastal waters. *Environ. Microbiol.* **11:**2434–2445.

Karner, M. B., E. F. DeLong, and D. M. Karl. 2001. Archaeal dominance in the mesopelagic zone of the Pacific Ocean. *Nature* **409:**507–5010.

Konneke, M., A. E. Berhnard, J. R. de la Torre,

C. B. Walker, J. B. Waterbury, and D. A. Stahl. 2005. Isolation of an autotrophic ammonia-oxidizing marine archaeon. *Nature* **437:**543–546.

Koops, H. P., U. Purkhold, A. Pommerening-Roser, G. Timmermann, and M. Wagner. 2006. The lithoautotrophic ammonia-oxidizing bacteria, p. 778–811. *In* M. Dworkin (ed.), *The Prokaryotes: an Evolving Electronic Resource for the Microbiological Community*. Springer-Verlag, New York, NY.

Krishnan, K. P., S. O. Fernandes, P. A. L. Bharathi, L. K. Kumari, S. Nair, A. K. Pratihary, and B. R. Rao. 2008. Anoxia over the western continental shelf of India: Bacterial indications of intrinsic nitrification feeding denitrification. *Mar. Environ. Res.* **65:**445–455.

Kuypers, M. M. M., A. O. Sliekers, G. Lavik, M. Schmid, B. B. Jorgensen, J. G. Kuenen, J. S. S. Damste, M. Strous, and M. S. M. Jetten. 2003. Anaerobic ammonium oxidation by anammox bacteria in the Black Sea. *Nature* **422:**608–611.

Kuypers, M. M., G. Lavik, D. Woebken, M. Schmid, B. M. Fuchs, R. Amann, B. B. Jørgensen, and M. S. M. Jetten. 2005. Massive nitrogen loss from the Benguela upwelling system through anaerobic ammonium oxidation. *Proc. Natl. Acad. Sci. USA* **102:**6478–6483.

Laanbroek, H. J., P. L. E. Bodelier, and S. Gerards. 1994. Oxygen-consumption kinetics of *Nitrosomonas europaea* and *Nitrobacter hamburgensis* grown in mixed continuous cultures at different oxygen concentrations. *Arch. Microbiol.* **161:**156–162.

Lam, P., J. P. Cowen, and R. D. Jones. 2004. Autotrophic ammonia oxidation in a deep-sea hydrothermal plume. *FEMS Microbiol. Ecol.* **47:**191–206.

Lam, P., M. M. Jensen, G. Lavik, D. F. McGinnis, B. Muller, C. J. Schubert, R. Amann, B. Thamdrup, and M. M. Kuypers. 2007. Linking crenarchaeal and bacterial nitrification to anammox in the Black Sea. *Proc. Natl. Acad. Sci. USA* **104:**7104–7109.

Laursen, A. E., and S. P. Seitzinger. 2002. The role of denitrification in nitrogen removal and carbon mineralization in Mid-Atlantic Bight sediments. *Cont. Shelf Res.* **22:**1397–1416.

Lilley, M., D. Butterfield, E. J. Olson, J. E. Lupton, S. A. Macko, and R. E. McDuff. 1993. Anomalous CH_4 and NH_4^+ concentrations at an unsedimented mid-ocean-ridge hydrothermal system. *Nature* **364**:45–47.

Lipschultz, F., O. C. Zafiriou, S. C. Wofsy, M. B. Melroy, F. W. Valois, and S. W. Watson. 1981. Production of NO and N_2O by soil nitrifying bacteria. *Nature* **294:**641–643.

Lipschultz, F., S. C. Wofsy, B. B. Ward, L. A. Codispoti, G. J. W. Friedrich, and J. W. Elkins. 1990. Bacterial transformations of inorganic nitrogen in the oxygen-deficient waters of the Eastern Tropical South Pacific Ocean. *Deep Sea Res.* **37:**1513–1541.

Lomas, M. W., and F. Lipschultz. 2006. Forming the primary nitrite maximum: nitrifiers or phytoplankton? *Limnol. Oceanogr.* **51:**2453–2467.

Martens-Habbena, W., P. M. Berube, H. Urakawa, J. R. de la Torre and D. A. Stahl. 2009. Ammonia oxidation kinetics determine niche separation of nitrifying Archaea and Bacteria. *Nature* **461:**976–979.

Martin, J. H., G. A. Knauer, D. M. Karl, and W. W. Broenkow. 1987. Vertex-carbon cycling in the northeast Pacific. *Deep Sea Res., Part A* **34:**267–285.

Meyer, R. L., D. E. Allen, and S. Schmidt. 2008. Nitrification and denitrification as sources of sediment nitrous oxide production: a microsensor approach. *Mar. Chem.* **110:**68–76.

Mincer, T. J., M. J. Church, L. T. Taylor, C. Preston, D. M. Kar, and E. F. DeLong. 2007. Quantitative distribution of presumptive archaeal and bacterial nitrifiers in Monterey Bay and the North Pacific Subtropical Gyre. *Environ. Microbiol.* **9:**1162–1175.

Molina, V., O. Ulloa, L. Farias, H. Urrutia, S. Ramirez, P. Junier, and K. P. Witzel. 2007. Ammonia-oxidizing beta-Proteobacteria from the oxygen minimum zone off northern Chile. *Appl. Environ. Microbiol.* **73:**3547–3555.

Morris, R. M., M. S. Rappe, S. A. Connon, K. L. Vergin, W. A. Siebold, C. A. Carlson, and S. J. Giovannoni. 2002. SAR11 clade dominates ocean surface bacterioplankton communities. *Nature* **420:**806–810.

Mosier, A. C., and C. A. Francis. 2008. Relative abundance and diversity of ammonia-oxidizing archaea and bacteria in the San Francisco Bay estuary. *Environ. Microbiol.* **10:**3002–3016.

Mulder, A., A. A. van de Graaf, L. A. Robertson, and J. G. Kuenen. 1995. Anaerobic ammonium oxidation discovered in a denitrifying fluidized-bed reactor. *FEMS Microbiol. Ecol.* **16:**177–183.

Muller-Neugluck, M., and H. Engel. 1961. Photoinaktivierung von *Nitrobacter winogradskyi* Buch. *Arch. Mikrobiol.* **39:**130–138.

Nevison, C., J. H. Butler, and J. W. Elkins. 2003. Global distribution of N_2O and the ΔN_2O-AOU yield in the subsurface ocean. *Glob. Biogeochem. Cycles* **17:**1119.

O'Mullan, G. D., and B. B. Ward. 2005. Relationship of temporal and spatial variabilities of ammonia-oxidizing bacteria to nitrification rates in Monterey Bay, California. *Appl. Environ. Microbiol.* **71:**697–705.

Olson, R. J. 1981a. ^{15}N tracer studies of the primary nitrite maximum. *J. Mar. Res.* **39:**203–226.

Olson, R. J. 1981b. Differential photoinhibition of marine nitrifying bacteria: a possible mechanism for the formation of the primary nitrite maximum. *J. Mar. Res.* **39:**227–238.

Ouverney, C. C., and J. A. Fuhrman. 2000. Marine planktonic Archaea take up amino acids. *Appl. Environ. Microbiol.* **66:**4829.

Prosser, J. I., and G. W. Nicol. 2008. Relative contribution of archaea and bacteria to aerobic ammonia oxidation in the environment. *Environ. Microbiol.* **10:**2931–2941.

Purkhold, U., M. Wagner, B. Timmermann, A. Pommerening-Roser, and H.-P. Koops. 2003. 16S rRNA and *amoA*-based phylogeny of 12 novel betaproteobacterial ammonia-oxidizing isolates: extension of the dataset and proposal of a new lineage within the nitrosomonads. *Int. J. Syst. Evol. Microbiol.* **53:**1485–1494.

Revsbech, N. P., N. Risgaard-Petersen, A. Schramm, and L. P. Nielsen. 2006. Nitrogen transformations in stratified aquatic microbial ecosystems. *Antonie Van Leeuwenhoek Int. J. Gen. Mol. Microbiol.* **90:**361–375.

Revsbech, N. P., L. H. Larsen, J. Gundersen, T. Dalsgaard, O. Ulloa, and B. Thamdrup. 2009. Determination of ultra-low oxygen concentrations in oxygen minimum zones by the STOX sensor. *Limnol. Oceanogr. Methods* **7:**371–381.

Richards, F. A. 1965. Anoxic basins and fjords, p. 611–645. *In* J. P. Riley and G. Skirrow (ed.), *Chemical Oceanography*, vol. 1. Academic Press, London, United Kingdom.

Richards, F. A., and W. W. Broenkow. 1971. Chemical changes, including nitrate reduction, in Darwin Bay, Galapagos Archipelago, over a 2-month period, 1969. *Limnol. Oceanogr.* **16:**758.

Santinelli, C., B. B. Manca, G. P. Gasparini, L. Nannicini, and A. Seritti. 2006. Vertical distribution of dissolved organic carbon (DOC) in the Mediterranean Sea. *Clim. Res.* **31:**205–216.

Schleper, C., G. Jurgens, and M. Jonuscheit. 2005. Genomic studies of uncultivated Archaea. *Nat. Rev. Microbiol.* **3:**479–488.

Shaw, L. J., G. W. Nicol, Z. Smith, J. Fear, J. I. Prosser, and E. M. Baggs. 2006. *Nitrosospira* spp. can produce nitrous oxide via a nitrifier denitrification pathway. *Environ. Microbiol.* **8:**214–222.

Sheridan, C. C., C. Lee, S. G. Wakeham, and J. K. B. Bishop. 2002. Suspended particle organic composition and cycling in surface and midwaters of the equatorial Pacific Ocean. *Deep Sea Res., Part I* **49:**1983–2008.

Sliekers, A. O., K. A. Third, W. Abma, J. G. Kuenen, and M. S. M. Jetten. 2003. CANON and anammox in a gas-lift reactor. *FEMS Microbiol. Lett.* **218:**339–344.

Smith, A. J., and D. S. Hoare. 1968. Acetate assimilation by *Nitrobacter agilis* in relation to its obligate autotrophy. *J. Bacteriol.* **95:**844.

Smith, Z., A. McCaig, J. Stephen, T. Embley, and J. I. Prosser. 2001. Species diversity of uncultured and cultured populations of soil and marine ammonia oxidizing bacteria. *Microb. Ecol.* **42:**228–237.

Starkenburg, S. R., P. S. G. Chain, L. A. Sayavedra-Soto, L. Hauser, M. L. Land, F. W. Larimer, S. A. Malfatti, M. G. Klotz, P. J. Bottomley, D. J. Arp, and W. J. Hickey. 2006. Genome sequence of the chemolithoautotrophic nitrite-oxidizing bacterium *Nitrobacter winogradskyi* Nb-255. *Appl. Environ. Microbiol.* **72:**2050–2063.

Starkenburg, S. R., F. W. Larimer, L. Y. Stein, M. G. Klotz, P. S. G. Chain, L. A. Sayavedra-Soto, A. T. Poret-Peterson, M. E. Gentry, D. J. Arp, B. Ward, and P. J. Bottomley. 2008. Complete genome sequence of *Nitrobacter hamburgensis* X14 and comparative genomic analysis of species within the genus *Nitrobacter. Appl. Environ. Microbiol.* **74:**2852–2863.

Steinberg, D. K., S. A. Goldthwait, and D. A. Hansell. 2002. Zooplankton vertical migration and the active transport of dissolved organic and inorganic nitrogen in the Sargasso Sea. *Deep Sea Res. Part I* **49:**1445–1461.

Stephen, J. R., A. E. McCaig, Z. Smith, J. I. Prosser and T. M. Embley. 1996. Molecular diversity of soil and marine 16S rRNA gene sequences related to beta-subgroup ammonia-oxidizing bacteria. *Appl. Environ. Microbiol.* **62:**4147–4154.

Suess, E. 1980. Particulate organic-carbon flux in the oceans—surface productivity and oxygen utilization. *Nature* **288:**260–263.

Suntharalingam, P. and J. L. Sarmiento. 2000. Factors governing the oceanic nitrous oxide distribution: Simulations with an ocean general circulation model. *Glob. Biogeochem. Cycles* **14:**429–454.

Sutka, R. L., N. E. Ostrom, P. H. Ostrom, and M. S. Phanikumar. 2004. Stable nitrogen isotope dynamics of dissolved nitrate in a transect from the North Pacific Subtropical Gyre to the Eastern Tropical North Pacific. *Geochim. Cosmochim. Acta* **68:**517–527.

Teira, E., P. Lebaron, H. van Aken, and G. J. Herndl. 2006. Distribution and activity of Bacteria and Archaea in the deep water masses of the North Atlantic. *Limnol. Oceanogr.* **51:**2131–2144.

Teske, A., E. Alm, J. M. Regan, S. Toze, B. E. Rittmann, and D. A. Stahl. 1994. Evolutionary relationships among ammonia- and nitrite-oxidizing bacteria. *J. Bacteriol.* **176:**6623–6630.

Treusch, A. H., S. Leininger, A. Kletzin, S. C. Schuster, H. P. Klenk, and C. Schleper. 2005. Novel genes for nitrite reductase and Amo-related proteins indicate a role of uncultivated mesophilic crenarchaeota in nitrogen cycling. *Environ. Microbiol.* **7:**1985–1995.

Usui, T., I. Koike, and N. Ogura. 2001. N_2O production, nitrification and denitrification in an estuarine sediment. *Estuar. Coast. Shelf Sci.* **52:**769–781.

van de Graaf, A. A., A. Mulder, P. Debruijn, M. S. M. Jetten, L. A. Robertson, and J. G. Kuenen. 1995. Anaerobic oxidation of ammonium is a biologically mediated process. *Appl. Environ. Microbiol.* **61:**1246–1251.

Venter, C. J., K. Remington, J. G. Heidelberg, A. L. Halpern, D. Rusch, J. A. Eisen, D. Wu, I. Paulsen, K. E. Nelson, W. Nelson, D. E. Fouts, S. Levy, A. H. Knap, M. W. Lomas, K. Nealson, O. White, J. Peterson, J. Hoffman, R. Parsons, H. Baden-Tillson, C. Pfannkoch, J.-H. Rogers, and H. O. Smith. 2004. Environmental genome shotgun sequencing of the Sargasso Sea. *Science* **304:**66–74.

Voss, M., J. W. Dippner, and J. P. Montoya. 2001. Nitrogen isotope patterns in the oxygen-deficient waters of the Eastern Tropical North Pacific Ocean. *Deep Sea Res., Part I* **48:**1905–1921.

Wakeham, S. G., C. Lee, J. I. Hedges, P. J. Hernes, and M. L. Peterson. 1997. Molecular indicators of diagenetic status in marine organic matter. *Geochim. Cosmochim. Acta* **61:**5363–5369.

Wankel, S. D., C. Kendall, J. T. Pennington, F. P. Chavez, and A. Paytan. 2007. Nitrification in the euphotic zone as evidenced by nitrate dual isotopic composition: observations from Monterey Bay, California. *Glob. Biogeochem. Cycles* **21•••.**

Ward, B. B. 1987a. Kinetic studies on ammonia and methane oxidation by *Nitrosococcus oceanus. Arch. Microbiol.* **147:**126–133.

Ward, B. B. 1987b. Nitrogen transformations in the Southern California Bight. *Deep Sea Res.* **34:**785–805.

Ward, B. B. 2005a. Molecular approaches to marine micobial ecology and the marine nitrogen cycle. *Annu. Rev. Earth Planet. Sci.* **33:**301–333.

Ward, B. B. 2005b. Temporal variability in nitrification rates and related biogeochemical factors in Monterey Bay, California, USA. *Mar. Ecol. Prog. Ser.* **292:**97–109.

Ward, B. B. 2008. Nitrification in marine systems, p. 199–261. *In* D. G. Capone, D. A. Bronk, M. R. Mulholland, and E. J. Carpenter (ed.), *Nitrogen in the Marine Environment*, 2nd ed. Academic Press, Burlington, MA.

Ward, B. B., and K. Kilpatrick. 1990. Relationship between substrate concentration and oxidation of ammonium and methane in a stratified water column. *Cont. Shelf Res.* **10:**1193–1208.

Ward, B. B., and G. D. O'Mullan. 2002. Worldwide distribution of *Nitrosococcus oceani*, a marine ammonia-oxidizing gamma-proteobacterium, detected by PCR and sequencing of 16S rRNA and *amoA* genes. *Appl. Environ. Microbiol.* **68:**4135–4157.

Ward, B. B., and O. C. Zafiriou. 1988. Nitrification and nitric oxide in the oxygen minimum of the eastern tropical North Pacific. *Deep Sea Res.* **35:**1127–1142.

Ward, B. B., M. C. Talbot, and M. J. Perry. 1984. Contributions of phytoplankton and nitrifying bacteria to ammonium and nitrite dynamics in coastal water. *Cont. Shelf Res.* **3:**383–398.

Ward, B. B., K. A. Kilpatrick, E. Renger, and R. W. Eppley. 1989. Biological nitrogen cycling in the nitracline. *Limnol. Oceanogr.* **34:**493–513.

Watson, S. W., and J. B. Waterbury. 1971. Characteristics of two marine nitrite oxidizing bacteria, *Nitrospina gracilis* nov. gen. nov. sp. and *Nitrococcus mobilis* nov. gen. nov. sp. *Arch. Microbiol.* **77:**203–230.

Watson, S. W., E. Bock, F. W. Valois, J. B. Waterbury, and U. Schlosser. 1986. *Nitrospira marina* gen. nov. sp. nov.: a chemolithotrophic nitrite-oxidizing bacterium. *Arch. Microbiol.* **144:**1–7.

Woese, C. R., L. J. Magrum, and G. E. Fox. 1978. Archaebacteria. *J. Mol. Evol.* **11:**245–252.

Wuchter, C., B. Abbas, M. J. L. Coolen, L. Herfort, J. van Bleijswijk, P. Timmers, M. Strous, E. Teira, G. H. Herndl, J. J. Middelburg, S. Schouten, and J. S. S. Damste. 2006. Archaeal nitrification in the ocean. *Proc. Natl. Acad. Sci. USA* **103:**12317–12322.

Yakimov, M. M., V. La Cono, and R. Denaro. 2009. A first insight into the occurrence and expression of functional amoA and accA genes of autotrophic and ammonia-oxidizing bathypelagic Crenarchaeota of Tyrrhenian Sea. *Deep Sea Res., Part Ii* **56:**748–754.

Yool, A., A. P. Martin, C. Fernandez, and D. R. Clark. 2007. The significance of nitrification for oceanic new production. *Nature* **447:**999–1002.

Zart, D., and E. Bock. 1998. High rate of aerobic nitrification and denitrification by *Nitrosomonas eutropha* grown in a fermentor with complete biomass retention in the presence of gaseous NO_2 or NO. Arch. Microbiol. **169:**282–286.

SOIL NITRIFIERS AND NITRIFICATION

James I. Prosser

14

INTRODUCTION

Ammonia is produced naturally in soil through the microbial degradation of organic matter and hydrolysis of urea. Anthropogenic input is of similar, if not greater, magnitude, comprising nitrogen fertilizers (100 Tg year^{-1}) and deposition of atmospheric nitrogen (25 Tg year^{-1}) (Gruber and Galloway, 2008), and outweighs input from nitrogen fixation (110 Tg year^{-1}). Conversion of soil ammonia to nitrite, nitrate, and nitrous oxide by nitrifiers is a central process within the soil nitrogen cycle. Nitrification determines the relative amounts of the two major inorganic nitrogen sources of plants, ammonium and nitrate, and controls total inorganic nitrogen availability in soil. Ammonium, as a positively charged ion, is retained within soil, which is dominated by negatively charged particles. Nitrate, in contrast, is readily leached from soil, polluting groundwaters. Nitrification of ammonia therefore significantly reduces the efficiency of ammonia-based fertilization, such that 70% of applied fertilizer can be lost from agricultural systems, and nitrate limits for drinking water are often exceeded in regions of intensive agriculture. Leaching losses are compounded by denitrification of soil nitrate to gaseous nitric and nitrous oxides and nitrogen, which are lost to the atmosphere. Nitrification leads to acidification of soil, increasing mobilization of toxic metals, particularly in poorly buffered soils. Ammonia oxidizers can also oxidize methane and can co-oxidize organic pollutants, providing a potential role in bioremediation in oligotrophic soil ecosystems.

Pasteur, in 1862, predicted a role for microorganisms in production of nitrate from ammonia, which was demonstrated by Schloesing and Muntz (1877a, 1877b, 1879) in a column containing sand and chalk perfused with sewage. Decreasing ammonia concentration in the effluent was matched by increasing nitrate concentration, and the process was reversed by addition of chloroform and reinstated when chloroform was removed. The organisms responsible for this process were cultured from soil by three independent groups (Frankland and Frankland, 1890; Winogradsky, 1890–1891; Warington, 1891), providing the foundation for biochemical and physiological studies of these organisms.

The 20th century saw major advances in our understanding of the process of soil nitrification and the physiology of nitrifying bacteria. Physiological studies demonstrated their metabolic diversity, but nitrifier community ecology was severely limited by difficulties in

James I. Prosser, Institute of Biological and Environmental Sciences, University of Aberdeen, Aberdeen AB24 3UU, United Kingdom.

Nitrification, Edited by Bess B. Ward, Daniel J. Arp, and Martin G. Klotz

isolating and identifying ammonia and nitrite oxidizers that persist today. The advent of cultivation-independent, molecular techniques for characterization of microbial communities in the 1990s focused attention on nitrifier diversity and community structure and led to the discovery of archaeal ammonia oxidizers (see Section III). These techniques have revolutionized our view of soil nitrifier communities, and attempts have been made to link soil nitrifier diversity with process studies, to determine the nature of diversity-ecosystem function relationships and to assess the impact of environmental change and soil management strategies on processes and communities.

This chapter provides an overview of soil nitrification, paying particular attention to these recent advances. The focus is on the specific factors associated with soil that influence the nitrification process and nitrifier communities (i.e., which characteristics of soil are important to nitrifiers). Thus, it will consider how soil nitrification differs from nitrification in pure cultures growing in liquid medium or in oceans, estuaries, and wastewater treatment plants. Information will be presented on the soil characteristics that drive nitrification rates and nitrifier growth and communities, relating findings, where possible, to physiological and genetic information described in earlier chapters.

COMMUNITY COMPOSITION OF SOIL NITRIFIERS

Ammonia Oxidizers

Prior to 1996, knowledge of soil ammonia oxidizer communities was based on characterization of a limited number of laboratory cultures of ammonia oxidizers, in particular *Nitrosomonas europaea*. Indeed, *N. europaea*, originally isolated from soil, was considered to be a typical soil ammonia oxidizer, although nitrosospiras had also been isolated. The development of 16S rRNA gene primers targeting betaproteobacterial ammonia oxidizers (Stephen et al., 1996) led to reassessment of soil ammonia oxidizer community composition. Analysis of a wide range of soils using 16S rRNA gene clone libraries and fingerprinting techniques, such as denaturing gradient gel electrophoresis (DGGE), showed that soil bacterial ammonia oxidizer communities are dominated by *Nitrosospira* strains. Although nitrosomonads are sometimes detected in soil, their relative abundance is generally low. Community analysis based on 16S rRNA genes was followed by analysis of *amoA* genes, with similar findings. Bacterial ammonia oxidizer phylogeny based on these two genes shows some congruence (Purkhold et al., 2000; Aakra et al., 2001), and *amoA* gene analysis slightly increases taxonomic resolution, particularly of nitrosospiras. The distribution of soil ammonia oxidizers is discussed in greater detail in chapter 3.

The links between phylogenetic diversity and physiological or functional diversity are discussed below in relation to soil characteristics and environmental factors. However, phylogenetic analysis based on a single gene is only likely to distinguish broad levels of physiological diversity. Certainly, in bacterial groups with many cultivated representatives, much of the diversity relevant to ecosystem function and ecology is only detectable using techniques such as multilocus sequence-type analysis. Consequently, although some broad generalizations may be made about physiological characteristics of different 16S rRNA- or *amoA*-defined groups, important properties (e.g., urease activity, sensitivity to high ammonium concentration) are found in phylogenetically distant groups, and the distribution of others (e.g., saturation constants, maximum specific growth rates, survival capability) is unknown. This applies particularly when viewing ammonia oxidizer phylogeny over broad geographical range and over many sites, limiting the ability to detect patterns in or mechanisms leading to distribution and community structure. Local studies are better able to demonstrate patterns and links to environmental factors. This assumes that such patterns exist, although application of neutral theory to other microbial systems indicates that much of the variability in bacterial communities is not due to environmental or physiological characteristics (Woodcock et al., 2007). Koops and Pommerening-Röser (2001) demonstrated some relationships between phylogeny of a number of ammonia oxidizer cultures, including several soil isolates, and a limited range of characteristics: urease activity, halophily, and ammonium inhibition. Because community composition is determined by diverse environmental factors, however, more detailed physiological studies are required with representatives of phylogenetic groups present in the soil (Smith et al., 2001). It is therefore likely that future advances will depend on cultivation-independent techniques. Substantial technical development is needed, however, to distinguish active organisms with high taxonomic resolution techniques and to determine physiological responses to changing conditions with greater quantitation.

Methodological caveats are required for all community studies. Primers may be biased toward different groups, although the dominance of soil nitrosospiras in both 16S rRNA and *amoA* gene libraries reduces concerns over primer or gene bias. Other biases, through differences in cell lysis efficiency and variability in gene copy number, may influence analyses, and studies performed in different laboratories, even with similar primers, may utilize different protocols that may influence results. This is particularly important for previous studies where techniques were being developed.

The discovery of colocation of an *amoA* gene homologue and a Group 1 crenarchaeal 16S rRNA gene on a soil metagenome fragment (Treusch et al., 2005) suggested the potential for ammonia oxidation by archaea, which was confirmed by isolation of a marine, autotrophic, crenarchaeal ammonia oxidizer (Könneke et al., 2005; Leininger et al., 2006) (see Section III). These findings, and subsequent studies, have seriously challenged the established view that betaproteobacteria perform the vast majority of ammonia oxidation in soil, with the possible exception of small contributions from heterotrophic nitrifiers (see Chapter 5). No soil crenarchaeon has yet been obtained in pure culture, but archaeal *amoA* genes are ubiquitous in soil, and the influence of temperature and pH on abundance, community structure, and transcriptional activity of archaeal and bacterial ammonia oxidizers is discussed in later sections. The relative roles of bacteria and archaeal ammonia oxidation are currently a subject of debate (see Section III) (Prosser and Nicol, 2008). The evidence is necessarily indirect and based on molecular techniques, as there is no known selective inhibitor that discriminates ammonia oxidation by the two groups.

Nitrite Oxidizers

Nitrite generally does not accumulate to high levels in soil. Nitrifier community studies have therefore focused on ammonia oxidizers, as the perceived "rate-limiting" organisms in nitrification. Cultivated nitrite oxidizers fall into five genera, *Nitrobacter*, *Nitrospira*, *Nitrococcus*, *Nitrospina*, and "Candidatus Nitrotoga arctica" (see Section V), of which only *Nitrobacter* and *Nitrotoga* have been isolated from soil. Molecular analysis is less straightforward than for ammonia oxidizers, because different 16S rRNA gene primer sets are required for each genus, but shows that both *Nitrobacter* and *Nitrospira* are present in soil. *Nitrospira* sequences from soil and wastewater treatment plants are highly diverse, with some links between clusters, environmental origin, and physiological diversity (Daims et al., 2000; Maixner et al., 2006), but there is less evidence of diversity within *Nitrobacter*. Analysis of nitrite oxidizers in grassland soils confirmed this view (Freitag et al., 2005), where DGGE analysis detected only one *Nitrobacter* band but several *Nitrospira* bands. Clone library analysis also suggested the existence of further diversity, with a new evolutionary group and novel subclusters within established groups. *Nitrobacter* grassland soil communities have also been investigated using the *nxrA* gene, encoding subunit A of nitrite oxidoreductase (Poly et al., 2008; Wertz et al., 2008), with evidence for diversity within *Nitrobacter* and the influence of grazing on community composition. This situation parallels that for ammonia oxidizers. Isolates indicate dominance by *Nitrosomonas* and *Nitrobacter* (the textbook nitrifiers), while molecular analysis suggests that these organisms are selected by laboratory cultivation conditions and that *Nitrosospira*, archaeal ammonia oxidizers, and *Nitrospira* may be more abundant and diverse in the soil.

Functional Redundancy and Resilience of Nitrifiers

The lack of cultivated representatives of natural communities restricts assessment of functional redundancy. However, experiments in which the diversity of natural soil communities has been manipulated provide evidence of high redundancy. Wertz et al. (2006) inoculated sterile soil microcosms with serial dilutions (over several orders of magnitude) of a suspension from the same, nonsterile soil, incubated microcosms to establish the original cell abundance, and determined diversity and function of all denitrifiers and ammonia oxidizers. Despite considerable reduction in diversity, ecosystem function of the three groups, including ammonia oxidizers, was not affected; the nitrification rate was unaffected despite a 1,000-fold reduction in diversity. A similar approach was used to determine the effect of diversity of denitrifier and nitrite oxidizer communities on resilience and resistance following heating to 42°C for 24 h (Wertz et al., 2007). Nitrite oxidizers were less resistant and resilient than denitrifiers, but the effects of heating and recovery from heating were not influenced by the diversity of the communities. Roux-Michollet et al. (2008) also showed an ability to recover following steaming of soil. Ammonia and nitrite oxidizer most probable number (MPN) counts were reduced by ~95%, but numbers recovered within 62 days, with effects greatest in the upper 0 to 2 cm layers of soil.

SURFACE ATTACHMENT

Many soil nitrifiers will be attached to particulate matter. This consists mainly of soil minerals and organic matter, the proportions and nature of which will vary with soil type, texture, and management history. Surface attachment has a number of consequences for soil nitrifier ecology, in comparison with that in other natural environments. Attachment reduces the likelihood of removal of cells by bulk flow of water through interstitial pathways and soil fissures. It therefore increases the stability of nitrifier communities, once established, and reduces transport when conditions become unfavorable. Soil nitrifiers must respond to changing conditions, and evolution and community structure may be driven by the ability to survive unfavorable conditions and respond rapidly when conditions improve. Surface attachment provides an additional mechanism for community stability, potentially increasing diversity.

Attachment to surfaces and establishment of biofilms modifies the physiological characteristics of nitrifiers, as for other microorganisms, with significant consequences for their ecology. This applies both to attachment to relatively inert particulate material and to charged mineral particles, which concentrate ions and nutrients. Soil particulate material has a net negative charge, increasing adsorption of ammonium and potentially favoring growth of attached ammonia oxidizers, but disadvantaging that of nitrite oxidizers. Concentration of substrate at surfaces increases colonization but not necessarily specific growth rate. Thus, ammonia and nitrite oxidizers colonize cation and anion exchange resin beads, respectively, to greater extents due to higher local concentrations of ammonia and nitrite (Prosser, 1989).

There is evidence that ammonia oxidizers are more strongly attached to soil particles than heterotrophs. Aakra et al. (2000) found that only 0.5% of indigenous ammonia oxidizers could be extracted from a clay loam soil using a dispersion-density-gradient centrifugation technique, but extraction efficiency increased to 8% after incubation with urea. They suggest that newly growing cells are less strongly attached and/or urea stimulated growth of strains that were less strongly attached.

The consequences of attachment have been observed in soil but are best studied using particulate material of defined composition, to eliminate the physicochemical complexity and heterogeneity of soil. Three important aspects of surface growth will be discussed here: effects on growth and inhibition, survival and recovery from starvation, and protection from effects of low pH.

Growth and Inhibition

The presence of glass slides during growth of suspended cells in batch cultures of *N. europaea* does not influence specific growth rate, determined through exponential increases in nitrite concentration (Powell and Prosser, 1992), but attached cell numbers increase at a faster exponential rate through attachment of free cells, surface growth, and detachment. To investigate the effects of clay minerals, which have much greater cation exchange capacity (CEC) than glass, Powell and Prosser (1991) determined growth of *N. europaea* in weakly buffered inorganic growth medium in the presence and absence of three clay minerals with increasing CEC:illite, vermiculite, and montmorillonite. Complete oxidation of ammonia was prevented by reduction in pH that occurred in the absence of clays and the presence of illite. In the presence of vermiculite and montmorillonite, initial growth was similar to that of suspended cells but was followed by a second, slower growth phase, reflecting growth of attached cells, and nitrite yield was greater. Surface attachment may also increase ammonia uptake through changes in cell physiology. Bollman et al. (2005) measured K_m values of 2.9 and 3.2 μM NH_3, respectively, for batch and continuous cultures of suspended cells of a community containing two nitrosospiras, *N. briensis* and *N. winogradskyi*, but values were significantly lower (1.8 μM NH_3) for cells attached to the vessel wall.

A mechanism for the apparent buffering effect of some clay minerals is suggested by growth of *N. europaea* in the presence of ammonia-treated vermiculite (ATV), in which ammonia is fixed to vermiculite at high temperature (Armstrong and Prosser, 1988). Growth in liquid culture was preceded by a lag phase and was incomplete, due to acidification of the medium. In the presence of ATV, the specific growth rate was slightly reduced, but the lag phase was abolished and pH did not decrease, leading to more prolonged growth and greater nitrite yield. This can be explained by localized buffering at the clay surface, in which ammonium is released and utilized by attached cells. H^+ ions produced during ammonia oxidation are then exchanged for ammonium ions, preventing reduction in the pH of the medium.

Both glass slides and clay minerals provide protection from inhibition of nitrifiers by nitrapyrin. Powell and Prosser (1992) found that inhibition of suspended cells by nitrapyrin (at 0.5 mg liter^{-1}), added prior to inoculation, was unaffected by glass slides, but inhibition of attached cells was reduced to 25%. During the early stages of colonization, surfaces provided some protection from inhibition, but continued supplementation of growth medium with ammonia led to establishment of mature biofilms, with clusters of cells often surrounded by extracellular material. The specific growth rate of these biofilms was only 65% that of cell suspensions, but they were not inhibited by 0.5 mg of nitrapyrin liter^{-1}. In addition, detached cells grew at the same specific rate as attached cells and were also not inhibited. The results suggest physiological changes during biofilm formation, possibly due to formation of extracellular material, which lead to protection of both biofilm and detached cells.

When grown in the presence of nitrapyrin, illite did not protect cells from inhibition (Powell and Prosser, 1991). In contrast, vermiculite and ATV completely protected cells from 0.5 μg of nitrapyrin ml^{-1}, while montmorillonite stimulated the early stages of growth and only slightly inhibited later growth. Growth in the presence of montmorillonite homoionic to aluminum was monophasic and gave no protection, suggesting that adsorption of ammonia was necessary for colonization. Biofilm formation on both glass slides and clays therefore protects cells from inhibition and is one explanation for reduced inhibition in soil (Rodgers and Ashworth, 1982; Powell and Prosser, 1986).

Recovery from Starvation

Extracellular polymeric material produced by nitrifiers may increase attachment, and production may increase during starvation (Stehr et al., 1995), leading to greater survival of attached ammonia oxidizers in soil (Allison and Prosser, 1991a). Abolition of the lag phase prior to growth on ATV, and other clay minerals, also reflects a potentially important additional advantage of surface growth, because ammonia supply will be intermittent, and competition for ammonia with other soil biota and plants will be strong (Verhagen and Laanbroek, 1991). Lag phases prior to recovery from starvation are common among bacteria, and the lag phase of suspended *N. europaea* cells increases with the starvation period (Batchelor et al., 1997). After starvation for 42 days, recovery did not occur for 153 h, which would obviously present serious problems in the competitive soil environment. In contrast, *N. europaea* biofilms colonizing sand or soil particles in continuous-flow, fixed column reactors exhibited no lag phase, even after starvation for 43 days. Surface growth and biofilm formation therefore give these organisms a significant ecological advantage in the soil. In many Gram-negative bacteria, cell density-dependent phenomena are mediated by *N*-acyl homoserine lactones, one of which [*N*-(3-oxohexanoyl)-L-homoserine lactone], significantly reduced the lag phase of suspended cells starved for 28 days. *N*-Acyl homoserine lactones may also be involved in interactions between nitrifiers, other soil microorganisms, and plants, particularly in the rhizosphere where they may accumulate due to high cell concentrations.

Protection from Effects of Low pH

Keen and Prosser (1987) investigated growth and activity of *Nitrobacter* colonizing anion-exchange resin beads in a nitrite-limited, airlift column fermentor. Prolonged growth at a range of dilution rates led to establishment of a mature biofilm, and steady states were then established as the pH of the inflowing medium was gradually reduced to 4.5, 1.5 pH units below the pH minimum for growth in liquid batch culture. To determine whether similar mechanisms exist for ammonia oxidizers, Allison and Prosser (1993) investigated the effect of pH on activity of *N. europaea*, which could not grow in liquid batch culture at a pH lower than 7. *N. europaea* was inoculated into packed columns of either sand or vermiculite, which were then supplied continuously with medium containing ammonium. For sand columns, medium was supplied at pH 8 until establishment of a steady-state effluent nitrite concentration, after which further steady states were established despite reductions in the pH of the medium to 7, 6.5, and 6. In vermiculite columns, steady states were established with medium down to pH 5.4, although the greater buffering capacity of vermiculite increased effluent pH to 6.3 and prevented detailed assessment of the effects of low pH on activity. Nevertheless, in both columns, ammonia oxidation occurred approximately 1 pH unit lower than in liquid culture. In addition, an enriched ammonia oxidizer culture, colonizing the wall of an ammonia-limited chemostat, was active at a pH of 5, and grew at 5.5.

Protection from the effects of low pH through high cell densities also occurs in soil through formation of cell aggregates. De Boer et al. (1991) observed aggregates in enrichment cultures containing cells morphologically similar to *Nitrosospira* strain AHB1, isolated from an acid soil. Aggregates were separated by filtration and could oxidize ammonia at pH 4, while free cells were inhibited. Growth at low pH in continuous culture was facilitated by coculture with an acidophilic nitrite oxidizer and by previous growth at pH 6, which enabled activity of freely suspended cells at pH 4. Maintenance of activity required high cell densities and may have resulted from removal of toxic nitrous acid and protection by extracellular polymeric material.

Low pH nitrification also occurs in mixed culture bioreactor systems (Green et al., 2006). High levels of nitrification activity and nitrifier growth were observed in biofilm and suspended-biomass reactors at pH values as low as 4.3 and 3.8, respectively. Ammonia oxidizer communities were dominated by the *Nitrosomonas oligotropha* group, rather than nitrosospiras, which dominate acid soils. In a second study, mixed culture biofilms on chalk particles and sintered glass could nitrify at pH 4, and ammonia oxidizer communities were dominated by *Nitrosospira* spp. and *N. oligotropha* and nitrite oxidizers by *Nitrospira*. Microelectrode measurements provided no evidence for high-pH microenvironments, even on chalk particles, and dominant ammonia oxidizer communities in both studies were characterized by sequence types whose cultured representatives have low saturation constants for ammonia. Bioreactors were supplied with oxygen, rather than air, potentially reducing CO_2 levels, which could limit nitrification. Coculture with nitrite oxidizers could also have reduced nitrous acid toxicity. Another feature of ammonia oxidizers in the reactor systems, in biofilms and when attached to soil particles, is that they are retained within the system during reduction in pH. Cells were therefore able to adapt to low ammonia availability, possibly inducing ammonium transport systems, which may have been repressed at higher pH.

SUBSTRATE SUPPLY

Ammonia supply is of obvious importance for soil nitrification, the major sources being degradation of organic matter and urea-containing animal waste, fertilizer nitrogen, and atmospheric deposition. Ammonia concentrations therefore vary temporally and spatially, and concentrations experienced by different components of the ammonia oxidizer community differ from total concentrations measured in bulk soil. The influence of ammonia on activity and growth rate of ammonia oxidizers is typical for substrates that are inhibitory at high concentration. Specific activity and specific growth rate increase with ammonia concentration at low concentrations following Michaelis-Menten or Monod kinetics, respectively. K_m and K_s values for ammonia are in the range 0.4 to 14 mM NH_4^+-N and 0.051 to 0.07 mM NH_4^+-N (Prosser, 1989). Few cultured ammonia oxidizers can grow at concentrations of >1 mg of NH_4^+-N ml^{-1}, but most grow at concentrations of up to 50 μg of NH_4^+-N ml^{-1}. Although bulk soil concentrations rarely reach the upper end of this concentration range, local concentrations will be high following, for example, animal urination, addition of solid fertilizer, or release from degrading organic material. Within soil aggregates, ammonia may be exhausted or at growth-limiting concentrations, even when bulk concentrations are relatively high. Tolerance to high ammonia concentration varies between strains and has been used as a taxonomic character (Koops and Pommerening-Röser, 2001), and K_s and K_m values also vary. Thus, long-term differences in ammonia supply and consequent differences in ammonia concentration are potentially important in determining nitrification rates, nitrifier abundance, and ammonia oxidizer community structure.

Effects of Soil on Ammonia Availability

The influence of ammonia availability on soil nitrification and nitrifiers is dependent on pH, which controls the NH_3:NH_4^+ equilibrium, reducing availability of the substrate for ammonia oxidizers, NH_3, as soil pH decreases. Many studies, of both laboratory cultures and soil, do not consider this distinction; saturation constants are quoted in terms of either soil NH_3 or NH_4^+, and comparison of values is not possible without additional information on pH. In addition, most soil particles have net negative charge, and NH_4^+ therefore exchanges with cations on mineral and organic material, reducing ammonium concentration in solution. The impact of this is significant, and information on extractable ammonium concentration may not be relevant if concentrations in solution determine cellular oxidation rates. This has consequences for choice of method for determination of nitrification rates. In soil slurries, ammonium concentration will be uniform and not significantly affected by adsorption. In intact soil, solution concentration will be significantly reduced through adsorption of NH_4^+.

The significance of these factors is exemplified by an investigation of the effect of soil on oxidation of ammonia and co-oxidation of ethylene, chloroethane, and 1,1,1-trichloroethane by *N. europaea* (Hommes et al., 1998). Nitrite production was determined in *N. europaea* cell suspensions in soil slurries containing a silt loam (CEC; 15 cmol kg of soil^{-1}) at up to 1 g of soil ml^{-1} and 10 mM ammonium. Soil significantly inhibited nitrite production, through reduction in pH and adsorption of ammonium. Adsorption reduced ammonia availability from 80 mM to 8 mM, within the range of K_s values reported for *N. europaea*. Activity was restored by increasing ammonia concentration to 50 mM. Similar effects were observed for oxidation of ethylene and chloroethane, but not 1,1,1-trichloroethane, which oxidizes much more slowly and therefore requires less reductant.

Measurement of Growth Parameters in Soil

If the effects of surface attachment on growth and effects of soil on ammonia availability and transport are taken into consideration, there is no a priori reason why ammonia oxidizers should grow differently in soil and liquid culture. This has been investigated by comparing growth kinetics of pure cultures in liquid medium and after inoculation into gamma-irradiated soil. For example, Taylor and Bottomley (2006) determined growth parameters of *N. europaea* and *Nitrosospira* NPAV after inoculation into three gamma-irradiated soils with different texture and different extractable NH_4^+ concentrations (2 to 11 mg of NH_4^+-N g of soil^{-1}). Nitrite was produced at a constant rate, suggesting that growth was not limited by reduction in pH or slow release of adsorbed ammonia. In soil, *Nitrosospira* NPAV had greater activity than *N. europaea*, and cell activities were 21% and 60% of those in liquid culture at 10 mM, respectively. *Nitrosospira* had the lower K_s value (0.14 mM NH_4^+ versus 1.9 mM) and the lower V_{max} (0.002 versus 0.007 pmol h^{-1} cell^{-1}). This was calculated to give *Nitrosospira* a competitive advantage at concentrations of <1 mM NH_4^+, while *Nitrosomonas* would have greater activity at concentrations of >2 to 2.5 mM NH_4^+. A calculated growth yield of 3.5 × 10^6 to 6 × 10^6 cells (mmol of NH_4^+)$^{-1}$ for *N. europaea* in soil was similar to values reported for growth in liquid culture.

Growth constants have also been determined for natural soil communities, effectively averaging characteristics of all members of the community. For example, ammonia oxidizer communities in soil slurries containing soil from under an oak woodland canopy and from adjacent open grassy spaces (Stark and Firestone, 1996) had a K_m of 15 μM NH_4^+ (equivalent to 0.012 μM NH_3) and were inhibited at 1.6 mM NH_4^+ (1.3 μM NH_3). Both values are significantly lower than previously reported, and K_m values were higher in enrichment cultures from these soils. It is not known whether these contained phylotypes dominant in natural communities, and the complexity of the soil system makes it difficult to explain these low values. However, they may be more relevant when predicting nitrification in these soils. Temperature optima were 31.8 and 35.9°C under trees and in open spaces, respectively, possibly due to temporal differences in temperature rather than mean temperatures, which were similar. Highest nitrification rates were found in March, and rates decreased with decreasing osmotic potential in both systems.

Cell activities, and other kinetic parameters, are being reassessed using cultivation-independent quantitative PCR (qPCR) techniques to assess cell abundance. Okano et al. (2004) used this approach to determine the influence of ammonia concentration on growth characteristics of soil communities in microcosms and in the field. In microcosms amended with 1.5 or 7.5 mM ammonium sulfate, bacterial ammonia oxidizer cell abundance increased from 4 × 10^6 to 35 and 66 × 10^6 cells g of dried soil^{-1}, respectively, with possible pH limitation at the higher concentration. Fertilization of a tomato field plot increased abundance from 8.9 × 10^6 to 38 × 10^6 cells g of dried soil^{-1} after 39 days. Doubling times calculated using these data were 28 and 52 h in low- and high-ammonia microcosms, respectively, and 373 h in the field. Cell activities decreased from initial values of 0.5 to 25 fmol of NH_4^+ h^{-1}, and growth yields were 5.6, 17.5, and 1.7 × 10^6 cells (mol of NH_4^+)$^{-1}$ in low- and high-ammonia microcosms and field soil, respectively. Values obtained in microcosms are often within the ranges of values for laboratory cultures, but the reasons for the longer generation time in field soil were not identified. Also, ammonia conversion was not complete, suggesting that other factors limited growth (e.g., acidification).

Cell activities and abundances have been used to assess the relative importance of bacterial and archaeal ammonia oxidizers in soil. Boyle-Yarward et al. (2008) investigated nitrification in two soils supporting Douglas fir and red alder. They estimated bacterial and archaeal ammonia oxidizer abundances required for observed nitrification potential using published activities of bacterial ammonia oxidizers (1×10^{-15} to 10×10^{-15} mol cell^{-1} h^{-1}) and a value for *N. maritimus* of 0.25×10^{-15} to 0.35×10^{-15} mol cell^{-1} h^{-1} calculated from Könneke et al. (2005). Bacterial activity could account for nitrification potential except in red alder soils at one of the sites. At one site, this required assumption of the maximum reported bacterial cell activity but archaeal *amoA* gene abundance was also sufficient to support nitrification, despite lower cell activity. Shauss et al. (2009) adopted a similar approach for two soils amended with pig manure containing the antibiotic sulfadiazine. Archaeal:bacterial *amoA* gene ratios were 7:1 and 73:1, and incorporation of gene abundances and estimated cell activities into a simple nitrification model predicted a requirement for archaeal ammonia oxidizer activity in one of the soils. Archaea also appeared to have greater tolerance to sulfadiazine. Changes in relative abundance of bacterial but not archaeal phylotypes, based on DGGE or restriction fragment length polymorphism profiles, are sometimes used to suggest greater bacterial activity. However, where archaeal:bacterial *amoA* gene ratios are ≥100, equivalent proportional changes in relative abundance will require 100-fold greater changes in absolute abundance of archaeal genes. This makes community changes much more difficult to detect. In addition, these calculations are limited by lack of knowledge of the proportions of ammonia oxidizers that are active in soil and ignorance of cell activity and yield of soil archaeal ammonia oxidizers, as data must be extrapolated from a single marine isolate, *N. maritimus*.

Influence of Ammonium Concentration on Abundance and Community Structure

Soil ammonia concentration will directly influence activity and specific growth rates of ammonia oxidizers and strain-specific differences in growth kinetics, and ammonia inhibition can lead to differences in community structure. However, ammonia oxidizer abundance will not depend on ammonia concentration per se, but on ammonia flux and rates of removal through death and predation. Similarly, changes in community structure through selection for strains that are tolerant of high ammonia concentration will only be evident if ammonia supply enables sufficient growth in selected strains for measurable differences in relative abundance. Although many studies have looked for positive correlations between ammonia concentration and abundance, there are frequent examples where the opposite might be expected. The most obvious of these is in N-saturated forest soils (see below) where ammonium concentrations are high, nitrification rates low, and ammonia oxidizers often are undetectable. Similarly, rates of ammonium supply and consequent integrated mean ammonium concentrations, rather than snapshot ammonium concentrations, will determine "bulk" community structure.

Plants vary in their preferences for ammonium and nitrate as inorganic nitrogen sources, influencing competition with ammonia oxidizers and availability of ammonium. Hawkes et al. (2005) found evidence of this in studying invasion of Californian grassland by exotic grasses. Invasion doubled nitrification rates, mainly through increased ammonia oxidizer abundance due to preference of the invading grasses for nitrate, rather than ammonium.

Heterogeneity of ammonium concentration will also lead to heterogeneity in community structure and, although prolonged fertilization is likely to change ammonia oxidizer communities, other management factors, such as tillage and liming, will complicate analysis and interpretation. There is evidence that plowing reduces spatial heterogeneity in AOB communities and reduces diversity (Webster et al., 2002), but this may be less important than changes in total biomass. Low cell concentrations and slow growth of soil ammonia oxidizers often mean that large changes in community structure are required for detection. The next sections describe the influence of general or localized input of ammonia-N, arising from soil management strategies, on nitrifier abundance and community structure.

Livestock Grazing

Excretion and urination by grazing animals effectively interconvert and redistribute nitrogen throughout a field, creating regions of high ammonia concentration, and grazing can influence competition between plants and ammonia oxidizers for nitrogen. Heterogeneity in ammonia concentration is a likely cause of the high variability in nitrification rates frequently observed within grasslands (White et al., 1987) through increased abundance of ammonia oxidizers and changes in community structure. The importance of such changes for grassland ecosystems was illustrated in a microcosm study (Webster et al., 2005) that demonstrated strong links between ammonia oxidizer community structure and dynamics, physiological diversity, and dynamics of soil nitrification. Soil microcosms containing unfertilized soil were amended with synthetic sheep urine (1 mg of urea-N g of soil^{-1}). Nitrate production varied significantly between microcosms, mainly through differences in the length of the lag phase prior to NO_3^- production. This variability was associated with differences in the composition of the ammonia oxidizer community. Soils were dominated by two subgroups of *Nitrosospira* cluster 3, clusters 3a and 3b. In microcosms with high relative abundance of cluster 3b, lag phases were short. In those dominated by cluster 3a, nitrate production was not apparent for several weeks and required increases in relative abundance of *Nitrosospira* cluster 3b strains, leading to their eventual dominance (Fig. 1). This link between ecosystem function and community structure appears to be due to differences in ammonia tolerance. Pure culture representatives of *Nitrosospira* clusters 3a and 3b, and enrichment cultures obtained from the same site, were sensitive and tolerant, respectively, to high ammonia concentration, typical of that found in sheep urine patches, explaining differences in lag phases in soils dominated by the different groups. In microcosms containing fertilized soil, long lag phases were not observed, and there was no clear trend in community structure changes during incubation. This is likely due to greater abundance of all ammonia oxidizer groups, through long-term fertilization, such that initial nitrification was not limited by low abundance of ammonia-tolerant strains.

Field studies have shown that grazing increases the nitrification rate (e.g., Groffman et al., 1993; Le Roux et al., 2003; Patra et al., 2005) and affects activity, abundance, and community structure of ammonia oxidizers (Webster et al., 2002; Patra et al., 2006). There is some evidence for links between community structure and plant species and for reduced complexity (increased dominance) of bacterial (rather than archaeal) ammonia oxidizer communities in grazed soils (Webster et al., 2002; Patra et al., 2006). To investigate temporal changes associated with grazing, Le Roux et al. (2008) determined nitrifier activity and ammonia oxidizer abundance in mesocosms 2 years after switches between grassland management to and from grazing. Different ammonia oxidizer communities were found in continuously grazed and ungrazed soils, and nitrifier activity and bacterial and archaeal ammonia oxidizer abundances were always greater in the former. Within 5 months of a switch to

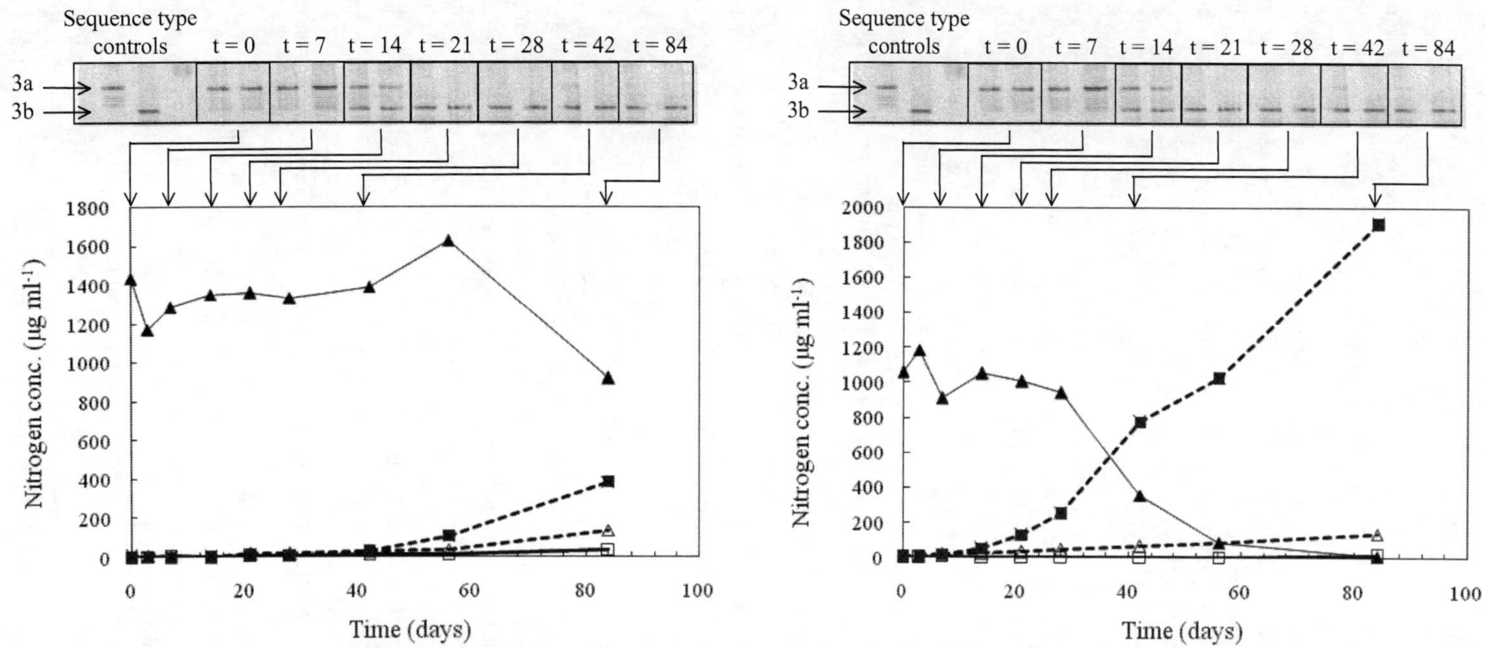

FIGURE 1 Ammonia oxidizer community structure influences nitrification kinetics following addition of synthetic sheep urine. Microcosms containing grassland soil were amended with synthetic sheep urine containing 1 mg of urea-N kg of soil^{-1}. Ammonium (filled and open triangles) and nitrite + nitrate (filled and open squares) concentrations and ammonia oxidizer sequence types were determined in control (open square and triangle) and in amended (filled square and triangle) microcosms. Ammonia oxidizer sequences were analyzed by DGGE. Soil was initially dominated by *Nitrosospira* cluster 3a (a) or cluster 3b (b). (From Webster et al. [2005], with permission.)

grazing, bacterial ammonia oxidizer community structure changed and nitrifier activity and abundance then increased. Increased nitrification activity therefore appeared to require a change in community structure, possibly through selection of strains tolerant to high localized ammonia concentration. Putative tolerant phylotypes were, however, not always the same as those found by others. Conversely, nitrifier activity and abundance decreased following cessation of grazing, and AOB community structure subsequently changed. This is presumably due to decreased abundance, with some groups decreasing faster than others. Nitrification activities in mesocosms that had been switched to grazed or ungrazed conditions were similar to those of equivalent unaltered soil within 12 months of switching.

Turf Grass

In highly managed turf grass systems, repeated application of high levels of nitrogen fertilizer increases nitrification potential (Shi et al., 2006) but may not reduce diversity. A turf grass chronosequence (1 to 95 years) contained a wide range of ammonia oxidizer phylotypes, with four *Nitrosospira* and two *Nitrosomonas* clusters, but their relative abundance did not change significantly with turf age, despite apparently strong selective pressure (Dell et al., 2008). This was explained by lack of disturbance and mixing in these nontilled soils, increased organic matter, and consequent improvements in soil structure, leading to greater spatial separation of potentially competing populations. Cantera et al. (2006) also found a relationship between ammonia concentration, nitrification activity, and ammonia oxidizer abundance in turf-covered soil irrigated with groundwater, highly saline river water, or wastewater and in sand traps. However, this positive effect was counterbalanced by a negative influence of salt concentration. Ammonia oxidizer communities were less diverse in turf-covered soils and were dominated by *Nitrosomonas* sequences, as often found in wastewater and other high-ammonia environments. Oved et al. (2001) also reported increases in *Nitrosomonas* sequences following irrigation of orchard soil with wastewater effluent, but not fertilizer-amended water, suggesting that factors specific to wastewater, including salinity, influence ammonia oxidizer communities.

Nitrogen Fertilization

Microcosm studies demonstrate clear impacts of ammonia-N or urea-N on ammonia oxidizers. Mahmood et al. (2006) showed distinct changes in communities of ammonia oxidizers at three different levels of nitrogen addition (100, 500, and 1,000 μg of urea-N g of soil^{-1}) to soil microcosms. Soils were dominated by *Nitrosospira* clusters 2 to 4. *Nitrosospira* cluster 2 increased in relative abundance at all applied N levels, while clusters 3 and 4 increased in relative abundance at low and high nitrogen, respectively. These effects of ammonia concentration were not seen in microcosms incubated at 4°C (Avrahami et al., 2002), where growth will be slow and measurable changes in relative abundance unlikely.

In general, there is evidence for greater abundance of ammonia oxidizers in fertilized soils, although this is not always apparent from MPN counts (Bruns et al., 1999; Phillips, et al., 2000). qPCR of 16S rRNA genes (Phillips et al., 2000a, 2000b) shows abundances 1 to 2 orders of magnitude greater in a range of continuously fertilized soils (Fig. 2). There is also evidence that disturbance of soil, through fertilizer application and tillage, leads to increased abundance, regardless of ammonia addition (Bruns et al., 1999; Phillips et al., 2000; Mendum and Hirsch, 2002). The influence of fertilization, liming, tillage, and herbicides on bacterial ammonia oxidizer community structure was reviewed by Kowalchuk and Stephen (2001). Generally, *Nitrosospira* cluster 3 dominates fertilized and managed systems, and cluster 4 dominates unmanaged or unimproved systems. *Nitrosomonas*-like sequences have also been reported in grassland soils following treatment with inorganic N (Webster et al., 2002) and compost (Kowalchuk et al., 1999; Avrahami et al., 2002). Although this has been suggested to indicate inhibition of *Nitrosospira*

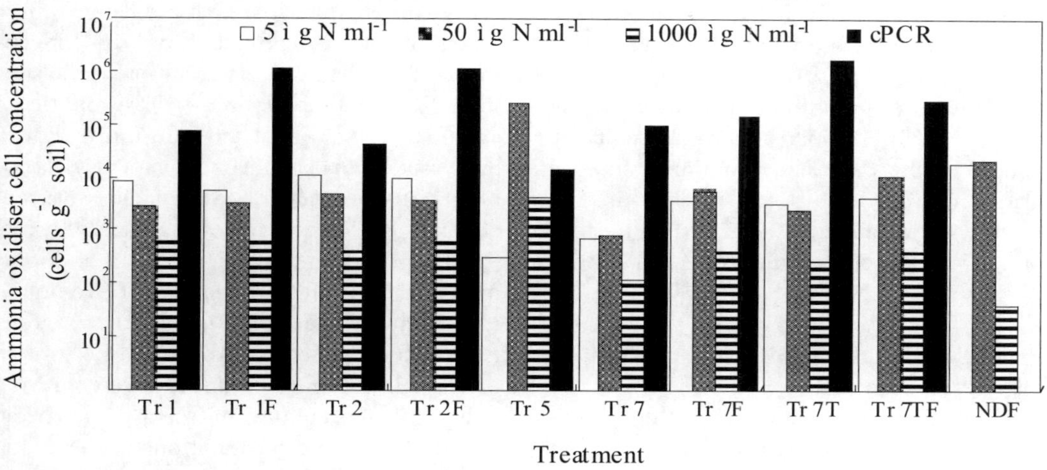

FIGURE 2 Bacterial ammonia oxidizer abundances in long-term ecological research plots subjected to different management regimes. Viable cell abundance was determined using the MPN method with mineral salts medium containing 5, 50, or 1,000 mg of NH_4^+-N ml^{-1}. Total bacterial ammonia oxidizer abundance was determined by competitive PCR amplification of 16S rRNA genes using primers targeting betaproteobacterial ammonia oxidizers. Treatments were as follows: Tr1, conventional tilling; Tr2, no tilling; Tr5, *Populus* perennial cover crop; Tr7, historically tilled. Suffixes T and F indicate tillage and fertilization, respectively, and NDF represents native deciduous forest. Error bars represent standard errors. (From Phillips et al. [2000], with permission.)

cluster 4 by high ammonium, cluster 4 isolates have been obtained on standard ammonia oxidizer growth medium (Mintie et al., 2003).

Mendum et al. (1999) found an increase in the nitrification rate within 3 days of ammonium nitrate fertilization of an arable soil and a decline after 6 weeks. Competitive PCR, targeting bacterial *amoA* and 16S rRNA genes, indicated growth from 10^4 to 10^6 cells g^{-1} and 10^5 to 10^8 cells g^{-1}, respectively. Although this suggests a lack of direct correlation between abundance and activity, the lack of intermediate sampling points makes interpretation difficult. In a subsequent study (Mendum and Hirsch, 2002), in which soil was fertilized with ammonium nitrate with or without manure, the nitrification rate again increased more rapidly than associated population changes. The effect of plowing on both nitrification rate and community structure differed with different fertilizer regimens. Plowing alone led to domination by *Nitrosospira* cluster 4 sequences, while cluster 3 sequences dominated plots that were plowed and fertilized with ammonium nitrate.

It is not clear how fertilization might influence nitrite oxidizer communities, unless it leads to accumulation of nitrite. Nevertheless, long-term ammonia fertilization and plowing influenced nitrite, but not ammonia oxidizers, in a grassland study (Freitag et al., 2005). *Nitrobacter* diversity was the same in all plots, *Nitrospira* communities varied with management regime, and novel *Nitrospira* groups were found. These changes are particularly difficult to explain given the paucity of knowledge of *Nitrospira* physiology and diversity.

SOIL pH

The pH of most soils ranges from 3.5 to 9 and appears to be a major factor influencing bacterial community structure (Fierer and Jackson, 2006). Soil with pH values >8 are relatively rare, but acid soils (pH <5) are common. Soil pH decreases to these values largely through the effects of microbial activity, associated with decomposition of organic matter, but anthropogenic activity can also decrease soil pH, through atmospheric deposition, including deposition of nitrogen. In contrast to marine and freshwater environments, where pH is close to neutral, the pH of many soils is unfavorable for nitrifier growth and activity. In many managed soils, pH is increased to neutrality by liming, but the existence and importance of nitrification in acid soils has led to laboratory and field studies aimed at determining mechanisms for acidophilic nitrification (see De Boer and Kowalchuk, 2001).

Growth and Activity of Laboratory Cultures at Low pH

pH is a major factor influencing the growth and activity of all microorganisms, and a number of mechanisms ensure pH homeostasis in response to alkaline or acid conditions. These mechanisms are designed to maintain intracellular pH close to neutral, even in many acidophiles (Baker-Austin and Dopson, 2007), but stress induced by acid or alkaline conditions increases energy demands and reduces activity and growth. For ammonia oxidizers, pH extremes have additional effects on substrate availability. The pK_a for the NH_3:NH_4^+ equilibrium is 9.25, and NH_3 concentration therefore changes approximately 10-fold for each unit change in pH. In alkaline soils, most will be present as NH_3, leading to significant losses through volatilization. At low pH, most will be in the form of NH_4^+, reducing the concentration of ammonia (NH_3), the substrate for ammonia monooxygenase.

The requirement for and energetics of NH_3 and NH_4^+ uptake by ammonia oxidizers are poorly characterized, and it is not clear whether reduced activity of pure cultures is due to increased energy demand, associated with transport into the cell, or reduced ammonia concentration. Suzuki et al. (1974) determined the combined effect of ammonia concentration and pH on oxygen consumption during ammonia oxidation by *N. europaea* cells, and cell extracts in the pH ranges 7 to 9.1 and 6.5 to 8.5, respectively. Maximum velocity (V_{max}) was independent of pH. K_m calculated for NH_3 + NH_4^+ decreased by factors of 26 and 83 with increasing pH, for whole cells and cell extracts, respectively, but changed little when calculated as a function of NH_3 concentration. This suggests that ammonia is the substrate for ammonia monooxygenase and that it enters the cell by facilitated diffusion, or at least by a process that is pH independent. Ammonium transport (if it occurs) demands sufficient energy to influence activity and growth rate significantly. Alternatively, uptake of ammonia may not be necessary, and the pH responses may be entirely due to ammonia concentration. Allison and Prosser (1991b) found no growth of several ammonia oxidizers below pH 6.5 in liquid batch culture, and Jiang and Bakken (1999a) found similar inhibition of growth and activity of four *Nitrosospira* strains isolated from soil. No strain grew below pH 6.5, but one was active at pH 5. Double reciprocal plots of activity versus NH_3 at different pH values were linear, and estimated half-saturation constants were similar to those of other strains. The influence of pH on growth did, however, differ between strains and reflected environmental origin, with greater acid tolerance by isolates from acid soils. Effects of pH on growth are therefore more complex than those on activity, possibly due to greater tolerance to pH per se, or to nitrite toxicity.

A number of factors, of relevance to soil nitrification, determine the influence of pH on ammonia oxidation. Many ammonia oxidizers can hydrolyze urea to ammonia, and ureolysis appears to be pH independent. For example, one ureolytic strain, *Nitrosospira* NPAV, which does not grow on ammonia below pH 7 in liquid batch culture, grows on urea at pH values as low as 4. When growing on urea, the maximum specific growth rate is independent of pH in the range 4 to 8 (Burton and Prosser, 2001), and nitrite production ceases as soon as urea hydrolysis is complete, even though ammonia is present in the medium (Fig. 3). The results suggest that urea uptake is independent of pH, that intracellular pH is favorable for both urea hydrolysis and ammonia oxidation, and that nitrite, and some ammonia, diffuse out of the cell. Once released, ammonia is ionized and, at low pH, is effectively unavail-

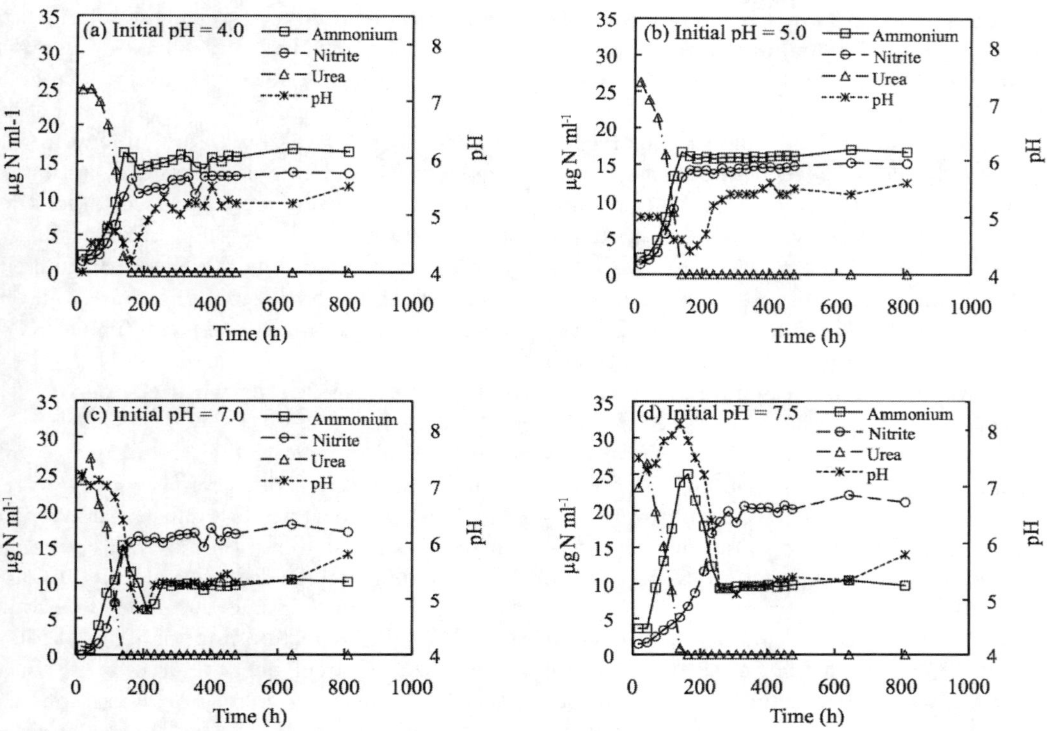

FIGURE 3 Growth of *Nitrosospira* NpAV, a ureolytic ammonia oxidizer, on urea in liquid batch culture at initial pH values of 4 (a), 5 (b), 7 (c), and 7.5 (d). Growth on ammonia was inhibited at pH <7. Growth was followed by measuring changes in urea, ammonium, and nitrite concentrations and pH. (From Burton and Prosser [2001], with permission.)

able. Urea hydrolysis may therefore be an important factor enabling nitrification in low pH soils, but requires that ammonia oxidizers gain access to urea prior to its conversion by other organisms, as ureolysis is a common characteristic of soil microbes. The influence of pH on growth and activity also depends on the physiological state of cells. Keen and Prosser (1987) showed that populations of *Nitrobacter* could not grow below pH 6 in liquid batch culture containing 50 μg of NO_2^--N ml^{-1} but grew at lower initial nitrite concentrations, or in nitrite-limited continuous culture in which the pH was reduced, gradually, from 6 to 5.5.

Nitrification in Acid Soils

Despite lack of acidophilic growth in batch culture, nitrification is common in acid soil (for review, see De Boer and Kowalchuk, 2001), even in soil with pH values as low as 3. The laboratory studies described above suggest that growth and activity of neutrophilic ammonia and nitrite oxidizers in acid soils can be explained by urease activity, supply of low concentrations of substrate, gradual reductions in pH, surface growth, and attachment and aggregate formation. Nitrification in acid soils may also result from heterotrophic nitrifiers that can grow and maintain activity at low pH (see Chapter 5). However, the cellular rates of nitrification by such organisms are low, and it remains difficult to determine their importance and activity in soil. In addition, soil is heterogeneous, and bulk pH measurements may not reveal neutral microenvironments (see below).

Although low pH inhibits ammonia oxidizers in laboratory culture, a metastudy of nitrification in almost 300 soils (Booth et al., 2005) provided no evidence for a significant effect of soil pH on nitrification (Fig. 4). Importantly, the study focused on gross rather than net nitrification, considering the combined effect of related soil processes, including mineralization and denitrification. Gross nitrification was controlled most by nitrogen mineralization, particularly at low mineralization rates: the proportion of mineralized nitrogen that was nitrified decreased from 63% at a mineralization rate of 1 mg of N kg of soil^{-1} day^{-1} to 28% and 19% at mineralization rates of 5 and 10 mg of N kg of soil^{-1} day^{-1}, respectively. Some of the highest nitrification rates were in low pH organic soils. These observations are not inconsistent with some of the proposed mechanisms for growth at low pH. Ammonia oxidizers growing on the surface of particulate organic matter, gradually releasing ammonia, will benefit from the advantages of surface growth, low ammonia concentration, and continuous growth, all of which enable growth at acid pH values in laboratory culture. Interestingly, for the grassland and agricultural soils, there was no detectable overall effect of N fertilization on nitrification rate, again suggesting that ammonium generated by mineralization may be a more important source of ammonia for nitrification. Ross et al. (2009) also found no correlation between nitrification rate and soil pH in ten forested watershed soils, but nitrification rate did correlate negatively with C/N ratio.

Nitrification in acid forest soils is particularly important, given the impact of atmospheric N deposition on these ecosystems, but can also be of importance in agricultural systems. For example, Kyveryga et al. (2004) investigated the influence of soil pH on nitrification following addition of fertilizer N to Corn Belt soils. Anhydrous ammonia is commonly used in such systems, and its application to some fields in fall and to others in spring is beneficial for management reasons. Rapid growth of corn does not occur until June, and nitrification inhibitors are therefore added to reduce fertilizer loss. Nitrification increased with pH in the relatively narrow range 6 to 8, but with significant variability, and the effectiveness of inhibition by nitrapyrin was reduced at higher pH. In general, inhibition is less effective under optimal conditions for nitrification, and knowledge of pH effects is necessary for efficient fertilizer management, suggesting that fertilizer application should be delayed until spring in soils with higher pH.

Ammonia Oxidizer Abundance in Acid Soils

MPN counts indicate that abundance of culturable ammonia oxidizers in low pH soils is 2 to 3 orders of magnitude less than that in typical agricultural soil. Often they are not detectable; for example, Klemedtsson et al. (1999) could not detect ammonia oxidizers in acid forest soils with a MPN detection limit of 500 g of soil^{-1}. MPN counts are likely to underestimate abundance. Media and cultivation conditions are selective, growth may not be detectable within the incubation period used (Matulewich et al., 1975), cells may be difficult to separate from soil particles, and some will be inhibited by other organisms present. Molecular detection of ammonia oxidizers eliminates these limitations, but abundances determined by qPCR of ammonia oxidizer 16S rRNA or *amoA* genes in acid soils are also low, although sometimes 1 to 2 orders greater than MPN counts. Jordan et al. (2005) reported abundances close to the detection limit of 10^4 gene copies g of soil^{-1} in soils with pH 3.7 to 4.9, and Nicol et al. (2008) measured 7.2×10^4 bacterial *amoA* gene copies g of soil^{-1} at pH 4.9. However, detection is often not possible, even when using nested PCR and fingerprinting approaches to determine community structure (Laverman et al., 2001; Bäckman et al., 2003; Jordan et al., 2005). Although detection limits are not always defined, this suggests abundances below 10^3 to 10^4 cells g of soil^{-1}. For example, Schmidt et al. (2007) could not amplify bacterial *amoA* genes from an acid heathland soil, even with a nested PCR approach. Nitrification in these soils was

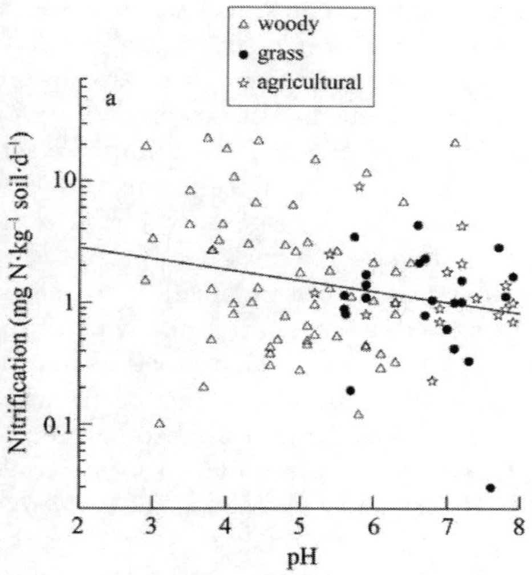

FIGURE 4 Relationships between soil pH and gross nitrification rate in mineral and organic soil layers from agricultural and woodland ecosystems. (Data kindly provided by J. M. Stark. Adapted from Booth et al. [2005].)

low and was assumed to be derived from heterotrophic nitrification, although this study did not target archaeal ammonia oxidizers, which may have been active at low pH.

Influence of pH on Soil Ammonia Oxidizer Communities

Enrichment of bacterial ammonia oxidizers from acid soils is relatively common, and several workers have successfully obtained pure cultures. Most isolates are nitrosospiras (see De Boer and Kowalchuk, 2001), although nitrosomonads are also found (Allison and Prosser, 1991b). Many, but not all, are urease positive, and some are tolerant of high-ammonia concentration. Attempts have been made to design media that select acidophilic strains, but there is only limited evidence that such strains exist. Smith et al. (2001) compared betaproteobacterial ammonia oxidizer 16S rRNA gene sequences in environmental clones and enrichment cultures from acid and neutral soil plots. Environmental clones from pH 4.2 soil were relatively evenly distributed between *Nitrosospira* clusters 2 to 4, with a minority of *Nitrosomonas* sequences. In contrast, enrichment cultures were dominated by *Nitrosospira* cluster 3, with only one nitrosomonad, suggesting selection by cultivation conditions. Interestingly, fewer enrichments were obtained from neutral soil than acid soil, none was obtained on low pH medium, and several clone sequences were identical to those of enrichments. This demonstrates the extent of selection by laboratory conditions but also indicates that strains that are abundant in soil can dominate enrichments.

Molecular analysis of communities provides quantitative data on the influence of pH. Although it remains dangerous to generalize, most studies indicate dominance of low pH soils by nitrosospiras, particularly *Nitrosospira* cluster 2 (Stephen et al., 1996, 1998; Kowalchuk et al., 2000; Laverman et al., 2001), although other *Nitrosospira* clusters are frequently obtained and *Nitrosomonas* sequences dominated DGGE profiles of ammonia oxidizers in an acid forest soil (Carnol et al., 2002). Unfortunately, with rare exceptions (Jiang and Bakken, 1999a; Burton and Prosser, 2001), physiological studies have focused on *Nitrosomonas* rather than *Nitrosospira*, and the physiological characteristics leading to greater abundance of nitrosospiras in soils and selection for *Nitrosospira* cluster 2 in acid soils are not known.

Selection for Ammonia Oxidizers in Long-Term pH Gradients

Long-term selection by soil pH and stability of communities has been investigated in a Scottish agricultural soil maintained for 36 years at pH values in the range 3.9 to 6 (Stephen et al., 1998). Low pH soils were dominated by *Nitrosospira* cluster 2 sequences, which decreased in relative abundance as pH increased, while *Nitrosospira* cluster 3 sequences increased in relative abundance. Nitrosospira cluster 4 and *Nitrosomonas* cluster 6a sequences were also detected, at lower relative abundance, and with less obvious pH trends. The stability of these long-term changes was confirmed by Nicol et al. (2008), who found similar distributions of bacterial ammonia oxidizers in the same plots after an additional 9 years. They also characterized both bacterial and archaeal *amoA* gene sequences, by DGGE, and quantified *amoA* gene and *amoA* gene transcript abundances. Relative abundance of bacterial and archaeal *amoA* gene sequence types changed with pH, suggesting that both groups contained lineages with different pH preferences. Bacterial and archaeal gene abundances showed contrasting behavior (Fig. 5). Archaeal *amoA* genes decreased with increasing pH but were always more abundant than bacterial *amoA* genes, which varied less with pH. Both archaeal and bacterial *amoA* gene transcripts were always less abundant than genes, suggesting that the majority of the community is inactive, although this could reflect methodological difficulties in quantifying transcripts. Archaeal *amoA* transcript abundance decreased with increasing pH, while that of bacteria increased. Gene:transcript ratio is a better measure of functional activity (Freitag and Prosser, 2009), and archaeal ratio decreased significantly with pH, but increased for bacteria (Fig. 5). Links between transcriptional activity and process rates are poorly understood, due to lack of physiological studies, but the data indicate significantly different pH preferences for archaeal and bacterial ammonia oxidizer communities in these soils, with archaea and bacteria apparently preferring acid conditions and neutral conditions, respectively. Archaeal *amoA* phylotypes that were dominant at low- and high-pH soils were also more transcriptionally active in short-term microcosm experiments with mixed pH 4.5 and 7 soils readjusted to either of these pH values (Nicol et al., 2008). This provides evidence that the long-term pH selection for different phylotypes is due to pH-related activities.

N Deposition and N-Saturated Soils

Atmospheric deposition of N has increased significantly since the industrial revolution and in terrestrial ecosystems leads to increased nitrification, leaching of nitrate, further production of nitrous oxide, and acidification of soil. Nitrogen saturation occurs when deposition rates exceed N demand from plants and soil microbial biomass. Forest ecosystems can act as a significant reservoir of reactive nitrogen, in part because nitrification rates are low, and conditions in such climax ecosystems are optimal for maintenance of nitrogen within the plant-soil system through minimal loss of nitrate. N deposition can lead to increased leaching, but liming is often used to reverse soil acidification. The subsequent effects on nitrification depend on the extent of N deposition and whether the soil is N saturated. At low levels, N is assimilated by plants, but nitrification can occur when demand is satisfied and soils become N saturated. Thus, although ammonia oxidizers could not be detected in a Swedish acidic coniferous forest soil (Klemedtsson et al., 1999; Bäckman et al., 2003), liming increased pH in the upper, organic horizons and led to detection of *Nitrosospira* cluster 2 and 4 sequences and to increased nitrification potential. Similar effects can result from clear-cutting, which increases nitrogen mineralization, nitrification potential, ammonia oxidizer abundance, and changes in community structure (Bäckman et al., 2004). Even in acid soils with low levels of nitrification and low ammonia oxidizer abundance, there can be significant potential for nitrification if ammonia concentrations are high and pH is increased. Jordan et al. (2005) performed similar studies on a Californian forest soil subjected to atmospheric N addition. Potential nitrification activity was detected and was greater in N-saturated soils, but nitrification was inhibited by acetylene, indicating heterotrophic activity. Nitrogen and soil pH did not influence community structure, and growth of *N. multiformis* was inhibited by soil from these sites, suggesting the presence of inhibitory compounds.

Another consideration is the way in which nitrification is measured. Stark and Hart (1997) found relatively high rates of gross nitrification in several forest soils. Nitrate uptake by heterotrophic microorganisms was also high, so that little nitrate was lost from the system and there was no significant relationship between gross and net nitrification rates. This emphasizes the need to consider nitrification within

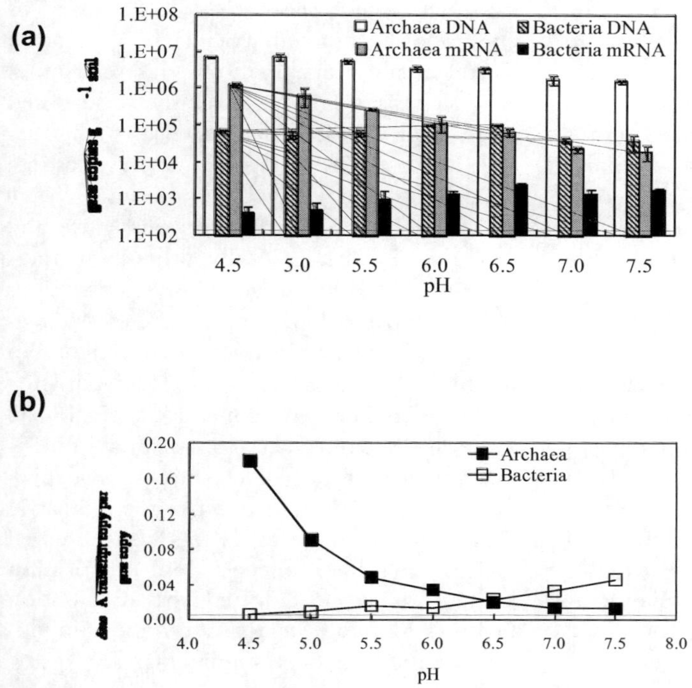

FIGURE 5 Abundance and transcriptional activity of crenarchaeal and bacterial ammonia oxidizers in long-term pH plots, maintained at pH values in the range 4.5 to 7.5, determined by quantification of *amoA* gene and gene transcripts (a) and by ratios of gene transcript:gene abundance (b). Error bars represent standard errors of replicate field samples at each soil pH. (From Prosser and Nicol [2008], with permission.)

the context of the soil carbon cycle and other nitrogen cycling processes, the need to measure gross nitrification rates, and the possibility of high ammonia oxidizer activity in forest soils. Internal cycling in these systems also has the potential to increase nitrous oxide production.

SOIL STRUCTURE, HETEROGENEITY, AND MICROENVIRONMENTS

The physical, chemical, and biological characteristics of soil are heterogeneous and vary between and within soils, giving different soils their distinctive properties. This in itself provides a wide range of local conditions and environments that influence soil microorganisms. Heterogeneity in particulate material also leads to heterogeneity in soluble and gaseous components. Diffusion of gaseous material will be low in water-saturated soils and, although greater at low-moisture content, will be restricted by soil tortuosity. Conversely, movement of soluble substrates and products by diffusion, and of motile cells, increases with moisture content, but again is limited by tortuosity, even in saturated soils. Transport of soluble nutrients and microorganisms increases with bulk flow of soil water, and "mixing" of all soil components is increased by root growth, burrowing by soil animals and predators, and large-scale events, such as plowing. These complex factors create a myriad of microenvironments with distinct physicochemical characteristics and different benefits and disadvantages for resident microbial communities. Complexity is further increased by the temporal changes in these characteristics through soil processes and seasonal influences. In addition to effects on ammonia transport and diffusion, water potential will influence nitrifier physiology, as dehydration will concentrate soil solutes, increasing osmotic pressure. Stark and Firestone (1995) found that substrate limitation by diffusion reduced nitrification at soil water potentials less than −20.6 MPa, while dehydration was more important at lower water potentials. Dehydration also increases ammonia concentration, which may influence nitrification rates more than increases in osmotic pressure (Low et al., 1997).

Attempts to understand the mechanisms driving soil microbial processes, community dynamics, and diversity must take this complexity into account. A major difficulty is the absence of techniques that reliably measure ecosystem process rates, community dynamics, and diversity at the scale required. Most measurements of soil nitrification are made at scales >1 g, which may have little relevance for organisms existing within a soil pore with a maximum width <50 μm. Some of the impacts of soil heterogeneity have already been considered within the section on surface growth, which significantly changes the activity of nitrifiers and their sensitivity to unfavorable conditions. This section deals with more general aspects of the impact of heterogeneity.

Consequences of Heterogeneity for Nitrification Kinetics

Potential nitrification is typically measured as the increase in product (nitrite + nitrate) concentration with time in soil or soil slurries. In the absence of ammonia or pH limitation, which are manipulated to maximize nitrification rates, short-term incubation leads to a linear increase in product concentration. Prolonged incubation leads to increases in product formation that are often described using the logistic equation. The logistic equation is commonly used to describe growth of animal populations and assumes that the specific growth rate is inversely proportional to population size, with a maximum value r, decreasing to zero at a population size, K, the carrying capacity of the environment. In applying this equation to microbial growth in liquid culture, it describes, in approximate terms, the acceleration, exponential, deceleration, and stationary phases. These growth phases, in turn, relate to initial substrate excess; reduced growth rate through reduction in substrate concentration and accumulation of metabolic byproducts or end products of metabolism; and cessation of growth, when substrate is fully utilized or conditions become unfavorable. Its application to soil nitrification kinetics therefore assumes a well-mixed, homogeneously distributed population of nitrifying bacteria, with each individual experiencing the same substrate concentration and growth conditions.

Molina (1985) challenged this view and proposed an alternative hypothesis: nitrifiers are distributed heterogeneously, growing as spatially separated clusters on discrete sources of ammonium within microenvironments. To test this, nitrification was investigated in microcosms, each of which contained a single soil aggregate. The time taken for cessation of nitrification in each aggregate, through acid production, was determined, and the cumulative distribution of these times followed observed increases in nitrate concentration. Nitrification kinetics therefore reflected the distribution of times required for activation and completion of nitrification in individual aggregates, and not the average kinetic constants (r and K) of the total community. This profoundly influences the way in which we consider soil nitrification kinetics. Activation following starvation may be more important than maximum specific growth rate, and cessation of nitrification will be due to acidification in individual aggregates, or microenvironments, and not due to low concentrations of ammonia.

Consequences of Heterogeneity for Nitrifier Communities

Spatial separation of soil communities in microenvironments is a potential mechanism maintaining high levels of diversity within soil nitrifier communities. Analysis of samples of ≥1 g prevents detection of small-scale diversity patterns but explains reduced diversity in managed soils, where plowing and regular addition of inorganic nitrogen fertilizer increase mixing and reduce heterogeneity in ammonia concentration. For example, Webster et al. (2002) found greater evenness in bacterial ammonia oxidizer communities in unfertilized grassland soils than in those subjected to high levels of inorganic fertilization. In addition, replicate 0.5-g soil samples from unfertilized soil showed greater heterogeneity than fertilized soils in terms of ammonia oxidizer community structure, ammonia concentration, and pH. Other studies have reported reduction in heterogeneity in community structure with soil management (Bruns et al., 1999) but, as discussed above, fertilization may not reduce heterogeneity if not accompanied by physical mixing of soil (Dell et al., 2008). Heterogeneity in ammonia oxidizer communities can also be seen at the cm-scale. For example, ammonia oxidizers in biological surface crust soils, found in arid regions, are restricted to a depth of 2 to 3 cm, where they are shielded from sunlight and oxygen concentration is sufficient (Johnson et al., 2005). Nitrogen is supplied to these soils by nitrogen fixation, but communities are limited by oxygen and not ammonium concentration.

Grundmann and Debouzie (2000) used a geostatistical approach to determine the structure and relationships between soil ammonia and nitrite oxidizer communities at the mm-scale. Analysis of samples taken at 1-mm intervals along a 10-cm transect exhibited spatial structure of ammonia and nitrite oxidizers at 4-mm and 2-mm scales, respectively, the shorter range for nitrite oxidizers reflecting their dependence on ammonia oxidizers for nitrite. Spatial distributions of the two groups were not independent, and only one of the six nitrite oxidizer serotypes detected exhibited spatial structure. Investigation of mechanisms driving nitrifier distribution and diversity in soil must therefore consider scales ≤1 mm, and structure is likely to be determined by soil pores, aggregates, and small roots. Grundmann et al. (2001) used an alternative approach, combining computer simulations with experimental data. The proportion of soil microsamples, ranging from 20 to 2,000 μm diameter, containing ammonia or nitrite oxidizers was determined experimentally, and comparison with computer simulations predicted that microhabitats (50-μm diameter) contained seven nitrite oxidizer cells (i.e., microcolonies were small). Patches colonized by ammonia and nitrite oxidizers were distributed randomly, and 50-μm microhabitats colonized by nitrite oxidizers were relatively distant (375 μm), potentially limiting nitrification rates through diffusion of nitrite produced by ammonia oxidizers. However, distribution of ammonia and nitrite oxidizers was not independent, suggesting a degree of colocalization facilitating interactions between these groups. Serotyping also showed colonization of 50-μm diameter microsamples by several nitrite oxidizer serotypes, implying high environmental heterogeneity at this microscale, with diversity similar to that found between reference strains from different geographical regions (Grundmann and Normand, 2000). These studies highlight the technical and conceptual difficulties in understanding and quantifying the forces and mechanisms driving nitrifier diversity and interactions in soil. Analysis of communities, rates, and environmental conditions is necessary at submillimeter scales, and analysis also depends on the phylogenetic resolution of methods used to determine diversity and the sampling effort (Grundmann, 2004).

Oxygen Diffusion and Limitation

Soils are heterogeneous with respect to oxygen concentration, rapidly become anaerobic when waterlogged, due to high microbial activity and availability of carbon substrates. At the other extreme, well-drained soils contain many regions with oxygen at atmospheric concentrations. Even in these soils, spatial heterogeneity will result in microsites in which nitrification will be oxygen limited. As soils dry, water is removed from increasingly small pores but will remain in some, providing the potential for anaerobic conditions where oxygen utilization exceeds supply, due to diffusional limitations. Conditions are most anaerobic where organic substrates and microbial activity are greatest, which is usually in the rhizosphere. However, roots of some plants, notably rice plants, release oxygen, making conditions favorable for nitrification and reversing typical oxygen gradients.

These processes are important for interactions between nitrifiers and denitrifiers, which rely on nitrifiers for nitrate and also require available organic carbon. Nitrifiers and denitrifiers will be active in different microsites within the soil, linked by nitrate diffusion, with the balance of the two processes dependent on the proportion of water-filled pore spaces and diffusion of oxygen and nitrate. Denitrifiers and nitrifiers are also linked through their ability to produce nitrogen oxides. Nitrous oxide is of particular concern because of its activity as a greenhouse gas. It has 310 times the global warming potential of carbon dioxide, atmospheric N_2O concentrations are increasing by 0.26% per year, and 10 to 50% of global anthropogenic N_2O emissions are produced by agricultural soils (Chen et al., 2008).

Ammonia oxidizers produce nitrogen oxides by two processes. In the first, N_2O is produced as a byproduct of hydroxylamine conversion to nitrite through chemical decomposition of intermediates. In the second process, nitrifier denitrification, nitrite is reduced to nitrous oxide, via nitric oxide, and then to nitrogen gas (Wrage et al., 2001; Stein and Yung, 2003). Nitrifiers also contribute indirectly to nitrous oxide production through provision of nitrate to denitrifiers. Nitrous oxide production is catalyzed by two mutually exclusive forms of nitrite reductase, encoded by *nirK* and *nirS*, both of which are present in *Nitrosomonas* and *Nitrosospira* and lead to production of nitric oxide. Nitric oxide reductase, which reduces nitric oxide to nitrous oxide, is encoded by *norB*, which has been found in *Nitrosomonas*, *Nitrosococcus* (Casciotti and Ward, 2005), and *Nitrosospira* (Garbeva et al., 2007). Production of both oxides therefore seems to be universal within bacterial ammonia oxidizers, and there is evidence from genomic studies that archaeal ammonia oxidizers have homologous genes (see Section III). Both *nirK* and *norB* are less well conserved in bacterial ammonia oxidizers than *amoA* and 16S rRNA genes, and lack of congruency in phylogenies suggests that they were acquired through lateral gene transfer (Garbeva et al., 2007; Cantera and Stein, 2007). Phylogeny is also not related to nitrous oxide production rates in a number of strains (Garbeva et al., 2007)

The process is readily demonstrated in pure culture, where N_2O constitutes 0.05 to 1% of nitrite produced by *Nitrosospira* (Garbeva et al., 2007; Jiang and Bakken, 1999b) with significantly higher yields for *N. europaea* (up to 1.95% [Remde and Conrad, 1990]). Production increases with decreased oxygen concentration (Goreau et al., 1980; Kester et al., 1997; Dundee and Hopkins, 2001) and is influenced by low pH and ammonia concentration (Jiang and Bakken, 1999b). Up to 54% of N_2O generated from added nitrite was converted by nitrifier denitrification in *Nitrosospira* 40KI (Shaw et al., 2006). Jiang and Bakken (1999b) found similar N_2O/NO_2^- ratios with ammonium or urea as substrate and an increased ratio at lower pH and under effective starvation conditions, suggesting greater production under starvation or acid-limiting conditions.

The significance of nitrifier-denitrification in soil is difficult to determine because of methodological difficulties. Traditional approaches involve analysis of correlation between N_2O production and nitrification rates (Sitaula et al., 2001) and incubation of samples after addition of $^{14}NH_4{}^{15}NO_3$ or $^{15}NH_4{}^{15}NO_3$ and in the presence or absence of inhibitors of nitrification or denitrification (Robertson and Tiedje, 1987; Webster and Hopkins, 1996a). Specific inhibitors of nitrification, but not denitrification, are available but denitrification can be suppressed by incubation under aerobic conditions. Such methods suffer from several limitations, including changes in substrate concentrations, disturbance, and turnover of label, such that nitrification will lead to production of $^{15}N\text{-}NO_3$, which can then be denitrified. The efficiency of inhibition will be reduced by diffusion limitations in soil, heterogeneous distribution and lack of specificity. There is also evidence that the sensitivity of nitrous oxide production to acetylene inhibition varies significantly between different ammonia oxidizers (Wrage et al., 2004b).

Better discrimination of sources of N_2O can be achieved using variations in both $^{15}N/^{14}N$ and $^{18}O/^{16}O$ isotope ratios at natural abundance levels (Webster and Hopkins, 1996b) and after incubation with artificially enriched compounds, including single or double ^{15}N-labeled ammonium nitrate and ^{18}O-labeled water (Wrage et al., 2005). $\delta^{15}N$ values in N_2O produced by nitrifiers or denitrifiers will depend on isotope ratios in ammonium and nitrate, respectively. $\delta^{18}O$ values will depend on those in oxygen and water. Ammonia oxidation to hydroxylamine utilizes O_2, while oxidation of hydroxylamine and nitrite utilize O in water. N_2O from nitrification, nitrifier-denitrification, and denitrification of nitrate produced by nitrification will therefore contain 100, 50, and 33% of $\delta^{18}O$ of that in oxygen. Complications arising from O-exchange between water and intermediates of the different pathways are discussed by Kool et al. (2007). An alternative approach is to quantify the distribution of isotopes between the central and terminal N atoms (isotopomers) within the N_2O molecule, expressed as site preference. Sutka et al. (2006) found that site preferences in nitrous oxide produced during ammonia and hydroxylamine oxidation by nitrifiers were around 33%, while that from nitrate and nitrite reduction by denitrifiers was approximately 0%. This overcomes inaccuracies of other stable isotope methods associated with variability in ammonia and nitrate isotope ratios. Ostrom et al. (2007), however, found that correction in site preference values may be required to allow for changes due to consumption of nitrous oxide before release from soil.

Despite these methodological difficulties, there is evidence that nitrifiers play an important role in terrestrial N_2O production, particularly in dry soils, where it can contribute up to 80% of total soil N_2O (Robertson and Tiedje, 1987; Webster and Hopkins, 1996a, 1996b; Gödde and Conrad, 1999; Wrage et al., 2005). There is no evidence for an effect of pH on N_2O production (Wrage et al., 2004a), but production increases following addition of high concentrations of artificial urine (Wrage et al., 2004a) and ammonium (Avrahami et al., 2002).

EFFECTS OF TEMPERATURE AND CARBON DIOXIDE ON NITRIFICATION

Jiang and Bakken (1999b) described the effect of temperature on activity of four *Nitrosospira* strains in terms of both the Arrhenius equation and the square root model, although the former gave good fit only in the range 3 to 21°C. Optima for three of the strains were in the range 26 to 29°C, while one strain, *Nitrosospira* strain AF, isolated from a Zambian soil (Utåker et al., 1996), had an optimum of 31 to 33°C and greatest activity over most of the range. Activation energies, calculated from the Arrhenius plots, were higher for the two vibrioid strains, which fell into a distinct 16S rRNA-gene phylogenetic group and had a lower pH minimum for activity. Activation energies were also higher than those found for *Nitrosomonas*.

Temperature kinetics in pure culture are reflected in soil, where accurate prediction of temperature responses is important in determining the timing of fertilizer application, potential losses, and nitrate leaching. For example, application of fertilizer in the autumn is possible in regions where nitrification rates over winter are small. The Arrhenius equation can be used to describe the effect of temperature on soil nitrification, but, as for pure cultures, is less reliable at higher temperatures (Stark, 1996). This equation is used in models of soil nitrification. Optimum temperature is greater in tropical regions, lower in northern regions, and varies with soil conditions. For example, temperature and pH will interact, as faster nitrification rates will increase the rate of acidification, and inhibition of nitrification and differential effects on ammonia and nitrite oxidizers can lead to accumulation of nitrite at temperatures <12°C (Russell et al., 2002). The relationships between nitrification rate and temperature are also influenced by water content, which can be described in terms of three parameters: maximum nitrification rate, optimal relative water content, and temperature (Grundmann et al., 1995). Interactions between these factors are related to their impact on respiration rate, oxygen diffusion, and aggregate structure.

Transect and transplant experiments have been used to study the influence of microclimatic conditions and soil management on ammonia oxidizers and nitrification. Mintie et al. (2003) found that ammonia oxidizer community structure was dominated by *Nitrosospira* cluster 4 sequences across two meadow-to-forest transects, despite differences in vegetation composition, plant nitrogen input and soil temperature, moisture, and pH. There was some evidence of increased relative abundance of *Nitrosospira* clusters 2 and 3 in meadow and transitional soils. Ammonia oxidizer isolates from these soils were dominated by *Nitrosospira* cluster 4 strains. Despite small differences in community structure, nitrification potential was significantly greater in meadow soils, possibly through increased abundance with no change in community structure. Alternative explanations are finer-scale discrimination of communities, functional redundancy, or differences in active, rather than "total" communities. The influence of environmental conditions on ammonia oxidizer community structure, abundance (MPN counts), and activity were then investigated by reciprocal transfer of soil cores between meadow and forest soils from each site (Bottomley et al., 2004). Within 2 years, net nitrification rates of forest soil cores increased to those of the meadow sites to which they had been transferred, and in some, but not all, cases, these increases were associated with increased nitrification potential and ammonia oxidizer abundance. Changes were explained in terms of climatic conditions, particularly increased temperature, rather than changes in community structure, which showed no consistent pattern.

Microcosm studies provide evidence for a relationship between phylogeny and temperature preferences. Tourna et al. (2008) found significant changes in archaeal, but not bacterial, 16S rRNA and *amoA* gene DGGE patterns and *amoA* gene transcripts with increasing temperature in soil microcosms (Fig. 6). Relationships with bacterial ammonia oxidizers are, however, not always simple and experimental studies with soil are often complicated by variation in other factors. *Nitrosospira* strains belonging to *amoA*-defined clusters 1, 2 and 4 are generally associated with low temperature environments (Avrahami and Conrad, 2005), while *Nitrosospira* strain AF and related sequences are found in warmer climates, although temperature responses interact with fertilizer and moisture in a complex manner (Avrahami and Bohannan, 2007). In long-term incubations of two low-temperature soils, *Nitrosospira amoA* clusters 3a, 3b, and 9 were most common at 30°C; cluster 4 was most common at 25°C; and cluster 1 was most common at <30°C (Avrahami and Conrad, 2003). The high-temperature soil was dominated by *Nitrosospira* cluster 3a at all temperatures, but with changes within this group. In an agricultural soil, nitrous oxide production increased with increasing temperature (4 to 37°C), while potential nitrification was greatest at intermediate temperatures (Avrahami et al., 2003). Ammonia oxidizer communities changed in response to both ammonium concentration and temperature in long-term experiments (16 weeks), with *Nitrosospira* cluster 1 dominant at low-temperature and temperature-related changes within *Nitrosospira* cluster 3 throughout the temperature range.

This complexity was also seen in experiments designed to investigate the impact of climate change (increased CO_2, temperature, and precipitation) on nitrification. Elevated atmospheric CO_2 concentration may not directly influence nitrification, as soil atmospheric CO_2 concentrations are severalfold greater than atmospheric concentrations. However, indirect effects, arising from predicted increases in soil water content, and readily available soil carbon and nitrogen could reduce nitrification. Carnol et al. (2002) found increased nitrification and N_2O production at

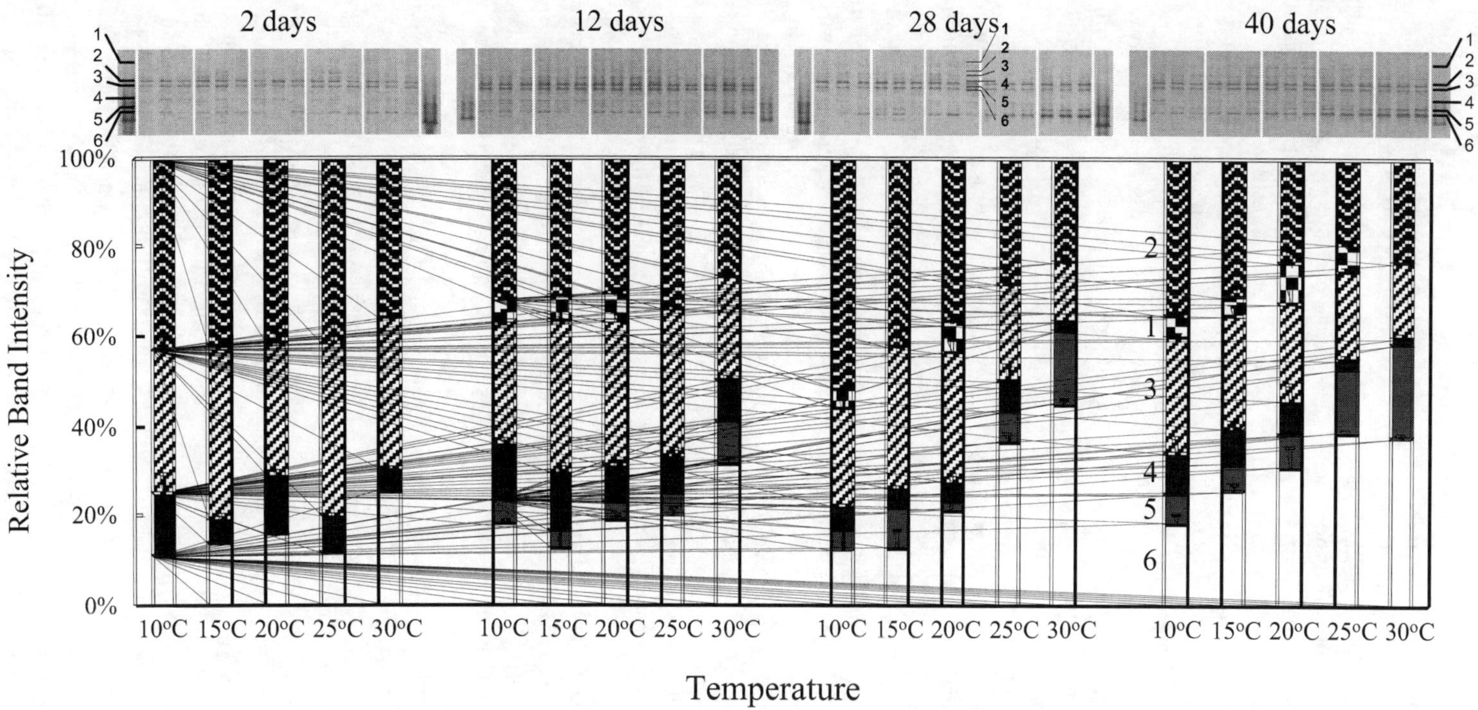

FIGURE 6 Influence of temperature and incubation time on relative abundance of crenarchaeal *amoA* gene transcripts of different phylotypes in soil microcosms incubated at temperatures in the range 10 to 30°C and sampled at 2, 12, 26, and 40 days. *AmoA* gene transcript sequences were analyzed by DGGE, and histograms represent relative intensity of bands within individual lanes from triplicate microcosms. Data are presented as means of triplicate values and standard errors. (From Tourna et al. [2008], with permission.)

elevated CO_2, explained through direct effects of high CO_2 concentration and/or indirect effects on ammonia availability. Barnard et al. (2004) found no effect of elevated CO_2 on nitrification activity, but Barnard et al. (2006) reported a negative effect when other factors (N addition, increased temperature) were considered. Similar complexity was found in responses of bacterial ammonia oxidizers to increased CO_2, temperature, nitrogen, and precipitation (Horz et al., 2004).

NITRIFICATION INHIBITORS

Inhibition of nitrification is of interest for several reasons. First, the chemical nature of inhibitors and the characteristics and kinetics of inhibition can give clues to biochemical mechanisms of ammonia and nitrite oxidation and have been used extensively in studies of nitrifier physiology (Arp and Stein, 2003). Second, specific inhibitors of nitrification are invaluable in discriminating between different processes contributing to changes in substrates and products (e.g., nitrous oxide) of nitrifiers and other groups. Third, natural and commercial inhibitors of nitrification may lead to retention of ammonia within soil, with economic and environmental benefits.

Chemically Synthesized Inhibitors

More than 50 chemicals have been characterized and investigated as inhibitors of soil nitrification with a view to their commercial use. These compounds and their mechanisms of action are described in detail in two recent reviews (Subbarao et al., 2006; Singh and Verma, 2007). The most commonly used are nitrapyrin (2-chloro-6-trichloromethyl pyridine; N-Serve), dicyandiamide, allylthiourea, carbon-disulfide-based inhibitors, 3,4-dimethylpyrazol-phosphate, 2-amino-4-chloro-6-methylpyrimidine, and acetylene. Inhibitors can be bactericidal or bacteriostatic, and mechanisms of inhibition include chelation with copper components of the ammonia monooxygenase, ligand binding, and suicide inhibition, notably by acetylene. Most inhibitors target ammonia oxidation, but chlorate has been used as a specific inhibitor of nitrite oxidation.

The enormous increases in application of N fertilizers, particularly ammonia-based fertilizers, and other modern farming practices have increased the risk of pollution from nitrification. For example, it is estimated that 67% of applied fertilizer is not taken up by plants, equivalent to a worldwide annual loss of $US15.9 billion (Raun and Johnson, 1999). In addition to the economic cost, this can increase levels of nitrate in groundwaters above regulatory standards, increase greenhouse gas production, through both nitrification and denitrification, and lead to eutrophication of rivers, through run-off of nitrate.

A number of strategies have been proposed and developed to increase fertilizer use efficiency, including use of slow-release fertilizers, more intelligent fertilizer application strategies, and precision farming. An alternative is the application of nitrification inhibitors with fertilizer, which is attracting increasing attention for mitigation of nitrous oxide production (De Klein and Ledgard, 2005; Subbarao et al., 2006; Singh and Verma, 2007). Where possible, inhibitors are added with ammonia- or urea-based fertilizers, although this may be restricted by the characteristics of the fertilizer and inhibitor, while some compounds act as both fertilizer and inhibitor (e.g., ammonium thiosulfate). Choice of inhibitor depends on effectiveness, specificity, volatility, ease of use, solubility, and degradation rate. Effectiveness will depend on a range of factors, including rates of inhibitor degradation, soil type, environmental conditions, and crop type. For example, inhibition will be most effective when nitrification rates are low (e.g., due to low temperature) and for crops with relatively high-ammonium requirements. It is also possible that different members of the nitrifier community may show differences in susceptibility to inhibition (Wrage et al., 2004b), although this has been little studied.

Wolt (2004) assessed long-term effectiveness of inhibitors using, as an example, nitrapyrin application to corn across the midwestern United States, fertilized with inorganic N or manure. Efficiency was assessed in terms of grain yield, maintenance of inorganic N in the root zone, nitrate leaching, and nitrous oxide production. Data from 158 locations indicated that nitrapyrin application led to a 7% increase in crop yield, a 28% increase in soil N retention, a 16% decrease in N leaching, and a 51% decrease in N_2O emissions. Nitrapyrin had a significant effect in 75% of studies analyzed.

"Natural" Inhibitors of Nitrification

In general, nitrification rates in managed agricultural systems are high, leading to low ammonium concentrations and nitrate concentrations. In grasslands and forest ecosystems, ammonium concentrations are high, and nitrate concentrations are low. Low rates may be due to increased demand for ammonium, following release of carbon from plant roots and consequent strong competition for ammonium from vegetation and heterotrophic microorganisms, or to high nitrate assimilation by plants (Stark and Hart, 1997). However, an alternative explanation is the production by plants of nitrification inhibitors in these climax systems. Subbarao et al. (2006) reviewed the extensive literature on potential allelopathic inhibitors of nitrification, particularly phenolics, terpenes, and flavonoids, produced by plants as root exudates or released during degradation of plant material and litter. These soils also have low pH, which itself inhibits nitrification, but there is evidence that plant species effects are independent of pH. The review highlights the many technical problems associated with chemical analysis of these complex compounds in soil and consequent difficulties in demonstrating inhibition in the laboratory and their conversion in soil.

Ammonia oxidation is considered to be a good measure of soil health and fertility (Ritz et al., 2009), and ammonia oxidizers are thought to be more sensitive than the majority of other organisms to toxic compounds. This sensitivity has been exploited in the development of a solid phase, ecotoxicity test involving a bioluminescent reporter strain of *N. europaea*, marked with *luxAB* genes (Brandt et al., 2002). This marker strain has been exploited in the search for inhibitors produced by 18 crops and grasses (Subbarao et al., 2007a). Many inhibitory compounds and root exudates from different genotypes of the tropical grass *Brachiaria humidicola* exhibited a range of inhibitory effects on the *lux*-marked sensor strain, while inhibition levels correlated with nitrification rates in the soils from which genotypes were obtained. Inhibition was also greater in root exudates of a wild relative of wheat (*Leymus racemosus*) than in cultivated wheat (*Triticum aestivum*) (Subbarao et al., 2007b). In both systems, production of inhibitors was stimulated by growth on ammonium, but not nitrate, and genetic improvement of genotypes to increase inhibition was proposed as a management strategy to increase nitrogen utilization efficiency.

CURRENT QUESTIONS AND FUTURE RESEARCH

The soil environment presents a number of challenges and benefits for ammonia oxidizers. The former include intermittent substrate and water supply, spatially and temporally heterogeneous physicochemical conditions, low pH, and strong competition for ammonia from plants and heterotrophic microorganisms. Benefits include considerable increases in anthropogenic supply of ammonia, and therefore nitrite, favorable temperature and oxygen supply, and attachment surfaces that reduce removal. Laboratory studies have increased understanding of the physiological mechanisms that enable nitrifiers to meet the challenges of life in the soil. The major recent advances have been in characterizing the diversity and community structure of ammonia and nitrite oxidizers and, to some extent, uncovering relationships between diversity and environmental distribution, ecosystem function, and physiological diversity. Such studies are, however, in their infancy. Patterns are developing, but we are far from being able to describe soil characteristics or land use management practices on the basis of nitrifier community structure, or to predict the influence of environmental change on nitrifier communities and their ecosystem function. Such understanding will hopefully come from improvements in enrichment and isolation techniques and study of laboratory cultures that are better representatives of natural soil communities. These will enable better ecophysiological studies informed by genomics and other "omics" approaches and by better quantitative data on cellular growth and activity parameters. This will hopefully be paralleled by the development of improved techniques for measuring in situ activity. An obvious requirement is the need to assess the relative importance of bacteria and archaeal ammonia oxidizers and their environmental niches, but it is equally important to assess within-group physiological diversity and, indeed, whether this diversity is truly important for understanding and predicting soil nitrification rates.

Of possibly greater importance is the requirement for conceptual and theoretical approaches that increase our understanding of the links between nitrifier ecology, the soil nitrification process and the soil environment. For the organisms, this requires consideration of the scale at which they interact with their environment and with other organisms. Several of the most important studies described above have advanced understanding by considering how the local microenvironment influences nitrifiers, rather than treating nitrifier populations as suspended cells growing in liquid medium. The extent of interactions among nitrifiers will influence their diversity, and better information is required on functional redundancy if we are to assess the importance of the high levels of soil nitrifier diversity for ecosystem function in changing environments.

Interactions with other functional groups and with other processes are also of enormous significance. This volume focuses on a single process within the nitrogen cycle, but soil nitrification is intimately linked to mineralization, soil organic matter decomposition, and denitrification. More indirectly, it is linked with other important biogeochemical cycles, soil physicochemical processes, and biological processes about which little is known (e.g., control of nitrifier populations by predation and phage). Understanding of nitrifier ecology and activity in microenvironments must be scaled up to the microcosm, mesocosm, and field scales for input into quantitative predictive models. This represents a major challenge in determining the importance of nitrification for atmospheric and groundwater pollution and fertilizer loss and our ability to control and reduce them.

METHODS

The "special" features of soil that influence nitrifier ecology and activity also influence methods used to investigate soil nitrification. The particulate nature of soil and its physicochemical properties make it more difficult to remove cells for isolation and analysis. They also introduce spatial heterogeneity that complicates measurement of activity and the factors that influence activity. Soil also reduces the ease with which modern molecular techniques can be applied, particularly those involving microscopy.

Enrichment and Isolation

Autotrophic ammonia and nitrite oxidizers are enriched by inoculation of mineral salts medium, supplemented with ammonia or nitrite, with soil. A pH indicator is often added to detect acid production by ammonia oxidizers, but growth should be confirmed by determination of ammonia, nitrite, and/or nitrate concentration. Growth usually occurs within several weeks. Isolation of pure cultures is achieved by dilution to extinction and further subculture in inorganic medium or subculturing of isolated colonies growing on solid medium. Isolation is difficult. Heterotrophic contaminants grow much more quickly than autotrophs, utilizing organic byproducts of autotroph growth and volatile organics. Growth is slow in liquid and on solid media, yields are low, attempts to increase yield by increasing ammonia or nitrite concentrations lead to substrate inhibition, and colonies produced after incubation on solid media are microscopic. Isolation can take several months, and cultures can be unstable, surviving only a limited number of subcultures. More importantly, enrichment and isolation processes are selective, and the majority of isolates are not representative of the dominant members of natural nitrifier communities. The extent of this problem has been highlighted by molecular analysis of enrichments, isolates, and soil DNA. In one study, 31 soil enrichments were dominated by the *Nitrosospira* cluster 3 strains, while sequences of 50 environmental clones from the same soil were distributed among *Nitrosospira* clusters 2 to 4 (Smith et al., 2001). Only 16% of soil enrichment culture sequences were identical to those in the sequenced clones. More importantly, no crenarchaeal ammonia oxidizer has yet been isolated from soil, despite significantly higher abundance of crenarchaeal *amoA* genes, and nitrite oxidizer isolates are dominated by *Nitrobacter*, despite indications of the importance of *Nitrospira* in soil (Freitag et al., 2005).

Abundance

Cultivation-based enumeration of nitrifiers uses the MPN method, as slow growth and formation of only microscopic colonies on solid medium prevent routine use of plate-counting methods. Dilutions of a soil suspension are used for multiple inoculation of liquid medium, and growth is determined qualitatively after incubation, typically for several weeks. This provides estimation of abundance of either, or both, ammonia or nitrite oxidizers, depending on the composition of the medium, but it suffers from a number of disadvantages. Calculation of MPN counts is based on probabilities of transferring cells during inoculation and has intrinsic statistical variation, in addition to experimental variability, although it can be reduced by reducing dilution factors and increasing replication of inocula. Other disadvantages have been discussed in earlier sections.

Increasingly, quantitative molecular methods are being used to enumerate nitrifiers. Soil ammonia oxidizers have been quantified by competitive PCR of 16S rRNA genes (Phillips et al., 2000; Hermansson et al., 2004) and real-time PCR of 16S rRNA and *amoA* genes (Okano et al., 2004; Leininger et al., 2006; Nicol et al., 2008). This removes major disadvantages of cultivation-based techniques: selective growth and inability to grow under laboratory conditions. Direct comparisons of MPN and qPCR counts indicate underestimation by the former by 1 to 2 orders of magnitude and selection for *Nitrosomonas* over *Nitrosospira* in medium containing higher ammonia concentrations. Combining MPN and qPCR can be used to quantify less abundant phylotypes, but this is best achieved using primers specific for different groups. There are no published reports of archaeal ammonia oxidizers in MPN enumeration media, but comparison of bacterial and archaeal *amoA* gene abundance has provided one of the major lines of evidence for the importance of archaeal ammonia oxidation in soil.

Although qPCR methods are gaining widespread acceptance for estimating gene abundance, they also have biases and limitations. Cells may contain several copies of the target gene; cell lysis and DNA extraction efficiencies will vary between different groups, and primer efficiencies will vary (Smith et al., 2006). It is also difficult to verify abundance data by direct microscopy. Although immunological and fluorescent in situ hybridization (FISH) techniques are available for nitrifying bacteria, the disadvantages of fluorescence-based techniques in soil and their low abundance per-unit surface area prevent routine quantification by microscopic methods in soil. This is unfortunate, given the important information that this approach has provided in activated sludge and marine studies of nitrification.

Community Composition

Diversity of nitrifiers is now determined routinely by molecular analysis of 16S rRNA or functional genes amplified from DNA or RNA extracted from soil. Primers are available for 16S rRNA genes of major groups of soil bacterial ammonia and nitrite oxidizers, and archaeal primers can often be used to assess crenarchaeal ammonia soil communities, because of low relative abundance of non-ammonia-oxidizing crenarchaea. Nitrifiers have also been characterized by analysis of the functional genes *amoA*, *amoB*, hydroxylamine oxidoreductase (*hao*), cytochrome c-554 (*hcy*) (Bruns et al., 1998) and *ureC* (Koper et al., 2004) (for ammonia oxidizers), and *nxrA* (nitrite oxidizers) (Wertz et al., 2008). Unfortunately, there are no 16S rRNA gene primers or functional gene primers that encompass all ammonia oxidizers (e.g., bacterial and archaeal) or all nitrite oxidizers (e.g., *Nitrobacter*, *Nitrospira*, and *Nitrotoga*). This complicates analysis of total functional group communities, unless reliable data are available on abundance of different functional subpopulations.

Detailed information on community composition is obtained by sequencing representatives of clone libraries of amplified genes and provides the basis for phylogenetic analysis of natural communities. Relative abundances of clones within different phylogenetic groups give some information on the influence of environmental factors on communities, but the requirement for sequencing large numbers of clones to achieve reasonable coverage and for analysis of replicate clone libraries have encouraged use of fingerprinting techniques. DGGE has been used most frequently, but terminal restriction fragment length polymorphism, temperature gradient gel electrophoresis, and SSCP methods have also been applied. These allow analysis of a greater proportion of sequence information, enable quantification of relative abundances of different sequence types, and are sufficiently cheap and rapid to enable necessary replication. Sequence data for DGGE bands may be obtained from excised bands, and fingerprinting can be used to analyze clone libraries and identify clones associated with specific phylotypes to infer sequence identity. Sequence data can be used to identify nitrifiers by comparison with database sequences. Databases contain considerable numbers of 16S rRNA gene sequences of betaproteobacterial ammonia oxidizers, but much less for nitrite oxidizers. The database for *amoA* gene sequences is also now large and expanding rapidly, following extensive environmental sequencing surveys after discovery of crenarchaeal ammonia oxidizers. Analysis of community structure will benefit from the new generation of molecular techniques. 16S rRNA gene and functional gene microarray systems include many nitrifier sequences, and high-throughput sequencing approaches will allow more in-depth characterization of community composition, potentially replacing fingerprinting approaches as costs decrease.

Nitrifier Activity

Growth parameters of soil ammonia oxidizers have been determined in laboratory culture, during batch or chemostat (substrate-limited) growth or in cell suspensions. Increasingly, molecular techniques are being used for "in situ physiological" studies. For example, combined measurement of nitrification rates and qPCR analysis of cell abundance can be used to determine in situ cellular activity. Use of rRNA- rather than DNA-targeted analysis of 16S rRNA genes indicates which ammonia oxidizers are active and responsive to changing environmental conditions. Quantification of *amoA* gene transcripts by qPCR and molecular analysis of *amoA* gene transcripts have been used to determine levels of transcriptional activity and to determine which ammonia oxidizers are active in soil. Stable isotope probing, involving incubation with ^{13}C-labeled CO_2 and subsequent molecular analysis of labeled and unlabeled nucleic acids, has been applied to estuarine nitrification (Freitag et al., 2006) and, more recently, to soil (Jia and Conrad, 2009). Other approaches are more difficult to transfer to the soil. For example, microautoradiography combined with FISH (MAR-FISH) and other FISH-based techniques suffer from difficulties associated with fluorescence microscopy of soil microorganisms.

Process Measurements

Potential nitrification rates can be determined relatively easily by measuring changes in concentrations of ammonia, nitrite, and nitrate at constant temperature and amending, if necessary, with ammonia. Denitrification is minimized by maintaining aerobic conditions (e.g., by using shaken soil slurries). In general, short-term incubations with nonlimiting substrate concentration and no significant growth lead to zero-order kinetics. Potential nitrification activity is the maximum nitrification rate possible for a particular soil and has traditionally been used as a measure of nitrifier biomass, although this represents a composite of biomass and activity. Kinetics are more complex if substrate concentrations are low or if extended incubation leads to nitrifier growth, generating Michaelis-Menten kinetics or exponential increases in product concentration, respectively. Use of logistic kinetics is described above.

Complexity is increased by links to other nitrogen cycle processes. Ammonia concentration will be increased by decomposition of organic matter, nitrate concentration will be reduced through denitrification, and both will be reduced through uptake by plants and assimilation by heterotrophic microorganisms. Assimilation will, in turn, depend on available organic carbon. Denitrification will be reduced under aerobic conditions used to measure nitrification, but all of these processes will interfere with rate measurements. Their effects are determined in measurements of net nitrification rate (nitrate production minus nitrate consumption) and gross nitrification rate (nitrate production plus nitrate consumption). This requires use of stable isotope methods, which are also required when ammonia concentrations are low, to increase sensitivity. The rate of production of $^{15}NO_3^-$ from added $^{15}NH_4^+$ may be measured, although ammonia addition may stimulate nitrification, and $^{15}NH_4^+$ may be diluted by mineralization (Hart et al., 1994). Alternatively, in the pool dilution method, nitrification leads to dilution of added $^{15}NO_3^-$ by unlabeled nitrate (Barraclough and Puri, 1995). Heterotrophic nitrification rates can be assessed by comparison of rates in the presence and absence of specific inhibitors of ammonia oxidation. Use of double-labeled compounds ($^{14}NH_4{}^{15}NO_3$ or $^{15}NH_4{}^{15}NO_3$), measurement of $^{15}N/^{14}N$ and $^{18}O/^{16}O$ isotope ratios, and isotopomer analysis have also been used to investigate other nitrogen cycle processes, particularly nitrous oxide production (Sutka et al., 2006; Ostrom et al., 2007). Stable isotope techniques are therefore used for accurate measurement of nitrification rates. They do not involve addition of significant amounts of substrate, enabling measurement of in situ rates. They can also distinguish between autotrophic heterotrophic nitrification and enable measurement of associated processes, such as nitrous oxide production and nitrifier denitrification.

Model Systems and Microcosms

Soil nitrification has been studied in a range of experimental systems, which attempt to mimic the soil system, but with greater control and monitoring, or to isolate particular environmental factors of interest. Chemostats and continuous flow systems containing suspended particles have been used to determine growth parameters and to study biofilm growth and activity of ammonia and nitrite oxidizers. Packed column reactors containing defined particulate material or soil have been used to investigate nitrification kinetics, effects of inhibitors, and the influence of specific environmental factors. These systems have also provided important information required for mathematical modeling and model parameterization.

REFERENCES

Aakra, A., M. Hesselsoe, and L. R. Bakken. 2000. Surface attachment of ammonia-oxidizing bacteria in soil. *Microb. Ecol.* **39:**222–235.

Aakra, A., J. B. Utåker, and I. F. Nes. 2001. Comparative phylogeny of the ammonia monooxygenase subunit A and 16S rRNA genes of ammonia-oxidizing bacteria. *FEMS Microbiol. Lett.* **205:**237–242.

Allison, S. M., and J. I. Prosser. 1991a. Survival of ammonia oxidising bacteria in air-dried soil. *FEMS Microbiol. Lett.* **79:**65–68.

Allison, S. M., and J. I. Prosser. 1991b. Urease activity in neutrophilic autotrophic ammonia-oxidizing bacteria isolated from acid soils. *Soil. Biol. Fertil.* **23:**45–51.

Allison, S. M., and J. I. Prosser. 1993. Ammonia oxidation at low pH by attached populations of nitrifying bacteria. *Soil Biol. Biochem.* **25:**935–941.

Armstrong, E. F., and J. I. Prosser. 1988. Growth of *Nitrosomonas europaea* on ammonia-treated vermiculite. *Soil Biol. Biochem.* **20:**409–411.

Arp, D. J., and L. Y. Stein. 2003. Metabolism of inorganic N compounds by ammonia-oxidizing bacteria. *Crit. Rev. Biochem. Mol. Biol.* **38:**471–495.

Avrahami, S., and B. J. M. Bohannan. 2007. Response of *Nitrosospira* sp. strain AF-like ammonia oxidizers to changes in temperature, soil moisture content, and fertilizer concentration. *Appl. Environ. Microbiol.* **73:**1166–1173.

Avrahami, S., and R. Conrad. 2003. Patterns of community change among ammonia oxidizers in meadow soils upon long-term incubation at different temperatures. *Appl. Environ. Microbiol.* **69:**6152–6164.

Avrahami, S., and R. Conrad. 2005. Cold-temperate climate: a factor for selection of ammonia oxidizers in upland soil? *Can. J. Microbiol.* **51:**709–714.

Avrahami, S., R. Conrad, and G. Braker. 2002. Effect of soil ammonium concentration on N_2O release and on the community structure of ammonia oxidizers and denitrifiers. *Appl. Environ. Microbiol.* **68:**5685–5692.

Avrahami, S., W. Liesack, and R. Conrad. 2003. Effects of temperature and fertilizer on activity and community structure of soil ammonia oxidizers. *Environ. Microbiol.* **5:**691–705.

Backman, J. S. K., A. Hermansson, C. C. Tebbe, and P. E. Lindgren. 2003. Liming induces growth of a diverse flora of ammonia-oxidising bacteria in acid spruce forest soil as determined by SSCP and DGGE. *Soil Biol. Biochem.* **35:**1337–1347.

Bäckman, J. S. K., A. K. Klemedtsson, L. Klemedtsson, and P. E. Lindgren. 2004. Clear-cutting affects the ammonia-oxidising community differently in limed and non-limed coniferous forest soils. *Biol. Fertil. Soils* **40:**260–267.

Baker-Austin, C., and M. Dopson. 2007. Life in acid: pH homeostasis in acidophiles. *Trends Microbiol.* **15**: 165–171.

Barnard, R., L. Barthes, X. Le Roux, and P. W. Leadley. 2004. Dynamics of nitrifying activities, denitrifying activities and nitrogen in grassland mesocosms as altered by elevated CO_2. *New Phytol.* **162:**365–376.

Barnard, R., X. Le Roux, B. A. Hungate, E. E. Cleland, J. C. Blankinship, L. Barthes, and P. W. Leadley. 2006. Several components of global change alter nitrifying and denitrifying activities in an annual grassland. *Funct. Ecol.* **20:**557–564.

Batchelor, S. E., M. Cooper, S. R. Chhabra, L. A. Glover, G. S. A. B. Stewart, P. Williams, and J. I. Prosser. 1997. Cell density-regulated recovery of starved biofilm populations of ammonia-oxidizing bacteria. *Appl. Environ. Microbiol.* **63:**2281–2286.

Bollmann, A., I. Schmidt, A. M. Saunders, and M. H. Nicolaisen. 2005. Influence of starvation on potential ammonia-oxidizing activity and *amoA* mRNA levels of *Nitrosospira briensis*. *Appl. Environ. Microbiol.* **71:**1276–1282.

Booth, M. S., J. M. Stark, and E. Rastetter. 2005. Controls on nitrogen cycling in terrestrial ecosystems: a synthetic analysis of literature data. *Ecol. Monogr.* **75:**139–157.

Bottomley, P. J., A. E. Taylor, S. A. Boyle, S. K. McMahon, J. J. Rich, K. Cromack Jr., and D. D. Myrold. 2004. Responses of nitrification and ammonia-oxidizing bacteria to reciprocal transfers of soil between adjacent coniferous forest and meadow vegetation in the Cascade Mountains of Oregon. *Microb. Ecol.* **48:**500–508.

Boyle-Yarwood, S. A., P. J. Bottomley, and D. D. Myrold. 2008. Community composition of ammonia-oxidizing bacteria and archaea in soils under stands of red alder and Douglas fir in Oregon. *Environ. Microbiol.* **10:**2956–2965.

Brandt, K. K., A. Pedersen, and J. Sorensen. 2002. Solid-phase contact assay that uses a lux-marked *Nitrosomonas europaea* reporter strain to estimate toxicity of bioavailable linear alkylbenzene sulfonate in soil. *Appl. Environ. Microbiol.* **68:**3502–3508.

Bruns, M. A., M. R. Fries, J. M. Tiedje, and E. A. Paul. 1998. Functional gene hybridization patterns of terrestrial ammonia oxidizing bacteria. *Microb. Ecol.* **36:**293–302.

Bruns, M. A., J. R. Stephen, G. A. Kowalchuk, J. I. Prosser, and E. A. Paul. 1999. Comparative diversity of ammonia oxidizer 16S rRNA gene sequences in native, tilled, and successional soils. *Appl. Environ. Microbiol.* **65:**2994–3000.

Burton, S. A. Q., and J. I. Prosser. 2001. Autotrophic ammonia oxidation at low pH through urea hydrolysis. *Appl. Environ. Microbiol.* **67:**2952–2957.

Cantera, J. J. L., F. L. Jordan, and L. Y. Stein. 2006. Effects of irrigation sources on ammonia-oxidizing bacterial communities in a managed turf-covered aridisol. *Biol. Fertil. Soils* **43:**247–255.

Cantera, J. J. L., and L. Y. Stein. 2007. Molecular diversity of nitrite reductase genes (*nirK*) in nitrifying bacteria. *Environ. Microbiol.* **9:**765–776.

Carnol, M., L. Hogenboom, M. E. Jach, J. Remacle, and R. Ceulemans. 2002. Elevated atmospheric CO_2 in open top chambers increases net nitrification and potential denitrification. *Global Change Biol.* **8:**590–598.

Carnol, M., G. A. Kowalchuk, and W. De Boer. 2002. *Nitrosomonas europaea*-like bacteria detected as the dominant beta-subclass proteobacteria ammonia oxidisers in reference and limed acid forest soils. *Soil Biol. Biochem.* **34:**1047–1050.

Casciotti, K. L., and B. B. Ward. 2005. Phylogenetic

analysis of nitric oxide reductase gene homologues from aerobic ammonia-oxidizing bacteria. *FEMS Microbiol. Ecol.* **52:**197–205.

Chen, D., Y. Li, P. Grace, and A. R. Mosier. 2008. N_2O emissions from agricultural lands: A synthesis of simulation approaches. *Plant Soil* **309:**169–189.

Daims, H., P. H. Nielsen, J. L. Nielsen, S. Juretschko, and M. Wagner. 2000. Novel *Nitrospira*-like bacteria as dominant nitrite-oxidizers in biofilms from wastewater treatment plants: Diversity and in situ physiology. *Water Sci. Technol.* **41:**85–90.

De Boer, W., and G. A. Kowalchuk. 2001. Nitrification in acid soils: micro-organisms and mechanisms. *Soil Biol. Biochem.* **33:**853–866.

De Boer, W., P. J. A. K. Gunnewiek, M. Veenhuis, E. Bock, and H. J. Laanbroek. 1991. Nitrification at low pH by aggregated chemolithotrophic bacteria. *Appl. Environ. Microbiol.* **57:**3600–3604.

De Klein, C. A. M., and S. F. Ledgard. 2005. Nitrous oxide emissions from New Zealand agriculture - key sources and mitigation strategies. *Nutr. Cycling Agroecosyst.* **72:**77–85.

Dell, E. A., D. Bowman, T. Rufty, and W. Shi. 2008. Intensive management affects composition of betaproteobacterial ammonia oxidizers in turfgrass systems. *Microbial Ecol.* **56:**178–190.

Dundee, L., and D. W. Hopkins. 2001. Different sensitivities to oxygen of nitrous oxide production by *Nitrosomonas europaea* and *Nitrosolobus multiformis*. *Soil Biol. Biochem.* **33:**1563–1565.

Fierer, N., and R. B. Jackson. 2006. The diversity and biogeography of soil bacterial communities. *Proc. Natl. Acad. Sci. USA* **103:**626–631.

Frankland, P. F., and G. Frankland. 1890. The nitrifying process and its specific ferment. *Trans. R. Soc. (London)* **B181**:107.

Freitag, T. E., and J. I. Prosser. 2009. Correlation of methane production and functional gene transcriptional activity in a peat soil. *Appl. Environ. Microbiol.* 75:6679–6687.

Freitag, T. E., L. Chang, C. D. Clegg, and J. I. Prosser. 2005. Influence of inorganic nitrogen-management regime on the diversity of nitrite oxidizing bacteria in agricultural grassland soils. *Appl. Environ. Microbiol.* **71:**8323–8334.

Freitag, T. E., L. Chang, and J. I. Prosser. 2006. Changes in the community structure and activity of betaproteobacterial ammonia-oxidizing sediment bacteria along a freshwater-marine gradient. *Environ. Microbiol.* **8:**684–696.

Garbeva, P., E. M. Baggs, and J. I. Prosser. 2007. Phylogeny of nitrite reductase (*nirK*) and nitric oxide reductase (*norB*) genes from *Nitrosospira* species isolated from soil. *FEMS Microbiol. Lett.* **266:**83–89.

Gödde, M., and R. Conrad. 1999. Immediate and adaptational temperature effects on nitric oxide production and nitrous oxide release from nitrification and denitrification in two soils. *Biol. Fertil. Soils* **30:**33–40.

Goreau, T. J., W. A. Kaplan, and S. C. Wofsy. 1980. Production of NO_2^- and N_2O by nitrifying bacteria at reduced concentrations of oxygen. *Appl. Environ. Microbiol.* **40:**526–532.

Green, M., M. Beliavski, N. Denekamp, A. Gieseke, D. De Beer, and S. Tarre. 2006. High nitrification rate at low pH in a fluidized bed reactor with either chalk or sintered glass as the biofilm carrier. *Israel J. Chem.* **46:**53–58.

Groffman, P. M., C. W. Rice, and J. M. Tiedje. 1993. Denitrification in a tallgrass prairie landscape. *Ecol.* **74:**855–862.

Gruber, N., and J. N. Galloway. 2008. An earth-system perspective of the global nitrogen cycle. *Nature* **451:**293–296.

Grundmann, G. L. 2004. Spatial scales of soil bacterial diversity - the size of a clone. *FEMS Microbiol. Ecol.* **48:**119–127.

Grundmann, G. L., and D. Debouzie. 2000. Geostatistical analysis of the distribution of NH_4^+- and NO_2^--oxidizing bacteria and serotypes at the millimeter scale along a soil transect. *FEMS Microbiol. Ecol.* **34:**57–62.

Grundmann, G. L., and P. Normand. 2000. Microscale diversity of the genus *Nitrobacter* in soil on the basis of analysis of genes encoding rRNA. *Appl. Environ. Microbiol.* **66:**4543-4546.

Grundmann, G. L., P. Renault, L. Rosso, and R. Bardin. 1995. Differential effects of soil water content and temperature on nitrification and aeration. *Soil Sci. Soc. Am. J.* **59:**1342–1349.

Grundmann, G. L., A. Dechesne, F. Bartoli, J. P. Flandrois, J. L. Chasse, and R. Kizungu. 2001. Spatial modeling of nitrifier microhabitats in soil. *Soil Sci. Soc. Am .J.* **65:**1709–1716.

Hawkes, C. V., I. F. Wren, D. J. Herman, and M. K. Firestone. 2005. Plant invasion alters nitrogen cycling by modifying the soil nitrifying community. *Ecol. Lett.* **8:**976–985.

Hermansson, A., J. S. K. Bäckman, B. H. Svensson, and P.-E. Lindgren. 2004. Quantification of ammonia-oxidising bacteria in limed and non-limed acidic coniferous forest soil using real-time PCR. *Soil. Biol. Fertil.* **36:**1935–1941.

Hommes, N. G., S. A. Russell, P. J. Bottomley, and D. J. Arp. 1998. Effects of soil on ammonia, ethylene, chloroethane, and 1,1,1-trichloroethane oxidation by *Nitrosomonas europaea*. *Appl. Environ. Microbiol.* **64:**1372–1378.

Horz, H. P., A. Barbrook, C. B. Field, and B. J. Bohannan. 2004. Ammonia-oxidizing bacteria respond to multifactorial global change. *Proc. Natl. Acad. Sci. USA* **101:**15136–15141.

Jia, Z. and R. Conrad. 2009. Bacteria rather than Archaea dominate microbial ammonia oxidation in an agricultural soil. *Environ. Microbiol.* **11:**1658–1671.

Jiang, Q. Q., and L. R. Bakken. 1999a. Comparison of *Nitrosospira* strains isolated from terrestrial environments. *FEMS Microbiol. Ecol.* **30:**171–186.

Jiang, Q. Q., and L. R. Bakken. 1999b. Nitrous oxide production and methane oxidation by different ammonia-oxidizing bacteria. *Appl. Environ. Microbiol.* **65:**2679–2684.

Johnson, S. L., C. R. Budinoff, J. Belnap, and F. Garcia-Pichel. 2005. Relevance of ammonium oxidation within biological soil crust communities. *Environ. Microbiol.* **7:**1–12.

Jordan, F. L., J. J. L. Cantera, M. E. Fenn, and L. Y. Stein. 2005. Autotrophic ammonia oxidizing bacteria contribute minimally to nitrification in a nitrogen-impacted forested ecosystem. *Appl. Environ. Microbiol.* **71:**197–206.

Keen, G. A., and J. I. Prosser. 1987. Interrelationship between pH and surface growth of *Nitrobacter*. *Soil. Biol. Fertil.* **19:**665–672.

Kester, R. A., W. De Boer, and H. J. Laanbroek. 1997. Production of NO and N_2O by pure cultures of nitrifying and denitrifying bacteria during changes in aeration. *Appl. Environ. Microbiol.* **63:**3872–3877.

Klemedtsson, L., Q. Jiang, A. K. Klemedtsson, and L. Bakken. 1999. Autotrophic ammonium-oxidising bacteria in Swedish mor humus. *Soil Biol. Biochem.* **31:**839–847.

Könneke, M., A. E. Bernhard, J. R. De La Torre, C. B. Walker, J. B. Waterbury, and D. A. Stahl. 2005. Isolation of an autotrophic ammonia-oxidizing marine archaeon. *Nature* **437:**543–546.

Kool, D. M., N. Wrage, O. Oenema, J. Dolfing, and J. W. Van Groenigen. 2007. Oxygen exchange between (de)nitrification intermediates and H_2O and its implications for source determination of NO_3^- and N_2O: a review. *Rapid Comm. Mass Spec.* **21:**3569–3578.

Koops, H. P., and A. Pommerening-Röser. 2001. Distribution and ecophysiology of the nitrifying bacteria emphasizing cultured species. *FEMS Microbiol. Ecol.* **37:**1–9.

Koper, T. E., A. F. El-Sheikh, J. M. Norton, and M. G. Klotz. 2004. Urease-encoding genes in ammonia-oxidizing bacteria. *Appl. Environ. Microbiol.* **70:**2342–2348.

Kowalchuk, G. A., and J. R. Stephen. 2001. Ammonia-oxidizing bacteria: A model for molecular microbial ecology. *Annu. Rev. Microbiol.* **55:**485–529.

Kowalchuk, G. A., Z. S. Naoumenko, P. J. L. Derikx, A. Felske, J. R. Stephen, and I. A. Arkhipchenko. 1999. Molecular analysis of ammonia-oxidizing bacteria of the beta subdivision of the class proteobacteria in compost and composted materials. *Appl. Environ. Microbiol.* **65:**396–403.

Kowalchuk, G. A., A. W. Stienstra, G. H. Heilig, J. R. Stephen, and J. W. Woldendorp. 2000. Changes in the community structure of ammonia-oxidizing bacteria during secondary succession of calcareous grasslands. *Environ. Microbiol.* **2:**99–110.

Kyveryga, P. M., A. M. Blackmer, J. W. Ellsworth, and R. Isla. 2004. Soil pH effects on nitrification of fall-applied anhydrous ammonia. *Soil Sci. Soc. Am. J.* **68:**545–551.

Laverman, A. M., A. G. C. L. Speksnijder, M. Braster, G. A. Kowalchuk, H. A. Verhoef, and H. W. van Verseveld. 2001. Spatiotemporal stability of an ammonia-oxidizing community in a nitrogen-saturated forest soil. *Microb. Ecol.* **42:**35–45.

Le Roux, X., M. Bardy, P. Loiseau, and F. Louault. 2003. Stimulation of soil nitrification and denitrification by grazing in grasslands: do changes in plant species composition matter? *Oecologia* **137:**417–425.

Le Roux, X., F. Poly, P. Currey, C. Commeaux, B. Hai, G. W. Nicol, J. I. Prosser, M. Schloter, E. Attard, and K. Klumpp. 2008. Effects of aboveground grazing on coupling among nitrifier activity, abundance and community structure. *ISME J.* **2:**221–232.

Leininger, S., T. Urich, M. Schloter, L. Schwark, J. Qi, G. W. Nicol, J. I. Prosser, S. C. Schuster, and C. Schleper. 2006. Archaea predominate among ammonia-oxidizing prokaryotes in soils. *Nature* **442:**806–809.

Low, A. P., J. M. Stark, and L. M. Dudley. 1997. Effects of soil osmotic potential on nitrification, ammonification, N-assimilation, and nitrous oxide production. *Soil Sci.* **162:**16–27.

Mahmood, S., and J. I. Prosser. 2006. The influence of synthetic sheep urine on ammonia oxidizing bacterial communities in grassland soil. *FEMS Microbiol. Ecol.* **56:**444–454.

Maixner, F., D. R. Noguera, B. Anneser, K. Stoecker, G. Wegl, M. Wagner, and H. Daims. 2006. Nitrite concentration influences the population structure of *Nitrospira*-like bacteria. *Environ. Microbiol.* **8:**1487–1495.

Matulewich, V. A., P. F. Strom, and M. S. Finstein. 1975. Length of incubation for enumerating nitrifying bacteria present in various environments. *Appl. Microbiol.* **29:**265–268.

Mendum, T. A., and P. R. Hirsch. 2002. Changes in the population structure of beta-group autotrophic ammonia oxidising bacteria in arable soils in response to agricultural practice. *Soil Biol. Biochem.* **34:**1479–1485.

Mendum, T. A., R. E. Sockett, and P. R. Hirsch. 1999. Use of molecular and isotopic techniques to monitor the response of autotrophic ammonia-oxidizing populations of the beta subdivision of the class proteobacteria in arable soils to nitrogen fertilizer. *Appl. Environ. Microbiol.* **65:**4155–4162.

Mintie, A. T., R. S. Heichen, K. Cromack, D. D. Myrold, and P. J. Bottomley. 2003. Ammonia-oxidizing bacteria along meadow-to-forest transects in the Oregon cascade mountains. *Appl. Environ. Microbiol.* **69:**3129–3136.

Molina, J. A. E. 1985. Components of rates of ammonium oxidation in soil. *Soil Soc. Am. J.* **49:**603–609.

Nicol, G. W., S. Leininger, C. Schleper, and J. I. Prosser. 2008. The influence of soil pH on the diversity, abundance and transcriptional activity of ammonia oxidizing archaea and bacteria. *Environ. Microbiol.* **10:**2966–2978.

Okano, Y., K. R. Hristova, C. M. Leutenegger, L. E. Jackson, R. F. Denison, B. Gebreyesus, D. Lebauer, and K. M. Scow. 2004. Application of real-time PCR to study effects of ammonium on population size of ammonia-oxidizing bacteria in soil. *Appl. Environ. Microbiol.* **70:**1008–1016.

Ostrom, N. E., A. Piit, R. Sutka, P. H. Ostrom, A. S. Grandy, K. M. Huizinga, and G. P. Robertson. 2007. Isotopologue effects during N_2O reduction in soils and in pure cultures of denitrifiers. *J. Geophys. Res.* **112.** doi:/10.1029/2006JG000287.

Oved, T., A. Shaviv, T. Goldrath, R. T. Mandelbaum, and D. Minz. 2001. Influence of effluent irrigation on community composition and function of ammonia-oxidizing bacteria in soil. *Appl. Environ. Microbiol.* **67:**3426–3433.

Patra, A. K., L. Abbadie, A. Clays-Josserand, V. Degrange, S. J. Grayston, P. Loiseau, F. Louault, S. Mahmood, S. Nazaret, L. Philippot, F. Poly, J.I. Prosser, A. Richaume, and X. Le Roux. 2005. Effects of grazing on microbial functional groups involved in soil N dynamics. *Ecol. Monogr.* **75:**65–80.

Patra, A. K., L. Abbadie, A. Clays-Josserand, V. Degrange, S. J. Grayston, N. Guillaumaud, P. Loiseau, F. Louault, S. Mahmood, S. Nazaret, L. Philippot, F. Poly, J. I. Prosser, and X. Le Roux. 2006. Effects of management regime and plant species on the enzyme activity and genetic structure of N-fixing, denitrifying and nitrifying bacterial communities in grassland soils. *Environ. Microbiol.* **8:**1005–1016.

Phillips, C. J., D. Harris, S. L. Dollhopf, K. L. Gross, J. I. Prosser, and E. A. Paul. 2000a. Effects of agronomic treatments on structure and function of ammonia-oxidizing communities. *Appl. Environ. Microbiol.* **66:**5410–5418.

Phillips, C. J., E. A. Paul, and J. I. Prosser. 2000b. Quantitative analysis of ammonia oxidising bacteria using competitive PCR. *FEMS Microbiol. Ecol.* **32:**167–175.

Poly, F., S. Wertz, E. Brothier, and V. Degrange. 2008. First exploration of *Nitrobacter* diversity in soils by a PCR cloning-sequencing approach targeting functional gene *nxrA*. *FEMS Microbiol. Ecol.* **63:**132–140.

Powell, S. J., and J. I. Prosser. 1986. Inhibition of ammonium oxidation by nitrapyrin in soil and liquid culture. *Appl. Environ. Microbiol.* **52:**782–787.

Powell, S. J., and J. I. Prosser. 1991. Protection of *Nitrosomonas europaea* colonizing clay minerals from inhibition by nitrapyrin. *J. Gen. Microbiol.* **137:**1923–1929.

Powell, S. J., and J. I. Prosser. 1992. Inhibition of biofilm populations of *Nitrosomonas europaea*. *Microb. Ecol.* **24:**43–50.

Prosser, J. I. 1989. Autotrophic nitrification in bacteria. *Adv. Microb. Physiol.* **30:**125–181.

Prosser, J. I., and G. W. Nicol. 2008. Relative contributions of archaea and bacteria to aerobic ammonia oxidation in the environment. *Environ. Microbiol.* **10:**2931–2941.

Purkhold, U., A. Pommerening-Röser, S. Juretschko, M. C. Schmid, H. P. Koops, and M. Wagner. 2000. Phylogeny of all recognized species of ammonia oxidizers based on comparative 16S rRNA and *amoA* sequence analysis: implications for molecular diversity surveys. *Appl. Environ. Microbiol.* **66:**5368–5382.

Raun, W. R., and G. V. Johnson. 1999. Improving nitrogen use efficiency for cereal production. *Agron. J.* **91**: 357–363.

Remde, A., and R. Conrad. 1991. Role of nitrification and denitrification for NO metabolism in soil. *Biogeochemistry* **12:**189–205.

Ritz, K., H. I. J. Black, C. D. Campbell, J. A. Harris, and C. Wood. 2009. Selecting biological indicators for monitoring soils: a framework for balancing scientific and technical opinion to assist policy development. *Ecol. Ind.* **9:**1212–1221.

Robertson, G. P., and J. M. Tiedje. 1987. Nitrous oxide sources in aerobic soils: nitrification, denitrification and other biological processes. *Soil. Biol. Fertil.* **19:**187–193.

Rodgers, G. A., and J. Ashworth. 1982. Bacteriostatic action of nitrification inhibitors. *Can. J. Microbiol.* **28:**1093–1100.

Ross, D. S., B. C. Wemple, A. E. Jamison, G. Fredriksen, J. B. Shanley, G. B. Lawrence, S. W. Bailey, and J. L. Campbell. 2009. A cross-site comparison of factors influencing soil nitrification rates in northeastern USA forested watersheds. *Ecosystems* **12:**158–178.

Roux-Michollet, D., S. Czarnes, B. Adam, D. Berry, C. Commeaux, N. Guillaumaud, X. Le Roux, and A. Clays-Josserand. 2008. Effects of steam disinfestation on community structure, abundance and activity of heterotrophic, denitrifying and nitrifying bacteria in an organic farming soil. *Soil. Biol. Fertil.* **40:**1836–1845.

Russell, C. A., I. R. P. Fillery, N. Bootsma, and K. J. McInnes. 2002. Effect of temperature and nitrogen source on nitrification in a sandy soil. *Commun. Soil Sci. Plant. Anal.* **33:**1975–1989.

Schauss, K., A. Focks, S. Leininger, A. Kotzerke, H. Heuer, S. Thiele-Bruhn, S. Sharma, B.-M. Wilke, M. Matthies, K. Smalla, J. C. Munch, W. Amelung, M. Kaupenjohann, M. Schloter, and C. Schleper. 2009. Dynamics and functional relevance of ammonia-oxidizing archaea in two agricultural soils. *Environ. Microbiol.* **11:**446–456.

Schloesing, J., and A. Muntz. 1877a. Sur la nitrification par les ferments organises. *Comptes Rend. Acad. Sci. Paris* **84:**301–303.

Schloesing, J., and A. Muntz. 1877b. Sur la nitrification par les ferments organises. *Comptes Rend. Acad. Sci. Paris* **85:**1018.

Schloesing, J., and A. Muntz. 1879. Sur la nitrification par les ferments organises. *Comptes Rend. Acad. Sci, Paris* **87:**1074.

Schmidt, C. S., K. A. Hultman, D. Robinson, K. Killham, and J. I. Prosser. 2007. PCR profiling of ammonia-oxidizer communities in acidic soils subjected to nitrogen and sulphur deposition. *FEMS Microbiol. Ecol.* **61:**305–316.

Shaw, L. J., G. W. Nicol, Z. Smith, J. Fear, J. I. Prosser, and E. M. Baggs. 2006. *Nitrosospira* spp. can produce nitrous oxide via a nitrifier denitrification pathway. *Environ. Microbiol.* **8:**214–222.

Shi, W., H. Yao, and D. Bowman. 2006. Soil microbial biomass, activity and nitrogen transformations in a turfgrass chronosequence. *Soil. Biol. Fertil.* **38:**311–319.

Singh, S. N., and A. Verma. 2007. The potential of nitrification inhibitors to manage the pollution effect of nitrogen fertilizers in agricultural and other soils: a review. *Environ. Pract.* **9:**266–279.

Sitaula, B. K., J. I. B. Sitaula, A. Aakra, and L. R. Bakken. 2001. Nitrification and methane oxidation in forest soil: acid deposition, nitrogen input and plant effects. *Water Air Soil Pollut.* **130:**1061–1066.

Smith, C. J., D. B. Nedwell, L. F. Dong, and A. M. Osborn. 2006. Evaluation of quantitative polymerase chain reaction-based approaches for determining gene copy and gene transcript numbers in environmental samples. *Environ. Microbiol.* **8:**804–815.

Smith, Z., A. E. McCaig, J. R. Stephen, T. M. Embley, and J. I. Prosser. 2001. Species diversity of uncultured and cultured populations of soil and marine ammonia oxidizing bacteria. *Microb. Ecol.* **42:**228–237.

Stark, J. M. 1996. Modeling the temperature response of nitrification. *Biogeochemistry* **35:**433–445.

Stark, J. M., and M. K. Firestone. 1995. Mechanisms for soil moisture effects on activity of nitrifying bacteria. *Appl. Environ. Microbiol.* **61:**218–221.

Stark, J. M., and M. K. Firestone. 1996. Kinetic characteristics of ammonium-oxidizer communities in a California oak woodland-annual grassland. *Soil. Biol. Fertil.* **28:**1307–1317.

Stark, J. M., and S. C. Hart. 1997. High rates of nitrification and nitrate turnover in undisturbed coniferous forests. *Nature* **385:**61–64.

Stehr, G., S. Zorner, B. Bottcher, and H. P. Koops. 1995. Exopolymers: an ecological characteristic of a floc-attached, ammonia-oxidizing bacterium. *Microb. Ecol.* **30:**115–126.

Stein, L. Y., and Y. L. Yung. 2003. Production, isotopic composition, and atmospheric fate of biologically produced nitrous oxide. *Annu. Rev. Earth Planet. Sci.* **31:**329–356.

Stephen, J. R., A. E. McCaig, Z. Smith, J. I. Prosser, and T. M. Embley. 1996. Molecular diversity of soil and marine 16S rRNA gene sequences related to beta-subgroup ammonia-oxidizing bacteria. *Appl. Environ. Microbiol.* **62:**4147–4154.

Stephen, J. R., G. A. Kowalchuk, M. A. V. Bruns, A. E. McCaig, C. J. Phillips, T. M. Embley, and J. I. Prosser. 1998. Analysis of beta-subgroup proteobacterial ammonia oxidizer populations in soil by denaturing gradient gel electrophoresis analysis and hierarchical phylogenetic probing. *Appl. Environ. Microbiol.* **64:**2958–2965.

Subbarao, G., O. Ito, K. Sahrawat, W. Berry, K. Nakahara, T. Ishikawa, T. Watanabe, K. Suenaga, M. Rondon, and I. Rao. 2006. Scope and strategies for regulation of nitrification in agricultural systems—challenges and opportunities. *Crit. Rev. Plant Sci.* **25:**303–335.

Subbarao, G. V., M. Rondon, O. Ito, T. Ishikawa, I. M. Rao, K. Nakahara, C. Lascano, and W. L. Berry. 2007a. Biological nitrification inhibition (BNI)—is it a widespread phenomenon? *Plant. Soil* **294:**5–18.

Subbarao, G. V., B. Tomohiro, K. Masahiro, I. Osamu, H. Samejima, H. Y. Wang, S. J. Pearse, S. Gopalakrishnan, K. Nakahara, A. K. M Zakir Hossain, H. Tsujimoto, and W.L. Berry. 2007b. Can biological nitrification inhibition (BNI) genes from perennial *Leymus racemosus (triticeae)* combat nitrification in wheat farming? *Plant. Soil* **299:**55–64.

Sutka, R. L., N. E. Ostrom, P. H. Ostrom, J. A. Breznak, H. Gandhi, A. J. Pitt, and F. Li. 2006. Distinguishing nitrous oxide production from nitrification and denitrification on the basis of isotopomer abundances. *Appl. Environ. Microbiol.* **72:**638–644.

Suzuki, I., U. Dular, and S. C. Kwok. 1974. Ammonia or ammonium ion as substrate for oxidation by *Nitrosomonas europaea* cells and extracts. *J. Bacteriol.* **120:**556–558.

Taylor, A. E., and P. J. Bottomley. 2006. Nitrite production by *Nitrosomonas europaea* and *Nitrosospira* sp. AV in soils at different solution concentrations of ammonium. *Soil. Biol. Fertil.* **38:**828–836.

Tourna, M., T. E. Freitag, G. W. Nicol, and J. I. Prosser. 2008. Growth, activity and temperature responses of ammonia-oxidizing archaea and bacteria in soil microcosms. *Environ. Microbiol.* **10:**1357–1364.

Treusch, A. H., S. Leininger, C. Schleper, A. Kietzin, H. -J. Klenk, and S. C. Schuster. 2005. Novel genes for nitrite reductase and *amo*-related proteins indicate a role of uncultivated mesophilic crenarchaeota in nitrogen cycling. *Environ. Microbiol.* **7:**1985–1995.

Utåker, J. B., L. Bakken, Q. Q. Jiang, and I. F. Nes. 1996. Phylogenetic analysis of seven new isolates of ammonia-oxidizing bacteria based on 16S rRNA gene sequences. *Syst. Appl. Microbiol.* **18:**549–559.

Verhagen, F. J. M., and H. J. Laanbroek. 1991. Competition for ammonium between nitrifying and heterotrophic bacteria in dual energy-limited chemostats. *Appl. Environ. Microbiol.* **57:**3255–3263.

Warington, R. 1891. On nitrification. *J. Chem. Soc.* **59:**484.

Webster, E. A., and D. W. Hopkins. 1996a. Contributions from different microbial processes to N_2O emission from soil under different moisture regimes. *Biol. Fertil. Soils* **22:**331–335.

Webster, E. A., and D. W. Hopkins. 1996b. Nitrogen and oxygen isotope ratios of nitrous oxide emitted from soil and produced by nitrifying and denitrifying bacteria. *Biol. Fertil. Soils* **22:**326–330.

Webster, G., T. M. Embley, and J. I. Prosser. 2002. Grassland management regimens reduce small-scale heterogeneity and species diversity of beta-proteobacterial ammonia oxidizer populations. *Appl. Environ. Microbiol.* **68:**20–30.

Webster, G., T. M. Embley, T. E. Freitag, Z. Smith, and J. I. Prosser. 2005. Links between ammonia oxidiser species composition, functional diversity and nitrification kinetics in grassland soils. *Environ. Microbiol.* **7:**676–684.

Wertz, S., V. Degrange, J. I. Prosser, F. Poly, C. Commeaux, T. Freitag, N. Guillaumaud, and X. Le Roux. 2006. Maintenance of soil functioning following erosion of microbial diversity. *Environ. Microbiol.* **8:**2162–2169.

Wertz, S., V. Degrange, J. I. Prosser, F. Poly, C. Commeaux, N. Guillaumaud, and X. Le Roux. 2007. Decline of soil microbial diversity does not influence the resistance and resilience of key soil microbial functional groups following a model disturbance. *Environ. Microbiol.* **9:**2211–2219.

Wertz, S., F. Poly, X. Le Roux, and V. Degrange. 2008. Development and application of a PCR-denaturing gradient gel electrophoresis tool to study the diversity of *Nitrobacter*-like *nxrA* sequences in soil. *FEMS Microbiol. Ecol.* **63:**261–271.

White, R. E., R. A. Haigh, and J. Macduff. 1987. Frequency distributions and spatially dependent variability of ammonium and nitrate concentrations in soil under grazed and ungrazed grassland. *Fertil. Res.* **11:**193–208.

Winogradsky, S. 1890–1891. Recherches sur les organismes de la nitrification. *Ann. Inst. Pasteur* **4:**213–231, 257–275, 760–771; **Ibid., 5:**92–100, 577–616.

Wolt, J. D. 2004. A meta-evaluation of nitrapyrin agronomic and environmental effectiveness with emphasis on corn production in the midwestern USA. *Nut. Cycl. Agroecosys.* **69:**23–41.

Woodcock, S., C. J. Van Der Gast, T. Bell, M. Lunn, T. P. Curtis, I. M. Head, and W. T. Sloan. 2007. Neutral assembly of bacterial communities. *FEMS Microbiol. Ecol.* **62:**171–180.

Wrage, N., G. L. Velthof, M. L. van Beusichem, and O. Oenema. 2001. Role of nitrifier denitrification in the production of nitrous oxide. *Soil Biol. Biochem.* **33:**1723–1732.

Wrage, N., G. L. Velthof, H. J. Laanbroek, and O. Oenema. 2004a. Nitrous oxide production in grassland soils: assessing the contribution of nitrifier denitrification. *Soil Biol. Biochem.* **36:**229–236.

Wrage, N., G. L. Velthof, O. Oenema, and H. J. Laanbroek. 2004b. Acetylene and oxygen as inhibitors of nitrous oxide production in *Nitrosomonas europaea* and *Nitrosospira briensis*: a cautionary tale. *FEMS Microbiol. Ecol.* **47:**13–18.

Wrage, N., J. W. Van Groenigen, O. Oenema, and E. M. Baggs. 2005. A novel dual-isotope labelling method for distinguishing between soil sources of N_2O. *Rapid Commun. Mass Spectr.* **19:**3298–3306.

NITRIFICATION IN INLAND WATERS

Hendrikus J. Laanbroek and Annette Bollmann

15

INTRODUCTION

According to the United Nations Millennium Ecosystem Assessment (Anonymous, 2007), the impact of nitrogen pollution on inland waters is very rapidly increasing. The nitrifying bacteria in lakes and rivers take advantage of the increasing amounts of ammonium to generate energy for growth and maintenance and to produce simultaneously more oxidized forms of inorganic nitrogen (i.e., nitrous and nitric oxides, nitrite, and nitrate), which all contribute to environmental and health problems. In addition, as nitrifying bacteria are aerobic organisms, they will add to the oxygen depletion of their surroundings. In his review in 1986 on nitrification in lakes, Hall (1986) collected all the information available at that time on the production of nitrate in lakes. He also described the microbiology involved. Since 1986, methods for detection of nitrifying bacteria have considerably improved, especially those methods based on genes. In addition, new metabolic pathways have been discovered with respect to the oxidation of ammonium under oxic as well as anoxic conditions.

This chapter aims at presenting the latest information on nitrification in inland waters. Most emphasis has been given to the ecology of the different lineages of ammonia-oxidizing bacteria (AOB) belonging to the betaproteobacteria, as these chemolithotrophic microorganisms seem to be the only performers of ammonia oxidation in freshwater environments. Until now, representatives of the AOB belonging to the gammaproteobacteria have never been encountered in freshwater ecosystems. The presence of crenarchaea with a possible role in ammonium oxidation has been shown in different habitats (Konneke et al., 2005; Schleper et al., 2005; Leininger et al., 2006) including lakes (Ye et al., 2009; E.W. Vissers and H. J. Laanbroek, unpublished results) and estuaries (Caffrey et al., 2003), but their role in the nitrogen cycle of inland waters is not clear. The same is true for the anaerobic, ammonia-oxidizing anammox bacteria that have been detected in sediment samples from geographically and biogeochemically distinct environments, including a hypereutrophic lake (Penton et al., 2006). Their role is well recognized under anoxic conditions in wastewater treatment plants and in marine ecosystems, (e.g., Van de Graaf et al., 1995; Strous et al., 2006).

Hendrikus J. Laanbroek, Department of Microbial Ecology, Netherlands Institute of Ecology (NIOO–KNAW), Nieuwersluis, The Netherlands. *Annette Bollmann,* Department of Microbiology, Miami University, Oxford, OH.

Nitrification, Edited by Bess B. Ward, Daniel J. Arp, and Martin G. Klotz

Aerobic nitrification is performed by AOB in combination with nitrite-oxidizing bacteria. Nevertheless, this chapter deals mostly with AOB as they are primarily important for the onset of the process of nitrification, although their activity might be influenced by the presence of active nitrite-oxidizing bacteria, especially after starvation for ammonium (Laanbroek and Bär-Gilissen, 2002). Also under conditions of oxygen limitation, when AOB and nitrite-oxidizing bacteria have to compete for this electron acceptor, the oxidation of ammonium and especially the accumulation of nitrite will be dependent on the specific combination of both types of nitrifying bacteria (Laanbroek and Gerards, 1993; Laanbroek et al., 1994). The presence or absence of active nitrite-oxidizing bacteria might also influence the emission of nitric and nitrous oxides during the oxidation of ammonium (Kester et al., 1997). The study of the ecology of AOB got an impetus by the introduction of gene-based analyses, which was facilitated by the monophyletic nature of this functional group of bacteria (Purkhold et al., 2000, 2003; Kowalchuk and Stephen, 2001; Koops et al., 2003). In contrast, nitrite-oxidizing bacteria belong to different classes of bacteria, and this fact hampers the study of their ecology. Nevertheless, nitrite-oxidizing bacteria have been studied, especially in rivers (e.g., Cebron et al., 2003; Freitag et al., 2006).

NITRIFICATION IN LAKES

Measurements in Lake Superior, one of the world's largest freshwater reservoirs, demonstrated, on average, a 1% increase in nitrate every year since 1960 (Sterner et al., 2007). In the 1980s, many other lakes in North America as well as in Europe also showed an increase in their nitrate concentrations (Stoddard et al., 1999). In the next decade, however, a small reversal in nitrate accumulation rates was observed in these lakes, except for the more oligotrophic alpine lakes as well as for Lake Superior (Stoddard et al., 1999; Skjelkvale et al., 2005; Sterner et al., 2007). In most of these studies as well as in others (e.g., Molot and Dillon, 1993; Kaste and Lyche-Solheim, 2005; Lepisto et al., 2006), changes in nitrate concentrations were ascribed to alterations in atmospheric nitrogen deposition, but none of them referred to the biological process of nitrification, except in the study by Sterner et al. (2007). Based on measurements of the nitrogen and oxygen isotopes of nitrate in Lake Superior, Finlay et al. (2007) came to the conclusion that microorganisms are responsible for 93 to 100% of the nitrate accumulation in the water column of this lake; atmospheric deposition and runoff by rivers were apparently less important with respect to nitrate. The annual increase in Lake Superior is due to the oligotrophic, phosphorus-limited character of this lake, which prevents the assimilation of larger amounts of organic matter and, consequently, the burial of organic nitrogen and denitrification of a surplus of nitrate in the sediment (Finlay et al., 2007; Sterner et al., 2007).

The relative contribution of internal nitrification to the total nitrate load of a lake will be dependent on the characteristics of the lake as well as on the season. Stewart et al. (1982) compared the internal and external nitrate loads of different eutrophic freshwater lakes in the United Kingdom and observed a large difference between the stratified Blelham Tarn in the English Lake District and the shallow Balgavies Loch in county Angus, in the east of Scotland. The latter lake forms part of a drainage system surrounded by intensively farmed agricultural land. Whereas in-lake nitrification amounted to 77% of the total nitrate load in Blelham Tarn, it was only 18% in Balgavies Loch. Hall (1986) observed a seasonal effect on the importance of internal nitrification of Lake Grasmere, also in the English Lake District. Whereas the annual average in-lake nitrification amounted to 31% of the total, contributions of more than 50% were noticed in spring and early summer. Notwithstanding the important role of nitrifying microorganisms in converting the reduced forms of inorganic nitrogen to nitrate in many freshwater systems, measurements of actual nitrification rates are rather limited. In addition, Hall (1986) concluded in his review that interpretation of the available data on nitrification in lakes is difficult due to the different methods used to estimate rates of nitrification and to the variety of the habitats involved.

Active nitrification in river sediments was demonstrated by Schwert and White (1974) and Curtis et al. (1975). It was suggested by Garland (1978) that planktonic nitrification in river water is related to scouring and resuspension of sediments. As concluded by Belser (1979) in his review on nitrification, biological nitrate production in aquatic ecosystems appears to be associated with the sediment rather than with the overlaying water. This behavior was demonstrated by Pauer and Auer (2000) in a study of microcosms filled with water and sediments from the hypereutrophic Lake Onondaga and its adjoining Seneca River located in metropolitan Syracuse, NY. In contrast to the sediment, initial nitrate production rates in the water compartment were zero. This difference in production rates could be explained by extremely low numbers of nitrifying bacteria in the water column compared to the sediment (i.e., a difference of four to five magnitudes). Spiking water with sediment particles gave rise to increased nitrate production rates in the water column after 4 days of incubation.

In lakes, the contribution of the sediment to the total in-lake nitrification will depend on the characteristics of the lake and on the season. In their comparison between the internal and external nitrate loads of Blelham Tarn and Balgavies Loch, Stewart et al. (1982) observed a large difference between the lakes with respect to the relative contribution of the water column to total in-lake nitrification: 56 and 5% for Blelham Tarn and Balgavies Loch, respectively. Hall (1986) also observed a seasonal effect on the contribution of the water column to the total nitrate load of Lake Grasmere. During the occasions of relatively high contributions of in-lake nitrification to the total nitrate load, which happened in spring and early summer, the water column contributed more than 70% to the total nitrification. Also in autumn, nitrification in the water column exceeded the production of nitrate in the sediment by almost a factor of 4, but the total amount of in-lake nitrate production was much lower in that season.

The interaction between water column and the sediment is determined by morphometric and climatic conditions. The same is true for the mixing of the lake. Due to seasonal warming of surface waters, deep lakes in the temperate zone usually exhibit thermal stratification, leading to an isolation of relatively cool bottom waters, the so-called hypolimnion, from the well-mixed, warmer surface waters, the epilimnion. The isolation of the hypolimnion from the oxygenated epilimnion may lead to the establishment of an oxygen gradient toward the sediment and sometimes to total anaerobiosis (Horne and Goldman, 1994). At the end of the warm season, the total water column gets mixed again, and the oxygen supply to deeper layers becomes restored. As stratification and mixing have an effect on the supply of oxygen, it can be imagined that the nature of the lake has an effect on the rate of aerobic nitrification. The release of ammonium from oxygen-limited sediments into the water column increases with the trophic status of the lake (Beutel, 2006). However, oxygen limitation in the hypolimnion will restrain the production of nitrate (Beutel, 2001). Nevertheless, the contributions of the oxygen-limited hypolimnion can be equally important or even more significant to the total nitrate production in the water column during stratification than the oxygenated epilimnion (Table 1).

Accumulation of nitrate beneath the thermocline is a general feature of aerobic lake water columns at all latitudes (Vincent and Downes, 1981). Rysgaard et al. (1994) found that rates of nitrification and coupled nitrification-denitrification increased with increasing dissolved oxygen in the water overlaying the sediment from a eutrophic lake. Ahlgren et al. (1994) observed the highest rates of coupled nitrification-denitrification in profundal sediments of a eutrophic Swedish lake just prior to and after thermal stratification, when the water column was oxygenated and contained some

TABLE 1 Observed rates of nitrification in epilimnion and hypolimnion of a selected number of lakes[a]

Method applied	Lake	Activity (μg of N liter^{-1} day $^{-1}$) in:		
		Epilimnion	Hypolimnion	Hypolimnion (% of total)
Nitrate produced in slurries	Buttermere (United Kingdom)	0–24	0–53	68.8
	Grasmere (United Kingdom)	0–25	0–22	46.8
	Esthwaite Water (United Kingdom)	0–92	0–159	63.3
	Balgavies Loch (United Kingdom)	3.65	1,100	99.7
$^{15}N-NH_4^+$	Mendota (United States)	1.7–5.0	4–26	81.5
	Hald (Denmark)	7	7	50.0
$^{13}C-CO_2$	Taupo (New Zealand)	0.5–4.0	0.5–4	50.0

[a]Modified from Hall (1986).

nitrate. On aeration of anoxic bottom water, nitrifying bacteria will become active again. In a microcosm experiment with sediment and water from two different Wisconsin hard-water lakes, nitrate production reappeared after a lag period of several days upon the release of anoxic conditions (Graetz et al., 1973). Chemolithotrophic nitrifying bacteria are able to survive periods of anaerobiosis, and bacteria from lake sediments resuscitated even earlier after exposure to oxic conditions than bacteria from terrestrial soils (Bodelier et al., 1996).

Isolation of bottom waters from the overlaying surface waters often also affected sediments in the deeper lakes. These so-called profundal sediments also experience more oxygen stress than the more shallow or littoral sediments. During a seasonal study of Esthwaite Water, a productive lake in the English Lake District, profundal sediments revealed smaller numbers of ammonia oxidizers and lower nitrification potentials than littoral sediment sites, especially when the lake was stratified in summer months (Hastings et al., 1998). This also suggests the importance of oxygen availability for the size of the ammonia-oxidizing community and the nitrification potential. The process of nitrification itself may be responsible for the oxygen deficit of the hypolimnion as shown by calculations (Hall, 1986) on the basis of published information. The median values given by Hall (1986) amounted to 30%, but contributions up to 100% have also been observed (Christofi et al., 1981). In this way, the hypolimnion of many lakes might be important as a sink of oxygen.

Although shallow lakes do not become stratified in summer, they often have submerged macrophytes that shape an additional habitat for AOB. These macrophytes offer additional structures and space for epiphytic bacteria to attach. Microcosm experiments with shoots of the macrophyte *Potemogeton pectinatus* have shown that these submerged plants can be important for the conversion of nitrogen in ammonium-rich freshwaters by stimulating nitrification through providing surfaces for attached nitrifying bacteria (Eriksson and Weisner, 1999; Eriksson, 2001). Also litter and dead stems from emergent macrophytes can offer surface area in freshwater wetlands and in the littoral zones of lakes, in that way providing a habitat for nitrifiers (Eriksson and Andersson, 1999). A study by these latter authors have demonstrated that the activity of attached nitrifying bacteria differs greatly between litter of different emergent macrophytes, suggesting that the spatial distribution of nitrification activity within wetland ecosystems is related to the species composition of the emergent vegetation. In addition, it indicates that compounds released from the emergent macrophytes during their decomposition may have positive or negative effects on the nitrification within macrophyte beds in wetlands and littoral zones.

In summary, nitrification in lakes takes place in the sediment as well as in the water column. Three factors have a strong influence on the nitrification: stratification of the water column, time of the year, and the presence of plants in shallow parts of lakes. All these factors control the oxygen and ammonium availability in the lakes and thereby also nitrification.

NITRIFICATION IN STREAMS AND RIVERS

As in lakes, nitrification in streams and rivers occurs primarily in the oxic surface layers of the sediment (Cooper, 1984; Delaune et al., 1991; Kemp and Dodds, 2001). By addition of a ^{15}N-ammonium tracer to a first-order deciduous forest stream in Tennessee, Mulholland et al. (2000) demonstrated that nitrification was an important sink for ammonium in stream water. Despite the low concentrations of ammonium and high demand for this anion by benthic organisms, nitrification rates were substantial and accounted for 19% of the total ammonium uptake rate. Applying the same ^{15}N-ammonium tracer technique, Peterson et al. (2001) found that, on average, 20 to 30% of ammonium removal from twelve different headwater streams across the United States was due to nitrification. The remainder was taken up by photosynthetic organisms, heterotrophic microorganisms, and sorption to sediments.

In a survey comprising 36 streams in northern Wisconsin and the upper peninsula of Michigan, nitrification rates appeared to be highly variable spanning over 2 orders of magnitude (Strauss et al., 2002). Of twelve environmental parameters measured, only the stream water pH was significantly correlated with nitrification rates. A multiple regression model containing stream temperature and pH, conductivity, dissolved organic carbon (DOC) concentrations, and total extractable ammonium explained 60% of the variation in nitrification rates. No single variable explained more than 20% of the total variation in nitrification rates measured in these 36 streams. On the basis of these correlations and additional experiments with nitrogen and carbon additions to stream microcosms, Strauss et al. (2002) concluded that nitrification in their streams is regulated by several variables, with ammonium availability and pH being the most important. Organic carbon was likely important at regulating nitrification only under high environmental C:N conditions and if most of the available carbon is relatively labile.

In a microcosm experiment aimed at studying the effect of variable ammonium, nitrate, and dissolved oxygen concentrations on a number of inorganic nitrogen-converting processes in a prairie stream, Kemp and Dodds (2002) demonstrated that the effect of addition of ammonium and oxygen was highly dependent on the substrata in the microcosms. All substrata showed a slight increase in nitrification rates after addition of nitrate, but a significant decrease was observed in a N_2 atmosphere, whereas the rates declined 100% after addition of nitrate under anoxic conditions. From their results, Kemp and Dodds (2002) concluded that the substrate concentration, the type of substrata present, and the relative abundance of those substrata types within the stream channel are important steering factors for nitrification and for nitrogen cycling, in general.

From estimating nitrification and nitrate uptake rates by short-term injections of ammonium into streams in a forested area, Bernhardt et al. (2002) concluded that in-stream nitrification was insufficient to explain the variation in nitrate concentrations among streams under the prevailing conditions of low ammonium concentrations in the stream water. At the same time, they suggested that nitrate may indirectly influence nitrification rates by mediating the competitive demand for ammonium between heterotrophic microorganisms and AOB. This implies that the heterotrophic organisms switch to nitrate assimilation before ammonium starts to become limiting for both heterotrophic and ammonia-oxidizing bacteria. This seems not very likely as ammonium assimilation is energetically preferable to nitrate assimilation and heterotrophic bacteria are the better competitors for limiting amounts of ammonium compared to AOB (Verhagen and Laanbroek, 1991; Verhagen et al., 1992). The experiments with pure cultures of ammonia-oxidizing and heterotrophic bacteria performed by Verhagen and coworkers (Verhagen and Laanbroek, 1991; Verhagen et al., 1992) was repeated by Straus and Lamberti (2000) with natural material from a third-order stream in northern Indiana flowing through an area of mixed land use. They also observed a repression of nitrification by the addition of carbon; however, the size of repression was dependent on the quality of the carbon source. Higher amounts of carbon in the form of more refractory leaf leachates were required to reach the same level of repression in comparison to glucose.

In larger rivers, because of the smaller surface:volume ratio compared to streams and small rivers, benthic nitrification might be insignificant, and most of the oxidation of ammonium may take place in the water column (Billen, 1975; Lipschultz et al., 1986). As presented by Admiraal and Botermans (1989), polluted rivers are characterized by large ammonium:nitrate ratios when compared to uncontaminated rivers. Such a high ratio may indicate repressed nitrification. Using multiyear data on the concentrations of dissolved inorganic nitrogen compounds in three branches of the lower River Rhine, Admiraal and Botermans (1989) reconstructed the nitrification rates, which yielded these dissolved inorganic nitrogen data. Irrespective of the river branch considered, ammonium concentrations declined strongly between 1972 and 1985, while simultaneously the nitrate and the oxygen saturation values increased. The sediment accounted for 90% of the total nitrification, and the authors argued that oxygen limitation in the sediment repressed nitrification. Relatively large differences were observed between the tributaries, which differ in physical characteristics. Highest rates were observed in the tributary with the highest values for water discharge, average flow rate, and intensity of shipping but the lowest value for water retention time. Such a combination of factors will determine the oxygen availability in the sediment and hence the overall nitrification rate. In contrast, Brion et al. (2000) explained the higher nitrification rates in the river branch with the most intensive shipping by increased turbulence and turbidity due to ship movements. In their own study of nitrification in the River Seine and its estuary, Brion et al. (2000) observed a slow nitrification in the freshwater part of this river but rapid nitrifying activities in the estuarine part. They assigned the difference in activity to the absence or presence of suspended particles. Due to strong tidal dynamics, particles in the water column are continuously resuspended, whereas they settle more in the river with its unidirectional discharge of water. Hence, nitrifying activity seems to be associated with particles. This was also shown by Helder and de Vries (1983) in the Ems-Dollard estuary on the border between Germany and The Netherlands and by Owens (1986), who showed the same phenomenon for the River Tamar estuary in the United Kingdom. To be attached to particles with a longer residence time than the water masses represents apparently a benefit for the functioning of these slowly growing microorganisms. Also in the Scheldt estuary, 57 to 86% of the nitrifying activity is associated with particles (De Bie et al., 2002b). During a 13-month survey on nitrification rates in the freshwater-brackish part of the estuary, a peak in nitrification rates was usually observed in the freshwater part of the estuary. Downstream of this peak, nitrification rates declined, presumably due to ammonium limitation. Year round, dissolved N_2O in the water column peaked at the same location as nitrification, which suggests that nitrification in the water column was the main source of N_2O (De Wilde and De Bie, 2000). A controlled laboratory experiment with natural bacterial communities from the Scheldt estuary showed that low oxygen concentrations trigger nitrous oxide production if ammonium is present in sufficient amounts (De Bie et al., 2002a).

In summary, nitrification in streams and rivers appears to be associated with particles, and the nature of the particles may determine the size of the nitrification rate. In streams, most activity is found in the benthic compartment, whereas as in larger rivers, the highest nitrifying activity is observed in the water column due to their smaller surface-to-volume ratios. But, also in the larger rivers, most activity is associated with particles. Since particles are not evenly distributed along the rivers, but tend to increase especially in the high turbidly zone in estuaries, nitrification often peaks at these zones. Also resuspension of sediment particles by, for example, intensive shipping may enhance nitrification. However, as in every ecosystem, nitrification activity is entirely dependent on the presence of ammonium, and where heterotrophic processes dominate due to the availability of labile carbon, nitrification will be repressed.

LINEAGES OF FRESHWATER AMMONIUM-OXIDIZING BACTERIA

Although less frequently than from other environments, aerobic AOB have been isolated from a number of inland waters (Koops et al., 2003). Most of these isolates appear to be related to the *Nitrosomonas europaea* lineage. This lineage is commonly known as a "sewage" ammonium oxidizer (Koops and Pommerening-Röser, 2001; Koops et al., 2003). Its members seem to be better adapted to most of the isolation procedures applied in the laboratory. Members of the *Nitrosomonas oligotropha* and the *Nitrosomonas communis* lineages have also been isolated from freshwater habitats, but in low numbers. According to Koops and Pommerening-Röser (2001) and Koops et al. (2003), most of the isolates of the *N. oligotropha* lineage have been isolated from oligotrophic freshwater environments, whereas most of the isolates of the *N. communis* lineage, and more precisely of the species *Nitrosomonas nitrosa*, originate from eutrophic freshwater ecosystems. However, this partitioning is probably not that strict as members of the *N. oligotropha* lineage have also been isolated from sediments of the eutrophic Lake Drontermeer, The Netherlands (Bollmann and Laanbroek, 2001). The isolation of a member of the *N. oligotropha* lineage from the root zone of the macrophyte *Glyceria maxima* in that lake was apparently facilitated by enrichment at low ammonium concentrations in continuous cultures. A member of the *Nitrosospira* lineage has once been isolated from an oligotrophic cave lake (Koops and Harms, 1985).

Isolation of ammonia-oxidization bacteria has always been hampered by their slow growth rates. The introduction of molecular techniques has considerably improved our insight in the composition of ammonia-oxidizing communities in inland waters (Table 2). The first analyses were based on the 16S rRNA gene. By using this gene in a first round of a nested PCR approach with a general primer set followed by a secondary amplification with primers specific for the *N. europaea–Nitrosomonas eutropha* lineage or for the *Nitrosospira* lineage, Hiorns et al. (1995) detected the presence of *Nitrosospira* DNA, but not of *Nitrosomonas* DNA, in water and sediment samples from Esthwaite Water. DNA from *Nitrosomonas* was only observed by this method after 2 weeks of incubation in the presence of ammonium. Hence, *Nitrosomonas* species were apparently present but required addition of ammonium to become more abundant. By applying the same technique during a seasonal study of Esthwaite Water, Hastings et al. (1998) obtained the same results: only *Nitrosospira* DNA was detected in water and sediment samples. However, a specific PCR amplification based on the ammonia monooxygenase (*amoA*) gene of *N. europaea* yielded positive results when applied directly to sediment and lake water samples. Furthermore, the presence of 16S rRNA genes related to the *N. europaea* lineage could be detected by specific oligonucleotide probing of enrichment cultures. Sediment and lake water samples collected periodically throughout the year from Buttermere, an oligotrophic freshwater lake in the English Lake District, showed also the predominance of *Nitrosospira* DNA when applying the same nested approach as Hiorns et al. in 1995 (Whitby et al., 1999). Only during the summer months did 16S rRNA genes related to the *N. europaea* lineage come to the fore. Surprisingly, partial 16S rRNA sequences related to *N. europaea* or *N. eutropha* segregated between littoral and profundal sediment samples, respectively. These data of Whitby et al. (1999) suggest that the different conditions at each site of the lake had selected for the different genotypes of *N. euro-*

TABLE 2 Distribution of ammonia-oxidizing betaproteobacteria in inland waters as detected by molecular analyses based on either the 16S rRNA or the *amoA* gene[a]

Habitat	*Nitrosospira* lineages					*Nitrosomonas* lineages						Gene	Reference
	0	1	2	3	4	5	6a	6b	7	8	9		
Constructed alpine wetland												16S rRNA	Gorra et al., 2007
Estuarine sediment				+				+				16S rRNA	Gorra et al., 2007
Estuarine sediment							+	+				16S rRNA	Coci et al., 2005
Estuarine sediment		+					+	+			+	16S rRNA	Freitag et al., 2006
Estuarine sediment				+							+	16S rRNA	Satoh et al., 2007
Estuarine sediment							+				+	16S rRNA	Urakawa et al., 2006
Estuarine water[b]							+		+			16S rRNA	Cebron et al., 2005
Estuarine water				+			+	+			+	16s rRNA	De Bie et al., 2001
Freshwater lake									+			16S rRNA	Whitby et al., 1999
Freshwater lake			+	+			+	+			+	16S rRNA	Kim et al., 2006
Freshwater lake				+			+					16S rRNA	Coci et al., 2008
Soda lake									+			16S rRNA	Carini and Joye, 2008
Tidal freshwater marsh	+			+		+	+					16S rRNA	Laanbroek and Speksnijder, 2008
Estuarine sediment[b]											+	*amoA*	Beman and Francis, 2006
Estuarine sediment								+				*amoA*	Francis et al., 2003
Estuarine sediment				+								*amoA*	Caffrey et al., 2003
Estuarine sediment							+				+	*amoA*	Bernhard et al., 2005
Estuarine sediment							+		+		+	*amoA*	Mosier and Francis, 2008
Freshwater lake	+		+	+			+		+			*amoA*	Kim et al., 2008
Soda lake									+			*amoA*	Hornek et al., 2006
Soda lake									+			*amoA*	Carini and Joye, 2008

[a]Classification according to Koops et al. (2003): clusters 0 to 4, all *Nitrosospira* lineage; cluster 5, *Nitrosomonas* lineage 5; cluster 6a, *N. oligotropha* lineage; cluster 6b, *N. marina* lineage; cluster 7, *N. europaea* lineage; cluster 8, *N. communis* lineage; cluster 9, *Nitrosomonas* Nm143 lineage.

[b]Contained also an undefined *Nitrosospira* sp.

paea and of *N. eutropha*. With stratification in summer, oxygen tension is assumed to be the most likely significant difference between the littoral and profundal sediment sites studied. In a study of eutrophic and oligotrophic basins of Lake Windemere, a large lake in the English Lake District, 16S rRNA gene fragments of the *Nitrosospira* lineage were readily detected in all samples, whereas DNA from the *N. europaea* lineage could only be detected in the oligotrophic basin, and more often in the sediment than in the water column of that basin (Whitby et al., 2001). These data suggested that ammonia-oxidizing communities might be physiologically distinguished between lake water and sediment and that species distribution in a single lake is not uniform.

The predominance of 16S rRNA genes related to the *Nitrosomonas oligotrophic* lineage was observed in water and sediment samples from a variety of freshwater habitats in The Netherlands (Speksnijder et al., 1998). In contrast to the studies in the British lakes mentioned above, Speksnijder et al. (1998) applied a more versatile 16S rRNA-based primer set in combination with denaturing gradient gel electrophoresis (DGGE) (Muyzer et al., 1993; Kowalchuk et al., 1997). Similar populations were observed in the water column and the sediment of the freshwater part of the Scheldt estuary (De Bie et al., 2001; Bollmann and Laanbroek, 2002; Coci et al., 2005). If present, 16S rRNA genes related to the *N. oligotropha* lineage have remained undetected in the above-mentioned studies on the British lakes when applying the *N. europaea* lineage-specific primer set of Hiorns et al. (1995). As inferred from 16S rRNA analysis, members of the *N. oligotropha* lineage were also the dominant AOB in the water column and the top sediment of Lake Drontermeer, The Netherlands (Speksnijder et al., 1998). Using the same primer set, 16S rRNA gene fragments belonging to the *Nitrosospira* lineage (clusters 3 and 4 [Gillan et al., 1998]) turned out to be the dominant representatives of the AOB inside and outside the root zone of the emergent and aerenchymatous macrophyte *G. maxima* (Kowalchuk et al., 1998). Although month-to-month differences were seen in the distribution of *Nitrosospira* clusters 3 and 4, such differences appeared random. Also, no consistent differences were detected between root zone and bare sediment samples. In contrast to the situation in the lake, 16S rRNA gene fragments related to the *N. oligotropha* lineage, and more precisely to the species *Nitrosomonas ureae*, were observed in dilution series inoculated with sediment. Whereas PCR-based techniques make no distinction between active and dormant cells, dilution series select for easily activated cells, which may account for the differences in dominant species observed with both methods.

On average, the communities from river and estuarine environments are not so different from those encountered in lakes (Table 2). The cluster comprising of the *N. oligotropha* and the *Nitrosomonas marina* lineages were, by far, the most numerous. Sequences belonging to cluster 2 of the *Nitrosospira* lineage were only observed in freshwater lakes and rivers, while sequences belonging to clusters 0 and 1 of the same lineage and to *Nitrosomonas* lineage 5 were only found in the estuarine environments.

The introduction of the 16S rRNA gene for the detection of AOB in freshwater habitats was rapidly followed by the application of the functional *amoA* gene in these habitats (e.g., Horz et al., 2000) (Table 2). This gene codes for the subunit A of the ammonium monooxygenase, the key enzyme in aerobic ammonium oxidation. The use of the *amoA* gene did not yield much difference with the application of the 16S rRNA gene. In both cases, members of the *N. oligotropha* and the *Nitrosomonas* Nm143 lineages appeared most frequent in the analyses. Members of cluster 1 of the *Nitrosospira* lineage and of the *Nitrosomonas* lineage 5 were not detected by the use of the *amoA* gene. However, these comparisons should be handled with care as the total number of molecular analyses based on both the 16S and the *amoA* gene is still small. Finally, soda lakes appeared to be rather limited in species richness as only sequences belonging to *Nitrosomonas halophila* were found in the analyses, irrespective of the gene used (Hornek et al., 2006; Carini and Joye, 2008).

In summary, whereas members of the *N. oligotropha* linage seems to dominate freshwater ecosystems according to molecular analyses based on both the 16S rRNA gene and the *amoA* gene, members of the *N. europaea* lineage appear more numerous in isolations, which may indicate that they are better adapted to the conditions applied in the isolation procedures. Hence, methods mimicking better the natural conditions, such as nutrient-limited chemostats, should be applied for isolating ecologically important strains.

LINEAGE SEGREGATION IN LAKES

Although all lineages of aerobic AOB have been detected in freshwater habitats, members of the *N. oligotropha* lineage appear mostly as the dominant fraction of bacterial communities involved in the oxidation of ammonium in inland waters. All the other lineages appear often as a minority in such communities. As detection on the level of DNA only suggests the presence of certain lineages, it does not specify the active part of the community. Hence, certain lineages might just be nonactive invaders from other habitats. Nevertheless, as different conditions will prevail in distinct parts of lakes such as sediment, water column, hypolimnion, and epilimnion, it is to be expected that different species or lineages of AOB will dominate. In a study on the distribution of AOB of the β-subclass of the *Proteobacteria* in stratified lakes in northern Germany, which differ mutually in their trophic status, Kim et al. (2006) observed a difference in community composition based on the 16S rRNA gene between the oxic epilimnion and the anoxic hypolimnion of the eutrophic Lake Plußsee. In mesotrophic Lake Schöhsee with an oxic hypolimnion, the communities were similar at all depths. The ammonia-oxidizing communities in the sediments were always different from those in the water column. Clone libraries of PCR products of 16S rRNA gene fragments from Lake Plußsee showed the presence of specific sequences in each habitat. Whereas members of the *N. oligotropha* lineage dominated in the first 1 m of the water column, members of the *Nitrosospira* lineage dominated in the sediment.

Since shallow lakes inhabit submerged macrophytes, which can promote nitrification activity as discussed above, it is interesting to know whether these plants present a niche for specific lineages or species of AOB. To study this, Coci et al. (2008) sampled the water column, the sediment and the epiphyton in a series of seven interconnected lakes in The Netherlands. Numbers in the water column and epiphyton were too low to yield PCR products after applying specific 16S rRNA-based as well as *amoA*-based primer sets in a direct way. Only a nested approach of a combination of a broad-range primer set for AOB (McCaig et al., 1994) and an AOB-specific amplification with the CTO primer set (Kowalchuk et al., 1997) generated products in all three compartments. The benthic communities were composed of members of both the *Nitrosospira* and the *N. oligotropha* lineages, whereas the pelagic communities contained only members of the *N. oligotropha* lineage. The epiphytic communities were mostly composed of members of the *Nitrosospira* lineage, but some communities also contained members of the *N. oligotropha* lineage. Epiphytic samples from Lake Gooimeer contained only members of the *N. oligotropha* lineage. The community analysis was repeated in three of the seven lakes in the following year, and the results were almost similar: the benthic and the epiphytic compartments contained both 16S rRNA gene fragments of the *Nitrosospira* and the *N. oligotropha* lineages, whereas the pelagic compartment had only members of the *N. oligotropha* lineage (Coci, 2007). A statistical test including all communities of AOB in the three lakes showed significant differences between the benthic and epiphytic communities on one side and the pelagic communities at the other side. The benthic and epiphytic communities were not significantly different from each other. To estimate numbers of AOB in different lake compartments, copy numbers of the 16S rRNA gene specific for AOB have been determined by quantitative PCR (Coci, 2007) following the method described by Hermansson and Lindgren (2001). Gene copy numbers per ml were usually 2 to 4 orders of magnitude larger in the upper 5 cm of the sediment compared to the other compartments, with the exception of the epiphyton on the macrophytes of Lake Gooimeer Fig. 1. The median value for the sediments were significantly ($P < 0.05$) larger than the medians of the pelagic and the epiphyton compartments. Between the latter two was no significant difference. The 16S rRNA-based analyses in the second year of study were accompanied by construction and analyses of clone libraries based on the *amoA* gene. The use of the latter gene led to quite different results: no *amoA* gene fragments related to the *Nitrosospira* lineage were found in any of the compartments; whereas the pelagic and the epiphytic compartments had only *amoA* gene fragments of the *N. oligotropha* lineage, the benthic compartments contained a mixture of fragments of the *N. oligotropha, N. europaea,* and *Nitrosomonas* sp. Nm143 lineages. Hence, both approaches gave rather different results with respect to the presence of members of the *Nitrosospira* lineage. Fluorescent in situ hybridization with *Nitrosospira*- and *Nitrosomonas*-specific probes did, however, confirm the presence of *Nitrosospira*-related bacterial cells on the leaves of the macrophyte *P. pectinatus* in Lakes Gooimeer and Vossemeer as well as in water samples of Lake Vossemeer (Coci, 2007). In contrast, in water samples of Lake Gooimeer, only cells

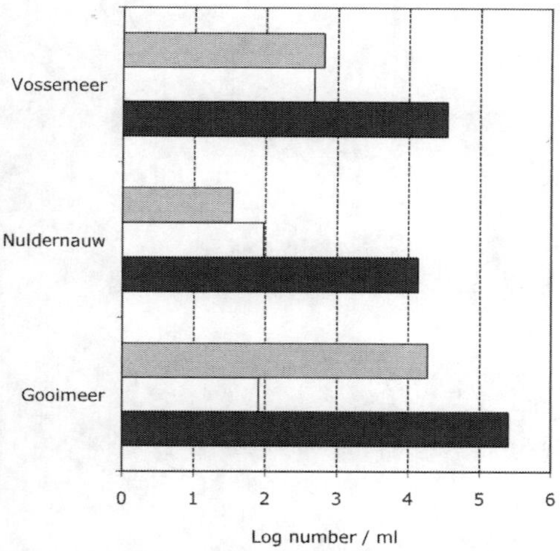

FIGURE 1 Numbers of *amoA* gene copy numbers obtained by quantitative PCR from the epiphyton on macrophytes (gray bars), the water column (white bars), and the sediment (black bars) from three eutrophic lakes in The Netherlands.

belonging to *Nitrosomonas* were observed. No ammonia-oxidizing cells were found in the water column of Lake Nuldernauw and on the leaves of the macroalga *Chara* sp. that specially inhabit this lake. This observation agreed with the low numbers of gene copies of AOB found in the epiphyton in Lake Nuldernauw (Fig. 1). *Chara* species are known to contain high concentrations of sulfur compounds (Anthoni et al., 1980) that may repress the activity of nitrifying bacteria (Joye and Hollibaugh, 1995) and prevent in this way the establishment of epiphytic communities of AOB.

As sediment and water column harbor often different lineages or species of AOB, it is interesting to know whether the epiphytic population originates either from the sediment or from the lake water. In a microcosm setup with sediment and lake water from Lake Vossemeer and with *P. pectinatus* as a model of a submerged macrophyte, Coci (2007) studied the colonization of plant leaves at low and high ammonium concentrations, which varied between 0.0 to 0.2 and 0.1 to 2.0, respectively. Due to the excess of ammonium, dense algal blooms developed at the highest level of ammonium concentrations and repressed significantly the growth of the macrophyte (Table 3). After 5 weeks of incubation, no 16S RNA gene fragments related to aerobic AOB of the betaproteobacteria could be detected in the epiphyton in the microcosms that lacked lake water but contained mineral medium suited for the growth of AOB. Of the three different types of gene fragments that originate from the lake sediment, two were also retrieved on the leaves of the macrophyte, but only in the presence of lake water. A second 16S rRNA gene fragment belonging to the *N. oligotropha* lineage was observed exclusively in the epiphytic community in the presence of lake water. This type could only stem from the water as it had never been found in the sediment. Hence, it seems that AOB in the epiphyton in the microcosms originate from the pelagic community and not from the sediment. This is in contrast to the observations in the lakes themselves, where epiphytic and benthic communities of AOB were more similar to each other and different from the pelagic community of AOB. In the lake itself, apparently, other factors are involved in the attachment of nitrifying bacteria on submerged macrophytes, such as, for example, the presence of suspended sediment particles in the water column.

In summary, a differentiation in niches is apparent for the ammonia-oxidizing betaproteobacteria in lakes. The environmental conditions of each compartment such as littoral or profundal sediment, submerged plants and the hypolimnion or hyperlimnion seem to select for specific species of bacteria. As with activity, the most important steering factors are likely the prevailing concentrations of ammonium and oxygen. The pH value seems less important with the exception of soda lakes that select for specific salt-tolerant species. If existing, a relationship between activity and community composition has still to be established.

LINEAGE SEGREGATION IN RIVERS

Not only in lakes, but also in rivers, particular species or lineages of AOB seem to find their specific habitats. One of the first studies

TABLE 3 Biomass of the submerged macrophyte *Potamogeton pectinatus*, seston weight, and community composition of aerobic AOB in microcosms incubated for 35 days at 20 to 23°C and a 12-h dark-light cycle (light intensity, 225 μmol s^{-1} m^{-2})[a]

16S rRNA genes in the:											
Epiphytic compartment				Benthic compartment							
Nitrosospira DGGE band 1	*Nitrosospira* DGGE band 2	*N. oligotropha* DGGE band 1	*N. oligotropha* DGGE band 2	*Nitrosospira* DGGE band 1	*Nitrosospira* DGGE band 2	*N. oligotropha* DGGE band 1	*N. oligotropha* DGGE band 2	Seston wt (g dry wt $liter^{-1}$)	Plant biomass (g dry wt m^{-2})	NH_4^+ concn (mM)	Absence or presence of lake water
+	-	+	+	+	-	+	-	1.5	72	0.0–0.2	Present
+	-	+	+	+	+	+	-	2.8	35	0.1–2.0	Present
-	-	-	-	-b	-	-	-	1.6	0.2	0.0–0.2	Absent
-	-	-	-	-	-	-	-	3.2	2	0.1–2.0	Absent

[a]Community composition was determined by 16S rRNA-based PCR-DGGE. The sediment originally contained 16S rRNA gene fragments related to the *Nitrosospira* lineage (DGGE bands 1 and 2) and to the *N. oligotropha* lineage (DGGE band 1). No 16S rRNA gene fragments could be detected in the lake water. Modified from Coci (2007).

[b]–, not detectable.

related to this topic was the survey by Stehr et al. (1995a) in the lower River Elbe, Germany. Of two *Nitrosomonas* strains isolated from this river, one freshwater strain belonging to the *N. oligotropha* lineage (Stehr et al., 1995b) had the capacity of flocculation by excretion of exopolymeric substances; the other strain that belonged to the *N. europaea* lineage did not have this ability. Consequently, cells of *N. oligotropha* in the River Elbe were found to occur predominantly attached to particles as demonstrated by immunofluorescence microscopy. This was less the case with cells of the *N. europaea* lineage. The amounts of exopolymers excreted by the cells were not significantly affected by changes in temperature, pH, or NaCl concentration, which are variable characteristics in the Elbe estuary. However, the ammonium concentration had an effect on the production of the exopolymers. More extracellular compounds were produced at low ammonium concentration, which led to less densely inhabited aggregates. The facilitation of attachment to particles will have survival value for slowly growing microorganisms such as the ammonia-oxidizing members of the betaproteobacteria in a river system with high current velocities and, consequently, low water retention times.

Members of the *N. oligotropha* lineage were the dominant AOB in the water column of the freshwater part of the Scheldt estuary as apparent from DGGE analyses based on 16S rRNA gene fragments (De Bie et al., 2001; Bollmann and Laanbroek, 2002). A shift in community composition was observed along the estuary within the region where gradients with respect to salinity, dissolved oxygen, and ammonia were sharpest and where ammonia oxidation was highest (De Wilde and De Bie, 2000). From these environmental factors, the salt concentration turned out to be most important with respect to selection of species of AOB in the Scheldt estuary (Bollmann and Laanbroek, 2002). This was not only demonstrated in ammonium-limited continuous cultures inoculated with samples from either the freshwater or the brackish part of the estuary, but also in batch cultures with excess ammonium. Irrespective of the origin of the inoculum, highest growth rates were observed in the medium composed of filter-sterilized freshwater from the river (Fig. 2). Addition of salt to the level of the brackish water sample decreased the growth rate; the same happened when the freshwater medium was replaced by mineral medium or by filter-sterilized brackish river water. DGGE analysis based on 16S rRNA gene fragments showed the predominance of the *N. marina* lineage in all enrichments, except in the enrichment of the freshwater inoculum in filter-sterilized river water from the same freshwater origin. In this latter enrichment, a representative of the *N. oligotropha* came to the fore. From these results, it could be concluded that members of the *N. marina* lineage are already present in the freshwater part of the Scheldt estuary, but some conditions inherent to the quality of the freshwater itself prevented them from becoming dominant in the river itself. However, in contrast to the enrichment experiments, they remain a minority in the brackish part of the estuary itself, just as representatives of the *Nitrosospira* lineage. The majority of 16S rRNA gene fragments in the brackish part of the estuary were related to the *Nitrosomonas* strain Nm143. The prevalence of this lineage under brackish conditions in estuaries has been confirmed in studies by Bernhard et al. (2005) in the Parker River estuary, on the east coast of Massachusetts, and by Freitag et al. (2006) in sediments of the Ythan estuary, on the east coast of Scotland.

The majority of genes from clone libraries based on the *amoA* gene and collected from the freshwater part of the River Seine, belonged also to the *N. oligotropha* lineage, while members of the *Nitrosospira* and the *N. europaea* lineages were present in smaller numbers (Cebron et al., 2003). Further downstream in the estuary, members of the *Nitrosospira* lineage became more important at the expense of the *N. oligotropha* lineage. A more detailed community analysis along the continuum of the River Seine estuary based on amplification of 16S

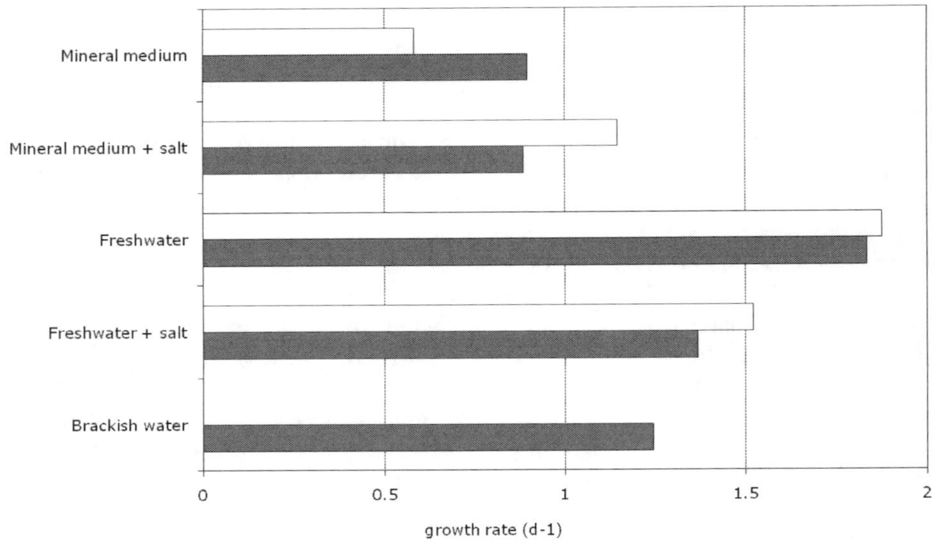

FIGURE 2 Growth rates (day^{-1}) of freshwater (white bars) and brackish (gray bars) inoculums containing AOB from the Scheldt estuary in medium of different compositions.

rRNA gene fragments in combination with DGGE demonstrated that a wastewater treatment plant just downstream of Paris inoculated the river with species of AOB belonging, again, in majority to the *N. oligotropha* lineage. They persisted in the river for a long stretch until they were replaced by other species under more estuarine conditions where lower concentrations of ammonium and increased amounts of suspended matter prevail (Cebron et al., 2004). Among these replacing species was again another member of the *N. oligotropha* lineage. This might well have been a strain with the ability to produce exopolymers as was shown by Stehr et al. (1995a) in the River Elbe.

As discussed above, members of the *N. oligotropha* lineage are sensitive to increased salt concentrations leading to their replacement by more salt-tolerant lineages of AOB. This was also shown to happen when intertidal freshwater sediments from the Scheldt estuary were flooded with brackish and marine waters in microcosms (Coci et al., 2005). As demonstrated by DGGE, a new 16S rRNA gene fragment appeared after 24 days in the microcosms flooded with marine water and after 35 days in the microcosms inundated with brackish water, in both cases, only at the top 1 cm of the sediment. The new AOB belonged to the *N. marina* lineage. However, not only in the brackish and marine microcosms but also in the microcosms flooded with freshwater, a change in community composition occurred. Already within 7 days, a second 16S rRNA gene fragment belonging to the *N. oligotropha* lineage appeared at the first cm of the sediment and after 14 days also in the layers below to a depth of 10 cm. Many indigenous worms of the class *Oligochaetes* kept the sediment well oxidized over the first 10 cm and facilitated in this way the growth of aerobic AOB. In the brackish and marine microcosms, the worms were killed, and only the top 1 cm was oxic and hence suitable for growth of salt-tolerant AOB. Replacement of the native strain of *N. oligotropha* by another strain of this lineage in the freshwater microcosms clearly shows that the laboratory conditions did not mimic the natural conditions in the intertidal sediments.

In the intertidal freshwater marsh just above the intertidal sediments described in the last paragraph, plants grow in distinct zones according to their position in relation to the low water line. Between these plants, nonvegetated zones occur. To determine whether specific plants species select for distinct lineages of aerobic AOB, samples were taken from the different vegetated and nonvegetated zones (Laanbroek and Speksnijder, 2008). According to DGGE analyses based on 16S rRNA gene fragments, the distribution of lineages of AOB was determined by the altitude on the marsh and not by the plant species. Hence, elevation determined both the distribution of plant species and of lineages of AOB. Gene fragments belonging to the *N. oligotropha* lineage were mostly found in the upper, most actively nitrifying layers of the sediment across the whole marsh, whereas members of the *Nitrosospira* lineage were encountered in the deeper layers of the sediment, where oxygen was likely limiting nitrification activity. 16S rRNA fragments of the environmental *Nitrosomonas* lineage 5 were only found in the deeper layers of the sediment closest to the low water line. It seems that members of *Nitrosomonas* lineage 5 and the *Nitrosospira* lineages are better adapted to conditions of starvation than members of the *N. oligotropha* lineage.

In summary, like in freshwater lakes, members of the *N. oligotropha* lineage appear to be the dominant ammonia-oxidizing betaproteobacteria in rivers and streams. Due to the effluent of sewage treatment plants or agricultural activities on the river forelands, the river-borne community might be exchanged or enriched by other AOB. Due to increased salt concentrations, the freshwater community is replaced by more salt-tolerant AOB downstream from the estuary.

CONCLUSIONS

Nowadays, inland water environments are often loaded with large amounts of ammonium, which raises one of the limitations for the occurrence of the process of nitrification. Under conditions of increased ammonium availability, nitrification in inland waters may still be repressed because of oxygen limitation and pH, salinity, or temperature constraints. The importance of anaerobic anammox bacteria for the oxidation of ammonium under limited oxygen availability still has to be demonstrated. Nitrification has been shown to occur in a large variety of inland water ranging from small first-order streams to large rivers and from shallow, vegetated ponds to deep lakes. In all these environments, the presence of surfaces appears to be a preferential place for AOB to be active. Hence, sediments, suspended particles, as well as submerged structures such as macrophytes are usually places of increased activity.

The total diversity of ammonia-oxidizing betaproteobacteria in inland waters is relatively large. Most lineages of AOB have been encountered by molecular analyses without large mutual differences between and among ecosystems. An exception is formed by soda lakes, which contain only sequences from the *N. europaea* lineage and more specifically from the *N. halophila* species. The differences in diversity between lakes and rivers are surprisingly rather small, although the latter may contain more salt-tolerant species in the brackish and marine parts. Overall, members of the *N. oligotropha* lineage are most numerous among the ammonia-oxidizing betaproteobacteria, both in rivers and lakes. After being detected in large numbers in soils and marine environments, the first observations of crenarchaea containing the *amoA* gene are published, but their role in nitrification in inland waters still has to be demonstrated. The same holds for anammox bacteria and their role in anaerobic ammonium oxidation.

REFERENCES

Admiraal, W. I. M., and Y. J. H. Botermans. 1989. Comparison of nitrification rates in 3 branches of the lower river Rhine. *Biogeochemistry* **8:**135–151.

Ahlgren, I., F. Sorensson, T. Waara, and K. Vrede. 1994. Nitrogen budgets in relation to microbial transformations in lakes. *Ambio* **23:**367–377.

Anonymous. 2007. United Nations Millennium Ecosystem Assessment. The Stationery Office, House of Commons, London, United Kingdom.

Anthoni, U., C. Christophersen, J. O. Madsen, S. Wiumandersen, and N. Jacobsen. 1980. Biologically-active sulfur-compounds from the green-alga *Chara globularis*. *Phytochemistry* **19:**1228–1229.

Belser, L. W. 1979. Population ecology of nitrifying bacteria. *Annu. Rev. Microbiol.* **33:**309–335.

Beman, J. M., and C. A. Francis. 2006 Diversity of ammonia-oxidizing archaea and bacteria in the sediments of a hypernutrified subtropical estuary: Bahia del Tobari, Mexico. *Appl. Environ. Microbiol.* **72:**7767–7777.

Bernhard, A. E., T. Donn, A. E. Giblin, and D. A. Stahl. 2005. Loss of diversity of ammonia-oxidizing bacteria correlates with increasing salinity in an estuary system. *Environ. Microbiol.* **7:**1289–1297.

Bernhardt, E.S., R. O. Hall, and G. E. Likens. 2002. Whole-system estimates of nitrification and nitrate uptake in streams of the Hubbard Brook Experimental Forest. *Ecosystems* **5:**419–430.

Beutel, M. W. 2001. Oxygen consumption and ammonia accumulation in the hypolimnion of Walker Lake, Nevada. *Hydrobiologia* **466:**107–117.

Beutel, M. W. 2006. Inhibition of ammonia release from anoxic profundal sediments in lakes using hypolimnetic oxygenation. *Ecol. Eng.* **28:** 271–279.

Billen, G. 1975. Nitrification in the scheldt estuary (Belgium and the Netherlands). *Estuar. Coas. Mar. Sci.* **9:**79–89.

Bodelier, P. L. E., J. A. Libochant, C. W. P. M. Blom, and H. J. Laanbroek. 1996. Dynamics of nitrification and denitrification in root-oxygenated sediments and adaptation of ammonia-oxidizing bacteria to low-oxygen or anoxic habitats. *App. Environ. Microbiol.* **62:**4100–4107.

Bollmann, A., and H. J. Laanbroek. 2001. Continuous culture enrichments of ammonia-oxidizing bacteria at low ammonium concentrations. *FEMS Microbiol. Ecol.* **37:**211–221.

Bollmann, A., and H. J. Laanbroek. 2002. Influence of oxygen partial pressure and salinity on the community composition of ammonia-oxidizing bacteria in the Schelde estuary. *Aquat. Microb. Ecol.* **28:**239–247.

Brion, N., G. Billen, L. Guezennec, and A. Ficht. 2000. Distribution of nitrifying activity in the Seine River (France) from Paris to the estuary. *Estuaries* **23:**669–682.

Caffrey, J. M., N. Harrington, I. Solem, and B. B. Ward. 2003. Biogeochemical processes in a small California estuary. 2. Nitrification activity, community structure and role in nitrogen budgets. *Mar. Ecol. Prog. Ser.* **248:**27–40.

Carini, S. A., and S. B. Joye. 2008. Nitrification in Mono Lake, California: activity and community composition during contrasting hydrological regimes. *Limnol. Oceanogr.* **53:**2546–2557.

Cebron, A., T. Berthe, and J. Garnier. 2003. Nitrification and nitrifying bacteria in the lower Seine River and estuary (France). *App. Environ. Microbiol.* **69:**7091–7100.

Cebron, A., M. Coci, J. Garnier, and H. J. Laanbroek. 2004. Denaturing gradient gel electrophoretic analysis of ammonia-oxidizing bacterial community structure in the lower Seine River: impact of Paris wastewater effluents. *Appl. Environ. Microbiol.* **70:**6726–6737.

Cebron, A., J. Garnier, and G. Billen. 2005. Nitrous oxide production and nitrification kinetics by natural bacterial communities of the lower Seine river (France). *Aquat. Microb. Ecol.* **41:**25–38.

Christofi, N., T. Preston, and W. D. P. Stewart. 1981. Endogenous nitrate production in an experimental enclosure during summer stratification. *Water Res.* **15:**343–349.

Coci, M. 2007. Niche differentiation between ammonia-oxidising bacteria in aquatic environments. Ph.D. thesis. Utrecht University, Utrecht, The Netherlands.

Coci, M., D. Riechmann, P. L. E. Bodelier, S. Stefani, G. Zwart, and H. J. Laanbroek. 2005. Effect of salinity on temporal and spatial dynamics of ammonia-oxidising bacteria from intertidal freshwater sediment. *FEMS Microbiol. Ecol.* **53:**359–368.

Coci, M., P. L. E. Bodelier, and H. J. Laanbroek. 2008. Epiphyton as a niche for ammonia-oxidizing bacteria: Detailed comparison with benthic and pelagic compartments in shallow freshwater lakes. *Appl. Environ. Microbiol.* **74:**1963–1971.

Cooper, A. B. 1984. Activities of benthic nitrifiers in streams and their role in oxygen consumption. *Microb. Ecol.* **10:**316–333.

Curtis, E. J. C., K. Durrant, and M. M. I. Harman. 1975. Nitrification in rivers in the Trent Basin. *Water Res.* **9:**255–268.

De Bie, M. J. M., A. G. C. L. Speksnijder, G. A. Kowalchuk, T. Schuurman, G. Zwart, J. R. Stephen, et al. 2001. Shifts in the dominant populations of ammonia-oxidizing beta-subclass Proteobacteria along the eutrophic Schelde estuary. *Aquat. Microb. Ecol.* **23:**225–236.

De Bie, M. J. M., J. J. Middelburg, M. Starink, and H. J. Laanbroek. 2002a. Factors controlling nitrous oxide at the microbial community and estuarine scale. *Mar. Ecol. Prog. Ser.* **240:**1–9.

De Bie, M. J. M., M. Starink, H. T. S. Boschker, J. J. Peene, and H. J. Laanbroek. 2002b. Nitrification in the Schelde estuary: methodological aspects and factors influencing its activity. *FEMS Microbiol. Ecol.* **42:**99–107.

De Wilde, H. P. J., and M. J. M. De Bie. 2000. Nitrous oxide in the Schelde estuary: production by nitrification and emission to the atmosphere. *Mar. Chem.* **69:**203–216.

Delaune, R. D., L. M. Salinas, R. S. Knox, M. N. Sarafyan, and C. J. Smith. 1991. Water quality of a coastal river receiving nutrient inputs—

ammonium nitrogen transformations. *J. Environ. Sci. Health A* **26:**1287–1302.

Eriksson, P. G. 2001. Interaction effects of flow velocity and oxygen metabolism on nitrification and denitrification in biofilms on submersed macrophytes. *Biogeochemistry* **55:**29–44.

Eriksson, P. G., and J. L. Andersson. 1999. Potential nitrification and cation exchange on litter of emergent, freshwater macrophytes. *Freshwater Biol.* **42:**479–486.

Eriksson, P. G., and S. E. B. Weisner. 1999. An experimental study on effects of submersed macrophytes on nitrification and denitrification in ammonium-rich aquatic systems. *Limnol. Oceanogr.* **44:**1993–1999.

Finlay, J. C., R. W. Sterner, and S. Kumar. (2007) Isotopic evidence for in-lake production of accumulating nitrate in lake superior. *Ecol. Applic.* **17:**2323–2332.

Francis, C. A., G. D. O'Mullan, and B. B. Ward. 2003. Diversity of ammonia monooxygenase (*amoA*) genes across environmental gradients in Chesapeake Bay sediments. *Geobiology* **1:**129–140.

Freitag, T. E., L. Chang, and J. I. Prosser. 2006. Changes in the community structure and activity of betaproteobacterial ammonia-oxidizing sediment bacteria along a freshwater-marine gradient. *Environ. Microbiol.* **8:**684–696.

Garland, J. H. N. 1978. Nitrification in the River Trent, p. 167. *In* A. James (ed.), *Mathematical Models in Water Pollution Control.* Wiley, Chichester, United Kingdom.

Gillan, D. C., A. Speksnijder, G. Zwart, and C. De Ridder. 1998. Genetic diversity of the biofilm covering *Montacuta ferruginosa* (Mollusca, Bivalvia) as evaluated by denaturing gradient gel electrophoresis analysis and cloning of PCR-amplified gene fragments coding for 16S rRNA. *Appl. Environ. Microbiol.* **64:**3464–3472.

Gorra, R., M. Coci, R. Ambrosoli, and H. J. Laanbroek. 2007. Effects of substratum on the diversity and stability of ammonia-oxidizing communities in a constructed wetland used for wastewater treatment. *J. Appl. Microbiol.* **103:**1442–1452.

Graetz, D. A., D. R. Keeney, and R. B. Aspiras. 1973. E_h status of lake sediment-water systems in relation to nitrogen transformations. *Limnol. Oceanogr.* **18:**908–917.

Hall, G. H. 1986. Nitrification in lakes, p. 127–156. *In* J. I. Prosser (ed.), *Nitrification.* IRL Press, Oxford, United Kingdom.

Hastings, R. C., J. R. Saunders, G. H. Hall, R. W. Pickup, and A. J. McCarthy. 1998. Application of molecular biological techniques to a seasonal study of ammonia oxidation in a eutrophic freshwater lake. *Appl. Environ. Microbiol.* **64:**3674–3682.

Helder, W., and R. T. P. Devries. 1983. Estuarine nitrite maxima and denitrifying bacteria (Ems-Dollard estuary). *Neth. J. Sea Res.* **17:**1–18.

Hermansson, A., and P. E. Lindgren. 2001. Quantification of ammonia-oxidizing bacteria in arable soil by real-time PCR. *Appl. Environ. Microbiol.* **67:**972–976.

Hiorns, W. D., R. C. Hastings, I. M. Head, A. J. McCarthy, J. R. Saunders, R. W. Pickup, and G. H. Hall. 1995. Amplification of 16S ribosomal RNA genes of autotrophic ammonia-oxidizing bacteria demonstrates the ubiquity of nitrosospiras in the environment. *Microbiology* **141:**2793–2800.

Horne, A. J., and C. R. Goldman (ed.). 1994. *Limnology.* McGraw-Hill Inc., New York, NY.

Hornek, R., A. Pommerening-Roser, H. P. Koops, A. H. Farnleitner, N. Kreuzinger, A. Kirschner, and R. L. Mach. 2006. Primers containing universal bases reduce multiple amoA gene specific DGGE band patterns when analysing the diversity of beta-ammonia oxidizers in the environment. *J. Microbiol. Methods* **66:**147–155.

Horz, H. P., J. H. Rotthauwe, T. Lukow, and W. Liesack. 2000. Identification of major subgroups of ammonia-oxidizing bacteria in environmental samples by T-RFLP analysis of *amoA* PCR products. *J. Microbiol. Methods* **39:**197–204.

Joye, S. B., and J. T. Hollibaugh. 1995. Influence of sulfide inhibition of nitrification on nitrogen regeneration in sediments. *Science* **270:**623–625.

Kaste, O., and A. Lyche-Solheim. 2005. Influence of moderate phosphate addition on nitrogen retention in an acidic boreal lake. *Can. J. Fish. Aquat. Sci.* **62:**312–321.

Kemp, M. J., and W. K. Dodds. 2001. Centimeter-scale patterns in dissolved oxygen and nitrification rates in a prairie stream. *J. North Am. Benthol. Soc.* **20:**347–357.

Kemp, M. J., and W. K. Dodds. 2002. The influence of ammonium, nitrate, and dissolved oxygen concentrations on uptake, nitrification, and denitrification rates associated with prairie stream substrata. *Limnol. Oceanogr.* **47:**1380–1393.

Kester, R. A., W. De Boer, and H. J. Laanbroek. 1997. Production of NO and N_2O by pure cultures of nitrifying and denitrifying bacteria during changes in aeration. *Appl. Environ. Microbiol.* **63:**3872–3877.

Kim, O. S., P. Junier, J. F. Imhoff, and K. P. Witzel. 2006. Comparative analysis of ammonia-oxidizing bacterial communities in two lakes in North Germany and the Baltic Sea. *Arch. Hydrobiol.* **167:**335–350.

Kim, O. S., P. Junier, J. F. Imhoff, and K. P. Witzel. 2008. Comparative analysis of ammonia monooxygenase (*amoA*) genes in the water column and sediment-water interface of two lakes and the Baltic Sea. *FEMS Microbiol. Ecol.* **66:**367–378.

Konneke, M., A. E. Bernhard, J. R. de la Torre, C. B. Walker, J. B. Waterbury, and D. A. Stahl. 2005. Isolation of an autotrophic ammonia-oxidizing marine archaeon. *Nature* **437:**543–546.

Koops, H. P., and H. Harms. 1985. Deoxyribonucleic-acid homologies among 96 strains of ammonia-oxidizing bacteria. *Arch. Microbiol.* **141:**214–218.

Koops, H. P., and A. Pommerening-Röser. 2001. Distribution and ecophysiology of the nitrifying bacteria emphasizing cultured species. *FEMS Microbiol. Ecol.* **37:**1–9.

Koops, H.-P., U. Purkhold, A. Pommerening-Röser, G. Timmermann, and M. Wagner. 2003. The lithotrophic ammonia-oxidizing bacteria. *In* M. Dworkin, S. Falcow, E. Rosenberg, K.-H. Schleifer, and E. Stackebrandt (ed.), *The Procaryotes, an Evolving Electronic Resource for the Microbiological Community*, 3rd ed. Springer, New York, NY.

Kowalchuk, G. A., and J. R. Stephen. 2001. Ammonia-oxidizing bacteria: a model for molecular microbial ecology. *Annu. Rev. Microbiol.* **55:**485–529.

Kowalchuk, G. A., J. R. Stephen, W. De Boer, J. I. Prosser, T. M. Embley, and J. W. Woldendorp. 1997. Analysis of ammonia-oxidizing bacteria of the beta subdivision of the class proteobacteria in coastal sand dunes by denaturing gradient gel electrophoresis and sequencing of PCR-amplified 16S ribosomal DNA fragments. *Appl. Environ. Microbiol.* **63:**1489–1497.

Kowalchuk, G. A., P. L. E. Bodelier, G. H. J. Heilig, J. R. Stephen, and H. J. Laanbroek. 1998. Community analysis of ammonia-oxidising bacteria, in relation to oxygen availability in soils and root-oxygenated sediments, using PCR, DGGE and oligonucleotide probe hybridisation. *FEMS Microbiol. Ecol.* **27:**339–350.

Laanbroek, H. J., and M.-J. Bär-Gilissen. 2002. Weakened activity of starved ammonia-oxidizing bacteria by the presence of pre-activated *Nitrobacter winogradskyi*. *Microb. Environ.* **17:**122–127.

Laanbroek, H. J., and S. Gerards. 1993. Competition for limiting amounts of oxygen between *Nitrosomonas europaea* and *Nitrobacter winogradskyi* grown in mixed continuous cultures. *Arch. Microbiol.* **159:**453–459.

Laanbroek, H. J., and A. Speksnijder. 2008. Niche separation of ammonia-oxidizing bacteria across a tidal freshwater marsh. *Environ. Microbiol.* **10:**3017–3025.

Laanbroek, H. ., P. L. E. Bodelier, and S. Gerards. 1994. Oxygen consumption kinetics of nitrosomonas europaea and nitrobacter hamburgensis grown in mixed continuous cultures at different oxygen concentrations. *Arch. Microbiol.* **161:**156–162.

Leininger, S., T. Urich, M. Schloter, L. Schwark, J. Qi, G. W. Nicol, et al. 2006. Archaea predominate among ammonia-oxidizing prokaryotes in soils. *Nature* **442:**806–809.

Lepisto, A., K. Granlund, P. Kortelainen, and A. Raike. 2006. Nitrogen in river basins: sources, retention in the surface waters and peatlands, and fluxes to estuaries in Finland. *Sci. Total Environ.* **365:**238–259.

Lipschultz, F., S. C. Wofsy, and L. E. Fox. 1986. Nitrogen-metabolism of the eutrophic Delaware reiver ecosystem. *Limnol. Oceanogr.* **31:**701–716.

McCaig, A. E., T. M. Embley, and J. I. Prosser. 1994. Molecular analysis of enrichment cultures of marine ammonia oxidisers. *FEMS Microbiol. Lett.* **120:**363–367.

Molot, L. A., and P. J. Dillon. 1993. Nitrogen mass balances and denitrification rates in central Ontario lakes. *Biogeochemistry* **20:**195–212.

Mosier, A. C., and C. A. Francis. 2008. Relative abundance and diversity of ammonia-oxidizing archaea and bacteria in the San Francisco Bay estuary. *Environ. Microbiol.* **10:**3002–3016.

Mulholland, P. J., J. L. Tank, D. M. Sanzone, W. M. Wollheim, B. J. Peterson, J. R. Webster, and J. L. Meyer. 2000. Nitrogen cycling in a forest stream determined by a N-15 tracer addition. *Ecol. Monogr.* **70:**471–493.

Muyzer, G., E. C. Dewaal, and A. G. Uitterlinden. 1993. Profiling of complex microbial populations by denaturing gradient gel-electrophoresis analysis of polymerase chain reaction-amplified genes coding for 16S ribosomal RNA. *Appl. Environ. Microbiol.* **59:**695–700.

Owens, N. J. P. 1986. Estuarine nitrification—a naturally occurring fluidized-bed reaction *Estuar. Coast. Shelf Sci.* **22:**31–44.

Pauer, J. J., and M. T. Auer. 2000. Nitrification in the water column and sediment of a hypereutrophic lake and adjoining river system. *Water Res.* **34:**1247–1254.

Penton, C. R., A. H. Devol, and J. M. Tiedje. 2006. Molecular evidence for the broad distribution of anaerobic ammonium-oxidizing bacteria in freshwater and marine sediments. *Appl. Environ. Microbiol.* **72:**6829–6832.

Peterson, B. J., W. M. Wollheim, P. J. Mulholland, J. R. Webster, J. L. Meyer, J. L. Tank, et al. 2001. Control of nitrogen export from watersheds by headwater streams. *Science* **292:**86–90.

Purkhold, U., A. Pommerening-Röser, S. Juretschko, M. C. Schmid, H. P. Koops, and M. Wagner. 2000. Phylogeny of all recognized species of ammonia oxidizers based on comparative 16S rRNA and *amoA* sequence analysis: implications for molecular diversity surveys. *Appl. Environ. Microbiol.* **66:**5368–5382.

Purkhold, U., M. Wagner, G. Timmermann, A. Pommerening-Roser, and H. P. Koops. 2003. 16S rRNA and *amoA*-based phylogeny of 12 novel betaproteobacterial ammonia-oxidizing isolates: extension of the dataset and proposal of a new lineage within the nitrosomonads. *Int. J. Syst. Evol. Microbiol.* **53:**1485–1494.

Rysgaard, S., N. Risgaardpetersen, N. P. Sloth, K. Jensen, and L. P. Nielsen. 1994. Oxygen regulation of nitrification and denitrification in sediments. *Limnol. Oceanogr.* **39:**1643–1652.

Satoh, K., C. Itoh, D. L. Kang, H. Sumida, R. Takahashi, K. Isobe, et al. 2007. Characteristics of newly isolated ammonia-oxidizing bacteria from acid sulfate soil and the rhizoplane of leucaena grown in that soil. *Soil Sci. Plant Nutr.* **53:**23–31.

Schleper, C., G. Jurgens, and M. Jonuscheit. 2005. Genomic studies of uncultivated archaea. *Nat. Rev. Microbiol.* **3:**479–488.

Schwert, D. P., and J. P. White. 1974. Method for in situ measurement of nitrification in a stream. *Appl. Environ. Microbiol.* **28:**1082–1083.

Skjelkvale, B. L., J. L. Stoddard, D. S. Jeffries, K. Torseth, T. Hogasen, J. Bowman, et al. 2005. Regional scale evidence for improvements in surface water chemistry 1990–2001. *Environ. Pollut.* **137:**165–176.

Speksnijder, A., G. A. Kowalchuk, K. Roest, and H. J. Laanbroek. 1998. Recovery of a *Nitrosomonas*-like 16S rDNA sequence group from freshwater habitats. *Syst. Appl. Microbiol.* **21:**321–330.

Stehr, G., S. Zorner, B. Bottcher, and H. P. Koops. 1995a. Exopolymers: an ecological characteristic of a floc-attached, ammonia-oxidizing bacterium. *Microb. Ecol.* **30:**115–126.

Stehr, G., B. Bottcher, P. Dittberner, G. Rath, and H. P. Koops. 1995b. The ammonia-oxidizing nitrifying population of the River Elbe estuary. *FEMS Microbiol. Ecol.* **17:**177–186.

Sterner, R. W., E. Anagnostou, S. Brovold, G. S. Bullerjahn, J. C. Finlay, S. Kumar, et al. 2007. Increasing stoichiometric imbalance in North America's largest lake: nitrification in Lake Superior. *Geophys. Res. Lett.* **34:**10406.

Stewart, W. D. P., T. Preston, H. G. Peterson, and N. Christofi. 1982. Nitrogen cycling in eutrophic freshwaters. *Philos. Trans. R. Soc. Lond. B* **296:**491–509.

Stoddard, J. L., D. S. Jeffries, A. Lukewille, T. A. Clair, P. J. Dillon, C. T. Driscoll, et al. 1999. Regional trends in aquatic recovery from acidification in North America and Europe. *Nature* **401:**575–578.

Strauss, E. A., and G. A. Lamberti. 2000. Regulation of nitrification in aquatic sediments by organic carbon. *Limnol. Oceanogr.* **45:**1854–1859.

Strauss, E. A., N. L. Mitchell, and G. A. Lamberti. 2002. Factors regulating nitrification in aquatic sediments: effects of organic carbon, nitrogen availability, and pH. *Can. J. Fish. Aquat. Sci.* **59:**554–563.

Strous, M., E. Pelletier, S. Mangenot, T. Rattei, A. Lehner, M. W. Taylor, et al. 2006. Deciphering the evolution and metabolism of an anammox bacterium from a community genome. *Nature* **440:**790–794.

Urakawa, H., S. Kurata, T. Fujiwara, D. Kuroiwa, H. Maki, S. Kawabata, et al. 2006. Characterization and quantification of ammonia-oxidizing bacteria in eutrophic coastal marine sediments using polyphasic molecular approaches and immunofluorescence staining. *Environ. Microbiol.* **8:**787–803.

Van de Graaf, A. A., A. Mulder, P. De Bruijn, M. S. M. Jetten, L. A. Robertson, and J. G. Kuenen. 1995. Anaerobic oxidation of ammonium is a biologically mediated process. *Appl. Environ. Microbiol.* **61:**1246–1251.

Verhagen, F. J. M., and H. J. Laanbroek. 1991. Competition for ammonium between nitrifying and heterotrophic bacteria in dual energy-limited chemostats. *Appl. Environ. Microbiol.* **57:**3255–3263.

Verhagen, F. J. M., H. Duyts, and H. J. Laanbroek. 1992. Competition for ammonium between nitrifying and heterotrophic bacteria in continuously percolated soil columns. *Appl. Environ. Microbiol.* **58:**3303–3311.

Vincent, W. F., and M. T. Downes. 1981. Nitrate accumulation in aerobic hypolimnia—relative importance of benthic and planktonic nitrifiers in an oligotrophic lake. *App. Environ. Microbiol.* **42:**565–573.

Whitby, C. B., J. R. Saunders, J. Rodriguez, R. W. Pickup, and A. McCarthy. 1999. Phylogenetic differentiation of two closely related *Nitrosomonas* spp. that inhabit different sediment environments in an oligotrophic freshwater lake. *Appl. Environ. Microbiol.* **65:**4855–4862.

Whitby, C. B., J. R. Saunders, R. W. Pickup, and A. J. McCarthy. 2001. A comparison of ammonia-oxidiser populations in eutrophic and oligotrophic basins of a large freshwater lake. *Antonie Van Leeuwenhoek Int. J. Gen. Mol. Microbiol.* **79:**179–188.

Ye, W. J., X. L. Liu, S. Q. Lin, J. Tan, J. L. Pan, D. T. Li, and H. Yang. 2009. The vertical distribution of bacterial and archaeal communities in the water and sediment of Lake Taihu. *FEMS Microbiol. Ecol.* **70:**263–276.

NITRIFICATION IN WASTEWATER TREATMENT

Satoshi Okabe, Yoshiteru Aoi, Hisashi Satoh, and Yuichi Suwa

16

INTRODUCTION

Importance of Nitrification in Wastewater Treatment Plants

Because of its oxygen demand and toxicity to aquatic life in receiving waters, ammonia nitrogen (NH_4^+-N) must be sufficiently removed from various wastewaters. Biological and physico-chemical treatments have been conventionally used for nitrogen removal. A variety of physico-chemical processes have been developed to treat especially high concentrations of NH_4^+-N, such as volatilization of nitrous acid (HNO_2), dinitrogen (N_2), and nitrous oxide (N_2O) (Udert et al., 2005), air stripping, ion exchange, membrane separation, and chemical precipitation to form magnesium ammonium phosphate hexahydrate (Zhang et al., 2009a). However, biological nitrogen removal processes are generally selected from an economic viewpoint. Engineered biological removal of ammonia and organic nitrogen is achieved by first oxidizing ammonia to nitrite/nitrate (via nitrification) that, in turn, is reduced to nitrogen gas (N_2) (via denitrification). Typically, the nitrite concentration found in the influent wastewater is low because the oxidation of ammonia to nitrite is limited. Microbial nitrification is, therefore, a necessary step in removing nitrogen from wastewaters via biological denitrification and is becoming more important due to strict regulations on nitrogen discharge. However, microbial nitrification is recognized as being difficult to maintain in practical wastewater treatment plants (WWTPs) owing to the lower kinetics, yields, and sensitivity of nitrifying bacteria to physical, chemical, and environmental disturbances as mentioned below, even though nitrification has been studied more than any other specific biochemical reactions occurring in wastewater treatment to date (Gujer, 2010).

Biological Nitrogen Removal Processes

Nitrifying bacteria, both ammonia-oxidizing bacteria (AOB) and nitrite-oxidizing bacteria (NOB), are autotrophs, chemolithotrophs, and obligate aerobes. Thus, they have much lower growth yields than do the aerobic heterotrophs that always coexist in activated sludge and biofilm systems. They are also known to be slow-growing bacteria and sensitive directly to various environmental factors (e.g., temperature, pH, dissolved oxygen [DO] concen-

Satoshi Okabe and Hisashi Satoh, Department of Urban and Environmental Engineering, Graduate School of Engineering, Hokkaido University, Sapporo 060-8628, Japan. *Yoshiteru Aoi*, Waseda Institute for Advanced Study, Tokyo 169-8050, Japan. *Yuichi Suwa*, Faculty of Science and Engineering, Chuo University, Tokyo 112-8551, Japan.

Nitrification, Edited by Bess B. Ward, Daniel J. Arp, and Martin G. Klotz

tration, alkalinity, chemical oxygen demand/total Kjeldahl nitrogen [COD:TKN] ratio, and presence of toxic chemicals). Effects of these factors on nitrification are discussed below in detail. (See "Factors affecting nitrifying activity in WWTP," below.)

In addition, the nitrifying bacteria have relatively high half-saturation constants for oxygen ($K_{A,O2}$) than do the heterotrophs. These features of nitrifying bacteria are the reason why they are usually outcompeted by the heterotrophs in the presence of organic carbon due to interspecies competition for oxygen and space, leading to deterioration or failure of process performance (Okabe et al., 1995, 1996; Satoh et al., 2000).

The nitrification process is undertaken in WWTPs predominantly as an activated sludge or as a biofilm-based process. Over the past few decades, a variety of process flow sheets for nitrogen removal have been proposed and studied. The flow sheets of the treatment plant depend on the characteristics of wastewater compositions. A successful nitrification process in both suspended growth or attached biofilm growth reactors is primarily dependent on solids (biomass) retention time (SRT), feeding pattern to the reactor (e.g., completely mixed reactor, plug-flow reactor, sequencing batch reactor, step feed, internal recycle, and so forth), aeration pattern in the reactor, and recycle ratio.

The SRT controls the concentrations of microorganisms in the system. A higher SRT contributes to a higher concentration of microorganisms. Biomass retention is achieved by separating the microbial flocs from the liquid by gravity sedimentation and recycling them in suspended growth reactors or by passing the liquid flow past the biofilm attached to the solid surfaces. When the suspended growth reactor is at steady state, SRT is defined as the inverse of the specific growth rate (μ) (Rittmann and McCarty, 2001). Hence, washout of nitrifiers occurs when the SRT is shorter than μ^{-1}. The maximum specific growth rate of nitrifiers is known to be much lower than that of heterotrophs; the maximum specific growth rate for heterotrophs is typically in the range of 4 to 13.2 day^{-1}, in contrast, that of nitrifiers is 0.62 to 0.92 day^{-1} (Rittmann and McCarty, 2001). In general, SRT of a nitrification tank is increased at the expense of that of a denitrification tank, especially at low temperatures.

For biofilm processes, a mass balance on active biomass is expressed as follows (Rittmann and McCarty, 2001):

$$[d(X_f dz)]/dt = Y(-R_{ut})dz - b'X_f dz \quad \textbf{(1)}$$

where X_f is a uniform biomass density, dz is the thickness of a differential section of biofilm, Y is the true yield for cell synthesis, R_{ut} is the substrate utilization rate, and b' is an overall biofilm specific loss rate.

At steady state, equation 1 is:

$$0 = YJ - b'X_f L_f \quad \textbf{(2)}$$

where J is the substrate flux into the biofilm and L_f is a uniform biofilm thickness.

Biomass density per unit area ($X_f L_f$) is obtained by dividing YJ by b':

$$X_f L_f = YJ/b' \quad \textbf{(3)}$$

Therefore, it is obvious that the substrate flux (J) and biofilm detachment rate (b') directly control biomass retention in the biofilm reactor (i.e., SRT) at steady state.

Nitrification is typically most efficient under aerobic conditions. On the other hand, chemo-organo-heterotrophic denitrification is typically most efficient under anoxic conditions and requires organic electron donors. Typical municipal wastewater is rich in both organic material and ammonia nitrogen (biochemical oxygen demand [BOD]/TKN ratio, 5 to 10) (Rittmann and McCarty, 2001). For removal of both ammonia and organic material, aerobic nitrification reactions must precede the anoxic denitrification reactions to generate nitrate, which is reduced to N_2 in the denitrifying tank. However, organic carbon concentrations must be low for nitrification to proceed due to the competition with heterotrophs for DO and space. Furthermore, denitrification is usually limited by the organic carbon source. Therefore, a portion of untreated wastewater is bypassed to anoxic denitrifying tank to supply organic carbon for denitrification reaction (Fig. 1A). Otherwise, addition of an exogenous carbon sources such as methanol, which is actually the least expensive among all commercially purchased external electron donors, is sometimes needed. In such a system, ammonia nitrogen present in the bypassed wastewater cannot be removed sufficiently, leading to a low maximum nitrogen removal rate (up to approximately 60%). An alternative process is the Bardenpho process (Barnard, 1975), in which denitrification is performed efficiently using untreated wastewater as the organic source (Fig. 1B). This system generally consists of oxic and anoxic tanks and requires a high recycle flow of nitrate produced in the oxic nitrifying tank to the anoxic denitrifying tank. The complication of the system sometimes makes it difficult to control the process performance. In addition, the operational cost (i.e., pumping cost) increases with the recycle ratio in this system.

Factors Affecting Nitrifying Activity in WWTP

Various investigations had been conducted to understand factors affecting nitrifying activity in WWTPs in the second half of the previous century; these contributed to fundamental information necessary for establishing stable biological nitrogen removal processes. An immense body of literature published up to the mid-1970s was thoroughly reviewed by such authors as Painter (1970, 1986), Focht and Chang (1975), and Sharma and Ahlert (1977). These most widely recognized review articles may not necessarily be outdated and are still quite informative as far as a lot of physicochemical and kinetic parameter values are systematically reviewed. Here, the authors refer to articles demonstrating recent progress in understanding physicochemical factors affecting nitrification activities, which may help in applying a nitrification process to a various types of wastewater.

EFFECT OF $NH_3 - NH_4^+$ CONCENTRATION AND pH

In a properly operated nitrification process, nitrification consumes significant alkalinity, and, in the absence of adequate control of pH, overall process failure can occur. In general, pH is controlled between 7.2 and 8.9 (Tchobanoglous et al., 2003). It has been recognized that NH_3 (free ammonia) rather than NH_4^+ (ionized form of the ammonium) is the energy substrate for *Nitrosomonas* and other chemolithotrophic aerobic AOB. pH is the key parameter governing $NH_3 - NH_4^+$ and $NO_2^- - HNO_2$ equilibria; NH_3 and HNO_2 concentrations are higher at higher and lower pH, respectively. Anthonisen et al. (1976) hypothesized that the nonionized forms of ammonium and of nitrite, NH_3, and HNO_2 inhibit nitrifying organisms. Based on this hypothesis, he created a diagram to specify which combination of pH and either total ammonium or total nitrite concentrations allow stable nitrification. Many researchers had supported this hypothesis (summarized by Sharma and Ahlert, 1977). This idea and his diagram still possess practical importance on operating and designing nitrification processes. Availability of CO_2 necessary for the growth of chemolithotrophic AOB and NOB is affected by pH as it dissolves more readily into water at higher pH. Considering availability of CO_2 and NH_3 and the potential adverse effect of NH_3 and HNO_2, weak alkaline pH around 7.5 would be the most favorable, especially for chemolithotrophic AOB. Sensitivity of ammonia also depends on the physiological nature of chemolithotrophic AOB. Suwa et al. (1994) found that predominant AOB in typical sewage sludges are often sensitive to a higher concentration of NH_4^+, while those in a reactor highly enriched with higher concentrations or loadings of NH_4^+ were more NH_4^+ tolerant. Both NH_4^+-sensitive and NH_4^+-tolerant strains were isolated. Each strain was grouped distinctively in distant lineages (Suwa et al., 1997). It was shown that values of half-saturation constant for NH_4^+, $K_{A,NH4}$, for sensitive strains were lower than those of tolerant strains (Suwa et al., 1994).

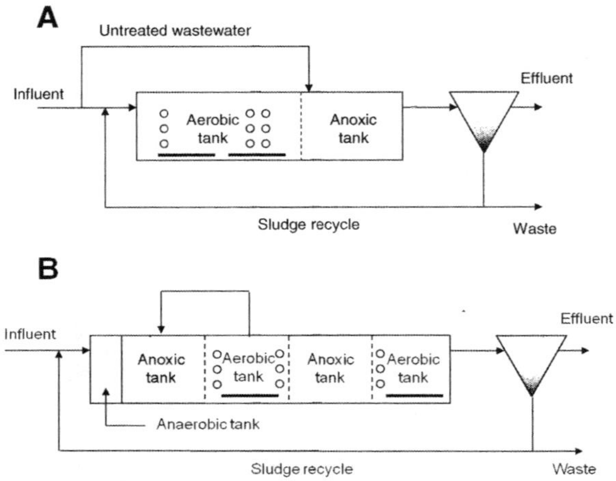

FIGURE 1 Typical process flow sheets for biological nitrogen removal. (A) A portion of the wastewater can be bypassed to the anoxic tank (denitrifying tank). (B) Bardenpho process.

Free ammonia inhibits not only ammonia oxidation but also nitrite oxidation (Anthonisen et al., 1976), and nitrite oxidation behaves often more sensitively than ammonia oxidation, which results in accumulation of nitrite. Concentrations in the range of 0.1 to 1.0 mg of NH_3 liter^{-1} is apparently inhibitory to nitrite oxidation, while appreciable inhibition of ammonia oxidation were observed on and higher than 7 to 10 mg of NH_3 liter^{-1} (Abeling and Seyfried, 1992). Nitrification processes of which the major product is nitrite have been developed as a key component in an energy-saving, short-circuit biological nitrogen removal system. Nitrification processes with nitrite accumulation were originally combined with the denitrification process and later incorporated with an anammox (anaerobic ammonia oxidation) process. In principle, nitrite accumulates when AOB is more active or grows faster than NOB. Such an unbalanced activity between AOB and NOB can be obtained in the presence of free ammonia at higher pH (Abeling and Seyfried, 1992; Isaka et al., 2007) as well as at higher temperature (Hellinga et al., 1998; van Dongen et al., 2001a, 2001b; Volcke et al., 2006) and at lower DO concentrations (Garrido et al., 1997; Bernat et al., 2001; Tokutomi, 2004). Inhibition of nitrite oxidation by free ammonia, of which level is controlled with combination of total NH_3 + NH_4^+ concentration and pH, has been demonstrated to be a realistic choice in developing a nitrification process with nitrite accumulation (Abeling and Seyfried, 1992; Isaka et al., 2007).

EFFECT OF DISSOLVED OXYGEN CONCENTRATION

As an oxidation process, nitrification significantly consumes oxygen, and dissolved oxygen (DO) concentration is a key factor for maintaining nitrification stably as well as pH. Nitrifying rates could readily be lowered at low DO concentrations, which could be explained by a relatively high half-saturation constant for oxygen ($K_{A,O}$) (Tchobanoglous et al., 2003). Thus, continuous operation with such a low DO level as below $K_{A,O}$ of nitrification may lead to washout of nitrifiers from the process and replacement of non-nitrifying organisms with lower $K_{A,O}$, which may result in failure of the process. In prolonged exposure to low DO conditions, physiological adaptation and/or population shift of nitrifying population to the given condition could not be ignored.

It is notable that differences in affinities to DO by AOB and NOB in a nitrification process have been observed. According to results obtained by Garrido et al. (1997), who have demonstrated appreciable nitrite accumulation in the process at low DO, $K_{A,O}$ for AOB population (nitritification) in a nitrifying biofilm process was about 0.5 mg of DO liter^{-1}, while that for NOB population (nitratification) was at least three times higher. This observation could be explained by the fact that AOB activity was not readily suppressed, while NOB activity was considerably lowered at low DO concentrations. Thus, DO may be another realistic controlling factor for establishing a nitrification process with nitrite accumulation (Garrido et al., 1997; Bernat et al., 2001; Tokutomi, 2004).

EFFECT OF TEMPERATURE

The growth rates of both AOB and NOB are greatly affected by temperature. The sludge retention time at which complete nitrification occurs progressively decreases with increasing temperature (Prosser, 1989). Applying and adapting a nitrification process to a lower temperature has been a major issue in northern climates, and a lot of fundamental studies have been undertaken mainly for upgrading a conventional biological BOD removal process to a nitrification process. Basically, these efforts could be made for extending SRT to retain nitrifiers by applying a membrane bioreactor that uses membrane separation instead of gravity sedimentation for obtaining treated water (Kishino et al., 1996), by applying granulated activated sludge biomass after acclimating it to low temperature (de Kreuk et al., 2005) and by adding support material (Hoilijoki et al., 2000). In other cases, the aeration period was extended to compensate for the low nitrification rate (Oleszkiewicz and Berquist, 1988). Frequent supplemental addition (i.e., bioaugmentation) of nitrifying biomass grown in a separate side-stream aeration tank for cultivating backup biomass of nitrifiers was also attempted to balance washout of nitrifiers from the process (Kos, 1998). These techniques were basically successful in maintaining nitrifying activity as low as 14°C, or at an even lower temperature, such as at 7°C.

Maximum specific growth rate (μ_{max}) of AOB often exceeds that of NOB at higher temperatures (van Dongen et al., 2001a, 2001b). Thus, by properly maintaining a higher temperature (30 to 35°C) and a shorter SRT (about 1 day), NOB can be eliminated exclusively from microbial consortia, while AOB is retained. This is the principle for maintaining AOB population in the single reactor system for high ammonium removal over nitrite (SHARON) process, in which nitrite is accumulated (Hellinga et al., 1998). By properly modifying operational conditions of the SHARON process, partial conversion of influent ammonium can be accomplished, and the most appropriate composition for the subsequent anammox process, 50% NH_4^+ + 50% NO_2^-, can be obtained (van Dongen et al., 2001a, 2001b; Volcke et al., 2006). Based on an observation that NOB in activated sludge is much more susceptible to heat shock than AOB, a feasibility study for developing a nitrification process with nitrite accumulation has also been attempted (Isaka et al., 2008).

SUBSTANCES INHIBITORY TO NITRIFYING ACTIVITY

Many compounds commonly found in wastewater, even volatile fatty acids, glucose, and soluble microbial products (SMPs) produced through microbial activities in activated sludge, have an adverse effect, more or less, on nitrification, directly and indirectly (Hanaki et al., 1990; Eilersen et al., 1994; Ichihashi et al., 2006). Thus, as nitrifiers are susceptible to various organic compounds, it would be better to eliminate BOD for stable nitrification.

Madoni et al. (1999) demonstrated that heavy metals Cd, Cu, Zn, Pb, and Cr were less toxic to nitrification, in this order. Among these heavy metals, the levels of inhibition by Cr^{6+} and Zn^{2+} were similar in both nitrification and heterotrophic oxygen uptake rate, while nitrifiers exhibited a lower sensitivity to Cd^{2+}, Cu^{2+}, and Pb^{2+} than did heterotrophs. Dahl et al. (1997) suggested that nitrification was more sensitive to heavy metals than denitrification. Toxicity of Cu^{2+} to *Nitrosomonas europaea* cell increased with increases in ammonia concentration (Sato et al., 1988), which was explained by the formation of copper-ammine complexes. Although Cd^{2+} has a strong inhibitory effect on nitrification, inhibition was almost completely recovered by the addition of EDTA, a good chelating agent to Cd^{2+} (Semerci and Cecen, 2007). EDTA prevented biosorption of Cd^{2+} onto a nitrifying bacterial cell, suggesting that the Cd-sensitive site may not exist inside of the cell but is possibly localized instead on the surface of the cell (Semerci and Cecen, 2007). It is reported that the presence of zeolite in the activated sludge process helped to recover from suppressed nitrifying activity caused by Zn, because zeolite adsorbs Zn (Park et al., 2003).

In an activated sludge process, 3% salt inhibited both the maximum utilization rate and the saturation constant, suggesting uncompetitive inhibition (Dinçer and Kargi, 2001). Actually, wastewater with high salinity at 1% or higher considerably alters the community structure of activated sludge (Chen et al., 2003) and also has a significant inhibitory effect (Furukawa et al., 1993; Chen et al., 2003). Wastewater containing high salinity (or substances at high concentrations) possesses high osmotic pressure. Addition of sodium sulfate to increase osmotic pressure up to 19.2×10^5 Pa, while the NH_4^+ concentration in influent and NH_4^+ loadings to nitrifying airlift reactor was unchanged, resulted in abrupt inhibition of nitrification (Jin et al., 2007). The inhibition to nitrification gradually recovered by lowering the osmotic pressure. This inhibitory effect of osmotic pressure on nitrification was partially relieved by the addition of potassium. The authors explained that potassium might help to properly control cytoplasmic water activity.

Phenol, cyanide, and thiocyanate in wastewater as a result of coke and steel processing and mining are typically powerful inhibitory substances for nitrification. However, phenol can be readily degraded in activated sludge, if it is properly acclimated to phenol, and does not inhibit nitrification as far as it is degraded to a low level (Amor et al., 2005). Contrary to the adverse effect to nitrification, phenol may work as an electron donor for subsequent denitrification, and nitrification and denitrification using phenol can be maintained in the same reactor (Yamagishi et al., 2001). Cyanide and thiocyanate can also be degraded microbiologically and produce ammonium, carbonate, and sulfate. A nitrifying microbial consortium capable of degrading thiocyanate can be established (Lay-Son and Drakides, 2008). Kim et al. (2008) demonstrated that thiocyanate at over 200 mg liter^{-1} apparently inhibited nitrification in activated sludge obtained from a full-scale process treating wastewater from a coke manufacturing plant; however, it is not due to toxicity of thiocyanate itself but due to increased loading of free ammonia, NH_3, produced via degradation. As thiocyanate degradation was inhibited by NH_3, pH, which controls NH_3-NH_4^+ equilibrium, would be a key factor to maintain nitrification and thiocyanate degradation concurrently. It is also notable that free cyanide is very toxic and inhibits nitrification at concentrations of 0.2 mg liter^{-1} or higher. On the other hand, its salt, ferric cyanide, is not so toxic, as it does not inhibit nitrification at concentrations of 100 mg liter^{-1} or lower (Kim et al., 2008).

Niche Separation

With the development of molecular techniques, the inherent biases in isolation and cultivation of microorganisms have been circumvented, and phylogenetic compositions of AOB and NOB have been determined. The phylogeny of aerobic AOB is simple with two monophyletic groups in the beta and gamma subclasses of the *Proteobacteria*, respectively. A continually expanding database of AOB 16S rRNA gene sequences has led to the description of distinct clusters within the betaproteobacterial AOB from the family "*Nitrosomonadaceae*," five within the genus *Nitrosomonas* and five within the genus *Nitrosospira*. All betaproteobacterial AOB form a phylogenetically coherent group within which all organisms exhibit the same primary physiology. Niche differentiation occurs based on the physiological characteristics of the groups.

In industrial and domestic wastewater treatment systems, there appears to be selection for either predominance of a single AOB population or several different AOB populations occur together (Mobarry et al., 1996; Dionisi et al., 2002; Adamczyk et al., 2003; Wittebolle et al., 2008; Wells et al., 2009). For example, *Nitrosococcus mobilis* (this species phylogenetically belongs to the genus *Nitrosomonas* and thus should be reclassified as *Nitrosomonas mobilis* [Head et al., 1993]) and *Nitrospira* sp. dominated in an industrial WWTP that receives extraordinarily high ammonia concentrations (up to 5,000 mg liter^{-1}) (Juretschko et al., 2002). A swine WWTP receiving more than 1,000 mg of NH_4^+-N liter^{-1} exhibited a predominance of AOB closely related to *Nitrosomonas* sp. clone 74 (S. Okabe, unpublished data), which was detected from the SHARON reactor treating an anaerobic digestion effluent with high concentrations of NH_4^+-N (Logemann et al., 1998).

In contrast, the coexistence of the different AOB populations related to the *N. europaea*, *Nitrosomonas oligotropha*, and *Nitrosospira* sp. clusters was found in a domestic WWTP and night soil treatment plant (the fourth aeration tank) with the middle NH_4^+-N concentration range (NH_4^+-N = 6.7 to 22 mg of N liter^{-1}) (S. Okabe, data unpublished data). Similarly, Gieseke et al. (2001) also reported the coexistence of three different AOB populations (*N. europaea/eutropha*, *N. mobilis*, and *N. oligotropha*) in a phosphate-removing biofilm from a sequencing batch biofilm reactor fed with artificial wastewater. Furthermore, fluorescence in situ hybridization (FISH) analysis revealed that *N. europaea* and *N. oligotropha* were detected at the surface biofilm, whereas only *N. oligotropha* dominated in the deeper biofilm layers. This separation in space is suggested as a mechanism that allows coexistence of three different AOB populations in the biofilm. Furthermore, four different AOB populations were found in a full-scale nitrifying trickling filter, with two *N. oligotropha* populations dominating at all depths of the filter and *N. europaea* only at 0.5 m (Lydmark et al., 2006). In the same NH_4^+-N concentration environments, *Nitrosospira* dominated 90% of AOB in the aquarium seawater purification system (S. Okabe, unpublished data). Thus, salinity is definitely one of the selective factors. However, a notably high diversity of AOB (*N. europaea*, *N. eutropha*, and *N. mobilis*) and NOB (*Nitrospira*-like and *Nitrobacter*) populations was found in a sequencing biofilm batch reactor receiving high ammonia and salt concentrations (Daims et al., 2001a). It is unusual that such an extreme condition generally selects a monoculture of AOB or NOB (Juretschko et al., 2002). This high diversity could be explained on the basis of a complex biofilm ecosystem where a different microenvironment was created within the biofilm due to various nutrient gradients (Okabe et al., 1999b).

Generally, the distribution patterns of the distinct species in engineering systems reflect the physiological properties such as affinity to NH_4^+ and DO so that the NH_4^+-N concentration influences the extent of AOB diversity. In high and low NH_4^+-N concentration environments, the level of AOB diversity was, in general, low. In contrast, a greater diversity was found in the middle NH_4^+-N concentration environments (e.g., domestic WWTP) (Juretschko et al., 1998; Okabe et al., 1999b; Daims et al., 2001a; Juretschko et al., 2002; Lydmark et al., 2006).

For NOB, *Nitrobacter* was traditionally considered to be the most important NOB in WWTPs. Using the full-cycle rRNA approach, the occurrence of yet uncultured *Nitrospira*-like NOB in nitrifying WWTPs often has been demonstrated (Juretschko et al., 1998; Okabe et al., 1999b, 2002; Daims et al., 2001a, 2001b; Gieseke et al., 2001; Kindaichi et al., 2004). Daims et al. (2001a) investigated the ecophysiology of the uncultured *Nitrospira*-like NOB in activated sludge by using microautoradiography combined with FISH (MAR-FISH). It has been suggested that *Nitrospira*-like NOB are probably *K*-strategists for oxygen and nitrite with high substrate affinities and low maximum activity or growth rates compared to the *r* strategists, such as *Nitrobacter* spp. (Blackburne et al., 2007; Ahn et al., 2008). Thus, *Nitrospira*-like NOB outcompete *Nitrobacter* under substrate-limiting conditions like WWTPs (Schramm et al., 1999a; Kim and Kim, 2006) or within the deeper part of the biofilm where the O_2 concentration is low (Okabe et al., 1999b). This hypothesis would also explain why *Nitrobacter* and *Nitrospira* coexist in sequencing batch biofilm reactors with temporarily high nitrite concentrations (Daims et al., 2001a). For a long time, the NOB isolated from activated sludge samples were the genus *Nitrobacter*. This is because *Nitrobacter* spp. grow better in pure cultures with high NO_2^- concentrations than do *Nitrospira* spp. and outcompete *Nitrospira* spp. during standard enrichment and isolation procedures. This is probably a reason why *Nitrospira* spp. have previously been overlooked in wastewater treatment processes.

Of course we must be cautious about the interpretation of these studies because the community structures of AOB and NOB would be greatly influenced by other environmental factors such as salinity, pH, and concentrations of DO and NO_2^-. In addition, environments like wastewater biofilms and microbial flocs are very heterogeneous, and thus it is difficult to define the numerical dominance of distinct species or groups of species in distinct environments. Hence, distribution patterns may overlap, even if clear differences in ecophysiological characteristics are recognizable among the species. In addition, the varying composition of the wastewater between different WWTPs, in combination with the different types of reactors used, makes it very difficult to draw general conclusions about the community structure of nitrifying bacteria.

The high level of AOB and NOB diversity found in the reactor might relate to the stability of the reactor performance. Hence, engineering a system with a greater diversity may improve the performance and stability by, for example, more efficient bioaugmentation (Rittmann and Whiteman, 1994; Satoh et al., 2003b).

ACTIVATED SLUDGE SYSTEMS

The activated sludge process is the most widely used system for nitrogen removal from both domestic and industrial wastewaters (Blackall and Burrell, 1999). Wastewater treatment in the activated sludge process is based on attachment and subsequent biological degradation of both organic and inorganic compounds by microorganisms retained in microbial aggregates, termed activated sludge flocs. Effective immobilization of nitrifying bacteria in the activated sludge flocs is one of the most important factors for efficient nitrogen removal in the activated sludge process. Activated sludge floc is a highly dense microbial aggregate that results in the development of heterogeneous microenvironments (e.g., stratification of electron donors and acceptors and steep gradients of pH and the oxidation-reduction potential) in a single floc (Lens et al., 1995; Schramm et al., 1999b; Satoh et al., 2003a; Li and Bishop, 2004) and makes chemical properties inside the floc quite different from that prevailing in the bulk liquid. In addition, planktonic bacteria growing at a low rate like nitrifying bacteria are likely to be washed out from a reactor (Larsen et al., 2008). Furthermore, the grazing impact exerted by predators is different from bacterial species (Jürgens and Matz, 2002). Therefore, microbial diversity in the floc becomes greater than that of planktonic cells in the bulk liquid (Pogue and Gilbride, 2007). Several excellent reviews offer detailed description of ecophysiology of nitrifying bacteria in activated sludge flocs (Blackall and Burrell, 1999; Schramm, 2003). Here, we will describe a few of the major features of community structures, spatial distribution, and population dynamics of nitrifying bacteria and their in situ activities within single activated sludge flocs.

Phylogeny and Spatial Distribution of Nitrifying Bacteria in Activated Sludge Systems

The 16S rRNA gene and/or ammonia monooxygenase subunit A (*amoA*) gene-based molecular approaches have revealed the genus *Nitrosomonas* is the most abundant AOB in activated sludges obtained from a full-scale municipal WWTP (Wittebolle et al., 2008), a continuously stirred tank reactor, a sequencing batch reactor (Mobarry et al., 1996), an aerated activated sludge bioreactor (Wells et al., 2009), nitrification tanks of industrial (*N. europaea* and *N. eutropha*) and municipal (*N. oligotropha*) WWTPs (Adamczyk et al., 2003), and aeration basins of municipal (*N. oligotropha*) and industrial (*N. nitrosa*) WWTPs (Dionisi et al., 2002). Predominance of the genus *Nitrosomonas* is due to their relatively higher growth rate than other AOB (Prosser, 1989).

Numerous reports have demonstrated that NH_4^+ oxidation is not restricted to autotrophic AOB. Heterotrophic nitrification was initially described as early as 1894 (Stutzer and Hartleb, 1894). Today, it is recognized to be a widespread phenomenon among different genera of fungi and heterotrophic bacteria, such as *Diaphorobacter* spp. (Khardenavis et al., 2007), *Alcaligenes faecalis* (Joo et al., 2006), and *Shinella zoogloeoides* (Bai et al., 2009). It was also recently discovered that NH_4^+ oxidation is not restricted to the domain *Bacteria* (Konneke et al., 2005). An ammonia-oxidizing archaeon, *Nitrosopumilus maritimus*, was isolated from the rocky substratum of a tropical marine aquarium tank, representing the first cultivated isolate of the ubiquitous marine group 1 of the phylum *Crenarchaeota* (Konneke et al., 2005). Ammonia-oxidizing archaea have been detected in activated sludge bioreactors (Park et al., 2006; You et al., 2009), a highly aerated activated sludge process (Wells et al., 2009), a laboratory-scale nitrogen removal bioreactor, and Hong Kong WWTPs treating saline or freshwater wastewater (Zhang et al., 2009b).

Concerning the NOB, recent findings obtained by 16S rRNA gene-based analysis demonstrated that the *Nitrospira* group dominated the *Nitrobacter* group in activated sludges. For example, in activated sludges obtained from a full-scale municipal WWTP, a sequential batch reactor, and a membrane bioreactor, the *Nitrobacter* group was not detected, whereas the *Nitrospira* group was present in high amounts (10^8 to 10^9 cells/mg of VSS) (Wittebolle et al., 2008). *Nitrobacter* were not detectable by FISH with probe NIT3, whereas *Nitrospira*-like bacteria were present in significant numbers (9% of the total bacterial counts) in an intermittently aerated nitrification-denitrification basin of an industrial WWTP (Juretschko et al., 1998). These results can be explained by the concept of *r*- and *K*-strategists (Kim and Kim, 2006; Blackburne et al., 2007; Ahn et al., 2008).

Microscope observations revealed that a floc consists of microorganisms, extracellular polymeric substances (EPSs), inert materials, and void space ranging from 30 to 80% (Schramm et al., 1999b) (Fig. 2). Figures 2B and C show images representing typical examples of an activated sludge floc structure and localization of the bacteria and AOB in a floc. AOB formed densely packed microcolonies. Filamentous microorganisms were found in the flocs, implying that could contribute to the stability of flocs (Bossier and Verstraete, 1996). Several studies provided evidence that NOB cells formed smaller clusters and associated with AOB microcolonies in the activated sludge flocs (Mobarry et al., 1996; Juretschko et al., 1998; Daims et al., 2006). This spatial organization may reflect the syntrophic association between AOB and NOB. AOB and NOB are partners in a mutualistic symbiosis. The AOB produce NO_2^- as the substrate of the NOB, whereas NO_2^- is toxic to the AOB. The NOB consume the NO_2^- that would otherwise inhibit the growth of AOB. These symbiotic interactions among microorganisms were successfully visualized and analyzed with novel three-dimensional image analysis software (Daims et al., 2006).

A number of studies have examined the influence of various environmental factors on AOB community structure in WWTPs (Tanaka et al., 2003; Park and Noguera, 2004). For example, Wells et al. (2009) comprehensively investigated correlations between AOB population dynamics and water quality represented by 20 operational parameters in an activated sludge bioreactor in a municipal WWTP by using nonmetric multidimensional scaling and redundancy analyses. Temperature was the most important variable affecting the AOB population dynamics; the *Nitrosospira* lineage showed strong negative correlations ($P < 0.001$) to temperature in the range of 18 to

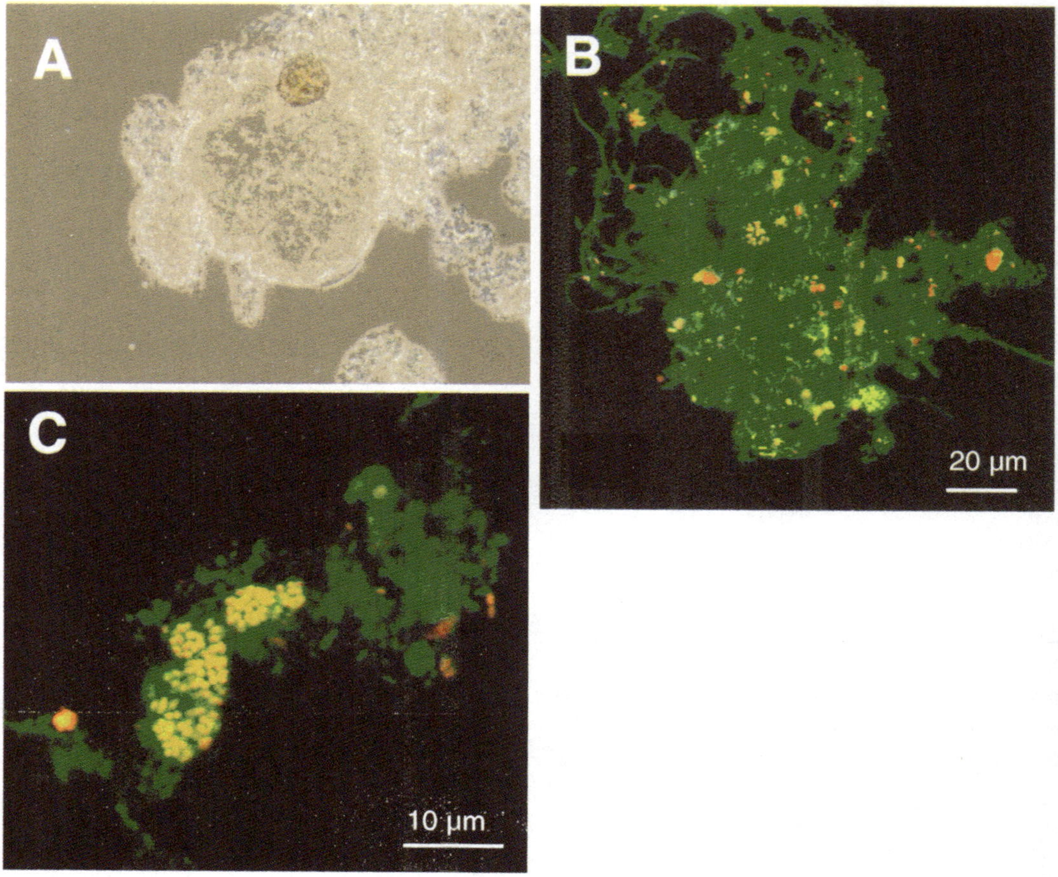

FIGURE 2 (A) Photomicrograph of an activated sludge floc. (B and C) Confocal laser-scanning microscope images of an activated sludge floc showing the in situ spatial organization of bacteria and AOB. FISH was performed using a fluorescein isothiocyanate-labeled EUB338-mixed probe and a tetramethylrhodamine 5-isothiocyanate-labeled Nso190 probe. The probe Nso190-stained AOB appear to be yellow because of binding of both probes, and bacterial cells are green.

25°C. DO was also linked to the AOB population dynamics and strongly negatively correlated ($P < 0.01$) especially with the *Nitrosospira*. Influent NO_2^-, chromium, and nickel influenced the AOB community structure, while correlations between other metals analyzed in this study and the AOB community structure were insignificant. These studies revealed how operational and environmental factors affect community dynamics within bioreactors. In addition, since such factors might elicit different responses from different nitrifying bacteria, their activities directly influence bioreactor performance due to their impacts on process efficiency and stability (Briones and Raskin, 2003). At present, although the relative importance of specific deterministic environmental factors to nitrifying bacteria in full-scale systems is uncertain, these studies might provide insight into parameters controlling nitrifying bacterial community structures for stable nitrification in activated sludge reactors.

Adhesion characteristics of nitrifying bacteria and their microcolonies should directly play an important role in the dynamics of nitrifying bacterial communities, because the ability of nitrifying bacteria to adhere to activated sludge flocs is a fundamental basis for stable nitrification of activated sludge, and poor floc formers are likely to be washed out with the effluent and are much more accessible to predators. Larsen et al. (2008) demonstrated AOB (*N. oligotropha*) and NOB (*Nitrospira* spp.) formed strong microcolonies that were very resistant to high shear force and different physicochemical treatments (pH, O_2, sulfide, and addition of EDTA and Triton X-100) compared with other bacteria. Only very little erosion of single cells took place. Microcolonies of *Nitrospira* spp. were generally slightly stronger than *N. oligotropha*. These results clearly showed that the nitrifying bacteria remained almost intact even under extreme physical and chemical conditions, such as in aeration tanks, settling tanks, and pumping.

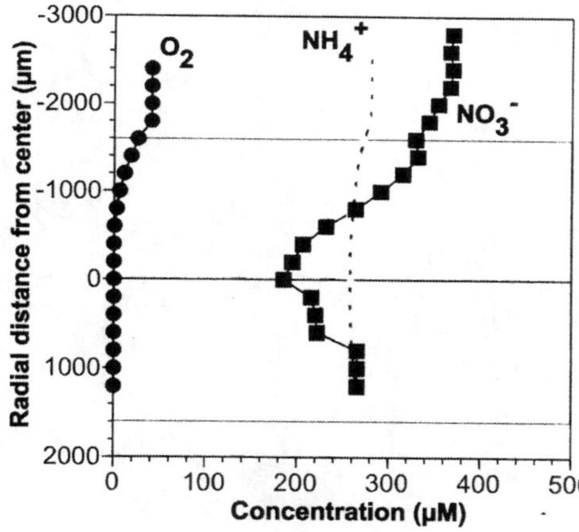

FIGURE 3 Typical concentration profiles of O_2, NH_4^+, and NO_3^- in an activated sludge floc at 45 µM O_2. The shaded area indicates the floc. The center of the floc is at a depth of 0 µm. (From Satoh et al. [2003a], with permission from *Biotechnology and Bioengineering*.)

In Situ Activity Measurement

A variety of molecular techniques described above have indeed provided valuable information on the community structure and diversity of nitrifying bacterial populations in the activated sludge process. In contrast, microsensor measurements have revealed spatial distribution of nitrifying activity in single flocs (Satoh et al., 2003a). O_2 penetrated the entire floc with a diameter of approximately 3,000 µm at an O_2 concentration of 195 µM. The NH_4^+, NO_3^-, and pH microprofiles showed that nitrification occurred throughout the floc. The O_2 penetration depth in the flocs decreased with decrease in the O_2 concentration in the bulk liquid. The O_2 penetration depths were 1,200 µm and only 200 µm at O_2 concentrations of 45 µM (Fig. 3) and 15 µM, respectively, and thus an anoxic zone was created in the flocs. Nitrification was restricted to an outer oxic zone of the floc, and denitrification occurred in an inner anoxic zone. Therefore, simultaneous nitrification and denitrification (SND) occurred at 45 µM O_2 in the bulk liquid, whereas only denitrification occurred at 15 µM of O_2. Similarly, the O_2 penetration in a floc was limited to the surface layer of the flocs, and SND occurred in the flocs with a diameter of approximately 1,000 µm at less than 1 mg liter of O_2^{-1} (Li and Bishop, 2004). Schramm et al. (1999b) detected anoxic zones in single flocs, where denitrification occurred but sulfate reduction could not be detected.

Batch experiments were performed to investigate the effect of bulk O_2 concentration on the rates of nitrification and denitrification of the activated sludges (Satoh et al., 2003a). Production of NH_4^+ at near 0 µM O_2 could be explained by biomass degradation and liberation of NH_4^+ adsorbed on biomass (Fig. 4). Nitrification occurred at >10 µM O_2. The nitrification rate of the activated sludges increased with increasing O_2 concentration in the bulk liquid up to 40 µM, indicating that O_2 was a limiting factor. Nitrification rates remained constant at more than 40 µM O_2. SND was observed at O_2 concentrations ranging between 10 and 35 µM with a maximum rate of 4 µmol g of MLSS^{-1} h^{-1} at 35 µM of O_2, although nitrification was incomplete. The absence of denitrification at O_2 concentrations near 0 or >35 µM could be explained by the absence of NO_3^- produced by nitrification and the inhibition of denitrification by O_2, respectively. The average O_2 concentration

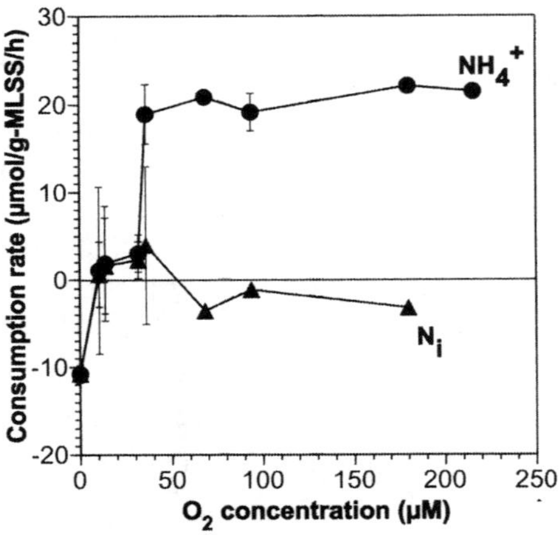

FIGURE 4 The consumption rates of NH_4^+ and inorganic nitrogen (Ni) defined as the sum of NH_4^+, NO_2^-, and NO_3^- as determined by the batch experiments at various O_2 concentrations in the bulk liquid. Rates shown are mean values, and error bars indicate standard deviations. (From Satoh et al. [2003a], with permission from *Biotechnology and Bioengineering*.)

in the aeration basin from which the activated sludge samples were obtained was in this range (15 μM), which could explain the occurrence of SND in the basin.

BIOFILM SYSTEMS

General Description of Biofilm Systems

Aerobic biofilm systems such as a biological aerated filter, trickling filter, and rotating biological contactor (Rowan, et al., 2003) allow slowly growing nitrifying bacteria to remain in the reactors by their attached growth and have been used for reliable nitrogen removal. The main drawbacks common to nitrifying biofilm processes are substrate transport limitation and interspecies competition for oxygen and space between nitrifying bacteria and heterotrophic bacteria in the biofilms (Okabe et al., 1996). In practice, various nitrifying biofilm reactors were operated at low surface organic loadings (in the range of 2 to 6 g BOD m^{-2} day^{-1}) to attain successful nitrification (Rittmann and McCarty, 2001). Keeping the organic loading rate below 2 to 6 g BOD m^{-2} day^{-1} controls the interspecies competition with heterotrophs. When the BOD:TKN ratio of wastewater is high (typical municipal wastewater has 5 to 10 g of BOD/g of N), the nitrifying bacteria are forced deeper into the biofilms, which incurs greater mass-transport resistance to the nitrifying bacteria in the long-term operation.

A better understanding of the microbiology, ecology, and population dynamics of nitrifying biofilms is essential for improving process performance and control. However, the investigation has been hampered by their slow growth rates and by the biases inherent in all culture-based techniques. Therefore, the in situ detection of nitrifying bacteria and their activity in biofilms is of great practical and scientific relevance. Especially, wastewater biofilms are complex multispecies biofilms, displaying considerable heterogeneity with respect to both the microorganisms present and their physicochemical microenvironments. Moreover, successive vertical zonations of predominant respiratory processes occurring simultaneously in close proximity have been found in wastewater biofilms with a typical thickness of only a few millimeters (Santegoeds et al., 1998; Okabe et al., 1999a). Therefore, several studies have been conducted on wastewater treatment biofilms by combining FISH and microsensor technology to link the spatial organization of microbial communities to physicochemical parameters in the biofilms (Schramm et al., 1996; Okabe et al., 1999b).

Interspecies Competition with Heterotrophs

In aerobic biofilm systems, competitive interactions of heterotrophic bacteria and nitrifying bacteria for DO, ammonia, and even space are well known. This problem is attributed to the presence of organic matter in wastewater and production of SMPs by nitrifying bacteria that further support heterotrophic growth (Rittmann et al., 1994; Kindaichi et al., 2004; Okabe et al., 2005). Nitrifying bacteria are usually outcompeted by the heterotrophic bacteria, because the $K_{A,O2}$ values are gener-

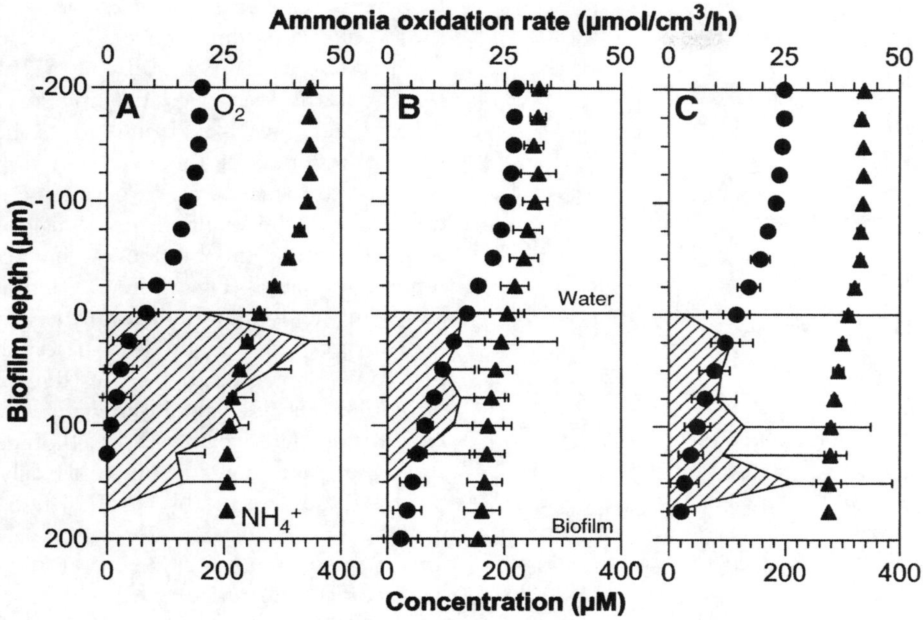

FIGURE 5 Steady-state concentration profiles of O_2 and NH_4^+ in the autotrophic nitrifying biofilm incubated in the media at C/N = 0 (A), C/N = 1 (B), and C/N = 3.4 (C), respectively. The modeled profiles are indicated by solid lines. The spatial distributions of NH_4^+ oxidation rates are indicated by stippled area. Surface is at a depth of 0 μm. (From Okabe et al. [2004b], with permission from *Biotechnology and Bioengineering*.)

ally higher than those of heterotrophs (0.1 mg liter^{-1} for heterotrophs but 0.5 mg liter^{-1} for nitrifying bacteria). Inhibition or elimination of nitrifying bacteria by the interspecies competition leads to a decrease in nitrification efficiency or even to a failure of the process. Therefore, a quantitative understanding of the effect of the substrate C/N ratio on the spatial distribution of nitrifying bacteria and their activities in biofilms is essential for improving process performance.

The effect of the substrate C/N ratio on the spatial distributions of AOB and their in situ activity was investigated by using microelectrodes with high spatial resolution and the FISH technique (Satoh et al., 2000). The volumetric NH_4^+ uptake rates in the surface of the biofilms markedly decreased with the addition of acetate from 22.6 μmol of NH_4^+ cm^{-3} h^{-1} at C/N = 0 to 2.6 μmol of NH_4^+ cm^{-3} h^{-1} at C/N = 3 (Fig. 5). In contrast, the rates were relatively unchanged (20 to 25 μmol of NH_4^+ cm^{-3} h^{-1}) at the bottom of the biofilm. The volumetric oxygen utilization rates were relatively constant in all C/N ratios tested (C/N = 0, 1, and 3). This experimental result clearly demonstrated that the addition of organic carbon immediately induced the interspecies competition for O_2 and space in the outer part of the biofilm, and AOB were outcompeted by heterotrophic bacteria due to high $K_{A,O2}$ (van Niel et al., 1993). As a result of the interspecies competition, the specific NH_4^+ oxidation rates decreased in the outer part of the biofilm, eliminating nitrifying bacteria in the outer part of biofilm and forcing the nitrifying bacteria deeper into the biofilms (Fig. 6), which further incurred greater mass-transport resistance to the nitrifying bacteria in the long-term operation. Furthermore, the average diameter of AOB microcolonies was relatively constant throughout the biofilm at C/N = 0 (Fig. 7). In contrast, the AOB microcolonies in the surface of the biofilm grown at C/N = 1 were

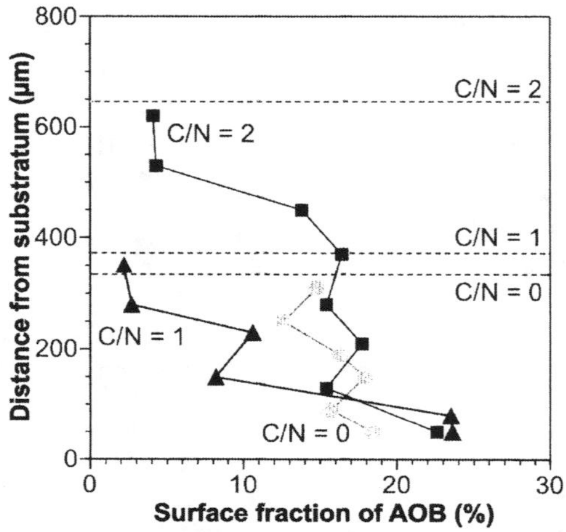

FIGURE 6 Spatial distributions of surface fractions of AOB in biofilms cultured in the media at C/N = 0, 1, and 2, respectively. The biofilm surfaces are indicated by the dotted lines.

significantly smaller than those found in the C/N = 0 and increased toward the bottom of the biofilm. These results clearly indicated that the presence of organic carbon significantly reduced the size of AOB microcolonies in the surface of the biofilm.

In addition, when the substrate C/N ratio increased, the vertical separation of active NH_4^+- and NO_2^--oxidizing zones became evident, resulting in the reduction on nitrifying performance. This experimental result clearly shows the stratified spatial distributions of AOB within the biofilms at higher substrate C/N ratios (Fig. 8). These findings are also significantly important to further improve mathematical models used to describe how the slow-growing AOB develop their niches in biofilms and how that configuration affects nitrification performance in the biofilm. It is also important to note that physiologically inactive AOB are detected by FISH as these AOB maintain high cellular ribosome contents under unfavorable conditions (Wagner et al., 1995). This is one of the difficult parts of the interpretation of FISH data. However, the physiologically active AOB can be detected when FISH analysis was combined with microautoradiography using ^{14}C-labeled bicarbonate as substrate (Lee et al., 1999; Okabe et al., 2004).

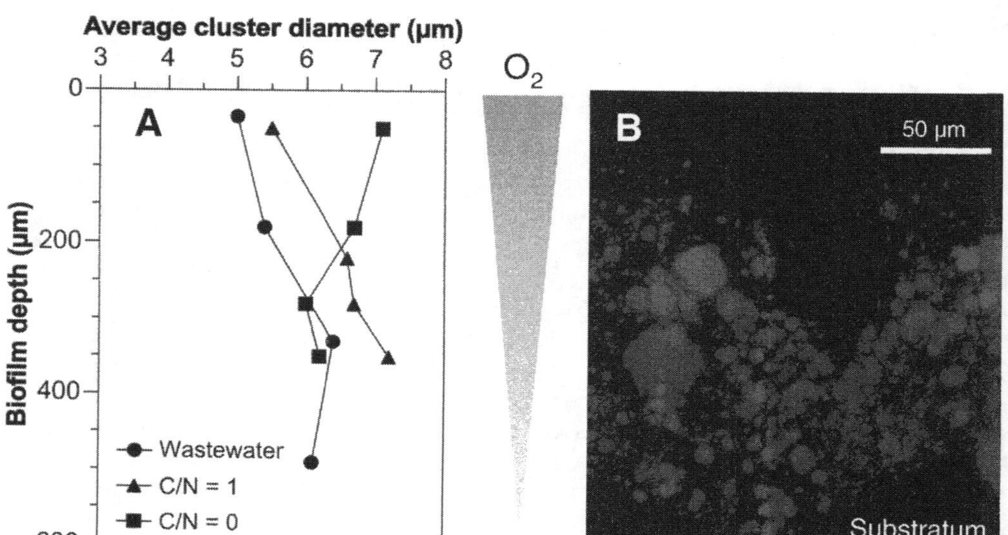

FIGURE 7 (A) Spatial distribution of average microbial cluster sizes of AOB hybridized with probe Nso190 in different biofilms cultured at C/N = 0 and 1. (B) Cross-section of biofilm cultured at C/N = 0.

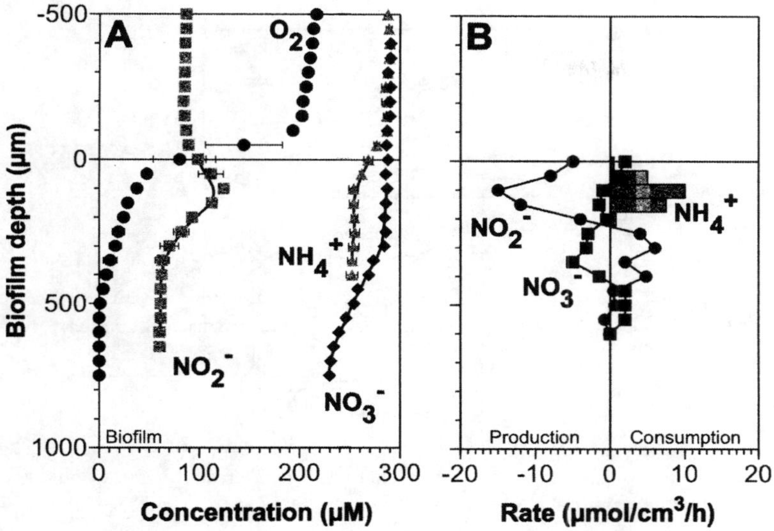

FIGURE 8 (A) Steady-state concentration profiles of O_2, NH_4^+, NO_2^-, and NO_3^- in a biofilm cultured at C/N = 2. (B) Spatial distribution of the estimated volumetric consumption and production rates of NH_4^+, NO_2^-, and NO_3^-. The biofilm surface is at a depth of 0 µm.

Ecophysiological Interaction with Heterotrophic Bacteria

As described above, in the presence of organic carbon, nitrifying bacteria are usually outcompeted by heterotrophs due to the low growth rate and growth yield. However, coexistence of a high level of heterotrophs with nitrifying bacteria has been found often in autotrophic nitrifying biofilms cultured without the external organic carbon supply (Okabe et al., 1999b, 2004). It is, therefore, speculated that they also interact through the exchange of organic matter, because autotrophic nitrifying bacteria reduce inorganic carbon to form organic carbon in cell mass and release SMPs from substrate metabolism (usually with biomass growth) and decay biomass. Ecophysiological interactions between nitrifiers and heterotrophic bacteria in a carbon-limited autotrophic nitrifying biofilm fed with only NH_4^+ as the energy source was investigated by a full cycle of 16S rRNA approach followed by MAR-FISH (Kindaichi et al., 2004; Okabe et al., 2005). The biofilm samples were first incubated with [^{14}C]bicarbonate to radiolabel only nitrifying bacteria (both AOB and NOB). After only nitrifying bacteria were labeled with [^{14}C]bicarbonate, the transfer of ^{14}C originally incorporated in nitrifying bacterial cells to heterotrophic bacteria was monitored with time by using MAR-FISH (Kindaichi et al., 2004; Okabe et al., 2005). MAR-FISH is a powerful technique to simultaneously examine the phylogenetic identity and the relative or actual specific activity of microorganisms within a complex microbial community at a single-cell level.

This autotrophic nitrifying biofilm comprised 50% of nitrifying bacteria (AOB and NOB) and 50% of heterotrophic bacteria. This result indicated that a pair of nitrifiers (AOB+NOB) supported a heterotrophic bacterium via production of SMPs. MAR-FISH analysis revealed that most phylogenetic groups of heterotrophic bacteria except β-*Proteobacteria* showed significant uptake of ^{14}C-labeled microbial products. Particularly, the members of the chloroflexi preferentially utilized microbial products derived from mainly biomass decay (Fig. 9). On the other hand, the members of the *Cytophaga-Flavobacterium* cluster gradu-

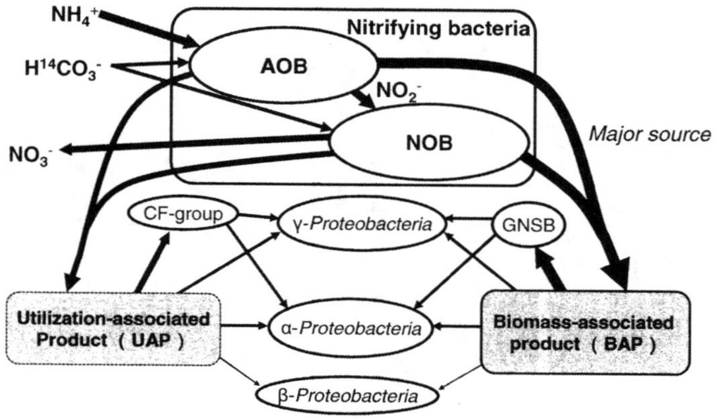

FIGURE 9 Proposed ecophysiological interactions between nitrifiers and heterotrophic bacteria in a carbon-limited autotrophic nitrifying biofilm fed with only NH_4^+ as energy source.

ally utilized ^{14}C-labeled products in the culture with NH_4^+ addition where nitrifying bacteria grew. These results suggested that they preferentially utilized substrate utilization-associated products (UAPs) of nitrifying bacteria and/or secondary metabolites of ^{14}C-labeled structural cell components. The heterotrophic bacterial community was composed of phylogenetically and metabolically diverse and, to some extent, metabolically redundant members, which ensures the stability of the biofilm ecosystem. These results clearly demonstrated that an efficient food web (carbon metabolism) existed in the autotrophic nitrifying biofilm community to assure the maximum utilization of SMP produced by nitrifiers and to prevent build-up of metabolites or waste materials of nitrifiers at significant levels.

MODELING

General Overview of Modeling Study for Nitrification

In the improvement of operation strategies and designs of WWTPs, mathematical models have become increasingly important. At the same time, the models of the involved processes are getting more and more complex as the knowledge of the underlying microbiology improves. Mathematical modeling in the science and engineering fields is used for generally following two purposes: "understanding" and "prediction" of phenomena in natural or engineered systems. The main engineering objective of modeling is prediction of processes to be investigated and control. The final goal is to optimize process performance and to improve system design. The scientific objective of modeling is better understanding of the system to be investigated through a verifying hypothesis because modeling provides an explanation of the system from a more theoretical point of view. Mathematical modeling is a powerful tool for better and general understanding of a complex system as well as for prediction and control of an engineering process, such as microbial activity and community in wastewater treatment processes.

Nitrification in Activated Sludge Model

Since nitrification is one key reaction in the biological wastewater treatment process, especially in a nitrogen removal process, it is an important target for modeling from the practical point of view, and thus nitrification is included in multiple types of mathematical models for wastewater treatment processes.

The Activated Sludge Model (ASM) is known as the most common and standard model for the wastewater treatment process; it was developed by a task group on Mathematical Modeling for Design and Operation of Activated Sludge Processes established by the International Water Association (IWA, formerly IAWQ and IAWPRC). They proposed basic models for design and operation of a biological wastewater treatment process with the aim of creating a world standard for the mathematical model. The model has been extended as proposed as ASM1, ASM2, ASM2d, and ASM3. ASM3, which is the most advanced model in ASMs, simulates removal of organic and nitrogen compound and sludge production through the calculation of biological reactions (nitrification, denitrification, and oxidation of organic compound; growth and decay of microorganisms, etc.) of an activated sludge process. The nitrification is a key process in all series of ASMs and is described as follows: ammonia is oxidized to nitrate via a single-step process resulting in production of biomass (nitrite oxidation step is not considered in the basic model, but it can be added by users if necessary). The nitrifying activity is modeled in ASM3 using Monod kinetics, as shown in the following equation (Gujer et al., 1999; Henze et al., 2000):

$$\frac{dS_{NH4}}{d_t} = \frac{\mu_1}{Y_A} \frac{S_{NH4}}{K_{A,NH4} + S_{NH4}} \frac{S_{O2}}{K_{A,O2} + S_{O2}} \frac{S_{ALK}}{K_{A,ALK} + S_{ALK}} X_4 \quad (4)$$

where μ_A is the maximum specific growth rate (day^{-1}), $K_{A,O2}$ is the oxygen half-saturation coefficient (mg of O_2 $liter^{-1}$), $K_{A,}$NH4 is the ammonium half-saturation coefficient (mg of N $liter^{-1}$), $K_{A,ALK}$ is the alkalinity half-saturation coefficient (mmol of HCO_3^- $liter^{-1}$), Y_A are nitrifying bacterial yields (g of COD g of N^{-1}), X_A is nitrifying bacterial biomass (mg of COD $liter^{-1}$), S_{NH4} is ammonium concentration (mg of N $liter^{-1}$), S_{O2} is DO concentration (mg of O_2 $liter^{-1}$), and S_{ALK} is alkalinity (mmol of HCO_3^- $liter^{-1}$).

In the above equation, μ_A, $K_{A,O2}$, $K_{A,NH4}$, $K_{A,ALK}$, and Y_A are parameters, while X_A, S_{NH4}, S_{O2}, and S_{ALK} are state variables.

The decay (cell death) process is also included in the model as a result of endogenous respiration.

The ASMs have been well received and widely used as a basis for further model development. To date, simulation of the activated sludge system is mature technology. Many types of software products are on the market with parameter sets for different purposes and can be applied to design, operation, and research of WWTPs (Gujer, 2006).

Biofilm Model

In the wastewater treatment process, microorganisms usually exist as biofilms on supporting materials or granules and the self-aggregated matrix called "flocs" in the activated sludge process. It helps to retain large numbers of valuable but slow-growing microorganisms such as ammonia and NOB in a reactor. Therefore, the model representing the biofilm system is necessary apart from the activated sludge model to address more complex spatial and microbial ecological structures and biological, physical, and chemical phenomena from both practical engineering and scientific viewpoints. Biofilm is a complex system, and the development includes many physical, chemical, and biological phenomena with a large number of micro- or macro-scale interactions.

Modeling of biofilm systems is to mathematically describe structure and activity of biofilm and to be able to predict biofilm structure from the initial environmental conditions. Perfect modeling of biofilm is to represent the development process of the following phenomena dynamically: (i) multidimensional heterogeneous morphology of biofilm; (ii) multidimensional spatial distribution of multiple species of microbial cells and their activities resulting from cell growth and decay; (iii) production of EPSs and their distributions; (iv) multidimensional spatial distribution of multiple soluble substrate compounds resulting from consumption and production by metabolic activity of microbes, and transportation by molecular diffusion and convection; (v) hydrodynamics that affect mass transport efficiency and the physical structure of the biofilm; and (vi) reactor performance (i.e., concentration of substrate in outlet or bulk liquid phase) resulting from the above phenomena.

The simplest model (earliest model) describes mass balances of biofilm as a result of transformation (biochemical reaction) and transportation (diffusion) and is shown as an equation described below.

$$\frac{\partial C}{\partial t} = D\left(\frac{\partial^2}{\partial_x^{\,2}}\right)C + r \qquad \textbf{(5)}$$

Transformation of a particular substrate (reaction rate r) can be represented by Monod kinetics as described above and diffusion by Fick's law with an effective diffusion coefficient D and substrate concentration C. In the advanced-type multidimensional biofilm model described below, diffusion direction in the model is simply increased to two or three dimensions as shown in equation 6.

$$\frac{\partial C}{\partial t} = D\left(\frac{\partial^2}{\partial_x^{\,2}}C + \frac{\partial^2}{\partial_y^{\,2}}C + \frac{\partial^2}{\partial_z^{\,2}}\right) + r \qquad \textbf{(6)}$$

The spatial distribution of microbial activity affects the substrate or product concentration. At the same time, the growth rate of a particular microorganism is influenced by the spatial distribution of particular substrates. Therefore, microbial growth rate varies through the biofilm because microorganisms are exposed to different substrate concentrations.

The early model described biofilm as uniform steady-state films consisting of a single species, in which one-dimensional mass transport and biochemical reactions occur to present the important phenomena that the substrate concentration decreased inside the biofilm. Then, the model extended to represent multisubstrate and nonuniform (layered) distribution of multispecies microorganisms. The core of the model is formed by a set of differential equations for the microbial species and substrates in the biofilm and in the bulk (Wanner and Gujer, 1986; Wanner and Reichert, 1996). Since the simple model often is sufficient for the practical purpose, this one-dimensional dynamic multispecies model implemented in AQUASM software is widely used for modeling of aquatic systems (Reichert, 1998).

More complex, bottom-up-type, two- or three-dimensional models describe the dynamics of multiple microbial populations by various approaches, such as grid-based biomass (e.g., cellular automata proposed or developed by Noguera et al. [1999], Picioreanu et al. [1999], and Bell et al. [2005]; continuum [Alpkvist and Klapper, 2007]). Furthermore, for the more realistic representation of biomass division and spreading, individual-based modeling (IbM) (Kreft et al., 2001; Picioreanu et al., 2004) and hybrid individual/continuum (Alpkvist et al., 2006) describe biomass (a bacterial cell or bacterial biomass) as spherical particles with positions in space defined by continuous coordinates. Each biomass particle (minimum units of bacterial biomass) contains active biomass of a single microbial type and inert material biomass and is surrounded by an EPS capsule produced by the biomass within the particle (Xavier et al., 2005). Each biomass particle grows and produces EPS according to a rate defined by concentrations of solutes and particulates. The biomass particles are divided into two daughter particles when their size exceeds a critical size. Each "type (species)" has its own set of variable parameters. Biofilm spreading as a result of biomass growth occurs by shoving the spheres when they get too close to each other. In this model, the pressure that builds up due to biomass growth is relaxed by minimizing the overlap of spheres (Fig. 10A). Since IbM, which treated bacterial cell as a minimum unit, requires more computer resources, biomass-based IbM using larger biomass particles (particles that are 2 to 20 μm in diameter) is more realistic for the general use, while still keeping the shoving or pushing principle for biomass redistribution. A more detailed two-dimensional model description is

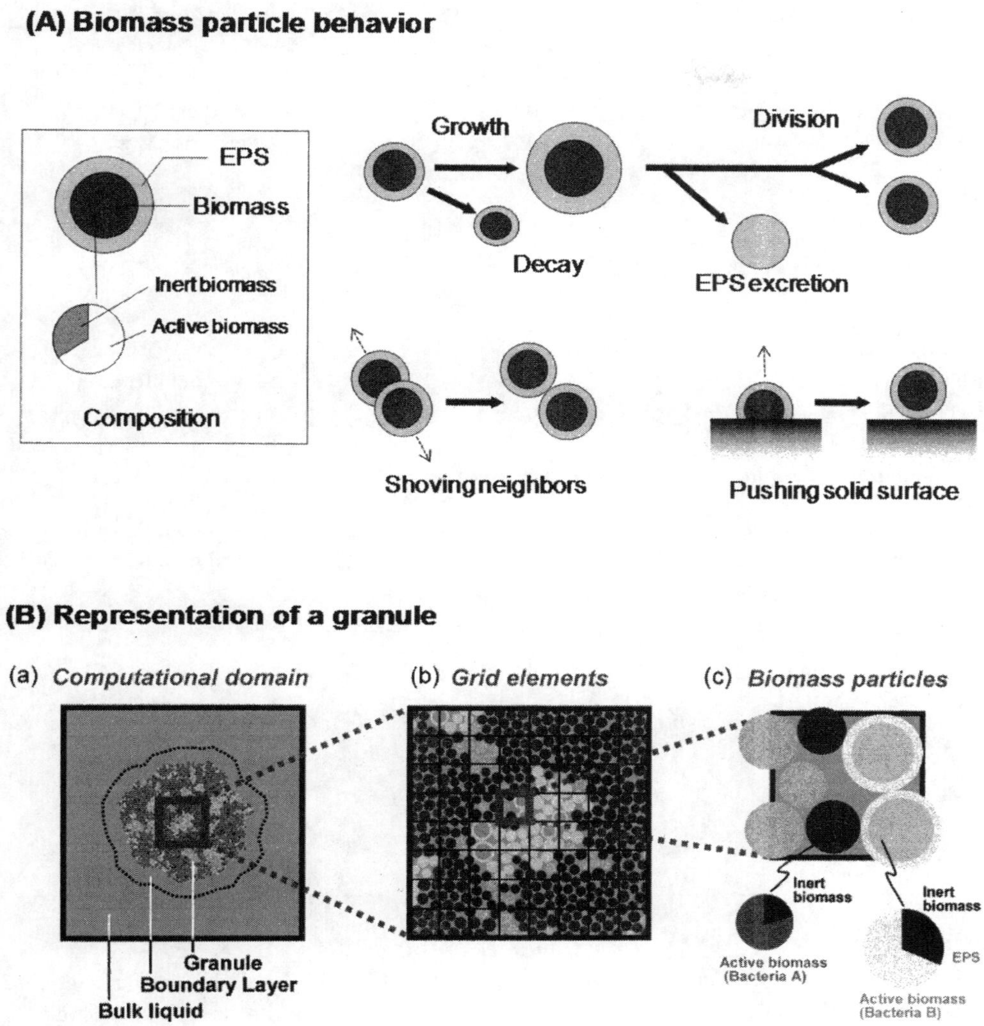

FIGURE 10 (A) Two-dimensional (2-d) IbM model description of actions occurring at the individual scale. (B) Spatial scales in the 2-d model of nitrifying granules. (a) Representative biomass granule comprised in a square computational domain; (b) the square grid elements discretizing the space, each containing several biomass particles; (c) individual biomass particles, of different possible biomass types. All biomass particles within a single grid element experience the same substrate concentrations. The biomass concentrations within a grid element are calculated from the mass of all individual biomass particles within the element volume (Xavier et al., 2005, Matsumoto et al., 2010).

shown by Picioreanu et al. (2004), Xavier et al. (2005), and Xavier et al. (2007).

Comparison of Modeling and Experimental Approach

Molecular techniques (in situ hybridization, etc.) combined with a confocal laser-scanning microscope and micro-sensor enabled investigating in situ microbial community structure and their activities. On the other hand, recent development of the multidimensional multispecies biofilm model features micro-scale distribution of various microbial cells and activities in the biofilm.

TABLE 1 Stoichiometric parameters for microbial reactions

Description	Symbol	Value	Unit	References
AOB				
Yield of biomass on substrate	Y_{AOB}	0.15	$g_{COD,X}\ g_N^{-1}$	Henze et al., 2000
UAP-formation coefficient	k_{UAP}^{AOB}	0.11	$g_{COD,P}\ g_N^{-1}$	Rittmann et al., 2002
NOB				
Yield of biomass on substrate	Y_{NOB}	0.123	$g_{COD,X}\ g_N^{-1}$	Adapted from Henze et al., 2000
UAP-formation coefficient	k_{UAP}^{NOB}	0.04	$g_{COD,P}\ g_N^{-1}$	Rittmann et al., 2002
UAP-utilizing heterotrophic bacteria (HetU)				
Yield of biomass on substrate	Y_{HetU}^{X}	0.25	$g_{COD,X}\ g_{COD,P}^{-1}$	Adapted from Rittmann et al., 2002
Yield of EPS on substrate	Y_{HetU}^{EPS}	0.35	$g_{COD,E}\ g_{COD,P}^{-1}$	Adapted from Rittmann et al., 2002
BAP-utilizing heterotrophic bacteria (HetB)				
Yield of biomass on substrate	Y_{HetB}^{X}	0.25	$g_{COD,X}\ g_{COD,P}^{-1}$	Adapted from Rittmann et al., 2002
Yield of EPS on substrate	Y_{HetB}^{EPS}	0.35	$g_{COD,E}\ g_{COD,P}^{-1}$	Adapted from Rittmann et al., 2002
Org-utilizing heterotrophic bacteria (HetO)				
Yield of biomass on substrate	Y_{HetO}^{X}	0.25	$g_{COD,X}\ g_{COD,P}^{-1}$	Adapted from Rittmann et al., 2002
Yield of EPS on substrate	Y_{HetO}^{EPS}	0.35	$g_{COD,E}\ g_{COD,P}^{-1}$	Adapted from Rittmann et al., 2002
Others				
Biodegradable fraction of active biomass	f_d	0.6	-	Alpkvist et al., 2000
Nitrogen content in active and inert biomass	i_{NXB}	0.087	$g_N\ g_{COD,X}^{-1}$	Rittmann et al., 2002

Here, the comparison of model simulation and experimental results targeting the nitrifying bacterial aggregate formation is shown as an example (Matsumoto et al., 2009). Granulation of nitrifying bacteria without using any carriers has been proposed for immobilizing nitrifying bacteria in an inorganic wastewater treatment processes fed with ammonia as the sole energy source (de Beer et al., 1997; Tay et al., 2002; Tsuneda et al., 2003). However, detailed information on granule formation and nitrogen conversion in primarily autotrophic systems is still lacking.

Recently, there were reports indicating ecophysiological interaction between nitrifying bacteria and heterotrophic bacteria in an autotrophic nitrifying reactor (Rittmann et al., 1994; Kindaichi et al., 2004; Okabe et al., 2005) grown without an external organic carbon source. Nitrifying bacteria are known to release soluble products (SMP) from substrate metabolism and biomass decay; these can provide a supplementary organic substrate for heterotrophic bacteria (Rittmann et al., 1994; Barker and Stuckey, 1999; Rittmann et al., 2002). There are a limited number of studies reporting about the organic substrate uptake pattern of heterotrophic bacteria detected in nitrifying granules. Okabe et al. (2005) reported that the members of the chloroflexi utilize the decaying nitrifying bacteria cell materials (i.e., biomass-associated products [BAP]), the members of the phylum *Cytophaga-Flavobacterium-Bacteroides* utilize substrate UAPs, and the members of the α-*Proteobacteria* and γ-*Proteobacteria* utilize low-molecular-weight organic matter (Org) produced by hydrolysis of EPSs.

In this section, the two-dimensional IbM simulation of nitrifying and heterotrophic bacterial interactions and microbial community structure during the development of nitrifying granules and its comparison with the experimental results are shown. A schematic diagram of the two-dimensional IbM spatial scales for representation of granule is shown in Figure 10B. A detailed model description is described elsewhere (Matsumoto et al., 2010). Kinetic parameters for microbial reaction are listed in Table 1.

IbM simulation results show that, as the granule size increases, an anoxic zone is created at the inner part of the granule. That condition is preferable for heterotrophic bacteria to denitrify, consuming the excreted organic material (SMP) by nitrifying bacteria as an electron donor. Therefore, in the later stage of granule formation, heterotrophic bacteria exist at the inner part of the granule, while nitrifying bacteria dominate at the outer edge. In the present model, based on above studies, we integrated three types of heterotrophic bacteria as HetU, HetB, and HetO, which utilize UAP, BAP, and Org, respectively. Detailed two-dimensional simulated distribution of heterotrophs in the granule is shown in Fig. 11. HetU mainly locates at the surface of nitrifying granule where sufficient UAP is produced by active nitrifying bacteria (Fig. 11a). HetB is mainly present in the inner parts of granule where BAP is produced from inactivated nitrifying bacteria because of depletion of oxygen (Fig. 11b). HetO are present throughout the granule because EPS produced by heterotrophic bacteria spread broadly in granule (Fig. 11c).

The spatial distributions of populations in the nitrifying granule were determined by results from fluorescent in situ hybridization with group-specific probes following quantitative two-dimensional image analysis of confocal image series (Fig. 12). Most bacteria at the outer edge of the granule were nitrifying bacteria; AOB dominated within first 200 μm below the granule surface, whereas NOB were most abundant in the layer between 200 and 300 μm below the granule surface. Bacteria related to CFB division and chloroflexi were detected mainly in the zone where nitrifying bacteria exist and in the inner part of the granules, respectively. Proteobacteria were present throughout the granules. It should be noted that absolute abundance of bacteria (Fig. 12, open circle) at the inner part of granules is much lower than at the surface of granules. Thus, even though chloroflexi occupy the

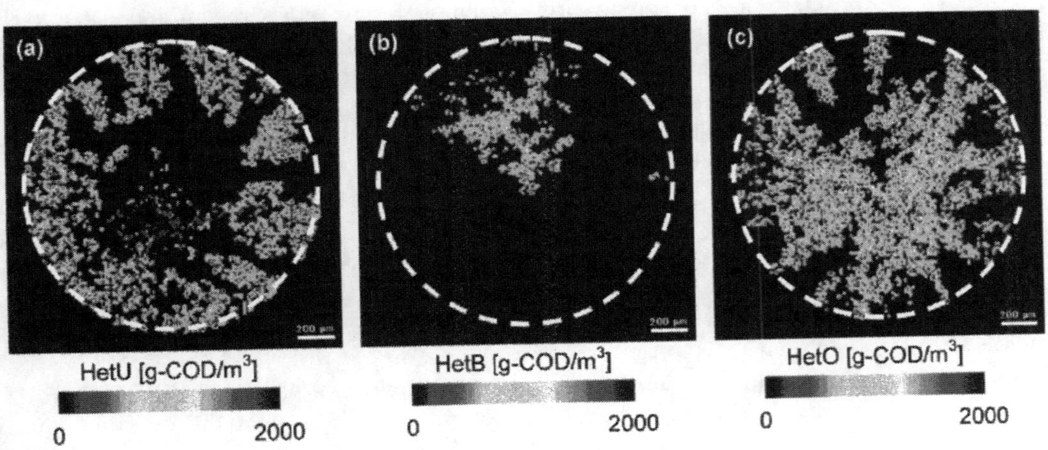

FIGURE 11 Detailed insight into the spatial localization of HetU (a), HetB (b), and HetO (c) provided by a two-dimensional IbM simulation, at day 100. White dotted line shows the granule surface (Matsumoto et al., 2010).

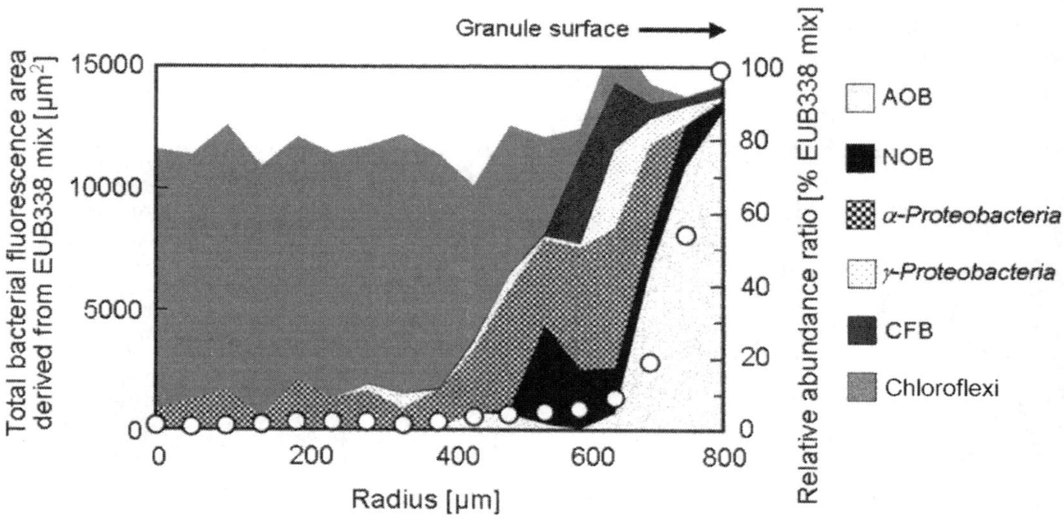

FIGURE 12 Spatial distributions of bacteria along the radius of a nitrifying granule as detected by FISH. The abundance ratio of each bacterium was quantified in 50-μm-thick shells starting at the granule surface (left). Abundance data presented are averages of six replicate measurements. The total bacterial volumetric occupancy in the granule is derived from fluorescence of EUB338 mix-tagged cells (open circles along the *R*-axis). The α-*Proteobacteria* constitute the bacterial group that hybridized with probe ALF1b, excluding the genus *Nitrobacter* that hybridized with probe NIT3. (The figure was reconstructed from the data of Matsumoto et al. [2010].)

inner part of the granules, the absolute amount of chloroflexi is much lower than that of AOB. Although the model simulations performed in this study were not intended to provide quantitative matches to the experimental data, the simulation results indicated that distributions of HetU, HetB, and HetO are consistent with those of CFB, chloroflexi, and *Proteobacteria*, respectively, obtained by FISH analysis (Fig. 12). Thus, it can be concluded that the implemented organic substrate uptake patterns of heterotrophic bacteria were plausible.

Next, the quantitative comparison between solute concentrations (ammonium, nitrite, nitrate, DO, and UAP, BAP, and Org) in the biofilm calculated by the two-dimensional model and those measured by microelectrodes is shown. The steady-state concentrations of main solute concentrations in the granule (ammonium, nitrite, nitrate, and oxygen) along the granule radius calculated with the model describe well the experimentally obtained microelectrode data (Fig. 13). Although it is impossible to experimentally measure each organic compound type, such as UAP and BAP, models simulated the microprofiles of these compounds inside the granule. Interestingly, microprofiles of these compounds dynamically changed along the granule radius and seemed to be related to the spatial distribution of microorganism types that consume or produce these compounds (Fig. 11 to 13).

Since community structure in the granule is determined by a complex interplay of various factors, including the concentration of chemical species, the presence of various types of bacteria, and their physiology, mathematical modeling provides a logical framework for the exploration of processes within granules.

CONCLUSION AND PERSPECTIVES

Simple models such as the one-dimensional model for the biofilm process generally seem to be sufficient for the prediction of the macroscopic phenomena such as overall nutrient

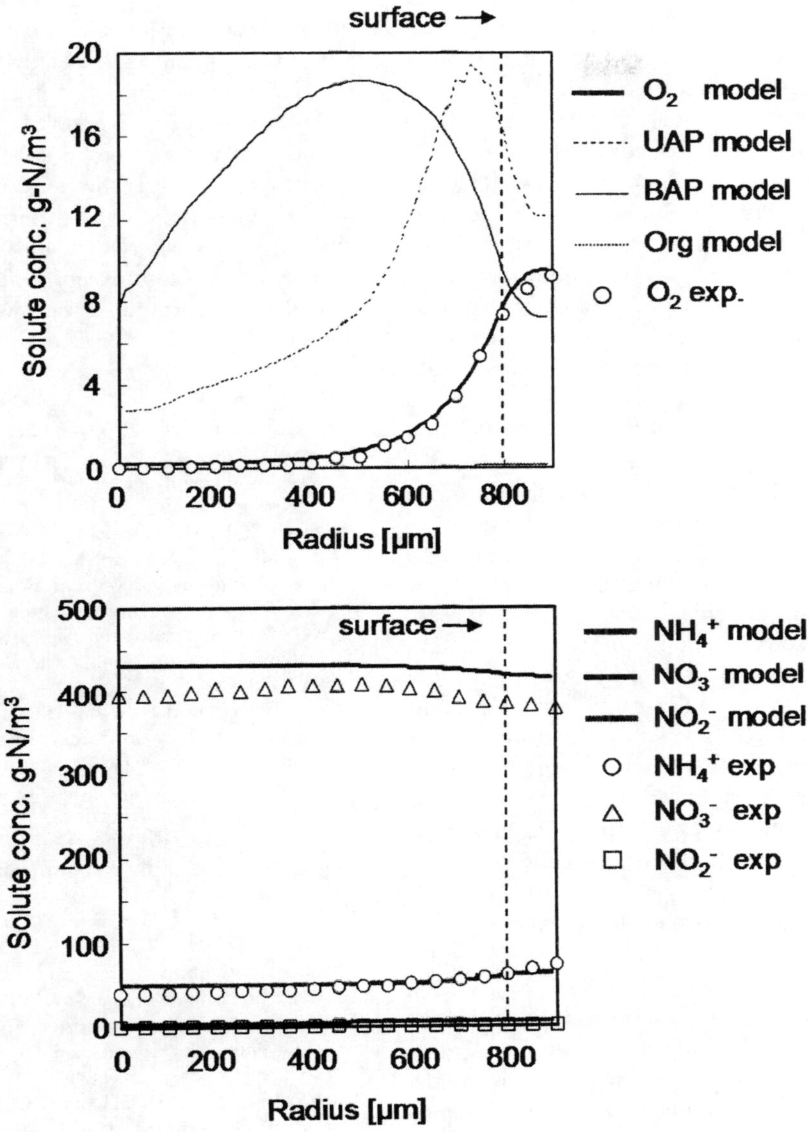

FIGURE 13 Comparison of the steady-state solute concentrations in the biofilm calculated in two-dimensional (2-d) IbM model simulations, with the experimental microelectrode data (open circles, triangles and squares for O_2, NH_4^+, NO_3^-, NO_2^-). The model results are at day 100. To obtain the comparable profiles along the radius, the 2-d concentration distributions were averaged in concentric shells with different radii. (The figure was reconstructed from the data of Matsumoto et al. [2010].)

flux, and thus it is preferable for an engineering purpose rather than a complex multidimensional model because of its simplicity. However, bottom-up type modeling such as "cellular automata" or IbM is very appealing to represent nonlinear natural systems. These model simulations have the potential to reveal interactions leading to the properties of a microbial ecological system that are generally not tractable by experimental approaches.

It is well acknowledged that a combination of laboratory experimental and mathematical modeling will be a strong tool for better understanding of complex systems such as a biological phenomenon (Eberal, 2003). However, there have been very few reports of experimental verification of the two- or three-dimensional multispecies biofilm model predictions (Xavier et al., 2005; Matsumoto et al., 2007), and there has been a single study that evaluated biofilm phenomena concerning microbial ecology by combining strategy of experimental and simulation analysis (Matsumoto et al., 2010). This is because there still remain problems to be solved. That is, (i) the modeling simulation is thought to be rather complex and not as familiar as experimental techniques such as molecular techniques for most microbiologists (necessity of the user-friendly software); (ii) present modeling algorithms should be improved further for different purposes; and (iii) it is difficult to obtain kinetic parameters, and this sometimes causes low creditbility of a modeling approach.

However, such attempts set a milestone by combining rather complex mathematical simulations with molecular methods to study microbial ecological systems as a new tool for investigating microbial ecological systems.

CONCLUDING REMARKS

Biological wastewater treatment is undoubtedly one of the most important biotechnological processes. Nitrification in wastewater treatment systems has been studied extensively. Despite their importance, knowledge about the identity and ecology of nitrifying bacteria carrying out nitrification in WWTPs has been scarce. Thus, biological nitrogen removal processes have been regarded as "a black box" in practice because the lack of fundamental microbiological understanding hampers knowledge-driven process design and operation. However, the recent application of molecular approaches has unveiled the entity of a complex microbial community in WWTPs and the spatial distribution of each member to address fundamental questions about microbial identities, functions, and spatial organization. This approach also revealed a remarkably vast microbial diversity including many yet uncultured species involved in nitrogen removal processes. The differences in nitrifying bacterial diversity may be related to the differences in the performance. The challenge now is to elucidate the mechanism underlying the community differences, which should be incorporated into WWTP design and operation.

REFERENCES

Abeling, U., and C. F. Seyfried. 1992. Anaerobic-aerobic treatment of high-strength ammonium waste-water—nitrogen removal via nitrite. *Water Sci. Technol.* **26:**1007–1015.

Adamczyk, J., M. Hesselsoe, N. Iversen, M. Horn, A. Lehner, P. Nielsen, M. Schloter, P. Roslev, and M. Wagner. 2003. The isotope array, a new tool that employs substrate-mediated labeling of rRNA for determination of microbial community structure and function. *Appl. Environ. Microbiol.* **69:**6875–6887.

Ahn, J., R. Yu, and K. Chandran. 2008. Distinctive microbial ecology and biokinetics of autotrophic ammonia and nitrite oxidation in a partial nitrification bioreactor. *Biotechnol. Bioeng.* **100:**1078–1087.

Alpkvist, E., and I. Klapper. 2007. A multidimensional multispecies continuum model for heterogeneous biofilm development. *Bull. Math. Biol.* **69:**765–789.

Alpkvist, E., C. Picioreanu, M. van Loosdrecht, and A. Heyden. 2006. Three-dimensional biofilm model with individual cells and continuum EPS matrix. *Biotechnol. Bioeng.* **94:**961–979.

Amor, L., M. Eiroa, C. Kennes, and M. C. Veiga. 2005. Phenol biodegradation and its effect on the nitrification process. *Water Res.* **39:**2915–2920.

Anthonisen, A. C., B. C. Loehr, T. B. S. Prakasam, and E. G. Srinath. 1976. Inhibition of nitrification by ammonia and nitrous acid. *J. Water Pollut. Control Fed.* **48:**835–852.

Bai, Y. H., Q. H. Sun, C. Zhao, D. H. Wen, and X. Y. Tang. 2009. Aerobic degradation of pyridine by a new bacterial strain, Shinella zoogloeoides BC026. *J. Ind. Microbiol. Biotechnol.* **36:**1391–1400.

Barker, D. J., and D. C. Stuckey. 1999. A review of soluble microbial products (SMP) in wastewater treatment systems. *Water Res.* **33:**3063–3082.

Barnard, J. L. 1975. Biological nutrient removal without the addition of chemicals. *Water Res.* **9:**485–490.

Bell, A., Y. Aoi, A. Terada, S. Tsuneda, and A.

Hirata. 2005. Comparison of spatial organization in top-down- and membrane-aerated biofilms: a numerical study. *Water Sci. Technol.* **52:**173–180.

Bernet, N., P. Dangcong, J. Delgenes, and R. Moletta. 2001. Nitrification at low oxygen concentration in biofilm reactor. *J. Environ. Eng.* **127:**266–271.

Blackall, L. L., and P. Burrell. 1999. The microbiology of nitrogen removal in activated sludge systems, p. 203–226. *In* R. J. Seviour and L. L. Blackall (ed.), *The Microbiology of Activated Sludge.* Kluwer Academic Publishers, Dordrecht, The Netherlands.

Blackburne, R., V. Vadivelu, Z. Yuan, and J. Keller. 2007. Kinetic characterisation of an enriched *Nitrospira* culture with comparison to *Nitrobacter. Water Res.* **41:**3033–3042.

Bossier, P., and W. Verstraete. 1996. Triggers for microbial aggregation in activated sludge? *Appl. Microbiol. Biotechnol.* **45:**1–6.

Briones, A., and L. Raskin. 2003. Diversity and dynamics of microbial communities in engineered environments and their implications for process stability. *Curr. Opin. Biotechnol.* **14:**270–276.

Chen, G., M. Wong, S. Okabe, and Y. Watanabe. 2003. Dynamic response of nitrifying activated sludge batch culture to increased chloride concentration. *Water Res.* **37:**3125–3135.

Dahl, C., C. Sund, G. H. Kristensen, and L. Vredenbregt. 1997. Combined biological nitrification and denitrification of high-salinity wastewater. *Water Sci. Technol.* **36:**345–352.

Daims, H., S. Lucker, and M. Wagner. 2006. daime, a novel image analysis program for microbial ecology and biofilm research. *Environ. Microbiol.* **8:**200–213.

Daims, H., J. L. Nielsen, P. H. Nielsen, K.-H. Schleifer, and M. Wagner. 2001a. In situ characterization of Nitrospira-like nitrite-oxidizing bacteria active in wastewater treatment plants. *Appl. Environ. Microbiol.* **67:**5273–5284.

Daims, H., U. Purkhold, L. Bjerrum, E. Arnold, P. A. Wilderer, and M. Wagner. 2001b. Nitrification in sequencing biofilm batch reactors: lessons from molecular approaches. *Water Sci. Technol.* **43:**9–18.

de Beer, D., A. Schramm, C. Santegoeds, and M. Kuhl. 1997. A nitrite microsensor for profiling environmental biofilms. *Appl. Environ. Microbiol.* **63:**973–977.

de Kreuk, M. K., M. Pronk, and M. C. M. van Loosdrecht. 2005. Formation of aerobic granules and conversion processes in an aerobic granular sludge reactor at moderate and low temperatures. *Water Res.* **39:**4476–4484.

Dinçer, A., and F. Kargi. 2001. Salt inhibition kinetics in nitrification of synthetic saline wastewater. *Enzyme Microb. Technol.* **28:**661–665.

Dionisi, H. M., A. C. Layton, G. Harms, I. R. Gregory, K. G. Robinson, and G. S. Sayler. 2002. Quantification of *Nitrosomonas oligotropha*-like ammonia-oxidizing bacteria and *Nitrospira* spp. from full-scale wastewater treatment plants by competitive PCR. *Appl. Environ. Microbiol.* **68:**245–253.

Eberal, H. J. 2003. What do biofilm models, mechanical ducks, and artificial life have in common? Mathematical modeling in biofilm research, p. 9–31. *In* S. Wuertz, P. L. Bishop, and P. A. Wilderer (ed.), *Biofilms in Wastewater Treatment: An Interdisciplinary Spproach.* IWA Publishing, London, United Kingdom.

Eilersen, A., M. Henze, and L. Kloft. 1994. Effect of volatile fatty acids and trimethylamine on nitrification in activated sludge. *Water Res.* **28:**1329–1336.

Focht, D. D., and A. C. Chang. 1975. Nitrification and denitrification processes related to wastewater treatment. *Adv. Appl. Microbiol.* **19:**153–186.

Furukawa, K., A. Ike, and M. Fujita. 1993. Preparation of marine nitrifying sludge. *J. Ferment. Bioeng.* **76:**134–139.

Garrido, J., W. van Benthum, M. vanLoosdrecht, and J. Heijnen. 1997. Influence of dissolved oxygen concentration on nitrite accumulation in a biofilm airlift suspension reactor. *Biotechnol. Bioeng.* **53:**168–178.

Gieseke, A., U. Purkhold, M. Wagner, R. Amann, and A. Schramm. 2001. Community structure and activity dynamics of nitrifying bacteria in a phosphate-removing biofilm. *Appl. Environ. Microbiol.* **67:**1351–1362.

Gujer, W. 2006. Activated sludge modelling: past, present and future. *Water Sci. Technol.* **53:**111–119.

Gujer, W. 2010 Nitrification and me—a subjective review. *Water Res.* **44:**1–19.

Gujer, W., M. Henze, T. Mino, and M. C. M. van Loosdrecht. 1999. Activated sludge model NO.3. *Water Sci. Technol* **39:**183–193.

Hanaki, K., C. Wantawin, and S. Ohgaki. 1990. Effects of the activity of heterotrophs on nitrification in a suspended-growth reactor. *Water Res.* **24:**289–296.

Head, I., W. Hiorns, T. Embley, A. McCarthy, and J. Saunders. 1993. The phylogeny of autotrophic ammonia-oxidizing bacteria as determined by analysis of 16S ribosomal RNA gene sequences. *J. Gen. Microbiol.* **139:**1147–1153.

Hellinga, C., A. Schellen, J. Mulder, M. van Loosdrecht, and J. Heijnen. 1998. The SHARON process: an innovative method for nitrogen removal from ammonium-rich waste water. *Water Sci. Technol.* **37:**135–142.

Henze, M., W. Gujer, T. Matsuo, and M.C.M. van Loosdrecht. 2000. Activated sludge models ASM1, ASM2, ASM2d and ASM3. *Scientific and Technical Reports.* IWA publishing, London, United Kingdom.

Hoilijoki, T., R. Kettunen, and J. Rintala. 2000. Nitrification of anaerobically pretreated municipal landfill leachate at low temperature. *Water Res.* **34:**1435–1446.

Ichihashi, O., H. Satoh, and T. Mino. 2006. Effect of soluble microbial products on microbial metabolisms related to nutrient removal. *Water Res.* **40:**1627–1633.

Isaka, K., S. Yoshie, T. Sumino, Y. Inamori, and S. Tsuneda. 2007. Nitrification of landfill leachate using immobilized nitrifying bacteria at low temperatures. *Biochem. Eng. J.* **37:**49–55.

Isaka, K., T. Sumino, and S. Tsuneda. 2008. Novel nitritation process using heat-shocked nitrifying bacteria entrapped in gel carriers. *Process Biochem.* **43:**265–270.

Jin, R., P. Zheng, Q. Mahmood, and B. Hu. 2007. Osmotic stress on nitrification in an airlift bioreactor. *J. Hazard. Mater.* **146:**148–54.

Joo, H. S., M. Hirai, and M. Shoda. 2006. Piggery wastewater treatment using *Alcaligenes faecalis* strain No. 4 with heterotrophic nitrification and aerobic denitrification. *Water Res.* **40:**3029–3036.

Juretschko, S., G. Timmermann, M. Schmid, K. H. Schleifer, A. Pommerening-Roser, H. P. Koops, and M. Wagner. 1998. Combined molecular and conventional analyses of nitrifying bacterium diversity in activated sludge: *Nitrosococcus mobilis* and *Nitrospira*-like bacteria as dominant populations. *Appl. Environ. Microbiol.* **64:**3042–3051.

Juretschko, S., A. Loy, A. Lehner, and M. Wagner. 2002. The microbial community composition of a nitrifying-denitrifying activated sludge from an industrial sewage treatment plant analyzed by the full-cycle rRNA approach. *Syst. Appl. Microbiol.* **25:**84–99.

Jürgens, K., and C. Matz. 2002. Predation as a shaping force for the phenotypic and genotypic composition of planktonic bacteria. *Antonie Van Leeuwenhoek Int. J. Gen. Mol. Microbiol.* **81:**413–434.

Khardenavis, A. A., A. Kapley, and H. J. Purohit. 2007. Simultaneous nitrification and denitrification by diverse *Diaphorobacter* sp. *Appl. Microbiol. Biotechnol.* **77:**403–409.

Kim, D. J., and S. H. Kim. 2006. Effect of nitrite concentration on the distribution and competition of nitrite-oxidizing bacteria in nitratation reactor systems and their kinetic characteristics. *Water Res.* **40:**887–894.

Kim, Y. M., D. Park, D. S. Lee, and J. M. Park. 2008. Inhibitory effects of toxic compounds on nitrification process for cokes wastewater treatment. *J. Hazard. Mater.* **152:**915–921.

Kindaichi, T., T. Ito, and S. Okabe. 2004. Ecophysiological interaction between nitrifying bacteria and heterotrophic bacteria in autotrophic nitrifying biofilms as determined by microautoradiography-fluorescence in situ hybridization. *Appl. Environ. Microbiol.* **70:**1641–1650.

Kishino, H., H. Ishida, H. Iwabu, and I. Nakano. 1996. Domestic wastewater reuse using a submerged membrane bioreactor. *Desalination* **106:**115–119.

Konneke, M., A. E. Bernhard, J. R. de la Torre, C. B. Walker, J. B. Waterbury, and D. A. Stahl. 2005. Isolation of an autotrophic ammonia-oxidizing marine archaeon. *Nature* **437:**543–546.

Kos, P. 1998. Short SRT (solids retention time) nitrification process/flowsheet. *Water Sci. Technol.* **38:**23–29.

Kreft, J. U., and J. W. T. Wimpenny. 2001. Effect of biofilm structure and function as revealed by an individual-based model of biofilm growth. *Water Sci. Technol.* **43:**135–141.

Larsen, P., J. Nielsen, T. Svendsen, and P. Nielsen. 2008. Adhesion characteristics of nitrifying bacteria in activated sludge. *Water Res.* **42:**2814–2826.

Lay-Son, M., and C. Drakides. 2008. New approach to optimize operational conditions for the biological treatment of a high-strength thiocyanate and ammonium waste: pH as key factor. *Water Res.* **42:**774–780.

Lee, N., P. Nielsen, K. Andreasen, S. Juretschko, J. Nielsen, K. Schleifer, and M. Wagner. 1999. Combination of fluorescent in situ hybridization and microautoradiography–a new tool for structure-function analyses in microbial ecology. *Appl. Environ. Microbiol.* **65:**1289–1297.

Lens, P., M. Depoorter, C. Cronenberg, and W. Verstraete. 1995. Sulfate-reducing and methane producing bacteria in aerobic waste-water treatment systems. *Water Res.* **29:**871–880.

Li, B. K., and P. L. Bishop. 2004. Micro-profiles of activated sludge floc determined using microelectrodes. *Water Res.* **38:**1248–1258.

Logemann, S., J. Schantl, S. Bijvank, M. van Loosdrecht, J. G. Kuenen, and M. Jetten. 1998. Molecular microbial diversity in a nitrifying reactor system without sludge retention. *FEMS Microbiol. Ecol.* **27:**239–249.

Lydmark, P., M. Lind, F. Sorensson, and M. Hermansson. 2006. Vertical distribution of nitrifying populations in bacterial biofilms from a full-scale nitrifying trickling filter. *Environ. Microbiol.* **8:**2036–2049.

Madoni, P., D. Davoli, and L. Guglielmi. 1999. Response of sOUR and AUR to heavy metal contamination in activated sludge. *Water Res.* **33:**2459–2464.

Matsumoto, S., A. Terada, Y. Aoi, S. Tsuneda, E. Alpkvist, C. Picioreanu, and M. van Loosdrecht. 2007. Experimental and simulation analysis of community structure of nitrifying bacteria in a membrane-aerated biofilm. *Water Sci. Technol.* **55:**283–290.

Matsumoto, S., M. Katoku, G. Saeki, A. Terada, Y. Aoi, S. Tsuneda, C. Picioreanu, and M. van Loosdrecht. 2010. Microbial community structure in autotrophic nitrifying granules characterized by experimental and simulation analyses. *Environ. Microbiol.* **12:**192–206.

Mobarry, B. K., M. Wagner, V. Urbain, B. E. Rittmann, and D. A. Stahl. 1996. Phylogenetic probes for analyzing abundance and spatial organization of nitrifying bacteria. *Appl. Environ. Microbiol.* **62:**2156–2162.

Noguera, D., G. Pizarro, D. Stahl, and B. Rittmann. 1999. Simulation of multispecies biofilm development in three dimensions. *Water Sci. Technol.* **39:**123–130.

Okabe S., Hirata K., and Watanabe Y. 1995. Dynamic changes in spatial microbial distribution in mixed-population biofilms: experimental results and model simulation. *Water Sci. Technol.* **32:** 67–74.

Okabe, S., K. Hiratia, Y. Ozawa, and Y. Watanabe. 1996. Spatial microbial distributions of nitrifiers and heterotrophs in mixed-population biofilms. *Biotechnol. Bioeng.* **50:**24–35.

Okabe, S, T. Itoh, H. Satoh, and Y. Watanabe. 1999a. Analyses of spatial distributions of sulfate–reducing bacteria and their activity in aerobic wastewater biofilms. *Appl. Environ. Microbiol.* **65:**5107–5116.

Okabe, S., H. Satoh, and Y. Watanabe. 1999b. *In situ* analysis of nitrifying biofilms as determined by in situ hybridization and the use of microelectrodes. *Appl. Environ. Microbiol.* **65:**3182–3191.

Okabe, S., H. Naitoh, H. Satoh, and Y. Watanabe. 2002. Structure and function of nitrifying biofilms as determined by molecular techniques and the use of microelectrodes. *Water Sci. Technol.* **46:**233–241.

Okabe, S., T., Kindaichi, and T. Ito. 2004. MAR-FISH – an ecophysiological approach to link phylogenetic affiliation and *in situ* metabolic activity of microorganisms at a single-cell resolution. *Microb. Environ.* **19:**83–98.

Okabe, S., T. Kindaichi, and T. Ito. 2005. Fate of ^{14}C-labeled microbial products derived from nitrifying bacteria in autotrophic nitrifying biofilms. *Appl. Environ. Microbiol.* **71:**3987–3994.

Oleszkiewicz, J., and S. Berquist. 1988. Low-temperature nitrogen removal in sequencing batch reactors. *Water Res.* **22:**1163–1171.

Painter, H. A. 1970. A review of literature on inorganic nitrogen metabolism in microorganisms. *Water Res.* **4:**393–450.

Painter, H. A. 1986. Nitrification in the treatment of sewage and waste-waters, p. 185–213. *In* J. I. Prosser (ed.), *Nitrification*, vol. 20. IRL Press Limited, Oxford, United Kingdom.

Park, H. D., and D. R. Noguera. 2004. Evaluating the effect of dissolved oxygen on ammonia-oxidizing bacterial communities in activated sludge. *Water Res.* **38:**3275–3286.

Park, S. J., J. W. Oh, and T. I. Yoon. 2003. The role of powdered zeolite and activated carbon carriers on nitrification in activated sludge with inhibitory materials. *Process Biochem.* **39:**211–219.

Park, H. D., G. F. Wells, H. Bae, C. S. Criddle, and C. A. Francis. 2006. Occurrence of ammonia-oxidizing archaea in wastewater treatment plant bioreactors. *Appl. Environ. Microbiol.* **72:**5643–5647.

Picioreanu, C., M. van Loosdrecht, and J. Heijnen. 1999. Discrete-differential modeling of biofilm structure. *Water Sci. Technol.* **39:**115–122.

Picioreanu, C., J. U. Kreft, and M. C. M. van Loosdrecht. 2004. Particle-based multidimensional multispecies biofilm model. *Appl. Environ. Microbiol.* **70:**3024–3040.

Pogue, A., and K. Gilbride. 2007. Impact of protozoan grazing on nitrification and the ammonia- and nitrite-oxidizing bacterial communities in activated sludge. *Can. J. Microbiol.* **53:**559–571.

Prosser, J. I. 1989. Autotrophic nitrification in bacteria. *Adv. Microb. Physiol.* **30:**125–181.

Reichert, P. 1998. AQUASIM 2.0—Computer program for the identification and simulation of aquatic systems, version 2.0. EAWAG, Dubendorf, Switzerland.

Rittmann, B. E., and P. L. McCarty. 2001. Environmental biotechnology: principles and applications. McGraw Hill, New York, NY.

Rittmann, B. E., and R. Whiteman. 1994. Bioaugmentation: a coming of age. *Water Qual. Int.* **1:**12–16.

Rittmann, B. E., J. M. Regan, and D. A. Stahl. 1994. Nitrification as a source of soluble organic substrate in biological treatment. *Water Sci. Technol.* **30:**1–8.

Rittmann, B., D. Stilwell, and A. Ohashi. 2002. The transient-state, multiple-species biofilm model for biofiltration processes. *Water Res.* **36:**2342–2356.

Rowan, A. K., J. R. Snape, D. Fearnside, M. R. Barer, T. P. Curtis, and I. M. Head. 2003. Composition and diversity of ammonia-oxidising bacterial communities in wastewater treatment reactors of different design treating identical wastewater. *FEMS Microbiol. Ecol.* **43:**195–206.

Santegoeds, C., A. Schramm, and D. de Beer. 1998. Microsensors as a tool to determine chemical microgradients and bacterial activity in wastewater biofilms and flocs. *Biodegradation* **9:**159–167.

Sato, C., S. W. Leung, and J. L. Schnoor. 1988. Toxic response of *Nitrosomonas europaea* to copper in inorganic medium and wastewater. *Water Res.* **22:**1117–1127.

Satoh, H., S. Okabe, N. Norimatsu, and Y. Wata-

nabe. 2000. Significance of substrate C/N ratio on structure and activity of nitrifying biofilms determined by in situ hybridization and the use of microelectrodes. *Water Sci. Technol.* **41:**317–321.

Satoh, H., Y. Nakamura, H. Ono, and S. Okabe. 2003a. Effect of oxygen concentration on nitrification and denitrification in single activated sludge flocs. *Biotechnol. Bioeng.* **83:**604–607.

Satoh, H., S. Okabe, Y. Yamaguchi, and Y. Watanabe. 2003b. Evaluation of the impact of bioaugmentation and biostimulation by in situ hybridization and microelectrode. *Water Res.* **37:**2206–2216.

Schramm, A. 2003. In situ analysis of structure and activity of the nitrifying community in biofilms, aggregates, and sediments. *Geomicrobiol. J.* **20:**313–333.

Schramm, A., L. Larsen, N. Revsbech, N. Ramsing, R. Amann, and K. Schleifer. 1996. Structure and function of a nitrifying biofilm as determined by *in situ* hybridization and the use of microelectrodes. *Appl. Environ. Microbiol.* **62:**4641–4647.

Schramm, A., D. de Beer, J. van den Heuvel, S. Ottengraf, and R. Amann. 1999a. Microscale distribution of populations and activities of *Nitrosospira* and *Nitrospira* spp. along a macroscale gradient in a nitrifying bioreactor: quantification by in situ hybridization and the use of microsensors. *Appl. Environ. Microbiol.* **65:**3690–3696.

Schramm, A., C. M. Santegoeds, H. K. Nielsen, H. Ploug, M. Wagner, M. Pribyl, J. Wanner, R. Amann, and D. De Beer. 1999b. On the occurrence of anoxic microniches, denitrification, and sulfate reduction in aerated activated sludge. *Appl. Environ. Microbiol.* **65:**4189–4196.

Semerci, N., and F. Çeçen. 2007. Importance of cadmium speciation in nitrification inhibition. *J. Hazard. Mater.* **147:**503–512.

Sharma, B., and R. C. Ahlert. 1977. Nitrification and nitrogen removal. *Water Res.* **11:**897–925.

Stutzer, H., and R. Hartleb. 1894. Uber Nitratbildung. *Zentralbl. Bakteriol. Parasitenkd. Infektionskr. Hyg. Abt. 1 Orig. Reihe A* **22:**701.

Suwa, Y., Y. Imamura, T. Suzuki, T. Tashiro, and Y. Urushigawa. 1994. Ammonia-oxidizing bacteria with different sensitivities to $(NH_4)_2SO_4$ in activated sludges. *Water Res.* **28:**1523–1532.

Suwa, Y., T. Sumino, and K. Noto. 1997. Phylogenetic relationships of activated sludge isolates of ammonia oxidizers with different sensitivities to ammonium sulfate. *J. Gen. Appl. Microbiol.* **43:**373–379.

Tanaka, J., K. Syutsubo, K. Watanabe, H. Izumida, and S. Harayama. 2003. Activity and population structure of nitrifying bacteria in an activated-sludge reactor containing polymer beads. *Environ. Microbiol.* **5:**278–286.

Tay, J., Q. Liu, and Y. Liu. 2002. Characteristics of aerobic granules grown on glucose and acetate in sequential aerobic sludge blanket reactors. *Environ. Technol.* **23:**931–936.

Tchobanoglous, G., F. L. Burton, and H. D. Stensel. 2003. *Wastewater Engineering: Treatment and Reuse*, 4th ed. McGraw-Hill, New York, NY.

Tokutomi, T. 2004. Operation of a nitrite-type airlift reactor at low DO concentration. *Water Sci. Technol.* **49:**81–88.

Tsuneda, S., T. Nagano, T. Hoshino, Y. Ejiri, N. Noda, and A. Hirata. 2003. Characterization of nitrifying granules produced in an aerobic upflow fluidized bed reactor. *Water Res.* **37:**4965–4973.

Udert, K. M., T. A. Larsen, and W. Gujer. 2005. Chemical nitrite oxidation in acid solutions as a consequence of microbial ammonium oxidation. *Environ. Sci. Technol.* **39:**4066–4075.

van Dongen, U., M. S. M. Jetten, and M. C. M. van Loosdrecht. 2001a. The SHARON-Anammox process for treatment of ammonium rich wastewater. *Water Sci. Technol.* **44:**153–160.

van Dongen, U., M. S. M. Jetten, and M. C. M. van Loosdrecht. 2001b. The combined SHARON/anammox process: a sustainable method for N-removal from sludge water, Stowa. IWA Publishing, London, United Kingdom.

van Niel, E. W. J., L. A. Robertson, and J. G. Kuenen. 1993. A mathematical description of the behaviour of mixed chemostat cultures of an autotrophic nitrifier and a heterotrophic nitrifier/aerobic denitrifier; a comparison with experimental data. *FEMS Microbiol. Ecol.* **102:**99–108.

Volcke, E., M. van Loosdrecht, and P. Vanrolleghem. 2006. Controlling the nitrite: ammonium ratio in a SHARON reactor in view of its coupling with an Anammox process. *Water Sci. Technol.* **53:**45–54.

Wagner, M., G. Rath, R. Amann, H. P. Koops, and K. H. Schleifer. 1995. *In situ* identification of ammonia-oxidizing bacteria. *Syst. Appl. Microbiol.* **18:**251–264.

Wanner, O., and W. Gujer. 1986. A multispecies biofilm model. *Biotechnol. Bioeng.* **28:**314–328.

Wanner, O., and P. Reichert. 1996. Mathematical modeling of mixed-culture biofilms. *Biotechnol. Bioeng.* **49:**172–184.

Wells, G., H. Park, C. Yeung, B. Eggleston, C. Francis, and C. Criddle. 2009. Ammonia-oxidizing communities in a highly aerated full-scale activated sludge bioreactor: betaproteobacterial dynamics and low relative abundance of Crenarchaea. *Environ. Microbiol.* **11:**2310–2328

Wittebolle, L., H. Vervaeren, W. Verstraete, and N. Boon. 2008. Quantifying community dynamics of nitrifiers in functionally stable reactors. *Appl.*

Environ. Microbiol. **74:**286–293.

Xavier, J., and K. Foster. 2007. Cooperation and conflict in microbial biofilms. *Proc. Natl. Acad. Sci. USA* **104:**876–881.

Xavier, J., C. Picioreanu, and M. van Loosdrecht. 2005. A framework for multidimensional modeling of activity and structure of multispecies biofilms. *Environ. Microbiol.* **7:**1085–1103.

Yamagishi, T., J. Leite, S. Ueda, F. Yamaguchi, and Y. Suwa. 2001. Simultaneous removal of phenol and ammonia by an activated sludge process with cross-flow filtration. *Water Res.* **35:**3089–3096.

You, J., A. Das, E. Dolan, and Z. Hu. 2009. Ammonia-oxidizing archaea involved in nitrogen removal. *Water Res.* **43:**1801–1809.

Zhang, T., L. Ding, H. Ren, and X. Xiong. 2009a. Ammonium nitrogen removal from coking wastewater by chemical precipitation recycle technology. *Water Res.* **43:**5209–5215.

Zhang, T., T. Jin, Q. Yan, M. Shao, G. Wells, C. Criddle, and H. H. P. Fang. 2009b. Occurrence of ammonia-oxidizing Archaea in activated sludges of a laboratory scale reactor and two wastewater treatment plants. *J. Appl. Microbiol.* **107:**970–977.

INDEX